Caldwell Environmental, Inc
(919) 781-1936

Coatings
Encyclopedic Dictionary

Edited by

Stanley LeSota

Federation of Societies for Coatings Technology
YE OLDE PAINT STONE 1737

This volume is an effort to improve communications and understanding in the coatings and interfacing industries, arts, and crafts.

Coatings
Encyclopedic Dictionary

Edited by

Stanley LeSota

Published by
Federation of Societies for Coatings Technology
492 Norristown Road, Blue Bell, PA 19422-2350

This document has been prepared with great care on the part of many people to eliminate possible errors. However, neither the editors nor the FSCT make any warranty, expressed or implied, or assume any legal liability or responsibility for the accuracy or completeness of any information contained herein. In addition, reference to any specific product, process or service by tradename, trademark, manufacturer or otherwise does not necessarily constitute or imply its endorsement, recommendation, or favor.

These definitions do not purport to address all the safety concerns, if any, associated with their use. It is the responsibility of the reader to establish appropriate safety and health practices and determine the applicability of regulatory limitations prior to use.

ISBN Number 0-934010-04-8 (Casebound)
ISBN Number 0-934010-05-6 (Perfect Bound)
Library of Congress Catalog Number 95-061644

492 Norristown Rd., Blue Bell, PA 19422-2350

Printed in the United States of America

Contents

Dedication

This *Coatings Encyclopedic Dictionary* is dedicated to the members of the original FSCT's Definition Committee that was responsible for compiling the *Paint/Coatings Dictionary,* a 13-year effort, which was the source of most of the definitions in this volume.

David J. Engler .... Retired
Ruth Johnston-Feller .... Retired
J. Richard Kiefer, Jr. .... Retired
Stanley LeSota .... Coatings Consultant
Seymour Mark .... Deceased
John Landis .... Whereabouts unknown
Michael L. McGroarty .... Columbus Coated Fabrics
Lothar Sander .... Retired
Francis J. Willard .... Deceased

Except for Ruth Johnston-Feller, all the above were members of the Philadelphia Society for Coatings Technology.

Preface

First published by the FSCT in 1978, the *FSCT Paint/Coatings Dictionary*[1] has become a standard among reference sources for the coatings industry. The *Coatings Encyclopedic Dictionary* is the expanded and encyclopedic version of the *Paint/Coatings Dictionary,* and it reflects the many changes experienced by the paint and coatings industry over the past 17 years.

Over 6000 definitions, including state-of-the-art technologies and regulatory terminology are featured in the *Coatings Encyclopedic Dictionary*. The broad scope of terms included reveals the usefulness of this volume to a wide audience, ranging from the layman, artists and artisans, legal and governmental agencies, to technicians in all coatings-related fields.

One of the most beneficial enhancements to this publication is the "encyclopedic" additions which are presented in *italic* type after many definitions. These succinct paragraphs relay significant information that the reader should know about the term defined. In many cases, ASTM Test Method references are provided, further enhancing the usefulness of the *Dictionary*.

Another unique feature is the classification of all terms into one or more of 80 categories (color, pigments, additives, methods of application, etc.). At the end of the definitions, numerical superscripts indicate the categories which are listed in the second section of this volume. This categorical listing of terms serves as a check list for key words which assist in the preparation of research papers, literature searches, etc.

To signal readers that a technical or non-commonplace word which is part of the definition of a term is also defined in another part of the *Coatings Encyclopedic Dictionary,* a different type style, called "CAP-SMALL CAP," is used. In some cases this type is used for emphasis.

To further benefit readers, there is a considerable amount of cross-referencing that refers the reader to other similar significant terms that may help in the understanding of definitions. There are also cf. (compare) references that indicate contrasting terms or antonyms. Synonyms (Syn:) follow the definition in many cases.

For many of the chemical definitions in this book, the chemical formula follows the chemical defined. Also, SI (System International d' Unites) units and spellings are used throughout.

Pigment terms have been classified by their Colour Index numbers (which immediately follow the pigment being defined) and pigment synonyms are extensively cross-referenced to a commonly accepted name.

After many definitions, there is a parenthetical reference such as ASTM, IUPAC, EPA, etc., to acknowledge their source. Descriptions of organizations, associations, and societies related to the coatings industry are featured in the *Coatings Encyclopedic Dictionary*, and addresses and phone numbers of many are included in the Appendix.

SAMPLE DEFINITIONS: — vocabulary entry

checking that phenomenon manifested in paint films by slight breaks in the film that do not penetrate to the underlying surface.[42,56,154] (ASTM) See COLD-CHECKING, and CHECKING RESISTANCE.[56,72] — source of definition; cross-reference to another entry

The break should be called a crack if the underlying surface is visible. Where precision is necessary in evaluating a paint film, checking may be described as visible (as seen with the naked eye) or as microscopic (as observed under a magnification of 10 diameters). — encyclopedic discussion

checking resistance the ability of a coating to resist slight breaks in its film that do not penetrate to the underlying surface.[56,72] cf. CHECKING[56,42,154] — cross-reference to another entry for comparison

The breaks should be called cracks if penetration extends through the underlying surface. See CRACKING RESISTANCE.[56,72] *Where precision is necessary in evaluating a coating film, checking may be described as visible (as seen by the naked eye) or as microscopic (as observed under minimum magnification of 10 diameters). The FSCT Pictorial Standards of Coating Defects contains photographic standards for the degree of checking of exterior paints.*

ASTM Test Method D660 has a different set of illustrated reference standards.

chelating agent (1) a sequestering or complexing agent that, in aqueous solution, renders a metallic ion inactive through the formation of an inner ring structure with the ion;[83] (2) ORGANIC COMPOUND that can remove many of the heavy metal cations from solution by forming soluble chelate compounds. Ethylenediamine tetraacetetic acid (EDTA) is typical.[83] Syn: sequestering agent, chelate — indicates word or phrase is also defined in this volume; indicates classification category, refers to listing in second part of *Dictionary*; synonyms

Chelating agents are also used to treat metal poisoning.

Philosophy of Consensus

If there was a governing philosophy for this effort, it was one of achieving consensus within the industry. Consensus dictated the choice of terms and their definitions. If they were ASTM, British Standards Institute (BSI), or International Union of Pure and Applied Chemistry (IUPAC) definitions, for example, they were already defined through consensus and, therefore, given preference and accepted intact. In some cases, a supplementary definition was added if it complemented the consensus definition.

Close liaison was maintained with groups such as the definitions subcommittee of the ASTM Committee D-1 on Paint and Coatings, the National Coil Coaters Association, the National Association of Corrosion Engineers, and particularly the FSCT's Inter-Society Color Council Committee. Most definitions were reviewed and approved by knowledgeable people in the industry, including representatives in Canada and Europe.

We are indebted to the many organizations who contributed to the preparation and production of this publication.

Classification System

To our knowledge, this is the first dictionary to contain classified definitions. Each definition is classified by a coded numerical superscript into one or more of 80 categories which, in turn, fall under nine hierarchies. The numbers appear as superscripts at

Table 1—Classification Categories and Numbers[a] for Definitions
(used as superscripts following each definition)

Category	No.	Category	No.
Additives	83	Monomers	141
Adhesion and Adhesives	79	Naval Stores	44
Analytical	37	Nomenclature	134
Appearance	35	Oils	63
Application, Methods of	77	Optical Properties	43
Architecture	82	Organizations	174
Art	81	Packaging	50
Biological	173	Paint Problems	42
Bituminous Products, Roofing	76	Physical Properties	142
Building Products, Materials	75	Pigments, Dyes	41
Catalysts	74	Plasticizers	170
Cementitious Products	73	Plastics	38
Chemicals; Chemistry	144	Polymerization	169
Coating Types	71	Polymers (Synthetic Resins)	130
Coil Coating	70	Printing Inks; Printing	168
Colloids	139	Processing, Process	131
Color	69	Regulatory	172
Corrosion	66	Resins (Natural)	167
Curing Chemistry, Cure	67	Resistance Properties	72
Decorating	148	Rheology	166
Degradation of Coatings	65	Safety and Toxicity	172
Dimensions, Scales, Values	128	Sealants, Caulks	165
Dispersion	64	Sealers, Fillers	147
Ecology, Pollution	171	Shellac	146
Elastomers	140	Solvents, Solvency, Thinners, Solubility	164
Electrical Properties	61	Specialty Coatings	163
Emulsions	60	Stability	161
Equipment, Apparatus, Machines	59	Statistical	36
Extenders, Fillers, Flatting Agents	58	Structure	133
Fabrication	57	Substrates	159
Film Defects	56	Surface Preparation and Pretreatments	158
Film Formation	55	Surfactants	127
Finishing Techniques	54	Test Methods	157
Furniture Finishing	175	Thermal Properties	176
Instruments	145	Varnishes	152
Lacquers	151	Vehicles	156
Manufacturing	49	Water-soluble Resins and Gums	155
Mechanical Properties	47	Waxes and Polishes	137
Metals, Metal Preparation	45	Weathering	154
Minerals	34	Wood, Wood Products	153

(a) The category numbers happen to be those on McBee Keysort cards originally used to complete and classify the definitions. They are in no significant order.

the end of each definition, and refer to the Categorical Listing of Terms which forms the second section of this volume.

Although this classification system was originally arbitrarily (and later empirically) developed, it classifies all the definitions into one or more of the categories listed in *Table* 1. The categories were further classified into the following nine hierarchies:

(1) *Related Products*—adhesives, elastomers, plastics, printing inks, etc.
(2) *Aesthetics*—architecture, art, color, decorating, finishing techniques, appearance.
(3) *Coating Constituents*—additives, pigments, resins, solvents, surfactants, minerals, etc.
(4) *Coating Problems*—corrosion, degradation, film defects, weathering, etc.
(5) *Coating Types*—lacquers, sealers, shellac, specialty coatings, varnishes, furniture finishes, etc.

(6) *Processing*—dispersion, polymerization, surface preparation, etc.
(7) *Regulatory*—ecology, pollution, safety, toxicity.
(8) *Test Methods and Properties*—mechanical properties, optical properties, physical properties, resistance properties, analytical, etc.
(9) *Miscellaneous*—dimensions, nomenclature, structure, substrates, biology, statistical, etc.

Examination of these groups and the categories reveals the broad scope of subjects in this volume. Further manipulation of these categories and groups might suggest a way of classifying the whole coatings industry.

Use of the Classification System

As mentioned, all the terms defined, along with synonyms and related terms, are included in one or more of the categorized listings. Keywords for articles in journals or reports, etc., can be conveniently chosen from these listings.

Table 2—Methods of Application, Category 77

aerosol coating
air brush
air spray
airless spraying
anionic electrodeposition
application
application rate
applicator
autodeposition
automatic electrostatic spray application
barrel finishing
barrel polish
barrelling
bell
bird applicator
brayer
brush
brushing
casting
cathodic electrodeposition
cloth distressing
coating powders
coating, dip
coating, spray
cobwebbing
cocooning
coil coating
cold flow
crossing
curtain coating
dabber
daub
deposition
dip coating
dipcoat
dipping
disk
distressing
doctor blade
doctor roll
drawdown
drawdown bar (blade)
drawout
EDP
electrocoating
electrodeposition
electrostatic spraying
Elpo
equalizer rod
fadding
film applicator
flame spray
flat screen printer
flock finish
flocking
flocking deposition
flow coating
fluidized bed coating
fresco
gilding
gouache
hand-held electrostatic spraying
hand-screening
hot spraying
hydraulic spraying
internal mix spray gun
knife coater
knife mark
lifting
lithography
Meier bar
metallizing
mini-bell
mitt
mohair paint roller cover
mopping
multinotch applicator
natural spreading rate
offset
padding
paint
paint brush
paint harling
paint mitt
paint pad
paint roller
painter's mitt
picking
plasma coating
plasma deposition technique
plasma flame
powder coating
rag rolled finish
rag rolling
ragging
reverse roll coater
roll coating
roller
roller application
roller coating
roller coating enamel
rotary screen printer
rotational electrostatic spray
roto-dipping
rumbling
screen painting
silk-screen printing
silk-screening
slushing
spin coating
spinning
sponging
sponging off
sponging on
spray coating
spraying
spreading rate
squeegee coat
stencilling
stipple
stippling
strip coating
thermospray
throwing power
transfer coating
tumbling
two-component gun
whirling
wire-bar applicator
wire-wound rod
wrap

In anticipation of literature searches, definitions were classified so that the reader could retrieve a vocabulary and a glossary for any of the categories. As an example, Methods of Application (category 77) could be retrieved and the interested party could study the definitions and encyclopedic sections and become an "instant expert" on this narrow subject area. *Table* 2 lists the surprisingly extensive vocabulary for this subject.

Pigment Classification

Another type of classification used in this dictionary is the Colour Index number following each common name of the defined pigment. This coding of pigments is patterned after the Colour Index (CI) system[2] established under the joint sponsorship of the American Association of Textile Chemists and Colorists (AATCC) in the United States and the Society of Dyers and Colourists in Great Britain. The system comprises a two-part code; the first part identifies the actual pigment (or colorant) and the second part (as applicable) identifies the method of preparation (or chemical nature). The Colour Index name consists of the general category, hue, and an identifying number (from a consecutive order). For example, the generic name for titanium dioxide would be Pig-

Table 3—Colour Index System for Coatings Pigments[2]

Example: Titanium Dioxide Rutile = C.I. Pigment White 6 (77891)

C.I. Generic Name — C.I. Pigment White 6: **General Category** (Pigment), **Hue** (White), **Consecutive Number** (6)

C.I. Constitution Number — (77891): **Chemical Class**

General Category	Hue	Chemical Class	
acid dyes	blue	nitroso	10000-10299
azoic components	black	nitro	10300-10999
and composition	brown	azo	
basic dyes	green	mono	11000-19999
developers	metal	ci	20000-29999
direct dyes	orange	tri	30000-34999
disperse dyes	violet	poly	35000-36999
fluorescent	red	azoic	37000-39999
brighteners	white	stilbene	40000-40999
food dyes	yellow	methane	
ingrain dyes		diphenyl	41000-41999
mordant dyes		triaryl	42000-44999
natural dyes		xanthene	45000-45999
oxidation bases		acridine	46000-46999
pigments		quinoline	47000-47999
reactive dyes		methine	48000-48999
reducing agents		thiazole	49000-49399
solvent dyes		indamine	
sulfur dyes-		and indophenol	49400-49999
condense sulfur dyes		azine	50000-50999
vat dyes		oxazine	51000-51999
		thiazine	52000-52999
		sulfur	53000-54999
		lactone	55000-55999
		ketone	
		amino	56000-56999
		hydroxy	57000-57999
		anthraquinone	58000-72999
		indigoid	73000-73999
		phthalocyanine	74000-74999
		natural organic	75000-75999
		oxidation bases	76000-76999
		inorganic colorants	77000-77999

STANLEY LE SOTA received his B.S. Degree in Chemistry from the City College of New York (CCNY). Associated with the coatings industry for over 50 years, he has served as Coatings Chemist for 32 years with Rohm and Haas Co.; 11 years with the Dutch Boy Division of NL Industries, and 4 years with the Mimex Company, a small pioneer in latex paint development. Since retiring from the Rohm and Haas Co., he has been a coatings consultant. He is Past- President of the Philadelphia Society for Coatings Technology and recipient of its Liberty Bell Award. For over 20 years, he has been serving as Definition Committee Chairman of both the Federation and ASTM D-1. Mr. LeSota has received many of the major awards from both Societies.

ment White 6. The Colour Index number is a five-digit number, 77891, which describes its chemical composition—an inorganic colorant (see *Table* 3).

Most of the pigment names are synonyms for a commonly accepted name as used in Volume I of the *Pigment Handbook*.[3] Typically, a pigment definition contains the common name; the chemical formula; the Colour Index number; a brief description of the pigment; a category superscript (pigment,[41] in the case below); synonyms; a cf. (compare) reference; sometimes a See_____ reference that points the reader to other significant terms; and frequently, as in this case, an encyclopedic section, which contains significant information about the pigment. An example follows:

> **titanium dioxide, rutile** TiO_2. Pigment White 6 (77891). A high-opacity, bright white pigment, nonchalking type, used as a prime pigment in paints, rubber, plastics.[41] cf. titanium oxide, anatase.[41] Syn: titania
>
> *Prepared from the mineral, ilmenite, or rutile ore. Properties: density, 3.9-4.2 g/cm³ (33-35 lb/gal); O.A., 16-48; particle size, 0.2-0.3 µm; refractive index, 2.76. Titanium dioxide in its rutile crystalline form has an exceptionally high refractive index and its opacity and tinting strength, when finely dispersed, exceed all other white pigments. It is the most widely used white pigment in the paint and coatings industry.*
>
> *In addition to giving opacity and whiteness to coatings, titanium dioxide absorbs or reflects harmful radiation, thereby protecting the substrate. In its pure form it can also act as a photoactive catalyst causing degradation of the binders, such as alkyds, and eventual loss of pigment from the surface—one of the mechanisms of chalking. To prevent or reduce this, titanium oxide particles are coated with, for example, alumina and silica. The coating of titanium dioxide is a highly developed process and has resulted in numerous grades intended for specific uses.*

Both the *Pigment Handbook*[3] and BSI's *Synonyms for Pigment Names*[4] served as sources of pigment synonyms—and there are many. For example, cobalt blue has 14. A considerable number of entries in this volume are synonyms for pigments.

Regulatory Terms and Safety Caveats

Safety was given a high priority in the *Coatings Encyclopedic Dictionary* and regulatory terms are comprehensively covered.[5-7] There are some safety caveats included in the encyclopedic sections. We strongly urge the readers to refer to the appropriate MSDS before using any chemical, pigment, or material listed in this volume.

Definitions of many toxic materials are presented in the past tense, and are included in this work for historic reference only. Although some materials may be obsolete or considered archaic to some of our more sophisticated readers, they are listed in the effort to present the most comprehensive reference source possible. Some of the raw materials listed that are no longer used in the United States may be indigenous to other countries.

Future Editions

Compiling a dictionary can be an endless task of polishing and refining definitions, searching for missing terms, making the entries more encyclopedic—as witness our 13-year stint on the original *Paint/Coatings Dictionary* and an additional three years on this current volume. This publication is still only a beginning. We believe that this should be a continuing effort—a living document. We ended the original *Paint/Coatings Dictionary* introduction with the phrase "The Beginning"; we end this one with . . ."To Be Continued."

Acknowledgment

We gratefully acknowledge the work done by the following:

Members of the Federation staff for their cooperation, encouragement, and assistance in editing and proofreading.

Leonard Schaeffer of the Leneta Company for his excellent "Condensed Index of ASTM Committee D-1 Standards" which we used for a checklist and source of ASTM D-1 Test Methods, as well as his help with the controversial "gloss" definition—about 20 years in the making.

— *Stanley LeSota*
Coatings Consultant,
720 Inverness Drive,
Horsham, PA 19044

References

(1) *Paint/Coatings Dictionary,* Federation of Societies for Coatings Technology, Philadelphia, PA, 1978.

(2) *Colour Index,* 3rd Ed., 5 volumes and revisions, The Society of Dyers and Colourists with the American Association of Textile Chemists and Colorants, Lund Humphries, London, 1971-75.

(3) *Pigment Handbook,* Patton, T.C. (Ed.), Vol. I, John Wiley and Sons, Inc., New York, 1973.

(4) *Synonyms for Pigment Names,* British Standards Institution, British Standards House, London, 1957.

(5) *Glossary for Air Pollution of Industrial Coatings Operation,* 2nd Ed., EPA-450/3-83-013 R, U.S. Environmental Protection Agency, N.C., Dec. 1983.

(6) Accrocco, J.O., *MSDS Pocket Dictionary,* Genium Publishing Corp., Schenectady, NY, 1990.

(7) Calvert, J.G., *IUPAC Glossary of Atmospheric Chemistry Terms,* Atmospheric Chemistry Division, Natural Center for Atmospheric Research, Boulder, CO, 1990.

Bibliography

Accrocco, J.O., *MSDS Pocket Dictionary,* Genium Publishing Corp., Schenectady, NY, 1990.

Artists' and Illustrators' Encyclopedia, Quick, J. (Ed.), McGraw-Hill Book Co., New York, 1969.

Calvert, J.G., *IUPAC Glossary of Atmospheric Chemistry Terms,* Atmospheric Chemistry Division, Natural Center for Atmospheric Research, Boulder, CO, 1990.

Colour Index, 3rd Ed., 5 volumes and revisions, The Society of Dyers and Colourists with the American Association of Textile Chemists and Colorants, Lund Humphries, London, 1971-75.

Compilation of ASTM Standard Definitions, American Society for Testing and Materials, Philadelphia, PA, 1976.

Concise Chemical and Technical Dictionary, 2nd Ed., Bennett, H. (Ed.), Chemical Publishing Co., Inc., New York, 1962.

Definitions of Terms Appearing in ISO Recommendations for Oils and Pigments (in English and French), International Organization for Standardization (ISO), Geneva, Switzerland, 1963.

Dictionaire du Batiment/Building Terms Dictionary, LeFebure, M. (Ed.), Les Editions Lemeac, Ottawa, Canada, 1965.

Dictionary of Architecture and Building Trade in Four Languages: Eng., Ger., Pol., Russ., Zboinski, A. and Tyszynski, L. (Eds.), Pergamon Press, Inc., New York, 1963.

Dictionary of Architecture and Construction, Harris, C.M. (Ed.), McGraw-Hill Book Co., New York, 1975.

A Dictionary of Art Terms & Techniques, Mayer, R. and Crowell, Thomas Y., (Eds.) New York, 1975.

A Dictionary of Building, Scott, J.S. (Ed.), Penguin Books Inc., Baltimore, MD, 1964.

Dictionary of Painting and Decorating Trade Terms, Goodier, J.M. (Ed.), Southerland Publishing Co., Southport, England, 1961.

Dictionary of Paint, Varnish and Lacquer Terms (English, French, German and Dutch), Raaff, J.J., Language Service, Inc., New York, 1967.

Dictionary of Science, Uvarov, E.B. and Chapman, D.R. (Ed.), Penguin Books, London, 1971.

Dictionary of the Varnish and Paint Industry (in Dutch, English, French, German and Italian), Merz, O., Wissenschaftliche Verlagsgesellschaft, Stuttgart, Germany, 1954.

Elsevier's Dictionary of the Printing and Allied Industries in Four Languages, F.J.M. Wijnekur Research Institute for the Graphic and Allied Industries, TNO, Amsterdam, Elsevier Publishing Co., New York, 1967.

Encyclopedia for the User of Petroleum Products, Humble Oil and Refining Co., Baltimore, MD, 1968.

Encyclopedia of Associations, Vol. I, National Organization of the U.S., Fisk, M. (Ed.), Gale Research Co., Detroit, MI, 1976.

An Encyclopedia of the Chemical Process Industries, Stewart, J.R. (Ed.), Chemical Publishing Co., Inc., New York, 1956.

Encyclopedia of Polymer Science and Technology: Resins, Plastics, Rubbers and Fibers, Wiley-Interscience, New York.

English/French Paints and Coatings Vocabulary, L'Association Québec Coise des Industries de la Peinture, Montreal, Canada, 1977.

Five Language Dictionary of Surface Coatings, Platings, Product Finishing, Plastics and Rubber, Santholzer, R.W. (Ed.), Pergamon Press, Inc., Elmsford, NY, 1969.

Glossary for Air Pollution of Industrial Coatings Operation, 2nd Ed., EPA-450/3-83-013 R, U.S. Environmental Protection Agency, N.C., Dec. 1983.

Glossary of Paint Terms, BS 2015, British Standards Institution, London, 1965.

Glossary of Paint, Varnish, Lacquer and Allied Terms, Martin, R.C., American Paint Journal Co., St. Louis, MO, 1937.

Glossary of Paint and Varnish Terms in Swedish, English, French, German and Finnish, Tekniska Nomenklaturcentralen, Stockholm, Sweden, 1967.

A Glossary of Pigments, Varnish and Lacquer Constituents, Martin, J.M. and Morgans, W.M., Chemical Publishing Co., Inc., New York, 1959.

"Glossary of Terms Used in Maintenance Painting," NACE Technical Committee Report, Materials Protection, 73-80, January 1965.

Glossary of Terms Used in the Paint, Varnish, and Allied Trades, Chatfield, H.W. (Ed.), Scott Greenwood and Son, Ltd., London, 1951.

Graystone, Jon, *The Care and Protection of Wood,* Published by ICI Paints Division, Slough SL2 5DS, England, 1985.

Guide to Pigments and to Varnish and Lacquer Constituents, Martin, J.H. and Morgans, W.M. (Eds.), L. Hill, London, 1954.

Kirk-Othmer Encyclopedia of Chemical Technology, Wiley-Interscience, New York.

Modern Plastics Encyclopedia, McGraw-Hill Book Co., New York.

The National Paint Dictionary, Stewart, J.R., Stewart Research Laboratory, Washington, D.C., 1948.

Paint/Coatings Dictionary, Federation of Societies for Coatings Technology, Philadelphia, PA, 1978.

The Paint Dictionary (in English, German and Japanese), Matsumoto, T. (Ed.), Gihodo, Tokyo, Japan, 1956.

Paint Dictionary in Four Languages, Van Goor Zonen (Ed.), The Hague, The Netherlands, 1968.

The Penguin Dictionary of Architecture, Fleming, J., Honour, H. and Pevsner, N. (Eds.), Penguin Books, Middlesex, England, 1966.

The Pocket Dictionary of Art Terms, Levy, M. (Ed.), N.Y. Graphic Society, Greenwich, CT, 1961.

Paint Materials, A Short Encyclopedia, Gettens, R.J. and Stout, G.L. (Eds.), Dover Publications, Inc., New York, 1966.

Pigment Handbook, Patton, T.C. (Ed.), Vol. I, John Wiley and Sons, Inc., New York, 1973.

Stewart's Scientific Dictionary, Stewart, J.R., Stewart Research Laboratories, Alexandria, VA, 1953.

Synonyms for Pigment Names, British Standards Institution, British Standards House, London, 1957.

Thesaurus of Paint and Allied Technology, Federation of Societies for Paint Technology, Philadelphia, PA, 1968.

Webster's Third New International Dictionary, G & C Merriam Co., Springfield, MA, 1963.

A

Å see ANGSTROM UNIT.[128,43]

a.i. abbreviation for ACTIVE INGREDIENT.

"a," Kubelka-Munk equation mathematical constant characteristic of a COLOR at complete hiding; dependent on the optical constants K and S: $a = \frac{1}{2}[1/R_\infty + R_\infty] = 1 + K/S$.[69,43]

"a" or "α" redness-greenness coordinate in certain transformed COLOR spaces, generally used as the Δa, or difference in "a" between a specimen and a standard reference color. If "a" or Δa is plus, there is more redness than greenness; if "a" or Δa is minus, there is more greenness than redness. It is normally used with b or ß as part of the chromaticity or chromaticity difference.[69] See UNIFORM CHROMATICITY COORDINATES.[69,46]

A-acid $NH_2C_6H_4$ COOH. Trade abbreviation for ANTHRANILIC ACID, an INTERMEDIATE used in the manufacture of the pigment, Lake Red D.[41,144]

AATCC abbreviation for the American Association of Textile Chemists and Colorists.[174]

abat-jour any beveled aperture, or a skylight, in a wall or in a roof to admit light from above.[82] (DAC)

abat-vent LOUVERS which are placed in an exterior wall opening to permit light and air to enter but which break the wind.[82] (DAC)

abatement action taken to reduce air POLLUTION which involves the use of control equipment or some new PROCESS.[172] (IUPAC)

This refers to a reduction or lessening as opposed to elimination of a type of discharge or pollutant.

Abbe Refractometer common form of REFRACTOMETER used for determining the REFRACTIVE INDEX of OILS and other liquids, or of grease-like products which are capable of liquefaction at moderate temperatures.[43,59,145]

Good accuracy is attainable in the range of 1.3 to 1.7, readings being given to the fourth decimal place. The prisms, which constitute the most important part of the instrument, and hence the liquid held between their faces, are capable of being maintained accurately at the temperature of the determination. With the use of special liquids to form an optical seal to the prisms and a special technique of viewing, it is also used for determining the refractive index of solids such as plastics cast in sheets with polished surfaces and edges.

abbozzo underpainting of an oil painting, either in MONOCHROME or COLOR.[69,81]

Sometimes called BOZZO or DEADCOLORING.

abhesive material that resists adhesion; applied to surfaces to prevent sticking, HEAT-SEALING, etc.[79,158]

abietates ESTERS or salts of ABIETIC ACID, a principal constituent of ordinary ROSIN from which the products of commerce are derived, no attempt being made to separate abietic acid from the other acids which rosin is likely to contain.[167,170,44] See ESTER GUM as an example of the GLYCEROL ESTER.[44,130]

Metallic abietates, as such, are rarely encountered under this name but generally as resinates. Esters of rosin, however, are commonly described as abietates and not as resinates. The ester abietates which have enjoyed some popularity are the methyl, ethyl, and benzyl derivatives. They are soft, resinous materials and are used chiefly as plasticizers.

abietic acid, commercial grade $C_{20}H_{30}O_2$. Product consisting chiefly of ROSIN ACIDS in substantially pure form, separated from rosin or TALL OIL commercially for specific purposes and in which ABIETIC acid and its ISOMERS are the principal components.[44,144] (ASTM) Syn: sylvic acid

ablation a self-regulating heat and mass transfer PROCESS in which incident THERMAL energy is expended by sacrificial loss of material.[65,72,176] (ASTM)

ablative coatings thick, mastic-like materials which absorb heat; they are designed to char and sacrifice themselves while protecting the metal substrate underneath.[71,163,176] See ABLATIVE PLASTIC.[38]

Used for missiles and re-entry rockets.

ablative plastic material which absorbs heat (while part of it is being consumed by heat) through a DECOMPOSITION PROCESS (pyrolysis) which takes place near the surface exposed to the heat.[38,176]

abrasiometer one of the many devices used to test ABRASION of a coating by using an air blast to drive an ABRASIVE against the test film, or by rotating a film submerged in an ABRASIVE, or by simply dropping a stream of abrasive onto the film.[47,145,157]

abrasion wearing away of a surface in service by action such as RUBBING, SCRAPING, or EROSION.[65]

Words presented in CAP/SMALL CAPS type indicate that the word is defined in another part of the *Coatings Encyclopedic Dictionary*.

Numerical superscripts classify terms in one or more of the categories listed in the second section of this volume.

abrasion coefficient method for reporting the result of an ABRASION TEST using the FALLING SAND ABRASION TESTER, in which it is assumed that the ABRASION RESISTANCE is proportional to the film thickness.

$$\text{Abrasion Coefficient} = \frac{(W_1 - W_2)}{T}$$

where:

W_1 = grams of abrasive and holder before test,
W_2 = grams of abrasive and holder after test, and
T = thickness of coating in mils (0.001 in.) (0.025 mm).[47,157]

abrasion cycle number of abrading revolutions to which a specimen is subjected in an abrasion test, as that made with a TABER ABRASER.[47,157]

abrasion resistance the ability of a coating to resist being worn away and to maintain its original APPEARANCE and structure as when subjected to RUBBING, SCRAPING, or EROSION.[56,72,47,157] (ASTM) See ABRASION,[65] WET ABRASION RESISTANCE,[72,157] SCRUB RESISTANCE.[72,157]

ASTM test methods for abrasion resistance include: air blast abrasive test, D658; falling sand/abrasive test, D968; Taber Abraser test, D4060; wear resistance, of traffic paint, D913; wet abrasion/scrubbing, D2486, D4213.

abrasive any material which, by a PROCESS of GRINDING down, tends to make a surface smooth or rough.[54,131,158]

abraum a red OCHER used to STAIN mahogany.[41,54] (DAC)

abridged spectrophotometer instrument which measures SPECTRAL TRANSMITTANCE or REFLECTANCE at a limited number of WAVELENGTHS, usually employing filters rather than a MONOCHROMATOR.[43,69,145] See FILTER SPECTROPHOTOMETER.[43,145]

ABS see ACRYLONITRILE-BUTADIENE-STYRENE.[38]

absolute adjective used to describe measurements in terms of fundamentally defined units.[134,128]

absolute alcohol pure anhydrous ETHYL ALCOHOL (ethanol).[144,164]

The term is used to distinguish it from the several varieties of alcohol which are available, and which contain varying amounts of water and/or other impurities.

absolute humidity see HUMIDITY, ABSOLUTE.[128]

absolute reflectance REFLECTANCE measured relative to the PERFECT DIFFUSER.[43,69]

absolute temperature see KELVIN TEMPERATURE SCALE.[128]

absolute viscosity (1) tangential force on unit area of either of two parallel planes at unit distance apart, when the space between the planes is filled with fluid (in question) and one of the planes moves with unit velocity in its own plane relative to the other; (2) FORCE required to move in opposite directions at a velocity of 1 m per s, two parallel planes of liquid, 1 m^2 in area and separated from each other by a distance of 1 m. The absolute viscosity is designated by the Greek letter η.[166,128] See POISE, VISCOSITY.[166]

absorbance the logarithm (must be specified as to base 10, lg, or base e, ln) of the reciprocal of transmittance [ln (I_o/I_t) or lg (I_o/I_t)], where I_o and I_t are the MONOCHROMATIC radiances (intensities) of light incident on and transmitted through, respectively, a sample which is usually contained in a sample cell.[43,69] (IUPAC) See BEER-BOUGUER LAW.[43,69]

absorber a device used commonly for sampling by ABSORPTION in which a gaseous or liquid material is removed from another gas or liquid by selection absorption.[59] (IUPAC)

These devices include scrubber, impinger, packed column, and spray chamber.

absorption the PROCESS of one material (ABSORBENT) being retained by another (ABSORBATE); this may be the physical solution of a gas, liquid, or solid in a liquid, attachment of molecules of a gas, vapor, liquid, or dissolved substance to a solid surface by physical forces, etc. The transfer of a component from one phase to another. (IUPAC) cf. ADSORPTION and SORPTION[142,43]

In SPECTROPHOTOMETRY, absorption of light at characteristic wavelengths is used to identify the chemical nature of molecules, atoms, or ions and to measure the concentrations of the species. (IUPAC)

absorption coefficient ABSORPTION of RADIANT ENERGY for a unit concentration through a unit pathlength for a specified wavelength and ANGLE OF INCIDENCE and viewing.[43,69,128] See BEER-BOUGUER LAW, KUBELKA-MUNK THEORY, MIE THEORY.[43,69]

absorption hygrometer any one of several types of HYGROMETERS containing a HYGROSCOPIC substance, the change in length, thickness, or mass of which is a measurable index of the humidity of the atmosphere.[59,145,157]

absorption, oil the amount of VEGETABLE DRYING OIL required to convert a given mass or volume of the dry powder (PIGMENT or EXTENDER) to a very stiff PUTTY-like paste, which does not break or separate.[41,63,157] See OIL ABSORPTION (VALUE).[63,157,64]

It is more usually expressed as the pounds of refined LINSEED OIL required for 100 pounds of pigment or g/100 g.

absorption spectrophotometry measurement of the amount of RADIANT ENERGY absorbed as a function of WAVELENGTH or frequency.[43,157]

ULTRAVIOLET radiant energy can be employed as the source of incident radiant energy, and it has been found that certain groupings of atoms in ORGANIC compounds influence the intensity and location of the absorption bands in the ultimate spectrum. A consideration of absorption SPECTRA of DRYING oils and their fatty acids has been used with great success in the elucidation, confirmation, and rectification of their chemical structures.

absorption tinting strength see TINTING STRENGTH, ABSORPTION.[69,41,43]

absorptivity, a, n the ABSORBANCE divided by the product of the concentration c of a substance and the sample optical pathlength b, a = A/bc. The units of b and c should be specified.[43] (ASTM)

ASTM Test Method E131 gives more details.

abstract art art in which the elements of FORM have been stressed in handling the subject matter—which may or may not be recognizable.[81]

Abstraction is a relative term; all art works exist on a continuum between total abstraction and full representation. Vassily Kandinsky is generally credited with having created the first purely abstract art form in 1910.

abut to adjoin at an end; to be contiguous.[82] (DAC)

acacia gum water-soluble GUM obtained from trees of the acacia species, as an EXUDATION from incisions in the bark.[79,155,83] Syn: gum arabic

It is water-soluble and is used as an ADHESIVE, THICKENING AGENT, and for TRANSPARENT paints.

academy board a board which is given a surface in preparation for painting, primarily oil painting.[81,159]

It is made of PAPER containing chalk and size and has a face of pale gray or white ground, oil, and chalk mixture.

ACB abbreviation for ASBESTOS-CEMENT BOARD.[73,75,159] (DAC)

accelerated aging any set of conditions designed to produce in a short time the results obtained under normal conditions of aging.[157]

In accelerated aging tests, the usual factors considered are heat, light, or oxygen, either separately or combined. Sometimes called ACCELERATED LIFE.

accelerated life see ACCELERATED AGING.[157]

accelerated testing tests, the conditions of which simulate those encountered in practice, but which have been accentuated artificially to provide performance results in shorter periods of time.[65,154,157]

Coatings do not necessarily behave under such tests exactly as they will under actual conditions, but any coating which gives good performance under test may be shown to have possibilities which are worthy of further considerations and experiments.

accelerated weathering machine-made tests designed to simulate, but at the same time to intensify and accelerate, the destructive action of natural outdoor WEATHERING on coatings films.[65,72,154,157] See WEATHER-OMETER[65,145,154,159] and QUV.[125]

The tests involve exposure to artificially produced components of natural weather, e.g., light, heat, cold, water vapor, rain, etc., which are arranged and repeated in a given cycle. There is no universally accepted test, and different investigators use different cycles.

accelerated weathering machine (or apparatus) device intended to accelerate the DETERIORATION of coatings by exposing them to controlled sources of RADIANT ENERGY, heat, water, or other factors that may be introduced.[154,65,157] See WEATHER-OMETER and ACCELERATED WEATHERING.[154,65,157,57,145]

accelerator (1) any substance used in small proportion which increases the speed of a chemical reaction. In the paint industry, the term usually indicates materials that hasten the CURING or CROSSLINKING of a RESIN system. In the POLYESTER RESIN field, it covers more specifically an ADDITIVE which accelerates the action of the CATALYST. See CROSSLINKING AGENT and ACTIVATOR;[67,83,74] (2) an ORGANIC or INORGANIC chemical which hastens the VULCANIZATION of RUBBER, natural or synthetic, causing it to take place in a shorter time or at a lower temperature. Accelerators, particularly ORGANIC, are not mere catalysts of vulcanization, however, because they produce different and generally beneficial states of cure and different degrees of STABILITY or resistance to chemical attack in the vulcanization.[83,67,140,74]

accra natural COPAL RESIN of African origin.[167]

accretion the PROCESS by which AEROSOLS grow in size by external addition of various chemical species; a form of AGGLOMERATION.[64] (IUPAC)

accroides resinous accumulation which occurs on the leaf and stem of the *Xanthorrhoea* species.[167]

It is native to Australia and Tasmania. It appears on the market in red and yellow forms, both of which are soluble in industrial ALCOHOL, and are used in spirit varnishes. The resin is also described as "black boy gum," "botany bay resin," "grass tree gum," "gum accaroid," "accaroid resin," "red gum," and "yacca" or "yacka" gum.

accumulator series of rolls which FESTOON strip metal on a continuous line both at the beginning and at the end.[59,70]

Words presented in CAP/SMALL CAPS type indicate that the word is defined in another part of the *Coatings Encyclopedic Dictionary.*

Numerical superscripts classify terms in one or more of the categories listed in the second section of this volume.

This allows the beginning or the end of the line to stop while the rest of the line is in operation. The accumulator actually accumulates a considerable length of strip, and gives a portion of its strip to whichever end is stopped.

accuracy the closeness of agreement between the true value and the measured value.[157,36] (IUPAC) cf. PRECISION[36]

Agreement between a single measured value, C_a, and the value of air quality characteristics, C_a, itself, or the accepted reference value; accuracy denotes extent to which bias and random errors are absent. (ISO)

acenaphthene $C_{10}H_6(CH_2)_2$. Solid with a melting point of 95°C, obtained from COAL TAR.[41,144]

It is a DYESTUFF INTERMEDIATE.

acetal $CH_3CH(OC_2H_5)_2$. (1) A colorless, FLAMMABLE liquid used in cosmetics and as a SOLVENT;[164] (2) any of a class of compounds formed from ALDEHYDES combined with ALCOHOL.[134]

acetal resins high MOLECULAR WEIGHT, stable, linear POLYMERS of FORMALDEHYDE; structurally, an oxygen atom joins the repeating units in an ether rather than ester-type link.[130]

These also include BUTYRALS.

acetaldehyde CH_3CHO. Low boiling liquid (21°C).[144]

acetaldehyde resin product of auto-CONDENSATION of ACETALDEHYDE.[130]

acetamide CH_3COHN_2. AMIDE of ACETIC ACID with a melting point of 81°C and boiling point of 222°C.[144,170]

Generally used as a PLASTICIZER for CELLULOSE ESTERS.

acetanilide $C_6H_5NHCOCH_3$. Its physical properties include: mp of 115°C; bp of 304°C; sp gr of 1.21; and flp of 165°C (329°F).[144] Syn: phenyl acetamide

acetate chromes LEAD CHROMATE PIGMENTS prepared from LEAD ACETATE or BASIC LEAD ACETATE.[41]

Were available as the lemon, primrose, medium, and orange shades.

acetate dope CELLULOSE ACETATE LACQUERS used for coating aircraft fabrics.[151]

acetate green range of BRUNSWICK or CHROME GREENS derived from mixtures of ACETATE CHROME and PRUSSIAN BLUE.[41]

These can be made wet or dry by mixing. The type made by wet mixing is preferred for paint making, as these greens have less tendency to partial separation in the film, a phenomenon known as "FLOATING."

acetates (1) metallic salts derived from ACETIC ACID, by interaction of the METALLIC OXIDE, hydroxide, or carbonate with the acid, or the esters derived by interaction of ALCOHOLS with acetic acid; (2) also used colloquially for CELLULOSE ACETATE PLASTICS.[144,38]

Typical metallic salts are lead, cobalt, and manganese acetates. Common esters are ethyl, propyl, isopropyl, butyl, and amyl acetates. Acetate salts have the formula CH_3COOMe, where Me is a monovalent metal. Divalent metals like lead, etc., obviously combine with two acid radicals. The formula given above for the acetate salts applies also to the esters, except that the Me becomes an alkyl radical.

acetic acid CH_3COOH. Monobasic acid.[144]

Its mp is 16°C and bp, 118°C. In the paint industry, its chief applications are in the manufacture of metallic acetates used for the production of DRIERS, and in the manufacture of acetate esters employed as SOLVENTS or PLASTICIZERS.

acetic anhydride $CH_3CO–O–OCCH_3$. Important REAGENT which has wide application in the manufacture of many raw materials and INTERMEDIATES for the paint trade, and also in analysis.[144]

It is used for the acetylation of hydroxyl groups as in the manufacture of acetyl ricinoleates, and in the manufacture of CELLULOSE acetate. It has a bp of 137°C.

acetic ester and acetic ether See ETHYL ACETATE.[144,164]

acetocopal product obtained when CONGO COPAL is reacted with ACETIC ANHYDRIDE.[130]

acetone CH_3COCH_3. Typical low-boiling KETONE. It is a liquid which flashes at ordinary room temperature, has a bp of 57.5°C, flp of -15°C (closed cup), and sp gr of 0.788 at 25°C.[164] Syn: dimethyl ketone or propanone

ASTM Test Method D1363 covers the detection in acetone and methanol of the presence of impurities that reduce potassium permanganate.

acetone resin SYNTHETIC RESIN formed by the POLYCONDENSATION of ACETONE with another compound such as PHENOL or FORMALDEHYDE.[130]

acetonyl acetone $CH_3COCH_2CH_2COCH_3$. SOLVENT containing two keto groups in each molecule.[164,144]

It has a boiling range of 188-193°C; sp gr of 0.973/15°C; refractive index of 1.449; flp of 85°C (185°F); and vapor pressure of less than 2 mm Hg at 30°C.

acetophenone $CH_3COC_6H_5$. SOLVENT with a bp of 202°C; mp of 20°C; sp gr of 1.023/25°C; flp of 83°C (180°F); and refractive index of 1.536.[164,144] Syn: phenyl methyl ketone or acetyl benzene

acetyl monovalent RADICAL CH_3CO-.[133]

acetyl benzene see ACETOPHENONE.[164,144]

acetyl number (value) number of milligrams of potassium hydroxide required to neutralize the ACETIC ACID set free from 1 g of an acetylated compound when the latter is subjected to HYDROLYSIS.[157,128]

acetyl ricinoleates PLASTICIZERS, such as butyl and ethyl acetyl ricinoleates.[170]

They can be regarded as esters of acetylated RICINOLEIC ACID. The hydroxy group in the ricinoleic chain has been acetylated or esterified, and the carboxylic group has also undergone esterification, but with an ALCOHOL.

acetylated congo see ACETOCOPAL.[167]

acetylated damar product obtained when DAMAR is reacted with ACETIC ANHYDRIDE.[167]

acetylation reaction wherein the HYDROGEN ATOM of a HYDROXYL GROUP is replaced by an acetyl RADICAL (CH_3CO). It can really be regarded as an ester formation, except that it is a specific ester, the ACETATE, which is formed.[131]

When the acetyl value of castor or other oils containing free hydroxyl groups, or of the reaction mixture of monoglycerides, or alkyds, is determined, ACETIC ANHYDRIDE is commonly used as the acetylating agent. This reacts with any free hydroxyl groups, whether they occur in the fatty acid chains of the vegetable oil or in unreacted polyhydric alcohols.

acetylene black particularly pure form of CARBON BLACK PIGMENT made by the controlled combustion of acetylene in air under pressure. It is graphitic in nature and has high electrical conductivity.[41]

ACGIH abbreviation for American Conference of Governmental Industrial Hygienists.[174]

achromatic having no distinguishable HUE; neutral. See GRAY SCALE.[69,43]

ACI abbreviation for American Concrete Institute.[73,174]

acicular needle-shaped.[133,41]

It is applied chiefly to describe the shape of pigment particles which are long and slender. To be acicular, the length of the particle must be at least three times the width.

acid a species which reacts in liquid water to generate hydrogen ions (conventionally represented as cations (H^+ or hydronium H_3O^+); anions (e.g., sulfate, SO_4^{2-}; nitrate, NO_3^-) which were associated with the H^+ in the acid are also released.[144,172] (IUPAC) See CARBOXYLIC ACID.[144]

acid catalysts ACIDS which may be either ORGANIC or INORGANIC, or SALTS from these acids which exhibit acidic characteristics. They are used to promote or ACCELERATE CHEMICAL REACTIONS, and find special applications in the manufacture and subsequent HARDENING of SYNTHETIC RESINS.[67,74,83,169]

Acid CATALYSTS have been employed in the manufacture of POLYMERIZED DRYING OILS, COUMARONE, UREA, PHENOL- and MELAMINE-FORMALDEHYDE RESINS, and in the cold-setting of compositions containing the last three named resins.

acid curing (hardening) (1) PROCESS of CURING or HARDENING RESINS through the use of ACID CATALYSTS. These are frequently employed with UREA and MELAMINE-FORMALDEHYDE RESINS;[67] (2) acid cure through use of sulfur monochloride in liquid or vapor form. This is a surface effect only suitable for thin articles such as fabric coatings or sheet gum, like dental dam, bathing caps, and similar articles. Often called a COLD CURE, since no heat beyond that required to vaporize the sulfur chloride is necessary.[140,169,67]

acid deposition the ACIDS deposited by transfer from the atmosphere either by precipitation (rain, fog, snow, etc.), called WET DEPOSITION, or by direct transfer to the surface of the earth induced by the flow of acid-containing air masses over the earth's surface, called DRY DEPOSITION.[172] (IUPAC)

acid dyes DYESTUFFS that possess acidic groupings, e.g., carboxy, sulfonic acid, or both. They form salts with heavy metals like barium and calcium, and this reaction is used in LAKE formation.[41]

It is usually an azo, triarylmethane, or anthraquinone DYE with acid substituents such as nitro, carboxy, or sulfonic acid. Includes such dyes as eosine, erioglaucine, fluorescein, naphthol yellow, ponceau, quinoline, tartrazine, etc. They precipitate with calcium, barium, or titanium glycero chlorides or other metallic salts, to give suitable insoluble pigment dyestuffs, or they may be precipitated on bases to form lakes.

acid groups FUNCTIONAL GROUPS having the properties of ACIDS.[130,134] (ASTM)

In CELLULOSE and its derivatives, these are usually CARBOXYL GROUPS.

acid number (value) number of milligrams of KOH required to neutralize the free acids in one gram of an OIL, RESIN, VARNISH, or other substance; generally reported on the nonvolatile.[157] (ASTM)

ASTM test methods for acid number (value) include: fatty acids, D1980; lac resins, test, D29; organic coating materials, D1639; pine tars and pine tar oils, test, D856; rosin oil, test, D1131; rosin test, D465; tall oil, test, D803; turpentine and pinene, test, D233.

Words presented in CAP/SMALL CAPS type indicate that the word is defined in another part of the *Coatings Encyclopedic Dictionary.*

Numerical superscripts classify terms in one or more of the categories listed in the second section of this volume.

acid rain rain with pH values < about 5; commonly results from ACIDS formed by POLLUTANTS.[171,172] (IUPAC)

"Pure" rain water equilibrated with atmospheric CO_2 and naturally occurring acids in relatively clean air usually has a pH > 5.

acid refined linseed oil LINSEED OIL which has been treated with ACID, usually sulfuric, to remove MUCILAGINOUS matter.[63,156]

acid resistance ability of materials to resist attack by ACIDS; generally, the acids concerned are MINERAL ACIDS.[72]

acid sludge residue which separates from mineral and related OILS when they are refined with sulfuric acid.[144]

acid value see ACID NUMBER (VALUE).[157]

acidity (1) measure of the free acid present; (2) in OILS, acidity denotes the presence of acid-type constituents whose concentrations are usually defined in terms of the NEUTRALIZATION NUMBER, called ACID NUMBER.[144,157]

acoustic coating coating which absorbs or deadens sound.[71,163] See ANTINOISE PAINTS.[71,163]

acoustical board a low-density, sound-absorbing structural INSULATING board having a factory-applied finish and a fissured, felted-fiber, slotted or perforated surface pattern provided to reduce sound reflection.[75,159,82] (ASTM)

These are usually supplied for use in the form of tiles.

acoustical ceiling board an ACOUSTICAL MATERIAL in board form, designed primarily for suspended ceiling application.[75,82] (DAC)

acoustical material any material considered in terms of its acoustical properties.[75,159] (ASTM)

Commonly, and especially, a material designed to absorb sound.

acoustical paint see ANTINOISE PAINTS and ACOUSTIC COATING.[71,163]

acoustical plaster a special low-density, sound absorptive plaster, applied in the form of a finish-coat, to provide a continuous finished surface.[73,82,75] (DAC)

acoustical tile an ACOUSTICAL MATERIAL in board form, usually having unit dimensions of 24 in. x 24 in. (approx. 61 cm x 61 cm) or less.[82,75] (DAC)

Usually used on ceilings but also may be applied to sidewalls.

acrolein $CH_2 = CHCHO$. UNSATURATED LIQUID ALDEHYDE with a bp of 52°C.[141,144]

It possesses a very pungent odor, and has strong lachrymatory properties.

acrolein polymers and resins HOMOPOLYMERS or COPOLYMERS of ACROLEIN.[130]

acrylate ester formed from ACRYLIC ACID.[71]

The term also applies to the metallic salts of this acid.

acrylate resins see ACRYLICS.[38,71,130]

acrylic acid $CH_2 = CHCOOH$. Propenoic acid, vinylformic acid. A MONOMER used in the manufacture of coatings and plastics.[141]

Properties: mol wt, 72.06; mp, 14°C; bp, 141.0°C; sp gr, 1.422.

acrylic ester ESTER of ACRYLIC ACID, or of a structural derivative of acrylic acid, such as METHACRYLIC ACID.[144]

acrylic latex AQUEOUS DISPERSION, THERMOPLASTIC or THERMOSETTING, of POLYMERS or COPOLYMERS of ACRYLIC ACID, methacrylic acid, esters of these acids, or ACRYLONITRILE.[130,156]

acrylic plastics THERMOPLASTIC or THERMOSETTING PLASTICS of POLYMERS or COPOLYMERS of ACRYLIC ACID, methacrylic acid, esters of these acids, or acrylonitrile.[38,130]

acrylics RESINS resulting from the POLYMERIZATION of derivatives of ACRYLIC ACIDS, including esters of acrylic acid, methacrylic acid, acrylonitrile, and their copolymers.[38,71,130] Syn: acrylic resins, and acrylate resins

acrylide maroon this group includes the AZO PIGMENTS based on acrylides of beta hydroxyl naphthoic acid (e.g., toluidine maroon); they are characterized by their excellent SOAP, ACID, and alkali resistance and good bake resistance.[41] See ALKALI-RESISTANT RED.[41]

Poor BLEED RESISTANCE in aromatic and ALCOHOL SOLVENTS, poor LIGHTFASTNESS in other than MASSTONE shades (including metallics), low HIDING POWER, and high cost discourage their use except where chemical resistance requirements demand; to this extent, they may be considered as specialty pigments.

acrylonitrile $CH_2 = CHCN$. A raw material for the manufacture of synthetic RESINS and RUBBERS.[141,144] Syn: vinyl cyanide

It is a liquid at room temperature, with a bp of 77°C and an flp of 0°C.

acrylonitrile-butadiene-styrene (ABS) ACRYLONITRILE and STYRENE liquids and BUTADIENE gas polymerized together in a variety of ratios to produce the family of ABS resins.[38,130]

ACS abbreviation for AMERICAN CHEMICAL SOCIETY.[174]

action level the exposure level (the material's concentration in air) at which OSHA regulations to protect employees takes effect; e.g., workplace air analysis, employee training, medical monitoring, and record keeping.[172]

Exposure at or above the action level is termed occupational exposure. Exposure below this level can also be harmful. This level is generally one half the TLV.

activate to put into a state of increased chemical activity.[74]

activated materials, specially treated to confer ABSORPTIVE, ADSORPTIVE, or CATALYTIC properties on them.[74,142]

Such substances include activated alumina, activated earths, and ACTIVATED CARBON.

activated carbon any form of CARBON characterized by high ADSORPTIVE capacity for gases, vapors, and colloidal solids.[37]

activator (1) a SUBSTANCE added in small quantity to another to promote its activity, e.g., traces of silver or copper can promote the LUMINESCENCE of ZINC SULFIDE PIGMENTS in LUMINOUS PAINTS, and peroxides which are added to polyesters, are sometimes called activators;[83,67] (2) a chemical which may act on the ACCELERATOR to change the rate of CURE and the properties of a vulcanizate; activators may be ORGANIC or INORGANIC.[83,67,140] (BSI) See CATALYST.[83,67,140,74]

They may be zinc oxide or other metallic oxides, fatty acid like stearic, ORGANIC COMPOUNDS which are mild accelerators if used alone.

active ingredient (a.i.) component of a material that produces the specific effect for which the material is designed.

active metal METAL in a condition of high chemical activity that is more susceptible to CORROSION.[66]

ACV abbreviation for ADAMS CHROMATIC VALUE.[69]

acyclic open chain, not ring formation.[133,134]

acyl, acyl groups when the OH group is removed from a FATTY ACID MOLECULE, the monovalent residue is described as an ACYL RADICAL.[133,134] (ASTM)

Examples of acyl radicals are CH_3CO–acetyl; C_2H_5CO–propionyl; and C_3H_7CO–butyryl. Radicals derived from carboxylic acids by removal of the hydroxyl group.

acylation formation or introduction of an ACYL RADICAL in or into an ORGANIC COMPOUND.[131]

Adams chromatic value color difference equation one of the transformations of CIE COLOR SPACE into another space which is perceived as visually more uniform.[43,69]

Used to describe the difference between two similar colors. It is based on the MUNSELL VALUE function:

$$Y_C/100 = 1.2219V_Y - 0.23111V_Y^2 + 0.23951V_Y^3 - 0.021009V_Y^4 + 0.0008404V_Y^5$$

For determining the V_X, substitute $X_C/0.9804$ for Y_C; for determining V_Z, substitute $Z_C/1.1810$ for Y_C in the above equation. The transformed space is then calculated according to the following equations:

$$a = V_X - V_Y$$

$$b = 0.4\,(V_Z - V_Y) \text{ or } 0.4\,(V_Y - V_Z)$$

$$L = 0.23V_Y$$

The total color difference, ΔE, is calculated as follows:

$$\Delta E = \sqrt{(0.23\Delta V_Y)^2 + [\Delta({}_X - V_Y)]^2 + [0.4\Delta(V_Z - V_Y)]^2}$$

Note that if b is calculated as $0.4\,(V_Z - V_Y)$, a plus number indicates blueness; if the b is calculated as $0.4\,(V_Y - V_Z)$, a plus number indicates yellowness. This equation is generally used with a normalizing constant placed outside the brackets. If the components L, a, and b are used individually they are also multiplied by the normalizing constant, f. The normalizing constant or factor, f, is used to convert the numbers obtained to the same magnitude as those obtained with the NBS equation. The factor used is generally 40, although it may vary between 40 and 50 and must therefore be specified. Unless specified otherwise, the value 40 is assumed to have been used.

Adams-Nickerson-Stultz color difference equation the normalizing factor, f, recommended in this equation is 41.78. See ADAMS CHROMATIC VALUE COLOR DIFFERENCE EQUATION.[69]

adaptation change in the sensitivity of the eye resulting from changes in the viewing stimuli, enabling the eye to meet the needs of sensitive vision in a very wide range of conditions.[69,173]

addition polymer polymer made by ADDITION POLYMERIZATION.[130,169] (IUPAC) cf. CONDENSATION POLYMER[130,169]

addition polymerization POLYMERIZATION in which MONOMERS are linked together without the splitting off of water or other simple molecules.[130,169] (IUPAC, ASTM)

Chemical reaction in which simple molecules (MONOMERS) are linked to each other to form long-chain molecules (POLYMERS) and no by-products are formed.[169]

additive any SUBSTANCE added in small quantities to another substance, usually to improve properties.[83]

Sometimes called MODIFIER.

Words presented in CAP/SMALL CAPS type indicate that the word is defined in another part of the *Coatings Encyclopedic Dictionary.*

Numerical superscripts classify terms in one or more of the categories listed in the second section of this volume.

additive color mixture COLOR which results when the same area of the retina of the eye is ILLUMINATED by LIGHTS of different SPECTRAL DISTRIBUTION, such as by two or more colored lights.[43,69]

Additive color mixture may result from addition of lights from two or more projectors, by visual averaging of small colored dots as on colored television screens, by visual averaging of spinning or flickering colors, or by different stimulation of each of the two retinas of a single observer (binocular fusion). Should be distinguished from SUBTRACTIVE COLORANT (PIGMENT or DYESTUFF) MIXTURE.

additive reaction chemical reaction in which two components join together to form a single reaction product.[133,134]

In a pure additive reaction, neither of the REACTANTS undergoes MOLECULAR fission or splitting, but attaches itself to the other reactant intact. In other additive reactions, one of the reactants may split into two separate parts, each of which attaches itself to the appropriate places of the other intact reactant. There is still, however, a single reaction product.

adduct a CHEMICAL addition product, such as the cyclic product of the addition of a diene with another UNSATURATED COMPOUND (as MALEIC ANHYDRIDE).[134]

adduct curing agent see CROSSLINKING AGENT.[67,83,74]

adeps lanae, anhydrous pharmaceutical name for LANOLIN or purified WOOL GREASE.[144,137]

adhere to cause two surfaces to be held together by ADHESION.[79] (ASTM)

adherend body which is held to another body by an ADHESIVE.[79]

adherometer instrument which measures the FORCE required to strip a coating from a metal surface.[145,79]

adhesion state in which two surfaces are held together by INTERFACIAL FORCES which may consist of valence forces or interlocking action, or both.[79] (ASTM)

ASTM test methods for adhesion include: cut tape test, D3359; Dillon dynometer test, D4796; HIPAC coatings, D3730; portable tester, pull-off strength, D4541; prepainted fabricated metal, D4145; scrape test on smooth surfaces, D2197; traffic marking paints, materials, D4796; zinc-rich primer on steel, D4146.

adhesion, mechanical ADHESION between surfaces in which the ADHESIVE holds the parts together by interlocking action.[79] (ASTM) See ADHESION, SPECIFIC.[79]

adhesion promoters materials built into the BINDER or added to the paint to form primary bonds to either the substrate or the previously applied coating with the specific aim of improving the dry or wet ADHESION, or both.[79,83] (ASTM)

adhesion, specific ADHESION between surfaces which are held together by valence forces of the same type as those which give rise to COHESION.[79] See also ADHESION, MECHANICAL.[79]

adhesive substance capable of holding materials together by surface attachment.[79] (ASTM)

Adhesive is the general term and includes among others CEMENT, GLUE, MUCILAGE, and PASTE. All of these terms are loosely used interchangeably. Various descriptive adjectives are applied to the term, "adhesive," to indicate certain characteristics as follows: physical form—liquid adhesive, tape adhesive; chemical type—silicate adhesive, resin adhesive; materials bonded—paper adhesive, metal-plastic adhesive, can label adhesive; and conditions of use—hot-setting adhesive.

adhesive, assembly ADHESIVE which can be used for bonding parts together, such as in the manufacture of a boat, airplane, furniture, and the like.[79,57,175] (ASTM)

The term, "assembly adhesive," is commonly used in the WOOD industry to distinguish such adhesives (formerly called "JOINT GLUES") from those used in making PLYWOOD (sometimes called "VENEER GLUES"). It is applied to adhesives used in fabricating finished structures or goods, or subassemblies thereof, as differentiated from adhesives used in the production of sheet materials for sale as such, for example, plywood or laminates.

adhesive, cold-setting ADHESIVE which sets at temperatures below 20°C.[79] (ASTM) cf. ADHESIVE, HOT-SETTING[79]

adhesive dispersion two-phase ADHESIVE system in which one phase is suspended in a liquid.[79]

adhesive, edge jointing ADHESIVE used to BOND strips of VENEER together by their edges in the formation of larger sheets.[57,79]

adhesive film SYNTHETIC RESIN ADHESIVE, usually of the THERMOSETTING type, in the form of a thin dry film of resin, used under heat and pressure as an interleaf in the production of laminated materials (particularly PLYWOOD and densified WOOD).[57,79]

adhesive, hot-setting ADHESIVE which requires a temperature at or above 100°C to set it.[79] (ASTM) cf. ADHESIVE, COLD-SETTING[79]

adhesive, pressure sensitive ADHESIVE made so as to adhere to a surface at room temperature by briefly applied PRESSURE alone.[79] (ASTM)

adhesive tape test see TAPE TEST.[157]

adipates esters of ADIPIC ACID.[144,170]

adipic acid HOOC–$(CH_2)_4$–COOH. Dibasic acid of ALIPHATIC type, used in the preparation of POLYESTERS.[144]

Properties: acid value, 767; molecular weight, 146.1; mp, 151°C; and bp, 216°C/15 mm Hg.

adobe brick large, roughly molded, sundried CLAY brick of varying sizes.[75,73] (DAC)

adronal see CYCLOHEXYL ACETATE.[144]

adsorbate (1) a molecular species of gas, dissolved substance, or liquid which adheres to or is ADSORBED in an extremely thin surface layer of solid substance; (2) the material accumulated on the surface of an ADSORBENT by the PROCESS of ADSORPTION.[139] (IUPAC)

adsorbed water water which is held on the surface of a material by electrochemical forces; its physical properties are substantially different from those of ABSORBED water or CHEMICALLY combined water at the same temperature and pressure.[142] (DAC)

adsorbent the solid material to which another species is adsorbed in a surface layer.[142] (IUPAC) cf. ABSORBENT

adsorption the PROCESS by which molecules of gas, dissolved substances, or liquids adhere to the SURFACE of solids through either weak physical forces (physical adsorption) or stronger chemical forces (chemical adsorption).[142,139] (IUPAC) cf. ABSORPTION

adulteration presence of inferior materials which reduce the quality of the standard.

advancing colors COLORS that give the illusion of being closer to the eye than their COMPLEMENTARY COLORS when both are adjacent and in the same plane; colors ranging from yellow-green to scarlet are of this character.[69]

aeration the PROCESS by which a volume filled with a liquid becomes permeated with air or another gas.[142] (IUPAC)

Aeration is often accomplished by spraying the liquid into the air, bubbling air through a liquid, or agitating the liquid to promote surface absorption of air. (IUPAC)

aeration cell ELECTROLYTIC CELL in which the driving force to cause CORROSION results from a difference in the amount of oxygen in solution at one point as compared to another. Corrosion is accelerated where the oxygen concentration is least or most.[66]

aerograph a SPRAY GUN for coatings, using air.[59]

aerosol mixtures of small particles (solid, liquid, or a mixed variety) and the carrier gas (usually air); owing to their size, these particles (usually less than 100 µm and greater than 0.01 µm in diameter) have a comparatively small settling velocity and hence exhibit some degree of stability in the earth's gravitational field.[64] (IUPAC)

An aerosol may be characterized by its chemical composition, its radioactivity, the particle size distribution, the electrical charge, and the optical properties. Aerosols are DISPERSIONS in gases. In aerosols, the particles often exceed the usual size limits for colloids. If the dispersed particles are solid, one speaks of "aerosols of solid particles," if they are liquid, of "aerosols of liquid particles." The use of terms "solid aerosol" and "liquid aerosols" is discouraged. An aerosol is neither "solid" nor "liquid," but if anything, gaseous. The term aerosol is recommended for general use in describing airborne particulate matter.[172] (IUPAC)

aerosol coating a conveniently packaged SPRAY coating in a sealed can. Pressure is supplied by compressed liquified gas.[71,77,163]

aerugo see VERDIGRIS.[41]

African ochre see PIGMENT YELLOW 42 (77492).[41]

after-bake see POST-CURE.[67]

after-tack FILM DEFECT in which the coated surface, having once reached a tack-free stage, subsequently develops a sticky condition.[47,56,168] (BSI)

The effect may be due to SYNERESIS (i.e., expulsion of liquid from a GEL). Also applies to PRINTING INKS.

aftercure a continuation of the PROCESS of CURING or VULCANIZATION after the cure has been carried to the desired degree and the source of heat removed, generally resulting in over-cure and a product less resistant to aging than properly cured products.[67,140] cf. OVERCURE

agar (or agar-agar) gelling hydrocolloid from sea ALGAE plants which are POLYMERS of galactose.[139,155]

age resistance RESISTANCE of DETERIORATION with time.[72]

agent a means by which something is done or caused; instrument.

agglomerate (1) cluster of individual PARTICLES;[64] (2) cluster of dry PIGMENT particles held together by surface forces. The spaces between the particles are filled with air; may also be present in liquid paints if the pigment has not been properly dispersed.[56,64,41] See AGGREGATE.[41,73,75]

Words presented in CAP/SMALL CAPS type indicate that the word is defined in another part of the *Coatings Encyclopedic Dictionary.*

Numerical superscripts classify terms in one or more of the categories listed in the second section of this volume.

agglomeration the growth of PARTICLES in size due to an attachment of other particles following collisions.[64] (IUPAC)

The term refers to AEROSOL particles, cloud droplets, etc. (IUPAC) Agglomeration may be loosely used to refer to undispersed material.

aggregate (1) a group of DRY PIGMENT PARTICLES held together by their surface FORCES; the spaces between the particles are filled with air;[41,56,64] (2) an inert granular material, such as natural sand, manufactured sand, gravel, crushed gravel, crushed stone, vermiculite, perlite, and air-cooled blast furnace slag, which when bound together into a conglomerate mass by a matrix, form concrete or mortar.[73,75,42] (DAC) See AGGLOMERATE.[56,64,41]

aggressive tack see TACK, DRY.[47,79]

aging (ageing) STORAGE of paints, varnishes, etc., under defined conditions of TEMPERATURE, RELATIVE HUMIDITY, etc. in suitable containers, or of dry films of these materials, for the purpose of subsequent tests.[157,161] See MATURING.[49,131]

agitation PROCESS of mixing or stirring to achieve homogeneity, but not necessarily DISPERSION.[49,131]

agitator MECHANICAL device used for mixing or stirring.[59,131]

AIA abbreviation for American Institute of Architects.[82,174]

AIC abbreviation for Association Internationale de la Couleur.[134,69]

AICE abbreviation for American Institute of Chemical Engineers.[174]

AIMA abbreviation for Acoustical and Insulating Materials Association.[82,174]

air assist thermoforming method of thermoforming in which air is employed to preform the sheet immediately prior to the final pulldown onto the mold, using vacuum.[57]

air blowing PROCESS employed for the production of low temperature BLOWN OILS or of high temperature isomerized oils.[63,131] cf. BLOWN BITUMEN,[76] BLOWN CASTOR OIL,[63,131] BLOWN OIL,[63,131] BLOWN STAND OIL.[63,131]

A large stream of air, sent through VEGETABLE OILS under suitable conditions, produces chemical changes in them. These changes are recognized by solubility characteristics, increased viscosity, and improved drying and polymerizing properties; but the changes which occur with a particular oil are governed by its chemical constitution.

air brush (1) very small SPRAY GUN, not much larger than a fountain pen, designed as an artist's tool.[77,81] (2) British term for spray gun.[59,82,168,77] (DAC)

A small tool used for the fine-spray application of paint, DYE, watercolor pigment, or INK by compressed air.

air bubble dry bubble in coating film caused by entrapped air.[56]

air bubble viscometer see BUBBLE TUBE VISCOMETER.[145,157]

air cap perforated housing for atomizing air at head of SPRAY GUN.[59]

air contamination FOREIGN SUBSTANCES introduced into the air which make the air impure.[171,172] See AIR POLLUTANT.[171,172] Syn: air pollution

air cure VULCANIZATION which takes place at ROOM TEMPERATURE with the use of ultrafast or fast-acting ACCELERATORS.[67,140,160]

air drying see DRYING.[67]

air ducts pipes that carry warm air or cold air to rooms and back to furnace or air conditioning system.[82]

air entraining agents natural WOOD RESINS, fats, INORGANIC materials, sulfonated compounds, and oils for AIR ENTRAPMENT in concrete.[73,83]

air entrapment inclusion of air bubbles in coating film or in other solids such as CONCRETE. See MICROVOIDS.[43,56]

air flotation PROCESS used to separate light from heavy pigment particles by a strong current of air.[49,131] Syn: air separation

The air stream is so arranged that the pigment particles are carried vertically from the grinding mill. By this means only the finest particles are carried away, the larger and heavier ones fall back to be re-ground.

air jet a type of sandblasting gun in which the ABRASIVE is conveyed to the gun by partial vacuum.[59,158]

air lock surface depression on a molded part, caused by trapped air between the mold surface and the PLASTIC material.[56,57]

air loss loss in MASS by a PLASTIC or coating on exposure to air at room temperature.[65]

air pollutant a substance, gaseous material, or AEROSOL which has been introduced into the air (either by human activity or by natural processes) in sufficient concentration to produce a measureable effect on humans, animals, vegetation, or materials (monuments, etc.): SO_2, NO_2, H_2S, CO, HYDROCARBONS, etc.[171,172] (IUPAC)

air pollutants, hazardous materials discharged into the ATMOSPHERE that have a proven relationship to increased human death rates.[171]

air pollution usually the presence of substances in the atmosphere, resulting either from human activity or natural processes, present in sufficient concentration, for a sufficient time and under circumstances such as to interfere with comfort, health, or welfare of persons or the environment.[171,172] (IUPAC) Syn: air contamination

air pollution control measures taken to control the purity of the air to some degree specified by law.[171,172] (IUPAC) See AIR QUALITY CONTROL REGIONS.[171,172]

air quality control regions geographical units of the country, as required by U.S. Law, reflecting common AIR POLLUTION problems, for purposes of reaching national standards.[171,172]

air sampling determining quantities and types of atmospheric contaminants by measuring and evaluating a representative sample of air.[171,172]

The most numerous environmental hazards are chemical, and can be conveniently divided into (1) the particulates and (2) the gases or vapors. Particulates are mixtures or dispersions of solid or liquid particles in air and include dust, smoke, mist, and similar materials.

air separation see AIR FLOTATION.[131]

air spray SPRAY COATING method in which the coating is atomized by mixing it with compressed air.[77] (EPA)

air vent small outlet to prevent entrapment of GASES.[59]

airless spraying PROCESS of atomization of paint by forcing it through an orifice at high pressure.[77,54] (BSI) See HYDRAULIC SPRAYING.[77]

This effect is often aided by the vaporization of the solvents, especially if the paint has been previously heated. The term is not generally applied to those electrostatic spraying processes which do not use air for atomization.

alabaster fine-grained, TRANSLUCENT variety of very pure GYPSUM, generally white or delicately shaded.[75,82] (DAC)

albino bitumen pale BITUMEN of PETROLEUM origin, similar in general physical characteristics to the black types, but distinguished by its deep golden brown COLOR.[76]

albite $NaAlSi_3O_8$. A widely distributed white FELDSPAR. One of the common rock-forming plagioclase group.[41,58,34]

albumin water-soluble PROTEIN derived from egg whites or animal blood, which is used in coatings and ADHESIVES.[155]

alburnum softer part of the WOOD between the innerpart and the HEARTWOOD.[153] Syn: sapwood

alcohol (1) any of the class of the CHEMICAL COMPOUNDS having the general formula R·OH, where R represents an alkyl group, and in systematic nomenclature indicated by -ol; (2) sometimes used to mean ETHYL ALCOHOL.[144,164]

ASTM test methods for alcohol include: amyl alcohol, D319; n-butyl alcohol (1-butanol), D304; sec-butyl alcohol (2-butanol), D1007; 2-ethylhexanol, D5008; isobutyl alcohol (isobutanol), D1719; isopropyl alcohol (isopropanol), D770; methyl isobutyl carbinol, D2635; methanol (methyl alcohol), D1152; n-propyl alcohol (n-propanol), D3622.

alcohol resistance ability of a DRIED or CURED coating to withstand the damaging effects of (ETHYL) ALCOHOL.[72]

alcohol-ethers SOLVENTS which are part ALCOHOL and part ETHER, made by etherifying one or more of the available hydroxyl groups of a polyhydric alcohol, but still leaving at least one hydroxyl group free.[164]

alcoholysis general CHEMICAL REACTION involving an ESTER exchange. The cleavage of a C–C bond by the addition of an ALCOHOL.[131]

alcove a small recessed space, opening directly into a larger room.[82] (DAC)

aldehyde a class of ORGANIC compound of general formula: R (C=O) H, where R is a hydrogen atom or hydrocarbyl group (e.g., CH_3, C_2H_5 ,etc.). These are products of HYDROCARBON oxidation.[134,144] (IUPAC)

The simplest ALDEHYDES are FORMALDEHYDE (methanal, CH_2O), ACETALDEHYDE (ethanal, CH_3CHO), and propionaldehyde (propanal, (CH_3CH_2CHO). Aldehydes absorb sunlight ($\lambda <$ 330nm) and generate radicals. They also react rapidly with HO-radicals in the atmosphere to produce acyl (RCO) and acylperoxyl ($RCOO_2$) radicals. Formaldehyde is the most abundant aldehyde in the atmosphere (formed from the natural hydrocarbon, CH_4, as well as other HYDROCARBONS) where its mixing ratios are often > 0.2 ppbv even in relatively clean air. Acetaldehyde is a major precursor to peroxyacetyl nitrate (PAN) in the troposphere.

aldehyde resin SYNTHETIC RESIN made by treating various ALDEHYDES with CONDENSATION agents.[130]

PHENOL, UREA, ANILINE, and MELAMINE react readily with ALDEHYDES, such as FORMALDEHYDE, which are also called aldehyde resins.

Words presented in CAP/SMALL CAPS type indicate that the word is defined in another part of the *Coatings Encyclopedic Dictionary.*

Numerical superscripts classify terms in one or more of the categories listed in the second section of this volume.

Aleurites botanical name for a species of shrub or tree which provides VEGETABLE OILS used in VARNISH manufacture.[63,153]

The various Aleurites include A. fordii, A. montana, A. cordata, A. trisperma, and A. moluccana. The first three yield TUNG OILS, the fourth, BAGILLUMBANG OIL, and the last, CANDLENUT OIL.

aleuritic acid $C_{15}H_{28}(OH_3)COOH$, DL-erythro-9,10,16-trihydroxyhexadecanoic acid. Acid obtained from the major constituent of the hard LAC portion of SHELLAC.[146]

It is present as an interester acid, of monocarboxylic type, which on HYDROLYSIS with caustic soda yields a mixture of two sodium salts, sodium aleuritate and sodium lacollate.

algae any of the various chiefly aquatic eukaryotic, photosynthetic organisms ranging in size from single-celled forms to the giant kelp.[173,171] See ANTI-FOULING COMPOSITION. cf. MILDEW

Algae (singular alga) were once considered to be plants but are now classified separately because they lack true roots, stems, leaves, and embryos. Algae should not be confused with mildew which is usually gray to black in color and can grow in sunny and drier areas. ASTM Test Method D3274 has photographic reference standards to provide a numerical basis for rating algal and FUNGAL growth or DIRT and soil accumulation on paint films.

algicide a CHEMICAL AGENT used to destroy ALGAE.[83,173]

algin soluble salts (sodium, ammonium, and potassium) of a polyuronic acid (mannuronic acid), a POLYMER obtained from brown sea plants.[139,155]

alginates see ALGIN.[144,139,155]

alginic acid polysaccharide composed of beta D mannuronic acid residues found in certain seaweeds.[139,155]

algistatic having the property of inhibiting ALGAL growth.[173]

aliphatic amino group $-NH_2$ RADICAL when attached to a chain.[144,134]

aliphatic compounds a class of ORGANIC COMPOUNDS which are composed of open chains of CARBON atoms. These include paraffins, olefins, etc.[134] cf. AROMATIC COMPOUNDS.[134]

aliphatic solvents HYDROCARBON SOLVENTS comprised primarily of paraffinic and cycloparaffinic (naphthenic) hydrocarbon compounds. Aromatic hydrocarbon content may range from less than 1% to about 35%.[164,134]

aliquot a representative portion of the whole. (IUPAC)

alizarin 1,2-dihydroxyanthraquinone obtained from madder roots, or made synthetically, used in the manufacture of LAKES.[41]

alizarin madder lake see ALIZARIN RED.[41]

alizarin red Pigment Red 83 (58000). Light maroon PIGMENT produced by treating ALIZARIN with calcium and aluminum salts in the presence of alumina. It is identical with MADDER LAKE.[41] Syn: alizarin madder lake, alizarin red B, dihydroxy anthraquinone lake

alkali broadly, any COMPOUND having highly basic properties; i.e., one that readily ionizes in aqueous solution to yield OH anions, with a PH of above 7. Alkalies are oxides and hydroxides of certain metals and belong to group IA of the periodic table (Li, Na, K, Rb, Cs, Fr). Ammonia and AMINES are also alkaline.[144] See ACID, and PH.[144] Syn: base and caustic

Common commercial alkalies are sodium carbonate (soda ash), lye (NaOH), LIME, potash, caustic soda, KOH, bicarbonate of soda, water glass, regular mortar, and PORTLAND CEMENT.

alkali blue complex ORGANIC blue TONER prepared by the phenylation of para rosaniline or fuchsine.[41]

alkali cellulose product obtained from CELLULOSE when it is treated with concentrated alkali at low temperatures.[155]

alkali resistance (1) the degree to which a coating resists reaction with alkaline materials such as LIME, CEMENT, PLASTER, soap, etc. On weathering over alkaline cementitious substrates, LATEX paints can result in premature film failure (CRACKING, CHECKING, CHALKING). This is most often due to the DEGRADATION of the BINDER (such as POLYVINYL ACETATE) by alkaline salts normally present in the substrate causing HYDROLYSIS, EFFLORESCENCE, and/or MOTTLING; (2) the degree to which a coating will resist attack by AQUEOUS ALKALINE SOLUTIONS.[72]

alkali-refined linseed oil RAW LINSEED OIL treated with alkali to reduce the free acidity by formation of water-soluble salts, which are subsequently removed by washing.[63,131]

alkali-resistant paint see ALKALI RESISTANCE.[71,72]

alkali-resistant red (12315, 12355). AZO PIGMENTS based on acrylides of beta hydroxy naphthoic acid and sometimes called naphthol reds; the distinction between them and the acrylide maroons is strictly tinctorial.[41]

For the most part, they have identical properties except the reds are somewhat easier grinding but poorer in bake resistance than the maroons. These reds find some application in LATEX paints, particularly for application over alkaline substrates; they are not sufficiently LIGHTFAST for exterior exposure.

alkali-soluble resins (ASR) these are generally lower molecular weight (than conventional latices) POLYMERS containing about 5 to 15% CARBOXYL GROUPS which require AMINE and/or COSOLVENT to solubilize them.[130,155]

These systems are generally DISPERSIONS of MICELLES rather than true solutions.

alkaline catalysts hydroxides of sodium, potassium, lithium, and ammonium, or salts derived from these metallic radicals, which exhibit alkaline characteristics. Gaseous ammonia can also be used, as well as a number of basic ORGANIC COMPOUNDS.[74]

The most important reactions in which alkaline catalysts are involved are in the CONDENSATION of PHENOLS with FORMALDEHYDE, the condensation of UREA with formaldehyde, and in the ISOMERIZATION of DRYING OILS.

alkyds SYNTHETIC RESINS formed by the CONDENSATION of polyhydric ALCOHOLS with polybasic acids. They may be regarded as complex polyesters. The most common polyhydric alcohol used is GLYCEROL, and the most common polybasic acid is PHTHALIC ANHYDRIDE.[71,130]

Modified alkyds are those in which the polybasic acid is substituted in part by a monobasic acid, of which the VEGETABLE OIL FATTY ACIDS are typical.

ASTM test methods for alkyds/alkyd resins include: carboxylic acids content, D2455; fatty acids content, D1398; isophthalic acid content, D2690; phthalic anhydride content, D563, D1306; polyhydric alcohol, D1615, D2456, D2998; rosin acid content, D1469; silicon content, D3733; test methods, selection of, D2689; unsaponifiable matter content, D1397.

alkyl groups monovalent ALIPHATIC RADICALS derived from aliphatic HYDROCARBONS by removal of a hydrogen.[134] (ASTM)

alkyl phenolic resin PHENOL-FORMALDEHYDE RESIN in which the phenol used has an alkyl group in the para position.[130]

In resins used in coatings, the most common are the tertiary butyl and tertiary amyl phenols.

all-over pattern the typical effect produced by a WALLCOVERING. A pattern in which the units of design are evenly distributed over a surface, without undue emphasis.[148]

alligatoring (1) a type of CRAZING or surface CRACKING of a definite pattern, as indicated by its name. The effect is often caused during weather AGING;[56,154] (2) the cracking of the surface BITUMEN on a built-up roof, producing a pattern of cracks similar to an alligator's hide; the cracks may not extend through the surface bitumen.[76] (ASTM)

allotonic ADDITIVE which changes the surface tension of water.[83,127]

allotropy property which an element or compound possesses, of existing in different forms which in themselves have different characteristics.[133]

These various forms are described as allotropic modifications. CARBON, for example, is found in an amorphous form as carbon black, and in the crystalline form as graphite, and as a diamond.

alloy a homogeneous MIXTURE or SOLID SOLUTION of two or more METALS, the ATOMS of one replacing or occupying interstitial positions between the atoms of the other.[45]

For example, brass is an alloy of copper and zinc.

allyl alcohol $CH_2 = CHCH_2OH$. UNSATURATED ALCOHOL with pungent odor. It is miscible with ETHER, water, and alcohol.[164]

allyl resin a SYNTHETIC RESIN formed by the POLYMERIZATION of chemical compounds containing the group $CH_2 = CH{-}CH_2{-}$.[130]

allyl starch soft, gummy mass prepared by the reaction of starch with allyl chloride in the presence of strong ALKALI.[139]

almaciga native Philippine name for MANILA COPAL.[167]

almond oil VEGETABLE NONDRYING OIL, with an iodine value of about 99, and a sp gr of 0.918/15°C.[63]

alpha cellulose very pure CELLULOSE used as a FILLER, particularly with AMINO PLASTIC RESINS.[153,167]

alpha-hydroxypropionic acid see LACTIC ACID.[144]

alpha-methylstyrene $C_6H_5C(CH_3) = CH_2$. Colorless liquid which can be POLYMERIZED by ACIDS to RESINOUS products and will COPOLYMERIZE with other MONOMERIC VINYL COMPOUNDS.[141,169]

Properties: bp of 165°C; sp gr of 0.914 at 20°C.

alpha-pinene see PINENE.[44,164]

alpha-protein PROTEIN obtained from SOYA BEANS which is chiefly glycinin.[155]

The process of extraction of the protein from the beans involves dissolution of the protein and separation from CELLULOSE, removal of CARBOHYDRATES, and HYDROLYSIS of the protein to reduce its chemical complexity. Its uses are similar to those of CASEIN, with which it has much in common, being employed in water paints, emulsions, and adhesives of all types.

alternating copolymer a COPOLYMER in which the two different monomeric types alternate along the

---

Words presented in CAP/SMALL CAPS type indicate that the word is defined in another part of the *Coatings Encyclopedic Dictionary.*

Numerical superscripts classify terms in one or more of the categories listed in the second section of this volume.

chain in an -A-B-A-B-A-B- manner.[133,134,169] See GRAFT, and BLOCK POLYMER.[169,130]

alternating strain amplitude analogous to ALTERNATING STRESS AMPLITUDE.[47]

alternating stress amplitude test parameter of a dynamic fatigue test; one-half the algebraic difference between the maximum and minimum stress in one cycle.[47]

alumina any of several forms of ALUMINUM OXIDE, Al_2O_3, occurring naturally as CORUNDUM, in a hydrated form of BAUXITE, and with various impurities as ruby, sapphire, and emery.[41,58] See ALUMINUM OXIDE ABRASIVE,[59,158] ALUMINUM OXIDE, HYDRATED,[41] and ALUMINUM OXIDE PAPER.[59,158]

alumina hydrate see ALUMINUM OXIDE, HYDRATED.[41,58]

alumina-blanc fixe composite COLOR base used in the manufacture of certain types of LAKES.[58,41]

It is made by first reacting aluminum sulfate with soda ash, and then adding barium chloride solution to form BARIUM SULFATE from the sodium sulfate derived from the first reaction.

aluminum (in U.K., aluminium) see ALUMINUM POWDER, ALUMINUM PASTE.[41]

aluminum arsenite admixture with ALUMINA used as a BASE for LAKES, its function being to improve the brilliance of the COLOR.[41] See ALUMINUM PHOSPHATE.[144]

aluminum hydrate see ALUMINUM OXIDE, HYDRATED.[41,58]

aluminum ink see SILVER INK.[168]

aluminum mixing varnish any of the vehicles for ALUMINUM PASTE or POWDER which have the property of causing the METALLIC FLAKES of the PIGMENT to "LEAF" or FLOAT, and that retain this property for an indefinite period of time.[156]

They are characterized by low acid number, usually less than 15; but in some specifications, the acid number is required to be less than 7. Ready mixed aluminum vehicles are those that: were made without lead drier, used in factory prepared paints, and capable of retention of excellent leafing properties for very long periods of time. Ready-to-mix aluminum vehicles are mixed with aluminum paste or powder, and used within a few hours at the job site. Vehicles for nonleafing aluminum paints are rarely referred to as aluminum mixing varnishes.

aluminum naphthenate pale brown, jelly-like product, containing approximately 4.0% of aluminum, although the precise metal content varies with the type of NAPHTHENIC ACID used in its manufacture.[83,74]

aluminum oxide see ALUMINA.[41,58]

aluminum oxide abrasive produced in electric furnaces by purifying BAUXITE to a crystalline form and adding various amounts of titanium to impart extra toughness.[158] See EMERY.[158,59]

The grain is a tough, durable abrasive characterized by the long life of its cutting edges. It is stable chemically and will not react with hydrochloric or nitric acids or with metals other than the alkaline metals.

aluminum oxide cloth extremely hard, sharp, and enduring coated ABRASIVE cloth, particularly suited for finishing metals, capable of the hardest service.[59,158] See ALUMINA.[41,58]

aluminum oxide, hydrated $Al(OH)_3$. Crystalline powder, balls, or granules. Used as a PIGMENT or a base for ORGANIC lakes and as an EXTENDER for INKS.[41] Syn: alumina hydroxide, alumina trihydrate, alumina hydrate, hydrated alumina, gibbsite, transparent white. See ALUMINA.[41,158]

Properties: density, 2.4 g/cm³ (19.8-20.2 lb/gal); O.A., 41-53 lb/100 lb. Particle size, 0.38-8.5 μm.

aluminum oxide paper extremely hard, sharp, and enduring coated ABRASIVE paper, it is particularly suited for hardwoods, LACQUER, and metals.[59,158] See ALUMINA.[41,58]

aluminum paint coating consisting of a mixture of METALLIC ALUMINUM PIGMENT in POWDER or PASTE form, dispersed in a suitable vehicle.[71]

aluminum paste METALLIC ALUMINUM FLAKE PIGMENT in PASTE form, consisting of ALUMINUM, SOLVENT, and various ADDITIVES.[41] See LEAFING.[55]

The metallic aluminum pigment can be in the form of very small, coated leaves or amorphous powder, known under the respective designations of "leafing" and "nonleafing."

ASTM Test Method D480 covers procedures for sampling, qualitative analysis, and physical testing of flaked aluminum powders and pastes (leafing and nonleafing) for coatings.

aluminum phosphate admixture with ALUMINA used as a BASE for LAKES, its function being to improve the brilliance of the COLOR.[144] See ALUMINUM ARSENITE.[41]

aluminum pigment see ALUMINUM POWDER.[41]

aluminum potassium silicate $3Al_2O_3{\cdot}K_2O{\cdot}6SiO_2{\cdot}2H_2O$. Pigment White 20 (77019). Generically known as MICA, a complex hydrous ALUMINUM SILICATE, based on several mineralogically related groups.[41,58,34] Syn: mica, ground muscovite, graphitic mica, sericite

The most common is muscovite. Mica crystals have well developed basic cleavage that permit splitting into thin flaky pigment particles. Widely used in paints, rubber, and sealants. Properties: density, 2.82 g/cm³ (23.5 lb/gal); O.A., 56.74. Particle size, 5-20 μm. Hardness (moh), 2.5.

aluminum powder POWDERED ALUMINUM METAL used as a PIGMENT. Available in LEAFING or NONLEAFING forms.[41]

Density, 2.73 g/cm^3 (22.5 lb/gal).

ASTM Test Method D480 covers procedures for sampling, qualitative analysis, and physical testing of flaked aluminum powders and pastes (leafing and nonleafing) for coatings.

aluminum primer (1) PRIMER specifically formulated for aluminum metal; (2) primer containing a proportion of aluminum PIGMENT, but distinguished from aluminum paints in which the aluminum is designed to float to the top of the film giving metallic brilliance, a feature undesirable in a primer.[71,158] (BSI)

Aluminum primers have been used on resinous TIMBER or timber which has been treated with oil-soluble wood preservatives.

aluminum resinate brown soft mass, insoluble in water; used for water-repellancy and as a DRIER in VARNISH.[83]

aluminum silicate (clay) $Al_2O_3{\cdot}2SiO_2{\cdot}2H_2O$. Pigment White 19 (77005). White inert PIGMENT of little color and opacity, obtained from certain natural deposits of CHINA CLAY, KAOLIN, FELDSPAR, and similar materials.[41,58] Syn: kaolin, hydrated aluminum silicate, china clay, pipe clay, white bole

Hydrated aluminum silicate, fine grain crystallized clay. Laminar structure with repeating alumina-silica configurations. Density, 2.58 g/cm^3 (21.5 lb/gal); O.A., 32.1-55 lb/100 gal. Particle size, 0.5-3.5 µm.

aluminum soaps aluminum compounds of FATTY ACIDS and naphthenic acid, etc.[127]

These materials are soluble generally in paint vehicles and THINNERS, and are employed as THICKENING, FLATTING, and WATERPROOFING agents.

aluminum stearate complex SALT or SOAP of ALUMINUM and STEARIC ACID. It is a white powder, which forms COLLOIDAL SOLUTIONS or gels with DRYING OILS and certain SOLVENTS.[58,83,139]

It is used as a FLATTING AGENT, ANTI-SETTLING AGENT for pigments, and to help prevent penetration.

amber a fossil RESIN found chiefly in the blue earth of East Prussia. It is primarily an exudation of *Pinus succinifera*.[167]

ambient usual surrounding environmental conditions, such as pressure or temperature.

ambient air the outdoor air in a particular location.[171,172] (IUPAC)

ambient air quality a general term used to describe the quality of the outside air.[171,172] (IUPAC)

Usually adjectives such as good, fair, bad, etc. are used by the media to describe this; often some form of AIR POLLUTION or air quality index is employed to determine the specific descriptive term to be used. These are very qualitative terms of little or no scientific value. (IUPAC)

ambient cure see SELF-CURING.[67,55]

American Chemical Society, The (ACS) is a nonprofit scientific and educational association of professional chemists and chemical engineers. Founded in New York City on April 6, 1876, with 133 charter members, the Society now has a membership of over 110,000 and is the world's largest organization devoted to a single science. It carries out a broad program of publications, meetings, and other activities in pursuit of the following objectives, as set forth in its National Charter granted by Congress in 1937: to encourage in the broadest and most liberal manner the advancement of chemistry in all its branches; the promotion of research in chemical science and industry; the improvement of the qualifications and usefulness of chemists through high standards of professional ethics, education, and attainments; the increase and diffusion of chemical knowledge; and by its meetings, professional contacts, reports, papers, discussions, and publications, to promote scientific interests and inquiry; thereby fostering public welfare and education, aiding the development of our country's industries, and adding to the material prosperity and happiness of our people.[144,174]

Its address is 1155 Sixteenth Street, N.W., Washington, D.C. 20036; Tel: (202) 872-4600.

American Conference of Governmental Industrial Hygienists (ACGIH) an organization of professionals in governmental agencies or educational institutions engaged in occupational safety and health programs. ACGIH develops and publishes recommended occupational exposure limits for chemical substances and physical agents.[174] See TLV.[172]

Its address is 6500 Glenway Ave., Bldg. D-7, Cincinnati, OH 45211; Tel.: (513) 661-7881.

American gallon see gallon, U.S.[128]

American National Standards Institute (ANSI) a privately funded organization that identifies industrial/public national consensus standards and coordinates their development. Many ANSI standards relate to safe design/performance of equipment and safe practices or procedures.[174]

Its address is: 1430 Broadway, New York, NY 10018; Tel.: (212) 642-4900.

American process zinc oxide zinc oxide PIGMENT made directly from zinc ores.[41,131] See ZINC OXIDE.[41]

Sometimes called "direct" process.

Words presented in CAP/SMALL CAPS type indicate that the word is defined in another part of the *Coatings Encyclopedic Dictionary.*

Numerical superscripts classify terms in one or more of the categories listed in the second section of this volume.

American Society for Testing and Materials (ASTM) is a not-for-profit organization that provides a forum for producers, users, ultimate consumers, and those having a general interest (representatives of government and academia) to meet on common ground and write standards for materials, products, systems, and services. Organized in 1898, ASTM has grown into one of the largest voluntary standards development systems in the world.[174] (ASTM)

From the work of 132 standards-writing committees, ASTM publishes standard test methods, specifications, practices, guides, classifications, and terminology. ASTM's standards development activities encompass metals, paints, plastics, textiles, petroleum, construction, energy, the environment, consumer products, medical services and devices, computerized systems, electronics, and many other areas. D01 is the technical committee on paint and related coatings and materials.

More than 9000 standards are published each year in the 68 volumes of the Annual Book of Standards.

Its address is: 100 Barr Harbor Dr., West Conshohocken, PA 19428-2959; Tel: (610) 832-9500.

American turpentine light-colored, volatile, ESSENTIAL OIL obtained from RESINOUS EXUDATES or RESINOUS WOOD associated with living or dead coniferous trees, particularly of the genus, *Pinus*, or more commonly pines.[44,164] See TURPENTINE.[41,144]

American vermillion a PIGMENT usually consisting of a lead molybdate or as a basic lead chromate (as chrome red).[41]

amides CARBOXYLIC ACIDS in which the HYDROXYL GROUP of the acid is replaced by an AMINO or AMIDO group (NH_2).[134,144]

Thus, acetamide (CH_3CONH_2) is obtained by substituting the hydroxyl (OH) of acetic acid with the NH_2 group. They may also be regarded as derivatives of ammonia (NH_3), in which one of the hydrogen atoms is replaced by an acyl group.

amido "amido" and "amino" apply to the same grouping, NH_2.[134,144]

The former term is usually applied to the NH_2 group when it occurs in an acid amide.

amine equivalent see AMINE VALUE.[157]

amine equivalent weight MOLECULAR WEIGHT of AMINE divided by the number of active HYDROGENS in the molecule.[128]

amine resin SYNTHETIC RESIN derived from the reaction of UREA, THIOUREA, MELAMINE, or allied compounds with ALDEHYDES, particularly FORMALDEHYDE.[130] See AMINO RESIN.[134]

amine value the number of milligrams of potassium hydroxide equivalent to the fatty amine basicity in 1 g of sample.[157] (ASTM) Syn: amine equivalent

amine-furfural resin see ANILINE-FURFURAL RESIN.[130]

amines ORGANIC BASES derived from the parent COMPOUND, ammonia (NH_3).[134,144]

The hydrogens of the ammonia may be substituted by alkyl groups, in which case the series of ALIPHATIC BASES is produced. Similarly, AROMATIC BASES are formed when the hydrogens are substituted with aryl groups. Primary, secondary, tertiary, and quaternary amines are formed as one, two, three, or four of the hydrogen atoms are substituted. Substitution of a fourth hydrogen atom is possible because it is considered available in the hypothetical compound, ammonium hydroxide (NH_4OH).

amino indicates the presence of an $-NH_2$ or $=NH$ group.[134]

amino plastics PLASTICS based on RESINS made by the CONDENSATION of amines, such as UREA and MELAMINE, with ALDEHYDES.[38] See AMINO RESIN.[156,130]

amino resins synthetic RESIN of the THERMOSETTING type made by the reaction of UREA, THIOUREA, MELAMINE, or allied compounds, usually with FORMALDEHYDE.[156,130] (BSI)

Amino resins are usually cured by baking and are blended with other resins (e.g., alkyds or epoxies). Amino resins are also cured by chemical means at normal air temperature, e.g., in wood finishes.

ASTM test methods for amino resins include: free formaldehyde content, D1979; test procedures, practice, D4277; nitrogen content, D1013; solvent tolerance test, D1198.

2-amino-2-methyl-1-propanol see AMP.[83,127]

ammonia cure a modification of a hot air pressure cure for RUBBER, often used for curing, in which ammonia gas is used to accelerate vulcanization and to prevent the deteriorating effect of air.[67,140]

ammonium caseinate CASEIN solubilized by ammonium hydroxide generally employing a hot water presoak.[83,127,139,155]

ammonium soaps SOAPS formed by reaction of AMMONIA with the higher molecular weight FATTY ACIDS, such as ammonium oleate, linoleate, stearate, etc., used in wetting or emulsifying agents.[127,83]

amoora oil semidrying VEGETABLE OIL with an iodine value of 135 and sp gr of 0.939/15°C.[63]

amorphous without definite form, not CRYSTALLIZED (noncrystalline).[133]

amorphous silica SiO_2. A naturally occurring or SYNTHETICALLY produced PIGMENT, characterized by the absence of pronounced crystalline structure, and which has no sharp peaks in its X-ray diffraction pattern. It may contain water of hydration or be an anhydrous type.[41,58] cf. SILICA, CRYSTALLINE.[41]

It is used as an EXTENDER PIGMENT, FLATTING AGENT, and as a desiccant in metal flake and metal powder coatings.

AMP abbreviation for 2-AMINO-2-METHYL-1-PROPANOL.[83,127]

Used as a PIGMENT dispersant and/or as a pH modifier.

ampere unit of current. One ampere (1 A) is produced when one volt is applied to a circuit with one ohm resistance.[61,128]

amphibole a group of ASBESTOS MINERALS.[41,58,34]

amphiphilic property of being both hydrophilic and hydrophobic.[127]

Surfactants are amphiphilic.

amphoteric compounds which, according to the circumstances, can behave either as normal METALLIC OXIDES or hydroxides and form normal SALTS, or as ACIDS and form salts with ALKALI METALS.[133]

amu abbreviation for ATOMIC MASS UNIT.[128,144]

amyl acetate $CH_3COOC_5H_{11}$. Medium boiling SOLVENT.[144,164] Syn: banana oil

The commercial product has a boiling range of 120-145°C, a sp gr of approx. 0.876, and an flp of 31°C (87°F).

amyl alcohol $C_5H_{11}OH$.[144,164]

Commercial ALCOHOL having a boiling range of 128-132°C; a sp gr of 0.816/15°C; an flp of 43°C (110°F); and a vapor pressure of 10 mm Hg at 20°C. It has some application in the manufacture of CELLULOSE lacquers as a high boiling DILUENT.

amyl benzoate $C_6H_5COOC_5H_{11}$.[164,170]

High boiling SOLVENT-PLASTICIZER, with a bp of 261°C, and sp gr of 0.994.

amyl borate $(C_5H_{11})_3BO_3$. High boiling SOLVENT-PLASTICIZER.[164,170]

The commercial liquid has a boiling range of 250-260°C.

amyl citrate see CITRATES.[144]

amyl ether $(C_5H_{11})_2O$. Amyl ether, or diamyl ether as it is correctly called, is a useful high boiling SOLVENT of the ether series, with a boiling range of 170-190°C.[164]

amyl formate $HCOOC_5H_{11}$. Medium boiling SOLVENT with a bp of 130°C, and an flp of 27°C.[164]

amyl lactate $CH_3CH(OH)COOC_5H_{11}$. Water white to pale yellow colored liquid used as an ester SOLVENT.[164]

It plasticizes in lacquers and has a bp of 210°C.

amyl propionate $CH_3CH_2COOC_5H_{11}$. Very strong SOLVENT, similar in properties and uses to AMYL ACETATE, but possessing a higher boiling range, 140-170°C and an flp of about 41°C.[164]

amyl stearate $C_{17}H_{35}COOC_5H_{11}$. Pale yellow colored substance used as a high boiling PLASTICIZER.[170]

It has an mp of 30°C and a bp of 360°C.

amyl tartrate PLASTICIZER with strong SOLVENT power and a bp of about 400°C.[170]

amylaceous pertaining to, or of the nature of, starch; starchy.[79] (ASTM)

amyrin $C_{30}H_{50}O$ for both α and ß modifications. Amyrins, isolated in α and ß forms, are constituents of ELEMI resin.[167]

They are believed to be alcoholic in character.

anacardic acid $C_{22}H_{32}O_3$. Ortho pentadecadienylsalicylic acid.[144]

Principal constituent of cashew nutshell liquid.

anacardol $C_{21}H_{32}O$. Monohydroxy phenol, which closely resembles the ANACARDIC ACID from which it is readily obtained by decarboxylation when heated.[144]

The substituent side chain is the same for both compounds.

anadonis green see CHROMIC OXIDE GREEN.[41]

anaerobic free of oxygen and/or air.[173]

Used in connection with BACTERIA, developing without air. Important for anaerobic CORROSION (tanks, ship bottoms).

anamorphosis distorted painting which appears normal when viewed from the side.[81]

A form of "trick" painting fashionable in the 16th and 17th centuries.

anatase see TITANIUM DIOXIDE, ANATASE.[41,133]

anchorage (1) property or profile of metal or WOOD SUBSTRATE to enhance the ADHESION of a coating;[158] (2) adhesion of RUBBER to fiber, fabric, metal, or other material to which the rubber compound is applied by calendering, welding, CEMENT spreading, or other means.[79,140]

Words presented in CAP/SMALL CAPS type indicate that the word is defined in another part of the *Coatings Encyclopedic Dictionary.*

Numerical superscripts classify terms in one or more of the categories listed in the second section of this volume.

anelasticity dependence of elastic strain on both stress and time. This can result in a lag of strain behind stress.[47]

In materials subjected to cyclic stress, the anelastic effect causes internal damping.

angle of contact associated with the phenomenon of WETTING.[127,142]

When a drop of liquid contacts a solid surface it can remain exactly spherical, in which case no wetting occurs; or it can spread out to a perfectly flat film, and in that latter case complete wetting occurs. The angle of contact is the angle between the tangent to the periphery of the drop at the point of contact with the solid, and the surface of the solid. When the drop spreads to a perfectly flat film, the angle of contact is zero. When it remains exactly spherical, the area of contact is a point only, and the angle of contact is 180°.

angle of incidence angle between the axis of an impinging light beam and the perpendicular to the object surface.[43,69]

angle of reflection angle between the axis of a reflected light beam and the perpendicular to the specimen surface.[43,69]

angle of viewing angle between the axis of a detected light beam and the perpendicular to the object surface.[43,69]

angocopalolic acid $C_{23}H_{36}O_3$. A constituent acid of ANGOLA COPAL. It is an unsaturated monobasic acid and has a melting point of 85°C.[144]

angocopaloresenes identified as both α and ß forms, these are constituents of ANGOLA COPAL.[144]

The former has an mp of about 64°C and the latter about 222°C; α angocopaloresene corresponds with $C_{30}H_{54}O_6$, and ß angocopaloresene with $C_{25}H_{38}O_4$.

Angola copal fossil COPAL of African origin.[167]

Angstrom unit a unit of linear measure named after A.J. Ångström. It is 1 x 10^{-10} metre; 1 µm = 10,000 A.[128,43]

It is generally abbreviated as A. in the United States; elsewhere, it is variously abbreviated A, Å., A.U., A., or AU. Its use is discouraged in favor of the nanometre = 10^9m.

anhydride COMPOUND derived from an ACID by the elimination of water. The acid may be either ORGANIC or INORGANIC.[134,144]

anhydrite the mineralogical name for native anhydrous calcium sulfate which is often associated in nature with calcium sulfate dihydrate or GYPSUM. See CALCIUM SULFATE, ANHYDROUS.[73,34,58]

It occurs occasionally as an impurity in gypsum and Plaster of Paris.

anhydrous entirely free from water.[134]

It is generally accepted to mean free from water of any kind, whether as such in the free state, or as water of crystallization or other tightly held water.

anilides acyl derivatives from ANILINE. Possibly the most common anilide in the paint trade is ACETANILIDE, $C_6H_5NHCOCH_3$.[144]

aniline, anilin a colorless, oily, poisonous BENZENE derivative, $C_6H_5NH_2$, used in the manufacture of RUBBER, DYES, RESINS, pharmaceuticals, and VARNISHES.[144]

aniline cloud point see ANILINE POINT.[128,157]

aniline dyes large class of synthetic DYES made from intermediates based upon or made from ANILINE.[41]

aniline inks solutions of COAL TAR DYESTUFFS in ORGANIC SOLVENTS. Modern inks employ PIGMENTS.[168] See FLEXOGRAPHIC INK.[168]

These are classified as spirit inks and emulsion inks.

aniline number see ANILINE POINT.[157]

aniline pigments made from DISTILLATION PRODUCTS OF COAL TAR, a by-product of coke and coal gas manufacture.[41] Syn: coal-tar colors

aniline point minimum equilibrium solution temperature for equal volumes of ANILINE and SOLVENT.[128,157] Syn: aniline cloud point, aniline number

aniline printing see FLEXOGRAPHY.[168]

aniline-formaldehyde resin SYNTHETIC RESIN formed by the POLYCONDENSATION of ANILINE with FORMALDEHYDE.[130]

aniline-furfural resin FURFURAL yields resinous COMPOUNDS not only with ANILINE but with aromatic amines generally.[130]

ROSIN MODIFIED RESINS have also been produced. Amine-furfural resins are dark in color. BLEACHING occurs on exposure to light. They possess the unique property of compatibility with CELLULOSE ACETATE.

animal blacks black PIGMENTS made by burning or calcining animal bones, ivory, etc. Three types are known: DROP BLACK, BONE BLACK, and IVORY BLACK.[41]

Drop black and bone black are produced by the calcination of bones, but ivory black in its original cake form is virtually extinct.

animal fats includes such products as lard, bone fat, tallow, and butter fat.[144]

They are semisolid at ordinary air temperature, and consist chiefly of the glycerides of oleic and stearic acids.

animal oils normally restricted to the animal foot oils and lard oil, and quite distinct from the fish oils.[63]

Neatsfoot oil is typical of the foot oils. The animal oils are characterized by the almost complete absence of acids more unsaturated than oleic acid, and by the presence of cholesterol. The presence of this latter compound enables them to be distinguished from the nondrying vegetable oils.

animal waxes obtained from a great variety of sources and have little in common, except absence of glycerides.[137]

animi fossil COPAL, sometimes described as "gooseskin" animi because of the characteristic markings on its surface.[167] Syn: gum animi, and mombassa gum

It originated in Zanzibar (now Tanzania) and hence its alternative name, Zanzibar copal.

anion ion having a negative charge.[61]

Anions in a liquid subjected to electric potential collect at the positive pole or anode.

anion exchange resins certain SYNTHETIC RESINS with the property of absorbing ANIONS from AQUEOUS SALT SOLUTIONS.[130] Syn: ion exchange resins

anionic electrodeposition ELECTRODEPOSITION in which the part being painted is wired to serve as the ANODE or electron donor.[77]

This was the first type of electrodeposition developed, but has been largely replaced by cathodic electrodeposition which gives better CORROSION protection.

anionic emulsion an EMULSION in which the emulsifying system establishes a predominance of negative charges on the DISCONTINUOUS PHASE.[60] (ASTM) cf. CATIONIC EMULSION[60]

anionic surfactants SURFACTANTS which give negatively charged ions in an aqueous solution.[127]

anisole $C_6H_5OCH_3$. Mixed AROMATIC-ALIPHATIC ETHER.[144,164]

It has good solvent properties and a bp of 154°C.

anisotropic not ISOTROPIC.[133]

anisotropy exhibiting different OPTICAL properties when tested along axes in different directions.[43,133] See ISOTROPIC.[43]

ANLAB color difference equation abbreviation used primarily in Britain for Adams-Nickerson, L, a, b color difference equation.[69] See ADAMS CHROMATIC VALUE COLOR DIFFERENCE EQUATION.[69]

The normalizing factor is generally 40 and b is calculated as $0.4(V_Z - V_Y)$.

annealing heat treatment of materials to increase the ductility or remove internal stresses, as of METAL, GLASS, OR PLASTICS.[38]

annual growth ring growth layer of a tree put on in a single growth year. This includes SPRINGWOOD and SUMMERWOOD.[153]

anode (1) the positive terminal of an electrical source to which electrons and negatively charged ions travel; (2) the electrode at which CORROSION (oxidation) occurs.[61,66]

anprolene see ETHYLENE OXIDE.[144,141]

ANS abbreviation for ADAMS-NICKERSON-STULTZ COLOR DIFFERENCE EQUATION.[69]

ANSI abbreviation for AMERICAN NATIONAL STANDARDS INSTITUTE, which is the U.S. member of ISO.[174]

anthracene $C_6H_4{:}(CH)_2{:}C_6H_4$. Yellow, crystalline solid obtained from the distillation of COAL TAR.[41,144]

anthracene oil fraction obtained from the distillation of COAL TAR, with a boiling range in the region of 270°C, and containing a substantial proportion of anthracene and similar HYDROCARBONS.[144]

anthranilic acid see A-ACID.[41,144]

antibiotic a substance derived from a living organism capable of killing or incapacitating another organism.[173]

antiblocking agent ADDITIVE used to prevent the undesirable ADHESION between touching layers of coated material such as occurs under moderate pressure (and sometimes heat and pressure) during storage, manufacture, or use.[83] cf. MOLD WASH[83,79,140]

anticondensation paint coating designed to minimize the effects of CONDENSATION of MOISTURE under intermittently dry and humid conditions. Such a material normally has a matte textured finish and frequently contains heat-insulating material as a filler.[71,163] (BSI)

anticorrosive paint or composition coating used for preventing the CORROSION of metals and, more particularly, specially formulated to prevent the RUSTING of iron, steel, and other metals.[66,71,72,163] (BSI)

antifloating agent ADDITIVE used to prevent FLOATING.[83]

Words presented in CAP/SMALL CAPS type indicate that the word is defined in another part of the *Coatings Encyclopedic Dictionary.*

Numerical superscripts classify terms in one or more of the categories listed in the second section of this volume.

antiflooding agent ADDITIVE used to prevent FLOODING.[83]

antifoaming agent ADDITIVE used to prevent formation of FOAM or added to the foam itself to break a foam already formed.[60,83] cf. DEFOAMER[83]

Long chain saturated alcohols containing from 6 to 18 carbon atoms and emulsions of certain silicone polymers are among the principal antifoams used.

antifouling paints coatings designed to prevent the growth of BARNACLES and other organisms on ships' bottoms.[71,163,173] (BSI)

It usually contains substances which are poisonous to such organisms in the early stages of growth. Such compositions are applied over protective paints, e.g., over anticorrosion paints in the case of steel ships.

ASTM test methods for antifouling paints include: submergence test, D3263; erosion test, high velocity water, D4938; erosion test, rotating drum, D4939; organotin release rate, D5108.

antigism interaction of discrete agents (such as drugs) such that the total effect is less than the sum of the individual effects; the antonym of SYNERGISM.

This is not a dictionary word, but is occasionally seen in print as the opposite of synergism.

antilivering agent ADDITIVE used to prevent the livering of a coating.[83,161] See LIVERING.[161,166,156,42]

antimony orange Sb_2S_3. Like ANTIMONY VERMILLION, which it resembles closely, it is a SULFIDE PIGMENT, and can be prepared by precipitation from an AQUEOUS antimony salt solution with hydrogen sulfide.[41]

antimony oxide Sb_2O_3. Pigment White 11 (77052). Antimony and oxygen combine to form trioxide, tetraoxide, and pentoxide.[41] Syn: antimony trioxide, antimony white, antimony sesquioxide, flower of antimony

The most common antimony trioxide pigment is used as a flame retardant in paints, plastics, and textiles. It contains, in addition, traces of the tetroxide (Sb_2O_4), and of the oxides of lead, arsenic, and iron. Density, 5.3-5.7 g/cm³ (44.0-47.5 lb/gal); O.A., 11-14; particle size, 1 µm.

antimony sesquioxide see ANTIMONY OXIDE.[41]

antimony trioxide see ANTIMONY OXIDE.[41]

antimony vermillion Sb_2S_3. PIGMENT varying in COLOR from pale orange to deep crimson, with good opacity and tinting strength. It is sometimes referred to as CRIMSON ANTIMONY.[41]

It is prepared either by adding a solution of sodium thiosulfate to a solution of tartar emetic and tartaric acid, or to a solution of another suitable antimony salt. The pigment is not resistant to ACIDS, ALKALIS, or heat, but, if properly made, is LIGHTFAST.

antimony white see ANTIMONY OXIDE.[41]

antimony yellow $Pb_3(SbO_4)_2$. Essentially, lead antimonate, which may be considered to be chemically combined lead and antimony oxides.[41] Syn: Naples yellow

It varies in COLOR from sulfur yellow to orange-yellow, depending upon the proportion of the two materials.

antinoise paint coating used for its sound deadening effect, on vibrating and naturally noisy machinery.[71,163] Syn: acoustical paint[71,163]

It relies for its effectiveness on its very rough surface, produced by the incorporation of cork and similar granular, or fibrous materials.

antioxidant compound which prevents oxygen from reacting with other compounds which are susceptible to OXIDATION.[67,83]

They are often themselves oxidized in the process of protecting the other compounds.

antiozonant substance that retards or prevents the action of OZONE on ELASTOMERS when exposed under tension, either statically or dynamically, to air containing ozone.[83]

antique finish usually applied to newly painted furniture or objects for the purpose of giving the appearance of mellowness of age.[54,148,163,175]

antiquing (1) treating or finishing of wood to make it look old;[54,148,163,175] (2) imparting a special coloring effect by the application of a BASECOAT followed by a transparent or semitransparent colored gaze, which is either applied by BRUSH, then partially wiped away or applied by wiping with a cloth. Parts of the undercoat are exposed by COMBING, GRAINING, or MARBLING. The effects obtained are: natural woods, aged coatings, or any number of special effects.[54,148,163] Syn: broken-color work

For added DURABILITY and control of final GLOSS, a clear SATIN FINISH is sometimes applied.

antisag agent ADDITIVE used to control SAGGING of a coating.[83,166]

antiseptic an agent used to destroy or restrain the growth of MICROORGANISMS.[173]

antiseptic wash see FUNGICIDAL WASH.[158]

antisettling agent substance incorporated into pigmented paint to retard SETTLING and to maintain uniform consistency during storage or, as in DIPPING paints, during painting operations.[83] (BSI) See SUSPENSION AID.[83,64,161]

These additives normally function by altering the rheological properties of the paint.

antishatter composition composition designed to prevent the fragmentation of glass and glass-like materials.[71,79,163]

The main use of these compositions is in the manufacture of splinterproof glass. They are applied as an adhesive, resilient sandwich between relatively thin sheets of glass. A violent blow may cause cracking of the glass, but the antishatter composition is designed to retain the various pieces in position. Properties include good permanent adhesion to the glass, a natural resilience such that it will absorb violent shocks without itself disintegrating, a permanent freedom from color, and absolute transparency.

antisilking agent ADDITIVE used to prevent SILKING, which is a special case of FLOATING, resulting in the formation of parallel hairlike striations of differing colors, running through the length of the painted surface.[83]

antiskid paint see ANTISLIP PAINT.[71] Syn: nonskid paint[71]

antiskinning agent any material added to a coating or printing INK to prevent or retard the processes of OXIDATION or POLYMERIZATION which result in the formation of an insoluble SKIN on the surface of the coating or printing ink in an unopened container.[83]

antislip paint used for APPLICATION to decks and other surfaces where the conditions conducive to slipping are present.[71,163] Syn: nonskid paint

It prevents slipping by providing a tough, rough surface. The roughness is usually achieved by the addition of sand or cork dust.

antistatic agents ADDITIVES which, when added to the molding material or applied on the surface of the molded object, make it less conducting (thus hindering the fixation of DUST).[61,83]

antitracking varnish when two points of high electrical potential difference are suitably located in contact with the surface of an ORGANIC insulating medium, the surface of the insulator may be degraded to CARBON under the imposed electrical stresses, and a carbon track formed between the two points.[71,152,163]

PHENOL-FORMALDEHYDE laminated boards are sometimes given a special coating of VARNISH at their exposed surfaces to prevent such carbonization. Varnishes used for such purposes are described as antitracking varnishes.

antiwrinkling agent material added to surface coating compositions to prevent the formation of wrinkles in films during drying.[56,83]

Antwerp blue essentially, zinc ferrocyanide with a pale greenish-blue COLOR.[41]

It is characterized by good opacity and almost complete absence of FLOATING tendency.

AOCS abbreviation for AMERICAN OIL CHEMISTS SOCIETY.[174]

APHA color scale abbreviation for American Public Health Association color scale developed for visually evaluating slight YELLOWNESS of solutions.[69,128,43,157] Syn: HAZEN COLOR SCALE.[69]

It is based on dilutions of a platinum-cobalt standard solution of 500 APHA. This single-number color scale was originally developed by Hazen.

API abbreviation for American Petroleum Institute.[174]

API gravity measure of SPECIFIC GRAVITY of PETROLEUM and petroleum products, defined by the following equation:[128]

$$\text{API Gravity, degrees} = \frac{141.5}{\text{Specific gravity } 60/60°F} - 131.5$$

apparent density mass per unit volume of loose, dry PIGMENT (or POWDER) including the air voids between the particles.[128] cf. BULKING VALUE[128] Syn: bulk density

Usually expressed as lbs/cubic foot or g/cm^3.

apparent viscosity (of a NON-NEWTONIAN fluid) the viscosity of a NEWTONIAN LIQUID that produces the same reading in the same apparatus under identical conditions. Avoidance of this artificial term is recommended.[166] (ASTM)

appearance manifestation of the nature of objects and materials through visual attributes such as size, shape, color, texture, glossiness, transparency, opacity, etc.[43,69,35]

appliance finish generally, THERMOSET coatings, which are characterized by their HARDNESS, MAR RESISTANCE, and good CHEMICAL RESISTANCE.[71,163]

application PROCESS by which surface coating compositions are transferred to a variety of surfaces, such as: BRUSHING; SPRAYING (cold and hot); DIPPING (simple immersion); DIPPING (assisted by application of vacuum and/or pressure); ROLLER COATING; FLUSHING, and SPREADING.[77,131]

application rate the quantity (mass, volume, or thickness) of material applied per unit area.[77] (ASTM)

applicator device to deposit a film of a specified thickness, such as DOCTOR BLADE, WIREWOUND ROD, DRAWDOWN BAR, etc.[77,59,157] See FILM APPLICATOR.[77,59,157]

applicator roll roll in a rollercoater which applies the paint to a continuous strip.[59,70]

Words presented in CAP/SMALL CAPS type indicate that the word is defined in another part of the *Coatings Encyclopedic Dictionary*.
Numerical superscripts classify terms in one or more of the categories listed in the second section of this volume.

applique (1) pattern or picture, formed by laying various materials usually on a FABRIC ground. Especially suitable for dressmaking and needlework;[81] (2) a design or ornament applied to another surface; (3) in WALLCOVERING, cut-outs applied to a plain, textured, or figured background.[148]

approximate value a value that is nearly but not exactly correct or accurate.[36,128] (ASTM)

appurtenance any built-in, nonstructural portion of a building, such as doors, windows, ventilators, electrical equipment, partitions, etc.[82] (DAC)

apricot kernel mixed fatty acids these have an mp of 12-14°C; an acid value of 194; and an iodine value of 103.[63,144]

apricot kernel oil NONDRYING OIL with iodine value of 96.[63,144]

apron a paved area, such as the juncture of a driveway with the street or with a garage entrance.[82]

aptitude, color see COLOR APTITUDE.[69]

A. pullulans see *AUREOBASIDIUM PULLULANS*.[173]

aquarelle (1) drawing or print which has been colored with TRANSPARENT WATER COLOR WASHES; (2) color that is made workable when mixed with water.[81]

aquatint method of ENGRAVING which, like etching, involves the use of a mordant acid for biting a METAL plate, but differing in that aquatint is used to render tonal instead of linear effects.[81]

aqueous water-containing or water-based.[155]

arabic see ACACIA GUM or GUM ARABIC.[79,155]

arachis oil see PEANUT OIL.[63]

Archangel pitch see PITCH, ARCHANGEL.[44,76]

architectural coatings ORGANIC coatings intended for on-site application at ambient temperatures to interior or exterior surfaces of residential, commercial, institutional, industrial, or government structures.[71] cf. PRODUCT COATINGS and SPECIAL-PURPOSE COATINGS[71,159] Syn: trade sales coatings

Architectural coatings include EXTERIOR and interior house paints, STAINS, UNDERCOATERS, PRIMERS and SEALERS. These coatings are distributed through wholesale or retail channels and are purchased by the general public, painters, building contractors, government agents, and others.

ASTM test methods for architectural paints include: block resistance, D4946; color permanence, white enamels, D1543; brushability (brush drag), D4958; efflorescence of wall paints, D1736; film failures, exterior latex paint, D1848; film porosity, D3258; freeze-thaw resistance, D2243; gloss or sheen uniformity, D3928; guides for testing architectural coatings: exterior paints, solvent-borne, D3323; waterborne, D3129; flat paints, solvent-borne, D3323; waterborne, D2431; floor paints, solvent-borne, D3383; waterborne, D3358; gloss and semigloss, solvent-borne, D3425; waterborne, D4540; high performance (HIPAC), D3730; solvent-borne (general), D5146; guide for purchasing: state and institutional, D3927; hiding power: brush application, visual, D344; drawdown, reflectometric, D2805;wet-to-dry hiding change, D5007; leveling, D4062; minimum film forming temperature, D2345; package stability, D1849; porosity of films, D3258; print resistance, D4207; roller spatter, resistance to, D4707; sag resistance, D4400; washability: soilant, mechanical test, D3450; practical multi-stain test, D4828; wet abrasion (scrub) resistance: scrub-to-failure test, D2486; weight-loss test, D4213.

architecture (1) the art and science of designing and building structures, or large groups of structures, in keeping with aesthetic and functional criteria; (2) structures built in accordance with such principles.[82] (DAC)

arcing swinging SPRAY GUN away from the perpendicular.[54]

Arco Microknife instrument designed for testing SCRATCH HARDNESS and ADHESION of coatings.[145,157,79]

A diamond point is weighted until it can penetrate a film to the metal substrate in two retracing steps. The mass necessary to achieve this cutting force for films of standard thickness is a measure of hardness.

argillaceous composed primarily of CLAY or SHALE; clayey.[41,134] (DAC)

Arnaudon's green see CHROMIUM OXIDE GREEN.[41]

aromatic compounds a class of ORGANIC COMPOUNDS which contain an UNSATURATED RING of CARBON atoms, including BENZENE, NAPHTHALENE, ANTHRACENE, and their derivatives.[134] cf. ALIPHATIC COMPOUNDS[134]

aromatic content percent AROMATIC HYDROCARBONS present in a SOLVENT mixture or in a COMPOUND.[134]

aromatic petroleum residues see DIOLEFIN RESINS.[130]

aromatic solvents HYDROCARBON SOLVENTS comprised wholly or primarily of aromatic HYDROCARBON COMPOUNDS.[164]

Aromatic solvents containing less than 80% aromatic compounds are frequently designated as partial aromatic solvents.

aromatization PROCESS of converting SATURATED or ALIPHATIC HYDROCARBONS to AROMATIC hydrocarbons.[131]

Arrhenius equation the equation, $k = Ae^{-E_a/RT}$, which describes the rate of coefficient (k) for an elementary reaction in terms of pre-exponential factor A, activation energy E_a, temperature T (K), and gas constant R.[157,161] (IUPAC)

Some accelerated tests such as the HEAT STABILITY TEST *are based on this equation.*

arsenic orange see REALGAR.[41]

artificial daylight light sources, frequently equipped with filters, which are claimed to reproduce the COLOR and spectral distribution of daylight.[43,69] See CORRELATED COLOR TEMPERATURE, COLOR RENDERING INDEX, and SPECTRAL POWER DISTRIBUTION CURVE.[43,69]

A more specific definition of the light source is to be preferred.

artificial illuminant a synthetic light source of SPECTRAL distribution as close as possible to that of the natural illuminant (usually daylight) to be duplicated.[43]

artificial stone special CONCRETES and tiles, artificially colored to simulate natural stone, obtained by mixing stone dust AGGREGATE and chips with PORTLAND CEMENT.[73,75]

artificial ultramarine synthetic ultramarine blue.[41] See ULTRAMARINE BLUE.[41]

artificial weathering EXPOSURE to cyclic laboratory conditions involving changes in temperature, RELATIVE HUMIDITY, and RADIANT ENERGY, with or without direct water spray, in an attempt to produce changes in the material similar to those observed after long-term continuous outdoor exposure.[154,157] (ASTM)

artists' colors (or paints) various paint media used by artists, such as OIL PAINTS, WATERCOLORS, GOUACHE, TEMPERA, ENCAUSTIC, FRESCO, SILICATE ESTERS, and LATEX.[71,81]

aryl pertaining to monovalent AROMATIC groups, e.g., C_6H_5, phenyl.[133,134]

asbestine see MAGNESIUM SILICATE, NONFIBROUS.[41,58]

asbestos a group of fibrous MINERALS which occur as small veins in the massive body of natural hydrous silicates of serpentine or amphibole and have heat-, fire-, and solvent-resistant properties.[41,83] (ASTM) See MAGNESIUM SILICATE, FIBROUS.

Typical types are chrysotile, amosite, crocidolite, tremolite, and actinolite. Two common forms of asbestos are serpentine and amphibole. The former is a hydrated variety of magnesium metasilicate, while the latter is chiefly a mixed calcium-magnesium metasilicate.

asbestos board see ASBESTOS-CEMENT BOARD.[73,159,75]

asbestos plaster was a fireproof insulating material generally composed of ASBESTOS with bentonite as the BINDER.[159,75] (DAC)

asbestos-cement board (ACB), asbestos-cement wallboard, asbestos sheeting a dense, rigid, board containing a high proportion of ASBESTOS fibers bonded with PORTLAND CEMENT; resistant to fire, flame, and weathering; has low resistance to heat flow.[73,75,159] (DAC) Syn: cement-asbestos board

Was used as a building material in sheet form and corrugated sheeting.

asbestosis a chronic lung disease caused by inhaling airborne ASBESTOS fibers.[172]

ash the nonvolatile INORGANIC matter of a COMPOUND which remains after subjecting it to a high decomposition temperature.[157]

aspartic acid $COOHCH_2CH(NH_2)COOH$. AMINO ACID constituent of PROTEINS of both animal and vegetable origin. In particular it is present in CASEIN. It is a dibasic acid.[144] (ASTM)

asphalt (1) black to dark-brown solid or semisolid CEMENTITIOUS material which gradually liquefies when heated. The predominating constituents are BITUMENS, all of which occur in the solid or semisolid form in nature or are obtained by refining PETROLEUM, or which are combinations of the bitumens mentioned with each other or with petroleum or derivatives thereof;[76] (2) a similar material obtained artificially in refining petroleum, used in BUILT-UP ROOFING SYSTEMS as a waterproofing agent; (3) a mixture of such substrates with an AGGREGATE for use in paving.[76] (ASTM) (DAC)

asphalt, air blown an ASPHALT produced by blowing air through molten asphalt at an elevated temperature to raise its softening point and modify other properties.[76] (ASTM)

asphalt cut back ASPHALT plus THINNER; asphalt solution; asphalt coating formed by dissolving asphalt.[76]

asphalt emulsion a SUSPENSION or EMULSION of ordinary ASPHALT in water.[76] Syn: emulsified asphalt[76]

Such emulsions have attained wide use because, unlike straight asphalt, they do not have to be heated to be applied. The suspended asphalt is spread on in the usual way; and after the water has evaporated, the asphalt hardens into a continuous mass. Uses: highways; cement waterproofing; roofing compounds; and the like.

asphalt paint a liquid ASPHALTIC product sometimes containing small amounts of other materials such

Words presented in CAP/SMALL CAPS type indicate that the word is defined in another part of the *Coatings Encyclopedic Dictionary.*

Numerical superscripts classify terms in one or more of the categories listed in the second section of this volume.

as lampblack, aluminum flakes, and mineral pigments.[71,76] (DAC)

asphalt paper a PAPER sheet material that has been coated, saturated, or laminated with ASPHALT to increase its toughness and its resistance to water.[75,76] (DAC)

asphalt rock (rock asphalt) a naturally occurring rock formation, usually limestone or sandstone, containing throughout its mass a minor amount of ASPHALT.[76] (ASTM)

asphalt seal coat BITUMINOUS coating, with or without AGGREGATE, applied to the surface of a pavement to waterproof and preserve the surface and to improve the texture of a previously applied bituminous surface.[76,163] (DAC)

asphalt, steam blown an ASPHALT produced by blowing steam through molten asphalt to modify its properties.[76] (ASTM) cf. ASPHALT, AIR BLOWN[76]

asphalt tile sealer RESIN or PLASTIC-type finish usually containing a water or ALCOHOL carrier.[147,163]

asphaltene a high MOLECULAR WEIGHT HYDROCARBON fraction precipitated from ASPHALT by a designated paraffinic naphtha solvent at a specified temperature and SOLVENT-asphalt ratio.[76] (ASTM)

The asphaltene fraction should be identified by the temperature and solvent-asphalt ratio used. (ASTM) Little is known of their chemical structure, but they are usually characterized by their insolubility in low boiling ALIPHATIC hydrocarbons. Their physical condition is solid or semisolid.

asphaltic bitumens may be either naturally occurring materials or otherwise obtained from the distillation of asphaltic base PETROLEUM; they are readily soluble in ALIPHATIC and AROMATIC HYDROCARBONS and have good film-forming properties.[76]

asphaltic pyrobitumens include elaterite, impsonite, and wurtzilite.[76]

They are characterized by their HYDROCARBON nature, infusibility, and insolubility in carbon disulfide. Oxygen-containing COMPOUNDS are present only in small amounts.

asphaltite a natural ASPHALT found below ground level.[76] (ASTM)

asphaltum U.S. paint industry's term for ASPHALT. In Britain, the term is reserved for natural asphalt.[76] See BITUMEN.[76]

ASR See ALKALI-SOLUBLE RESINS.[130,155]

assembly group of materials or parts, including ADHESIVE, which has been placed together for bonding or which has been bonded together.[79] (ASTM)

assembly time elapsed time after the ADHESIVE is spread and until the pressure becomes effective.[79]

associative thickener (AT) water-soluble or water-swellable POLYMERS that have chemically attached HYDROPHOBIC groups that are capable of nonspecific hydrophobic association similar to those of conventional SURFACTANTS that elevate viscosity presumably by association between thickener particles or thickener and dispersed particles that may be present in the LATEX particles rather than through high molecular weight or chain stiffness of the thickener particles themselves.[83,166] See RHEOLOGY MODIFIER, HEUR, and HASE.[83,166] cf. THICKENER Syn: rheology modifier

Associative thickeners impart many desirable properties to a latex paint. They provide: high shear viscosity which increases film BUILD; ROLLER SPATTER RESISTANCE; can provide essentially perfect LEVELING; offer the highest GLOSS potential; tend to increase OPEN TIME of the paint film (particularly the HEUR types); microbial degradation resistance; enhances the corrosion resistance by allowing the formulation of nonflocculated paints that form tighter films and provide lower permeability to undesirable corrosive ions.

A-stage early stage in the reaction of certain THEMOSETTING RESINS in which the material is fusible and still soluble in certain liquids.[67,79] (ASTM) Syn: RESOL (ASTM) cf. B-STAGE and C-STAGE[67,79]

ASTM abbreviation for AMERICAN SOCIETY FOR TESTING AND MATERIALS.[174]

AT abbreviation for ASSOCIATIVE THICKENER.[83,166]

atactic polymer a polymer in which there is complete randomness with regard to the configurations at all main chain sites of steric isomerism.[133,130,169] (IUPAC) cf. SYNDIOTACTIC POLYMER and ISOTACTIC POLYMER[133,130]

atmosphere a unit of pressure measurement which has been employed in describing the pressure of gases in storage tanks. It is roughly related to the average pressure of the atmosphere at sea level. One atmosphere = 1.01325×10^5 N m^{-2} (Pa) = 1013.25 mbar – 760 Torr.[128] (IUPAC)

The use of atm unit of pressure is no longer recommended by scientific bodies; it has been recommended that standard pressure be redefined as 10^5 Pa. (IUPAC)

atom a unit of matter, the smallest unit of an element, having all the characteristics of that element and consisting of a dense, central, positively charged nucleus surrounded by a system of electrons. The entire structure has an approximate diameter of 10^{-8} centimetre and characteristically remains undivided in CHEMICAL REACTIONS except for limited removal, transfer, or exchange of certain electrons.

atomic mass unit (amu) the unit of measure of the MASS of a particle (atom, molecule, ion, etc.), defined as 1/12 of the mass of carbon-12.[128,144] (ASTM)

The numerical value of the mass of a particle in terms of amu is identical with the older ATOMIC WEIGHT. (ASTM)

atomic weight see ATOMIC MASS UNIT (AMU).[128,144]

atomize to subdivide a liquid into very small particles.[131]

attapulgite $(Mg,Al)_5Si_8O_{20}(OH)_2 \cdot 8H_2O$. A CLAY MINERAL with the ideal formula in which there is considerable replacement of magnesium by aluminum.[41,58,34] See HYDRATED MAGNESIUM ALUMINUM SILICATE.[41,58] cf. KAOLINITE,[41,58,34] MONTMORRILONITE,[41,58,34] HECTORITE,[41,58,34] BENTONITE[41,139]

Electron micrographs show single laths or needles and bundles of laths or needles oriented in random fashion, which structure is generally believed to account for the high degree of porosity and absorptivity characteristic of this mineral. The name is derived from the place name, Attapulgus, GA.

attribute distinguishing characteristic of a sensation, perception, or mode of appearance.[43,35]

Distinction is made between chromatic and geometric appearance attributes.

A.U. see ANGSTROM UNIT.[128]

Aureobasidium pullulans a common ubiquitous, omniverous, and highly polymorphic species of FUNGUS which is often the primary organism in defacement of paint films.[173]

Often referred to as a "black yeast." A. pullulans is heterocaryotic since cells of most strains contain one to eight nuclei. It is now thought that Pullularia pullulans and A. pullulans are one and the same organism.

aureolin potassium cobaltinitrite yellow PIGMENT.[41] Syn: cobalt yellow

Austrian cinnabar see CHROME ORANGE.[41]

autoclave closed vessel for conducting a chemical reaction, sterilization, or other operation under pressure and heat.[59]

autodeposition a single-step immersion METAL FINISHING PROCESS in which an ORGANIC coating is applied by means of unique surface chemical reactions carried out in an AQUEOUS LATEX DISPERSION, also referred to as CHEMIPHORESIS. Components within the bath give rise to chemical reactions that slightly solubilize the metallic surface and lead to destabilization, DEPOSITION, and COALESCENCE of the dispersed LATEX particles at that surface.[77] (ASTM)

The deposition rate of the latex is controlled by the rate of surface solubilization. The process does not require any pretreatments such as phosphating, needs no external energy input, and gives rise to deposition wherever the solution wets the substrate. Irregularly shaped parts can be uniformly coated.

autohesion term which refers to the ability of two contiguous surfaces of the same material to form a strong bond which prevents their separation at the place of contact.[79] Syn: self-adhesion[79]

autoignition temperature the minimum temperature at which a substance ignites without application of a flame or spark.[172]

There are safety caveats recommending that materials not be heated to greater than 80% of this temperature.

automatic electrostatic spray application ELECTROSTATIC SPRAY applied by a robot or other self-acting mechanism.[77] (EPA)

Automatic electrostatic spray can be a very efficient way to transfer coatings to the SUBSTRATE thus reducing POLLUTION which would otherwise result from wasted paint OVERSPRAY.

automatic press hydraulic press for compression molding or an injection machine which operates continuously, being controlled mechanically, electrically, hydraulically, or by a combination of any of these methods.[59]

automatic turning sander a machine designed to use coated ABRASIVES to polish chair legs and posts and other irregular shapes that are produced from a turning lathe.[59,158] Syn: brush backed sander

The turnings are rotated against a bank of abrasive papers or cloths which are backed up and supported by BRISTLE BRUSHES.

automation as used by engineers, term applied to several types of technical developments, including: (a) a continuous-flow production process which integrates various mechanisms to produce a finished item with relatively few or no worker operations, usually through electronic control; (b) self-regulating machines that can perform highly precise operations in sequence; and (c) electronic computing machines.[131]

In common usage, however, the term is often used in reference to any type of advanced mechanization or as a synonym for technological change.

automotive finish complex number of coatings specifically designed for automobiles, which typically includes ZINC-RICH PHOSPHATE PRETREATED clean STEEL coated with an ELECTRODEPOSITION PRIMER, then

---

Words presented in CAP/SMALL CAPS type indicate that the word is defined in another part of the *Coatings Encyclopedic Dictionary.*

Numerical superscripts classify terms in one or more of the categories listed in the second section of this volume.

a SPRAY PRIMER followed with a COLOR COAT and finally a CLEARCOAT.[71,163]

auto-oxidation the breakdown of COMPOUNDS such as FATTY ACIDS into peroxides and hydroperoxides upon addition of molecular oxygen across double bonds during the STORAGE of OIL.[65]

auxochrome see CHROMOPHORE.[69,41]

average molecular weight (viscosity method) molecular weight of POLYMERIC materials determined by the VISCOSITY of the polymer in solution at a specific temperature.[166,169]

This gives an average molecular weight of the molecular chains in the polymer independent of specific chain length.

azelaic acid $HOOC(CH_2)_7COOH$. Dibasic ACID of ALIPHATIC type.[144]

Properties: acid value, 595.5; mol wt, 188.2; mp, 106°C; bp, 237°C/15 mm Hg.

azeotrope a liquid mixture that is characterized by a constant minimum or maximum boiling point which is lower or higher than any of the components, and that distills without change in composition.[164]

azeotropic copolymer a COPOLYMER in which the relative numbers of the different kinds of units (MERS) are the same as in the mixture of MONOMERS from which it was obtained.[130]

azo dyes (1) derivatives of azo-BENZENE formed by the interaction of diazonium salts with phenols or tertiary AMINES; (2) DYES that have the –N = N– group as a CHROMOPHORE group in their molecular structure and are produced from amino compounds by the processes of diazotization and coupling.[41,69] See DYES and DYESTUFFS.[41]

Over half of the commercial dyestuffs are in this general category. By varying the chemical composition, it is possible to produce acid, basic, direct, or mordant dyes. This general group is subdivided as monoazo, disazo, trisazo, and tetrazo according to number of –N = N– groups in the molecule. Examples of these dyes are chrysodine Y, bismarck brown 2R, and direct green B.

azo pigment see BENZIDINE YELLOWS.[41]

azure blue see COBALT BLUE.[41]

azurite natural blue PIGMENT derived from the MINERAL, azurite, a basic copper carbonate.[41,34] Syn: mountain blue

The mineral occurs in various parts of the world in secondary copper ore deposits where it is frequently associated with malachite, a green basic carbonate of copper.

B

"b," Kubelka-Munk mathematical constant characteristic of a COLOR at complete OPACITY; dependent on the optical constants K and S: $b = [2(K/S) + (K/S)^2]^{1/2}$.[69,43] See "A," KUBELKA-MUNK EQUATION.[69]

"b" or "ß" YELLOWNESS-blueness coordinate in certain COLOR SPACES, generally used as Δb, the difference in "b," between a specimen and a standard reference color, normally used with "a" or "α" as part of the chromaticity difference. Generally, if "b" is plus, there is more yellowness than blueness; if "b" is minus, there is more blueness than yellowness. The exception is Adams Chromatic Value.[69,43] See UNIFORM CHROMATICITY SCALE DIAGRAM and COLOR DIFFERENCE EQUATIONS.[69,43]

back paint see BACK PRIMING.[71,54]

back priming the PROCESS of applying a coating to the back SURFACE of construction materials prior to installation, usually for protection against the weather.[71,54]

back putty see BED PUTTY.[165]

backing away from fountain a condition in which an INK lacking in FLOW will not keep in contact with the fountain ROLLER so that the latter can transfer INK to the ductor roller.[168]

Eventually, the prints become uneven, streaky, and weak.

backing coat in COIL COATING, the FINISH on the back side of continuous strip.[70,71]

bacteria any of the numerous microscopic, spherical, rod-shaped, or spiral organisms of the class, *Schizomycetes*.[173]

bacterial corrosion a CORROSION which results from substances (e.g., ammonia or sulfuric acid) produced by the activity of certain BACTERIA.[66,173] (DAC)

bacterial resistance see BIODETERIORATION.[173]

bactericidal capable of causing the death of BACTERIA.[173]

bactericidal paint coating which discourages the multiplication of BACTERIA.[71,163,173]

bactericide material used in small percentages to kill BACTERIA.[83,173]

bacteriostat substance which prevents or slows down the growth of BACTERIA.[83,161,173]

bacteriostatic inhibiting the growth of BACTERIA, but not BACTERICIDAL.[173]

baddeleyite see ZIRCONIUM OXIDE.[41,34]

baffle device used for directing a fluid current in a desired direction.[59,70,131]

A metal piece so placed in a rollercoater pan to direct the stream or flow of paint to the pick-up roller. Also used to direct air currents in an oven.

bag molding method of molding or bonding involving the application of fluid pressure, usually by means of air, steam, water, or vacuum, to a flexible cover which, sometimes in conjunction with the rigid die, completely encloses the material to be bonded.[131,57] (ASTM)

bagillumbang oil DRYING OIL obtained from the species *Aleurites trisperma*, which grows in the Philippines.[63]

bail semicircular handle of a kettle, pail, or can.[50]

bailing machine mechanical device designed to automatically attach BAILS to containers.[50,59]

bake hardness increased HARDNESS on heating due to elimination of retained SOLVENT in THERMOPLASTIC coatings and thermal CROSSLINKING in THERMOSETTING coatings.[47,67,176]

bake system set of PRIME, INTERMEDIATE, and/or TOPCOATS which require baking to effect a cured or dried film, and which together have been determined to yield the desired properties for a particular purpose.[55,67,176]

Bakelite® proprietary name for PHENOLIC materials which were produced by Union Carbide Corp., but often used indiscriminately to describe any phenolic material.[38,130]

The name was derived from the Belgian chemist, Baekeland, who developed PHENOL-FORMALDEHYDE RESINS.

baking PROCESS of drying or CURING a coating by the application of heat in excess of 65°C (150°F).[67,131,176] Syn: stoving. When below this temperature, the process is referred to as FORCED DRYING.

baking finish paint or VARNISH that requires baking at temperatures above 65°C (150°F), for the development of desired properties.[67,71,176] (ASTM)

Words presented in CAP/SMALL CAPS type indicate that the word is defined in another part of the *Coatings Encyclopedic Dictionary.*

Numerical superscripts classify terms in one or more of the categories listed in the second section of this volume.

baking schedule set of related baking temperatures and baking times which will yield a CURED or dried film.[55,67,176]

The schedule will start with the lowest temperature, above 65°C (150°F), and longest time and progress through higher temperatures and corresponding shorter times, stopping short of the temperature at which the defects of overbake will occur.

baking temperature temperature above 65°C (150°F) at a specified time in a baking schedule which has been determined to produce a CURED or dried film having optimum desired properties for the coating.[67,55,176]

baking time time at a specified temperature, above 65°C (150°F), in a BAKING SCHEDULE which has been determined to produce a CURED or dried film having optimum desired properties for the coating.[67,55,176]

baking vehicle (varnish) vehicle specifically formulated for coatings which are intended to be baked in order to effect a CURED or dried film.[156,152,176]

balance, Westphal type see WESTPHAL BALANCE and MOHR BALANCE.[145]

balanced reaction a reaction in which a state of equilibrium has been reached, and it can be made to proceed in one direction or another by adjusting the conditions.[131]

Conditions which affect the direction of a reaction are the concentrations of the reactants involved and the temperature and the pressure.

balanced solvents combination of SOLVENTS designed to give a specified performance.[164]

balata dried juice of a West Indian tree (*Mimusops globosa*).[140]

It has rubbery characteristics and is sometimes used as a substitute for GUTTA-PERCHA. It resembles both this and ordinary rubber in that it consists of ISOPRENE molecules (C_5H_8), but POLYMERIZED in its own peculiar way.

bald spot area or patch, usually in a WRINKLE FINISH FILM, which has failed to wrinkle or give the desired optical effect.[56]

bale quantity of compressible articles or materials assembled into a shaped unit and bound with cord or metal ties.[50]

baling machine device used to form and fasten compressible materials into a shaped unit (BALE).[50,59]

ball when RUBBING DOWN paint, VARNISH films, or the like, with ABRASIVE PAPER, the material removed by the abrasive action may be either in the form of a dry powder or, if it has a tendency to softness or stickiness, will collect as relatively large lumps or balls. A film which exhibits this latter phenomenon is said to ball.[56,54]

ball and ring method system of testing the MELTING and SOFTENING temperatures of ASPHALT, WAXES, RESINS, and PARAFFINS.[130,167,137,157] See SOFTENING POINT.[130,167,137,157,176]

A ring 15.875 mm in diameter is filled with the substance to be tested, and a steel ball 9.5 mm in diameter is placed upon the substance in the ring. The end point is the temperature at which the substance softens sufficiently to allow the ball to fall through the ring to the bottom of the beaker.

ball charge volume of PORCELAIN or STEEL BALLS loaded in a BALL MILL.[49,59,64,131]

It is generally one third the total volume of the ball mill.

ball mill a rotating cylinder wherein CERAMIC materials are wet or dry ground, generally using pebbles or porcelain balls as GRINDING media.[49,59,64] (ASTM)

For paint, steel balls are frequently used.

ball milling a method of GRINDING and mixing material, with or without liquid, in a rotating cylinder or conical mill partially filled with grinding media such as balls or pebbles.[49,59,64] (ASTM)

ball punch impact test measure of the resistance of a coating to impact or shock. A ball punch is used for this test.[47,157]

ballotini GLASS beads used in reflective paints.[83] See BEADED PAINT[71,163] and TRAFFIC PAINT.[71]

balsam OLEORESINOUS exudations from plants which are characterized by a softness or a semiliquid consistency.[167]

They consist of mixtures of RESIN, ESSENTIAL OILS, and other compounds. Typical balsams are Canada, gurjun, Peru, tolu, and storax balsams.

baluster (1) one of a number of short vertical members, often circular in section, used to support a stair handrail or a coping; (2) a post in a BALUSTRADE.[82] (DAC) Syn: banister[82]

balustrade an entire railing system (as along the edge of a balcony) including a top rail and its BALUSTERS, and sometimes a bottom rail.[82] (DAC)

banana oil synonym for AMYL ACETATE.[144,164]

banbury compounding apparatus composed of a pair of contrarotating blades which masticate the materials to form a homogeneous blend.[49,59,64] See INTERNAL MIXER.[59,131]

An internal type heavy duty mixer.

band spectrum lines produced by MOLECULAR VIBRATIONS, which are so close together that they appear to be continuous.[43]

band wire a wire attached between dispensing and collecting vessels in order to equalize the electrical potential between them to dissipate electrostatic charge.[61,172]

banister see BALUSTER.[82]

bank, rubber mill a relatively small amount of unvulcanized RUBBER COMPOUND rolling in the space between two MILL rolls, while the major portion of the compound is bonded around the first roll.[140,57]

A PENCIL BANK is a small, smooth rolling amount of compound between a CALENDER roll and fabric surface being surface or friction coated.

Barcol hardness HARDNESS value obtained by measuring the RESISTANCE to PENETRATION of a sharp steel point under a spring load.[47,145,157,128]

The instrument, called the Barcol Impressor, gives a direct reading on a 0-100 scale. The hardness value is often used as a measure of the degree of cure of plastics; not suitable for paints.

bare glass GLASS (yarns, rovings, fabrics) from which the sizing or finish has been removed: also, such glass before the application of sizing or finish.

bargeboard a sloping board along a gable, covering the ends of roof TIMBERS.[82] Syn: vergeboard, gableboard

barium carbonate $BaCO_3$. A white compound, insoluble in water, occurring as a mineral or made by direct PRECIPITATION.[41,58,34] Syn: witherite

It has an sp gr of 4.275 and an mp of 1360°C. Used in CERAMICS, FILLERS, EXTENDERS.

barium chromate see BARIUM YELLOW.[41]

barium chrome see BARIUM YELLOW.[41]

barium ion stability the resistance of a LATEX to coagulation or FLOCCULATION in the presence of barium ions.[161]

The use of barium metaborate pigments with certain latices requires that they have stability to barium ions.

barium lithol red see LITHOL RED.[41]

barium lithol toner barium salt of 2-amino-1-naphthalene sulfonic acid coupled to 2-naphthol.[41]

barium metaborate, modified $BaB_2O_4{\cdot}H_2O$. White crystalline PIGMENT prepared by PRECIPITATION from aqueous solution; sp gr of 3.3, density of 27.5 lb/gal, refractive index of 1.55-1.60, O.A. of 30 lb/100 lb, and pH of 9.8 - 10.3.[41] See BARIUM ION STABILITY.[161]

barium naphthenate metallic NAPHTHENATE used as a WETTING AGENT for certain PIGMENTS, a HARDENER for some ALKYDS, and a THICKENER for OILS.[83,127]

barium oleate barium salt of OLEIC ACID.[83,127]

Used for the prevention of CHALKING, the maintenance of COLOR of titanium paints, and the DISPERSION of PIGMENTS in media during MILLING.

barium plaster a special MILL-mixed GYPSUM PLASTER containing barium salts; used to plaster walls of X-ray rooms.[75,159] (DAC)

barium stearate $Ba(C_{18}H_{35}O_2)_2$. White crystalline solid; insoluble in water or ALCOHOL.[83]

It has an mp of 160°C, and sp gr of 1.145. Used as a light and heat stabilizer in plastics.

barium sulfate $BaSO_4$. Natural Pigment White 22 (77120). Mineral consisting essentially of barium sulfate.[41,58,34] Syn: barytes, heavy spar, permanent white, barium white. cf. BARYTES[41,58]

It may contain, in addition, small amounts of sulfates of calcium and strontium, common chalk, calcium fluoride, silica, and iron oxide. Used in paints as a filler or EXTENDER. Density, 4.5 g/cm³ (37.5 lb/gal); O.A., 9; particle size, 2-30 µm.

barium sulfate $BaSO_4$. Synthetic Pigment White 22 (77120). Fine particle size white EXTENDER PIGMENT of high specific gravity, insoluble in water, soluble in concentrated sulfuric acid.[41,58] Syn: blanc fixe, precipitated barium sulfate, process white, barium white, constant white

Used as extender pigment in primers, coatings, leather, and paper. Density, 4.3-4.5 g/cm³ (35.9-37.5 lb/gal); O.A., 11.6-19.7; particle size, 0.5-1.0 µm.

barium white see BARIUM SULFATE, SYNTHETIC AND NATURAL.[41,58]

barium yellow essentially BARIUM CHROMATE made by adding a solution of sodium dichromate to barium chloride solution.[41] Syn: barium chromate, barium chrome, or lemon yellow

barium-extended titanium dioxide analogous to calcium-extended TITANIUM DIOXIDE, prepared by coprecipitating the $BaSO_4$ with TiO_2 or alternately made by a physical blending. In either case, the product contained 30% TiO_2 and 70% $BaSO_4$. It is no longer commercially available.[41]

barnacle any of certain crustaceans of the group *cirripedia*, as the goose barnacles, the stalked species which cling to ship bottoms and floating TIMBER; and the rock barnacles, the species which attach themselves to marine rocks.[42,173] SEE ANTI-FOULING PAINTS.[71,163,173]

Words presented in CAP/SMALL CAPS type indicate that the word is defined in another part of the *Coatings Encyclopedic Dictionary.*

Numerical superscripts classify terms in one or more of the categories listed in the second section of this volume.

barrel standard unit of liquid volume in the PETROLEUM industry.[50,128]

It is equal to 42 U.S. gallons, approximately 35 Imperial gallons; approximately 160 litres.

barrel finishing see TUMBLING.[77,131]

barrel polish see TUMBLING.[77,131]

barrelling see TUMBLING.[77]

barrier coat coating used to isolate a paint system from the surface to which it is applied in order to prevent chemical or physical interaction between them, e.g., to prevent the paint solvent attacking the underlying paint or to prevent BLEEDING from underlying paint or material.[71,72,158] (BSI) See TIE COAT, and TRANSITION PRIMER.[71] Syn: block coat

barytes see BARIUM SULFATE, NATURAL.[41,58]

bas-relief (basso-rilieva) sculpture in low relief.[81]

base (1) a species which when dissolved in water generates hydroxide (OH^-) ions or is capable of reacting with an acid to form a salt.[144] (IUPAC) *Brønsted base — a molecular entity capable of accepting a proton from an acid (i.e., a "proton acceptor") or the corresponding chemical species. For example: OH^-, H_2O, $CH_3CO_2^-$, HSO_4^-, SO_4^{2-}. A Lewis base — a molecular entity (and the corresponding species) that is able to provide a pair of electrons and thus capable of coordination to a Lewis acid, thereby producing a Lewis adduct, (IUPAC)* (2) reinforcing material (glass fiber, paper, cotton, ASBESTOS, etc.) which is impregnated with resin in the forming of laminates and is used as an insulating support for an electrical printed pattern;[38] (3) metallic salt upon which coloring matter is precipitated to form the insoluble pigments called LAKES;[41] (4) a paste or liquid which is to be thinned or TINTED;[41,58] (5) in INK manufacture, a dispersion containing usually only one coloring matter, pigment or DYE, properly dispersed in a vehicle. This is subsequently used for mixing to produce the desired end product.[168] (6) see BASEBOARD;[82] (7) see SUBSTRATE and GROUND.[82,159]

base box in can coating, the standard unit equal to 31,360 square inches (20.23 square metres).[128]

baseboard (1) board along the base of a wall usually finished with MOLDINGS;[82] (2) a molding that conceals the joint between an interior wall and floor.[82] Syn: base, skirting board, base plate, mop board, scrub board, kick board, washboard

basecoat (1) the first coat applied to a surface, as paint; a PRIME coat; (2) an initial coat applied to a wood surface before staining or otherwise finishing it; (3) all PLASTER applied before the finish coat; may be a single coat or a scratch coat and a brown coat.[75,71] (DAC) See BASECOAT/CLEARCOAT.[71]

basecoat/clearcoat a two step FINISH coat system in which a highly pigmented, often metallic, basecoat is followed by a thicker clearcoat.[54,71] (EPA)

It reportedly results in a finish with high gloss characteristics. (EPA)

base plate see BASEBOARD.[82]

base unit the smallest possible repeating unit of a POLYMER.[169,130] (IUPAC)

basic of an alkaline nature. Capable of uniting with an acid to form a salt.[134]

basic carbonate of lead see CARBONATE WHITE LEAD.[41]

basic dye having a slightly basic property due to the presence of ANILINE or a similar group.[41]

basic fuchsin magenta see FUCHSIN.[41]

basic lead acetate compound derived from lead oxide (LITHARGE) by treatment with insufficient ACETIC ACID for complete neutralization.[41,]

It was used in the manufacture of ACETATE CHROMES.

basic lead carbonate see CARBONATE WHITE LEAD.[41]

basic lead chromate $PbCrO_4 \cdot PbO$. Normal lead chromates are represented by the formula $PbCrO_4$, but if these are treated with alkalis, the COLOR can be changed from the original bright yellow to an orange color, and basic chromates were produced.[41]

basic lead nitrate $Pb(NO_3)_2$. Compound was prepared in a similar manner to BASIC LEAD ACETATE, but nitric acid was used in place of ACETIC ACID to produce nitrate chromes.[41]

basic lead silicate $3\ PbO \cdot 2\ SiO_2 \cdot H_2O$. Pigment White 16 (77625). PIGMENT consisting of an adherent surface layer of basic lead silicate "cemented" to silica. Compositions and constants varied depending on commercial grade.[41]

basic lead silico chromate calcined BASIC LEAD CHROMATE–BASIC LEAD SILICATE complex on a silica core, used as a CORROSION-INHIBITING PIGMENT.[41,66]

Properties: sp gr of 4.1; particle size of 7 μm; O.A. of 10-18 g/100 g; CrO_3—5.4 wt%; SiO_2—47.6 wt%; PbO—47.0 wt%.

basic lead sulfate $PbSO_4$, Pigment White 2 (77633). PIGMENT containing normal lead sulfate ($PbSO_4$) in "combination" with a proportion of basic lead which is taken to be an oxide (PbO).[41] Syn: sublimed white lead

It is a white pigment which was used in considerable amounts in weather-resisting undercoats and finishes, usually in conjunction with carbonate white lead and/or zinc oxide. O.A. of is 12-18 g/100 g and sp gr of 6.4.

basic pigment any PIGMENT which is capable of reacting with FATTY ACIDS to form SOAPS.[41]

basic zinc chromate $ZnCrO_4 \cdot 4\ Zn(OH)_2$. Yellow pigment used primarily for its CORROSION-INHIBITING properties.[41,66] Syn: zinc tetroxy chromate

basswood oil SEMIDRYING OIL.[63]

Sp gr of 0.938/15°C; iodine value of 111.0.

bastard sawn HARDWOOD LUMBER in which the annual rings make angles of 30 to 60° with the surface of the piece.[75,153]

Batavia dammar or damar GUM or RESIN exported from Batavia (now called Jakarta), Indonesia.[167]

Batavia dammar has two subgrades: padang or sumatra.

batch industrial unit or quantity of production used in one complete operation.[128,49] cf. LOT.[128,49]

The volume or mass which constitutes a batch is flexible and varies with the size of the plant and its facilities for converting the raw materials into the finished products.

batch adjustment record a record of materials added to a paint BATCH to adjust viscosity, color and such.[49]

batch process any PROCESS in which the charge is added intermittently in definite portions, or BATCHES, the operations of the process and the removal of products being completed on each portion before the addition of the next; as opposed to continuous process.[49,131]

batch record a record of all materials and proportions used to produce a BATCH of paint.[49]

batt INSULATION in the form of a blanket, rather than loose filling.[75]

batten small thin strips covering joints between wider boards on exterior building surfaces.[82] See BOARD AND BATTEN.[75,82,153]

batu East Indian semifossil RESIN, resembling the DAMMARS.[167]

Soluble in aromatic or mixed AROMATIC-ALIPHATIC HYDROCARBONS. Dissolves readily in warm VEGETABLE OIL. It is almost neutral. Acid value averages about 30.

baudouin test specific test for SESAME OIL, which gives a strong carmine coloration when a dilute ALCOHOLic solution of FURFURAL is added to a mixture of the oil and hydrochloric acid, and the whole violently shaken.[157,63]

Baumé (Bé) a hydrometer designed by the French scientist Antoine Baumé. Used for determining the SPECIFIC GRAVITIES of liquids, engraved, not directly in units of specific gravity, but in an arbitrary scale consisting of Baumé degrees.[145,157]

These Baumé degrees (often written °Bé) can be converted to true specific gravities by reference to conversion tables.

bauxite $Al_2O_3 \cdot 2H_2O$. Aluminum mineral, consisting chiefly of a hydrated aluminum oxide. The COLOR of native bauxite may be white, brownish red, or gray.[41,58,144,34]

Used as the source of many aluminum compounds. Crude powdered bauxite has been used to some extent as a FILLER for paints.

bay space between columns or supports of a building.[82]

A bay window is placed between such supports and usually projects outward. If curved or semicircular, it is called a bow window.

bead (1) heavy accumulation of a coating which occurs at the lower edge of a panel or other vertical surface as the result of excessive flowing;[42] (2) solidified droplet of an oil-resin mix, withdrawn from a varnish or resin mix, for testing purposes. Instructions frequently include a statement that the oil-resin mix shall be heated until a hard or tough bead is obtained;[157] (3) GLASS spheres. See BALLOTINI.[43,83]

bead and butt framed work in which the panel is flush with the framing and has a bead run on two edges in the direction of the grain; the ends are left plain.[75,82] Syn: bead, butt, and bead butt work

bead, butt and square similar to BEAD and BUTT, but having the panels flush on the beaded face only, and showing square reveals on the other.[75,82] (DAC)

beaded paint traffic or marking paint with reflecting beads (BALLOTINI) applied after striping or manufactured with BEADS and applied.[71,163]

beads, reflective see BEADED PAINT,[71,163] and BALLOTINI.[43,83]

beam TIMBER or girder supporting a floor, roof, or ceiling, usually supported at each end by a pier, wall, post, or girder.[82,75] (DAC)

A structural member whose prime function is to carry transverse loads, as a joist, girder, rafter, or purlin. [45,82,153,75]

beam ceiling (1) a ceiling, usually of WOOD, made in imitation of exposed floor beams with the flooring showing between; (2) the underside of a floor, showing the actual beams, and finished to form a ceiling.[82,153,75] (DAC)

beaumontage a RESIN, BEESWAX, and SHELLAC mixture used for filling small holes or cracks in WOOD or METAL.[147] (DAC)

Words presented in CAP/SMALL CAPS type indicate that the word is defined in another part of the *Coatings Encyclopedic Dictionary.*

Numerical superscripts classify terms in one or more of the categories listed in the second section of this volume.

Becchi-Millian test sensitive test for COTTONSEED OIL.[157,63]

If cottonseed fatty acids are dissolved in 90% ETHYL ALCOHOL and a little dilute silver nitrate solution added, they separate, colored brownish or black, as the ALCOHOL is boiled away on heating. It is claimed that as little as 1% of cottonseed oil is detected in a mixture by this test.

bed putty the glazier's PUTTY under GLASS, on which the glass is bedded. See FACE PUTTY.[165] Syn: back putty

beechnut oil a NONDRYING OIL.[63]

Its iodine value averages 110, and its sp gr, 0.922/15 °C.

Beer's Law this law describes the proportionality between the ABSORPTION of RADIANT ENERGY at any WAVELENGTH by a nonscattering material and the concentration of the absorbing material: $-\text{Log } T_i/c = K$, where T_i is the internal transmittance, c is the concentration of the absorbing material, and K is the unit ABSORPTION COEFFICIENT. Unit thickness is assumed. See BEER-BOUGUER LAW.[43,69]

Beer-Bouguer Law (Beer-Lambert Law) an equation useful in the description of light ABSORPTION by a compound of concentration $\underline{[A]}$ (often in mol L^{-1}) and with an optical path $\underline{l}$ (often in cm): $\lg (I_o/I_t) = \underline{[A]}\underline{l}\varepsilon$; here I_o and I_t represent the RADIANCE (intensity) of the incident and the transmitted light, respectively, and e is the decadic molar ABSORPTION COEFFICIENT with common units of L mol^{-1} cm^{-1} coherent SI units mol^{-1} m^2. For a collimated beam having a unique pathlength through the absorbing medium, this law applies to irradiance as well.[43,69] (IUPAC) See ABSORPTION COEFFICIENT.[43,69,128]

beeswax mixture of cured CEROTIC ACID and myricin separated from honey.[137]

It is a soft natural wax, with an mp of 63°C, an acid value usually below 20, a saponification value of 90-95, and a very low iodine value of 6-13.

behenic acid $CH_3(CH_2)_{20}COOH$. Long chain, ALIPHATIC ACID.[144]

It has an mp of 81°C, a bp of 306°C/60 mm Hg and an acid value of 164.8.

bell an ELECTROSTATIC SPRAY device in which the paint applicator is shaped like a bell.[77] (EPA)

This bell revolves rapidly so that paint, metered to the center of the bell, is thrown by centrifugal force off of the edge as a fine mist. High electrostatic voltage maintained on the bell also contributes to keeping the paint spray a fine mist. This fine mist allows electrostatic forces to draw the paint to the part being coated very efficiently with very little OVERSPRAY. The bell is one of the most efficient pieces of spray equipment with transfer efficiencies commonly over 90%. When high bell speeds are used (over 30,000 rpm) then high-solids coatings may be sprayed more easily than with other types of spray equipment. (EPA)

belly benzoin BENZOIN obtained as an exudation from the tree in the later years of its life.[167] See BENZOIN.[167]

Belly benzoin is usually regarded as an inferior grade, obtained during the nine years following the tenth year of the tree's life.

ben oil nondrying VEGETABLE OIL.[63]

The physical properties recorded vary widely from different samples; for example, iodine values have been reported from 72 to 112.

Bénard (vortex) cell hexagon-shaped cell that is produced by the vortex action of SOLVENT EVAPORATION in thin films.[42,164,55]

During drying, all solvent paints and many VARNISHES exhibit Bénard cell formation, i.e., roughly hexagonal cells, generally with a well-marked center. The whole cell is in movement with currents streaming up the center and flowing down the walls. These currents, because they affect pigments differently, lead to segregation and deposition of different components in different parts of the film. This phenomenon manifests itself in color and surface irregularities of the film.

bend test see FLEXIBILITY TEST.[47,157]

Benedict's test special test designed to detect the presence of reducing sugars, and it may also be applied to the products of HYDROLYSIS of certain GUMS, starch, etc.[157]

Benedict's reagent consists of a solution of sodium citrate and sodium carbonate in water, to which a small quantity of copper sulfate is subsequently added. An orange precipitate is obtained when a few ml of the reagent is added to the solution of the suspected compound, if a reducing sugar is present.

bengucopalic acid $C_{19}H_{30}O_2$. Monobasic acid which has been isolated from BENGUELA COPAL.[167,141]

It constitutes about 44% of the copal. The acid is said to resemble the congocopalic acid of CONGO COPAL.

bengucopalresene constituent of BENGUELA COPAL.[144,167]

Present in two forms, the α and ß varieties. The former is present to the extent of about 5% and the latter to the extent of 15%.

bengucopalolic acid $C_{21}H_{32}O_3$. Major constituent acid of BENGUELA COPAL, being present to the extent of about 25% of the whole.[144,167]

It has an mp of 115°C.

Benguela copal fairly hard natural COPAL obtained from West Africa.[167]

Sometimes known as Lisbon copal, and, like other similar copals from this part of West Africa, gives satisfactory oleoresinous varnishes after RUNNING.

beni (Japanese) a delicate pink or red PIGMENT of vegetable origin.[41,81]

beni-ye (Japanese) a print in which BENI is the chief COLOR used.[81]

Generally employed to describe all those two-COLOR prints which immediately preceded the invention of polychrome printing.

benne oil see SESAME OIL.[63]

bentonite a distinct type of fine-grained CLAY containing not less than 85% MONTMORILLONITE clay having the formula $(OH)_4Si_8Al_4O_{20}nH_2O$ and composed of units made up of two silica tetrahedral sheets with a central alumina octahedral sheet.[41,133,139,34] (ASTM) Syn: Wilkinite. cf. KAOLINITE, ATTAPULGITE, and HECTORITE[41,58]

It is a mixture and not a definite mineral type. It is derived from volcanic ash. Aqueous dispersions of bentonite, even at low concentrations, are very thixotropic. Bentonite is used in drilling muds for oil wells to support the walls of the drilled holes.

bentonite clay see BENTONITE.[41,133,139]

benzene C_6H_6. Aromatic HYDROCARBON.[164]

Pure benzene has a bp of 80°C and an mp of 5.5°C. Its flp is below normal air temperature and its vapor pressure is about 118 mm Hg at 30°C. It is a carcinogen.

benzene ring basis of all AROMATIC COMPOUNDS.[134,133]

It consists of six CH groups joined together to form a six-membered ring. Each carbon atom has one bond, and a shifting bond, or it is assumed that a double bond occurs between every alternate pair of carbon atoms.

benzenoid benzenoid compound is one which has a structural resemblance to BENZENE.[133,134]

benzidine yellow benzidine or diarylide yellows are disazo PIGMENT DYESTUFFS.[41]

They are approximately twice as strong, much more bleed and heat resistant, and markedly inferior in LIGHTFASTNESS versus hansa yellows. Two forms of benzidine yellow are made in quantity: (1) the transparent aceto acetanilide (AAA) type; and (2) the more opaque orthotoluidide (OT) type. Of the two types, the orthotoluidide type is the more lightfast.

benzilic acid $(C_6H_5)_2C(OH)COOH$. It has a mol wt of 228.17 and a mp of 150°C.[144] Syn: diphenyl glycolic acid

benzine (deprecated) (1) this term is archaic and misleading and should not be used;[164] (ASTM) (2) European term for gasoline.

benzofuran see COUMARONE.[141,167]

benzoguanamine resins RESINS based on benzoguanamine.[130]

benzoic acid C_6H_5COOH. It has an mp of 121°C, a bp of 249°C, and an acid value of 460.[144]

It is a useful PEPTIZING AGENT for METALLIC SOAPS, either in their solid form or in hydrocarbon solutions. It is sometimes effective in reducing SEEDING, especially in spirit enamel containing small amounts of basic pigments.

benzoin GUM benzoin, as it is more commonly known, is a spirit-soluble RESIN which appears on the market as Siam, Palembang, Padang, Sumatra, and Penang benzoins.[167] See BELLY BENZOIN.[167]

It has a very pleasant "vanilla" smell, and it is used partly on account of this, and partly as a PLASTICIZING RESIN in spirit varnishes. Benzoin is obtained as an exudation from a tree as a result of deliberate incision.

benzol or benzole commercial SOLVENT derived from the distillation of coal.[164]

It is a crude mixture of HYDROCARBONS, consisting chiefly of BENZENE, although it may contain in addition some toluene, xylene, carbon disulfide, thiophene, pyridine, acetonitrile, and paraffinic hydrocarbons. A 90%, or 90°, benzol is one from which 90% by volume can be distilled below 100°C.

benzoline (deprecated) synonym for PETROLEUM BENZINE.[164]

benzoperoxide see BENZOYL PEROXIDE.[74,144]

benzophenone an AROMATIC KETONE having excellent general compatibility.[144] Syn: diphenyl ketone

It is a solid at ordinary temperature, has an mp of 48°C, and a bp of 305°C. It is insoluble in water.

benzoyl C_6H_5CO–. Monovalent aryl RADICAL.[134]

benzoyl peroxide $(C_6H_5CO)_2O_2$. Its natural state is colorless crystals of mp of 103°C.[74,144] Syn: benzoperoxide

Among its principal uses is that of a CATALYST in many types of POLYMERIZATION reaction, including the manufacture of POLYSTYRENE, VINYL ACETATE, VINYL CHLORIDE, POLYESTER, and ACRYLIC RESINS.

benzyl monovalent RADICAL $C_6H_5CH_2$–.[134]

---

benzyl abietate very high boiling PLASTICIZER derived from ROSIN.[44,170] Syn: benzyl resinate

It is compatible with many other VARNISH and LACQUER constituents. As a rosin derivative, it is susceptible to atmospheric oxidation and yellowing.

benzyl alcohol $C_6H_5CH_2OH$. It has excellent SOLVENT properties for a wide range of VARNISH and LACQUER constituents.[164,144]

It has some application as a high boiling solvent, more particularly in baking finishes. It has a bp of 205°C, an flp of 300°C, and a vp of less than 1 mm Hg at ordinary temperature.

benzyl benzoate $C_6H_5COOCH_2C_6H_5$. A PLASTICIZER with an mp of 19°C, and a bp of 323°C.[164,170]

benzyl butyl phthalate high melting SOLID, with a bp of 208-288°C/20 mm Hg.[170]

benzyl Cellosolve® $C_6H_5CH_2OCH_2CH_2OH$. High boiling SOLVENT with a bp of 256°C and sp gr of 1.070/20°C.[164]

benzyl cellulose CELLULOSE ETHER prepared by reacting benzyl chloride on alkali cellulose; soluble in aromatic HYDROCARBON SOLVENTS.[151]

Used as LACQUER material.

benzyl phthalate $C_{22}H_{18}O_4$. A solid PLASTICIZER: mp of 42°C, and bp of 277°C/15 mm Hg.[170] Syn: dibenzyl phthalate

benzyl resinate see BENZYL ABIETATE.[44,170]

ber one of the host trees to which the LAC insect affixes itself.[146]

Berlin blacks low-cost form of air-drying blacks based on solutions of BITUMENS to which black pigments or fillers are added.[76,156,163]

They possess excellent opacity, but reduced gloss. Some Berlin blacks are semimatte.

Berlin blue see IRON BLUE.[41]

BESA abbreviation for British Engineering Standards Association.[174]

BET abbreviation for BRUNAUER, EMMETT, AND TELLER, applied to an equation and method for determining the surface area of an absorbent, such as CARBON.[157]

beta gauge (beta-ray gauge) gauge consisting of two facing elements, a ß-ray-emitting source and a ß-ray detector.[145]

When a sheet material is passed between the elements, some of the ß-rays are absorbed, according to the areal density or the thickness of the sheet.

beta-pinene see PINENE.[44,164]

beton a kind of CONCRETE; a mixture of LIME, sand, and gravel.[73] (DAC)

bevel siding see CLAPBOARD.[75,82,153,]

bicyclic consisting of two rings.[134]

biddiblack a particular type of MINERAL BLACK, mined near Bideford, Devonshire, England.[41,34]

bidirectional in two directions: (1) the term may be applied to illumination angles in three-dimensional space. Thus, the incident angle may be 45° relative to the material surface, but this designation does not describe the portion of the circumferential 45° angle possible. The term "bidirectional illumination" generally implies two incident beams separated circumferentially by 180°; (2) the term may also be applied to GONIOPHOTOMETRIC measurements, to the illumination and to the viewing angle.[69,43]

In either case, the angular distance is expressed as the degrees (or radians) from the perpendicular (normal) to the surface as 0°. Thus, the grazing angle approaches 90°.

bidirectional reflectance distribution function (BRDF) the ratio of radiance per unit irradiance, used for describing the geometrical reflectance properties of the surface.[43] See BIDIRECTIONAL.[69,43]

biff see MONKEY.[49,131,152]

bimodal distribution the occurrence of two maxima in a frequency distribution.[145] (IUPAC)

bin cure RUBBER term for partial or complete VULCANIZATION of a mixed COMPOUND while stored in a bin or pile waiting for molding or further processing.[67,140] Syn: pile burning or premature vulcanization

binder (1) NONVOLATILE portion of the LIQUID VEHICLE of a coating. It binds or cements the PIGMENT particles together and the paint film as a whole to the material to which it is applied;[156](BSI) cf. VEHICLE[156] (2) component of an ADHESIVE composition which is primarily responsible for the adhesive forces which hold two bodies together;[79] (ASTM) (3) RESIN or cementing constituent of a PLASTIC compound which holds the other components together. The agent applied to GLASS mat or preforms to bond the fibers prior to laminating or molding;[38] (4) the components in an INK film which hold the pigment to the printed surface.[168]

binder demand that amount of BINDER needed to completely wet a pigment by displacing the air voids.[156,157,41]

This is determined primarily by: the particle size, shape, chemical composition, and density of the pigment; and the particle size, degree of POLYMERIZATION, and WETTING properties of the binder. (Whereas the binder demand refers to a particular pigment-vehicle system, OIL ABSORPTION *is a numerical value assigned to a pigment based on a specific test method.)*

binder ratio see PIGMENT/BINDER RATIO.[41,156]

binding power index is the binding efficiency of a LATEX. It is the ratio of LINSEED OIL to latex volume solids needed to bind a given volume of pigment.[156,157,41] cf. BINDER DEMAND[156,157,41]

It can be calculated from the experimentally determined CPVC of the latex with a given pigmentation, and the CPVC of linseed oil with the same pigmentation calculated from the oil adsorption. Oil adsorption values are often given by pigment and extender manufacturers, and may be found in several coatings handbooks. Since extenders especially are not closely controlled in particle size and surface area, it is preferable to determine the oil adsorption using the same lot of pigment used in determining the CPVC with the latex binder.

binding varnish a viscous VARNISH in PRINTING INK that is used to toughen the dried film.[168]

Bingham body material displaying PLASTIC FLOW.[166](PTM)

Bingham liquid liquid exhibiting PLASTIC FLOW.[47,166]

bioavailability the extent that a substance can be inhaled or absorbed in a biologically active form.[172,173] (ASTM)

biocide a chemical agent capable of killing organisms responsible for microbiological DEGRADATION.[173]

biodegradable an ORGANIC material's capacity for decomposition as a result of attack by microorganisms.[173]

Sewage-treatment routines are dependent on this property. Phosphates and chlorinated HYDROCARBONS *(such as DDT) are not biodegradable.*

biodegradable surfactant SURFACTANT which may be decomposed by biological action.[127,173]

biodegradation PROCESS by which matter may be broken down by living organisms.[65,173]

biodeterioration any undesirable change in material properties brought about by the activities of microorganisms.[173]

ASTM test methods for biodeterioration include: emulsion paints in container, D2574; paint films: discoloration—exterior exposure, D3456; mold—environmental chamber, D3273; removal of fungal/algal growth, D4610; soil/dirt/fungal accumulation, D3274.

biological monitoring periodic examination of body substances, such as blood or urine, to determine the extent of hazardous material ABSORPTION as opposed to mere exposure.[172,173]

biopolymer a POLYMER directly produced by living or once-living cells or cellular components.[173]

bipolymer a COPOLYMER in which there is irregularity with regard to the placement (relative locations within the POLYMER chain, in a linear copolymer) of two or three or more chemically different types of units.[130]

These units may be MERS, *in a product of* ADDITION POLYMERIZATION, *or residues of the condensing small molecules in a* POLYCONDENSATE.

bird applicator device used for the laboratory application of coating of a prescribed thickness to test panels.[145,157,77] See APPLICATOR.[77,59]

It is a machined steel bar with a fixed clearance having a beveled undercut to guide the coating material.

bird's-eye small localized areas in WOOD with the FIBERS indented and otherwise contorted to form few to many small circular or elliptical figures remotely resembling birds' eyes on the tangential surface.[153]

Common in sugar maple and used for decorative purposes; rare in other HARDWOOD *species.*

birefringence property of ANISOTROPIC materials, those which possess different refractive indices according to the direction of vibration of light passing through a crystal; usually observed with the aid of a polarizing microscope.[43,69]

The degree of birefringence is expressed numerically as the difference between the highest and lowest indices of refraction, even though there may be more than two different indices of refraction.

bisacki grade of seed LAC.[146]

bismarck brown brown-colored DYESTUFF, which chemically is triaminoazobenzene.[41]

It dissolves in ALCOHOL *and water, to yield reddish-brown solutions. It has some application in the manufacture of mahogany stains.*

bisphenol A and F dihydroxydiphenyldimethylmethane, mol wt 224.1.[130,144]

Insoluble in water. Used in the manufacture of PHENOLIC *and* EPOXY RESINS.[130,144] CONDENSATION PRODUCT *formed by reaction of two (bis) molecules of* PHENOL *with acetone. This polyhydric phenol is a standard resin* INTERMEDIATE, *along with* EPICHLOROHYDRIN, *in the production of epoxy resins.*

Words presented in CAP/SMALL CAPS type indicate that the word is defined in another part of the *Coatings Encyclopedic Dictionary.*

Numerical superscripts classify terms in one or more of the categories listed in the second section of this volume.

BISRA abbreviation for British Iron and Steel Research Association.[174]

bistre brown water COLOR PIGMENT which is derived from the tarry soot of burned, resinous wood and beechwood.[76]

It is similar to ASPHALTUM in color and composition. The color varies from saffron yellow to brown-black, depending upon the source and treatment of the raw material.

bite ability of a coating to PENETRATE or SOFTEN a previous coating or SUBSTRATE.[164]

bittiness presence of material that hinders the APPEARANCE of a smooth and uniform coating film.[42,56,35]

bitty said of coatings containing bits of SKIN or foreign matter of any type which project above the coating film and hinder the formation of a smooth and uniform film.[56,42] See NIBS.[56] Syn: peppery

bitumen (1) a class of AMORPHOUS, black or dark-colored (solid, semisolid, or viscous) CEMENTITIOUS substances, natural or manufactured, composed principally of high molecular weight HYDROCARBONS, soluble in carbon disulfide, and found in ASPHALT, TARS, PITCHES, and ASPHALTITES; (2) a generic term used to denote any material composed principally of bitumen.[76] (ASTM)

bituminous cement a black substance available in solid, semisolid, or liquid states at normal temperatures; composed of mixed indeterminate HYDROCARBONS; appreciably soluble only in carbon disulfide or other volatile liquid HYDROCARBON; esp. used in sealing BUILT-UP ROOFING and between joints and in cracks of CONCRETE pavements.[76,147] (DAC)

bituminous coating an ASPHALT or TAR COMPOUND used to provide a protective finish for a surface.[76] (DAC)

bituminous emulsion a SUSPENSION of minute globules of BITUMINOUS material in water or of minute globules of water in a liquid bituminous material; used as a protective coating against weather, esp. where appearance is not important.[76,71] (DAC)

bituminous grout a mixture of BITUMINOUS material and fine sand that will flow into place without mechanical manipulation when heated.[76,165] (ASTM) See GROUT.[165]

bituminous paints (1) originally, the class of paints consisting essentially of natural BITUMENS dissolved in ORGANIC SOLVENTS: they may or may not contain softening agents, PIGMENTS, and INORGANIC FILLERS.[76,71] *They are usually black or dark in color. Within recent years the term "bituminous" has, by common usage, come to include bitumen-like products such as PETROLEUM ASPHALT;* (2) a low-cost paint containing asphalt or COAL TAR, a thinner, and drying oils; used to waterproof CONCRETE and to protect piping where bleeding of the asphalt is not a problem.[76,71,163] (DAC)

bituminous varnish see VARNISH.[168,152]

black ideally, the complete ABSORPTION of incident light; the absence of any reflection.[69,43]

In the practical sense, any COLOR which is close to this ideal in a relative viewing situation, i.e., of very low SATURATION and of low LUMINANCE.

black boy gum see ACCROIDES.[167]

black chalk a bluish black CLAY containing CARBON.[41]

black dammar see BLACK EAST INDIA.[167]

black East India dark colored RESIN collected in the East Indies.[167] Syn: black dammar

It resembles the DAMMARS in several respects, notably in solubility and free acid content.

black iron oxide $FeO{\cdot}Fe_2O_3$ or Fe_3O_4. Pigment Black 11 (77499). Two black oxides of iron are used, namely, the natural mineral, MAGNETITE, and the artificially produced oxide. Both types are magnetic. Neither possesses outstanding opacity.[41] Syn: pure black iron oxide, synthetic magnetite, black magnetic oxide

Magnetite consists of about 95% Fe_3O_4. Artificial black oxide consists of a mixture of ferric oxide (Fe_2O_3) and ferrous oxide (FeO). The artificial oxide is made by the precipitation of a soluble iron salt with an alkaline carbonate under oxidizing conditions. Properties: density, 4.95-5.98 g/cm^3 (41.2-43.2 lb/gal); O.A., 18-28; particle size, 0.1 to 0.01 μm.

black japan black VARNISH-like material in which the higher grades of ASPHALT or other resinous compositions are employed, blended with suitable DRYING OILS to produce hard, glossy black coatings which become insoluble in mineral spirits.[71,76,152] See JAPAN.[71,76]

Properly formulated black japan should show no tendency to BLEED. No pigments are incorporated.

black lead synonym for GRAPHITE.[41]

black light (deprecated) popular term for ULTRAVIOLET RADIATION without any visible radiation.[43,69] Syn: ultraviolet radiation

This is an incorrect term because light is defined as visible radiation and ultraviolet radiation is not detected by the visual mechanism of the eye.

black magnetic oxide see BLACK IRON OXIDE.[41]

black varnish any VARNISH in which the RESIN component is substituted completely, or in part, by a PETROLEUM BITUMEN, or natural ASPHALTUM.[76,152]

blackboard paint rapid drying flat paint capable of permitting writing on it by chalk; usually black or green in COLOR.[163,71]

blackbody theoretically perfect absorber of all INCIDENT RADIATION.[43,69]

blackout paint OPAQUE PAINT, usually black or dark in COLOR, used during war time, on windows, skylights, etc., to prevent the passage of light to the outside of a building.[71,163]

blanc fixe see BARIUM SULFATE, SYNTHETIC.[41,58]

blanch (1) to make WHITE; (2) as distinct from BLUSH or BLOOM, a pale milky cast on a coating film.[42,55,56]

blank a zero control in an experiment—generally a treatment in which the variable of interest (such as conditions or ingredients) are not changed.[157]

blank reading or blank value instrument reading for a zero sample. A reading originating from the matrix, reagents and any residual bias in the measurement instrument or process, but not from the analyte.[145,157] (IUPAC)

This is generally a treatment in which the variable of interest (such as conditions or ingredients) are not changed.

blast cleaning cleaning and roughening of a surface (particularly steel) by the use of natural or artificial GRIT or fine metal shot (usually steel), which is projected against a surface by compressed air.[158] Syn: power cleaning

bleach refers to the method of measuring the TINCTORIAL STRENGTH of an INK or TONER.[168]

Usually accomplished by mixing a small portion of the ink (or toner) with a large amount of white base and evaluating the tinctorial strength of the ink versus a control standard.

bleached oil oil which has been refined by the ACID, ALKALI, or MECHANICAL PROCESS and in which the refining includes a treatment at about 88°C (190°F) with an adsorbent such as FULLER'S EARTH or activated charcoal, after which the oil is filtered.[63,131]

bleached shellac substantially colorless product obtained by BLEACHING natural orange gum LAC with chlorine or chlorine-containing agents, such as sodium or calcium hypochlorites. Syn: bone dry bleached shellac

bleaching (1) uniform loss of COLOR of a paint or VARNISH. This may be due to internal chemical or physical action in the paint itself, to influences from the surface on which it is applied or to weathering or contamination from the atmosphere;[65,43,69] (2) intentional lightening of the color of a material such as WOOD, VEGETABLE OILS, varnishes, etc.[54,158,131] (BSI) See FADING, and WHITENING IN THE GRAIN.[42,43,56,69,154]

bleaching clay CLAY that possesses decolorizing characteristics for use in refining of mineral, petroleum, vegetable, and animal oils.[58,131]

bleed DISCOLORATION resulting from the DIFFUSION of coloring matter from a previously painted SUBSTRATE; also, discoloration of a painted surface by the action of a solvent on a substate.[42,69,56,35] See BLEEDING.[42,41]

bleed characteristics see BLEEDING.[41,42]

bleeding (1) the DIFFUSION of COLORANTS through a coating from a previously painted SUBSTRATE due to the action of the VEHICLE or SOLVENT or both. The action is dependent on the pigments, vehicles, and solvents of the systems;[56,69,42] (2) diffusion of a soluble colored substance from, into, and through a coating from beneath, thus producing an undesirable STAINING or DISCOLORATION.[56,42,153,35] *Materials which give rise to this effect are TANNINS or DYES in some types of wood; WOOD PRESERVATIVES; BITUMINOUS COATINGS; PIGMENT DYESTUFFS; and STAINS;* (3) diffusion of coloring matter from the substrate; also, the discoloration arising from such diffusion. In the case of printing ink, the spreading or running of a pigment color by the action of a solvent such as water or ALCOHOL;[56,42,168] (ASTM) (4) migration to the RUBBER SURFACE of an OIL, WAX, or PLASTICIZER as a film or in drops, sometimes called sweating. Also a term applied to ORGANIC PIGMENT colors if they migrate into an adjacent stock of a different color, or when they are removable at the surface by water or other solvent;[56,140] (5) the spreading or migration of an INK component into an unwanted area.[168]

bleeding knots usually circular-shaped DISCOLORATION in stained or painted siding.[42,153] See KNOT SEALER.[153,71,42]

It is generally caused by RESIN, dissolved by solvent in the paint, that worked its way to the surface from improperly sealed knots in the wood. It can be remedied by stripping and sanding down to the knots or PITCH POCKETS in the wood and applying a SEALER or SHELLAC to the knots; then priming and repainting.

bleeding pigment a PIGMENT in a coating which is partially soluble in the SOLVENT or vehicle portion of subsequent coats applied to it.[41,42,56]

blemish any surface imperfection of a coating or SUBSTRATE.[56,35]

blender (1) a round, softish BRUSH of badger hair or similar material with a blunt tip, used for blending colors and removing brush marks left by coarser brushes; [81,54,59] Syn: softener (2) a small

---

Words presented in CAP/SMALL CAPS type indicate that the word is defined in another part of the *Coatings Encyclopedic Dictionary.*

Numerical superscripts classify terms in one or more of the categories listed in the second section of this volume.

laboratory mixer used to disperse pigment in a vehicle.[59,64]

blending mixing two or more materials to obtain a particular kind or quality.[49]

blinding of lithographic plates loss of INK-receptivity in the plates' image areas.[168]

blister undesirable rounded elevation of the coating.[56,42]

Usually dome shaped. The formation of blisters results from localized loss of ADHESION. Paints are most susceptible to blistering early in their exposure life. This is induced by adverse climactic conditions involving high surface moisture levels associated with rain and heavy dew. Blisters may recover but may leave surface cosmetic imperfections. In extreme cases, massive loss of adhesion may result causing the destruction of the paint film in sheets of larger areas.

blister house house-like structure used as an ACCELERATED TEST for BLISTER RESISTANCE of coatings.[42,157]

Under controlled conditions, it attempts to simulate or accelerate BLISTERING which occurs on buildings.

blister resistance the ability of a coating to resist the formation of dome-shaped, liquid- or gas-filled projections in its film resulting from local loss of ADHESION and lifting of the film from the underlying surface or coating.[56,72] See BLISTERING.[56,42]

Blistering of exterior paints applied to wood siding is often due to high humidity within a structure. A driving force that causes blistering is humidity originating behind the substrate.

There are various tests that simulate this particular situation. ASTM Test Method D1735 uses a water fog apparatus similar to that used for salt spray testing. ASTM Test Method D4585 uses controlled CONDENSATION. ASTM Test Method D714 contains photographic standards for the degree of blistering size and frequency. The FSCT Pictorial Standards of Coatings Defects also has these same photographic standards.

blistering formation of dome-shaped hollow projections in paints or VARNISH films resulting from local loss of ADHESION and lifting of the film from the underlying surface.[56,42] (BSI) cf. BUBBLING,[42] and POPPING.[56,42] See BLISTER RESISTANCE.[56,72]

A phenomenon peculiar to painted surfaces is the formation of blisters relative to some system weakness.

ASTM Test Method D714 employs photographic reference standards to evaluate the degree of blistering. The FSCT Pictorial Standards of Coating Defects contains the same photographic reference standards. ASTM test methods for blistering (blister resistance) include: pictorial standards/evaluation, D714; water condensation test, D4585; water fog test, D1735.

block coat see TIE COAT, TRANSITION PRIMER, and BARRIER COAT.[71,158]

block copolymer (1) a POLYMER consisting of molecules in which two or more POLYMERIC segments of different chemical composition are attached end-to-end;[130,169] (2) an essentially linear COPOLYMER in which there are repeated segments of different chemical structure.[130,169] (IUPAC) (ASTM)

block cutter a craftsman who, by hand, hammers into a roller the brass strips and FELT from which the WALLPAPER will be printed.[148]

block filler a pigmented coating of heavy CONSISTENCY used to fill VOID spaces in CONCRETE or CINDER BLOCKS, prior to the application of the TOPCOAT, in order to produce a smooth surface.[71,158]

block flooring end GRAIN blocks GLUED down in DECORATIVE patterns.[153]

block polymer see BLOCK COPOLYMER.[130,169]

block printing a PROCESS of printing with blocks on which the unit of design stands out in relief.[81,148,168] Syn: hand-blocking

Pear wood is traditionally used, and a different block is needed for each color.

block resistance the resistance to BLOCKING.[72,42,56] cf. PRINT RESISTANCE[56]

Interior paint often comes in contact with itself, especially in window and door areas, and sometimes sticks to itself (blocks). Blocking depends on the HARDNESS of the coating, the pressure, temperature, humidity, dry time, solvent evaporation, and duration of time the surfaces are in contact.

Block resistance of architectural paints can be determined by using ASTM Test Method D4946. This is not to be confused with block resistance ASTM Test Method D3003 which is concerned with blocking of industrial coatings on metal substrates, nor with ASTM Test Method D2793 which is concerned specifically with wood product finishes.

blocking (1) the undesirable sticking together of two painted surfaces when pressed together under normal conditions or under specified conditions of temperature, pressure, and relative humidity;[157,42,56] (2) undesired ADHESION between touching layers of a material, such as occurs under moderate pressure and sometimes pressure and heat, during STORAGE or FABRICATION.[42,79] (ASTM)

ASTM test methods for blocking (block resistance) include: of architectural coatings, D4946; lacquers on metal substrates, D3003; on WOOD SUBSTRATES, D2793.

blocking point lowest temperature at which two coated surfaces in mutual contact will stick together sufficiently to injure the surfaces permanently and/or prevent easy separation.[137,157,128]

Blocking point is a major quality criterion for evaluating coating-grade PETROLEUM WAXES.

blocking test procedure for determining the tendency of painted surfaces to stick together (BLOCK) when stacked or placed in contact with each other under a weighted load.[157] See BLOCK RESISTANCE.[145,168]

bloom (1) a HAZINESS which develops on HIGH-GLOSS SURFACES resulting in scattering of the surface reflectance. One mechanism is by EXUDATION of a component such as a PLASTICIZER out of the paint film;[56,42,35] (2) undesirable deposit which sometimes forms on a glossy coating, resulting in WHITENING or loss of GLOSS and lightening of the color;[42,56,35] (3) similar to BLEEDING in that it is MIGRATION of liquids or solids to the surface of a RUBBER compound to cause a change of appearance in color or cloudiness at the surface. WAXES used in excess of their solubility point in rubber come to the surface as a wax BLOOM, as does sulfur that remains as an excess over the amount actually chemically combined with the rubber;[42,140,56,137] (4) undesirable EXUDATION of PENTACHLOROPHENOL on the surface of wood that has been treated with a PRESERVATIVE solution;[153] (5) in the PLASTICS industry, bloom is a visible exudation on the surface of a plastic, generally caused by lubricant, plasticizer, etc.;[38,170] (6) the FLUORESCENCE of MINERAL OIL; the blue or purple cast apparent when the oil is spread in a thin film.[63] See BLUSH.[56,42] Syn: fog. See OIL, DEBLOOMED.[63]

bloom oil one of the liquid DECOMPOSITION products obtained from the destructive distillation of ROSIN.[44]

It can be regarded as a rosin oil, and has a bp of about 270°C.

blotching see MOTTLE.[56,42]

blow molding (1) method of producing hollow objects (bottles, etc.) by injecting a parison (blob) of hot melt into a hollow, two-part mold, then inflating the parison against the cool mold surface where it "freezes" into shape; (2) method of fabrication in which a hollow tube is forced into the shape of the mold cavity by internal pressure.[57,131]

blowing POROSITY or sponginess occurring during CURE, either deliberately through use of a gas releasing material to form sponge or expanded RUBBER, or inadvertently due to entrapped moisture to cause undesirable porosity.[67,140]

blowing agents agents added to PLASTICS and RUBBERS that generate inert gases on heating, causing the RESIN to assume a cellular or FOAM STRUCTURE.[83] Syn: foaming agents

blowing (of plaster) APPEARANCE on the SURFACE of the PLASTER of conical hollows (pops or blows).[35] (BSI)

They are due to the presence of PARTICLES of REACTIVE MATERIAL which expand, after the plaster has SET, with sufficient force to push out the plaster in front of the particles. They may occur in undercoats or finishing coats of plaster, or both.

blown bitumen generally regarded as the BLOWN products of PETROLEUM BITUMENS, which have been previously obtained as residues from the DISTILLATION of crude petroleum.[76] See BITUMEN, and AIR BLOWING.[76,131]

The petroleum bitumens are subjected to high temperature blowing with air, which has profound effects on the properties of the material. MELTING POINT is raised considerably, and the product becomes RUBBERY. For bitumens of the same melting point, the blown product has a higher penetration figure.

blown castor oil product of very high VISCOSITY and reddish-brown COLOR, obtained by BLOWING RAW CASTOR OIL.[63,131,156]

One of its chief applications is as a PLASTICIZER or SOFTENER for CELLULOSIC compositions.

blown oil VEGETABLE or FISH OIL which has been partially OXIDIZED by blowing with a current of air while at an elevated temperature.[63,67,156,169] (BSI)

The characteristics of the oil such as its increased VISCOSITY and degree of oxidation can be controlled by time, temperature, and amount of air.

blown stand oil HEAT POLYMERIZED or STAND OILS, which are subsequently blown.[63,131,156]

blue basic lead sulfate (blue lead) a variant of WHITE BASIC LEAD SULFATE containing 78% monobasic lead sulfate.[41]

It was not actually a blue pigment, but had a dark gray or slate COLOR.

blue stone see COPPER SULFATE.[41]

blue toner PIGMENT or DYE used in small quantities to neutralize the yellow cast of certain white pigments or coatings, increasing apparent WHITENESS; also used to make black coatings more jet black.[41]

blue verditer basic COPPER CARBONATE.[41] Syn: azurite

blue vitriol see COPPER SULFATE.[41]

blueing neutralizing the yellow cast of certain white pigments or paints by adding a trace of blue, thereby increasing apparent WHITENESS.[43,69] See BLUE TONER.[41]

Words presented in CAP/SMALL CAPS type indicate that the word is defined in another part of the *Coatings Encyclopedic Dictionary*.

Numerical superscripts classify terms in one or more of the categories listed in the second section of this volume.

blush, blushing (1) FILM DEFECT which appears as a MILKY OPALESCENCE as the film dries; can be a temporary or permanent condition. It is generally caused by rapid EVAPORATION, MOISTURE, or INCOMPATIBILITY;[56,42,35] See BLOOM.[63] Syn: fog (2) milky opalescence which sometimes develops as a film of LACQUER dries, and is due to the deposition of moisture from the air and/or PRECIPITATION of one or more of the solid constituents of the lacquer; usually confined to lacquers which dry solely by evaporation of solvent. Also applies to PRINTING INKS;[43,55,56,168] (BSI) (3) a white or grayish cast on high gloss paint; results from the precipitation of BINDER solids owing to incompatibility with water, oil, or SOLVENT.[42,56,35] (DAC)

board LUMBER less than 2 in. (6 cm) thick and between 4 in. (10 cm) and 12 in. (30 cm) in width; a board less than 4 in. (10 cm) wide may be classified as a STRIP.[75,82,153] (DAC)

board and batten a type of WALL cladding for wood-frame houses; closely spaced, applied boards or sheets of PLYWOOD, the joints of which are covered by narrow (usually vertical) WOOD STRIPS.[75,82,153] (DAC) See BATTEN.[82]

board and brace a type of carpentry work consisting of BOARDS which are grooved along both edges and have thinner boards fitted between them.[75,82,153] (DAC)

board foot a unit of cubic content, used in measuring LUMBER; equal in volume to a piece 1 ft. square and 1 in. thick, or 2360 cm^3.[75,82,128,153] (DAC)

bodied oil in the strictest sense, this is a DRYING OIL which has been HEAT POLYMERIZED.[63,131,169,156]

The term is also used more broadly to describe drying oils which have viscosities greater than those of the oils in their natural condition, irrespective of how the increased viscosity is attained, e.g., BLOWN OIL or STAND OIL.

body (1) apparent CONSISTENCY or VISCOSITY of a paint as assessed subjectively. A practical term widely used to give a qualitative picture of consistency. For NEWTONIAN liquids, body is the same as viscosity;[166] (2) usually descriptive of liquids, e.g., a heat-bodied oil (LINSEED OIL which has been polymerized by heating).[166]

body gum (no. 8 varnish) in PRINTING INKS, LINSEED OIL that has been HEAT POLYMERIZED to a heavy, gummy state. It is commonly used as a BODYING AGENT.[63,168]

body putty (body solder) paste-like mixture of PLASTIC RESIN (POLYESTER) and TALC, used in repair of METAL SURFACES such as auto bodies.[38,165]

bodying increase in the APPARENT VISCOSITY of a paint, VARNISH, RESIN, or LACQUER, which occurs either deliberately during manufacture or adventitously during STORAGE.[161,166] (BSI)

bodying agent see THICKENER.[83,166]

boea one of the classes of MANILA COPAL.[167]

boiled oil LINSEED OIL, and sometimes SOYA OIL, which has been treated in any of the following ways: (a) heated for a short time to slightly increase viscosity and shorten drying time; (b) heated with compounds of lead and/or cobalt and/or manganese forming fatty acid salts of the metal(s), thus shortening the drying time; (c) by adding soluble driers with or without heating. If the driers are added without heating, the product is sometimes referred to as "BUNG HOLE OIL," from the practice of adding the driers simply by pouring through the bung opening of a drum.[61,131,63]

boilers group name for NITROCELLULOSE and LACQUER SOLVENTS.[164]

boiling point the temperature at which the VAPOR PRESSURE of a liquid is just slightly greater than the total pressure of the surroundings. The liquid as a consequence is rapidly converted from the liquid state into a vapor.[142,128]

boiling range range between initial and final boiling temperatures of a SOLVENT.[142]

bold term applied to natural COPALS and other NATURAL RESINS, and used as a means of indicating that the pieces of resin in question are fairly large lumps and not dust or chips.[128,167]

bole a mixture of gilder's CLAY and rabbit skin glue, used as an undercoating for water-gilded finishes in applications of real GOLD LEAF.[148,158,81]

BON acronym for betaoxy-naphthoic acid which can be coupled with a variety of AMINE COMPOUNDS to form a soluble AZO DYE, which is then PRECIPITATED by any of several METALS to form "BON" reds or maroons.[41]

BON reds and maroons Pigment Red 48 (15865). BON PIGMENTS are monoazo couplings of diazotized aromatic AMINES with betaoxy-naphthoic acid precipitated with metals such as calcium, barium, and manganese.[41]

A range of colors can be obtained extending from the toluidine red shade through maroon, by proper selection of the first component of the metallic precipitant. LIGHTFASTNESS in tints ranges from good to poor; increased MASSTONE depth is accompanied by poorer tint lightfastness.

bond (1) chemical—force of attraction between atoms in a molecule;[133] (2) ADHESIVE—adhesion between two materials, such as between an adhesive and a given surface;[79] (3) coatings—adhesion between a coating and a SUBSTRATE to which it has been applied.[79]

bond coat coating used to improve the ADHERENCE of succeeding coats.[71,79]

bond plaster, concrete bond plaster a MILL-mixed GYPSUM PLASTER containing a small percentage of LIME; just before application it is mixed with water only and then applied at a maximum thickness of 1/4 in. (0.6 cm) over prepared concrete surfaces.[73,79,82,75] (DAC)

It serves as the BOND between the CONCRETE and succeeding coats of gypsum plaster.

bond strength amount of ADHESION between bonded surfaces; a measure of the stress required to separate a layer of material from the base to which it is bonded.[47,57,79]

bonding action see BOND STRENGTH.[47,57,79]

bone ash see BONE WHITE.[41]

bone black PIGMENT made by charring animal bones in closed retorts.[41] cf. BONE BROWN, BONE WHITE, AND IVORY BLACK.[41] Syn: drop black

Bone black is blue-black in color and is fairly smooth in texture. It contains CARBON, CALCIUM PHOSPHATE, and CALCIUM CARBONATE.

bone brown PIGMENT related to BONE BLACK. Both are obtained from bones by CALCINATION, but with bone brown the temperature of calcination is much lower and less CARBONIZATION occurs.[41]

The pigment is not widely used.

bone dry bleached shellac a BLEACHED, light-colored orange GUM SHELLAC.[146] See BLEACHED SHELLAC.[146]

bone glue impure GELATIN prepared from bones.[79] See GELATINGLUE.[79]

bone oil FATTY OIL obtained by the DRY DISTILLATION of bones.[63]

bone white PIGMENT made by CALCINING animal bones. Composed chiefly of tricalcium phosphate.[41] cf. BONE BLACK, and BONE BROWN[41] Syn: bone ash

CALCIUM CARBONATE and minor constituents make up the rest. Bone white is a grayish white and slightly gritty powder.

bookbinders' varnish rapid drying VARNISH based on SHELLAC, SANDARAC, or GUM MASTIC, or mixtures of these RESINS.[71,152,156]

boomerang a single magnet, direct reading, dry film thickness gage.[145,157]

boot-legging separation of plies progressively occurring in conveyor, transmission, or other belts made of separate plies of RUBBER and fabric.[140]

booth see SPRAY BOOTH.[59]

borax $Na_2B_4O_7 \cdot 10H_2O$. Natural, hydrated SODIUM BORATE, found in SALT LAKES and ALKALI soils.[144]

Also, the commercial name for sodium borate. It is used in some FIRE-RETARDANT PAINTS, and to dissolve CASEIN in distemper.

bordeaux turpentine natural EXUDATION from the maritime pine which grows in Southern Europe.[44,164] See TURPENTINE.[44,164] Syn: galipot

border a narrow strip around an edge. See SOFFIT.[148]

A border WALLCOVERING is used for trimming, or as a FRIEZE, generally just under the ceiling.

borneol $C_{10}H_{18}O$. Constituent of pine oil. A TERPENIC ALCOHOL.[44]

It is a solid at ordinary temperature, with an mp of 203°C (397.4°F), bp of 212°C (413.6°F), sp gr of 1.011. Borneol is sometimes described as bornyl alcohol or borneo camphor. SOLUBILITY characteristics are good, and it has some application as a PLASTICIZER.

boron resins ESTERS derived from boric ACIDS and polyhydric ALCOHOLS. Characterized by SOLUBILITY in water.[144,167]

Boston Stone, The originally a PAINT MILL imported around 1700 by a painter, Thomas Child. Later used as a landmark or central point from which distances from Boston were measured. It is the earliest known implement of the paint industry in America. The round stone was rolled back and forth in the trough of a larger flat stone underneath, in which OIL and PIGMENT were DISPERSED. The Boston Stone is the official insignia of the FEDERATION OF SOCIETIES FOR COATINGS TECHNOLOGY.[59,64]

botany bay resin see ACCROIDES.[167]

bottom (1) ship's bottom;[159] (2) hard deposit in paint cans.[42]

bottom-drying DRYING of a coating film from the bottom towards the top, e.g., LEAD NAPHTHENATE is used as a bottom drier.[67]

bottoms see TAILINGS.[49]

Bouguer's (Lambert's) Law describes the proportionality between the ABSORPTION of RADIANT ENERGY by a nonscattering material and its optical path length: $-\text{Log } T_i/l = K$, where T_i is the internal transmittance, l is the path length, and K is the unit ABSORPTION COEFFICIENT.[43,69] See BEER-BOUGUER LAW.[43,69]

bounce back a SPRAY rebound.[42]

Words presented in CAP/SMALL CAPS type indicate that the word is defined in another part of the *Coatings Encyclopedic Dictionary.*

Numerical superscripts classify terms in one or more of the categories listed in the second section of this volume.

bow distortion of a board in which the face is convex or concave longitudinally.[153]

bow window see BAY.[82]

boxed heart where the pith falls entirely within the four faces anywhere in the length of a piece of WOOD.[153]

boxing combining of two or more separate BATCHES to one uniform batch.[131]

bozzo see ABBOZZO.[69,81]

BPI abbreviation for BINDING POWER INDEX.[156,157,41]

brake lining medium a medium for bonding together the chosen filling agent for the brake lining.[79]

branched refers to side chains attached to the main chain in molecular structure of POLYMERS, as opposed to linear arrangement.[130,169,133]

branched polymers two-dimensional POLYMERS having relatively short side chains branching from the main backbone.[130,169]

brand name see TRADE NAME.

brashness condition of WOOD characterized by low RESISTANCE to shock and by an abrupt failure across the GRAIN without splintering.[153]

brayer a small hand ROLLER for distributing INK.[168,59,77]

Brazil wax a name sometimes applied to CARNAUBA WAX.[137]

Brazil-wood a natural red DYE from the WOOD of *Caesalpinia braziliensis*.[41]

break (of oils) the separation of MUCILAGINOUS product which occurs when certain unrefined VEGETABLE OILS are heated. This appears as a hazy SUSPENSION which may COAGULATE into a spawn-like mass. The separated material generally remains insoluble, and cannot be redissolved in the oil, even on prolonged heating. When separation occurs, the oil is said to "break." The insoluble matter is also referred to as the "break" or "MUCILAGE."[63] (BSI)

break test (oils) break in DRYING OILS is a measure of the materials rendered insoluble in CARBON TETRACHLORIDE under the conditions of test by a prescribed method.[157,63]

breakdown to soften or PLASTICIZE RUBBER by working it on a rubber MILL or in an internal mixer. Same as to mill or MASTICATE.[140]

breakdown voltage the voltage required, under specific conditions, to cause the failure of an INSULATING VARNISH or other insulating material.[61,157]

breakfree oil OIL which has been refined mechanically by heating to about 88°C (190°F), adding an adsorbent such as FULLER'S EARTH or activated charcoal, agitating, and filtering.[63,131] Syn: nonbreak oil

breaking separation of an EMULSION or LATEX into two phases.[60,139]

breaking extension ELONGATION necessary to cause rupture of a test specimen; the tensile strain at the moment of rupture.[47,157]

breathing (1) PERMEABILITY of a coating, such as a LATEX paint film which permits WATER VAPOR to pass through the film without detrimental effects;[142] (2) flow of air or gas into and out of a partially filled storage tank, caused by changes of temperature.[59]

breathing-type coating a coating that is sufficiently PERMEABLE to permit transmission of water vapor without detrimental effect to itself.[71]

brewer's pitch see PITCH, BREWER'S.[44]

brick (1) U.S.A.—a solid or hollow MASONRY unit of CLAY or SHALE, molded into a rectangular shape while PLASTIC, and then burnt in a kiln; (2) Brit.—a walling unit not exceeding 33.75 cm in length, 22.5 cm in width, or 11.25 cm in height.[82,75] (DAC)

bridging covering over of an unfilled gap such as a crack or corner with a coating.[54] (BSI)

bridle series of rolls that keeps the tension on the strip of a continuous line.[59,70]

bright blast see WHITE BLAST.[158]

brightener, optical FLUORESCENT DYE or PIGMENT which absorbs UV RADIATION and re-emits light of a violet or bluish HUE. Used to increase the luminance factor and to remove the yellowishness of white or off-white materials.[41,43,69,83] Syn: brightening agent

brightening agent see BRIGHTENER, OPTICAL.[41,43,69,83]

brightness (deprecated) the dimension of COLOR that is referred to an ACHROMATIC scale, ranging from black to white, also called LIGHTNESS or LUMINOUS REFLECTANCE or TRANSMITTANCE.[69]

Because of confusion with SATURATION, the use of this term should be discouraged.

brightness flop change in LIGHTNESS with direction of observation.[43]

brilliance a subjective term referring to: (1) CLARITY or freedom from visible suspended matter or OPALESCENCE in clear coatings;[43] (2) cleanness or lack of a muddy or dirty tone in pigmented coatings;[43,69]

(3) the combined effect of LIGHTNESS and strength, or purity of tone of PRINTING INKS.[168,43,69]

brilliant ultramarine synthetic ULTRAMARINE BLUE.[41]

brine CONCENTRATED AQUEOUS SALT SOLUTION.[144]

The term can be applied to solutions of any salt, but it is commonly applied to solutions of sodium chloride.

bring forward in repainting, to repair local defective areas with the appropriate paints so as to bring them into conformity with the surrounding areas before applying the FINISHING COATS.[54,158] (BSI)

bristle material used in certain BRUSHES. It is frequently hog hair.[59]

British gum see DEXTRIN.[79,155]

British Standards Institution (BSI) a national organization (corresponding to the AMERICAN NATIONAL STANDARDS INSTITUTE and the AMERICAN SOCIETY FOR TESTING AND MATERIALS) which establishes and publishes standard specifications and codes of practice.[174]

British thermal unit (BTU) the quantity of heat required to raise the temperature of one pound of water one degree Fahrenheit at atmospheric pressure.[128,176] cf. CALORIE[128]

brittle easily broken when bent rapidly or scratched as with a knife blade or the finger nail; the opposite of tough.[47,56]

brittle point, brittle temperature lowest temperature at which a material withstands an IMPACT TEST under standardized conditions.[157,176]

brittleness tendency to crack or snap when subjected to DEFORMATION.[47,56]

broadcast to apply or sprinkle solid particles on an UNCURED coating SURFACE.[54]

broken color general MULTICOLORED effect brought about by the automatic merging of wet paints of various colors or by manipulation which produce apparently accidental effects.[42,43,54] (BSI)

broken white OFF-WHITE in which the initial pure WHITENESS of the principal PIGMENT has been very slightly "broken" or modified in a yellow or brown direction.[69,43] (BSI)

broken-color work see ANTIQUING.[54,163,148,175]

bronze an APPEARANCE characteristic of some printed films in which the apparent COLOR of the print depends upon the angles of viewing and illumination.[43,168,69,35] See BRONZING.[69,56]

bronze blue one of the names applied to the complex FERRIC FERROCYANIDE or iron blues.[41] See IRON BLUE.[41]

bronze dusting see BRONZING.[168]

bronze gold powders METALLIC POWDERS made from alloys of copper. They range in COLOR from bronze gold to green yellow.[41,45]

bronze paint paint incorporating bronze gold powder in a BRONZING LIQUID.[71,163] See BRONZING MEDIUM OR LIQUID.[41,45]

bronze paste paste consisting of fine METALLIC FLAKES in a volatile medium. The metal may be aluminum, copper, or alloys of copper.[41] (BSI)

bronze powders METAL FLAKE POWDERS made commonly from aluminum, copper, or copper-zinc alloys. [41,45,168]

Used in the manufacture of "gold paint" and PRINTING INKS. Mixed with a VARNISH, it makes a BRONZING MEDIUM or liquid that is then painted on a surface to produce a dull gold-colored finish. The finish will tarnish if it is not coated.

bronze scarlet see LAKE RED C.[41]

bronzing (1) a subjective, descriptive, APPEARANCE term applied to the METAL-like REFLECTANCE which sometimes appears at the surface of nonmetallic colored materials. It is perceived at the specular angle, by observing the image of a white light source, for example, and is characterized by a distinct HUE of different DOMINANT WAVELENGTH than the hue of the paint film itself. The origin of the selective SPECULAR REFLECTANCE observed is generally considered to be reflectance from very small particle size pigments partially separated from surrounding vehicle at or near the surface;[69,43,56,35] (2) characteristic metallic luster shown by certain highly colored pigments in full strength, e.g., certain iron and PHTHALOCYANINE BLUES. May also occur with BLEEDING type pigments in baking finishes;[41,42] (3) application of imitation gold or other metals either in powder form or in leaves;[148] (4) PRINTING with a sticky size and dusting same with finely powdered METAL PARTICLES or FLAKES to give the appearance of metallic printing. [168] (BSI) Syn: bronze dusting

bronzing fluid see BRONZING MEDIUM OR LIQUID.[71,152]

bronzing lacquer TRANSPARENT LACQUER for APPLICATION to bright METAL in order to preserve LUSTER and enrich COLOR.[71,151,163] (BSI)

bronzing medium or liquid vehicle, either a VARNISH or a LACQUER, which is specially formulated

---

Words presented in CAP/SMALL CAPS type indicate that the word is defined in another part of the *Coatings Encyclopedic Dictionary.*
Numerical superscripts classify terms in one or more of the categories listed in the second section of this volume.

for use as a BINDER with aluminum or bronze gold powders.[71,152] (BSI)

Brookfield® viscometer a VISCOMETER in which a cylinder or disk is rotated in a TEST LIQUID, and the torque necessary to overcome VISCOUS RESISTANCE is measured.[166,157,59,145]

The cylinder or disk element (spindle) is electrically driven through a beryllium copper spring. The degree to which the spring is wound (i.e., the angular distance by which the element lags) as a result of viscous drag is indicated by a pointer on the instrument dial while the test is being made, and is proportional to the viscosity of the fluid for any given speed and spindle. Low shear rate viscosities (ranging from 0.1 to 50 s^{-1}), like those measured by the Brookfield viscometer, give an indication of the flow and pumping properties of materials; of how easily it can be pumped into a mixing vessel.

The Brookfield viscometer measures the viscosity by measuring the force required to rotate a spindle in a liquid. The forces to which it responds are extremely small, and optimum performance of the instrument depends on the elimination of any extraneous friction (such as that supplied by the walls of the container) which may disturb the sensitivity of its measurements.

ASTM Test Method D2196 covers the determination of the APPARENT VISCOSITY and the SHEAR THINNING and THIXOTROPIC properties of NON-NEWTONIAN materials in the shear range of 0.1 to 50 s^{-1}.

brookite RUTILE form of TITANIUM DIOXIDE characterized by its orthorhombic form. Its color varies from black to reddish brown. It is not used directly as a PIGMENT.[41,34]

brown coat the coat of roughly finished PLASTER beneath the FINISH COAT; in three-coat work, the second coat of PLASTER, applied over a SCRATCH COAT and covered by the finish coat; in two-coat work, the BASECOAT plaster applied over LATH or MASONRY; may contain a greater proportion of AGGREGATE than the scratch coat.[73,75] (DAC) Syn: floating coat

brown hard varnish SPIRIT VARNISH prepared from SHELLAC dissolved in industrial ALCOHOL.[146,152]

Cheaper grades may consist of shellac-rosin mixtures, spirit manila, spirit manila-rosin mixtures, or alcohol soluble rosin-modified phenolic resins.

brown iron oxide pigment Pigment Brown 6 (77491, 77492, 77499). (1) Natural—mined, classified, and processed to a definite COLOR and Fe_2O_3 content; (2) Synthetic—(a) mixtures of red, black, and yellow oxides formulated in proportions dependent on the color required; (b) precipitated to color requirements.[41] Syn: metallic brown and mineral brown

brown madder PIGMENT made by the gentle charring of MADDER LAKE or ALIZARIN to give a dull, brownish red color.[41]

brown pink a deep variety of QUERCITRON LAKE.[41]

brown rot see DRY ROT.[65,154,173]

browne heat test HEAT POLYMERIZATION test designed to determine the purity of TUNG and similar OILS. The result is reported as the number of minutes required to cause GELATION at the specified temperature (282°C).[157,169,176]

Brownian motion the movement of particles in a COLLOIDAL system such as an AEROSOL caused by collision with the molecules in the fluid in which the particles are imbedded.[139] (IUPAC)

The accepted explanation is that movement of the particles is due to bombardment by molecules of the fluid. The average length of path L followed by a particle in any given direction at a given time t is directly proportional to the square root of the absolute temperature T and the time t; it is also inversely proportional to the square root of the viscosity η of the medium and to the diameter d of the particle (Avogadro's number N and the gas constant R enter in as constants):

$$L = \frac{2R}{N}\sqrt{\frac{T \cdot t}{\eta d 3\pi}}$$

Brunswick black solution consisting of GILSONITE, PETROLEUM PITCH, or similar material in SPIRIT or AROMATIC HYDROCARBONS.[76,163] (BSI)

Brunswick blue see IRON BLUE.[41]

Brunswick green a light green POWDER; a PIGMENT consisting of LEAD CHROMATE and IRON BLUE pigments.[41] See CHROME GREENS.[41] Syn: lead chrome green

brush a tool composed of BRISTLES set into a handle; most often used to apply coatings. Bristles may be synthetic or natural.[59,77]

brush backed sander see AUTOMATIC TURNING SANDER.[59,158]

brush binder tough product for cementing the hairs or bristles of BRUSHES in the handles.[79]

brush drag see DRAG, AND PULLING.[42,166]

ASTM test methods for brushability (brush drag) include: manual application test, D4958; high shear (ICI) viscosity, D4287.

brush graining an imitation effect of WOOD GRAIN; produced by drawing a clean dry BRUSH through a dark liquid STAIN, applied over a dry, light BASECOAT.[54,163] (DAC)

brush mark a small ridge or valley produced in a paint film by the combing action of the BRISTLES of a BRUSH.[42,56] (DAC)

brush (or roller) cleaner a combination of chemicals in which a BRUSH or rollers may be soaked in order to permit cleaning with soap and water.[127]

brush-off blast lowest blast cleaning standard. Nace No. 4.[158]

brushability the ability or ease with which a coating can be brushed.[157,166]

ASTM test methods for brushability (brush drag) include: manual application test, D4958; high shear (ICI) viscosity, D4287.

brushing application of a coating by means of a BRUSH.[77]

brushing consistency rheological conditions at which a coating can be properly applied to a substrate with a BRUSH.[166]

brushout the APPLICATION of paint on a small SURFACE for testing.[157] (DAC)

BS see BRITISH STANDARD.[174]

BSCP abbreviation for BRITISH STANDARD CODE OF PRACTICE.[174]

BSI abbreviation for BRITISH STANDARDS INSTITUTION.[174]

B-stage an intermediate stage in the reaction of certain THERMOSETTING RESINS in which the material softens when heated and swells when in contact with certain liquids, but may not entirely fuse or dissolve.[67,79,176] (ASTM) cf. A-STAGE, C-STAGE[67,79,176]

The resin in an uncured thermosetting adhesive is usually in this stage. Sometimes referred to as RESITOL.

BTU abbreviation for BRITISH THERMAL UNIT.[128]

bubble coating coating wherein the LIGHT SCATTERING characteristics are derived partially or wholly from small bubbles or MICROVOIDS in a TRANSPARENT BINDER. The air-filled or VAPOR-SATURATED VOIDS having diameters in the range of the WAVELENGTH OF LIGHT, are trapped in a water soluble medium such as soya protein.[71,43] See MICROVOIDS.[43]

This type of coating has a low specific weight and high LIGHTNESS.

bubble tube viscometer one of the bubble VISCOMETERS which is used by comparing the speed of a rising air bubble in a liquid of unknown VISCOSITY with that of a liquid of known viscosity, both liquids being in corked glass tubes of identical size, and at the same temperature.[157,166,145] Syn: air bubble viscometer

bubbling FILM DEFECT, temporary or permanent, in which bubbles of air or SOLVENT VAPOR, or both, are present in the APPLIED FILM.[42,56] cf. BLISTERING[42,56]

buffer a substance that reduces the change in hydrogen ion concentration otherwise produced by adding ACIDS or BASES to a solution.[144] See pH.[128,157]

buffer coat a finishing material applied over an old coating to protect it from SOLVENT action of the new FINISH.[158,71]

buffing to polish or GRIND down a SURFACE.[45,131,158]

Buhr mill stone MILL in which a viscous PIGMENT-medium paste is dispersed between the flat surfaces of two large stones, one of which rotates.[49,59]

Buhrstone type of POROUS FLINT, found in France, used for MILLSTONES.[64]

build real or apparent thickness, fullness, or depth of a dried film.[133,166,35] See HIGH BUILD.[54,35]

This term is used to describe the real or apparent FILM THICKNESS of a coating and is an indication of the way the coating disguises or covers any surface TEXTURE. Build can be directly related to the film thickness of the final coating system but glossy products usually appear to have a higher build than flatter ones even though the thickness may be the same.

build of wet film the amount of wet paint applied by BRUSHING onto a smooth impervious surface.[128] See COVERAGE.[168,128]

Wet film build is expressed as mass of wet paint per unit area of surface covered. The function is a gravimetric inverse of SPREADING RATE, which is noted volumetrically.

builder a substance, usually a POLYPHOSPHATE, added to a detergent formulation, to improve detergency.[139]

Its action is probably multiple; it acts to sequester ions contributing to HARDNESS, it may affect surface charges (and thus stabilize soil particles) and it maintains an alkaline pH, which is required for optimum detergency. (Note: in recent years, concern for the ecological effects of the discharge of phosphate-containing detergents into surface waters have led to a reduction in the quantity employed, and the use of other types of materials, e.g., sequestering agents, as a substitute.)

building a more or less enclosed and permanent STRUCTURE for housing, commerce, industry, etc., distinguished from mobile structures and those not intended for occupancy.[82] (DAC)

Words presented in CAP/SMALL CAPS type indicate that the word is defined in another part of the *Coatings Encyclopedic Dictionary.*

Numerical superscripts classify terms in one or more of the categories listed in the second section of this volume.

building block a rectangular MASONRY unit, other than a brick, made of burnt CLAY, CEMENT, CONCRETE, GLASS, GYPSUM, or any other material suitable for use in building construction.[73,75,82] (DAC)

building board any sheet of building material, often faced with paper or VINYL; suitable for use as a finished SURFACE on walls, ceilings, etc.[75,82] (DAC)

building coats coats used to build up a SURFACE or surfaces before RUBBING.[54]

building code a collection of rules and regulations adopted by authorities having appropriate jurisdiction to control the design and construction of buildings, alteration, repair, quality of materials, use and occupancy, and related factors of buildings within their jurisdiction; contains minimum architectural, structural, and mechanical standards for sanitation, public health, welfare, safety, and the provision of light and air.[82,171,172] (DAC)

building paper a heavy, relatively cheap, durable PAPER, such as ASPHALT paper, used in building construction, especially in frame construction, to improve thermal insulation and weather protection and to act as a vapor barrier.[75,76,82] (DAC)

Special types are: sheathing paper, used between sheathing and siding; and floor lining paper, used between rough and finish floors.

building stone any stone which may be used in building construction: granite, limestone, marble, etc.[75,82,73] (DAC)

built-up roofing (BUR) a continuous, semiflexible membrane consisting of PLIES of SATURATED FELTS, coated felts, fabrics, or mats assembled in place with alternate layers of BITUMEN, and surfaced with mineral AGGREGATE, bituminous materials, or a granule surfaced sheet.[76] (ASTM)

bulk density see APPARENT DENSITY.[128,142]

bulk modulus relationship of applied STRESS and volumetric STRAIN produced when a stress is applied uniformly to all sides of a body.[47]

bulking in the process of formulating coatings, the step wherein ingredient weights are converted to their volume equivalents.[128]

bulking value reciprocal of APPARENT DENSITY. Solid volume of a unit weight of material, usually expressed as gallons per pound or litres/kg.[128] (ASTM) Syn: specific volume

For practical purposes, this is 0.120 divided by the specific gravity.

bunghole orifice, usually in a steel drum, a barrel, or a keg, which is stopped by a "bung" or large stopper.[50]

bunghole oil see BOILED OIL.[63]

BUR abbreviation for BUILT-UP ROOFING.[76]

burgundy pitch see PITCH, BURGUNDY.[44]

burl swirl or twist in the GRAIN of WOOD that usually occurs near a knot, but does not contain a knot.[153]

burlap see JUTE.

burned finish WOOD FINISH in which the hard portion of the GRAIN stands out in relief, the effect being produced by using a blowtorch and a stiff-bristled BRUSH.[54] Syn: fiery finish

burning off removal of paint by a process in which the paint is softened by heat, e.g., from a flame, and then scraped off while still soft.[158]

burning-in PROCESS of repairing SCRATCHES and other damaged places in a finish, by melting compounds (stick SHELLAC) into the DEFECT with a heated tool.[54]

burnish to polish or rub to a smoother or glossier surface.[42,43]

burnish resistance RESISTANCE of a COATING to an increase in GLOSS or SHEEN due to polishing or RUBBING.[43,142,72]

Paints may become burnished in areas where clothing or upholstered furniture rub against a wall or in areas where scrubbing or cleaning have taken place. Burnish problems occur more often in FLAT and sheen paints than in SEMIGLOSS or gloss paints.

burnt island red (Brit.) NATURAL IRON OXIDE.[41]

burnt lime see CALCIUM OXIDE.[41,58]

burnt sienna see IRON OXIDE, NATURAL.[41]

burnt umber naturally occurring FERRIC OXIDE containing manganese dioxide and varying amounts of SILICA and ALUMINA. Characteristic COLOR is a rich, dark brown.[41] See UMBER.[41]

Ferric oxide range, 45-65%; manganese dioxide range, 3-20%. Produced by calcining raw umber.

bush kauri copal special grade of the KAURI COPAL, which has fossilized on trees at the point of exudation and above ground level.[167]

butadiene $CH_2{=}CH{-}CH{=}CH_2$. Gas, insoluble in water, but soluble in ALCOHOL and ether, obtained from the cracking of PETROLEUM, from COAL TAR BENZENE or from acetylene.[141]

It is widely used in the formation of copolymers with styrene, acrylonitrile, vinyl chloride, and other monomeric substances, where it imparts FLEXIBILITY.

butadiene styrene latex a SYNTHETIC LATEX similar

to synthetic RUBBER; used for latex paints.[140,130,156] Syn: styrene butadiene

1-butanol see N-BUTYL ALCOHOL.[164]

butt to meet without overlapping.[54,82]

butt veneer VENEER having a strong curly figure, caused by roots coming into the trunk at all angles.[153]

button lac LAC refined in the form of buttons, which are usually about 3 in. (7.5 cm) in diameter and 1/4 in. (6 mm) in thickness.[146]

n-butyl acetate $CH_3COOC_4H_9$. Limpid, colorless liquid with fruity odor.[164]

Prepared by heating and distillation, after contact of n-butyl alcohol with acetic acid in the presence of a CATALYST such as SULFURIC ACID. SOLVENT used in production of LACQUERS, natural gums, and SYNTHETIC RESINS.

n-butyl acrylate $CH_2{=}CHCOOC_4H_9$. Colorless liquid. A MONOMER used in the manufacture of SYNTHETIC RESINS and as a SOLVENT for CELLULOSE ESTERS.[141]

Properties: sp gr., 0.898; bp, 145°C

n-butyl alcohol $CH_3(CH_2)_2CH_2OH$. Colorless liquid, with vinous, irritating odor which causes coughing.[164] Syn: 1-butanol and propyl carbinol

SOLVENT used for RESINS, coatings, SHELLAC, and butylated melamine resins.

Butyl Carbitol® $HOCH_2CH_2OCH_2CH_2OC_4H_9$. DIETHYLENE GLYCOL MONOBUTYL ETHER, butyldiglycol, 2-[2-(1-butoxy)ethoxy] ethanol. Practically odorless OIL, SOLVENT, water miscible liquid.[164]

Properties: bp, 230.4°C; mp, -68.1°C; sp gr, 0.9536.

Butyl Carbitol® acetate $CH_3COO(CH_2CH_2O)_2C_4H_9$. DIETHYLENE GLYCOL MONOBUTYL ETHER ACETATE.[164]

Butyl Cellosolve® $HOCH_2CH_2OC_4H_9$. ETHYLENE GLYCOL MONOBUTYL ETHER, 2-butoxyethanol. Liquid soluble in 20 parts water; soluble in most ORGANIC SOLVENTS.[164]

Butyl Cellosolve® acetate $CH_3COOC_2H_5OC_4H_9$. ETHYLENE GLYCOL MONOBUTYL ETHER ACETATE.[164]

butyl citrate colorless or pale yellow, stable, odorless, nonvolatile liquid. PLASTICIZER, ANTIFOAM agent, SOLVENT for CELLULOSE NITRATE.[144,164] See CITRATES.[144,170] Syn: tributyl citrate

butyl lactate $CH_3CH(OH)COO(CH_2)_3CH_3$. Slow evaporating liquid which is miscible with most of the ORGANIC SOLVENTS.[144,164]

It is a solvent for NITROCELLULOSE, ETHYL CELLULOSE, OILS, DYES, NATURAL GUMS, and SYNTHETIC RESINS.

butyl phthalate see PHTHALATE ESTERS.[170]

butyl ricinoleate $C_{17}H_{32}(OH)\ COOC_4H_9$. Yellow to colorless oleaginous liquid derived from castor oil fatty acid (RICINOLEIC ACID) and butyl ALCOHOL.[170]

It is used as a PLASTICIZER or lubricant.

butyl rubber (1) generic name for VULCANIZABLE ELASTIC COPOLYMERS of isobutylene and small amounts of diolefins; (2) mixture of isobutylene, 98% and butadiene or isoprene, 2%.[140]

Butyl (tetra) titanate $Ti(OC_4H_9)_4$. Used for ADHESION PROMOTER, CATALYST, and in HEAT-RESISTANT PAINTS.[74]

butyraldehyde $CH_3(CH_2)_2CHO$. Colorless liquid. Syn: butaldehyde, butyl aldehyde, butyric aldehyde

Used in the manufacture of RUBBER and SYNTHETIC RESINS. Sp gr, 0.817; bp, 75.7°C.[144]

butyrates ESTERS of butyric acid, such as CELLULOSE butyrate, ethyl butyrate, etc.[144]

butyric alcohol see N-BUTYL ALCOHOL.[164]

Words presented in CAP/SMALL CAPS type indicate that the word is defined in another part of the *Coatings Encyclopedic Dictionary.*
Numerical superscripts classify terms in one or more of the categories listed in the second section of this volume.

C

C degrees Celsius (previously called centigrade). It is the S.I. temperature scale on which 0 (zero) C equals the freezing point of water and 100 C equals the boiling point of water.[128] cf. F[128]

°C degree Celsius.[128] cf. °F[128]

C or Δc abbreviations for chromaticity or chromaticity difference, respectively.[69] See COLOR DIFFERENCE EQUATIONS.[69]

c or ca. circa; about, approximate.

CAA abbreviation for the Clean Air Act.[172] (EPA)

cabinet finish a VARNISH or polished HARDWOOD interior FINISH as distinguished from a painted SOFTWOOD finish.[71,175] (DAC)

cabinet making the art or craft of making fine articles of wooden FURNITURE.[175]

cable varnishes and lacquers employed for impregnating the cloth or similar covering of cables. Normally, OIL VARNISHES are used.[151,152,163]

cacahuananche oil this oil is obtained from the nuts of the tree, *Licania arborea*.[63] (PTM) cf. OITICICA OIL.[63] Syn: Mexican oiticica oil

So far as the usual laboratory tests are concerned, this oil and Brazilian oiticica oil are much alike. The raw oil becomes lard-like on aging but may be permanently liquefied by heat. The raw and lightly heat-treated oil wrinkles as it dries, similarly to oiticica and tung oils.

cadmium green Pigment Green 18. Composed of 93-94% hydrated chromium oxide, the balance being CADMIUM YELLOW.[41]

cadmium lithopone Pigment Red 108 (77202). Range of yellow, orange, and red colors consisting of either CADMIUM SULFIDE or CADMIUM RED reduced on barium sulfate. They are usually produced by coprecipitation followed by CALCINATION.[41] See CADMIUM RED.[41]

cadmium/mercury lithopones orange, red, and maroon PIGMENTS consisting of calcined coprecipitations of CADMIUM SULFIDE and mercury sulfide, with barium sulfate.[41]

cadmium/mercury sulfides CdS·HgS. Pigment Orange 23 (77201). A series of cadmium/mercury sulfides of mixed crystal composition. Colors vary according to CdS/HgS ratio from deep orange to maroon.[41] Syn: Mercadium® orange, mercury cadmium red/orange, mercury/cadmium sulfide

Known for high HEAT STABILITY, SOLVENT INSOLUBILITY, good chemical resistance and excellent LIGHTFASTNESS. CdS/HgS ratio, 89.1/10.9. Density, 4.8-5.2 g/cm³ (40-43 lb/gal); O.A., 33.2; particle size, 0.1-1.0 μm.

cadmium orange mixture or coprecipitate of CADMIUM SULFIDE and CADMIUM SELENIDE used as a PIGMENT.[41] See CADMIUM SULFIDE.[41]

cadmium plating ELECTRODEPOSITION of cadmium on iron wire, steel articles, etc., to make them relatively rust resistant.[45,158]

This surface treatment makes the substrate more suitable for subsequent painting.

cadmium red Pigment Red 108 (77202). PIGMENTS of calcined coprecipitations of COMPOUNDS of cadmium, sulfur, and selenium; a brilliant red pigment, opaque, with good staining power, fast to light, unaffected by EXPOSURE to sulfur fumes; considerable resistance to heat.[41]

Properties: density, 4.9 g/cm³ (41 lb/gal); O.A., 20 lb/100 lb.

cadmium selenide see CADMIUM RED and CADMIUM ORANGE.[41]

cadmium sulfide (Cd/Zn)S or CdS Pigment Yellow 37 (77199). Class of yellow, orange, and red pigments that include pure cadmium sulfides and blends of cadmium sulfides with ZnS and CdSe and used with $BaSO_4$ to form LITHOPONE type PIGMENTS.[41] Syn: cadmium yellow, cadmium orange, Jaure brilliant

They show excellent heat and good alkali resistance especially for baking finishes. Properties: density, 4.6-4.7 g/cm³ (38-39 lb/gal); O.A., 18-22 lb/100 lbs.

cadmium yellow see CADMIUM SULFIDE.[41]

caisson recessed compartment in a vault or ceiling; recessed panel in a ceiling or SOFFIT.[82]

cajeputene see DIPENTINE.[164,44]

cake, press the thick cake of wet PIGMENT that is withdrawn from the filter press.[41,131]

caking (1) settling of PIGMENT PARTICLES of a paint into a compact mass which is not easily redispersed by stirring;[42] (2) in PRINTING INKS, caking is the collecting of pigment upon plates, rollers, or blankets caused primarily by the inability of the vehicle to hold the pigment in suspension.[168] See PILING. (BSI)[42]

calcareous any material containing calcium or calcium compounds.[134]

calcicoater (1) VEGETABLE OIL product that has been reacted with LIME to form a heavy gel-like soap, which has almost infinite tolerance for mineral spirits; used in flat wall paints, PRIMERS, and SEALERS;[71,156] (2) PIGMENTED FLAT PAINTS using such heavy gel-like soap vehicles.[71]

calcimine also spelled "kalsomine." Essentially, chalk and glue ready to mix with water. Used as a decoration for interior surfaces. It will not withstand washing. In Britain, it is referred to as powdered distemper.[71]

calcination (1) PROCESS of heating or roasting a material to a high temperature, but below its fusing point, to cause it to lose moisture or other volatile material or to be oxidized or reduced. Originally a heat process for the production of LIME (CaO) from limestone;[131] (2) the process of subjecting a sorptive mineral to prolonged heating at fairly high temperature, resulting in the removal of water, and an increase in the HARDNESS, physical stability, and absorbent properties of the material.[131] (ASTM)

calcined clay CHINA CLAY (KAOLIN) that has been heated until the combined water is removed and the plastic character of the clay is destroyed. This produces an air-solid interface within the particle which increases hiding in the resulting coating.[41,58,131]

calcite naturally occurring form of CALCIUM CARBONATE. It is an essential material of LIMESTONE, marble, and CHALK.[41,58]

calcium carbonate, natural $CaCO_3$,$CaMg(CO_3)_2$. Pigment White 18 (77220). White EXTENDER PIGMENT derived from natural CHALK, LIMESTONE, or DOLOMITE, consisting of CALCIUM CARBONATE with up to about 44% MAGNESIUM CARBONATE.[41,58] Syn: calcite, limestone, whiting, marble flour, Paris white, chalk, ground oyster shells, Iceland spar, Spanish white

Properties: density, 2.71 g/cm³ (22.6 lb/gal); O.A., 6-15; particle size, 1.5-12 µm.

The ASTM specification for calcium carbonate is D1199.

calcium carbonate, synthetic CaCO3. Pigment White 18 (77220). CALCIUM CARBONATE manufactured by a PRECIPITATION PROCESS in order to obtain a finer or more uniform particle size range. Four commercial processes are known.[41,58] Syn: precipitated calcium carbonate

Properties: density, 2.65 g/cm³ (22.07 lb/gal); O.A., 28-58; particle size, 0.6 to 3.0 µm.

calcium chloride $CaCl_2$. SALT used in the manufacture of some LAKES and TONERS from ACID DYESTUFFS, fireproof paints, sizing compounds, wood preservatives, snow melter, and antifreeze.[83,144]

calcium hydroxide $Ca(OH)_2$. Used in MORTAR and PLASTER.[144] Syn: slaked lime

calcium ion stability the RESISTANCE of a LATEX to COAGULATION or FLOCCULATION in the presence of calcium ions.[161]

Due to the increased concentration of calcium ions in "hard" water and their presence in some PIGMENTS, EXTENDERS, and SUBSTRATES (such as spackled areas, plaster walls, etc.), it is necessary to test latexes for resistance to calcium and other multivalent ions (see BARIUM ION STABILITY). In addition, multivalent ion stability is necessary for latices which are used for cement or other masonry substrate.

calcium linoleate white amorphous POWDER soluble in alcohol and ether; insoluble in water.[127]

Used for waterproofing compounds, emulsifying agents, and as a stabilizer for oleoresinous paints.

calcium lithol red see LITHOL RED.[41]

calcium metasilicate see CALCIUM SILICATE, NATURAL.[41]

calcium naphthenate calcium salt of naphthenic acids.[64,74,67]

Used as an auxiliary DRIER, dispersing and stabilizing aid.

calcium octoate calcium salt of 2-ethyl hexoic acid.[64,74,67]

Used as an auxiliary DRIER, dispersing and stabilizing aid.

calcium oxide CaO.[144] Synonym for LIME or quicklime.

calcium resinate calcium salt of ROSIN used as an auxiliary DRIER and dispersing and stabilizing aid.[64,74,67]

Commonly known as LIMED ROSIN, which is really rosin, the acidity of which has been substantially neutralized.

calcium silicate, natural $CaSiO_3$. It has white COLOR and acicular particles. It is characterized by high FLATTING action, combined with low OIL ABSORPTION, and is used as an EXTENDER in paints, CERAMICS, and PLASTICS.[41] Syn: wollastonite, calcium metasilicate

Properties: density, 2.9 g/cm³ (24.2 lb/gal); O.A., 25-30; particle size, 7 µm (fine particle grade); hardness (mohs), 4.5.

calcium silicate, synthetic $CaSiO_3 \cdot nH_2O$. EXTENDER PIGMENT with some DRY HIDING OPACITY.[41,58]

Words presented in CAP/SMALL CAPS type indicate that the word is defined in another part of the *Coatings Encyclopedic Dictionary.*

Numerical superscripts classify terms in one or more of the categories listed in the second section of this volume.

Properties: density, 2.26 g/cm³ (18.8 lb/gal); O.A., 280; particle size, 10-12 μm.

calcium stearate calcium salt of stearic acid. A METALLIC SOAP used in paints as WETTING aid for PIGMENTS, FLATTING AGENTS, and ADDITIVE for improving sanding sealers.[64,127,83]

calcium sulfate $CaSO_4 \cdot 2H_2O$. Obtained from GYPSUM or as a byproduct from other PRECIPITATION processes.[41,58] Syn: terra alba and gypsum

Notwithstanding its slight solubility in water, it is used to some extent in DISTEMPERS and water paints. It forms the base of certain LAKES and other PIGMENTS. PLASTER OF PARIS is a partially calcined gypsum.

calcium sulfate, anhydrous $CaSO_4$. Manufactured by calcining natural GYPSUM ($CaSO_4 \cdot 2H_2O$) or as byproducts from other PRECIPITATION processes.[41,58] Syn: anhydrite, dead burned calcium sulfate, dehydrated gypsum

Notwithstanding its slight water solubility, it is used in distempers and water paints. It also forms the base of certain LAKES and other PIGMENTS. PLASTER OF PARIS is partially calcined gypsum. Properties: density, 2.96 g/cm³ (24.7 lb/gal); O.A., 20; particle size, 1.8-2.2 μm.

calcium sulfate hemihydrate $CaSO_4 \cdot {}^1/_2H_2O$. See GYPSUM (CALCINED), and PLASTER OF PARIS.[58]

calcium tallate calcium salt of tall oil acids used as an auxiliary DRIER, DISPERSING, and STABILIZING aid.[64,74,67]

calender (1) to produce or process sheets of material by pressing between a series of revolving heated rolls; (2) the machine used in this process.[59,131]

calendered cloth basis of varnished cloths for electrical insulation.[131]

The process of calendering cloths involves the passing of moist cloth between heavy rollers, resulting in a smooth or glazed appearance.

calendered papers WALLPAPERS with hard coatings.[148]

calibrate to check, adjust, or determine by comparison with a standard.[37,157,145]

calk see CAULK.[165]

calking see CAULKING COMPOUND.[165]

calorie unit of heat. It is the amount of heat required to elevate the temperature of 1 g of water 1°C.[128] cf. BRITISH THERMAL UNIT[128]

camber strip under tension, as on a coating line that has a tendency to deviate from the horizontal or to slope to one side.[70]

cambogia see GAMBOGE.[167]

camel hair inaccurately applied to a fine haired BRUSH, the hair of which is obtained from squirrels native to Russia and Siberia.[59,81]

camphol see OXANILIDE.[170]

camphor $C_{10}H_{16}O$. Solid KETONIC COMPOUND with an mp of 175°C and a bp of 204°C. Used as a PLASTICIZER for CELLULOID, CELLULOSE NITRATE, and LACQUERS.[38,170]

can coating used on a wide variety of decorated METALS for packaging purposes. They are invariably THERMOSETTING coatings.[71,50]

These include cans for food, paint, etc., aerosols, collapsible tubes, caps and closures of many kinds. Generally, the paint is applied to flat sheet stock, followed by printing, stacked up for storage, then stamping and forming. The most common substrates are tin-plated steel, aluminum, and to a smaller extent, directly chromated steel.

can preservative see IN-CAN PRESERVATIVE.[83,173]

can stability see SHELF LIFE, and STORAGE STABILITY.[157,161]

Canada balsam oleoresin which exudes naturally from *Pinus balsamea*, the Canadian balsam pine.[167]

It is essentially a RESIN dissolved in an essential oil. When freshly exuded it is a viscous liquid, but it hardens on EXPOSURE. Its chief use is for cementing lenses and other GLASS objects, because its refractive index, 1.53, is near that for glass. It is also used in the manufacture of fine lacquers.

Canada turpentine see CANADA BALSAM.[167]

candelilla wax natural WAX obtained from the stems of the *Eurorbia antisiphylitica* and *Pedilanthus paroms* plant which grows in Mexico and the Southwest U.S.[137]

It varies in color from pale yellow to a yellow or greenish-brown. It is one of the harder waxes, although not as hard as carnauba wax. Used in varnishes, electric insulation compositions, waterproofing and insect proofing, and paint removers.

candle (unit of luminous intensity) candle power is a measure of intensity of a source of light as compared with a standard candle.[128,43] See FOOTCANDLE.[128,43]

candlenut oil OIL derived from the nuts of the candlenut tree (*Aleurites moluccana*) of tropical Asia and Polynesia.[63]

canopy ceiling a decoration composed of a ceiling paper, or a sidewall paper such as a stripe, in which the strips are cut in triangles and hung so that the apex will terminate in the center of the ceiling to produce a striking domed effect.[148]

canvas a coarse cloth made from cotton, hemp, or flax. It may be used for artist's canvas or a picture painted on canvas.[159,81]

canvas board a paper board with primed canvas fastened to one face.[81,159]

caoutchouc synonym for pure natural (raw) rubber.[140]

capillarity force of attraction between like and dissimilar substances.[142,157]

It is exhibited, for example, by the rise of a liquid up a capillary tube, or by the wetting of solids by liquids.

capillary viscometer see VISCOMETER.[145,157,166]

caprolactam cyclic amide-type COMPOUND, containing six carbon atoms.[130]

When the ring is opened, caprolactam is polymerizable into a nylon RESIN known as type-6 nylon or polycaprolactam. Used in preparation of PLASTICS, PLASTICIZERS, coatings, and POLYURETHANE CROSSLINKERS.

captive production production of raw materials or intermediates by a company for its own use.

carbamide see UREA.[133,144]

carbazole dioxazine violet Pigment Violet 23 (51319). These PIGMENTS are manufactured in a variety of shades from red-shade to blue-shade violets. The latter appear more popular for TONING (reddening) PHTHALOCYANINE BLUES.[41]

These pigments have excellent fastness properties even at high degrees of dilution but exhibit poorer color retention in metallized (aluminum flake) finishes than in light TiO_2 pastels. They are produced in both TONER and LAKE (alumina hydrate) forms.

carbenes constituents of bituminous type products which are insoluble in the CHLORINATED HYDROCARBONS, carbon tetrachloride, and chloroform, but soluble in carbon disulfide.[76]

carbinol (1) $-CH_2OH$. Monovalent primary alcohol radical. It may be part of an aliphatic or an aromatic alcohol. For example, ETHYL ALCOHOL (C_2H_5OH) can be described as methyl carbinol, or benzyl alcohol ($C_6H_5CH_2OH$) as phenyl carbinol; (2) it is sometimes used as a synonym for METHANOL.[164]

Carbitol® $C_2H_5O(CH_2)_2O(CH_2)_2OH$. Monoethyl ether of diethylene glycol. A water-miscible solvent.[164]

Properties: bp, 195°C; sp gr, 0.990/20°C; flp, 96°C (205°F); vp, less than 0.1 mm Hg/20°C. If the ethyl (C_2H_5) group is substituted by methyl (CH_3), butyl (C_4H_9) groups, etc., the corresponding methyl and butyl carbitols are obtained.

carbohydrates COMPOUNDS containing carbon and the elements of water, and thus their only elements are carbon, hydrogen, and oxygen.[134,144]

They are all variations of the same basic formula: $C_m(H_2O)_n$. The carbohydrates include the mono-, di-, tri-, tetra-, and poly-saccharides, as well as the conjugated saccharides.

carbolic acid common name for phenol.[130,144]

carbon black Pigment Black 7 (77266). Finely divided CARBON formed by any one of the following processes: (1) incomplete combustion of natural gas in burners under moving channel irons (channel carbon black); (2) incomplete combustion of natural gas and petroleum in large, closed furnaces (furnace carbon black); (3) decomposition of gases in large converters filled with hot refractory brick checkerwork.[41] Syn: gas black

These carbon blacks vary in particle size and some of them may be surface treated.

carbon dioxide (CO_2) colorless, odorless gas which is the major carbon-containing product of combustion of carbon compounds.[144,171] (IUPAC) Syn: carbonic acid gas

Its removal in the atmosphere involves the photosynthetic cycle in plants which convert CO_2 into carbohydrate and other organic molecules important in plant growth, as oxygen is released. It is also removed by the oceans where it is converted to carbonate deposits. Its concentration in the atmosphere varies with position on the earth, season, time of the day, etc. However, its average mixing ratio in pristine air has increased from 315 to 340 ppmv during the last 25 years, presumably as a result of increased burning of fossil fuels. Concern has been expressed in the possible effects of this increase on global climate change; see GREENHOUSE EFFECT. (IUPAC) In the solid state it is called dry ice.

carbon monoxide (CO) colorless, odorless gas which is highly poisonous even at the level of 0.2% in air.[171,172] (IUPAC)

It is a product of the incomplete combustion of hydrocarbon fuels, and it is also formed in the clean atmosphere as a result of the oxidation of methane and its oxidation product FORMALDEHYDE (CH_2O); mixing ratios of CO in pristine air masses are usually of the order of 100 ppbv near sea level. (IUPAC)

carbon tetrachloride CCl_4. Clear liquid with an odor similar to that of chloroform.[164] Syn: tetrachloromethane, perchloromethane

It has a sp gr of 1.629 and is nonflammable. It is miscible with most organic solvents. Solvent action on oils and similar products and soft

---

Words presented in CAP/SMALL CAPS type indicate that the word is defined in another part of the *Coatings Encyclopedic Dictionary.*

Numerical superscripts classify terms in one or more of the categories listed in the second section of this volume.

resins is good. It was frequently used to reduce fire hazards. Very toxic. (No flash point or fire point).

carbonate white lead $Pb_3O_8C_2H_2$. Pigment White 1 (77597). Basic carbonate of LEAD, the composition of which approximates to the formula $Pb(OH)_2 \cdot 2PbCO_3$.[41] Syn: basic carbonate of lead, basic lead carbonate, cremnitz white, kremnitz white, white lead

It was the normal white lead of industry and was often given the above description to differentiate it from white lead sulfate which is also used as a pigment. It was marketed in several grades designated according to the method of manufacture, e.g., stack, chamber, electrolytic, or precipitated. The latter was further subdivided into high, medium, and low stain varieties.

carbonific chemical COMPOUND which, upon DECOMPOSITION, produces a mass of carbon which frequently occupies a volume much greater than the original unburned materials.[172]

Carbonifics function to produce the insulating, relatively incombustible properties of INTUMESCENT COATINGS.

carbonium a positively charged organic ion such as H_3C, having one less electron than a corresponding free radical and behaving chemically as if the positive charge were localized on the carbon atom.[134]

carbonization PROCESS of degrading organic matter to elemental carbon.[56,65]

The term is used chiefly in describing the DECOMPOSITION of VARNISH FILMS and the like, when subjected to elevated temperatures. Carbonization of films on stoving is always associated with pronounced darkening.

carbonyl group the bivalent radical, =CO, especially as it occurs in aldehydes or ketones.[134] (ASTM)

carbonyl value keto group =CO, is sometimes referred to as the carbonyl radical. Thus the amount of keto groups in a compound is its carbonyl value.[134]

Carborundum® a trademark used for an abrasive of silicon carbide crystals.[158]

carboxyl group the radical –COOH, characteristic of most organic acids.[134] (ASTM)

carboxylic acid organic acid which possesses one or more carboxyl groups.[144]

The simplest member of the series is formic acid, H·COOH.

carboxymethyl cellulose (CMC) cellulose ether of glycolic acid. It is usually marketed as a water-soluble sodium salt, more properly called sodium carboxymethyl cellulose.[83,139,155] (ASTM) cf. CELLULOSE GUM[83,155]

In the early literature, it is sometimes called cellulose glycolate or cellulose glycolic acid.

carcinogen a material that either causes cancer in humans, or, because it causes cancer in animals, is considered capable of causing cancer in humans.[172]

Findings are based on feeding of large quantities of a material to test animals or by application of concentrated solutions to the animal's skin. A material is considered a carcinogen if (1) the INTERNATIONAL AGENCY FOR RESEARCH ON CANCER (IARC) has evaluated it and found it to be a carcinogen or potential carcinogen; (2) the National Toxicological Program's (NTP) "Annual Report on Carcinogens" lists it as a carcinogen or potential carcinogen; (3) OSHA regulates it as a carcinogen; or (4) one positive study has been published.

cardboard synonym for paperboard and pasteboard.[81]

cardol one of the original constituents of cashew nutshell liquid, occurring to the extent of about 10%.[144]

It is a dihydroxy phenol, containing a side chain with two double bonds.

carene TERPENE HYDROCARBON, which is a constituent of certain TURPENTINES.[164]

Carene has been reported in German, Indian, Russian, and Finnish turpentines.

carmine ALUMINUM LAKE of a PIGMENT from COCHINEAL.[41]

carmine lake Natural Red 4 (75470). Barium lake of the DYESTUFF produced by coupling 7-amino-1-naphthalenesulfonic acid with R-sale (2-naphthol-3:6-disulfonic acid).[41]

carmine vermillion see MERCURIC SULFIDE.[41]

carminic acid see COCHINEAL.[41]

carnauba wax extremely hard WAX obtained from the palm leaves of the cerara palm (*Copernia cerifera*) a Brazilian tree.[137] Syn: Brazil wax

It appears on the market in several grades, such as bleached, yellow, fatty gray, and chalky gray. The wax has the following approximate constants: mp, 84°C; sp gr., 0.998; acid value, 2; saponification value, 80; iodine value, 13. It is a constituent of wax polishes, and has some application in matte, overprinting, and baking varnishes. Its solubility is of a very low order in most varnish solvents and other constituents.

carnaubic acid $C_{23}H_{47}COOH$. One of the constituent acids of CARNAUBA WAX; monocarboxylic acid. Mp, 72°C.[137]

carnaubyl alcohol $C_{24}H_{49}OH$. Alcoholic constituent of CARNAUBA WAX. Mp, 69°C.[144]

carotene (carotin) $C_{40}H_{56}$. HYDROCARBON found in LINSEED and other VEGETABLE OILS where it is believed to act as an ANTIOXIDANT.[144]

carrageen (sometimes spelled "carrageenan" or "carrageenin.") Hydrocolloid obtained from a group of seaplants related to *Chondrus crispas*, colloquially called Irish moss or moss.[139,155] Syn: Irish moss, Irish gum, pearl moss, pig wrack, rock salt moss

It is a complex carbohydrate made up of galactose, dextrose, and levecose residue. Useful for mucilaginous and gelatizing qualities.

Carter lead WHITE LEAD which used to be manufactured in U.S.A. by what is known as the Carter process.[41]

This was somewhat similar to the chamber process except that the metallic lead is used in finely divided state instead of sheets. The process is therefore quicker, but the resulting product was very similar to chamber white lead.

cartoon (1) preliminary sketch or detailed drawing for a painting, mural, etc., either to scale or actual size. The design is transferred from the paper to the working surface;[81] (2) also used to describe a comic drawing or an animated film based on a succession of comic, grotesque drawings.[81]

carvone KETONE derived from the terpene dipentine.[144]

CAS Registration Number (CAS; CAS RN.) an assigned number used to identify a material. CAS stands for Chemical Abstracts Service, an organization that indexes information published in *Chemical Abstracts* by the AMERICAN CHEMICAL SOCIETY and that provides index guides by which information about particular substances may be located in the abstracts. Sequentially assigned CAS numbers identify specific chemicals. The numbers have no significance. The CAS number is a concise, unique means of material identification and is often used in MSDSs.[172]

Contact: Chemical Abstracts Service, Division of the American Chemical Society, Box 3012, Columbus, OH 43210; (614) 421-3600.

case hardening SURFACE HARDENING without thorough drying of the film.[55]

casein naturally occurring protein obtained from cow's milk, either by PRECIPITATION by ACID or by coagulation with rennet.[173,156] See PROTEIN.[173]

In the paint industry its chief application is in DISTEMPERS, where it is applied as a COLLOIDAL SUSPENSION of CALCIUM CASEINATE, which becomes insolubilized by drying out and by CARBONATION. For other end uses, casein is insolubilized by FORMALDEHYDE.

casein finish see CASEIN PAINT, and DISTEMPER.[163]

casein paint coatings in which CASEIN replaces the ordinary DRYING OILS, or is used as an EMULSIFYING AGENT in emulsion paints. Both types may be thinned with water.[163]

caseinates METALLIC SALTS derived from CASEIN.

casement a window sash that opens on hinges at the vertical edge.[82] See CASEMENT WINDOW.[82]

casement window window hinged at the side to swing in or out.[82]

cashew nut shell liquid resin RESIN derived from the liquid obtained from the shells of the nuts of the species, *Anacardium occidentale*, which grows chiefly in India and South America.[167]

The naturally occurring liquid is a mixture of a dihydroxy phenol CARDOL and ANACARDIC ACID. This acid readily decarboxylates on heating to yield a monohydroxy phenol with an unsaturated side chain, anacardol.

cashew nut shell oil natural OIL from the shells of the nuts of the species, *Anacardium occidentale*, containing a high proportion of CARDOL—a substituted phenol.[63,167]

By reaction of this oil with FORMALDEHYDE, a RESIN is produced which gives films with good chemical resistance.

casing knife in paperhanging, a knife used to trim WALLPAPER around casings, at moldings, baseboards, etc.[59] (DAC)

cassel brown see VANDYKE BROWN.[41]

cassel earth see VANDYKE BROWN.[41]

cassel yellow $PbCl_2{\cdot}7PbO$. Lead oxychloride PIGMENT.[41]

cassone painting pictures used to decorate the sides of an Italian marriage chest (cassone).[81,175]

The most common subjects were scriptural, chivalrous, mythological, and heraldic.

cast (1) a tinge; a subjective term used in combination with a directional indicator to describe a slight deviation from the norm in COLOR or appearance. Most often applied to color, and refers to a small difference in HUE. For example, four pure hues are generally recognized psychologically, red, green, blue, and yellow. Each is unique and contains no quality of the others. If a blue hue seems to contain some element of green, that

---

Words presented in CAP/SMALL CAPS type indicate that the word is defined in another part of the *Coatings Encyclopedic Dictionary.*

Numerical superscripts classify terms in one or more of the categories listed in the second section of this volume.

blue is said to have a green "cast"; (2) may also be applied to appearance aspects other than color. For example, a material which is normally expected to be transparent (nonscattering) but which is slightly cloudy (due to a small amount of light scattering) is said to have a "milky cast" or a "cloudy cast."[43,69,35]

cast coating a FILM made by depositing a layer of PLASTIC material in solution, in a DISPERSION, or in a molten state onto a (usually) smooth solid surface. The plastic is solidified and removed from the surface as a film having the same type of surface as the solid surface on which it was formed.[38,71] (EPA)

cast film a FILM made by depositing a layer of plastic, either molten, in solution, or in a dispersion, onto a surface, solidifying and removing the film from the surface. The film then has the same type of surface as the solid surface on which it was formed.[38] (ASTM)

cast film extrusion see CHILL ROLL EXTRUSION.[57,59]

cast-coated paper a paper or board, the coating of which is allowed to harden or set while in contact with a finished casting surface. Cast-coated papers have a high gloss.[168]

castanha de cotia kernel oil a DRYING OIL from a tree indigenous to Brazil.[63]

It is believed to contain a substantial proportion of eleostearic acid, or acid of similar type. Its rate of POLYMERIZATION is much slower than that of TUNG OIL. Its iodine value is 154.

casting see DRAWDOWN.[77]

casting plaster see GYPSUM, CALCINED.[58]

casting resin RESIN which can be CAST and hardened in a MOLD to form a shaped article.[57,169]

castor oil VEGETABLE OIL consisting of FATTY ACID ESTERS of glycerol, minor quantities of plant sterol, phosphatides, etc. It is unique in that its principal fatty acid is a HYDROXY ACID, RICINOLEIC ACID.[63] (PTM)

This oil is obtained from the seed of Ricinus communis. Its principal characteristics are light COLOR, relatively high SPECIFIC GRAVITY and VISCOSITY, and its SOLUBILITY in ALCOHOL. It differs from other oils in that its composition is mostly hydroxy fatty acids. It is essentially a nondrying oil, but it may be converted to a drying oil by "chemical dehydration" by which a hydroxy group and an adjacent hydrogen atom are removed as water to form a drying oil fatty acid ester with two double bonds, one of them being conjugated. This dehydration yields what commonly is known as dehydrated castor oil. In its original undehydrated form, castor oil is well known for its use in RESINS and as a PLASTICIZER for CELLULOSE ESTER LACQUERS.

castor oil, dehydrated CASTOR OIL from which some chemically combined water has been removed to improve its DRYING properties.[63,131] See DCO.[63]

cat eyes undissolved globules of RUBBER in a CEMENT generally made from only rubber and SOLVENT.[79,140] Syn: FISH EYES

Expression also used for PVC sheet and paint films, etc.

catalysis the alteration of the rate of a chemical reaction (usually increase in the rate) by the addition of some substance (the CATALYST) which does not undergo a net chemical change.[74] (IUPAC)

Cases occur with certain reactants in which the addition of a substance reduces the rate of a particular reaction, for example, the addition of an inhibitor in a chain reaction or a poison in a catalytic reaction. The term "negative catalysis" has been used for these phenomena, but this usage is not recommended; terms such as inhibition or poisoning are preferred.

catalyst substance whose presence increases the rate of a chemical reaction.[67,74,83]

In some cases the catalyst functions by being consumed and regenerated; in other cases the catalyst seems not to enter the reaction and functions by virtue of surface characteristics of some kind. A negative catalyst (inhibitor, retarder) slows down a chemical reaction. See HARDENER, INHIBITOR, CURING AGENT, ACID CATALYSTS.[67,74,83]

catalytic curing mechanism by which a coating is CROSSLINKED by the action of a CATALYST as opposed to OXIDATION, etc. Examples of such systems are TWO-PART (POT) EPOXIES and POLYURETHANES.[67,74]

catalytic incinerator a control device which oxidizes VOC by using a CATALYST to promote the combustion process.[172,74] (EPA)

The catalyst allows the combustion process to proceed at lower temperature (usually around 600°F to 800°F) than a conventional thermal incinerator would require (1100°F to 1400°F), resulting in fuel savings and lower cost incineration.

cataphoresis see ELECTROPHORESIS.[61,139]

catch-up in LITHOGRAPHY this is the PRINTING of nonimage areas of the plate and is overcome generally by increasing the amount of fountain solution applied to it.[168]

It may occur as a run is started before the dampening adjustment is set correctly.

cathode negative terminal of an electrical cell, vacuum tube, or other electrical device. cf. ANODE[61,145]

Also used in connection with CATHODIC PROTECTION.

cathodic corrosion CORROSION due to the development of ALKALINITY by a reaction at a cathode—

sometimes experienced with ALUMINUM and LEAD.[61,66] See CELL, ELECTROLYTIC.[61,66]

Alkalinity from this source may also affect paints vulnerable to alkali attack.

cathodic electrodeposition see ANIONIC ELECTRODEPOSITION and ELECTRODEPOSITION.[77]

cathodic polarization portion of the reduction in the initial potential of a CORROSION cell that occurs at the CATHODE.[66]

cathodic protection reduction or elimination of CORROSION of a METAL achieved by making current FLOW to it from a solution by connecting it to the negative pole of some source of current. The source of the protective current may be a SACRIFICIAL METAL, such as ZINC, MAGNESIUM, or ALUMINUM. The current may also be derived from a rectifier, generator, or battery applied through an appropriate anode which may be consumed by the applied current, as in the case of steel, or remain substantially unaffected by the current, as in the case of graphite or platinum.[66]

cation positively charged atom or radical, which moves to the negative electrode or cathode during electrolysis.[61,66]

Metallic ions, such as iron and copper, etc., are cations.

cationic detergent a DETERGENT that produces positively charged colloidal ions in solution.[127] (ASTM)

cationic emulsion an EMULSION in which the emulsifying system establishes a predominance of positive charges on the discontinuous phase.[60] (ASTM) cf. ANIONIC EMULSION[60]

cationic polymerization PROCESS in which the active end of the growing POLYMER molecule is a positive ion. If the ion is a carbonium ion, it is referred to as carbonium ion polymerization.[169]

cationic surfactant SURFACTANT which gives a positively charged ion in aqueous solution.[127]

caul see PLATEN.[59,168]

caulk (calk) to fill voids with PLASTIC or semiplastic materials; to fill crevices in the adherend surface with adhesive materials; to provide a seal against moisture or solvent intrusion.[75,165] cf. SEALANT[165]

caulking see CAULKING COMPOUND.[165]

caulking cartridge an expendable container made of PLASTIC, FIBERBOARD, or METAL, filled with CAULKING COMPOUND, for use in a CAULKING GUN.[59,165] (DAC)

A common type is 2 in. (5 cm) in diameter, approx. 8 in. (20 cm) long, and fitted with a plastic nozzle.

caulking compound a soft, PLASTIC, PUTTY-like material, consisting of PIGMENT and VEHICLE, used for SEALING JOINTS in buildings and other structures where normal structural movement may occur, or for preventing leakage.[75,165]

Caulking compound retains its PLASTICITY for an extended period after APPLICATION. It is usually available in two consistencies: "gun grade," for use with a caulking gun, and "knife grade," for application with a putty knife; extruded preformed shapes are also available.

caulking gun a device for applying CAULKING COMPOUND by extrusion.[59,165] (DAC)

In a hand gun, the required pressure is supplied mechanically by hand; in a pressure gun, the pressure required usually is greater and is supplied pneumatically.

caustic a strong chemical base.[144]

caustic potash (KOH). Archaic name for potassium hydroxide. A strong base.[144]

caustic soda (NaOH). Archaic name for sodium hydroxide. A strong base.[144]

cave painting the art produced in the form of paintings on the walls of caves from the beginning of the Old Stone Age to the end of the New Stone Age, around 3000 B.C.[81] See PREHISTORIC ART.[81]

The entire period covers a span of some 200,000 years. The Paleolithic cave paintings at Altamira (Spain) and those at Lascaux (France) are believed to date between 40,000 and 10,000 B.C.

ceara wax see CARNAUBA WAX.

cedar a durable SOFTWOOD generally noted for decay resistance; includes western red cedar, incense cedar, eastern red cedar.[153] (DAC)

cedar nut oil *Pinus cembra*, from the seeds of which this OIL is derived, grows prolifically in several parts of the world. The main constituent acids are LINOLEIC, LINOLENIC, and OLEIC ACIDS, and the oil possesses useful DRYING properties. Iodine values up to 160 have been reported.[63]

ceiling the overhead surface of a room, usually a covering or decorative treatment used to conceal the floor above or the roof.[82] (DAC)

celestial blue pigment IRON BLUE PRECIPITATED on BARYTES.[41]

Usually only a few percent of blue is used.

celestite mineral which is chiefly strontium sulfate.[41,58,34]

Words presented in CAP/SMALL CAPS type indicate that the word is defined in another part of the *Coatings Encyclopedic Dictionary.*

Numerical superscripts classify terms in one or more of the categories listed in the second section of this volume.

When ground, it has some application as a FILLER, being used as an alternative to BARYTES. The ground mineral is also known as STRONTIUM WHITE.

cell a small etched depression in a gravure cylinder that carries the INK.[168]

cell, electrolytic source of electrical current that is responsible for CORROSION, consisting of an anode and a cathode immersed in an electrolyte and electrically bonded together.[61,66]

The anode and cathode may be separate metals or dissimilar areas on the same metal. They will develop a difference in potential that causes current to FLOW and corrosion at the anode when the electrodes are in electrical contact with each other.

cellophane thin TRANSPARENT film consisting of a base sheet of CELLULOSE, regenerated from viscose, containing variable amounts of water and softener, and coated on one or both sides to render it moisture resistant and capable of being sealed with heat or solvent.[38]

Cellosolve® $C_2H_5OCH_2CH_2OH$. Proprietary name for the monoethyl ether of ethylene glycol, ethoxyethanol.[164]

It is also the generic name for a comprehensive series of ethers of similar type such as methyl cellosolve, butyl cellosolve, etc. Cellosolve is water miscible. Properties: bp, 135°C; sp gr, 0.931/20°C; flp, 42°C (107°F); vp, 4 mm Hg/20°C.

Cellosolve® acetate $CH_3COOCH_2CH_2OCH_2CH_3$. Medium high boiling solvent (ethylene glycol monoethyl ether acetate).[164]

Properties: bp, 156°C (313°F); sp gr, 0.975/120°C; flp, about 53°C (128°F); and vp, 1.5 mm Hg/20°C.

cellulase enzyme that attacks and breaks down CELLULOSIC SUBSTRATES such as wood, CELLULOSIC THICKENERS in paint, etc.[173]

Water soluble cellulose-ethers are widely used as THICKENERS for LATEX paints. These are readily degraded to smaller molecules by cellulase produced by BACTERIAL and FUNGAL contaminants in the raw materials and finished products, resulting in a paint viscosity loss. A biocide may be very effective in controlling microbial contamination in paints, but not necessarily effective for inhibiting cellulases which may have already been present in the raw materials and become blended into the finished paints. An in-can preservatiave that both kills microorganisms and inhibits cellulases should be added early in the manufacture of the paint to prevent this and to contribute to plant hygiene.

celluloid THERMOPLASTIC material which is a solid solution of CELLULOSE NITRATE in camphor or some other PLASTICIZER.[38]

ALCOHOL, KETONES, or ESTERS are normally employed as volatile SOLVENTS to assist PLASTICIZATION.

cellulose $(C_6H_{10}O_5)_n$. A naturally occurring POLYSACCHARIDE made up solely of glucose units and found in most plants; the main constituent of dried WOODS.[153] (ASTM)

Chief constituent of the fiber of plants; cotton is the purest natural form, containing about 90%; it occurs to a small extent in the animal kingdom. (Chemically, cellulose is 1-4 glucan of high degree of polymerization. It is desirable to apply "cellulose" to this material only and to designate the predominantly cellulosic residue obtained by subjecting woody tissues to various pulping processes as "cellulosic residues," "cellulosic pulp," or the like.)

cellulose acetate (CA) ACETIC ACID ESTER of CELLULOSE.[130]

It is obtained by the action, under rigidly controlled conditions, of acetic acid and acetic anhydride on purified cellulose usually obtained from cotton linters. When compounded with suitable plasticizers, it forms a tough thermoplastic material.

cellulose acetate butyrate (CAB) A mixed ESTER of CELLULOSE containing both acetate and butyrate groups.[38,130] (ASTM)

cellulose acetate propionate (CAP) Mixed CELLULOSE ESTER, similar to the acetate butyrate except that the butyryl radical is replaced by the propionyl.[38,130]

In properties and applications it resembles the acetate butyrate rather than the straight acetate.

cellulose ester CELLULOSE in which the free hydroxyl groups have been replaced wholly or in part by acidic groups.[38,130]

cellulose ethers derivatives of CELLULOSE in which one or more of the hydroxyl hydrogens have been replaced by alkyl groups.[83,155] (ASTM)

cellulose glycolate see CARBOXYMETHYL CELLULOSE.[83,155]

cellulose gum see CARBOXYMETHYL CELLULOSE.[83,155]

cellulose lacquer see LACQUER.[38,151]

cellulose nitrate nitric ACID ESTER of CELLULOSE manufactured by the action of a mixture of sulfuric and nitric acids on cellulose, such as purified cotton linters.[151,38] See NITROCELLULOSE and PYROXYLIN.[151,144]

cellulose plastics PLASTICS based on derivatives of CELLULOSE, such as ESTERS (cellulose acetate) and ETHERS (ethyl cellulose).[38]

cellulose propionate ESTER of CELLULOSE and PROPIONIC ACID.[38]

cellulosic coating one in which a large portion of the solids are made up of chemically treated CELLULOSE.[71,175] (EPA)

Cellulose lacquers are widely used in the WOOD FURNITURE *industry for* TOPCOATS.

cellulosic thickeners AQUEOUS THICKENERS based on CELLULOSE COMPOUNDS such as carboxymethyl cellulose, hydroxyethyl cellulose, and methyl cellulose.[83,155]

Celsius scale see CENTIGRADE.[128] cf. KELVIN and FAHRENHEIT SCALE[128]

cement a material or a mixture of materials (without aggregate) which, when in a PLASTIC state, possesses ADHESIVE and COHESIVE properties and HARDENS in place.[73,75,79] (DAC)

Frequently, the term is used incorrectly for CONCRETE, *e.g., a "cement" block for concrete block.*

cement coating see CEMENT PAINT.[73,71] cf. MASONRY PAINT[71,73]

cement colorants COLORS with sufficient TINCTORIAL STRENGTH and ALKALI RESISTANCE to be suitable for coloring PORTLAND CEMENT or CONCRETE.[41,73]

The natural earth colors, synthetic iron oxide colors, chromium oxide, ultramarine, and some of the ORGANIC PIGMENTS *are used for this purpose.*

cement paint paint supplied in DRY POWDER form, based essentially on PORTLAND CEMENT, to which PIGMENTS are sometimes added for DECORATIVE purposes.[71,73]

This dry powder paint is mixed with water immediately before use.

cement plaster (1) PLASTER with PORTLAND CEMENT as the BINDER; SAND and LIME are added on job. Used for exterior work or in wet or high humidity areas; (2) in some regions, GYPSUM plaster.[73,75] (DAC)

cement, Portland see PORTLAND CEMENT.[73]

cement, rubber an ADHESIVE that is a dispersion or solution of raw or COMPOUNDED RUBBER, or both, in a suitable liquid.[79] (ASTM)

cement-asbestos board a dense, rigid, noncombustible board containing a high proportion of ASBESTOS FIBERS which are bonded with PORTLAND CEMENT; highly resistant to WEATHERING.[73,75] (DAC) Syn: asbestos-cement board

cementitious having cementing properties.[134] (DAC)

cenobrium see MERCURIC SULFIDE.[41]

center of interest in a room, the principal focal point, architecturally speaking, such as a fireplace.[82,148]

In a WALLCOVERING *design, the dominant motif, usually hung at eye level in the central area of a room to establish a starting point for hanging the rest of the wallcovering.*

centigrade one hundredth of the thermometer scale, divided into 100 degrees, in which 0°C is the freezing point of water and 100°C is the boiling point at 1.013 bar.[128]

The preferred name according to the International System of Units (SI) is degree Celsius (°C).

centimetre (cm) a measure of length equal to a hundredth part of a metre, or 0.3937 in.; an inch equals 2.54 cm.[128] (DAC)

centipoise (cP) one hundredth of a poise (unit of measurement for viscosity). Liquids of low viscosity are usually given in centipoise units; for example, water at 20.2°C has a viscosity of 1.0000 centipoise. In SI units 1 cP = 10^{-3} Pa.s.[128,166]

centistokes (cST) one one-hundredth of a stokes, which is the unit of kinematic viscosity:

$$v = \frac{\eta}{\rho}$$

where: η= viscosity ρ = density.[128,166]
In SI units 1 cST = 1 mm^2/s.

centrifugal clarifier apparatus somewhat similar to a cream separator, used for clarifying clear and colored solutions by throwing out solid particles by means of centrifugal force.[59]

Clarifiers of this type are operated at high speed. The bulky and apparently undissolved particles are thrown aside into a collector.

centrifuge device for separating items of various masses by inducing gravitational forces/actions through the use of high rotational velocities.[145] (ASTM)

It exploits centrifugal force as a means of separating solid or semisolid particles from liquids.

cera alba bleached or white BEESWAX.[137]

cera flava yellow or unbleached BEESWAX.[137]

ceramic technology of producing fired CLAY and porcelain articles, their glazes, pigments, and modifiers.[49,75,131]

ceramic coating an inorganic, essentially NONMETALLIC COATING, on METAL.[71] (ASTM)

CERCLA abbreviation for the COMPREHENSIVE ENVI-

Words presented in CAP/SMALL CAPS type indicate that the word is defined in another part of the *Coatings Encyclopedic Dictionary.*

Numerical superscripts classify terms in one or more of the categories listed in the second section of this volume.

RONMENTAL RESPONSE, COMPENSATION, AND LIABILITY ACT.[171]

ceresine noncrystalline WAX made by refining ozokerite. Mp, 60-71°C (140-160°F).[137]

cerium naphthenate rare earth DRIER for air drying and baking finishes.[74,83,67]

cermet composite materials consisting of two components, one being either an oxide, carbide, boride, or similar INORGANIC COMPOUND and the other a METALLIC BINDER.[131]

cerotic acid a common name for either heptacosanoic acid or hexacosanoic acid.[137,144]

The former has the formula: $C_{26}H_{53}COOH$, and an mp of 82°C; the latter, the formula: $C_{25}H_{51}COOH$, and an mp of 88°C. Cerotic acid is a constituent of natural WAXES, in which it occurs as the cerotate ESTERS. Neocerotic acid, formula, $C_{24}H_{49}COOH$, mp of 77.8°C, is a known constituent acid of BEESWAX.

cerulean blue very complex PIGMENT consisting essentially of a combination of cobalt and tin oxides.[41]

It is bluish-green in color and very stable.

cerussite Pb CO_3. Natural LEAD CARBONATE, found in the upper zone of lead deposits. Colorless, white, and gray. Sp gr of 6.55. Effervesces in nitric acid.[41]

ceryl alcohol $C_{26}H_{53}OH$. ALCOHOL obtained from CHINESE WAX.[144]

cetoleic acid $CH_3(CH_2)_9CH{=}CH(CH_2)_9COOH$. One of the constituent fatty acids of many fish oils. It has a single double bond.[144]

cetraria see ICELAND MOSS.[139,155]

CFR abbreviation for CODE OF FEDERAL REGULATIONS.[172]

c.g.s.-emu system of units the system for measuring physical quantities in which the basic units are the centimetre, gram, and second, and the numerical value of the magnetic constant, tm, is unity. It is being replaced by the SI UNITS.[128] (ASTM) See F.P.M. SYSTEM OF UNITS and SI UNITS AND M.K.S. SYSTEM.[128]

c.g.s. system centimetre-gram-second system. Now replaced by SI UNITS.[128]

chain length the total length of a chain molecule, measured from atom to atom along the chain.[130,169] (IUPAC)

This term should not be used for the direct distance between the ends of the molecule.

chair rail the topmost molding of a dado, placed on the wall at the height of a chairback as a protection.[82]

chalk (1) soft naturally occurring form of calcium carbonate. The whiter varieties are powdered, washed, and dried. The product is known as whiting; the finer varieties as Paris white; (2) the product of chalking.[41,58,154] (BSI)

chalk kauri soft, white, powdery form of kauri COPAL.[167]

chalk masking the color change due to chalking and not due to a change in the colorant.[42,65,56]

chalk, precipitated obtained as a by-product in water softening by sodium carbonate. It consists chiefly of calcium carbonate, but is often contaminated by appreciable quantities of calcium sulfate.[58]

chalking (1) formation of a friable powder (pigment rich) on the surface of a paint film caused by the disintegration of the binding medium due to disruptive factors during weathering. The chalking of a paint film can be considerably affected by the choice and concentration of the pigment.(BSI) It can also be affected by the choice of the binding medium;[41,154] (2) a condition of a printing INK in which the pigment is not properly bound to the paper and can be easily rubbed off as a powder;[168] (3) formation of a powdery surface condition due to oxidation of the surface of rubber and release of pigments and fillers at the surface. Not to be confused with bloom, which looks similar.[154,140]

The FSCT Pictorial Standards of Coating Defects contains photographic standards for the degree of chalking of exterior paints. ASTM Test Method D4214 also has these same photographic standards as well as a second set of photographic standards.

chalking resistance the ability of a coating to resist the formation of a friable powder on the surface of its film caused by the disintegration of the binding medium due to degradative weather factors.[56,72]

The chalking of a coating can be considerably affected by the choice and concentration of pigment and binding medium.

chamber white lead particular type of white lead PIGMENT, which was made by hanging strips or straps of lead in chambers, and subjecting them to the corrosive action of moist air, carbon dioxide, and acetic acid vapor. Chamber white lead differed from the stack form, and was usually finer and of a better color.[41] See CARTER LEAD.[41]

channel black form of CARBON BLACK made from natural gas by the channel combustion process.[41]

The gas is burned with insufficient air in jets, and the flames are allowed to impinge on a cool, channeled metallic surface. The deposited carbon is then scraped from the channel after a certain period of burning.

charcoal a carbonaceous solid residue which is formed in destructive distillation of wood.[153] (IUPAC)

It has a very high surface area per unit weight which makes it especially useful as an absorber of various gases and vapors.

charcoal black black PIGMENT obtained from charcoal.[41]

charge measurement or weight of material used to load a mill, kettle, or other batch processing equipment at one time or during one cycle.

chaser MILL for DISPERSING PIGMENTS in BINDER to form very stiff PASTES or PUTTY.[59,131]

It consists of a circular pan in which one or two massive rollers run about central axes.

chatki stone MILL for grinding SHELLAC.[59,131,146]

chatter transverse roll marks on roll coat-painted strip with varying film thicknesses which are usually due to roll coater vibration or a nonconcentric roller.[57]

checking that phenomenon manifested in paint films by slight breaks in the film that do not penetrate to the underlying surface.[42,56,154] (ASTM) See COLD-CHECKING, and CHECKING RESISTANCE.[56,72]

The break should be called a crack if the underlying surface is visible. Where precision is necessary in evaluating a paint film, checking may be described as visible (as seen with the naked eye) or as microscopic (as observed under a magnification of 10 diameters).

checking resistance the ability of a coating to resist slight breaks in its film that do not penetrate to the underlying surface.[56,72] cf. CHECKING[56,42,154]

The breaks should be called cracks if penetration extends through the underlying surface. See CRACKING RESISTANCE.[56,72] Where precision is necessary in evaluating a coating film, checking may be described as visible (as seen by the naked eye) or as microscopic (as observed under minimum magnification of 10 diameters). The FSCT Pictorial Standards of Coating Defects contains photographic standards for the degree of checking of exterior paints.

ASTM Test Method D660 has a different set of illustrated reference standards.

checks (1) rough surface due to fine cracks from WEATHERING; (2) roughness formed on calendered sheet when temperature of calender rolls is too low or when the sheet is chilled too suddenly.[154,140]

cheesy character of a paint or VARNISH film which, although dry, is mechanically weak and rather soft.[42,47,56] (BSI)

chelating agent (1) a sequestering or complexing agent that, in aqueous solution, renders a metallic ion inactive through the formation of an inner ring structure with the ion;[83] (2) ORGANIC COMPOUND that can remove many of the heavy metal cations from solution by forming soluble chelate compounds. Ethylenediaminetetraacetetic acid (EDTA) is typical.[83] Syn: sequestering agent, chelate

Chelating agents are also used to treat metal poisoning.

chelation reversible reaction of a METALLIC ION with a molecule or ion to form a complex molecule which does not have all or most of the characteristics of the original metallic ion.[131] Syn: sequestration

chemical a substance with a distinct molecular composition that is produced by or used in a chemical process.[144]

chemical conversion coating a treatment, either chemical or electrochemical, of the metal surface to convert it to another chemical form which provides an insulating barrier of exceedingly low solubility between the metal and its environment, but which is an integral part of the metallic substrate.[45,66,158]

It provides greater CORROSION RESISTANCE to the metal and increased adhesion of coatings applied to the metal. Examples are phosphate coatings on steel or zinc and chromate coatings on aluminum.

chemical family a group of single elements or COMPOUNDS with a common general name, e.g., acetone, METHYL ETHYL KETONE (MEK), and METHYL ISOBUTYL KETONE (MIBK) are of the ketone family; ACROLEIN, FURFURAL, and ACETALDEHYDE are of the aldehyde family.[144]

chemical name the scientific designation of a chemical or name that clearly identifies the chemical for hazard evaluation purposes.[172,144]

chemical pretreatment see CHEMICAL CONVERSION COATING.[45,66]

chemical resistance that property of coating materials to withstand the attack of specified chemicals.[72]

ASTM test methods for chemical resistance include: acid and mortar, of coated aluminum, D3260; alkali, of varnish films, D1647; alcohol,

---

Words presented in CAP/SMALL CAPS type indicate that the word is defined in another part of the *Coatings Encyclopedic Dictionary*.

Numerical superscripts classify terms in one or more of the categories listed in the second section of this volume.

of wood furniture lacquers, D2571; household chemicals, of coatings, D1308.

chemical stability RESISTANCE to chemical damage due to internal or external reactions.[65,72]

Chemical Transportation Emergency Center (CHEMTREC) established in Washington, D.C., by the Chemical Manufacturers Association (CMA) to provide emergency information on materials involved in transportation accidents.[172]

chemical vapor deposition (CVD) chemical coating processes carried out in the gas phase.[45]

These are generally operated in closed chambers where the product is heated to a high temperature at which the constituents of the gas decompose or react to form metallic coatings. Tungsten and nickel coatings are produced this way.

chemical-resistant paint see CHEMICAL RESISTANCE.[71]

chemiluminescence see LUMINESCENCE.[43]

chemisorption (or chemical adsorption) adsorption which results from chemical bond formation (strong interaction) between the absorbent and absorbate in a monolayer on the surface.[139] (IUPAC)

Adsorption in which the forces involved are valence forces of the same kind as those operating in the formation of chemical compounds. If the molecule is adsorbed without fragmentation, the adsorption process is nondissociative. Adsorption of carbon monoxide is frequently of this type.

chemist a professional who possesses an earned bachelor's or higher degree with a major in a science from an accredited institution and who develops, applies, or communicates the principles of chemistry and exercises independent judgment and discretion in conceiving, planning, coordinating, or executing chemical projects, or who has experience in so doing. (ACS)

The chemical sciences deal with the composition, structure, and properties of substances and of the transformations they undergo.

chemistry the study of the elements and the compounds they form.[144]

Chemistry is mainly concerned with effects that depend on the outer electrons in atoms.

CHEMTREC abbreviation for the Chemical Transportation Emergency Center.[172]

cherry an even-textured, moderately high-density wood of the eastern U.S.A., rich red-brown in COLOR; takes a high LUSTER; used for cabinetwork and paneling.[153,175] (DAC)

chia oil Mexican OIL obtained from *Salvia hispanica.* It has excellent DRYING properties.[63]

A prominent characteristic of the oil is its high surface tension, which causes it to "crawl"; cooking at 260°C (500°F) for a short time destroys this property. Its main constituent acids may make up more than 90% of the total acids. Properties: sp gr, 0.9338/15°C; refractive index, 1.4855; iodine value, 196.3; saponification value, 192.2.

chill roll in EXTRUSION coating, a cored roll, usually cooled with circulating water, which cools the coated SUBSTRATE before winding. The chill roll and the pressure roll form a nip that combines the coating with the substrate by the application of pressure.[57,59]

chill roll extrusion cooling of an extruded film by drawing around two or more highly polished chill rolls.[57,59] Syn: cast film extrusion

chill-back a material capable of reducing the temperature of a batch of VARNISH or RESIN, either because of a considerable temperature difference between itself and the hot material, or because it absorbs heat in melting also.[83]

chilling (1) in VARNISH or RESIN manufacture, the deliberate and rapid reduction of the temperature of the charge in a varnish or resin pot.[131] Syn: quenching; (2) on a painted or varnished surface, a clouding of the surface or a reduction of LUSTER as a result of the movement of cold air over the drying surface.[42,56,35] (BSI) (DAC)

chimb see CHIME.[50]

chime the top ring of a coating container into which the friction top is pressed in order to effect an air tight seal.[50] Syn: chimb, chine

chimney an incombustible vertical structure containing one or more flues to provide draft for fireplaces, and to carry off gaseous products of combustion to the outside air from fireplaces, furnaces, or boilers.[82] (DAC) See FLUE.[82]

china clay $Al_2O_3 \cdot 2SiO_2 \cdot 2H_2O$. Pigment White 19 (77005). A complex hydrated ALUMINUM SILICATE produced by the breakdown of the mineral FELDSPAR. The finest grades are used by the COLOR and paint trades—often under proprietary names—and are utilized as a base for LAKES and as an EXTENDER and ANTISETTLING agent in paints.[58,41] See ALUMINUM SILICATE.[41] Syn: kaolin and kaolinite

china wood oil see TUNG OIL.[63,156]

chine see CHIME.[50]

Chinese blue often applied to better grades of PRUSSIAN BLUE.[41] See IRON BLUE.[41]

Chinese ink see INDIA INK.[168]

Chinese lac resinous exudation of the wild Chinese tree, *Angia sinensis* or *Tssichau*, from which it is obtained by making deliberate incisions in the trunk.[167]

It is very susceptible to EXPOSURE, *changing in* COLOR *gradually, through red to black. The resinous exudation has very good* SOLUBILITY, *blending satisfactorily with* TURPENTINE *and* ALCOHOL.

Chinese red see CHROME ORANGE.[41]

Chinese vegetable tallow fatty substance from the kernels of the Chinese vegetable tallow tree, *Stillingia sebifera.* [137]

This is the same tree which yields stillingia oil.

Chinese vermillion see MERCURIC SULFIDE.[41]

Chinese wax an insect secretion, collected from certain evergreens which are indigenous to China.[137] cf. SHELLAC[146] Syn: insect wax

Properties: mp, 81°C; sp gr, 0.970; saponification value, 92; and a negligible iodine value.

Chinese white (1) a paint using ZINC OXIDE as the principal PIGMENT;[71,41](DAC) (2) synonym for ZINC OXIDE.[41]

chinoiserie originally, European designs "in the Chinese taste."[148,81]

Now loosely applied to almost any oriental form of decoration.

chintz WALLPAPERS resembling the printed cotton materials from India once known as "chints," featuring brightly colored flowers.[148]

chip board paperboard generally made from reclaimed paper stock and frequently used as a facing for partitions.[75]

The board is of moderately low density and should not be confused with PARTICLE BOARD.

chip resistance the ability of a coating or layers of coatings to resist total or partial removal, usually in small pieces, resulting from impact by hard objects or from wear during service.[56,72] See CHIPPING.[54,148,47]

ASTM test methods for chipping (chip resistance) include: of coatings, D3170; of traffic paint, D913.

chipping (1) removal of paint, or rust and scale, by mechanical means;[158] (2) total or partial removal of a dried paint film in flakes by accidental damage or wear during service; in traffic paints, this failure is usually characterized by sharp edges and definite demarcation of the base area;[65,42] (3) spreading of white or colored chips or flakes during the application of seamless flooring. This procedure is used to impart a decorative effect to the flooring.[54,148] See CHIP RESISTANCE.[56,72]

ASTM test methods for chipping (chip resistance) include: of coatings, D3170; of traffic paint, D913.

chipping rating standards a set of photographic standards depicting the size and number of chips in each of several categories, as described in ASTM Test Method D3170.[157]

chips (1) the size of RESIN particles. It represents a form INTERMEDIATE between both lumps and dust;[128] (2) dry combination of BINDER and PIGMENT which in SOLVENT forms paint, or which can be added to COLOR PLASTICS (e.g., vinyl chips and nitrocellulose chips).[41]

chlorazol dyes range of soluble DYESTUFFS which are used in the manufacture of colored inks.[41]

chlorinated diphenyls (PCB). Range of CHLORINATED HYDROCARBONS varying in properties from very liquid plasticizers, through thick syrups to solid resins. They are characterized by excellent compatibility, stability, absence of free acidity, high toxicity, and nonflammability.[170]

chlorinated hydrocarbons powerful SOLVENTS which include such members as chloroform, carbon tetrachloride, ethylene dichloride, methylene chloride, tetrachlorethane, trichlorethylene, etc.[164]

Generally speaking, they are toxic. Their main applications include nonflammable paint removers, cleaning solutions, and special finishes where presence of residual solvent in the film is a disadvantage. Many are now prohibited in most countries.

chlorinated para nitraniline red see CHLORINATED PARA REDS.[41]

chlorinated para reds Pigment Red 4 (12085). There are two varieties of chlorinated para red: (1) ortho-chlor-para nitraniline; and (2) para-chlor-ortho nitraniline; both are diazotized and coupled to beta naphthol. Both are light yellow shade reds of the toluidine red orange.[41]

chlorinated paraffin wax ordinary PARAFFIN WAX can be chlorinated, under certain conditions, to yield products which no longer resemble the parent wax, but which have assumed definite RESINOUS characteristics.[137,170]

The amount of chlorine present determines the physical properties of the resultant waxes. Those with smaller amounts of chlorine are liquids, whereas those with about 70% chlorine are brittle resins. Their applications are in FIRE-RETARDING *compositions and as plasticizers.*

Words presented in CAP/SMALL CAPS type indicate that the word is defined in another part of the *Coatings Encyclopedic Dictionary.*
Numerical superscripts classify terms in one or more of the categories listed in the second section of this volume.

chlorinated polyether POLYMER obtained from PENTAERYTHRITOL by preparing a chlorinated oxetane and polymerizing it to a polyether by means of opening the ring structure.[130]

chlorinated rubber RESIN formed by the reaction of RUBBER with chlorine.[130]

Unlike rubber, the resulting product is readily soluble and yields solutions of low viscosity. It is sold as white powder, fibers, or as blocks. Commercial products generally contain about 65% chlorine. It has good chemical resistance properties. It tends to cobweb when sprayed. Now, mostly chlorinated polymers are used, as 1-butene, polyethylene, etc.

chloroform $CHCl_3$. Trichloromethane. Clear, colorless, volatile liquid.[144,164]

Properties: sp gr, 1.485 (20/20°C); bp, 61.2°C; fp, 63.5°C; wt/gal, 12.29 lb (25°C); refractive index, 1.4422.

chroma one of the three terms used in MUNSELL NOTATIONS. It correlates approximately with the psychological dimensions of SATURATION.[43,69,35]

chromate of lead see CHROME YELLOW.[41]

chromatic perceived as having HUE, not white, gray, or black.[69,35]

chromatic attributes those attributes associated with the spectral distribution of light; HUE and saturation.[69]

chromaticity dimensions of a COLOR stimulus expressed in terms of HUE and SATURATION, or redness-greenness, and yellowness-blueness, excluding the LUMINOUS intensity; generally expressed as a point in a plane of constant LUMINANCE.[69,43] Syn: chromaticness

chromaticity coordinates, CIE the ratios of each of the three TRISTIMULUS VALUES, X, Y, and Z in relation to the sum of the three; designated as x, y, and z, respectively.[69]

They are sometimes referred to as the TRICHROMATIC COEFFICIENTS. When written without subscripts they are assumed to have been calculated for ILLUMINANT C and the 2° (1931) STANDARD OBSERVER unless specified otherwise. If they have been obtained for other illuminants or observers, a subscript describing the observer or illuminant should be used. For example, x_{10} and y_{10} are chromaticity coordinates for the 10° observer and Illuminant C.

chromaticity coordinates, general the two dimensions of any COLOR ORDER SYSTEM which exclude the LIGHTNESS dimension and describe the chromaticity.[69] Syn: color coordinates

Thus, x and y, or a and b, or u and v, or YB and RG, may be considered as chromaticity coordinates in various color spaces. (Unless otherwise specified, the term chromaticity coordinates is assumed to refer to the CIE coordinates x, y, and z for ILLUMINANT C and the 2° (1931) Standard Observer.)

chromaticity diagram, CIE a two-dimensional graph of the CHROMATICITY COORDINATES, x as the abscissa and y as the ordinate, which shows the spectrum locus (chromaticity coordinates of MONOCHROMATIC LIGHT, 380-770 nm).[69]

It has many useful properties for comparing colors of both luminous and nonluminous materials.

chromaticity diagram, general plane diagram formed by plotting one of the CHROMATICITY COORDINATES against the other.[69]

chromaticity difference diagrams plane diagram formed by plotting the differences of one of the CHROMATICITY COORDINATES against the differences of the other chromaticity coordinate, the differences being taken from the neutral point, or, for a specific COLOR, from the chromaticity coordinates for the standard of that color.[69]

chromaticness combined HUE and SATURATION of a COLOR, with no reference to its LIGHTNESS.[69] Syn: chromaticity

chromatogram in chromatography, a graphic presentation of detector signal against time, having baseline and peaks.[37] (IUPAC)

chromatographic analysis (chromatography) a method of separation of different chemical species based upon selective adsorption. GAS CHROMATOGRAPHY involves the separation of components (gases or vapors) in a gaseous stream (carrier gas) on a column of some suitable sorbent (stationary phase).[37] (IUPAC)

Nonvolatile components are often separated by liquid chromatography where a solution of the components is separated by partition between a liquid mobile phase and a "solid" stationary phase. In both cases, the stationary phase may either be a liquid (often chemically bonded to the support) in which the components selectively dissolve, or a solid on which they are selectively adsorbed. ION CHROMATOGRAPHY and ion exchange chromatography are useful for the separation and analysis of the ions present in rainwater or cloud-water. The stationary phase may be packed in a column, spread as a layer, or distributed as a film, etc.; in these definitions chromatographic bed is used as a general term to denote any of the different forms in which a stationary phase may be used. The mobile phase may be gaseous or liquid.

chrome green $PbCrO_4xPbSO_4yFeNH_4Fe(CN)_6$. Pigment Green 15 (77510/77603). Chrome greens are blends of chrome yellows and iron blues ranging from light green to dark green.[41] Syn: chromium green, lead chrome green, brunswick green, milori green

Variation can be achieved by pigment blends with EXTENDERS. Noted for good HIDING and TINTING STRENGTH, poor ALKALI RESISTANCE. Properties: density, 4.1-5.4 g/cm³ (34.2-45.0 lb/gal); O.A., 13-25; particle size, 0.2-1.2 µm.

chrome ochre see CHROMIUM OXIDE GREEN.[41]

chrome orange, light and deep $PbCrO_4 \cdot PbO$. Pigment Orange 21 (77601). Chrome orange is chemically BASIC LEAD CHROMATE. The HUE ranges from light to deep orange, and the COLOR is related to particle size.[41] Syn: orange chrome, Austrian cinnabar, Persian red, Chinese red, derby red, American vermillion, chrome red

A pigment characterized by good opacity, fair LIGHTFASTNESS, good GLOSS RETENTION and DURABILITY. Disadvantages are poor alkali and acid resistance and poor high temperature stability. Properties: density, 6.6-7.1 g/cm³ (55.1-58.8 lb/gal); O.A., 9-12; particle size, 0.1-1.0 µm.

chrome oxide green see CHROMIUM OXIDE GREEN.[41] Syn: Schnitzer's green

chrome red see BASIC LEAD CHROMATE, and CHROME ORANGE, LIGHT and DEEP.[41]

chrome vermillion see MOLYBDATE ORANGE.[41]

chrome yellow, light and primrose $xPbCrO_4 \cdot yPbSO_4$. Pigment Yellow 34 (77603). Yellow pigments based on combinations of LEAD CHROMATE and lead sulfate, the HUE and TINT STRENGTH being controlled by the ratio of chromate to sulfate during precipitation. In the light and lemon hues, the approximate ratio of x/y is 2.5/1 ($2.5\ PbCrO_4 \cdot PbSO_4$); in the primrose hues, the x/y ratio is approximately 3.2/1. Syn: lemon chrome, golden chrome, middle chrome, Paris yellow, leipsic yellow, cologne yellow

The pigments are characterized by good OPACITY, fair LIGHTFASTNESS, good GLOSS RETENTION and DURABILITY, and poor resistance to alkali, acid, and high temperature.[41] Properties: density, 5.44-6.09 g/cm³ (17-39 lb/gal); O.A., 17-39; particle size, 0.1 to 0.8 µm.

chrome yellow, medium $PbCrO_4$. Pigment Yellow 34 (77600). Approaches chemically pure LEAD CHROMATE.[41]

Characterized by good OPACITY, fair to good LIGHTFASTNESS, good GLOSS RETENTION and DURABILITY, poor alkali resistance. Properties: density, 5.58-6.04 g/cm³ (46.6-50.3 lb/gal); O.A., 16-35; particle size, 0.15-1 µm.

chromic hydrate see CHROMIUM HYDROXIDE.[41]

chromic oxide see CHROMIUM OXIDE GREEN.[41]

chromium green see CHROME GREEN.[41]

chromium hydrate see CHROMIUM HYDROXIDE.[41]

chromium hydroxide $Cr(OH)_3$. Green PIGMENT, manufactured by adding a solution of ammonium hydroxide to the solution of a chromium salt.[41] Syn: chromic hydrate, chromium hydrate

chromium oxide green Cr_2O_3. Pigment Green 17 (77288). Most permanent green PIGMENT, almost pure CHROMIUM SESQUIOXIDE. May be manufactured by reducing a chromate (e.g., sodium dichromate) with sulfur or carbonaceous materials.[41] Syn: Arnaudon's green, chromium sesquioxide, chrome oxide green, vert emeraude (dull), anadonis green, chrome ochre, green cinnabar, green rouge, leaf green, oil green, Dingler's green, Schnitzler's green

Outstanding LIGHTFASTNESS and resistance to acid, alkali, and high temperatures; often used as a COLORANT in CEMENTITIOUS products. Properties: density, 5.1-5.4 g/cm³ (42.5-45.0 lb/gal); O.A., 12-24; particle size, 2-6 µm.

chromium oxide, hydrated see HYDRATED CHROMIUM OXIDE.[41]

chromium sesquioxide see CHROMIUM OXIDE GREEN.[41]

chromophore certain groups of atoms such as, –C=C–, –C=N–, –N=N–, –N=O, when present in an organic molecule, can give rise to colored COMPOUNDS.[69,41]

Groups acting in this way are called CHROMOPHORES, although compounds containing chromophores are not necessarily colored. The HUE is affected by the presence of other groups known as "AUXOCHROMES," such as $-NH_2$, $-OH$, $-NO_2$, $-CH_3$ and halogens –Cl, –Br, and –I.

chrysotile see MAGNESIUM SILICATE, FIBROUS.[69,41]

CI abbreviation for COLOUR INDEX.[41]

CIE abbreviation for Commission Internationale d'Eclairage, the French name for the International Commission on Illumination.[174]

In older publications, the abbreviation ICI was used. It is the main international organization concerned with problems of COLOR and color measurement.

Words presented in CAP/SMALL CAPS type indicate that the word is defined in another part of the *Coatings Encyclopedic Dictionary.*

Numerical superscripts classify terms in one or more of the categories listed in the second section of this volume.

CIE 1976 (L*a*b*) color difference equation (from *J. Opt. Soc. Am.*, 64, 896 (1974) (Modified Adams-Nickerson).

$$\Delta E_{CIE}(L^*,a^*,b^*) = [(\Delta L^*)^2 + (\Delta a^*)^2 + (\Delta b^*)^2]^{1/2}$$

where

$$L^* = 25\left(\frac{100Y}{Y_0}\right)^{1/3} - 16 \quad (1 \le Y \le 100)$$

$$a^* = 500\left[\left(\frac{X}{X_0}\right)^{1/3} - \left(\frac{Y}{Y_0}\right)^{1/3}\right]$$

$$b^* = 200\left[\left(\frac{Y}{Y_0}\right)^{1/3} - \left(\frac{Z}{Z_0}\right)^{1/3}\right]$$

X, Y, and Z are the TRISTIMULUS VALUES of the sample. X_0, Y_0, and Z_0 define the COLOR of the nominally white object color stimulus (the ILLUMINANT). $\Delta L^* = L^*$ for the sample —L^* for the standard; $\Delta a^* = a^*$ for the sample —a^* for the standard; $\Delta b^* = b^*$ for sample —b^* for the standard.[69,43]

CIE 1976 (L*u*v*) color difference equation (from *J. Opt. Soc. Am.*, 64 (1974).

$$\Delta E_{CIE}(L^*u^*v^*) = [(\Delta L^*)^2 + (\Delta u^*)^2 + (\Delta v^*)^2]^{1/2}$$

where

$$L^* = 25\left(\frac{100\,Y}{Y_0}\right)^{1/3} - 16 \quad (1 \le Y \le 100)$$

$$u^* = 13L^*(u' - u'_0)$$

$$v^* = 13L^*(v' - v'_0)$$

with

$$u' = \frac{4X}{X + 15Y + 3Z}$$

$$v' = \frac{9Y}{X + 15Y + 3Z}$$

$$u'_0 = \frac{4X_0}{X_0 + 15Y_0 + 3Z_0}$$

$$v'_0 = \frac{9Y_0}{X_0 + 15Y_0 + 3Z_0}$$

X,Y, and Z are the TRISTIMULUS VALUES of the sample. X_0, Y_0, and Z_0 are the tristimulus values for the ILLUMINANT. $\Delta L^* = L^*$ for the sample —L^* for the standard; $\Delta u^* = u^*$ for the sample —u^* for the standard; $\Delta v^* = v^*$ for sample —v^* for the standard.[69,43]

CIE chromaticity coordinates see CHROMATICITY COORDINATES, CIE.[69]

CIE color difference equation may refer to the color difference equation provisionally adopted by the CIE in 1964, one of four then recommended for study. It incorporates a linear transformation into a more nearly uniform COLOR SPACE. The following are the equations used:

$$U^* = 13W^*(u - u_0)$$

$$V^* = 13W^*(v - v_0)$$

$$W^* = 25Y^{1/3} - 17 \quad (1 \le Y \le 100)$$

$$u = \frac{4X}{X + 15Y + 3Z} \quad \text{or} \quad u = \frac{4x}{-2x + 12y + 3}$$

$$v = \frac{6Y}{X + 15Y + 3Z} \quad \text{or} \quad v = \frac{6y}{-2x + 12y + 3}$$

u_o and v_o refer to a nominally achromatic color. For the 1931 STANDARD OBSERVER and ILLUMINANT C, $u_o = 0.2009$ and $v_o = 0.3073$. The total color difference is calculated as:

$$\Delta E = [(\Delta U^*)^2 + (\Delta V^*)^2 + (\Delta W^*)^2]^{1/2}$$

$\Delta U^* = U^*$ sample $-$ U^* standard, indicating redder if positive and greener if negative.

$\Delta V^* = V^*$ sample $-$ V^* standard, indicating yellower if positive and bluer if negative.

This equation is properly titled the 1964 CIE U*, V*, W* equation.[69]

CIE Color Notation System COLORIMETRIC specification system based on stimulus-response characteristics adopted by the CIE in 1931.[69]

The current recommendations for the system may be obtained from the official publication, CIE Publication No. 15 (E-1.3.1) 1971, "Colorimetry Official Recommendation of the International Commission on Illumination," available from National Bureau of Standards, Washington, D.C. 20234.

ASTM Test Method E308 covers the Color Notation System.

CIE Luminosity Curve see LUMINOSITY CURVE, CIE.[69]

CIE Standard Observer the observer data adopted by the Commission Internationale d'Eclairage to represent the response of the average human eye, when light-adapted, to an equal-energy spectrum.[43,69]

Unless otherwise specified, the term applies to the data adopted in 1931 for a 2° field of vision. The data adopted in 1964, sometimes called the 1964 observer, were obtained for a 10°, annular field which excludes the 2° field of the 1931 observer functions.

CIE tristimulus values see TRISTIMULUS VALUES, CIE.[69,43]

CIEF abbreviation for COATINGS INDUSTRY EDUCATION FOUNDATION.[174]

cill British term for sill.[82] (DAC)

cimene see DIPENTINE.[164,44]

cinaper an obsolete form of CINNABAR.

cinder block a lightweight MASONRY unit made of cinder concrete; widely used for interior partitions.[75] See CLINKER BLOCK[75] and SINTERING.[73,75,159] cf. CONCRETE BLOCK[73,75,159]

cinnabar see MERCURIC SULFIDE.[41]

cinoper an obsolete form of CINNABAR.[41]

cissing (1) a slight shrinkage of a GLOSSY paint coat resulting in small cracks through which the undercoat may be seen; a mild form of CRAWLING.[56] British syn for crawling; (2) a process for preparing a wood surface for GRAINING by WETTING with a sponge.[158] (DAC) Also spelled "sissing."

citrates ESTERS derived from CITRIC ACID. The triethyl, -butyl, and -amyl citrates form a series of PLASTICIZERS.[144,170]

citric acid $C_3H_4(OH)(COOH)_3$. TRICARBOXYLIC, MONOHYDROXY ACID. One molecule of water is normally included in the formula.[144]

It has an mp of 154°C, but decomposes on further heating.

citronella oil ESSENTIAL OIL, known also as oil of lemongrass. It possesses a strong lemon-like smell, and is used to a small extent as a deodorant.[83]

CLA abbreviation for center line average (BS 1134); indication of surface roughness.[158]

clapboard a WOOD SIDING commonly used as an exterior covering on a building of frame construction; applied horizontally and overlapped, with the grain running lengthwise; usually thicker along the lower edge than along the upper.[75,82,153] (DAC) Syn: bevel siding, and lap siding

claret red AZO PIGMENT produced by coupling Tobias acid (2-naphthylamine-1-sulfonic acid) with 3-oxynaphthoic acid, followed by conversion to the calcium LAKE.[41]

clarity the characteristic of a TRANSPARENT material whereby distinct images may be observed through it.[43,35]

class I, II, or III liquids groupings of FLAMMABLE and COMBUSTIBLE liquids arbitrarily classified by boiling points and closed cup FLASH POINT methods, in order to specify particular procedures and types of electrical equipment to be used when handling them.[172]

clay see ALUMINUM SILICATE, KAOLIN, KAOLINITE, CHINA CLAY.[41,58]

clean a complete lack of any visible nonuniformity sometimes referred to as SEEDS, when viewed in thin films by any macroscopic or microscopic use of visible light.[157,43,35] (ASTM)

Clean Air Act (CAA) as amended, it provides the foundation for EPA's efforts to improve air quality. Public law under EPA jurisdiction, it is the regulatory vehicle that sets and monitors airborne pollution hazardous to public health or natural resources. The EPA sets national ambient air-quality standards. Enforcement and issuance of discharge permits are carried out by the state and are called state implementation plans.The Clean Air Act, building on earlier legislation in 1970, was amended in 1977. Further amendments proposed to the original rule in 1990 were made into law in 1991 and lowered the allowable emission levels for certain pollutants and mandated that additional pollutants be monitored.[172]

Clean Water Act (CWA) regulates the discharge of nontoxic and toxic pollutants into surface waters.[172]

Its ultimate goal is to eliminate all discharges into surface waters. Its interim goal is to make surface waters usable for fishing, swimming, etc. EPA sets guidelines, and states issue permits (NPDES, Natural Pollution Discharge System permit) specifying the types of control equipment and discharges for each facility.

cleaner (1) DETERGENT, ALKALI, ACID, or other cleaning material; usually water or steamborne;[127] (2) SOLVENT for cleaning paint equipment.[164]

clear a complete lack of any VISIBLE nonuniformity when viewed in mass, in bottles or test tubes, by strong TRANSMITTED LIGHT.[157,43] (ASTM)

clear coating TRANSPARENT protective and/or DECORATIVE film.[71]

clear lacquer see CLEAR COATING.[71,151]

clear liquid separation (CLR) often referred to as SYNERESIS, is the spontaneous appearance of a transparent liquid over an opaque liquid paint during storage.[161,42] See PHASE SEPARATION.

Often this occurs after months or years of aging, but can happen in as little as several hours. Although this problem has been observed in latex paints containing cellulosic or urethane type associative thickeners (HEUR), it is particularly prevalent when both thickener types are combined.

clear water repellent coating a TRANSPARENT coating formulated for the purpose of protecting porous SUBSTRATES by preventing the PENETRATION of liquid water.[71,165] (ASTM) See CLEAR COATING.

Water is known to penetrate WOOD and other substrates causing dimensional instability and eventual deterioration. Clear water repellent coat-

---

Words presented in CAP/SMALL CAPS type indicate that the word is defined in another part of the *Coatings Encyclopedic Dictionary.*
Numerical superscripts classify terms in one or more of the categories listed in the second section of this volume.

ings are designed to protect these substrates from the damaging effects of water.

clearcoat a PIGMENT-free coating used to protect the pigmented TOPCOAT CAR FINISH.[71,163]

These prevent car topcoats from superficial scratches, marring, and ETCHING *by environmental fallout such as* ACID RAIN.

clearcole, clairecolle (1) GLUE SIZE in appropriate DILUTION with the addition of a small quantity of WHITING. It is used to reduce POROSITY of ceiling and other surfaces prior to the application of size-bound DISTEMPER;[147] (BSI) (2) a primer consisting of glue, water, and whiting;[71] (3) a CLEAR COATING used in application of gold leaf.[71] (DAC)

cleavage (1) breaking of a LAMINATE due to the separation of the strata; (2) portion of the material adhering to the side of a container after the major portion has been drained.[50]

Cleveland condensing humidity cabinet (QCT) an ACCELERATED WEATHERING apparatus which operates on a CONDENSATION type of water EXPOSURE at elevated temperature.[59,157]

Cleveland open cup device used in determining FLASH and FIRE POINTS of PETROLEUM products.[145,157]

Cleve's acid old name for α-naphthol-5-sulfonic acid. It was also known as L acid and is used in the manufacture of Helio Bordeaux BL.[144]

climate cabinet any enclosure used to simulate selected climatic conditions.[154,157]

clinker block British term for CINDER BLOCK.[75] (DAC)

clockspring slipping action in COIL COATING, usually at the center of a coil of the strip during the recoiling operation.[49,57,70]

cloisonne a surface DECORATION in which differently colored ENAMELS or GLAZES are separated by fillets applied to the design outline.[81,148] (DAC)

For porcelain enamel, the fillets are wire (gold, silver, or copper) secured to the metal body; for tile and pottery, the fillets are made of ceramic paste, squeezed through a small-diameter orifice.

close drying material is said to be close drying when it does not show much fullness or body on the substrate.[55,67]

Often, the true fullness is not brought out until after rubbing. This is especially true with lacquer finishes.

close grain WOOD having narrow and inconspicuous annual growth rings.[153] (ASTM)

The term is sometimes used to designate wood having small and closely spaced pores, but in this sense the term "fine textured" is more often used.

closed coat in coated ABRASIVES, when the abrasive grains completely cover coatside surface of the backings.[158] See OPEN COAT.[158]

Closed coats are designed for severe service and are used for most applications.

cloth distressing see RAGGING.[77,148] cf. DISTRESSING[54,148,175]

cloud point point at which a definite lack of CLARITY (CLOUDINESS) appears when a liquid is subject to adulteration or when it is mixed with another substance, or the temperature at which a liquid becomes cloudy when it is cooled.[157,161]

cloudiness the lack of CLARITY or TRANSPARENCY in a paint or VARNISH film.[56,42] (DAC)

clouding development in a clear VARNISH or LACQUER film or liquid of an opalescence or CLOUDINESS caused by the PRECIPITATION of insoluble matter or IMMISCIBILITY of components.[161] (BSI)

CLR abbreviation for CLEAR LIQUID SEPARATION.[161,42]

clupanodonic acid acid found in many FISH OILS. The acid possesses an unusually high IODINE VALUE—well in excess of 300.[63] Syn: docosapentanoic acid

CMC (1)abbreviation for CARBOXYMETHYL CELLULOSE[83,139,155] and (2) CRITICAL MICELLE CONCENTRATION.[127]

CO_2 chemical formula for CARBON DIOXIDE.

coacervate an AGGREGATE of COLLOIDAL droplets held together by electrostatic attractive forces.[139]

coacervation (1) reversible collection of emulsoid particles into liquid droplets preceding FLOCCULATION. An INTERMEDIATE stage between SOL and GEL formation; (2) the separation of a solution (in a single solvent) containing two different macromolecular (or of one macromolecular and one micromolecular) species, of equal sign or charge (or of no charge), into two phases, one of which is rich in dissolved solute molecules of one species, while the other is enriched in the other dissolved solute.[139] cf. COAGULATION[42,131,161]

coagulation (1) PROCESS whereby a fluid liquid is changed into a thickened, curdled, or congealed mass;[42,131,161] (2) irreversible AGGLOMERATION of PARTICLES originally dispersed in a RUBBER LATEX.[42,131,161] (ASTM) cf. FLOCCULATION[42,64,139,161]

coagulum an AGGLOMERATE of PARTICLES.[42,161,139] (ASTM)

coal tar a dark brown to black CEMENTITIOUS material produced by the DESTRUCTIVE DISTILLATION of BITUMINOUS COAL.[76] (ASTM)

coal tar colors see ANILINE PIGMENTS.[41]

coal tar-epoxy coating coating in which BINDER or vehicle is a combination of COAL TAR with EPOXY RESIN.[71]

coal tar hydrocarbons AROMATIC HYDROCARBONS derived from COAL TAR, including BENZENE, TOLUENE, XYLENE, NAPHTHA, etc.[164]

coal tar pitch DISTILLATION residue from COAL TAR.[76]

It varies considerably from a very soft to a very hard product. Fusion points vary from as low as 27°C (80°F) to as high as 232°C (450°F).

coal tar resins see COUMARONE-INDENE RESINS.[167]

coal tar-urethane coating coating in which BINDER or VEHICLE is a combination of COAL TAR with a POLYURETHANE RESIN.[71]

coalescence the formation of a film of resinous or polymeric material when water evaporates from an emulsion or latex system, permitting contact and fusion of adjacent latex particles. Action of the joining of particles into a film as the volatile evaporates.[55] (DAC)

ASTM Test Method D3793 covers the coalescence of latex paints.

coalescent (coalescing agent) SOLVENT with a high boiling point which, when added to a coating, aids in FILM FORMATION via temporary PLASTICIZATION (softening) of the vehicle.[55,83,164]

coarse grain WOOD with wide and conspicuous annual rings having considerable difference between SPRINGWOOD and SUMMERWOOD.[153] (ASTM)

The term is sometimes used to designate wood with large pores, such as oak, ash, chestnut, and walnut, but in this sense the term "coarse texture" is more often used.

coarse texture see COARSE GRAIN.[153]

coat paint, VARNISH, or LACQUER applied to a SURFACE in a single APPLICATION (one layer) to form a properly distributed film when dry.[54,55]

A coating system usually consists of a number of coats separately applied in a predetermined order at suitable intervals to allow for DRYING or CURING. It is possible with certain types of material to build up coating systems of adequate THICKNESS and OPACITY by a more or less continuous process of application, e.g., WET-ON-WET spraying. In this case, no part of the system can be defined as a separate coat in the above sense.

coated abrasive a FLEXIBLE-type backing upon which a film of ADHESIVE holds and supports a coating of ABRASIVE GRAINS.[59]

The backing may be paper, cloth, vulcanized fiber, or a combination of these materials. Various types of RESIN and hide GLUES are used as adhesives. The abrasives used are flint, emery, crocus, garnet, aluminum oxide, and silicon carbide.

coated paper a PAPER coated with CLAY, other white pigments, and a suitable binder.[168]

coater apparatus which applies paint.[59,70]

coating (1) generic term for paints, LACQUERS, ENAMELS, PRINTING INKS, etc;[71] (2) a liquid, liquefiable or mastic composition which is converted to a solid protective, decorative, or functional ADHERENT FILM after APPLICATION as a thin layer. It is also used to refer to films applied to PAPER, PLASTICS, or FOILS.[71] (ASTM) See FINISH.[54,71]

coating, dip see DIP COATING.[77]

Coatings Industry Education Foundation (CIEF) Foundation which has as its goal to advance, through education and research, the chemical, physical, and mathematical sciences relating to the technology of protective coatings, and to aid in the dissemination of the results of such research and education to the public, through scientific publications and lectures. The Board of Directors of the FEDERATION OF SOCIETIES FOR COATINGS TECHNOLOGY constitute the stockholders or members of the CIEF. It is the duty of the Trustees to manage the business affairs of CIEF, including allocating funds for educational purposes.[174]

coating powders (1) finely divided, solid PLASTIC materials which are heat fusible and form relatively smooth, tough, electrical insulating coatings upon APPLICATION to METAL surfaces;[71] (ASTM) (2) finely divided particles of ORGANIC POLYMER, either THERMOPLASTIC or THERMOSETTING, which generally contain pigments, FILLERS, and ADDITIVES and which remain finely divided during STORAGE under suitable conditions.[71] (ASTM) See POWDER COATINGS.[71,77]

coating solids the part of the coating which remains after the coating is DRIED or CURED.[128] (EPA) See SOLIDS, NONVOLATILE MATTER, SOLIDS BY VOLUME, and SOLIDS BY WEIGHT.[128]

coating, spray the PROCESS in which a SUBSTRATE is sprayed with the coating material.[77] (ASTM)

coating system see COAT.[54]

coatings remover see PAINT and VARNISH REMOVER.[158,164]

Coatings Societies International (CSI) an international organization of coatings technical associations whose purpose is the advancement of international relations and technical communications. Member organizations include FEDERATION D'ASSOCIATION DES TECHNICIANS DES PEINTURES,

Words presented in CAP/SMALL CAPS type indicate that the word is defined in another part of the *Coatings Encyclopedic Dictionary.*

Numerical superscripts classify terms in one or more of the categories listed in the second section of this volume.

VERNIS, EMAUX, ET D'IMPRIMERIE DE L'EUROPE CONTINENTALE (FATIPEC); FEDERATION OF SOCIETIES FOR COATINGS TECHNOLOGY (FSCT); JAPAN SOCIETY OF COLOUR MATERIAL (JSCM); OIL & COLOUR CHEMISTS' ASSOCIATION (OCCA); SKANDINAVISKA LACKTEKNIKERS FORBUND (SLF); SURFACE COATINGS ASSOCIATION AUSTRALIA (SCAA); SURFACE COATINGS ASSOCIATION NEW ZEALAND (SCANZ).[174]

Founded in Paris on June 23, 1979 as the International Committee to Coordinate Activities of Technical Groups in the Coatings Industry (ICCATCI), the organization was reformed in 1993 to create CSI. Delegates from member organizations meet annually at the national conventions of member organizations to review local activities and to discuss areas of mutual support and cooperation.

cobalt aluminate see COBALT BLUE.[41]

cobalt blue $CoO{\cdot}Al_2O_3$. Pigment Blue 28 (77346). Bright blue PIGMENT, which is a complex product derived from oxides of cobalt and ALUMINUM.[41] Syn: cobalt aluminate, King's blue, Thenard's blue, azure blue, cobalt ultramarine, Dumont's blue, enamel blue, Gahn's ultramarine, Hungary blue, Leithner's blue, Leyden blue, Vienna blue, Wentzel's blue, zaffre

It has excellent LIGHTFASTNESS and good CHEMICAL and HEAT RESISTANCE and with low OPACITY. Properties: density, 4.2-4.3 g/cm³ (35.0-35.8 lb/gal); O.A., 27-36; particle size, 0.5-1.0 µm.

cobalt drier one of the many ORGANIC COBALT SALTS (COBALT NAPHTHENATE, COBALT OCTOATE, etc.) which are soluble in paints and VARNISHES and used to speed the DRYING and HARDENING of the OIL VEHICLE.[83,67,74]

Also used as an INITIATOR in POLYESTERS.

cobalt green similar to COBALT BLUE, except that ZINC OXIDE replaces wholly or partly the aluminum oxide in the latter.[41] Syn: Rinmann's green or zinc green

It does not contribute to good HIDING POWER.

cobalt naphthenate see COBALT DRIER.[83,64,74]

cobalt octoate see COBALT DRIER.[83,64,74]

cobalt ultramarine red shade COBALT BLUE.[41]

cobalt yellow see AUREOLIN.[41]

cobweb coating see COBWEBBING.[71]

cobwebbing production of fine FILAMENTS instead of the normal atomized particles when some coatings are SPRAYED.[54,42,77] (BSI)

Although generally considered a defect in ordinary LACQUERS, use is made of this property to provide a textured coating or a protective covering for equipment during STORAGE. A COCOON is formed around the article by the pronounced cobwebbing action.

cochin oil refined grade of COCONUT OIL.[63]

cochineal natural ORGANIC DYESTUFF made from the bodies of the female insect, *Coccus cacti*, which lives on plants in Central and South America.[41]

The coloring principle is known as CARMINIC ACID and is generally LAKED. It is not fast to light.

coconut oil expressed from the nut kernels of the coconut palm, the OIL is composed of GLYCERIDES of LAURIC ACID, capric acid, myristic, palmitic, and oleic acids. Its main use is in nonyellowing alkyds.[63]

Properties: sp gr, 0.9190/25°C; refractive index, 1.4545/40°C; iodine value, 9; and saponification value, 255.

cocooning a cocoon formed around an article by the controlled COBWEBBING action of a specially formulated coating.[55,77]

A durable continuous coating is then sprayed over the support cocoon to protect the equipment.

Code of Federal Regulations a collection of the regulations established by law.[172]

For copies, contact the agency that issued the regulation for details, interpretations, etc. They are sold by the Superintendent of Documents, Government Printing Office, Washington, D.C. 20402; Tele. (202) 783-3238.

coefficient of cubical expansion see COEFFICIENT OF EXPANSION.[142]

coefficient of elasticity reciprocal of YOUNG'S MODULUS in a tension test.[47]

coefficient of expansion ratio of increase in length, area, or volume of a substance for a 1°C rise in temperature.[142] Syn: coefficient of cubical expansion

coefficient of friction measure of the RESISTANCE of sliding of one SURFACE in contact with another surface.

Coefficient of friction is an intrinsic property of two interfacing, interacting surfaces and serves as a measure of their micro- and macro-roughness, inter- and intra-molecular forces of attraction and repulsion, and their viscoelastic (POLYMER DEFORMATION) properties.

coefficient of plastic viscosity see VISCOSITY, PLASTIC.[166]

coefficient of scatter see SCATTER, COEFFICIENT OF.[43]

coefficient of viscosity see VISCOSITY (2).[166]

cohesion propensity of a single substance to ADHERE to itself; the internal attraction of molecular particles toward each other; the ability to resist partition from the mass; internal adhesion; the force holding a single substance together.[79]

coil coating PROCESS wherein a continuous coil of metal is unwound, cleaned, surface-treated, coated, heat-cured, cooled, and rewound in one operation.[70,77,131] Syn: strip coating

The coated coil is subsequently unwound and formed into any number of products, such as house siding, venetian blinds, and automotive and appliance parts.

ASTM test methods for coil coatings include: wire-wound bar application, D4717; and guide for testing, D3794.

coiling soup in coating or spreading operations, a DEFECT caused by the curling or turning motion of a bank of LACQUER or other "soup" COMPOUND in front of a DOCTOR KNIFE, or similar application device, wherein streaks are formed on the surface of the coated fabric due to uneven APPLICATION of the soup.[42,56] Syn: curling

coin marking see METAL MARKING.[70,56]

coking in the RUNNING of COPALS, or SWEATING of ASPHALTUMS, and other materials which naturally contain a proportion of infusible dirt or mineral matter, this extraneous matter is likely to escape the action of the stirrer. [76,131]

As a consequence, any organic matter associated with it is liable to be overheated, and carbonization occurs. This ultimately results in the formation of carbonaceous lumps, and it is then said to have coked.

cold DULL, FLAT SURFACE where a bright, lustrous one is desired.[43]

cold cut dissolving a RESIN or other material in a suitable SOLVENT by mechanical agitation without the application of heat.[131]

cold cycle low temperature cycle during many ACCELERATED TESTS.[157]

cold flex temperature the lowest temperature at which the test strip can be twisted through a 200° arc without breaking.[157] (BS 2571, app. L.).[157]

cold flow continuing DISTORTION, DEFORMATION, or DIMENSIONAL change which takes place in materials under continuous load (constant stress) at AMBIENT TEMPERATURES. See CREEP.[166] Syn: compression set

cold pressing (1) assembly method in which the bonded structures are held in place by pressure without the application of heat or drying air until the ADHESIVE INTERFACE has solidified and reached proper SHEAR proportions;[57,131] (2) an EXTRACTION method for OILS such as fish oil.[131]

cold stretch pulling operation, usually on EXTRUDED FILAMENTS, to improve TENSILE properties.[47,57]

cold test refers to POUR POINT. A low cold test means that the product has a low temperature pour point.[157,166]

cold water paint paint in which the BINDER or vehicle portion is composed of SOYBEAN or other VEGETABLE PROTEIN, GLUE, RESIN EMULSION, or other similar material dispersed in water.[71,155]

cold-bend test used to assess the FLEXIBILITY of coatings at low temperatures, by bending a standard specimen to a specific radius while maintained at a specific subnormal temperature.[157]

cold-checking SURFACE DEFECT on LACQUER, especially FURNITURE FINISHES.[56,175] See CHECKING, and TEMPERATURE CHECKING.[56,42,154]

It appears as a pattern of fine cracks developing on film surface, when the finished article is subjected to alternate hot and cold conditions. ASTM Test Method D1211 covers cold-checking.

cold-cracking CRAZING and CRACKING of a coating subjected to low temperature or cold/ambient cycling.[56,157]

cold-curing PROCESS of CURING at normal atmospheric temperature (i.e., AIR DRY).[67] (BSI)

cold-process roofing a BITUMINOUS ROOFING MEMBRANE which consists of layers of coated FELTS that have been bonded with cold-applied asphalt roof CEMENT and surfaced with an emulsified or cutback asphalt roof coating.[75,76] (DAC)

cold-rolled steel low-carbon, cold-reduced sheet steel.[159]

cold-setting RESIN or LACQUER products which, under the influence of suitable CATALYSTS, are able to form suitable film properties without the application of heat.[67,169]

cold-setting adhesive SYNTHETIC RESIN ADHESIVE capable of HARDENING at normal room temperature in the presence of a HARDENER.[67,79]

cold-setting inks solid INKS which must be melted and applied on a hot press. They solidify again on contact with unheated paper.[168]

collage (Fr.) Literally, pasting.[81,148]

Employed to describe the practice of some 20th-century artists in introducing PAPER *or other nonpigmented materials in painting.[81] Basically,*

Words presented in CAP/SMALL CAPS type indicate that the word is defined in another part of the *Coatings Encyclopedic Dictionary.*

Numerical superscripts classify terms in one or more of the categories listed in the second section of this volume.

a pasting technique whereby pictorial images or patterns, and pieces of colored, textured material are superimposed on each other.

collector (1) a device for removing and retaining contaminants from air and other gases; often applied to cleaning devices in exhaust systems.[131,145,172] *Also used to designate a device for removing and retaining samples from media in different environmental compartments to be investigated. A collector is sometimes used to describe a scavenger;* (2) a solid substance added to or formed within a solution to collect a micro- or macro-component.[83] (IUPAC)

collet rigid lateral container for the mold-forming material; a dam, a restriction box; the drive wheel that pulls GLASS FIBERS from the bushing.[57,59]

A forming tube is placed on the collet and a package of strand (forming cake) is wound up on the tube.

colloid state of subdivision of matter which comprises either single large molecules or AGGREGATIONS of smaller molecules.[139,64]

Size range, 1×10^{-7} to 1×10^{-5} cm.

colloid, linear any COLLOID with FIBROUS PARTICLES or macromolecules.[64,139]

The long, asymmetrical particles may be coiled up or, in some instances, they may be BRANCHED.

colloid, micellar COLLOID composed of MICELLES.[64,139]

A micellar colloidal is an AGGREGATION of molecules, usually arranged in a definite order, whose dimensions are less than 0.05 micrometres. Example: a colloid solution of SOAP in water.

colloid mill machine consisting of a high-speed rotor and stator in which the emulsifying action takes place between the opposing faces.[59,131]

It is used primarily to produce COLLOIDAL DISPERSIONS and EMULSIONS.

colloid, molecular any COLLOID in which the COLLOIDAL PARTICLES are MACROMOLECULES.[64,139]

colloid, protective see PROTECTIVE COLLOID.[139]

colloid, reversible COLLOID SUBSTANCE that can be converted from the GEL to the SOL form without the expenditure of chemical energy, that is, by simply warming the gel form or redissolving the dried substance in water.[64,139]

It is a colloid which forms a HYDROPHILIC sol, such as gelatin, agar, GUM ARABIC, starch; a protective colloid.

colloid, sphere COLLOID possessing more or less symmetrically shaped, corpuscular, and relatively compact particles.[64,139]

colloidal pertaining to COLLOID.[64,139]

colloidal dispersion see COLLOID.[139]

colloidal particle large molecules or AGGREGATES of smaller molecules in a size range of 1 to 500 nanometres (10^{-9} m).[64,139]

colloidal solution any solution containing COLLOIDAL PARTICLES.[64,139]

Colloidal solutions are sometimes called SOLS.

colloidal state particular state in which any substance may exist under the proper conditions, determined by fineness of particle subdivision.[64]

The colloidal state is defined by a more or less well-marked ultramicroscopic zone in the scale of subdivision, the lower extreme of the zone approaching molecular dimensions, and the upper end gradually passing over into molecular aggregates (SUSPENSIONS) visible under the ordinary microscope.

collotype (1) a PRINTING PROCESS utilizing a GLASS plate with a GELATIN surface carrying the image to be reproduced;[168] (2) a print made by this process.[168] Syn: photogelatin process

cologne earth, cologne brown a PIGMENT, sometimes referred to erroneously as VANDYKE BROWN, made from roasted American CLAYS which contain OCHRE and BITUMINOUS MATTER.

cologne spirits see ETHYL ALCOHOL.[164]

cologne yellow see CHROME YELLOW, LIGHT and PRIMROSE.[41]

colonial spirit synonym for METHYL ALCOHOL.[164]

colophony, colophonium old name for ROSIN; used in Europe.[167]

color one aspect of APPEARANCE; a stimulus based on VISUAL response to light, and consisting of the three dimensions of HUE, SATURATION, and LIGHTNESS.[43,69,35]

ASTM test methods for color include: CIE color system, E308; color difference, instrumental, D2244; Colour Index, artist pigments, D4302; defining and evaluating tolerances, D3134; evaluating color change, gray scale, D2616; metamerism, visual evaluation, D4086; Munsell system, D1535; tristimulus values, D2244; visual evaluation of differences, D1729.

color acceptance see COLOR UNIFORMITY and TINTABILITY.[69,42]

color aptitude ability to work with COLOR, involving inborn factors as well as acquired abilities.[69]

color aptitude test test designed to examine any of the factors involved in COLOR APTITUDE.[69]

Most widely used was the ISCC Color-matching Aptitude Test, or CAT, based on differences in chroma.

color blindness (deprecated) An incorrect term applied to defective COLOR VISION.[69,173]

An extremely small number of persons have complete lack of color response (are achromatopes). Most so-called color-blind persons are anomalous trichromats, seeing all three primaries but having responses which are weaker than normal to one of the primaries. Persons who lack response to one primary are called dichromats, seeing only two primaries.

color burn-out an objectionable change in the COLOR of a PRINTING INK which may occur either in bulk or on the printed sheet.[168]

In the former case it is associated primarily with tints and is caused by chemical reaction between certain components in the ink formulation. In the latter case it is generally caused by heat generated in a pile of printed material during the drying of an oxidizing type of ink.

color code a system of coloring piping and parts of equipment with various precoded colors for identification purposes.[172]

color constancy relative independence of perceived object COLOR to changes in color of the light source.[69,43,35] See FLAIR[43,69,45] and METAMERISM.[69,43,35]

color coordinates see CHROMATICITY COORDINATES, GENERAL.[69]

color development The extent to which the COLORANT has achieved its full tinting potential, as evidenced by the COLOR change or lack thereof, when the tinted paint is subjected to very strong shear stress.[43,69] (ASTM) See TINTABILITY. cf. COLOR ACCEPTANCE[43,69]

Incomplete color development in tinted paints is undesirable from the standpoint of color uniformity and color efficiency. Color development can be caused by varying shearing forces associated with different methods of APPLICATION, tending to bring out the undeveloped color in varying degrees. For example, ROLLER APPLICATION (ca. 2000 s^{-1} shear rate) will show less color development than BRUSH APPLICATION (ca., 10,000 s^{-1} shear rate), the latter being at a higher shear rate. Thus, if both roller and brush are used on a wall, for example, there may be areas that show different depth of color.

Tinted paints are commonly tested for color development using ASTM Test Method D5326. A colorant sometimes fails to disperse completely in a tint base paint due to poor compatibility, which can be the fault of the colorant, the paint, or both. This will result in poor color development, which is readily manifested by the common procedure of applying the paint with a DOCTOR BLADE and subjecting the DRAWDOWN to high shear stress by finger-rubbing a small area of the partially dry film. (See RUB UP.[69,157,42]) This tends to disperse undeveloped colorant, if any, and produces a color variation between the unsheared and sheared areas of the paint film. The variation can be measured colorimetrically to give a numerical color difference value that is a measure of the color development of the original paint, the smaller the difference the better the color development and vice versa. Poor color development can be a problem in the production of paints, and in their performance in the field. In production it causes a loss of colorant monetary value, and unpredictable tinting results. In field performance it results in color variations in the applied paint film due to the varying shear forces to which the paint is subjected at different stages or by different modes of application.

Although poor color development is primarily and most often related to the colorant portion of a tinted paint, the white pigment in the base paint can also be poorly developed due to flocculation or other causes. In the latter case, shear dispersion can make the paint film lighter and less colorful, rather than the reverse. Then too, the colorant and the white might be poorly developed, and the color change due to the shear stress would then be the combined effect of both.

ASTM Test Method D5326 covers a method for measuring color development in tinted latex paints for the purpose of determining the efficiency of colorants and the potential for poor color uniformity of applied paint films.

color difference magnitude and character of the difference between two colors under specified conditions.[69]

color difference equations equations that transform CIE CHROMATICITY COORDINATES into a more uniform matrix such that a specified distance between two colors is more nearly proportional to the magnitude of an observed difference between them regardless of their HUE. The total color difference is generally designated as ΔE, the lightness difference as ΔL, the total chromaticity difference as ΔC, the redness-greenness difference as Δa, $\Delta \alpha$, or ΔRG, and the yellowness-blueness difference as Δb, $\Delta ß$, or ΔYB. The directional differences, such as ΔL, ΔRG, and ΔYB are frequently designated as the components of ΔE. Color difference equations may also be based on MUNSELL COLOR NOTATIONS, although such equations are seldom used today.[69]

color floating see FLOATING.[43,56]

color gamut see GAMUT, COLOR.[69]

Color Harmony Manual orderly array of colors spaced (approximately) according to the OSTWALD COLOR SYSTEM.[69]

Several editions were issued, each slightly different. It is no longer available.

Words presented in CAP/SMALL CAPS type indicate that the word is defined in another part of the *Coatings Encyclopedic Dictionary.*

Numerical superscripts classify terms in one or more of the categories listed in the second section of this volume.

color in oil see COLORANT.[41]

color in varnish see COLORANT.[41]

Color Index Name see COLOUR INDEX NAME.[69,134]

Color Index Number see COLOUR INDEX NUMBER.[69,134]

color match pair of colors exhibiting no perceptible difference when observed under specified conditions.[69,157]

The quality of an attempted match is described by the closeness to this ideal match.

color matching see MATCHING, COLOR.[69]

color matching functions relative amounts of three additive primaries required to match each WAVELENGTH of light.[69] See OBSERVER, STANDARD.[69,46,134]

The term is generally used to refer to the CIE STANDARD OBSERVER COLOR MATCHING functions designated $\bar{x} + \bar{y} + \bar{z}$.

color measurement physical measurement of LIGHT RADIATED, TRANSMITTED, or REFLECTED by a specimen under specified conditions, and mathematically transformed into standardized colorimetric terms which can be correlated with visual evaluations of colors relative to one another.[69]

Although the term "color measurement" is normally used, color itself cannot be measured.

color mixture, additive see ADDITIVE COLOR MIXTURE.[69]

color notation orderly system of numbers, letters, or a combination of both, which serves to describe the relationship of colors in three-dimensional space. Thus, three dimensions must be included; for example, HUE, VALUE, and CHROMA of the MUNSELL SYSTEM. Single-dimensional notations, such as a yellowness scale, can be used only if the other two dimensions are fixed or described; two-dimensional notations can be used only if the third is fixed or described.[69] See COLOR ORDER SYSTEMS.[69]

color of an object the aspect of the APPEARANCE of an object dependent upon the spectral composition of the INCIDENT LIGHT, the SPECTRAL REFLECTANCE or TRANSMITTANCE of the object, and the spectral response of the observer. HUE—the attribute by which a perceived color is distinguished as red, yellow, blue, purple, or a combination of these. (White, gray, black colors possess no hue.) LIGHTNESS—the attribute by which a perceived color is judged to be equivalent to a member of the continuous series of grays ranging from black to white. SATURATION—the attribute by which a perceived color is judged to depart from gray of equal lightness toward a pure hue.[43,69,35] (ASTM)

color order systems systems used to describe an orderly three-dimensional arrangement of colors. Three bases can be used for ordering colors: (1) an appearance basis, i.e., a psychological basis; in terms of HUE, SATURATION, and LIGHTNESS—an example is the MUNSELL SYSTEM; (2) an orderly additive color mixture basis, i.e., a psychophysical basis—examples are the CIE SYSTEM and the OSTWALD SYSTEM; and (3) an orderly subtractive color mixture basis—an example is the PLOCHERE COLOR SYSTEM, based on an orderly mixture of INKS.[69,35]

color rendering index method for describing the effect that a particular light source has on the COLOR APPEARANCE of objects, in comparison with the color appearance under a reference light source.[69,128,35]

The Illuminating Engineering Society of the U.S. and CIE Committee 1.3.2 have developed a method. A color rendering index of 100 indicates perfect agreement between the test source and the reference source.

color retention see LIGHTFASTNESS.[43]

color space three-dimensional solid enclosing all possible colors.[69] See COLOR ORDER, SYSTEMS and UNIFORM COLOR SPACE.[69]

The dimensions may be described in various geometries giving rise to various spacings within the solid.

color specification (1) the notation for a COLOR standard using one of the color order systems; or (2) the allowable tolerance of samples from a standard in terms of color difference units.[69]

For (1), the best specification is a sample of the color desired.

color strength in PRINTING INK, the effective concentration of coloring material per volume.[69,168] Syn: tinctorial strength

color temperature the COLOR of a PLANCKIAN (BLACKBODY) RADIATOR.[43,69]

The color is expressed in CHROMATICITY COORDINATES, x and y, determined from theoretical SPECTRAL POWER DISTRIBUTION CURVES, and the temperature is expressed in the ABSOLUTE (KELVIN) SCALE. This term is sometimes used incorrectly to describe the color of "white" light sources. The correct term to describe the color of light sources is "CORRELATED COLOR TEMPERATURE."

color tolerance limit of COLOR DIFFERENCE from a STANDARD which is acceptable.[69]

It is generally expressed in terms of a particular color difference equation, which must be described, and may consist of a total color difference, ΔE, or directional color differences, or both.

color uniformity an APPEARANCE property characterized by the unvarying COLOR of the tinted paint when applied by various techniques.[69,42,35] See TINTABILITY[69,42] and COLOR DEVELOPMENT.[69,43]

color vision, anomalous "anomalous" is used to imply the defective vision of an anomalous TRICHROMAT, an observer who sees the primaries but has weaker than normal response to one.[69,173] See COLOR BLINDNESS.[69,173]

The degree of the anomaly varies from slight to severe.

color vision, defective general term used to describe abnormal vision.[69,173] See COLOR BLINDNESS, and COLOR VISION, ANOMALOUS.[69,173]

This term is preferred to the common term "color blindness."

color vision, normal vision of observers requiring mixtures of three independent primary colors to COLOR MATCH all colors, using quantities of the primaries sufficiently close to those required for the average observer.[69]

The observer with normal vision is referred to as a normal TRICHROMAT.

color wash earth pigments, with or without WHITING, lightly bound in GLUE SIZE so as to facilitate ready removal since frequent removal is necessary, e.g., tinted LIME WASH.[41,71] (BSI)

colorant (1) any substance that imparts COLOR to another material or mixture. Colorants can be either DYES or PIGMENTS;[41,69] (2) concentrated pigment DISPERSIONS (or PASTE) in WATER, SOLVENT, or BINDER (SOLVENT or AQUEOUS). They are generally classified as aqueous, solvent-based, or UNIVERSAL. These are added to coatings during production, in paint stores or on-site to produce the desired colors.[41]

colorant dispenser mechanical device used to dispense precisely measured volume amounts (usually) of COLORANTS for the purpose of TINTING or SHADING coatings BASES.[59]

colorant match COLOR match made by using the same COLORANTS in the match as were used in the STANDARD.[69]

colorant mixture mixture of COLORANTS containing PIGMENTS, or DYESTUFFS, or both. The color of a mixture may be predicted by subtractive colorant mixture theories.[69] See SUBTRACTIVE COLORANT MIXTURE, COMPLEX and SIMPLE.[69]

colored cement a CEMENT to which COLOR PIGMENT has been added.[73] (DAC)

colorfastness see LIGHTFASTNESS.[69]

colorimeter an instrument used for COLOR measurement based on optical comparison with standard colors.[69,145] (IUPAC)

Two general types of colorimeters are commonly used: (1) for measuring concentrations of colored materials for analytical purposes; and (2) for measuring quantities which can be correlated with a psychophysical description of color. Generally, the second type should properly be referred to as a TRISTIMULUS COLORIMETER.

colorimeter, tristimulus see TRISTIMULUS COLORIMETER.[69,43,145]

colorimetric adjective used to refer to measurements converted to psychophysical terms describing COLOR or color relationships.[69]

colorimetric purity see PURITY, COLORIMETRIC.[69]

colorimetry light measurements converted to a psychophysical description or notation which can be correlated with visual evaluations of COLOR and COLOR DIFFERENCES.[69,145,157]

colorist a person skilled in the art of COLOR MATCHING (colorant formulation) and knowledgable concerning the behavior of COLORANTS in a particular material; a TINTER (in the American usage) or a shader.[69]

The word "colorist" is of European origin.

coloristic properties originating in Europe they are derived from the characteristics which a COLORIST sees and describes.[69,43]

coloristically the nature of the comparison of two COLORANTS in terms as used by a COLORIST. It may imply HUE, SATURATION, or STRENGTH or any combination of these, as well as TRANSPARENCY, UNDERTONE, etc.[69]

colour British spelling of "COLOR"; used in English-speaking countries of the old British Empire, including Canada, Australia, New Zealand, etc.[69]

Colour Index (C.I.) a publication of the Society of Dyers and Colourists (Great Britain) and the American Association of Textile Chemists and Colorists which includes periodic additions and amendments.[69,134] See preface under pigment classification.

It provides C.I. generic names for classifying commercial DYES and PIGMENTS with respect to usage and C.I. numbers for classifying them with respect to chemical composition. It gives basic chemical and usage information, performance characteristics, and manufacturers of them.

Colour Index Name consists of the category, HUE, and an identifying number.[69,134] See PREFACE FOR DETAILS.

For example, one PHTHALOCYANINE GREEN PIGMENT is C.I. Pigment Green 7.

Words presented in CAP/SMALL CAPS type indicate that the word is defined in another part of the *Coatings Encyclopedic Dictionary.*

Numerical superscripts classify terms in one or more of the categories listed in the second section of this volume.

Colour Index Number a five digit number which describes the chemical constitution.[69,134] See Table 3 in Preface for example.

For example, PHTHALOCYANINE GREEN, *C.I. Pigment Green 7, is C.I. 74260. (All phthalocyanines are, or will be, numbered 74000-74999.)*

colza oil another name for RAPESEED OIL, usually reserved for refined grades.[63]

comb thin spring-steel tool used in GRAINING or COMBING.[54,145]

combined water the water chemically held, as water of CRYSTALLIZATION, by the calcium sulfate dihydrate or hemihydrate crystal.[134] (ASTM)

combined-oxide formula $3Na_2O{\cdot}2CaO{\cdot}3Al_2O_3{\cdot}2CO_2{\cdot}6SiO_2$, $Na_2O{\cdot}Al_2O_3{\cdot}4SiO_2{\cdot}2H_2O$. A formula which represents the constituents as "OXIDES," with the metallic oxides preceding the ACID ANHYDRIDES, each arranged in the order of increasing valence.[134] (ASTM)

Water of composition appears last. Examples: cancrinite and analcite.

combing act of partially removing a coat of wet paint with combs to imitate the GRAIN of WOOD or other pattern.[54] (BSI)

combustible liquid any liquid having a flash point (flp) at or above 37.8°C (100°F).[172] See FLAMMABLE LIQUID.[172]

Combustible liquids are divided into two classes: Class II and Class III. Class II liquids—Those with an flp at or above 37.8°C (100°F) and below 60°C (140°F), except a mixture having components with flp of 93.3°C (200°F) or higher, the volume of which make up 99% or more of the total volume of the mixture. Class III liquids—Those with flp at or above 60°C (140°F). Class III is subdivided into two classes: Class IIIA—Those with flp at or above 60°C (140°F) and below 93.3°C (200°F), except a mixture having components with flp of 93.3°C (200°F) or higher, the volume of which make up 99% or more of the total volume of the mixture; Class IIIB—Those with flp at or above 93.3°C (200°F). When a combustible liquid is heated for use to within 16.7°C (30°F) of its flp, it is handled in accordance with the requirements for the next lower class of liquids.

comic inks see NEWS INKS.[168]

commercial blast see NACE No. 3.[158]

commercial xylene see XYLENE.[164]

comminution PROCESS by which AGGREGATES are reduced to small size.[131]

common name a designation for a material other than its chemical name, such as a code name, or code number or trade, brand, or generic name.

compacting compacting of a PLASTIC material and the forcing of it through an orifice in more or less continuous fashion.[38,131] See EXTRUSION.[57,151]

compatibility ability of two or more substances mixed with each other to form a stable homogeneous mixture.[161]

complementary colors two colors which, when mixed together in the proper proportions, result in a neutral color.[69]

Colored lights which are complementary, when mixed additively, form white light and follow the laws of additive color mixture. COLORANTS *which are complementary, when mixed together, form black or gray and follow the laws of subtractive colorant mixture. With the exception of* SPECTROPHOTOMETRIC *complementaries, which exist only in theory, colors which are complementary depend on the* ILLUMINANT CHROMATICITY *considered as the neutral point.*

complete hiding see HIDING, COMPLETE.[43,69]

complex dielectric constant vectorial sum of the DIELECTRIC CONSTANT and the loss factor; analogous to complex shear modulus and to complex YOUNG'S MODULUS.[61]

complex shear modulus vectorial sum of the shear modulus and the loss modulus; analogous to complex DIELECTRIC CONSTANT.

complex subtractive colorant mixture see SUBTRACTIVE COLORANT MIXTURE, COMPLEX.[69]

complex Young's modulus vectorial sum of YOUNG'S MODULUS and the loss modulus; analogous to the complex DIELECTRIC CONSTANT.[47]

compliance tensile compliance: the reciprocal of YOUNG'S MODULUS; shear compliance: the reciprocal of SHEAR MODULUS.[47]

compliance coating one whose VOC does not exceed that allowed by regulation. Compliance coatings may be WATERBORNE, low SOLVENT (HIGHER SOLIDS) or POWDER.[71,172] (EPA)

compliance schedule a negotiated agreement between a pollution source and a government agency that specifies dates and procedures by which a source will reduce emissions and, thereby, comply with a regulation.[172] (EPA)

composite homogeneous material created by the synthetic assembly of two or more materials.[133]

composite pigment PIGMENT usually made by the mechanical operation of intermixing or blending two or more pigments.[41]

composition leaf see DUTCH METAL.[148]

composition rollers PRINTING PRESS rollers made primarily of GLUE and GLYCERIN.[168]

composition roofing see BUILT-UP ROOFING.[75,76,82]

composition siding see HARDBOARD.[75,153]

compound (1) a pure, macroscopically homogeneous SUBSTANCE consisting of atoms or ions of two or more different elements that cannot be separated by physical means.[144] *A compound usually has properties unlike those of its constituent elements;* (2) a mixture of a POLYMER with other ingredients such as FILLERS, STABILIZERS, CATALYSTS, PROCESSING AIDS, MODIFIERS, PIGMENTS, or CURING AGENTS.[144] (ASTM)

compounder's modulus STIFFNESS measurement extensively used by RUBBER technologists, expressed as "modulus at 300%" or "300% modulus" (any other percent elongation may be indicated, but 300% is commonly used).[47] See MODULUS OF ELASTICITY.[47]

By this is meant the tensile stress at the indicated ELONGATION.

compregnated wood consolidation of the term "compressed-impregnated wood," referring usually to an assembly of layers of veneer impregnated with a liquid RESIN and bonded under very high pressures.[57,153] Syn: compreg

Comprehensive Environmental Response, Compensation, and Liability Act (CERCLA) The Superfund Law, Public Law PL 96-510 is under EPA jurisdiction. It provides for the identification and the cleanup of the hazardous materials that have been released over the land and into the air, waterways, and groundwater. It covers areas affected by newly released materials and older leaking or abandoned dump sites. Report releases of hazardous materials to the National Response Center, (800) 424-8802. CERCLA established the Superfund, a trust fund to help pay for the cleanup of hazardous material sites. The EPA has authority to collect the cleanup costs from those who release the waste material. Cleanup funds come from fines and penalties, from taxes on chemical/petrochemical feed stocks, and the U.S. Treasury Department. A separate fund collects taxes on active disposal sites to finance their monitoring after they are closed.[171]

CERCLA is a result of the serious problems that arose from the release of hazardous materials at the Love Canal area near Niagara Falls, NY, in August 1978. This law holds all parties that contributed to the landfill in any way, including the waste generators, trucking companies that deposited at the site, and owner of the land, with the financial responsibility of cleaning it up.

compression molding method of forming an object from granular, powdered, or preformed PLASTICS by applying HEAT and PRESSURE to the material in a chamber with an enclosing piston-type ram. Also, the product formed by this method.[57]

compression set sometimes used to mean CREEP.[47,166] See COLD FLOW.[47,166]

compressive strength maximum compressive stress which a material is capable of developing.[47]

concentration (1) amount of a substance expressed in relationship to the whole; (2) act or PROCESS of increasing the amount of a given substance in relationship to the whole;[131] and (3) a method of SOLVENT RECOVERY in which the vaporized solvent is liquefied generally by cooling.[172]

concentration cell ELECTROLYTIC CELL consisting of an ELECTROLYTE and two electrodes of the same metal or alloy that develop a difference in potential as a result of a difference in concentration of ions (most often metal ions) or oxygen at different points in a SOLUTION.[61,66]

concordance an alphabetized index of all the principal words in a book or in all the works of a single author, together with their context.[134] (ASTM)

concrete a composite material which consists essentially of a binding medium within which are embedded particles or fragments of AGGREGATE; in PORTLAND CEMENT CONCRETE, the binder is a mixture of Portland cement and water.[73] (DAC) See BETON.[73] cf. CEMENT[73,75,79]

concrete block a hollow or solid CONCRETE MASONRY unit consisting of PORTLAND CEMENT and suitable AGGREGATES combined with water. (DAC) Sometimes incorrectly called CEMENT BLOCK.[73,75,159] cf. CINDER BLOCK[73,75,159]

LIME, fly ash, air-entraining agents, or other admixtures may be included.

concrete bond plaster see BOND PLASTER, CONCRETE BOND PLASTER.[73,79,82,75]

concrete gun a SPRAY GUN used in applying freshly mixed CONCRETE; compressed air forces the concrete along a flexible hose and through a nozzle.[73,59]

condensate product obtained by cooling a vapor, such that it is converted either to a liquid or a solid.[131]

condensation (1) the physical PROCESS of converting a material from the gaseous or vapor phase to a liquid or solid phase; this commonly results when the temperature is lowered and/or the vapor pressure of the material is increased.[131] *The tendency exists for CONDENSATION to occur when*

Words presented in CAP/SMALL CAPS type indicate that the word is defined in another part of the *Coatings Encyclopedic Dictionary.*
Numerical superscripts classify terms in one or more of the categories listed in the second section of this volume.

the partial pressure of a given component of a gaseous mixture at a given temperature exceeds the vapor pressure of the liquid or solid form of that component at the given temperature; (2) chemical reaction in which two or more molecules combine with the separation of water or some other simple substance. If a POLYMER is formed, the process is called POLYCONDENSATION.[169] (IUPAC)

condensation polymer a POLYMER made by CONDENSATION POLYMERIZATION.[169,130] (ASTM) cf. ADDITION POLYMER[169,130]

condensation polymerization POLYMERIZATION in which MONOMERS are linked together with the splitting off of water or other simple molecules.[130,169] (ASTM)

conditional match a pair of COLORS which appear to match only under limited conditions, such as under a particular light source and a particular observer; a METAMERIC MATCH.[69,157,35,43]

conductance the reciprocal of RESISTANCE.[61]

conductivity the current transferred across per unit potential gradient. K = amperes/cm^2 divided by volts/cm. Reciprocal of RESISTIVITY.[61,128] cf. SPECIFIC RESISTANCE (RESISTIVITY)[61] Syn: specific conductance

cone mill old type of mill used for dispersing PIGMENTS in media.[59,131,64]

They usually have conical hoppers, and the grinding surfaces are also inclined like the surface of a cone.

congeal to change from a liquid or soft state to a solid or rigid state.[63,133,169]

Congo copal most important of the COPALS used for OIL varnishes.[167]

It originates in the Belgian Congo (now Zaire). It becomes soluble in VEGETABLE OILS after RUNNING.

Congo gum fossilized GUM RESIN obtained from the Congo region of Africa (now Zaire).[167]

Used mostly in manufacturing VARNISHES.

Congocopalic acid see BENGUCOPALIC ACID.[167]

Congolene bicyclic HYDROCARBON, obtained from the oily layer resulting from the running of CONGO COPAL.[167]

coniferous wood WOOD produced by coniferous trees; same as softwood or nonporous wood.[153]

conjugated double bond –CH = CH–CH = CH–. Two pairs of carbon atoms joined by double bonds unite the arrangement.[133]

consensus standard a standard developed according to a consensus agreement or general opinion among representatives of various interested or affected organizations and individuals.[157]

consistency the property of a material or composition which is evidenced by its RESISTANCE to FLOW, represented by an undefined composite of properties, each measurable from the complete, force-rate flow curve as PLASTIC VISCOSITY, YIELD VALUE, and THIXOTROPY.[166] cf. VISCOSITY[166]

The term is applied to a variety of materials. For NEWTONIAN LIQUIDS, consistency is simply viscosity. While "consistency" is an accepted RHEOLOGICAL term, it has qualitative meaning only, and is used with qualifying adjectives as "buttery," "thin," "high," etc., in describing plastic flow. Usually measured in an empirical manner and in arbitrary units. Often used incorrectly as a synonym for "viscosity."

constant white alternative name for BLANC FIXE.[58]

constitutional formula device used to illustrate the composition of a chemical COMPOUND by displaying the individual atoms and radicals, joined together by valency linkages.[133,134]

container lining protective coating applied to the inner walls of a container to prevent interaction between the contents and the material of construction of the container or to make the walls of the container impervious to the contents.[50,71]

continuous phase the medium or continuum in which the dispersed phase is contained.[139] Syn: external phase. cf. DISCONTINUOUS PHASE[139]

contrast ratio (1) ratio of the REFLECTANCE of a dry paint film over a black substrate of 5% or less reflectance, to the reflectance of the same paint, equivalently applied and dried, over a substrate of 80% reflectance; (ASTM) (2) the ratio of the LUMINOUS REFLECTANCE, Y, measured on a film over a black substrate, to the Y measured on the same film over a white substrate.[43,69,157]

The Y of the white and black must be specified. The contrast ratio will vary depending on the thickness of the film and on the CONCENTRATIONS OF COLORANTS.

ASTM Test Method D2805 covers hiding power by reflectometry.

control in the AIR POLLUTION field, this means the abatement of pollutants which might be exhausted into the atmosphere.[172] (EPA)

It often refers to the collection or destruction efficiency of an add-on device such as an incinerator or carbon adsorber as opposed to capture of the pollutants into the device.

control device any equipment which reduces the quantity of pollutant that is emitted to the air. The device may destroy or secure the pollutant for subsequent recovery.[172] (EPA)

Examples are incinerators, carbon adsorbers, condensers.

control device efficiency the ratio of the pollution released by a control device and the pollution introduced to the control device expressed as a percentage.[172] (EPA)

control technique guidelines (CTG) a series of documents prepared by EPA to assist states in defining reasonable available control technology (RACT) for major sources of VOC.[172] (EPA)

The documents provide information on the economic and technological feasibility of available techniques; and, in some cases, suggest limits on VOC emissions.

convection (as applied to air motion) vertical motion of the air induced by the expansion of the air heated by the air's surface and its resulting buoyancy.[171,172] (IUPAC)

conventional base unit of a polymer base unit, defined without regard to steric ISOMERISM.[130,169] (IUPAC)

conversion coating see CHEMICAL CONVERSION COATING.[45,66]

conversion varnish a short oil alkyd RESIN formulated with a high excess hydroxyl content and blended with an AMINO RESIN. It is cured by addition of a CATALYST producing a hard durable clean finish for FURNITURE and cabinetry at ambient temperature.[152,175]

convertible coating irreversible transformation of a coating after its film formation to a film insoluble in the solvent from which it was deposited.[55,71] cf. NONCONVERTIBLE COATING.[55,71] This should not be confused with chemical conversion coating.

This can be effected by OXIDATION, THERMAL CROSSLINKING, or CATALYTIC CURING.

copal esters these are normally regarded as the GLYCERIDE ESTERS of RUN CONGO COPAL, although the term can also be applied to esters of other copals.[167]

copal oil another name for the condensed FUMES obtained during the RUNNING of natural COPALS.[167]

copal varnish a HIGH-GLOSS VARNISH made with a DRYING OIL, such as LINSEED OIL, and COPAL.[152,156] (DAC)

copals GUM RESINS exuded from living plants and fossilized in the ground. Some of the copals are: Zanzibar, amber, kauri, Manila, and CONGO.[167]

coping a protective cap, top, or cover of a WALL, PARAPET, PILASTER, or CHIMNEY; often of stone, TERRA-COTTA, CONCRETE, or WOOD.[82] (DAC)

It is usually sloped to carry off water.

copolymer a POLYMER consisting of molecules containing large numbers of units of two or more chemically different types in irregular sequence.[130,169] (IUPAC) See BIPOLYMER and TERPOLYMER.[130]

If the sequence is unknown, it is assumed to be irregular.

copolymerization POLYMERIZATION in which a COPOLYMER is formed.[130,131,169] (IUPAC)

copper acetoarsenite synonym for Paris green, emerald green, and Schweinfurt green.[41]

copper blue CuS. Synonym for cupric sulfide (copper (II) sulfide).[41] Syn: covellite

copper bronze a range of metallic powders ranging in COLOR from "pale gold" to the deep "coin bronze."[41] See COPPER POWDER.[41]

They consist generally of alloys of copper, but dyed aluminum powders of similar shades are also used.

copper carbonate $CuCO_3$. Material frequently used as a poison in antifouling paints.[83]

copper driers (deprecated) Copper salts of ACIDS generally used for DRIERS, such as NAPHTHENIC and 2-ethyl hexoic, which are used as FUNGICIDES and preservative ADDITIVES in coatings, but do not function as driers.[83]

They are deep green and cannot be used in whites or light colors.

copper powder METALLIC COPPER and copper bronze powders have some application in ANTIFOULING paints, and in special PRIMERS.[41]

Copper bronzes of various colors are produced according to the amount of alloy metal present. In producing finishes from copper and copper bronze powders, great care in formulation is necessary, for these pigments are apt to cause inhibition of DRYING, DISCOLORATION OF MEDIA, and GELATION.

ASTM Test Method D283 covers the analysis of copper powder and ASTM D964, the specification for copper powder.

copper resinate green COMPOUND formed by dissolving copper acetate, VERDIGRIS or other copper salt in Venice turpentine, balsam, or similar resinous solutions.[83]

copper soap combination of copper with a FATTY ACID.[83,127]

copper staining usually caused by CORROSION of copper screens, gutters, or downspouts washing down on painted surfaces.[42]

Words presented in CAP/SMALL CAPS type indicate that the word is defined in another part of the *Coatings Encyclopedic Dictionary.*

Numerical superscripts classify terms in one or more of the categories listed in the second section of this volume.

copper sub-oxide Cu_2O. Synonym for CUPROUS OXIDE [copper (I) oxide].[41,83]

copper sulfate $CuSO_4 \cdot 5H_2O$. Used occasionally in PIGMENT manufacture, e.g., para brown.[41] Syn: blue vitriol and blue stone

copperas see FERROUS SULFATE.[41]

corduroy in WALLPAPER, a very narrow stripe imitating the fabric.[148]

The name is derived from the French "corde du roi" or king's cord.

cork bark of the cork oak, *Quercus suber*, which is indigenous to those countries that form the shores of the Western Mediterranean.[50,83]

cork dust very finely divided CORK which is used in ANTICONDENSATION paints.[83]

corn oil OIL obtained from the kernels of Indian corn, maize, Zea mays.[63] (PTM)

It is SEMIDRYING, lying between COTTONSEED and SOYBEAN OILS.

cornice (1) a horizontal molding or combination of moldings to FINISH the top of a wall; (2) PAPER simulating the same.[148,82]

corona CRACKS on the surface caused by ozone, usually appearing at right angles to a stress, CHECKING or cutting.[65]

corona discharge (1) FLOW of electrical energy from a conductor at high potential to the surrounding media; (2) CURING method involving the bombardment of organic vapors with high-energy electrons at very low pressures in contact with the substrate to be coated.[61,67] Syn: glow discharge

corona, internal in electrical cable, fault due to ionization of air between conductor and insulation.[61]

corona resistance ability of a material to withstand the effects of corona discharge.[61]

correlated different types of merchandise systematically related in COLOR and design, as WALLPAPER with a fabric, or a group of PAPERS designed to be used together.[148]

correlated color temperature the COLOR of "white" light sources. Specifically, it is the temperature of the PLANCKIAN (BLACK-BODY) RADIATOR which produces the CHROMATICITY most similar to that produced by the light source in question.[69,43] See COLOR TEMPERATURE.[69,43]

The temperature is expressed in degrees on the ABSOLUTE or KELVIN SCALE, or in MIREDS (micro-reciprocal degrees), $10^6/T$.

correlation relationship, degree of association, or index of prediction between two scores or sets of data. Measures the tendency of one score or set of data to vary concomitantly with the other (e.g., the tendency of students with high IQ to be above average in reading ability).[157,36]

The existence of a strong relationship—i.e., a high correlation—between two variables does not necessarily indicate that one has any causal influence on the other. Usually expressed as a decimal coefficient between -1.00 and +1.00 (Pearson v), where -1.00 indicates a perfect negative relationship, 0 indicates no relationship, and +1.00 indicates a perfect positive relationship. The Pearson v coefficient of correlation can assume any value on a continuum between -1.00 and +1.00.

corrosion the DETERIORATION of METAL or of CONCRETE by chemical or electrochemical reaction resulting from EXPOSURE to WEATHERING, MOISTURE, CHEMICALS, or other agents in the environment in which it is placed.[66] (DAC)

corrosion coating see CHEMICAL CONVERSION COATING.[158]

corrosion fatigue reduction by CORROSION of the ability of a METAL to withstand cyclic or repeated STRESSES.[66]

corrosion inhibitor any of a number of materials used to prevent the OXIDATION of METALS; may be a coating applied to the surface, a paint UNDERCOAT, an ADDITIVE, or an element alloyed with the metal.[66,83] (DAC)

corrosion potential potential that a freely corroding METAL or ALLOY exhibits in a particular solution.[66]

corrosion rate speed at which a METAL or ALLOY is wasted away because of CORROSION.[66]

It is expressed in inches of steel per year; accompanied by temperature.

corrosion resistance ability of a SUBSTANCE to resist DETERIORATION because of a reaction with its environment.[66]

corrosion-inhibiting paint see ANTICORROSIVE PAINT or COMPOSITION.[66,71,72,163]

corrosion-inhibitive pigment a PIGMENT which when made into a paint has the property of minimizing CORROSION of the SUBSTRATE to which it is applied.[41,66]

corrosive (1) having the capability or tendency to cause CORROSION: a corrosive acid;[66] (2) a chemical that causes visible destruction of or irreversible alterations in living tissue by chemical action at the site of contact; (3) a waste that exhibits a "characteristic of corrosivity (40 CFR 261.22)," as defined by RCRA, may be regulated by EPA as a hazardous waste.[172]

corundum natural ALUMINUM OXIDE generally used for ABRASIVES because of its HARDNESS.[158]

cosanic acids a series of FATTY ACIDS beginning with eicosanic acid ($C_{20}H_{40}O_2$) and ending with nonacosanic acid ($C_{29}H_{58}O_2$), each acid differing from its predecessor by one CH_2 group.[134]

cosolvent (1) a nonsolvent which becomes a good solvent on the addition of a relatively small amount of another solvent; (2) a SOLVENT which will cause two immiscible liquids to mix; (3) a functional additive for LATEX paints (see below).[164,83] Syn: mutual solvent coupling agent, latent solvent

They are frequently used in latex paints to prevent freezing, slow down drying rates by extending OPEN TIME, aid in latex coalescence, help wet substrates, and since the performance of ASSOCIATIVE THICKENERS is affected by cosolvents, they may be used to finetune the rheology of paints thickened with associative thickeners.

cotton-free see DUST-FREE.[67,157]

cotton-free dry see DUST-FREE.[67,157]

cottonseed oil a SEMIDRYING OIL obtained from the seeds of many types of plants of the genus, *Gossypium*. Its main constituent acids are LINOLEIC (46%), PALMITIC (29%), and oleic (24%).[63] See BECCHI-MILLIAN TEST.[157]

As oil, it is rarely used in paint, but its FATTY ACIDS are used in the manufacture of ALKYD RESINS.

cottonseed pitch see STEARIN PITCH.[63]

coulomb the metre-kilogram-second unit of electrical charge equal to the quantity of charge transferred in one second by a steady current of one ampere.[61,128]

coulometry analytical method for determining the amount of substance released during electrolysis in which the number of coulombs used is measured.[37,61]

coumarone C_8H_6O. Bicyclic ring COMPOUND.[141,167] See COUMARONE-INDENE RESINS.[167] Syn: benzofuran, and cumarone[141,169]

Parent substance for the coumarone resins. Properties: colorless liquid, aromatic odor; sp gr of 1.078; mp of –18°C (–0.4°F); bp of 177°C (351°F); insoluble in water; soluble in alcohol and ether; derived from the coal tar naphtha fraction boiling between 150-200°C (302-392°F).

coumarone-indene resins COAL TAR RESINS; INDENE resins; POLYCOUMARONE resins; POLYINDENE resins. Resins obtained by heating mixtures of coumarone and indene (such as those that occur in the light-oil fraction from coal-tar refining) with sulfuric acid, so as to cause POLYMERIZATION to THERMOPLASTIC materials with softening points of up to about 150°C (302°F).[167]

Properties: these vary from fairly viscous liquids to hard resins; color—pale yellow to nearly black; sp gr of 1.05-1.10; soluble in HYDROCARBON solvents, pyridine, acetone, carbon disulfide, and carbon tetrachloride; insoluble in water and alcohol. Used as components in aluminum paints, concrete curing compounds, pipe oils, rubber compounding, adhesives, chewing gum, printing inks, floor tile binding, and phonograph records.

counter-current PROCESS used in many industries in which material to be DRIED or EXTRACTED is caused to FLOW against the stream of drying agent or extracting liquid.[131]

A typical example is the tunnel drying of pigments. Here the wet filter PRESS CAKES and the stream of hot air enter the tunnel from opposite ends. The driest air meets the practically dry pigment so that all MOISTURE is removed. By the time the air has reached the other end, contact with the wet pigment insures that the maximum amount of moisture has been removed by the air stream.

counterflashing formed METAL or ELASTOMERIC sheeting secured on or into the wall, curb, pipe, rooftop unit, or other surface, to cover and protect the upper edge of a base flashing and its associated fasteners.[82,76] (ASTM) See FLASHING (4).[82]

country tar see PINE TAR, KILN BURNED.[44]

couple two or more dissimilar METALS or ALLOYS in electrical contact with each other so that they can act as the electrodes of a cell if they are immersed in an ELECTROLYTE.[61]

coupling the reaction which takes place between DIAZOTIZED AMINES on the one hand and PHENOLIC or AMINO COMPOUNDS on the other to give the COLORED AZO DYES.[41]

coupling agent SOLVENT which will cause two IMMISCIBLE liquids to mix.[83,164] See COSOLVENT.[83,164]

cove ceiling a ceiling which is rounded where it meets the wall.[148]

covellite see COPPER BLUE.[41]

coverage (1) spreading rate generally expressed in ft^3/gal or $metres^2$/litre. In pigmented coatings, it is related to HIDING POWER. In clear coatings, it refers to the area coated at a desired film THICKNESS; (2) description of the amount of paint per unit area which must be applied to achieve a specified con-

Words presented in CAP/SMALL CAPS type indicate that the word is defined in another part of the *Coatings Encyclopedic Dictionary.*

Numerical superscripts classify terms in one or more of the categories listed in the second section of this volume.

trast ratio;[43,69,128] (3) the surface area covered by a given quantity of INK or coating material.[128,168] See HIDING POWER.[43,69,157]

coverage rate is an ambiguous term that is used in some instances to refer to HIDING POWER and others to mean SPREADING RATE.[43]

The precise terms hiding power and spreading rate are preferred.

covering power (1) used occasionally in the paint industry as synonymous with HIDING POWER but which actually has no precise meaning;[43] (2) the ability of an INK to hide the material beneath (substrate) and to produce a uniform, opaque surface.[168]

cowtail small marks of paint applied to WOOD FURNITURE to give an antique effect.[54,175] (EPA)

CPVC abbreviation for CRITICAL PIGMENT VOLUME CONCENTRATION.[41]

cracking (1) generally, the SPLITTING of a dry paint or VARNISH film, usually as a result of AGING. The following terms are used to denote the nature and extent of this defect: HAIR-CRACKING: fine cracks which do not penetrate the TOPCOAT; they occur erratically and at random. CHECKING: fine cracks which do not penetrate the topcoat and are distributed over the surface, giving the semblance of a small pattern. CRACKING: specifically, a breakdown in which the cracks penetrate at least one coat and which may be expected to result ultimately in complete failure. See CRAQUELURE.[81] CRAZING: resembles checking, but the cracks are deeper and broader. CROCODILING OR ALLIGATORING: a drastic type of crazing, producing a pattern resembling the hide of a crocodile;[42,56,65](BSI) (2) the PROCESS of breaking down certain HYDROCARBONS into simpler ones of lower boiling points, by means of excess heat, distillation under pressure, etc., in order to give a greater yield of low boiling products than could be obtained by simple distillation;[56,131,164,42] (3) cracking is also the treatment of RUBBER, uncured and cured, by passing it through moving corrugated rolls, as in preparing tires and other vulcanized rubber for reclaiming.[154,140]

ASTM test methods for cracking (cracking resistance) include: evaluating exterior paints, D661; mandrel bend test, D522.

crack resistance the ability of a coating to resist breaks of the film where the breaks extend through to the surface painted and the underlying surface is visible.[56,72] (ASTM) cf. CHECKING RESISTANCE[154,140]

The use of a minimum magnification of 10 diameters is recommended in cases where it is difficult to differentiate between cracking and checking. The FSCT Pictorial Standards of Coatings Defects contains photographic standards for the degree of cracking of exterior paints.

ASTM Test Method D661 also has these same photographic reference standards. ASTM test methods for cracking (cracking resistance) include: evaluating exterior paints, D661; mandrel bend test, D522.

crackle finish APPEARANCE resulting from applying a TOPCOAT designed to shrink and crack and expose a more flexible UNDERCOAT, usually of a different COLOR.[54,35]

crackle varnish clear protective TOPCOAT applied over a crackle FINISH.[54]

craquelure the network of hairline cracks in old VARNISHES and paintings. (It is French for CRACKING.)[81]

Craquelure is commonly faked on intentionally deceptive reproductions.

cratering the formation in a wet coating film of small bowl-shaped depressions that persist after drying.[56]

craters small bowl-shaped depressions frequently having drops or bands of material at their centers and raised circular edges in a coating film.[56] See CRATERING.

crawl shrinkage of milled and calendered stock after removal from rolls.[140]

crawl space (1) a shallow, unfinished space beneath the first floor of a house which has no basement, used for visual inspection and access to pipes and ducts; (2) a shallow space in the attic, immediately under the roof.[82]

crawling (1) DEFECT in which a wet paint or VARNISH film recedes from small areas of the surface leaving them uncoated;[42,55,56] British Syn: CISSING; (2) the contraction of an INK film into drops after printing on a surface which the ink does not wet completely.[168] Syn: retraction

This occurs when the surface energy of the coating or ink to be applied is higher than the substrate to be coated or printed.

crayon a small stick of PIGMENT in OIL or WAX.[81]

Usually covered with paper for ease of handling. Certain crayons contain water-soluble dyes and are prepared in an aqueous medium.

crazing fine, random cracks or fissures in a network on or under a surface of PLASTER, CEMENT, MORTAR, CONCRETE, CERAMIC coating, or paint film; caused by SHRINKAGE.[56] (DAC) Syn: cracking and craze cracks

creaming of emulsion separation of an emulsion into two layers of different concentration.[60]

The more concentrated top layer, i.e., the layer containing the greatest number of dispersed droplets per volume, has a creamy appearance. Gentle agitation of the two layers often effects uniformity of the emulsion.

creep dimensional change with time of a material under load, following the initial instantaneous ELASTIC or rapid DEFORMATION.[47,166,140] (ASTM) Syn: drift, COLD FLOW, compression set, tension set, and strain relaxation

Creep at room temprature is sometimes called cold flow. Deformation occurring with lapse of time, in both cured and uncured rubber, in a body under stress in addition to the immediate elastic deformation.

creeping spontaneous SPREADING of a liquid on a SURFACE. In the case of an applied film of paint, VARNISH, or LACQUER it refers to the spread of the wet film beyond the area to which it was applied.[56] (BSI)

cremnitz white see CARBONATE WHITE LEAD.[41]

creosote heavy, high-boiling oil obtained from COAL and WOOD TARS.[76] Syn: dead oil, pitch oil

A major use for creosote is as a wood preservative, or as a BASE for the same.

cresol three cresols are possible, namely: (a) o-cresol, mp, 30°C; bp, 191°C; (b) m-cresol, mp, 4°C; bp, 205°C; (c) p-cresol, mp, 36°C; bp, 201°C, and these are found together in the crude cresylic acid from COAL TAR.[76] Syn: cresylic acid

The main use of the cresols is in the manufacture of cresol-FORMALDEHYDE resins, and cresylic acid, rich in the meta isomer, is usually chosen for this purpose.

cresylic acid see CRESOL.[76]

crevice corrosion CORROSION which occurs within or adjacent to a crevice formed by contact with another piece of the same or another METAL or with a nonmetallic material.[66]

When this occurs, the intensity of attack is usually more severe than on surrounding areas of the same surface.

cricket a relatively small, elevated area of a roof constructed to divert water from a horizontal intersection of the roof with a chimney, wall, expansion joint, or other projection.[82] (ASTM)

crimson antimony Syn: antimony vermillion[41]

crimson lake PIGMENT derived from COCHINEAL by PRECIPITATING the extract with ALUMINUM and TIN SALTS. It has a deep blue-red COLOR.[41] Syn: florentine lake

crimson toner AZO PIGMENT produced by coupling 4-aminotoluene-3-sulfonic acid with ß-oxynaphthoic acid, followed by conversion to the calcium LAKE.[41]

crinkle finish Syn: ripple finish[163]

crinkling see WRINKLING.[54,56,42]

criteria pollutant a pollutant for which a criteria document has been issued as described by Section 108 of the Clean Air Act.[172] (EPA)

Criteria pollutants are nitrogen oxides, sulfur oxides, ozone, particulate matter, and carbon monoxide. A Natural Ambient Air Quality Standards (NAAQS) exists for each criteria pollutant.

critical micelle concentration (CMC) the point beyond which the CONCENTRATION of single molecules of surfactants remains relatively constant.[127]

Much more surfactant may be dissolved to produce clear solutions, but the added increments form micelles in the solution instead of appearing as individual molecules.

critical pigment volume concentration (CPVC) that level of PIGMENTATION, PVC VALUE in the dry paint, where just sufficient BINDER is present to fill the voids between the PIGMENT PARTICLES.[41]

At this level, a sharp break occurs in film properties such as SCRUB RESISTANCE, HIDING, CORROSION RESISTANCE, ease of stain removal, etc. Different requirements for each product would dictate different PVC or CPVC ratios. Ceiling paints, for instance, are not required to be very washable and can be formulated at or above CPVC, whereas gloss paints and many exterior formulations are designed well below their CPVC, where CPVC has no significance. CPVC has significance only in flat paints.

critical point the temperature at which the liquid and vapor intensive properties (density, heat capacity, etc.) become equal. It is the highest temperature (critical temperature) and pressure (critical pressure) at which both a gaseous and a liquid phase of a given compound can coexist.[142] (IUPAC)

crocidolite blue ASBESTOS (mineral riebeckite).[41,34]

crocking (1) removal of COLOR on ABRASION or RUBBING;[56,42] (2) smudging, or rubbing off, of INK;[168] (3) color PIGMENT in RUBBER which may not appear on the surface as a bloom but which will rub off and discolor an adjacent surface; (4) staining of a white cloth by rubbing lightly over a colored surface.[56,140]

crocodiling see CRACKING.[56] (BSI)

crocoisite, crocolite, or crocoite native lead chromate mineral of red color.[41,34]

crocus abrasive either SYNTHETIC or NATURAL IRON OXIDE, crocus is the basis of the rouge used in many fine polishing and buffing operations.[158]

It is very soft, approximately 6 on the Mohs scale, bright red, and contains a small amount of silicon dioxide.

Words presented in CAP/SMALL CAPS type indicate that the word is defined in another part of the *Coatings Encyclopedic Dictionary.*

Numerical superscripts classify terms in one or more of the categories listed in the second section of this volume.

crocus cloth an IRON-OXIDE coated ABRASIVE cloth, used as a polishing agent after most of the work has been done with emery or aluminum oxide, capable of giving a mirror-like FINISH to metal.[59,158]

cross coating APPLICATION of a coat of paint by a series of strokes or spray passes, each at right angles to the previous series.[54] cf. crossing[77]

cross grain a pattern in which the FIBERS and other longitudinal elements deviate from a line parallel to the sides of the piece.[153] (ASTM) See DIAGONAL GRAIN.[153]

Applies to either diagonal or spiral grain or a combination of the two.

cross-cut (cross-hatch) tape test (1) an ADHESION TEST in which a cross-cut (X-cut) is made in a paint film to the SUBSTRATE, pressure-sensitive tape is applied over the cut and then removed, and adhesion is assessed qualitatively on a 0 to 5 scale; (2) a lattice pattern with either six or 11 cuts in each direction is made in the film to the substrate, pressure-sensitive tape is applied over the lattice and then removed, and the adhesion is evaluated by comparison with descriptions and illustrations.[79,157,128]

ASTM Test Method D3359 covers these two methods for assessing the adhesion of coating films to metallic substrates by applying and removing pressure-sensitive tape over cuts made in the film.

Test method (1) is primarily intended for use at job sites while test method (2) is more suitable for use in the laboratory. Also, test method (2) is not considered suitable for films thicker than 5 mils (125 μm).

crossing method of obtaining even distribution of paint by means of a BRUSH whereby the direction of brushing each series of strokes lies at right angles to that of the previous series.[77] (BSI)

crosslinking applied to POLYMER molecules, the setting up of chemical links between the molecular chains to form a three-dimensional or network polymer generally by COVALENT BONDING.[67,169]

When extensive, as in most THERMOSETTING resins, crosslinking makes one infusible larger molecule of all the linked chains. Crosslinking generally toughens and stiffens coatings. Thermosetting materials crosslink under the influence of heat and CATALYSIS. It may also be inducted by ELECTRON CURING.

crosslinking agent (or crosslinker) substance which will react chemically with the molecular chains of a THERMOPLASTIC material, and by linking them together create a more rigid structure resulting in a more or less infusible product.[67,74,83] (BSI)

crowsfooting (1) type of film DEFECT where small wrinkles occur in a pattern resembling that of a crow's foot; (BSI) See WRINKLING.[54,52,42] (2) type of CRYSTALLIZATION on the surface of a varnish or paint film after EXPOSURE to the gas test.[56] See GAS CHECKING.[56,63,157]

crude resinous liquid see TALL OIL.[63]

cryptometer instrument for measuring the HIDING POWER or OPACITY of a pigmented composition.[145,157,43]

It functions on the principle that a film or wedge of the paint is obtained with a uniformly varying thickness. By adjusting the instrument, the thickness of film necessary to obliterate can be determined.

crystalline silica see SILICA, CRYSTALLINE.[41]

crystallization (1) a condition in which a dried INK film repels a second ink which must be printed on top of it;[168] (2) LACQUERS containing materials which CRYSTALLIZE out from the medium as the solvent evaporates;[151] (3) materials in which advantage is taken of the tendency of certain DRYING OILS, notably TUNG OIL, to "FROST" or "crystallize" when dried under certain conditions.[63,148]

CSI see COATINGS SOCIETIES INTERNATIONAL.[174]

CSMA abbreviation for Chemical Specialties Manufacturers Association.[174]

C-stage final stage in the reaction of certain thermosetting resins in which the material is relatively insoluble and infusible.[67,79] (ASTM) See A-STAGE, and B-STAGE.[67,79]

Certain thermosetting resins in a fully cured adhesive layer are in this stage. Sometimes referred to as resite.

CTG see CONTROL TECHNIQUE GUIDELINE.

cube root color difference equation Specific COLOR DIFFERENCE EQUATION. The following equations are those recommended by the CIE in 1967 for study:

$L = 25.29G^{1/3} - 18.38$

where $G = 0.0010X + 1.005Y + 0.0004Z$

$a = K_a(R^{1/3} - G^{1/3})$

where $R = 1.1084X + 0.0852Y - 0.1454Z$

and $K_a = 105.0$ for $R<G$; $K_a = 125.0$ for $R>G$

$b = K_b(G^{1/3} - B^{1/3})$

where $B = -0.0062X + 0.0394Y + 0.8192Z$

and $K_b = 30.5$ for $B<G$; $K_b = 53.6$ for $B>A$

$\Delta E = [(\Delta L)^2 + (\Delta a)^2 + (\Delta b)^2]^{1/2}$

The differences (Δ's) are calculated as sample minus standard.[69]

culture (1) PROCESS of securing the growth of FUNGI or other microorganisms upon artificial media; (2) the organisms resulting from the culturing process.[173]

cumar gum a SYNTHETIC RESIN, used in VARNISHES to provide ALKALI-RESISTANT properties.[130] (DAC)

cumarone see COUMARONE.[141]

cumene $C_6H_5 \cdot C_3H_7$. A HYDROCARBON of good SOLVENT power. [164] Syn: isopropyl benzene

Properties: bp, 153°C; sp gr, 0.862/20°C; refractive index, 1.506.

cumulative sample a sample collected over an extended period of time.[157] (IUPAC)

cup efflux cup.[145] See FORD CUP.[59,157,166]

cup, weight per gallon brass cup with a volume of exactly 83.3 ml used for quickly determining the weight per gal (or DENSITY) of a finished liquid product.[145] cf. PYCNOMETER[145]

cupric sulfide see COPPER BLUE.[41]

cuprous (copper) oxide Cu_2O. Occurs in nature as the mineral, cuprite.[34,41]

Prepared commercially by furnace-reduction of a mixture of cupric oxide and copper. Used as a toxin in ANTIFOULING compositions.

curcas oil a VEGETABLE OIL obtained from the SEEDS of *Jatropha curcas*, grown in Central America, and in the Cape Verde Isles, the Comores Isles, and in some of the former Portuguese colonies, Siam and the East Indies. [63]

The percentage fatty acids as triglycerides are oleic, 62%; linoleic, 19%, and saturated fatty acids, the remainder.

curdle See COAGULATION, and CONGEAL.[161] Syn: coagulation

cure (1) to change the properties of a POLYMERIC system into a final, more stable, usable condition by the use of HEAT, RADIATION, or reaction with chemical ADDITIVES;[67] (ASTM) (2) synonymous with VULCANIZE. It includes time and temperature of VULCANIZATION.[67] (ASTM)

cure cycle the schedule of time periods at specified conditions to which a reacting THERMOSETTING PLASTIC or RUBBER composition is subjected to reach a specified property level.[67] (ASTM)

cure stroke time the time required for a RESIN being held at a specified temperature on a hot plate and being stroked with a spatula, to adhere to the hot plate and not the spatula.[67] (ASTM) cf. CURE TIME[67]

ASTM Test Method D4640 covers the determination of the cure rate at any specified temperature for a thermosetting phenol-formaldehyde resin. The time at that temperature required to make a fluid mixture change into a hard immovable mass indicates the speed of cure.

cure time the period of time that a reacting THERMOSETTING PLASTIC is exposed to specific conditions to reach a specified property level.[67] (ASTM)

cure volatiles REACTION products which are emitted during the chemical reaction which takes place in some coating films at the cure temperature. These emissions are other than those from the solvents in the coating.[67] (EPA)

curing the toughening or HARDENING of a coating film as a result of elevated temperature or significant reaction other than simple oxidation.[67]

curing agent ADDITIVE which promotes the curing of a film.[67,74,83] (BSI) See CATALYST,[74] and CROSSLINKING.[67] Syn: hardener

curing agent blush a BLUSHING, BLOOMING, or SWEATING caused by applying coatings such as amine cured epoxies under conditions of high humidity.[42,56]

curing oven a heating chamber where wet paint solidifies either by drying or by chemical CROSSLINKING of the paint film.[59,67] (EPA)

curing temperature see TEMPERATURE, CURING.[67]

curing time time necessary for curing.[67]

curling (1) see COILING SOUP;[42,56] (2) curling also refers to excessive warping of sheet goods, or distortion by uneven shrinkage.[55,56] See CRAWL.[140]

current density magnitude of current per unit area of a metal surface, usually expressed in millampere per sq ft (mA/ft^2 or mA/m^2).[61]

curtain coating method of APPLYING paint to an object by moving it through a failing curtain of paint.[77] (BSI)

curtaining See SAG or SAGGING.[56]

custom color special colors made by adding COLORANT to paint or by intermixing colors, which permits the retailer to match a color selected by the consumer.[41]

cut (1) to dilute an INK, LACQUER, or VARNISH with SOLVENTS or with clear BASE; to thin;[164] (2) a TYPOGRAPHIC PRINTING PLATE;[168] (3) the proportion of SHELLAC GUM in ALCOHOL. A shellac varnish is referred to as a four pound or five pound cut, or weight, which means 4 lbs or 5 lbs of shellac gum to 1 gal of alcohol.[146]

cutback SOLVENT-THINNED BITUMEN used in COLD-PROCESS ROOFING ADHESIVES, flashing cements, and roof coatings.[76] (ASTM)

Words presented in CAP/SMALL CAPS type indicate that the word is defined in another part of the *Coatings Encyclopedic Dictionary.*

Numerical superscripts classify terms in one or more of the categories listed in the second section of this volume.

cutting in painting of a SURFACE adjacent to another surface which must not be painted, for example, painting the frames of a window and avoiding painting the GLASS, or the painting of an area on a previously painted surface.[54] (BSI)

CVD abbreviation for CHEMICAL VAPOR DEPOSITION.[45]

cyclic all cyclic COMPOUNDS are recognized by their ring structure.[134]

Benzene is the simplest and most well-known cyclic compound, consisting of a single ring. It is monocyclic. When two or more rings are involved as with naphthalene, for example, the compound is bicyclic, tricyclic, etc. Cyclic compounds may contain from three to six or more carbon atoms in the ring, e.g., cyclopropane . . . 3 atoms; cyclobutane . . . 4 atoms; cyclopentane . . . 5 atoms; cyclohexane . . . 6 atoms.

cyclic diolefin resins POLYMERIZED products obtained from the cracked DISTILLATES of PETROLEUM.[130]

They are UNSATURATED products, with excellent solubility, acid, alkali, and water resistance. A special application is in METALLIC paint media.

cyclic ketone resins CYCLOHEXANONE is the primary constituent of these RESINS, and it is condensed either by the action of ALKALIS or ACIDS.[130]

Mixed CONDENSATION products of cyclohexanone and ALDEHYDES are also resinous. The cyclic ketone resins which are available commercially are notable chiefly by reason of their extremely pale colors and characteristic smell. They are usually very readily soluble in VEGETABLE OILS.

cyclized rubber RUBBER in which the natural chain structure of the molecules has been partly converted to a ring structure. This change is accompanied by a reduction in unsaturation.[167]

Cyclized rubber is made by heating a rubber solution with stannic chloride. These rubbers vary from elastic products to fairly brittle resins, and are soluble in HYDROCARBONS.

cyclohexadiene ordinary BENZENE is UNSATURATED to the extent of three double bonds. When these unsaturated groupings are partly saturated, a new class of compound, the cyclohexadienes, is formed. Thus, $\Delta^{3.5}$ cyclohexadiene.[144,164]

cyclohexan-hexol a CYCLIC COMPOUND with six hydroxyl groups. It has a melting point of 200°C, at which temperature it decomposes. The presence of six hydroxyl groups makes it an interesting polyhydroxyl alcohol for the manufacture of alkyd and ESTER resins.[144] Syn: inositol

cyclohexane C_6H_{12}. A fully saturated HYDROCARBON. It is a constituent of certain crude PETROLEUMS.[144,164]

Properties: mp, 6.4°C; bp, 81°C; sp gr., 0.780; vp, 77 mm Hg/20°C; flp –15°C (–4°F).

cyclohexanol $C_6H_{11}OH$. A medium boiling SOLVENT.[144,164] Syn: hexahydrophenol, and hexalin

Properties: bp, 158°C; mp, 15°C; sp gr, 0.945/15°C; flp, 58°C (136°F) (average); vp, 2 mm Hg/30°C.

cyclohexanone $C_6H_{10}O$. Valuable KETONIC SOLVENT.[164] Syn: sextone, hexanone, and anon

Properties: bp, 155°C; sp gr, 0.945; flp, 44°C (111°F) (average); vp, 7 mm Hg/30°C.

cyclohexyl $-C_6H_{11}$. The monovalent RADICAL, derived from cyclohexane.[144,164]

cyclohexyl acetate a powerful high boiling ESTER SOLVENT.[164] Syn: hexalin acetate, and adronal

Properties: bp, 170-176°C; sp gr, 0.949; flp, 67°C (153°F); vp, 7 mm Hg/30°C.

cyclohexyl oxalate $(COO)_2(C_6H_{11})_2$. Solid PLASTICIZER.[170]

Properties: mp, 40°C; bp, 190°C/10 mm Hg.

cyclohexyl phthalate $C_6H_4(COO)_2(C_6H_{11})_2$. PLASTICIZER of the phthalate class, distinguished by its PERMANENCY and COLOR RETENTION.[170]

It is a very thick liquid at normal temperatures and has a high boiling point. Its compatibility is excellent, being used with CELLULOSE ESTERS and ethers, urea combinations, and SHELLAC. Sp gr, 1.075.

cyclohexyl stearate $CH_3(CH_2)_{16}COOC_6H_{11}$. SATURATED semisolid PLASTICIZER.[170]

Properties: mp, 25°C; bp, 220-240°C/4 mm Hg. Its general compatibility is somewhat restricted by the presence of the STEARIC ACID RADICAL, but it is a substantially inert product.

cycloparaffins ring COMPOUNDS of SATURATED HYDROCARBON type based on groupings of methylene RADICALS (CH_2).[164] Syn: the naphthenes

Typical cycloparaffins are cyclopropane, cyclobutane, cycloheptane, etc. The cycloparaffins have very good SOLVENT properties, and are constituents of crude PETROLEUM.

cynoper see MERCURIC SULFIDE.[41]

D

D_{6500} or D_{65} refers to the daylight SPECTRAL POWER DISTRIBUTION CURVE with a correlated color temperature of 6500 K.[69] See DAYLIGHT ILLUMINANTS, CIE.[69]

dabber dome-shaped BRUSH of soft hair for applying SPIRIT VARNISH or for POLISHING and finishing GILDING.[59,77]

dado (1) lower part of an interior wall when paneled or decorated customarily of chair rail height;[54,82,148] (2) PAPER that provides this architectural division.[82,148] See WAINSCOT.[82,75,153]

dagincoleic acid $C_{22}H_{44}O_4$. ACID constituent of BORNEO DAMMAR. Mp, 170°C.[167]

dagingenoleic acid $C_{13}H_{26}O_3$. ACIDIC constituent of BORNEO DAMMAR. Mp, 126°C.[167] (BSI)

dalamar yellow Pigment Yellow 74 and 65 (11741). An azo coupling of m-nitro-o-anisidine with acetoacet-o-anisidine; it is one of the growing group of close chemical relatives of HANSA YELLOW G (an azo coupling of m-nitro-p-toluidine with acetoacetanilide).[41]

The high strength TONER is superior in TINTING STRENGTH and FASTNESS properties to hansa yellow G.

dammar (damar) natural RESINOUS EXUDATION from trees of the *Dipterocarpaceae* family, which grow chiefly in the East Indies (now Indonesia) and Malaya. Very pale in color (practically colorless to deep yellow).[167]

Soluble in HYDROCARBONS without the necessity of RUNNING. Average acid value of 30.

dammar wax see RESENE, ß.[137]

dammarolic acid $C_{56}H_{80}O_8$. ACIDIC constituent of DAMMAR.[167]

dammarylic acid $C_{36}H_{60}O_3$. ACIDIC constituent of DAMMAR.[167]

dampening solution see FOUNTAIN SOLUTION.[168]

daturic acid $CH_3(CH_2)_{15}COOH$. Synonym for MARGARIC ACID. A straight chain ALIPHATIC ACID. Mp, 59.5°C.[144]

daub to apply a coating by crude unskillful strokes.[54,77,148]

daylight natural ILLUMINATION that is the result of various mixtures of direct sunlight and skylight covering a wide range of correlated COLOR temperatures.[69,43]

For precision in definitions, see NORTH LIGHT, ARTIFICIAL DAYLIGHT, and DAYLIGHT ILLUMINANTS, CIE.[69,43]

Daylight Illuminants, CIE series of ILLUMINANT SPECTRAL POWER distribution curves based on measurements of natural daylight and recommended by the CIE in 1965.[69]

Values are defined for the wavelength region 300 to 830 nm. They are described in terms of the CORRELATED COLOR TEMPERATURE. The most important is D_{6500} because of the closeness of its correlated color temperature to that of ILLUMINANT C, 6774 K. D_{7500}, bluer than D_{6500}, and D_{5500}, yellower than D_{6500}, are also used.

DCO nondrying CASTOR OIL which has been converted into a DRYING OIL by the CATALYTIC removal of water from its principal FATTY ACID (80% RICINOLEIC ACID), forming approximately equal quantities with 9-11 and 9-12 unsaturation.[63]

The term DCO rather than "DEHYDRATED CASTOR OIL" has been suggested and is being used for this oil in order to reduce the tendency to confuse castor oil with DCO, which is entirely different in its properties from the oil from which it is derived.

dead burned calcium sulfate see CALCIUM SULFATE, ANHYDROUS.[41,58]

dead flat said of a coating having no GLOSS or SHEEN.[43] See GLOSS.[43,35] Syn: lusterless

dead level absolutely horizontal, or zero SLOPE.[82] (ASTM) See SLOPE.[82]

dead oil see CREOSOTE.[76]

deadcoloring see ABBOZZO.[69,81]

deairentrainment any material that will release entrained air from the system.[83] See DEFOAMER and ANTIFOAMING AGENT.[83]

Entrained air is within the bulk liquid of the system as opposed to foam which is generally viewed at the surface.

debloomed oil see OIL, DEBLOOMED.[63]

decahydronaphthalene $C_{10}H_{18}$. HYDROCARBON SOLVENT with a boiling range for the commercial product (175-205°C), slightly higher than that for MINERAL SPIRITS.[164]

Words presented in CAP/SMALL CAPS type indicate that the word is defined in another part of the *Coatings Encyclopedic Dictionary.*

Numerical superscripts classify terms in one or more of the categories listed in the second section of this volume.

Properties: sp gr, 0.894/20°C; flp, 60°C (140°F); refractive index, 1.467; and vp, 2 mm Hg/30°C. In Europe, decaline.

decalcomania, decal prepared designs that are transferred from PAPER to china, GLASS, furniture, etc.[54,148,175]

decantation pouring or siphoning off the liquid from a PRECIPITATE or SEDIMENT or the upper layer of two IMMISCIBLE liquids as a partial means of separating the phases.[131,157]

decapping removal of GILDING from antiques, frames, etc.[148,158,175]

decarboxylation removal of CARBOXYL GROUPS from ORGANIC ACIDS, usually as the result of heating.

deck the structural surface to which the ROOFING or waterproofing system (including insulation) is applied.[159,82] (ASTM)

deck paint a paint having a high degree of RESISTANCE to mechanical wear; esp. used on surfaces such as porch floors and ships' decks.[71] (DAC)

decomposition chemical breakdown of a substance into two or more simpler substances which differ from each other and from the original substance.[56,65,134] See DEGRADATION.[56,154]

decor the combination of materials, furnishings, and objects used in interior DECORATING to create an atmosphere or style.[148] (DAC)

decorated adorned, embellished, or made more attractive by means of COLOR or surface detail.[148]

decorative painting ARCHITECTURAL or aesthetical painting.[54,148]

decoupage modern technique of "burying" a picture design, which has been affixed to a plain surface, in multiple coats of a CLEAR, VARNISH-TYPE coating. Many coats are applied until the surface no longer reveals the thickness of the design.[54,81,148]

This art form is said to have developed from an old French peasant craft consisting of the DECORATION of a plain surface with cut-out designs of cloth or paper, and then coating it with SHELLAC, VARNISH, or CLEAR LACQUER. Another version is that the art of decoupage was originated by craftsmen of 18th century Europe in an effort to duplicate Chinese and Japanese lacquer-ware.

deep having depth (in COLOR); often used improperly for dark.[69,43]

defect any irregularity or imperfection whose indications do not meet specified acceptance criteria.[56]

definition a statement of the meaning of a word, phrase, or TERM, as in a DICTIONARY entry.[134]

deflocculant an ADDITIVE that prevents PIGMENTS in SUSPENSION from COALESCING to form FLOCS. See DISPERSANT (DISPERSING AGENT).[83,64] Syn: deflocculating agent

deflocculation state or condition of a DISPERSION of a solid in a liquid in which each solid particle remains independent and unassociated with adjacent particles.[166,64]

A deflocculated SUSPENSION shows zero or very low YIELD VALUE.

defoamer ADDITIVE used to reduce or eliminate FOAM formed in a coating or a coating constituent.[83] cf. ANTIFOAMING AGENT[83]

deformation any change of form or shape produced in a body by a STRESS or force.[55,142]

degradation change in characteristics or quality of the original substance through chemical breakdown or physical wear.[56,154] See DECOMPOSITION.[56,65,134]

degras natural dark WOOL GREASE from which the various grades of LANOLIN are prepared.[137]

degreaser SOLVENT or compounded material for removing OILS, FATS, or GREASE from a SUBSTRATE. Also the apparatus in which this operation is carried out.[45,158] See VAPOR DEGREASER.[158,59]

degreasing removal of GREASE, OIL, and other fatty matter (from METALS, fabrics, etc.) by the use of SOLVENTS or chemical cleaners, electro, or heat processes.[158]

degree of efflorescence (1) FLATTING and GLOSSING—a nonuniform decrease or increase in gloss noticed when the surface is illuminated and viewed at near grazing angles. There is no apparent change in COLOR when viewed perpendicular to the surface; (2) fine efflorescence—a barely discernible WHITENING of the surface when viewed perpendicularly; (3) medium efflorescence—a readily noticeable whitening of the surface without a marked masking of the color; (4) heavy efflorescence—a white deposit sufficient to mask the color.[42,56] (ASTM)

degree of metamerism see METAMERISM, DEGREE OF.[69]

degree of polymerization (DP) in general, the average number of base units, or of MONOMERIC units per molecule in linear POLYMERS. Specifically, the average number of anhydroglucose units (or derivative units) per molecule of CELLULOSE (or cellulose derivative).[169]

The type of average obtained depends upon the method used for the determination. Hence, the method must always be specified.

dehydrated castor oil see DCO.[63]

dehydrated gypsum see CALCIUM SULFATE, ANHYDROUS.[41,58]

dehydration removal of WATER from a COMPOUND.[131]

Generally applied to the removal of combined water as, for instance, from hydrates.

dehydroabietic acid $C_{20}H_{28}O_2$. ROSIN derivative present in HYDROGENATED rosin.[167]

dehydrogenated oils SEMIDRYING OILS which have been converted into DRYING OILS by first treating them with a halogen (usually chlorine), then eliminating hydrogen halide. This results in CONJUGATED DOUBLE BONDS in the fatty acids chain.[63]

dehydrogenation PROCESS whereby hydrogen is removed from COMPOUNDS by chemical means.[131]

deionization the removal of ions from a solution by ion exchange.[131] (ASTM)

delaminate to separate existing layers.[131]

delaminated clay EXTENDER PIGMENT produced by mechanically separating the platelets of natural KAOLIN so that particles of approximately 1/4 micrometre thickness and up to 10 micrometres in diameter are produced.[41,58]

This extender pigment is used to increase the opacifying efficiency of TITANIUM DIOXIDE by keeping the titanium dioxide particles separated by the thickness of the CALCINED CLAY particle. This spacing allows each titanium dioxide particle to intercept a wavelength of visible light and thereby results in maximum LIGHT-SCATTERING efficiency.

delamination separation of the WOOD surfaces at the bonded joints caused by a COHESIVE failure in the ADHESIVE or a failure of the adhesive at the SURFACE.[79,153] (ASTM)

When the separation occurs in the wood, even though very close to the bonded joint, the separation is termed wood failure or checking rather than delamination. Magnification is often necessary to determine whether the failure is in the bond or in the wood. The term also applies to failure occurring between successive coatings.

deleafing loss of LEAFING of METALLIC PIGMENTS in paints, giving rise to reduced metallic LUSTER.[41,43,56] (BSI)

deliquescence the PROCESS that occurs when the vapor pressure of the saturated aqueous solution of a substance is less than the vapor pressure of water in the ambient air. When water vapor is collected by the pure solid compound, a mixture of the solid and liquid, or an aqueous solution of the compound forms until the substance is dissolved and is in equilibrium with its environment; at this time the vapor pressure of water over the aqueous solution will equal the partial pressure of water in the atmosphere in contact with it.[142] (IUPAC) Deliquescence is the reverse of EFFLORESCENCE. See HYGROSCOPIC; HYDROPHILIC.[139,152,155]

A crystalline SALT AEROSOL particle will deliquesce in the atmosphere when the relative humidity surpasses a characteristic value, the so-called DELIQUESCENCE POINT.

delta E, ΔE total COLOR DIFFERENCE computed by use of a color difference equation. It is generally calculated as the square root of the sum of the squares of the CHROMATICITY difference, ΔC, and the LIGHTNESS difference, ΔL: $\Delta E = [(\Delta C)^2 + (\Delta L)^2]^{1/2}$.[69] See COLOR DIFFERENCE EQUATIONS.[69]

denaturation addition of unpleasant or TOXIC substances to a product to make it unfit for human consumption.[164,83]

denatured alcohol ETHYL ALCOHOL to which a denaturant, e.g., METHYL ALCOHOL, PYRIDINE, etc., has been added.[164]

The denaturant chosen is one designed to discourage the use of the alcohol in beverages, but not to interfere with its use as a CHEMICAL INTERMEDIATE or SOLVENT.

densitometer instrument to measure the OPTICAL DENSITY of a transmitting material, or an instrument to measure the negative log of the REFLECTANCE of a reflecting material.[43,145,157,168]

Such instruments do not measure COLOR. They are widely used in the graphic arts and photographic industries for process control. In SPECTROGRAPHIC and dispersive X-ray analyses utilizing photography, they are adapted for scanning and plotting the density of the photographed transparency as a function of the WAVELENGTH.

density the mass per unit volume of a gas, liquid, or solid under specified conditions (temperature, pressure, etc.).[128] (IUPAC) See SPECIFIC GRAVITY, and BULKING VALUE.[128]

Usually expressed in units such as g/ml or lb/gal or lb/cu ft.

ASTM coating test methods for density include: dipentene/terpene solvents, D801; industrial aromatics, D2935; paint liquids, D1963; pigments, tests, D153; pine oil, D802; pine tars and oils, D856; turpentine and pinene, D233.

density, apparent see APPARENT DENSITY.[128]

density, number the number of atoms, molecules, or AEROSOL particles per unit volume of gas (commonly in units of molecules cm^{-3}).[128] (IUPAC)

density, optical negative log of the TRANSMITTANCE.[43,69] See BEER-LAMBERT LAW, and ABSORPTION COEFFICIENT.[43,69]

Words presented in CAP/SMALL CAPS type indicate that the word is defined in another part of the *Coatings Encyclopedic Dictionary*.

Numerical superscripts classify terms in one or more of the categories listed in the second section of this volume.

deodorizer equipment for the removal of noxious gases and odors, which may consist of combustion, ABSORPTION, or adsorption units.[59] (IUPAC)

Department of Transportation, U.S. (DOT) regulates transportation of materials to protect the public as well as fire, law, and other emergency-response personnel.[174,172]

DOT classifications specify appropriate warnings such as Oxidizing Agent or Flammable Liquid, that must be used.

Its address is 400 7th St., S.W., Washington, D.C. 20590.

depolymerization splitting of POLYMERS into molecules or polymers of lower molecular weight.[65,169]

depolymerized rubber RUBBER which has been extensively MILLED or otherwise treated so that it has sufficient solubility in AROMATIC HYDROCARBONS for the resulting solution to be used as a VARNISH or varnish constituent.[152,169]

deposit attack excessive CORROSION that may occur under solid substances that may be deposited on a METAL SURFACE and thus shield it from ready access to oxygen or ions in the SOLUTION.[66]

This is a form of concentration cell corrosion.

deposition (1) usually referred to in METAL treatment as the amount of treatment by weight (mg/sq ft); (2) PROCESS of applying a material to a base by means of vacuum, electrical, chemical, screening, or vapor methods.[70,77,158]

depth of finish the APPEARANCE phenomenon of a coating having an "apparent" depth.[43,35] Syn: "seeing into the film"

depth of shade concept which embodies the subjective or visual equality of CONCENTRATION of COLORANTS, DYES, or PIGMENTS.[41,69,35] See DEPTH OF SHADE, STANDARD.[69,35]

depth of shade, standard an arbitrarily chosen DEPTH OF SHADE for all HUES, from which a uniform depth of shade may be determined for purposes of comparison.[69]

Depth of shade can then be described as a multiple or fraction of standard depth. (Developed primarily, and used most widely, in Europe.)

derby red BASIC LEAD CHROMATE.[41] See CHROME ORANGE, LIGHT and DEEP.[41]

derived high polymer a POLYMER which has been produced by chemical alteration of a primary high polymer or a natural high polymer.[130] See PRIMARY HIGH POLYMER.[130]

dermatitis inflammation or irritation of the skin. Industrial dermatitis is an occupational skin disease.[172]

There are two general types of skin reactions: primary irritation dermatitis and sensitization dermatitis.

descaling removal of MILL SCALE or caked RUST from steel by chemical and/or mechanical means, sometimes assisted by FLAME CLEANING.[66,158]

desensitize to treat nonimage areas of a water LITHOGRAPHIC PLATE to make them water-receptive and INK-repellent.[168]

desiccant a substance such as CALCIUM OXIDE (LIME) or SILICA GEL, that has a high affinity for water and is used as a DRYING AGENT.[83] Syn: drying agent[83]

desiccator enclosure containing DRYING AGENTS, in which substances can be kept in a controlled dry atmosphere.[59]

designated pollutant a pollutant which is not a CRITERIA POLLUTANT (as described in Section 108 of the CLEAN AIR ACT) but for which new source performance standards exist.[172] (EPA)

States are required to control these pollutants under Section 111d of the Clean Air Act. Designated pollutants acid mist, total reduced sulfur (TRS), and fluorides.[172]

designers colors OPAQUE WATERCOLOR PAINTS that dry FLAT and are not necessarily permanent.[81,163]

desorption the removal of a substance from a surface on which it became absorbed.[139,131] (IUPAC) cf. ABSORPTION

Desorption is accelerated by heating and EXPOSURE of the surface to a low pressure (or vacuum). It is the converse of ADSORPTION, i.e., the decrease in the amount of adsorbed substance.

destructive distillation DECOMPOSITION of ORGANIC COMPOUNDS by heat out of contact with air.[131]

destructively distilled pine oil see PINE OIL.[44,164]

destructively distilled wood turpentine see TURPENTINE.[44,164]

detection limit (threshold) the minimum concentration of substance which produces an observable response.[128] (IUPAC)

detergent a SURFACE-ACTIVE AGENT that possesses the ability to clean soiled surfaces.[127,83] (ASTM) See SURFACTANTS.[127,83]

Anionic detergent—a detergent that produces AGGREGATES of negatively charged ions with COLLOIDAL properties. Cationic detergent—a detergent that produces aggregates of positively charged ions with colloidal properties. Nonionic detergent—a detergent that produces aggregates of electrically neutral molecules with colloidal properties.

deterioration a permanent impairment of the physical properties.[65] (ASTM) See DEGRADATION.[56,154]

developing ink a nondrying greasy INK composition, specifically formulated for use in initial fixing or subsequent renewal of the image on a LITHOGRAPHIC PLATE.[168]

deviation variation from a specified dimension or design requirement, usually defining upper and lower limits.[36,157] (ASTM) See TOLERANCE.[36,157]

dew point that temperature at which water pressure present in the atmosphere is just sufficient to saturate it.[128] (IUPAC)

When air is cooled below the dew point, the excess of water vapor appears as tiny droplets or crystals of ice depending on the temperature of the air mass. This term is also applicable to acid gases as in the problem of acid corrosion and acid smut formation.

dewaxing PROCESS of removing WAXES from natural RESINS (DAMMARS, SHELLACS, etc.).[131,137]

dextrin water-soluble ADHESIVE or THICKENER prepared by the incomplete HYDROLYSIS of starch by thermal, diastatic (use of diastase), or acid methods.[79,155,83] Syn: British gum

di - compounds see under other headings, such as ethyl oxalate for di-ethyl oxalate, etc.[133,134]

diacetone alcohol $CH_3COCH_2C(CH_3)_2OH$. Colorless, flammable liquid.[164] Syn: diacetone

Properties: mol wt, 116.16; sp gr, 0.9406/20°C; bp, 169.1°C (760 mm Hg); flp, 52°C (125°F).

diagonal grain grain in which the longitudinal elements form an angle with the axis of the piece as a result of sawing at an angle with the bark of the tree or log; a form of CROSS GRAIN.[153] (ASTM)

diagonal-grained wood wood in which the fibers are at an angle with (that is, diagonal to) the axis of a piece as a result of sawing at an angle with the axis of the tree.[153] (ASTM)

diallyl esters series of UNSATURATED ESTERS which POLYMERIZE rapidly to yield CROSSLINKED RESINS in the presence of peroxides at relatively low temperatures (e.g., 40-100°C).[130,144]

diarylide yellow see BENZIDINE YELLOW.[41]

diatomaceous calcite 2.4 $CaCO_3 \cdot SiO_2$ (approx.). Unique blend of $CaCO_3$ and diatomaceous SILICA that is mined from a single large deposit located in Kansas.[41,58]

Used as an extender. Sp gr, 2.54; density, 21.2 lb/gal; O.A., 25 lb/100 lb; particle size, 10 µm.

diatomaceous earth see DIATOMACEOUS SILICA.[41,58]

diatomaceous silica $SiO_2 \cdot 7H_2O$. Pigment White 27 (77811). A form of hydrous SILICA, processed from natural diatomite, a sedimentary rock of varying degrees of consolidation that is composed essentially of the fossilized siliceous skeletal remains of single-cell aquatic plant organisms called diatoms.[41,58] Syn: diatomaceous earth, diatomite, and the obsolete names of kieselguhr, infusorial earth, tripoli

Refractive index, 1.42; density, 2.00 g/cm³ (16.7 lb/gal); particle size, 6 to 10 µm.

diazo –N = N– grouping.[41]

Diazo compounds are INTERMEDIATES for the manufacture of DYESTUFFS.

diazo solution solution resulting from treatment of an AROMATIC primary amine with sodium nitrite and MINERAL ACID.[41,144]

Such solutions are stable only below 5°C and are employed as one of the ingredients in the manufacture of AZO COLORS.

dibenzyl phthalate see BENZYL PHTHALATE.[170]

dibutyl phthalate (DBP) $C_6H_4\,(COOC_4H_9)_2$. PLASTICIZER in NITROCELLULOSE LACQUERS, PLASTICS, SYNTHETIC RESINS, RESIN SOLVENT. Colorless, odorless, nonvolatile, nontoxic, stable, oily liquid.[144,170]

Properties: sp gr, 1.0484 (20/20°C); flp, Cleveland Open Cup, 171°C (340°F); bp, 340°C.

1,1-dichloroethylene see VINYLIDENE CHLORIDE.[141]

dichroism special case of PLEOCHROISM for crystals or minerals in the rhombohedral or tetragonal systems that have two distinctly different ABSORPTION COEFFICIENTS, one for vibration of light parallel to the longitudinal axis (extraordinary ray) and the other for directions perpendicular to it (ordinary ray).[69,43,34]

dichromism displaying different COLORS as a result of changes in CONCENTRATION or THICKNESS.[69,43,35]

Such changes do not result from changes in the ABSORPTION COEFFICIENTS for the material, but rather on the change in perceived color resulting from more or less absorption across the visible spectrum as the concentration is changed. The term "dichroism" is frequently used (incorrectly) to describe this effect. Dichromism is predictable on the basis of the laws of SUBTRACTIVE COLORANT MIXTURE.

dictionary a language-focused collection of terms, organized alphabetically, with etymological and other characteristics provided. Each term is treated separately with individual DEFINITIONS.[134] (ASTM)

---

die stamping see INTAGLIO.[168]

dielectric constant specific property of an INSULATING material that is both defined and measured by the ratio of the electrical capacity of a condenser having that material as the dielectric to the capacity of the same condenser having air as the dielectric.[61] Syn: specific inductive capacity

The dielectric constant of air or vacuum is unity.

dielectric dissipation factor the cotangent of the dielectric phase angle or the tangent of the dielectric phase angle or the tangent of the dielectric loss angle, represented by the symbol δ.[61]

dielectric heating method of heating a material between metallic plates and thereby subjecting it to a strong ELECTROSTATIC field produced by high-frequency voltage; the heat is developed within the material rather than being brought to it from the outside, and hence the material is heated more uniformly throughout.[67]

dielectric strength INSULATING power expressed as volts at which an insulator breaks down, divided by the thickness in mils (0.001 in.).[61]

Diels-Alder reaction very general and important ORGANIC REACTION for the SYNTHESIS of the six-membered rings. The reaction consists of the 1,4- addition of an ETHYLENIC COMPOUND to a CONJUGATED DIENE.

diene value or number amount of MALEIC ANHYDRIDE (expressed as equivalents of iodine) which will react with 100 parts of oil under specific conditions. It is a measure of the CONJUGATED DOUBLE BONDS in the OIL.[63,128] Syn: the maleic anhydride value or maleic value

diethyl oxalate see ETHYL OXALATE.[164]

diethylene glycol monobutyl ether see BUTYL CARBITOL®.[164]

diethylene glycol monobutyl ether acetate see BUTYL CARBITOL® ACETATE.[164]

diffraction a modification which light (or electron, neutron beams, etc.) undergoes in passing by the edges of OPAQUE bodies or through narrow slits or in being reflected from ruled surfaces (or crystalline materials).[43,69] (IUPAC) See DISPERSION, DIFFRACTION GRATING.[43,69]

The light waves, owing to their wave-like nature, appear to be deflected and produce fringes of parallel light and dark bands corresponding to regions of constrictive reinforcement or destructive interference, respectively, of the waves.

diffraction analysis the application of diffraction techniques (X-rays, electrons, neutrons) which are sometimes used to identify the presence of certain solid AEROSOLS and dust particles through the characteristic DIFFRACTION patterns which result from each unique crystal structure.[37] (IUPAC)

diffraction grating device used to disperse a beam of ELECTROMAGNETIC RADIATION into its constituent WAVELENGTHS, i.e., for producing a spectrum.[43]

Grating may be prepared by ruling equidistant parallel lines onto a GLASS surface (TRANSMISSION grating) or a METAL surface (REFLECTION grating).

diffuse reflectance see REFLECTANCE, DIFFUSE, and SPECULAR REFLECTANCE EXCLUDED.[43,69]

diffuse transmittance see TRANSMITTANCE, DIFFUSE.[43]

diffused light nondirectional light.[43]

diffusion the spreading or scattering of a gaseous or liquid material.[142] (IUPAC)

EDDY DIFFUSION in the atmosphere is the PROCESS of transport of gases due to turbulent mixing in the presence of a composition gradient. Molecular diffusion is the net transport of molecules which results from their molecular motions alone in the absence of turbulent mixing; it occurs when the concentration gradient of a particular gas in a mixture differs from its equilibrium value. Eddy diffusion is the most important mixing process in the lower atmosphere, while molecular diffusion becomes significant at the lower pressures of the upper atmosphere.

digester jacketed REACTION vessel used to decompose, soften, or cook substances at high pressure and temperature.[59,131]

dihydroxy anthraquinone lake see ALIZARIN RED.[41]

dilatancy (1) property of certain SUSPENSIONS in which the resistance to FLOW increases at a greater rate than the increase in rate of flow. Dilatancy in a PIGMENT-VEHICLE SYSTEM or PIGMENT DISPERSIONS commonly results from high pigment loadings; (2) apparent viscosity increases with increased shear.[166] See SHEAR THICKENING.[166] cf. PSEUDO-PLASTICITY

dilatant flow (shear thickening) a type of FLOW characterized by an increase in VISCOSITY as SHEAR STRESS is increased.[166] (ASTM)

The curve of the plot of shear stress versus SHEAR RATE is nonlinear with shear stress increasing faster than the shear rate.

diluent (1) a VOLATILE liquid which, while not a SOLVENT for the NONVOLATILE CONSTITUENTS of a coating or PRINTING INK, may yet be used in conjunction with the true solvent, without causing PRECIPITATION;[164] (2) an inert solid or liquid used to increase the bulk of another substance; (3) an adulterant.[164] cf. REACTIVE DILUENT[164,171]

dilute to make THINNER or less CONCENTRATED by adding a liquid such as water or SOLVENT.[164] See LETDOWN.[131,49]

dilution ratio ratio by which a given SOLVENT or SOLUTION may be diluted without adverse effects.[151,164]

dilution ventilation see GENERAL VENTILATION.[172]

dimension a geometric element in a design, such as length or angle, or the magnitude of such a quantity.[157,128] (ASTM)

dimension stone stone that is selected, trimmed, or cut to desired shapes and/or sizes for such uses as building stone, markers, paving blocks, flagging, or curbing.[75]

dimensional stability ability of a substance (WOOD, PLASTIC, etc.) to retain its shape when subjected to varying degrees of temperature, moisture, pressure, or other stress.[161]

dimer CONDENSATION product or POLYMER of two molecules.[169] See DIMERIZATION.[169]

dimer acids liquid polycarboxylic acids produced by POLYMERIZATION of UNSATURATED FATTY ACIDS. Pure dimer acid is a C_{36} dicarboxylic acid.[144]

Commercial dimer acids are composed of predominantly C_{36} dicarboxylic acid with minor amounts of C_{18} monocarboxylic acid and varying proportions of C_{54} tricarboxylic acid.

dimerization state of POLYMERIZATION when two similar molecules have united.[169]

The product is described as a dimer.

dimethyl ketone see ACETONE.[164]

dimethyl phthalate $C_6H_4(COOCH_3)_2$. PLASTICIZER for NITROCELLULOSE and CELLULOSE ACETATE, RESINS, LACQUERS, PLASTICS. Colorless, odorless, lightfast, stable, nontoxic liquid.[144,170]

Properties: sp gr, 1.189; bp, 282°C; flp, 149°C (300°F).

dimethyl urea $(CH_3NH)_2CO$. Primary CONDENSATION product of UREA and FORMALDEHYDE.[130]

When condensed in the presence of ALCOHOLS, it forms oil-soluble resins. With mildly ACID SALTS, it may be used as an ADHESIVE.

dimethylolurea reaction product formed by the interaction of one molecule of UREA with two molecules of FORMALDEHYDE.[144]

dimorphous material which can exist in two distinct CRYSTALLINE forms, having different MELTING POINTS.[133]

DIN (1) abbreviation for Deutsche Industrie Norm (German Industry Standard); (2) also abbreviation for Deutsches Institut für Normung (formerly DNA).[157]

DIN Color Difference Equation COLOR DIFFERENCE EQUATION based on the DIN COLOR ORDER SYSTEM.[69]

It is described in English by Richter in J. Opt. Soc. Am., 45, 223 (1955).

DIN Color System official German Standard Color System (DIN 6164).[69]

The color solid is described in terms of HUE (DIN - Farbton), abbreviated F or T, SATURATION (DIN-Sättigung), abbreviated S, and relative lightness (DIN - Dunkelstufe), abbreviated D.

Dingler's green see CHROMIUM OXIDE GREEN.[41]

dinitraniline BASE which, when diazotized and coupled with ß-naphthol, forms Permanent Red 2G.[41]

dinitraniline orange Pigment Orange 5 (12075). A monoazo PIGMENT DYESTUFF prepared by coupling DIAZOTIZED dinitro aniline to beta naphthol. It BLEEDS in ORGANIC SOLVENTS and has fair HEAT RESISTANCE.[41]

Its LIGHTFASTNESS in MASSTONE is good. The good chemical resistance, freedom from toxic metals, and high COLOR intensity of dinitraniline orange encourage its use over the less expensive chrome and molybdate oranges in some specialized finishes not requiring good bake or BLEED RESISTANCE.

diol compound containing two ALCOHOLIC HYDROXYL GROUPS.[133,134]

ETHYLENE GLYCOL is a typical diol.

diolefin resins more generally known as AROMATIC PETROLEUM RESIDUES.[130]

By-products from the CRACKING of PETROLEUM.

dioxan (dioxane) cyclic 1.4 diethylene dioxide. $CH_2CH_2OCH_2CH_2O$.[164]

Bp, 101°C; sp gr, 1.035/20°C; fp, 11°C (52°F); refractive index, 1.423; vp, 26 mm Hg/20°C. A WATER-MISCIBLE SOLVENT for CELLULOSE ESTERS.

dioxin colloquial (short) name of a toxic by-product (and sometimes contaminant) of herbicides; the full name of this species is: 2,3,7,8-tetrachlorodibenzo[b,e] [1,4] dioxin.[172] (IUPAC)

dip coating a method of applying a coating in which a substrate is dipped into a tank of coating and then withdrawn.[77] (EPA) cf. ELECTRODEPOSITION[77]

dip type viscosity cup see ZAHN CUP.[145,157,166]

Words presented in CAP/SMALL CAPS type indicate that the word is defined in another part of the *Coatings Encyclopedic Dictionary.*

Numerical superscripts classify terms in one or more of the categories listed in the second section of this volume.

dip-grained wood WOOD which has single waves or undulations of the fibers, such as occur around knots and pitch pockets.[153] (ASTM)

dipcoat a paint or PLASTIC coating which is applied by completely immersing an article in a tank of the coating.[77]

dipentene optically inactive form of the monocyclic TERPENE HYDROCARBON limonene.[44,164]

Commercial dipentenes contain substantial portions of other monocyclic and bicyclic, as well as some oxygenated terpenes having closely related boiling ranges. They are generally obtained by FRACTIONAL DISTILLATION from the crude OILS recovered in the several commercial methods of utilizing pine wood, also by isomerization during the chemical processing of TERPENES. The four kinds of commercial dipentene are: (1) Steam-distilled dipentene—from the crude OLEORESINOUS extract used for the processing of related steam-distilled wood naval stores; (2) sulfate dipentene—from the crude condensate of the vapors generated in the digestion of wood in the sulfate PAPER PULP PROCESS; (3) destructively distilled dipentene— from the lighter portions of the oil, recovered during the destructive distillation of pine wood; (4) chemically processed dipentene—recovered as a by-product in connection with the chemical treatment and conversion of other terpenes. Used as a SOLVENT, but also has ANTISKINNING properties. Sp gr, 0.850 (15.6/15.6°C); flash point, 52°C (Cleveland Open Cup); boiling range, 175°C (347°F) to 188°C (370°F).

dipentene (pure compound) $C_{10}H_{16}$.dl-1,8(9)-p-menthadiene.[164,44] Syn: dl-limonene, cajeputene, cimene, kautschin

Colorless liquid; sp gr, 0.84; bp, 176°C; molecular weight, 136.12, insoluble in water, soluble in ETHYL ALCOHOL.

diphenyl liquid whose vapor is used as a heat transfer agent in chemical processes.[144,83]

At low pressures, the temperatures obtainable with diphenyl vapor are higher than those obtainable with steam.

diphenyl ether $C_6H_5OC_6H_5$. A PLASTICIZER.[170]

Properties: bp, 252°C; mp, 28°C.

diphenyl ketone see BENZOPHENONE.[164]

diphenyl phosphate (DPP) a permanent PLASTICIZER.[170]

diphenylamine $C_6H_5NHC_6H_5$. A weak secondary BASE used as a DYESTUFF INTERMEDIATE and also as an indicator.[41,144]

diphenylmethane diisocyanate (MDI) an AROMATIC ISOCYANATE MONOMER used as an INTERMEDIATE in urethane coatings.[141] See ISOCYANATE RESINS, POLYURETHANES, POLYURETHANE FINISH, and URETHANE COATINGS.[156,71,130,163,67] cf. HEXAMETHYLENE DIISOCYANATE (HDI), ISOPHORONE DIISOCYANATE (IPDI), TOLUENE DIISOCYANATE (TDI), and TETRAMETHYLXYLENE DIISOCYANATE (TMXDI)[141]

dipole combination of two electrically or magnetically charged particles, or parts of a structure of opposite sign which are separated by a finite distance.[61]

dipping method of APPLICATION in which the complete immersion of an article in the coating is followed by draining after removal.[77] See COATING, DIP.[77]

Paints, LACQUERS, and STAINS designed for such use are usually designated "dipping." This process may be carried out either at ordinary or elevated temperatures.

dipropyl ketone $C_3H_7COC_3H_7$. SOLVENT.[164]

Properties: bp, 143°C; flp, 49°C (120°F); vp, 5 mm Hg/20°C.

dipropyl oxalate high boiling SOLVENT.[164]

Bp, 213°C.

dipropyl succinate $(CH_2)_2(COOC_3H_7)_2$. PLASTICIZER.[170]

Properties: bp, 246°C; sp gr, 1.016/4°C.

direct lighting lighting which is so controlled by a REFLECTOR that the major portion of the light directly reaches the spot to be ILLUMINATED without redirection, DIFFUSION, or REFLECTION.[43]

directional reflectance see REFLECTANCE, DIRECTIONAL.[43]

dirt disfiguring foreign material other than microorganisms on or embedded in a dried coating.[56,158,42] Syn: soil

dirt collection the deposit of foreign matter consisting of dirt, soot, or stain present on the surface of exposed exterior coated panels.[42,65] (ASTM) Syn: dirt pick-up

Exterior coatings often experience the accumulation of dirt not entirely removable from the paint film and result in change of COLOR or appearance. Generally, harder and/or more hydrophobic films are more resistant to dirt collection.

ASTM Test Method D3719 quantifies dirt collection on coated exterior panels. This test method measures total dirt collection. Loose dirt may be washed and removed from the specimens if only imbedded dirt collection is to be measured.

dirt pick-up see DIRT COLLECTION.[42,65]

dirt/soil resistance the ability of a coating to resist soiling by foreign material, other than microorganisms, deposited on or embedded in the dried coating.[56,72] See SOILANT.[42,157]

The FSCT Pictorial Standards for Coating Defects contains photographic standards for evaluating the degree of surface disfigurement of paint films by dirt and soil accumulation or by FUNGAL *growth.*

ASTM Test Method D3274 also has these same photographic reference standards to provide a numerical basis for rating the degree of dirt and soil accumulation or fungal and ALGAL GROWTH.

ASTM test methods for dirt/soil resistance include: exterior white coatings, D3719; practical washability, D4828; washability, mechanical test, D3450.

discolor to become altered or spoiled in COLOR.[69]

discoloration STAINING; also changing or darkening in COLOR from the standard or original.[42,56,157]

discontinuities see HOLIDAYS.[56]

discontinuous phase see DISPERSED PHASE.[60] cf. CONTINUOUS PHASE.[139]

dished having a symmetrical concave depression in the surface. [56]

Opposite of DOMED.

disk an ELECTROSTATIC SPRAY APPLICATION method which consists of a rapidly spinning metal disk onto which paint is fed. The paint is spun off the edge of the rotating disk by centrifugal force and the paint is formed into a fine mist.[59,77] (EPA)

This device applies paint with HIGH TRANSFER EFFICIENCY.

dispersant ADDITIVE that increases the stability of a SUSPENSION of POWDERS (PIGMENTS) in a liquid medium.[64,83,127] Syn: dispersing agent

Dispersants are used to prevent FLOCCULATION, *stabilize properties (like* VISCOSITY *or* COLOR*), and/or reduce processing time. Using dispersants reduces the tendency of some* PIGMENTS *to overgrind, resulting in color inconsistency from* BATCH *to batch. In addition, they reduce the viscosity of the dispersion, which allows flowable compositions to be made at a higher-solids content.*

Dispersants function by introducing BARRIERS *to flocculation. These barriers are usually explained in terms of either* ELECTROSTATIC *or* STERIC STABILIZATION *mechanisms. In electrostatic stabilization, one portion of the dispersant molecule is adsorbed on the particle surface while an ionic portion of the dispersant is exposed to the liquid phase. This creates a charged surface that repels other like-charged particles. In steric stabilization, a long soluble portion of the dispersant molecule extends into the liquid phase, creating a stabilizing solvate layer.*

In both cases, the dispersant attaches itself to the particle by adsorption. A second, nonadsorbed portion of the dispersant molecule interacts with the liquid phase. Often, one portion is polar and the other is nonpolar. Two polarity, amphipathic molecules are used. Materials of this type include surfactants and certain specially designed polymers.

disperse system any system in which particles (of any size and state) are DISPERSED in a homogeneous medium.[64]

A COLLOIDAL SOLUTION *represents a special kind of disperse system.*

dispersed finely divided or colloidal in nature.[139,64]

dispersed phase that phase in an EMULSION or SUSPENSION which is broken down into droplets or discrete particles and DISPERSED throughout the other or CONTINUOUS PHASE.[60,64] Syn: discontinuous phase

dispersing agent see DISPERSANT.[83,127]

dispersion (1) any composition, PROCESS, or state, relating to heterogeneous systems of solids, gases, liquids, or immiscible liquids. *This term is used loosely in many ways with special meanings which are often implied but not defined. Since the term is decidedly generic and includes all types of* COLLOIDAL *systems, it should not be used on the assumption that it refers to a desirable feature of one type of* DISPERSION. *For example, in paint and printing ink technology a "good" dispersion usually means that a pigment is finely divided and* DEFLOCCULATED *in the vehicle. Where* AGGREGATES *are present in the dispersed phase, the material is considered without exception a "poor" dispersion regardless of the state of* FLOCCULATION. *Where the pigment particles are finely divided but flocculated, the material has been called variously a "good" and a "poor" dispersion;*[166,64] (2) process of dispersing a dry POWDER (or PIGMENT) in a liquid medium in such a way that the individual particles of the powder become separated from one another and are reasonably evenly distributed throughout the entire liquid medium; (3) two-phase system in which one phase, called the DISPERSED PHASE, is permanently distributed as small particles through the second phase called the CONTINUOUS PHASE;[64] (BSI) (4) the dilution of a pollutant by spreading in the atmosphere due to DIFFUSION or turbulent action (EDDY DIFFUSION).[172] (IUPAC)

ASTM test methods for dispersion (fineness of grind) include: coarse particles by sieve test, D185; dispersion stability, phthalo blue, D963; in pigment-vehicle systems, D1210; in printing inks, NPIRI grindometer, D1316; in titanium dioxide slurries, D3926.

dispersion curve a graph of the REFRACTIVE INDEX (ordinate) as a function of WAVELENGTH of RADIATION (abscissa).[69,43]

Words presented in CAP/SMALL CAPS type indicate that the word is defined in another part of the *Coatings Encyclopedic Dictionary.*

Numerical superscripts classify terms in one or more of the categories listed in the second section of this volume.

dispersion, degree of quantity varying reciprocally with AGGREGATE size.[41,64]

The larger the aggregates, the lower the degree of DISPERSION.

dispersion, light the separation of light into separate WAVELENGTHS by means of a DISPERSION device such as a prism or DIFFRACTION GRATING.[69,43] See DISPERSION CURVE.[69,43]

Fundamentally, dispersion is related to the speed of different wavelengths of light in a given medium.

dispersion, pigment SUSPENSION of PIGMENT particles uniformly in a medium such as a paint vehicle, plastic matrix, etc.[41,64,131]

The PROCESS of dispersing the pigment involves the separation of individual pigment particles, and coating them with the medium.

dispersion, resin see ORGANOSOL, and PLASTISOL.[64,71]

dispersoid particles of a DISPERSION.[64]

distemper heavily pigmented, matte drying composition, capable of being thinned with water, in which the binding medium consists essentially of either GLUE or CASEIN or similar sizing material.[71,81,155]

distemper brush a flat BRUSH, five to ten inches wide, well-packed, with long BRISTLES.[59]

distemper colors COLORANTS suitable for TINTING DISTEMPERS.[81]

The two chief requirements are LIGHTFASTNESS in reduced form and fastness to ALKALI.

distemper, oil-bound see OIL-BOUND DISTEMPER.[60,71]

distensibility ability to be stretched. See ELONGATION.[47]

distillation PROCESS of separation consisting of vaporizing a liquid and collecting the vapor, which is then usually condensed to a liquid.[131]

ASTM test methods for distillation include: dipentene/terpene solvents, D801; distillation range, D1078; ethyl acetate, spec., D4614; in vacuum, solvent-type paints, D3272; petroleum products, D86; pine oil, D802; pine tars and tar oils, D856; turpentine and pinene, D233.

distillation range temperature range over which a mixture of liquids is distilled.[157]

distinctness-of-image gloss the sharpness with which image outlines are REFLECTED by the surface of an object.[43]

distressing in ANTIQUING, the FURNITURE PROCESS designed to produce a more authentic look of age or craftsmanship.[54,148,175] cf. CLOTH DISTRESSING[77]

It attempts to simulate the small visible marks of abuse picked up by a piece of furniture during a long period of use. In simulating natural MARRING or aging of some kind, it may involve the use of gouges, hammers, chains, blunt or pointed instruments such as common car keys, a belt buckle, or sharp pencil point. Other effects may be achieved with a fine BRUSH or a pointed WOOD INSTRUMENT dipped into ink or glaze and placed on a surface. If no natural counterpart to an effect exists, it is not authentically aged and should be considered a novelty finish or accent piece.

DIY abbreviation for "do it yourself."

dl-limonene see DIPENTINE.[44,164]

DLVO theory a DISPERSION STABILIZING theory in which ZETA POTENTIAL is used to explain that as two PARTICLES approach one another, their ionic atmospheres begin to overlap and a repulsion force is developed.[64]

This theory for electrical stabilization was first established by Derjaguin and Landau of the then Soviet Union and independently by Verway and Overbeek of the Netherlands, thus the DLVO. According to this theory, stability is determined by a balance of van der Waals' attractive energy, V_A, and the electrical repulsive energy, V^{el}_R, arising from the repulsion between the electric double layers on the particles. Since V_A and V^{el}_R operate independently, the total potential energy, V_T, for the system is given by the sum of the two terms: $V_T = V_A + V^{el}_R$.

doctor blade (1) device for spreading a thin film of even thickness on base material. *In the paint industry, it is used to prepare paint and VARNISH films of even and predetermined thickness;*[59,77] (2) in GRAVURE PRINTING and coating, a blade that scrapes off the excess INK or LACQUER from the surface of the etched cylindrical roll just prior to printing, leaving the "cells" filled with ink or lacquer.[168,145]

doctor mark or streak streak or ridge in coated FABRICS caused by a damaged DOCTOR BLADE.[56] Syn: knife mark

doctor roll ROLLER mechanism which is revolving at a different surface speed, or in an opposite direction, resulting in a wiping action for regulating the ADHESIVE supplied to the spreader roll.[59,77,79] (ASTM)

doctor test method for detecting the presence of mercaptan sulfur in gasolines, naphthas, and kerosenes.[157]

dodecanoic acid see LAURIC ACID.[144,63]

dogskin see ORANGE PEEL (1).[56]

dolomite $CaCO_3 \cdot MgCO_3$. A mineral having the

empirical composition, 1 mole of CALCIUM CARBONATE and 1 mole of MAGNESIUM CARBONATE.[41,58,34] See CALCIUM CARBONATE, NATURAL.[41]

Any calcium carbonate rock containing 20% or more of magnesium carbonate.

domed having a symmetrical convex protrusion in the surface.[56]

Opposite of DISHED.

dominant wavelength, CIE WAVELENGTH of SPECTRALLY pure light which, when additively mixed with the ILLUMINANT and adjusted for LUMINANCE, would match the COLOR.[43,69]

It is expressed in wavelength units, nm (or µm), and is determined graphically from the CHROMATICITY DIAGRAM or by suitable computation. The dominant wavelength of the nonspectral colors, the purples, is described in terms of the dominant wavelength of the green, to which they are the additive complements, followed by the letter c. Thus, a dominant wavelength designation of 520c describes a purple which is additively complementary to a green having a dominant wavelength of 520 nm. Dominant wavelength describes the HUE of a color, although the visual evaluation of constant hue does not necessarily coincide with constant dominant wavelength.

domino early French WALLPAPER made in small sheets, originally imitating marble, later patterned.[148]

doorframe an assembly built into a wall consisting of two upright members (JAMBS) and a head (LINTEL) over the doorway; encloses the doorway and provides support on which to hang the door.[82,75] (DAC)

dope a composition, usually a CELLULOSE LACQUER, for application on textiles and leathers. Also a very viscous crude reaction product, as ACETYLATION DOPE.[71,151] (ASTM)

dope cotton designation of high VISCOSITY NITROCELLULOSE, i.e., 20-30 second, 40 second, and 70 second nitrocellulose.[151,144]

dormer GABLE-like projection or a window set upright in a sloping roof, or the roofed projection in which this window is set.[82]

DOT abbreviation for DEPARTMENT OF TRANSPORTATION, U.S.[174,172]

DOT Identification Numbers four-digit numbers [preceded by UN (United Nations) or NA (North America)] used to identify particular materials for regulation of their transportation.[172]

See DOT publications that describe the regulations (49 CFR 172.102). These numbers are called product identification numbers (PINs) under the Canadian Transportation of Dangerous Goods Regulation.

double bond elements possess a definite number of VALENCY linkages or BONDS, and these are employed in uniting with other elements. When all these bonds are completely employed in such unions, SATURATED COMPOUNDS result. If, however, a residual bond or linkage remains, this may be involved with another in forming a REACTIVE or UNSATURATED DOUBLE BOND. If a hydrogen atom is removed from each of the carbon atoms, two spare linkages occur, and an unsaturated double bond results. Double bonds are centers of great reactivity, and are able to accept addition of suitable elements such as hydrogen, chlorine, etc., to yield fully saturated compounds. The reactivity of double bonds in DRYING OILS is responsible for their DRYING and POLYMERIZING properties.[133,134]

double rub the act of rubbing a SOLVENT saturated cloth in one complete forward and backward motion over the coated SURFACE.[157,164] (ASTM)

double seam can can formed by interlocking the edges of both the end and the body, double folding and rolling firmly together.[50]

double spread see SPREAD.[79]

double-hung sash window divided into two sections, one lowering from the top and the other rising from the bottom.[82]

doubletone ink a type of PRINTING INK which produces the effect of two-COLOR printing with a single impression.[168]

These inks contain a soluble toner which BLEEDS out to produce a secondary color.

Dowtherm trademark for eutectic mixture of phenyl ether and DIPHENYL.

Bp, 258°C. Used for heat transfer.

DPP abbreviation for DIPHENYL PHOSPHATE.[170]

drag (brush drag) RESISTANCE encountered when applying a coating by BRUSH.[166]

The British term for this condition is "GUMMY." As the brush drag of a paint increases, any natural tendency of the painter is to overspread the paint is reduced. When all other factors are held constant, increased brush drag will result in greater FILM THICKNESS with consequent improvement in DURABILITY and HIDING. Conversely, sometimes it may be preferred to have a lesser degree of drag for easier application (that is, the amount of time and effort in applying a paint to a specific area is obviously reduced with a lesser degree of brush drag).

The brush drag of paints is directly related to their high-shear viscosity (ca. 10,000 s^{-1}). There

Words presented in CAP/SMALL CAPS type indicate that the word is defined in another part of the *Coatings Encyclopedic Dictionary.*

Numerical superscripts classify terms in one or more of the categories listed in the second section of this volume.

is generally good rank order agreement between results obtained by brushout procedure for comparing the brush drag of architectural type LATEX *paints (ASTM Test Method D4958) and the ICI Cone/Plate viscometer readings (ASTM Test Method D4287). The sensitivity of this ASTM brushout method has been found sufficient to distinguish between* BRUSHABILITIES *corresponding to high-shear viscosity differences not lower than 0.3 poise (0.03 Pa.s). Round robin data show that rank order agreement between the brushout and viscometric methods is poor when latex and solvent paints are part of the same comparison group. This is the result of these two paint types having markedly different* RHEOLOGICAL *properties that affect the relative perception of brush drag. (ASTM)*

dragon's blood dark red RESINOUS EXUDATION, obtained from the surface of the fruit of the *Calamus draco* and similar species.[167]

Used to some extent as a COLORANT *in coatings.*

drain a piping system used to collect and carry off surface and ground water.[82] (ASTM)

drawdown (1) a film of INK deposited on PAPER by a smooth edged blade to evaluate the UNDERTONE and MASSTONE of the ink;[157] (2) a layer of paint deposited on a SUBSTRATE by use of DRAWDOWN BAR to evaluate the characteristics of the paint.[157,77] (ASTM) Syn: pulldown and casting

drawdown bar (blade) METAL bar designed to deposit a specified thickness of wet coating film on test panels or other SUBSTRATES.[77]

drawout a method of APPLICATION in which the coating is applied, usually on PAPER, by spreading the PIGMENT-vehicle mass evenly with a broad knife.[77,157]

Useful for examination of MASSTONE, UNDERTONE, OPACITY, PRINTING STRENGTH, *and in some cases,* FADEOMETER *ratings of* PIGMENTS.

dressed lumber LUMBER having one or more of its faces planed smooth.[75,153] (DAC) Syn: dressed stuff and surfaced lumber

dressing compound BITUMINOUS liquid used hot or cold for dressing the exposed surface of ROOFING FELT.[76,163]

drier a composition which accelerates the DRYING of OIL, PAINT, PRINTING INK, or VARNISH.[74,67,83] (ASTM) Syn: siccative

Driers are usually metallic compositions and are available in both solid and liquid forms.

ASTM test methods for driers include: calcium/zinc content, EDTA, D2613; cerium content, test, D3970; clarity/cleanliness, visual, D2190; cobalt content, EDTA method, D2373; iron content, EDTA method, D2374; manganese content, EDTA method, D2375; rare earths content, EDTA method, D3989; selection of test methods, D564; specification, D600; vanadium content, EDTA method, D3988; volatile/nonvolatile content, D4140; zirconium content, EDTA method, D3969.

drier absorption see DRIER DISSIPATION.[74,161,42,168,67]

drier dissipation a loss in CATALYTIC power of a DRIER due to a physical ABSORPTION or a chemical reaction with certain PIGMENTS.[74,42,161,67,168]

drift see OVERSPRAY.[42,54]

drips the small drops of coating that collect on the edge of a coated work.[42,56] (ASTM)

drop one vertical descent of a scaffold.[59]

drop black see BONE BLACK.[41]

drop ceiling a form of DECORATION in which the ceiling PAPER is brought down to a suitable depth on the walls of a room and divided from the sidewall by a border or molding.[148,82]

Gives the illusion of a lower ceiling.

drop cloth cover used to protect floors and FURNITURE, etc., from paint spillage and droppings.[59]

drop siding an exterior wall cladding of WOODEN boards (or strips of other material, such as aluminum or vinyl), which are tongued and grooved or rabbeted and overlapped so that the lower edge of each board interlocks with a groove in the board immediately below it.[82,153,75] (DAC) Syn: novelty siding and rustic siding

drum (cylinder) sander a machine employing one or more revolving cylinders, each wrapped with coated ABRASIVES, used for GRINDING and sanding surfaces to a desired degree of smoothness.[59,158]

With multiple-drum sanders, the first drum may have a coarse grain for a heavy cut while subsequent drums are set up with progressively finer grades of coated abrasives to achieve a fine FINISH, *with only one pass through the machine. The original purpose to sand wood or finish wood is in preparation for painting.*

dry a film is considered dry when it feels firm to the finger, using moderate pressure.[47,157]

dry back the change in COLOR or FINISH of an INK film as it dries.[168]

dry blending a PROCESS for POWDER coatings manufacture where materials are blended without melting.[131]

dry brush technique in which PAINT or INK is APPLIED sparingly with a semidry BRUSH.[54,148]

dry bulk see APPARENT DENSITY.[128]

dry color dry colored PIGMENTS. A concentrated PIGMENT or coloring matter suitable for GRINDING or DISPERSING and use in coatings.[41]

dry deposition see SCAVENGING.[171,172]

dry hiding the increment of increasing SCATTERING due to the larger difference in REFRACTIVE INDEX between air and PIGMENT vs. pigment vs. BINDER and pigment.[43,55] cf. WET HIDING. See MICROVOIDS.[43]

It is the increase in the HIDING POWER of a paint which occurs in the drying process. It is most significant in nonglossy paints, which, after drying, have pigments or EXTENDERS protruding from the surface of the paint film which scatter the incident light, adding to hiding.

Hiding of paints changes during the drying process for two principal reasons. Initially, as the water and solvents leave the film, the pigment particles become more concentrated (crowded), thus lowering the hiding. Once the paint has dried enough that all the binder and pigment particles are packed together, air replaces the remaining water as it evaporates. Since the refractive index of air (1.0) is lower than that of water (1.3), hiding increases with the development of the air voids.

The dry hiding contribution to SCATTERING can be estimated from S_{dry} - S_{wet}, the scattering value for the dry films minus the scattering value for the films after BRUSHING with high boiling ALIPHATIC SOLVENT which replaces the air and eliminates dry hiding. Above CPVC, paints have a large component of dry hiding, below CPVC, paints do not. In addition, S_{wet} values are sometimes used as an indicator of the hiding of wet paints when they are first applied to the substrate.

Many ARCHITECTURAL PAINTS, particularly white and light tints, change significantly in FILM OPACITY as they dry. An increase in hiding is sometimes associated with porosity and poor FILM INTEGRITY with conventionally formulated coatings.

ASTM Test Method D5007 covers the determination of the change in hiding of white and tinted architectural coatings during drying, by visual evaluation of the wet and dry film. ASTM Test Method D2805 covers the determination of hiding power by reflectometry.

dry hydrate CALCIUM HYDROXIDE; HYDRATED LIME POWDER.[41,58]

dry ice solidified CARBON DIOXIDE.[144]

dry offset an indirect LETTERPRESS PROCESS in which the INK is transferred from a relief plate to a blanket and then to the stock.[168]

dry point during a DISTILLATION TEST, the temperature indicated at the instant the last drop of liquid EVAPORATES from the lowest point in the DISTILLATION flask.[157,164] (ASTM)

dry rot special type of brown ROT, causing underspread damage in buildings.[65,154,173]

In the U.S., the causal organism is Portia incrassata.

dry spray a condition caused by the partial drying of the liquid coating prior to reaching the SURFACE to be coated. The paint particles do not FLOW out to form a smooth film.[42,56]

Dry spray results in a rough, sandpaper texture on the surface. It often is not a continuous film and can result in early CORROSION or loss of ADHESION between coats. Dry spray should be removed by sanding and the coating reapplied. The most common causes of dry spray are holding the gun too far from the surface, using too much atomizing air, and spraying when the wind velocity is too high.

dry strippable paper see STRIPPABLE COATING.[148]

dry tack see TACK, DRY.[79]

dry wall (1) an interior wall, constructed with a dry-wall FINISH material such as GYPSUM BOARD or PLYWOOD; (2) in MASONRY construction, a self-supporting rubble or ashlar (squared building stone) wall built without MORTAR.[82,159,75] (DAC)

dry weight usually applied to the mass of the ovendry material, but is ambiguous unless the method of drying is specified.[37,157]

dry-hard film is considered dry-hard when any mark left by the thumb, exerting maximum downward pressure on a film, is completely removed by a light polishing with a soft cloth.[47,157]

dry-through film is considered dry-through when no loosening, detachment, wrinkling, or other distortion of the film occurs when the thumb is borne downward while simultaneously turning the thumb through an angle of 90° in the plane of the film.[157,67]

The arm of the operator is kept in a vertical straight line from the wrist to the shoulder and maximum pressure is exerted by the arm.

dry-to-handle time time interval between APPLICATION and ability to pick up without damage.[157] See DRYING TIME.[55]

dry-to-recoat time time interval between the APPLICATION of the coating and its ability to receive the next coat satisfactorily.[55]

dry-to-sand that stage of DRYING when a coating can be SANDED without the sandpaper sticking or clogging.[158]

Words presented in CAP/SMALL CAPS type indicate that the word is defined in another part of the *Coatings Encyclopedic Dictionary.*

Numerical superscripts classify terms in one or more of the categories listed in the second section of this volume.

dry-to-touch time interval between APPLICATION and tack-free time.[157,67,55] See SET-TO-TOUCH TIME.[67]

drying (1) PROCESS of change of coatings from the liquid to the solid state, due to EVAPORATION of the solvent, physico-chemical reactions of the binding medium, or a combination of these causes;[55] (2) process of removing moisture from pigments. It is carried out either in a current of hot air or in a vacuum.[67,74,131]

drying agent a substance which removes water (liquid or gaseous).[83] (IUPAC) Syn: desiccant

drying of ink the conversion of an INK film to a solid state.[168]

This can be accomplished by any of the following means, either singly or in combination: OXIDATION, EVAPORATION, POLYMERIZATION, PENETRATION, GELATION, PRECIPITATION.

drying oil an OIL which possesses to a marked degree the property of readily taking up oxygen from the air and changing to a relatively hard, tough, elastic substance when exposed in a thin film to the air.[63] (ASTM)

drying time (1) time required for an applied film of a coating to reach the desired stage of CURE, HARDNESS, or nontackiness. (See DRY, DRY-HARD, DUST-FREE TIME, SET-TO-TOUCH TIME, SAND-DRY, THROUGH-DRY, DRY-TO-HANDLE TIME, TOUCH-DRY, TACK-FREE, SURFACE DRYING, DRY-TO-RECOAT TIME, DRY-TO-SAND);[47,157,55] (2) the time required for an INK to form a tack-free surface after being applied to the paper, or other printed surface.[168,67]

dubbing out (1) filling in hollow and irregular surfaces and leveling walls with PLASTER before regular plasterwork; (2) forming, very roughly, a plaster cornice, before the final plaster coat is applied.[158] (DAC)

ductility tendency of a material to undergo DEFORMATION and extension without CRACKING or SPLITTING.[47]

ductor roller the ROLLER which is in intermittent contact with the FOUNTAIN ROLLER and transfers INK to the distribution system of the press.[168]

On a LITHOGRAPHIC PRESS, it is also the roller which transfers the FOUNTAIN SOLUTION to the dampening rollers.

dull rubbing rubbing a DRIED FILM of a coating to DULL FINISH, with an ABRASIVE PAPER, PUMICE, or STEEL WOOL and OIL or water.[54]

dullness (1) lack of LUSTER or GLOSS;[43,35] (2) COLORS of low CHROMA or SATURATION.[43,69]

dumbbell a piece of RUBBER cut in the shape of a dumbbell used for physical testing.[47,140,157]

Dumont's blue see COBALT BLUE.[41]

DuNouy Ring a method for measuring SURFACE TENSION under static or equilibrium conditions.[157,127]

duplex paper WALLPAPER which consists of two separate papers pasted together used to create a highly embossed effect.[148]

durability the degree to which coatings can withstand the destructive effect of the conditions to which they are subjected and how long they retain an acceptable APPEARANCE and continue to protect the SUBSTRATE.[35,154]

It is also important that the surface of the weathered coating should be suitable for RECOATING. A serious failure, such as FLAKING or CRACKING, requires more preparation for redecoration than an EROSION-type failure.

Durability can be assessed by physical tests, ACCELERATED WEATHERING and EXPOSURE panels, but most convincing evidence is provided by user trials carried out under site conditions.

No single property can be isolated as the key to good durability, but MECHANICAL PROPERTIES, ADHESION, PERMEABILITY, and retention of APPEARANCE are among those which are important.

durometer instrument for measuring the SURFACE HARDNESS of RUBBER, PLASTIC, or protective coatings.[47,145]

dust (1) small, dry solid particles projected into the air by natural forces, such as wind, volcanic eruption, and by mechanical or manmade processes such as crushing, GRINDING, MILLING, drilling, demolition, shoveling, conveying, screening, bagging, and sweeping.[64,171,139] (IUPAC) *Dust particles are usually in the size range from 1 to 100 μm in diameter, and they settle slowly under the influence of gravity*; (2) one of the size gradings of various gums, being composed of very small pieces, almost the size of dust.[128]

dust collector (1) a device for monitoring DUST emissions;[145,172] (IUPAC) (2) the equipment used to remove and collect dust from process exhaust gases; this may employ simply sedimentation (dustfall jars, coated slides, paper, etc.), inertial separation (cyclones, impactors, impingers, etc.), PRECIPITATION (thermal and electrostatic), or FILTRATION.[59,172] (IUPAC)

dust fall solid particles in the air which fall to the ground under the influence of gravity.[171,172] (IUPAC)

dust-free (1) descriptive of the stage in the DRYING of a paint or VARNISH film at which dust will no longer stick to the SURFACE;[67] (2) that stage of drying when cotton fibers, that have been dropped onto the film from a height of one inch, can be removed by blowing lightly across the surface of the film.[157] (DAC) Syn: cotton-free or cotton-free dry

dust-free time the time required for a freshly applied paint or compound to form a SKIN on its surface so that DUST will not adhere to it.[157] (DAC)

dusting bronzes see BRONZING.[168]

Dutch gold see STRIPING.[148,175]

Dutch metal leaves of bright brass which are used for overlaying in the same manner in which gold leaf is applied.[148] Syn: metal leaf and composition leaf

Dutch pink tint of red.[41]

The name is also used for certain yellow LAKES, prepared from QUERCITRON, Persian berries, or similar natural yellow coloring matters.

Dutch process old name for stack process formerly used for manufacture of WHITE LEAD.[41,131]

dye COLORANT which does not SCATTER LIGHT, but which absorbs certain WAVELENGTHS and TRANSMITS others.[41,69] See DYE PIGMENTS.[41]

Dyes are generally ORGANIC in nature and generally soluble in water or some other SOLVENT system; or they may exist in such a finely DISPERSED state that they do not scatter light and behave as though they were in SOLUTION. The dividing line between a DYE and a PIGMENT may, therefore, be indefinite and dependent on the particular total system involved.

dye, acid see ACID DYES.[41]

dye, basic see BASIC DYE.[41]

dye, mordant a range of DYESTUFFS, the COLORS of which are developed when PRECIPITATED onto bases to form LAKES.[41]

Often a salt-forming reaction takes place between the DYESTUFF and the base.

dye pigments DYES that by nature are insoluble in water and can be used directly as PIGMENTS without any chemical transformation.[41]

dye, spirit-soluble see SPIRIT-SOLUBLE DYE.[41]

dyes and dyestuffs ORGANIC COLORING substances.[41] (BSI)

PIGMENT DYESTUFFS are virtually insoluble in water and in the usual paint and varnish vehicles. Some examples of pigment dyestuffs are TOLUIDINE RED, PHTHALOCYANINE BLUE, and ARYLAMIDE YELLOW. Some dyes are soluble in water and organic solvents and are classed according to their solubilities, for example, water-soluble dyes, spirit-soluble dyes, and oil-soluble dyes.

dyestuff COLORANT which can be fixed firmly to a material to be dyed so as to be more or less fast to light, water, soap, etc.[41]

dynamic fatigue test see ALTERNATING STRESS AMPLITUDE.[47]

dyne absolute unit of force in the old CGS SYSTEM which, acting upon a mass of 1 g, will impart to it an acceleration of 1 cm per second squared.[128]

E

ΔE, Δe the total COLOR DIFFERENCE computed with a COLOR DIFFERENCE EQUATION.

It is generally calculated as the square root of the sum of the squares of the CHROMATICITY difference, ΔC, and the LIGHTNESS difference, ΔL: $\Delta E = [(\Delta C)^2 + (\Delta L)^2]^{1/2}$.

earth pigment class of PIGMENTS which are usually mined directly from the earth, dried, generally CALCINED and GROUND.[41] Syn: mineral pigments, natural pigments

Typical examples are red and yellow oxides of iron, yellow ochre, raw and burnt siennas, and raw and burnt umbers.

eave lower part of a roof projecting beyond the wall underneath.[82] See SOFFIT.[82,75]

EB see ELECTRON BEAM (EB) CURING.[67,55]

ebonize to blacken with paint or stain to look like ebony.[148] (DAC)

ecology the science of the relationship between organisms and their environment.[171]

eddy in turbulent fluid motion, a blob of the fluid that has some definitive character and moves in some ways differently from the main flow.[171,172] (IUPAC)

eddy diffusion see DIFFUSION.[142]

edge grain WOOD or VENEER so sawed that the annual rings form an angle of 45° to 90° with the surface of the piece.[153] (ASTM) Syn: vertical grain

edge runner mill MILL consisting of a horizontal, or substantially horizontal, circular pan which may be of cast iron or stone-lined.[59,131] Sometimes referred to as a putty chaser.

One or more circular rollers, which may also be of cast iron or suitable stone, are made to rotate edgewise around the pan. The material to be GROUND, which may be DRY PIGMENT or other POWDER or PIGMENT PASTE, is disintegrated between the surfaces of the rollers and the pan.

edgewise plane of the LAMINATE perpendicular to the laminations.[57]

EDP see ELECTRODEPOSITION.[61,77]

EDTA ethylenediamine tetraacetic acid or any of its salts; used as a chelating (sequestering) agent.[83]

effect finish any automotive FINISH which exhibits a significant change in any color attributes as a function of viewing or illumination geometry.[71,43,163]

Metallic paints, for example, refers to an effect finish containing metallic flake PIGMENT.

effervescence the vigorous evolution of gas which accompanies some chemical reactions, e.g., treatment of a CARBONATE with certain mineral acids.[144]

efflorescence (1) the reverse of DELIQUESCENCE: the drying of a salt solution when the vapor pressure of water in the saturated solution of a substrate is greater than the partial pressure of water in ambient air. Also refers to the loss water of crystallization from a solid salt such as $Na_2CO_3 \cdot 10H_2O$;[42,56,73] (IUPAC) (2) (for interior wall paints) a condition that occurs when soluble salts in the dried paint film or in the substrate migrate to the film surface during EXPOSURE.[42,56,73] (ASTM). *It should not be confused with other causes of nonuniform appearance or WHITENING that may occur at the time of application. Efflorescence is seen as either a light, medium, or heavy deposit of crystals. It exists both in isolated patches and over wide areas. ASTM Test Method D1736 can be used for this problem. Some studies have shown that efflorescence of interior paints is due to excessive sodium salts in the paint formulation and that this can be overcome by replacing the sodium salts with potassium salts or overwhelming existing sodium salts with three times the amount of potassium salts;*[42,56] (ASTM) (3) (for exterior paints) an incrustation of soluble salts, commonly white, deposited on the surface of coatings, stone, brick, plaster, or mortar; usually caused by salts or free alkalies leached from mortar or adjacent concrete as moisture moves through it. Syn: LAITANCE. *Efflorescence can affect the GLOSS and/or whiten the coating. Water-based coatings are most susceptible to this film defect. Since efflorescence results in the formation of white deposits on the film surface, it is more readily observed in TINTED coatings, especially deep tones.*

ASTM Test Method D1848 covers descriptions of properties that may be observed in weathered latex paint films. Efflorescence is included. The others are MOTTLING (or BLOTCHING), WATER SPOTTING, CRAWLING, PINHOLING, and MUDCRACKING.[42,56]

Efflorescence cannot be readily removed by wiping or with water only but can be removed with dilute mineral acid. ASTM Test Method D1848 can be used for this problem.

efflorescence/mottling nonuniform discoloration or whitening of a film, typically over masonry or cementitious substrates.[42,56] cf. EFFLORESCENCE and MOTTLING[42,56]

The basic cause may be leaching of soluble salts from the substrate (efflorescence) or an alkaline reaction with the vehicle and/or colorant (mottling). These two properties are grouped together because there is no practical way to distinguish them from each other. Mottling due to HYDROLYSIS *of the vehicle usually leads to adhesion type failures* (BLISTERING, FLAKING, *etc.).*

effluent any spent liquors or other waste material which are emitted by a source (waste from plating shops, pickling tanks, sewage treatment plants, chemical manufacturing plants, etc.).[171,172] (IUPAC)

effluent limitations any restrictions, established by the government or by management, on quantities, rates, and concentrations of chemical, physical, biological, and other constituents which are discharged from paint sources.[171]

efflux viscometer a cup type VISCOMETER containing an orifice.[166] See FORD CUP.[166]

Whatever PRECISION *and* ACCURACY *an efflux instrument may have is essentially dependent upon the dimension of the orifice. The closer an orifice resembles a* CAPILLARY, *the more accurate is the instrument.*

eggshell see GLOSS.[43,35,69] Associative terms: pearl, platinum, suede.

eggshell finish (1) low SHEEN (semimatte) surface which exhibits surface REFLECTANCE (GLOSS) similar to that from an EGGSHELL. Thus it is a gloss lying between FLAT and SEMIGLOSS and is generally thought to be between 20 and 35, as determined with a 60° GLOSSMETER; (2) paint which exhibits eggshell gloss.[43,35]

Egyptian asphaltum see SYRIAN ASPHALTUM.[76]

Egyptian blue blue PIGMENT, of ancient origin, which is substantially a copper silicate, obtained by CALCINATION.[41]

eicosanic acid see COSANIC ACIDS.[134]

elaidic acid $CH_3(CH_2)_7CH{=}CH(CH_2)_7COOH$. Properties: mp, 51°C; bp, 225°C/10 mm Hg; iodine value, 90.1.[144]

elastic ability of a substance to return to its approximate original shape or volume after distorting force on the substance has been removed.[47,140] See ELASTICITY.[47,140]

elastic limit maximum STRESS a material can withstand and still return to its original size and shape after the force is removed.[47,140]

elastic solid solid in which, for all values of the SHEARING STRESS below the rupture stress (SHEAR STRENGTH), the STRAIN is fully determined by the stress regardless of whether the stress is increasing or decreasing.[47,140]

elasticity property by which a material tends to recover its original size and shape immediately after removal of the force causing deformation.[47,140]

elastomer a natural or SYNTHETIC POLYMER which, at room temperature, can be stretched repeatedly to at least twice its original length and which after removal of the TENSILE load will immediately and forcibly return to approximately its original length.[47,140] (ASTM) See HYSTERESIS.[166]

elastomeric rubberlike; relating to or having the properties of ELASTOMERS.[47,140]

elaterite infusible ASPHALTIC pyrobitumen of PETROLEUM origin, sometimes described as mineral rubber or mineral CAOUTCHOUC.[76,167]

electric (electrical) of, relating to, or operated by electricity: electric current, an electric motor.[61,142]

electric strength Syn: dielectric strength

electricity the physical phenomena arising from the behavior of electrons and protons that is caused by the attraction of PARTICLES with opposite charges and repulsion of particles with the same charge.[61,142]

electrochemical methods of detection (analysis) methods in which either current or potential is measured during an electrochemical reaction.[37] (IUPAC)

The gas or liquid containing the trace impurity to be analyzed is sent through an electrochemical cell containing a liquid or solid ELECTROLYTE *and in which an electrochemical reaction specific to the impurity takes place.*

electrocoating see ELECTRODEPOSITION.[77]

electrocuring PROCESS which uses an ELECTRON BEAM to cure ORGANIC coatings applied to commercial products on a continuous production line.[67]

electrode potential difference in potential between an ELECTRODE and the immediately adjacent electrolyte referred to some standard electrode as zero potential.[61] (ASTM)

electrodeposition a form of DIP COATING in which an electric field is used to promote the DEPOSITION of a coating material onto the part. The part being painted acts as an electrode which is oppositely charged from the particles of paint in the dip

---

tank.[61,77] (EPA) See ANIONIC ELECTRODEPOSITION. cf. AUTODEPOSITION

This method of paint application in which an article to be coated which is an electric conductor is made one of the electrodes in a tank of water-thinned paint. The other electrode is generally a metal such as copper. The two electrodes are connected to a source of electric power, the polarity of the article to be coated being of the opposite sign to that of the particles in the liquid paint in the tank. The charged particles move towards the articles under the influence of the electric field, and when they give up their charge at the electrode (article) they are deposited and ultimately form a continuous film of paint. (BSI)

electrokinetic potential see ZETA POTENTIAL.[61,139]

electrolysis chemical change, especially DECOMPOSITION, produced in an ELECTROLYTE by an electric current.[61,144]

electrolyte a nonmetallic substance that dissociates into ions (rather than electrons) in solution or when fused, thereby becoming electrically conducting.[61]

electrolytic cell see CELL, ELECTROLYTIC.[61,66]

electrolytic white lead form of WHITE LEAD, chemically similar to the stack and CHAMBER types, which was made by the electrolysis of a solution of a lead salt followed by treatment with CARBON DIOXIDE gas.[41]

electromagnetic waves a wave of energy having a frequency within the electromagnetic spectrum and propagated as a periodic disturbance of the electromagnetic field when an electric charge oscillates or accelerates.[61]

electron a stable subatomic particle in the lepton family having a rest mass of 9.1066×10^{-28} gram and a unit negative electric charge of approximately 1.602×10^{-19} COULOMB.[61,133]

electron beam a method of CURING coatings which uses high energy radiation to CROSSLINK POLYMERS.[67,43] (EPA)

electron beam (EB) curing conversion of a coating from its application state to its final use state by means of a mechanism initiated by ELECTRON BEAM radiation generated by equipment designed for that purpose.[67,55] (ASTM) See CURE.[67]

electrophoresis the motion of charged particles, especially COLLOIDAL PARTICLES, through a relatively stationary liquid under the influence of an applied electrical field provided, in general, by immersed electrode.[61,139] Syn: cataphoresis

electrostatic of or relating to electric charges at rest.[61]

electrostatic detearing PROCESS of removing blobs and the thick edges of paint from an article which has been coated by DIPPING.[131,61] (BSI)

The process consists of passing the dipped article, after a limited period of draining, over a grid at a high electrical potential. The blobs and thick edges of paint are removed from the article by attraction to the grid.

electrostatic precipitation the removal of very fine particles suspended in a gas and subsequent PRECIPITATION onto a collector in a strong electric field.[61]

electrostatic spraying methods of APPLICATION SPRAYING in which an ELECTROSTATIC potential is created between the article to be coated and the atomized PAINT PARTICLES.[77] (BSI) cf. HAND-HELD ELECTROSTATIC SPRAYING[77]

The charged particles of paint are attracted to the article being painted and are there deposited and discharged. The electrostatic potential is used in some processes to aid the atomization of the paint.

Electrostatic air spray guns are similar in operation to conventional air SPRAY GUNS. These guns are made so that the paint will be negatively charged during atomization. The charged paint is attracted to any positively charged or grounded surface, preferably the part being painted. Electrostatic painting reduces OVERSPRAY and increases TRANSFER EFFICIENCY. HYDROCARBON emissions can be reduced through such overspray reduction and paint transfer improvements.

element a substance composed of atoms having an identical number of protons in each nucleus. Elements cannot be reduced to simpler substances by normal chemical means.[144,142]

See the PERIODIC TABLE OF THE ELEMENTS in this volume.

elemi gum any of various natural OLEORESINS derived from certain tropical trees, especially *Canarium luzonicum* of the Philippines, and used in making VARNISHES and INKS.[167,63]

Soluble in most ORGANIC SOLVENTS, and used chiefly to impart ELASTICITY and ADHESION to LACQUERS and varnishes.

eleomargaric acid synonym for ELEOSTEARIC ACID.[144]

eleostearic acid $CH_3(CH_2)_3(CH{=}CH)_3(CH_2)_7COOH$. Principal constituent ACID of TUNG OIL, characterized by the presence of three conjugated double bonds.[144] Syn. eleomargaric acid

Two forms are known, namely, a and ß types, with mps of 49°C and 71°C, respectively.

elongation the increase in specimen length from the point of initial load application to the point of film rupture in a tension test.[47,157] (ASTM)

Elongation measurements are of value in studying the behavior of coatings subjected to environmental stresses, such as those produced by aging and weathering. They may vary with specimen thickness, method of preparation, GAGE length, rate of load application, TENSILE TESTER response, and types of grips used.

ASTM Test Method D2370 covers the measurement of elongation as well as TENSILE STRENGTH and STIFFNESS (modulus of elasticity) of organic coatings when tested as free films.

Elpo see ELECTRODEPOSITION.[61,77]

elute to extract (one material) from another, usually by means of a fluid called the ELUENT.[164]

elutriation the PROCESS of separating the lighter particles of a powder from the heavier ones by means of an upward directed stream of fluid (gas or liquid).[131] (IUPAC)

embossed hardboard HARDBOARD that is manufactured with a TEXTURED surface.[75,153,159]

Wood-like and stucco patterns are examples of typical embossed hardboard surfaces.

embossed paper WALLPAPER run through ROLLERS with raised areas, to provide a light relief effect.[148]

embossing method of DECORATION in which the material is stamped or molded so that the pattern protrudes above the surface.[54,148]

emerald green applied to two distinctly different green PIGMENTS: complex copper acetoarsenite (Syn: Paris green, Schweinfurt green) and CHROMIUM HYDROXIDE.[41] See HYDRATED CHROMIUM OXIDE.[41]

emerald oxide of chromium see HYDRATED CHROMIUM OXIDE.[41]

emery a granular form of impure CARBORUNDUM®; used as an ABRASIVE for GRINDING and polishing GLASS, stone, and metal surfaces.[59,158] (DAC)

emery abrasive a natural composition of corundum and IRON OXIDE, found in large deposits both in the United States and the Near East.[59,158]

It is magnetic and partly soluble in hydrochloric acid. The grains are blocky, cut slowly, and tend to polish the material being abraded.

emery cloth coated ABRASIVE cloth used for light polishing of METAL or for removing RUST spots or similar light work; not recommended for large metal surfaces or tough metal, where ALUMINUM OXIDE CLOTH is much more satisfactory.[59,158]

emission the total rate at which a solid, liquid, or gaseous pollutant is emitted into the atmosphere from a given source; usually expressed as mass per unit time.[171,172,43,69] (IUPAC)

Primary emissions are those substances which are emitted directly to the atmosphere (e.g., NO, SO_2, etc.), while secondary emissions are formed from the primary emissions through thermal or photochemical reactions (e.g., ozone, ALDEHYDES, ketones, sulfuric acid, nitric acid, etc.). The point or area from which the discharge takes place is called the SOURCE; the area in which the emission or its transformed products (e.g., in the case of AEROSOLS, acidic deposition, etc.) may be deposited is called the receptor area or sink. Emission may be applied to noise, heat, etc., as well as pollutants.

emission control equipment air pollution control equipment which either converts the pollutant chemically to a nonpolluting substance or collects the pollutant by some means including gravity settling chambers, inertial separators, cyclonic separators, filters, electrical precipitators or scrubbers (spray towers, jet scrubbers, Venturi scrubbers, inertial scrubbers, mechanical scrubbers, and packed scrubbers).[59,171] (IUPAC)

Certain gases and bubble cap or sieve plate contactors, and packed towers, and adsorption units (packed beds or fluidized beds).

emission reduction the decrease in VOC emitted when (1) a low solvent coating is used in place of a higher solvent coating or (2) an add-on CONTROL DEVICE (such as CARBON adsorber or incinerator) is used on a process.[172] (EPA)

Emission reduction is often expressed as a percentage.

empirical formula type of formula which indicates the number and kind of atoms in a given COMPOUND, without providing information on the grouping of the atoms.[134,144]

For example, the empirical formula of OLEIC ACID is $C_{18}H_{34}O_2$.

emulgator see EMULSIFIER.[60,127,83]

emulsification the PROCESS of DISPERSING one liquid in another (the liquids being mutually insoluble or sparingly soluble in each other).[60,131]

When water is one of the liquids, two types of emulsions are possible: oil-in-water (water is the continuous state), and water-in-oil. The term "oil" describes any organic liquid sparingly soluble in water.

emulsified asphalt see ASPHALT EMULSION.[76]

emulsifier substance that intimately mixes, modifies the SURFACE TENSION of COLLOIDAL droplets, and DISPERSES dissimilar materials ordinarily IMMISCIBLE, such as oil and water, to produce a stable EMULSION.[60,127,83] Syn: emulsifying agent, emulgator

The emulsifier has the double task of promoting the emulsification and of stabilizing the finished product.

Words presented in CAP/SMALL CAPS type indicate that the word is defined in another part of the *Coatings Encyclopedic Dictionary.*

Numerical superscripts classify terms in one or more of the categories listed in the second section of this volume.

emulsifying agent see EMULSIFIER.[60,127,83]

emulsion two-phase liquid system in which small droplets of one liquid (the INTERNAL PHASE) are IMMISCIBLE in, and are DISPERSED uniformly throughout, a second CONTINUOUS LIQUID PHASE (the EXTERNAL PHASE).[60] cf. LATEX[130,155]

emulsion paint paint, the vehicle of which is an EMULSION of BINDER in water.[71,155]

The binder may be oil, OLEORESINOUS VARNISH, RESIN, or other emulsifiable binder. Not to be confused with a LATEX paint in which the VEHICLE is a LATEX.

emulsion polymerization PROCESS of POLYMERIZING a MONOMERIC material when it is present as the DISCONTINUOUS PHASE of an EMULSION.[60,169]

This process is to be distinguished from SOLUTION POLYMERIZATION and BULK POLYMERIZATION.

emulsion stabilizer see PROTECTIVE COLLOID.[60,83,139,161]

emulsoid colloidal particle which takes up water.[130,60] Syn: hydrophile

enamel (1) TOPCOAT which is characterized by its ability to form a smooth surface; originally associated with a high GLOSS, but may also include lower degrees of gloss, i.e., FLAT ENAMELS; (2) a class of substances having similar composition to GLASS with the addition of stannic oxide, SnO_2, or other infusible substances to render the enamel opaque.[43,71]

enamel blue see COBALT BLUE.[41]

enamel holdout the ability of a glossy ENAMEL (TOPCOAT) to dry to its normal uniform GLOSS when applied over a somewhat porous or nonuniform surface.[72] See HOLDOUT.[72,159] Syn: topcoat holdout

enantiotropic COMPOUND or an element able to exist in two distinct CRYSTALLINE forms, depending upon its maintenance above or below a certain transition temperature.[133]

encaustic (1) painted with a mixture of a paint solution and WAX which, after application, is set by heat; (2) COLORS which have been applied to brick, GLASS, porcelain, and tile and set by the application of heat.[71,81,137] (DAC)

encyclopedia where the emphasis is on concepts and meaning, and where general descriptions (rather than succinct DEFINITIONS) are presented.[134] (ASTM)

end grain the surface of TIMBER exposed when a tree is felled or when timber is cross-cut in any other way.[153,159]

end groups FUNCTIONAL GROUPS at either end of a POLYMERIC MOLECULE.[130]

These are normally different from the repeating unit group or groups.

end point (1) maximum DISTILLATION TEMPERATURE when a substance is distilled; (2) STOICHIOMETRIC point as shown by an indicator, potentiometer, or other means, in a titration; (3) required values of VISCOSITY, ACID VALUES, etc., the attainment of which indicates the conclusion of a PROCESS in RESIN or VARNISH manufacturing.[135,143,168]

end sealing compound a viscous coating used to form a seal or gasket between the end of a can and the can body.[163,71] (EPA)

endothermic designating or pertaining to a reaction which occurs with ABSORPTION of heat; opposite of EXOTHERMIC.[144]

energy at break see TOUGHNESS.[47,157]

English gallon see IMPERIAL GALLON.[128]

English vermillion see MERCURIC SULFIDE.[41]

engraving machine-printing (of WALLPAPER) with etched-out rollers to obtain subtle and fine effects not possible with surface printing.[148]

entrainment the act of forming a mist or fog droplets of a liquid carried off by the vapors of a boiling liquid or from a liquid through which bubbles of gas or vapor are passing rapidly.[142] (IUPAC)

Environmental Protection Agency (EPA) a Federal agency with environmental protection regulatory and enforcement authority.[174,172]

It administers the CAA, CWA, RCRA, TSCA, and other Federal laws and was created to permit coordinated and effective governmental action on behalf of the environment. It endeavors to abate and control pollution systematically, by proper integration of a variety of research, monitoring, standard setting, and enforcement activities. As a complement to its other activities, the agency coordinates and supports research and antipollution activities by State and local governments, private and public groups, individuals, and educational institutions. It also reinforces efforts among other Federal agencies with respect to the impact of their operations on the environment, and it is specifically charged with publishing its determinations when those hold a proposal is unsatisfactory from the standpoint of public health or welfare or environmental quality. In all, the EPA is designed to serve as the public's advocate for a livable environment.

Its address is 400 M St., S.W., Washington, D.C. 20460; Tel: (202) 382-2090.

enzyme any of numerous proteins or CONJUGATED PROTEINS produced by, and functioning as, biochemical CATALYSTS in living organisms.[173,74]

eosin red acid DYE used in the manufacture of LAKE PIGMENTS.[41]

Sodium salt of brominated FLUORESCEIN.

EPA abbreviation for ENVIRONMENTAL PROTECTION AGENCY.[174,172]

epichlorohydrin 1-Chloro-2,-3-epoxypropane. C_3H_5O Cl. It has a molecular weight of 92.51. INTERMEDIATE in the manufacture of RESINS.[83,141]

Also used as a STABILIZER in the formulation of CHLORINATED COMPOUNDS. Acts as an acid scavenger when used as stabilizer for chlorinated compounds.

epoxide a ring-shaped CHEMICAL STRUCTURE consisting of an oxygen atom bonded to two other atoms, usually CARBON, that are already bonded to each other.[67,133] See OXIRANE.[67,133] Syn: epoxy

epoxy group having the OXIRANE structure.[67,133,134]

epoxy adduct RESIN having all the required AMINE incorporated but requiring additional EPOXY RESIN for CURING.[130,156]

epoxy ester an EPOXY RESIN partially esterified with FATTY ACIDS, ROSIN, etc.; single package epoxy.[130,156]

epoxy paint paint based on an EPOXY RESIN.[71] (BSI)

epoxy plastics PLASTICS based on RESINS made by the reaction of epoxides or OXIRANES with other materials such as AMINES, ALCOHOLS, PHENOLS, CARBOXYLIC ACIDS, ACID ANHYDRIDES, and UNSATURATED COMPOUNDS.[38] (ASTM)

epoxy resins CROSSLINKING RESINS based on the reactivity of the EPOXIDE group. One common type is the RESIN made from EPICHLOROHYDRIN and BISPHENOL A.[71,133,156,130]

ALIPHATIC polyols such as GLYCEROL may be used instead of the aromatic bisphenol A or bisphenol F.

epsom salts $MgSO_4 \cdot 7H_2O$; Magnesium sulfate. Water-soluble magnesium salt for the preparation of PRECIPITATED SOAPS. Rhombic, colored or white; sp gr, 1.68.[144]

equalizer rod see WIRE-WOUND ROD.[145,157,77]

equilibrium a dynamic or static balance between opposing forces or actions.[144] (IUPAC)

A dynamic equilibrium exists between two different phases of a given pure compound (liquid-gas, gas-solid, solid-solid) when there is no net change in the amount of either phase when the two remain in contact; this occurs at a given temperature when the vapor pressures of each of the two phases of the compound are equal. A dynamic equilibrium temperature and pressure, when the reactions generating and destroying each reactant and product continue to occur, but there is no net change in the amounts of the reactants or products present.

ergonomics the study of human characteristics for the appropriate design of living and work environments.[172,173]

It involves people and their occupations; the equipment they use, their work environment, and the application of anatomical, physiological, and psychological knowledge. It is basically user-centered. Its aim is to improve the whole situation at work and play.

erioglaucine dye see ACID DYES.[41]

erlangen blue Syn: Prussian blue

erode see ETCH.[66]

erosion (1) wearing away of the top coating of a painted surface, e.g., by CHALKING, or by the ABRASIVE action of windborne particles of GRIT, which may result in EXPOSURE of the underlying surface; (2) phenomenon manifested in paint films by the wearing away of the FINISH to expose the SUBSTRATE or UNDERCOAT. The degree of failure is dependent on the amount of substrate or undercoat visible.[65,154,56,42]

erosion resistance the ability of a coating to withstand being gradually worn away by CHALKING or by the ABRASIVE action of water or windborne particles of GRIT.[56,72] See EROSION.[65,154,56,42]

The degree of RESISTANCE is dependent upon the amount of coating retained. The <u>FSCT Pictorial Standards of Coatings Defects</u> contains photographic standards for the degree of checking of exterior paints. ASTM Test Method D662 also has these same photographic reference standards.

erucic acid $CH_3(CH_2)_7CH{=}CH(CH_2)_{11}COOH$. UNSATURATED FATTY ACID found in many VEGETABLE OILS.[144] Syn: cis-13-do-cosenoic acid

Properties: mp, 33°C; bp, 281°C/30 mm Hg; sp gr, 0.860/55°C; and iodine value, 75.2.

erythrene 1-3 butadiene.[141]

erythritol $H(CHOH)_4H$. Tetrahydric ALCOHOL.[144]

Properties: mp, 112°C; bp, 330°C.

esparto wax hard vegetable WAX obtained from esparto grass (*Stipa tenacissma*), indigenous to North Africa and parts of Spain.[137]

It is used to some extent as a substitute for CARNAUBA, CANDELLILA, and similar hard waxes. Properties: mp, approx. 71°C; acid value, 30; saponification value, 75.

essential fatty acids the fatty acids which are synthesized in the human body.[173]

Words presented in CAP/SMALL CAPS type indicate that the word is defined in another part of the *Coatings Encyclopedic Dictionary.*

Numerical superscripts classify terms in one or more of the categories listed in the second section of this volume.

essential oils volatile OILS or essences derived from vegetation and characterized by distinctive odors and a substantial measure of RESISTANCE to HYDROLYSIS.[63]

Chemically, essential oils are often principally TERPENES. Some essential oils are nearly pure single COMPOUNDS. Some contain RESINS in solution and are then called OLEORESINS or BALSAMS.

ester ORGANIC SALT formed from an ALCOHOL (BASE) and an ORGANIC ACID by elimination of water.[134]

ester gum RESIN made from ROSIN or rosin acids and a polyhydric ALCOHOL, such as GLYCEROL or PENTAERYTHRITOL.[44,130] (ASTM)

ester interchange heating together of two or more ESTERS, under suitable conditions, causing an exchange of ACID units from one molecule to another.[144,131]

ester number difference between the SAPONIFICATION VALUE and the ACID NUMBER.[157,128]

Therefore, it is the number of mg of potassium hydroxide necessary to saponify the neutral ESTERS in one g of FAT, WAX, or OIL.

esterification PROCESS involving the interaction of a COMPOUND possessing a HYDROXYL GROUP with an ACID, with the elimination of water.[131,144]

eta (η) greek letter; symbol for ABSOLUTE VISCOSITY.[166]

etch wear away or roughen a surface with, or as if with, an ACID. Erode is sometimes used as a synonym.[158]

etching (1) in LITHOGRAPHY, the use of acidic substances to produce a surface in the nonprinting areas that is receptive to the FOUNTAIN SOLUTION but not to the INK; (2) in ENGRAVING, a treatment with ACID or by mechanical means to make certain areas considerably lower than the surface of the engraving.[168]

etching primer see WASH PRIMER.[71,158]

ethanite SYNTHETIC RUBBER made by the interaction of POTASSIUM POLYSULFIDE and ETHYLENE DICHLORIDE.[140]

ethanol see ETHYL ALCOHOL.[164]

ethanolamine $NH_2(CH_2)_2OH$. Bp, 171°C/150 mm Hg.[144] Syn: monoethanolamine, MEA, colamine, 2-amino ethanol, betahydroxyamine

ether ORGANIC COMPOUND in which two HYDROCARBON RADICALS are joined through an atom of oxygen, e.g., diethyl ether: C_2H_5–O–C_2H_5.[134,144]

etherified urea resins UREA RESINS in which the METHYLOL groups have been etherified with suitable ALCOHOLS.[130]

Alcohols commonly used include butyl and octyl alcohols. The etherified resins are characterized by improved HYDROCARBON tolerance, some types permitting substantial additions of ALIPHATIC HYDROCARBONS.

ethyl –C_2H_5. Monovalent alkyl RADICAL.[134]

ethyl abietate $C_{19}H_{29}COOC_2H_5$. Amber colored VISCOUS LIQUID which hardens upon OXIDATION.[144,151,167,170]

Properties: bp, 350°C; sp gr, 1.03; flp, 171°C (340°F).

n-ethyl acetanilide $C_6H_5N(C_2H_5)COCH_3$. Substitute for CAMPHOR in the NITROCELLULOSE industries.[144,170]

Properties: bp, 258°C, flp, 124°C (255°F).

ethyl acetate $CH_3COOC_2H_5$. SOLVENT for NITROCELLULOSE and for PHENOLFORMALDEHYDE RESINS in industrial ALCOHOL.[151,164] Syn: acetic ether and acetic ESTER

Properties: bp, 77°C; sp gr, 0.901/20°C; flp, 0.56°C (31°F); refractive index, 1.373; vp, 77 mm Hg/20°C.

ethyl acetyl ricinoleate $C_{17}H_{32}(OCOCH_3)COOC_2H_5$. Properties: bp, 400°C; sp gr, 0.931.[144,170]

ethyl acrylate CH_2:$CHCOOCH_2CH_3$. A colorless liquid, insoluble in water; MISCIBLE with most ORGANIC SOLVENTS.[141,144]

Used in the manufacture of SYNTHETIC RESINS. Properties: sp gr, 0.9283; bp, 101°C.

ethyl alcohol C_2H_5OH. A colorless, volatile liquid used as a SOLVENT in certain coatings such as SHELLAC.[144,164] Syn: alcohol, grain alcohol, ethanol, fermentation alcohol, spirit of wine, ethyl hydroxide, Cologne spirits, EtOH

Properties: bp, 78°C; sp gr, 0.7938/15°C; refractive index, 1.367.

ethyl carbamate see URETHANE.[141]

ethyl cellulose the ethyl ether of CELLULOSE; it is a white granular THERMOPLASTIC RESIN soluble in ORGANIC SOLVENTS.[130,151,156]

ethyl hydroxide see ETHYL ALCOHOL.[164]

ethyl lactate $CH_3CH(OH)COOC_2H_5$. Used as a SOLVENT for NITROCELLULOSE and CELLULOSE ACETATE. Colorless liquid with a mold odor.[144,164]

Properties: mol wt, 118; bp, 154°C; sp gr, 1.024/20°C; flp, 45°C (113°F); refractive index, 1.4111; vp, 5 mm Hg/33°C.

ethyl oxalate $(COOC_2H_5)_2$. Colorless, unstable, oily, aromatic liquid. Solvent for CELLULOSE ESTERS and ETHERS, many natural and synthetic RESINS.[164] Syn: diethyl oxalate

Properties: sp gr, 1.09 20/20°C; bp, 186°C; mp, -40.6°C; fp, 75.6°C (168°F).

ethyl propionate $CH_3CH_2COOC_2H_5$. Colorless liquid, fruity odor. Medium boiling solvent for NITROCELLULOSE.[151,164]

Properties: bp, 98°C; sp gr, 0.912/0°C; flp, 12°C (43.6°F); vp, 43 mm Hg/30°C.

ethyl succinate $(CH_2)_2(COO)_2(C_2H_5)_2$. Saturated PLASTICIZER.[144,170] Syn: diethyl succinate

Properties: bp, 216°C; sp gr, 1.049.

ethylbenzene $C_6H_5C_2H_5$. Colorless liquid with an aromatic odor.[144,164]

Used as an INTERMEDIATE in STYRENE production and as a SOLVENT. Properties: mol wt, 106.16; bp, 136°C; sp gr, 0.8673.

ethylene dichloride $ClCH_2CH_2Cl$. Colorless, oily liquid used for METAL DEGREASING; paint and VARNISH REMOVER.[164] Syn: 1,2 dichlorethane

Properties: bp, 83°C; sp gr, 1.252/20°C; vp, 65 mm Hg/20°C; mol wt, 99.

ethylene (ethene) $CH_2=CH_2$. A colorless gas of a faint, pleasant odor, obtained from petroleum sources.[144]

Properties: mol wt, 28; bp, -1.025°C; sp gr of liquid, 0.610/0°C.

ethylene glycol CH_2OHCH_2OH. A type of ALCOHOL, completely MISCIBLE in water, was used in LATEX and WATER-BASED PAINTS to provide STABILITY when frozen; used in heating and cooling systems as a fluid for TRANSFERING heat.[164,83] (DAC) Syn: ethylene alcohol

Used as a DIOL in manufacturing RESINS. Properties: bp, 197°C; sp gr, 1.12/4°C; flp, 111°C (232°F).

ethylene oxide C_2H_2O. Colorless, flammable gas at ordinary room temperature and pressure. It is used in organic syntheses, especially in the production of ETHYLENE GLYCOL and it is the starting material for the manufacture of ACRYLONITRILE and NONIONIC SURFACTANTS.[144,141] Syn: oxirane, anprolene

Properties: bp, 10.7°C; sp gr, 0.869.

ethylenediamine tetraacetic acid see EDTA.[83]

EtOH see ETHYL ALCOHOL.[164]

eupolymer POLYMER with a MOLECULAR WEIGHT over 10,000.[130]

evaporation the physical PROCESS by which a liquid substance is converted to a gas or vapor.[142] (IUPAC)

This may occur at or below the normal boiling point of the liquid (the temperature at which a liquid boils at 1 atmosphere pressure), and the process is ENDOTHERMIC.

evaporation rate a measure of the length of time required for a given amount of substance to evaporate (vaporize, volatilize), compared with time required for an equal amount of known material to evaporate.[157,164]

The known material is usually normal BUTYL ACETATE (N-BuAc or n-BuAc), with an evaporation rate designated as 1.0. Evaporation rates of other solvents or materials are then classified as (1) Fast evaporating if greater than 3.0, e.g., METHYL ETHYL KETONE (MEK), 3.8; ACETONE, 5.6; HEXANE, 8.3; (2) Medium evaporating if 0.8 to 3.0, e.g., 190-proof (95%) ETHYL ALCOHOL, 1.4; VM&P NAPHTHA, 1.4; MIBK, 1.6; (3) Slow evaporating if less than 0.8, e.g., XYLENE, 0.6; ISOBUTYL ALCOHOL, 0.6; NORMAL BUTYL ALCOHOL, 0.4; water, 0.3; MINERAL SPIRITS, 0.1.

evaporation rate, final time interval for complete EVAPORATION of all SOLVENT.[157,164]

evaporation rate, initial time interval during which low boiling SOLVENT EVAPORATES completely.[157,164]

evaporometer INSTRUMENT for measuring the EVAPORATION RATE of a liquid.[145,157]

excitation purity see PURITY, EXCITATION.[43,69]

exempt solvent any SOLVENT that has not been declared PHOTOCHEMICALLY REACTIVE by any of several regulatory agencies, most notably, the Los Angeles Air Pollution Control District.[164,171] See VOC.[164,172]

exfoliate SCALING from a SURFACE in FLAKES or layers.[42,45,65]

exothermic characterized by, or formed with, evolution of heat; opposite of ENDOTHERMIC.[144]

expansion joint a structural separation between two building elements that allows free movement between the elements without damage to the roofing or waterproofing system.[82] (ASTM)

explosive limits when COMBUSTIBLE VAPOR is mixed with air in the proper proportions, ignition will produce an explosion. This proper proportion is called the explosive range. The explosive range includes all concentrations of a mixture of flammable vapor or gas in air, in which a flash will occur or a flame will travel if the mixture is ignited. The lowest percentage at which this occurs is the lower explosive limit; and the highest percentage, the upper explosive limit.[172]

Explosive limits are expressed in percent by volume of vapor in air and, unless otherwise specified, under normal conditions of temperature and pressure.

Words presented in CAP/SMALL CAPS type indicate that the word is defined in another part of the *Coatings Encyclopedic Dictionary.*

Numerical superscripts classify terms in one or more of the categories listed in the second section of this volume.

exposure limits the concentration in workplace air of a chemical thought acceptable.[172] See MAXIMUM ALLOWABLE CONCENTRATION.[172]

This means that most workers can be exposed at the given levels or lower without harmful effects. The exposure limits in common use are (1) TLV-TWA (threshold limit value - time-weighted average); (2) STEL (short-term exposure limit) or STEV (short-term exposure value); (3) C (ceiling value).

exposure rack a frame on which test panels are exposed for DURABILITY TESTS.[154,157] See TEST FENCE.[154,157]

exposure tests tests which are conducted to evaluate the DURABILITY of a coating or film.[157]

They include exposure to ultraviolet light, moisture, cold, heat, salt water, mildew, etc. They can be generated either naturally or artificially.

ASTM test methods for exposure tests include: house paints on new wood, D1006; paints on steel surfaces, D1014, D5065; quantifying dirt collection, D3719; recording results on standard forms, D1150; wood panel substrates, D358.

expression removal of a liquid from a solid by pressing, as in the manufacture of VEGETABLE OILS from meal cakes.[63,131]

extender (1) see EXTENDER (PIGMENT).[58,41] (2) A TRANSPARENT or SEMITRANSPARENT white PIGMENT or a VARNISH that is used to alter the COLOR STRENGTH and working properties of an INK, without affecting its HUE. However, in FLAT paints, when used properly they can impart DRY HIDING and help spacing of TITANIUM DIOXIDE. Also, air voids in CALCINED CLAYS help hiding.[168]

The word EXTENDER has a pejorative connotation, however, it is true that they can be used to cheapen coatings. Their use is, however, invaluable in imparting desirable specific properties: aid SANDING, control SHEEN levels, improve ABRASION and BURNISHING resistance, affect RHEOLOGY, reinforce the film and increase BUILD and filling, and, as described above, contribute to hiding of flat paints.

Extenders can be conveniently divided into four groups: SULFATES—barytes and blanc fixe (both barium sulfates), GYPSUM (calcium sulfate); CARBONATES—whiting (calcium carbonate); SILICATES—CLAY (aluminum silicate), wollastonite (calcium silicate), TALC (magnesium silicate), MICA (aluminum potassium silicate); and OXIDES—silica, alumina.

extender (pigment) a specific group of ACHROMATIC pigments of low REFRACTIVE INDEX (between 1.45 and 1.70) incorporated into a vehicle system whose refractive index is in a range of 1.5 to 1.6.[58,41]

exterior basecoat a coating applied to the outside of a beverage can to provide both corrosion resistance and as a background for LITHOGRAPHY or PRINTING.[163,71] (EPA)

exterior finishes coatings which are expected to possess reasonable DURABILITY when exposed to natural WEATHERING.[71] See EXTERIOR PAINTS and VARNISHES.[71,152]

exterior paints and varnishes material formulated for use in conditions exposed to the weather.[71,152]

exterior type plywood PLYWOOD BONDED with a fully water-resistant glueline.[153,159,75]

external mix SPRAY EQUIPMENT in which fluid and air join outside of aircap.[59]

external phase of an emulsion synonym for the continuous phase.[60]

external plasticizer postadded PLASTICIZER as opposed to PLASTICIZATION by means of internally combined groups, such as COPOLYMERIZATION.[170] cf. INTERNAL PLASTICIZERS[170]

extinction coefficient an older synonym for ABSORPTION COEFFICIENT.[69,43]

extraction method by which the soluble parts of a substance are separated from the substance by a liquid.[157,131]

extrusion method whereby heated or unheated material forced through a shaping orifice becomes one continuously formed piece.[57,151] See COMPACTING and STRAINER.[38,131]

extrusion mark in EXTRUDED items, a cleft, gash, slit, or notch.[38]

extrusion moldings MOLDINGS which are made from plastic material by forcing it through a shaped orifice by means of pressure.[38,57]

exudation the MIGRATION of a substance to the SURFACE, such as RESIN from WOOD, PLASTICIZER from films.[42,56,154] cf. BLEEDING and SWEATING[69,42]

F

°F degrees Fahrenheit.[128] cf. °C

F.P.S. system of units the foot-pound-second system of units.[128] cf. C.G.S. EMU SYSTEM OF UNITS, SI UNITS and M.K.S. SYSTEM[128]

The British system of physical units derived from the three fundamental units of length, mass, and time, i.e., the foot, pound, mass, and the second.

f/cc fibers per cubic centimeter of air.[172]

fabric a cloth produced especially by knitting, weaving, or felting FIBERS.[159]

fabric coating a PROCESS which applies a uniform layer of POLYMERIC RESIN on a supporting FABRIC SUBSTRATE.[163,71] (EPA)

Typical coatings are RUBBERS, POLYURETHANES, VINYLS, and ACRYLICS. (EPA)

fabric printing a PROCESS which applies a DECORATIVE, usually multicolored, pattern on a FABRIC SUBSTRATE.[148] (EPA)

fabricate to work a material into a finished form by machining, forming, or other mechanical operations.[57]

facade one of the faces of a building, usually that containing the main entrance.[82]

facade paint a DECORATIVE and protective coating for exterior MASONRY SURFACES—usually for buildings and walls.[163]

This is a term more commonly used in Europe.

face color the COLOR observed on paints (particularly METALLIC paints) when they are viewed near to the normal (near to the perpendicular).[69]

It may also be referred to as the "TOP COLOR." It is generally used in contrast to the term "FLOP COLOR," which is the color observed at an angle near to the grazing angle (parallel to the surface). (EPA)

face putty triangular fillet of glazier's PUTTY on the exposed surface of GLASS.[165]

face velocity the velocity of air through a SPRAY BOOTH.[172] (EPA)

Where people are working, health regulations usually require this velocity to be at least 100 feet per minute. When ELECTROSTATIC SPRAYING is used a face velocity of 60 feet per minute is permitted. (EPA)

factice ELASTOMERIC products made by reaction of sulfur or sulfuric chloride with VEGETABLE OILS.[140]

factitious ultramarine synthetic ULTRAMARINE BLUE.[41]

fadding APPLICATION of SHELLAC LACQUERS by means of a PAD known as a "fad."[77,146,151] (BSI)

fadeometer an apparatus for determining the RESISTANCE of RESINS and other materials to FADING.[59,154] (DAC) Syn: fugitometer

It accelerates the fading by subjecting the article to high-intensity ULTRAVIOLET WAVELENGTHS similar to those found in sunlight.

fading subjective term used to describe the lightening of the COLOR of a PIGMENTED paint following EXPOSURE to the effects of light, heat, time, temperature, chemicals, etc.[42,43,56,69,154,35]

The observed fading may result from DETERIORATION of the pigment, from deterioration of the VEHICLE, or from a decrease in GLOSS. A separation of the vehicle from the pigment particle in the interior of the film, with the subsequent introduction of MICROVOIDS which SCATTER LIGHT, may also be interpreted visually as fading.

Fahrenheit scale temperature scale on which the freezing point of water is represented by 32°F (0°C) and the boiling point by 212°F (100°C) under normal pressure at sea level.[128]

falling sand abrasion test ABRASION RESISTANCE of coatings of paint, VARNISH, LACQUER, and related products is determined by the amount of abrasive (Ottawa) sand required to wear through a unit THICKNESS of the coating, when the sand falls against it at a specified angle from a specified height through a guide tube.[157] (ASTM)

fallout a measurement of air contamination consisting of the mass rate at which solid particles deposit from the atmosphere.[171,172] (IUPAC)

false body THIXOTROPIC FLOW property of a SUSPENSION or DISPERSION.[166]

When a composition "thins down" on stirring and "builds up" on standing it is said to exhibit false body. (See THIXOTROPY).[49,166] The term "false body" is also used in practice for "buttery" materials which are characterized by a relatively low VISCOSITY and high YIELD VALUE.

Words presented in CAP/SMALL CAPS type indicate that the word is defined in another part of the *Coatings Encyclopedic Dictionary.*

Numerical superscripts classify terms in one or more of the categories listed in the second section of this volume.

Farnsworth-Munsell 100 hue test test devised to check for both defective COLOR VISION and color discrimination ability.[69]

The test requires the subject to arrange four sets of randomized colors, contained in plastic caps, in order of HUE change. The test can be administered in about 15 minutes.

fast solvent SOLVENT that EVAPORATES rapidly under atmospheric conditions.[164]

fast to light see FASTNESS, FAST.[43,41,154]

fastness, fast the ability of a PIGMENTED or DYED material to resist COLOR change following EXPOSURE to the deleterious elements.[41,154] See LIGHTFASTNESS and COLORFASTNESS.[41,154,142] Syn: fast to light

fastopake inks compositions for PRINTING on WAXED SURFACES.[168]

fat edge accumulation of paint in the form of a ridge at the edge of a painted surface.[56]

fat turpentine oxidized TURPENTINE.[44,164]

Turpentine thickened by EXPOSURE to air at elevated temperature or for long periods at ambient temperature.

FATIPEC see FEDERATION D'ASSOCIATIONS DES TECHNICIENS DES PEINTURES, VERNIS, EMAUX ET D'IMPRIMERIE DE L'EUROPE CONTINENTALE.[174]

fats GLYCERYL ESTERS of higher FATTY ACIDS, such as STEARIC and PALMITIC.

fattening increase in CONSISTENCY of paint on STORAGE, not necessarily to such an extent as to make it unusable.[161,166] (BSI)

fatty acid ORGANIC ACID of the general formula $C_nH_{2n}O_2$, e.g., butyric acid and stearic acid; organic acids of ALIPHATIC or open chain structure.[134]

These are generally classified as SATURATED or UNSATURATED. The saturated fatty acids range from C_4 to C_{26}.

fatty acid pitches PITCH-like residues obtained from the distillation of FATS and OILS of animal or vegetable origin, or of their FATTY ACIDS.[144]

Fatty acid pitches include cottonseed pitch, palm oil pitch, wool grease pitch, etc. They have comparatively high acid and SAPONIFICATION VALUES.

fatty nitrogen products nitrogen-containing COMPOUNDS derived from vegetable and animal FATTY ACIDS.[134] (ASTM)

fatty paint paint which has thickened because of OXIDATION and POLYMERIZATION of the DRYING OIL VEHICLE during STORAGE.[42] (DAC)

faux hois (Fr.) painted DECORATION simulating marble, tortoise shell, or WOOD GRAIN.[54,148]

FDA abbreviation for FOOD AND DRUG ADMINISTRATION, an agency established under the Department of Health, Education, and Welfare, an executive branch of the Federal Government, whose function is to enforce the Federal Food, Drug and Cosmetic Act.[172,174]

feather edging reducing the THICKNESS of the edge of a dry paint film, e.g., the edge of a damaged area, prior to repainting.[54] (BSI)

feathering (1) operation of tapering off the edges of a coat of paint by laying off with a comparatively DRY BRUSH or ROLLER;[54] (2) PRINTING with irregular edges to the design;[168] (3) a ragged or feather edge which shows at the edge of type or cuts. It may be caused by poor INK distribution, bad impression, excessive ink, or an ink not suitable for the paper.[168] (BSI) Syn: laying off

feculose a mixture of ESTERS prepared by treating starch with glacial ACETIC ACID and washing the product free of uncombined acid.

Federal Register, The (FR) a daily publication that lists and discusses Federal regulations.[172]

Available from the Government Printing Office.

Federation d'Association des Technicians des Peintures, Vernis, Emaux, et d'Imprimerie de l'Europe Continentale (FATIPEC) [Federation of Associations of Technicians of Paint, Varnish, Lacquer and Printing Ink Industries of Continental Europe] an organization composed of the coatings technical associations of Belgium, France, Germany, Hungary, Italy, Netherlands, and Switzerland under a General Secretariat located in Paris.[174]

Founded in 1950, FATIPEC serves to promote and disseminate scientific and technical knowledge concerning the manufacture and use of paints, VARNISHES, ENAMELS and PRINTING INKS and to promote research and the interchange of ideas. FATIPEC is a member organization of the COATINGS SOCIETIES INTERNATIONAL.

Federation of Societies for Coatings Technology (FSCT) an organization composed of 26 Constituent regional Societies (22 in the U.S., two in Canada, one in England, and one in Mexico) whose members are the research and development, laboratory, technical, and production and manufacturing personnel employed in the protective and decorative coatings manufacturing industry.[174]

Founded in 1922, the FSCT was previously named the Federation of Paint and Varnish Production Clubs and, subsequently, the Federation of Societies for Paint Technology; in 1975 it assumed its current name.

The FSCT serves to provide materials and services for the technical education and professional development of its members and the industry at large. Acting through and with its Constituent Societies, and as a federation, the FSCT sponsors conventions, trade shows, seminars, symposia and committee activities which serve to gather and disseminate information on the technical and manufacturing aspects of the coatings industry.

The FSCT also produces a variety of educational publications and electronic media, including the monthly Journal of Coatings Technology.

Its address is: 492 Norristown Rd., Blue Bell, PA 19422; Tel: (610) 940-0777.

The FSCT is a member organization of the COATINGS SOCIETIES INTERNATIONAL.

feeding British term for LIVERING.[166,161]

feel a journeyman's term for the working qualities of a paint.[142]

feldspar $K_2O{\cdot}Al_2O_3{\cdot}6SiO_2$. Natural potassium aluminum silicate which occurs in granite.[58,41] Brit. Syn: felspar

The gradual DECOMPOSITION *of feldspar by natural agencies yields* CHINA CLAY *and* SILICA.

felt a flexible sheet manufactured by interlocking of fibers with a binder or through a combination of mechanical work, moisture, and heat.[75] (ASTM)

Felts are manufactured principally from VEGETABLE FIBERS *(organic felts),* ASBESTOS *fibers (asbestos felts), or* GLASS *fibers (glass fiber felts); other fibers may be present in each type. (ASTM) (It is doubtful whether asbestos fibers are still being manufactured.)*

felt and gravel roofing see BUILT-UP ROOFING.[75,76,82]

felting down operation of FLATTING down a dry film of VARNISH or paint by means of a PAD made of FELT or similar material, charged with a very fine ABRASIVE POWDER and lubricated with water or other suitable liquid.[54] (BSI)

fenchol $C_{10}H_{17}OH$. High boiling ALCOHOL present in TERPENE SOLVENTS.[164] Syn: fenchyl alcohol

Properties: bp of 201°C, sp gr of 0.964/20°C, and mp of 38°C.

fenchone $C_9H_{16}C{=}O$. Ketonic constituent of TERPENE SOLVENTS.[164]

Properties: bp of 193°C, sp gr of 0.945, refractive index of 1.4625, mp of 5°C.

fenchyl alcohol see FENCHOL.[164]

fermentation alcohol see ETHYL ALCOHOL.[164]

ferric oxide Fe_2O_3. Iron (III) OXIDE. Stable anhydrous oxide of IRON.[41] See IRON OXIDES.[41]

This constitutes the major ingredient in the chemically prepared synthetic reds, Indian reds, iron oxides, etc., but in the natural red oxides the proportion of ferric oxide varies considerably and is frequently associated with hydrated forms of the oxide.

ferriferro cyanide see IRON BLUE.[41]

ferrite yellow $Fe_2O_3{\cdot}xH_2O$. Pigment Yellow 42 (77492). Synthetic type IRON OXIDE with a COLOR range from bright yellow to dark orange. Manufactured by oxidizing ferrous hydroxide to the desired shade of hydrated ferric oxide.[41] See IRON OXIDES, SYNTHETIC.[41] Syn: yellow iron oxide

ferrous sulfate $FeSO_4{\cdot}7H_2O$.[41] Syn: copperas, and green vitriol

ferrule METAL portion of the BRUSH holding BRISTLES, hair, feathers, or other fibrous and flexible materials.[59,81]

festooning method of DRYING employed for heavy FABRICS, impregnated with VARNISH or OILS, which involves hanging the treated material over horizontal rods or poles in large drying rooms.[131]

While the impregnant is still wet, the fabric is moved gradually to avoid excessive accumulations of impregnant in the bottom of the folds. PROCESS *by which* WALLPAPER *is hung and dried after being* PRINTED. *Sticks carry the paper, in long loops, through a drying chamber.*

fiber any material in a form such that it has a minimum ratio of length to average maximum transverse dimension of 10 to 1, a maximum cross-sectional area of 7.85 x 10^{-5} in.^{2} corresponding to a circular cross section of 0.01 in. (0.254 mm in diameter) and a maximum transverse dimension of 0.010 in.[130] (ASTM)

*These include animal (wool); vegetable (cotton); mineral (*ASBESTOS*);* METAL *(*STEEL*); and synthetic (rayon, carbon, high* POLYMERS*).*

fiberboard building material composed of wood or other plant FIBERS BONDED together and COMPRESSED into rigid sheets.[75]

fibrous asbestos see MAGNESIUM SILICATE, FIBROUS.[41]

fibrous magnesium silicate see MAGNESIUM SILICATE, FIBROUS.[41]

field coat the coat(s) applied at the site of erection or FABRICATION.[54]

field painting SURFACE PREPARATION and painting operations of STRUCTURAL STEEL or other materials conducted at the project site.[158]

Words presented in CAP/SMALL CAPS type indicate that the word is defined in another part of the *Coatings Encyclopedic Dictionary*.

Numerical superscripts classify terms in one or more of the categories listed in the second section of this volume.

fiery finish see BURNED FINISH.[54]

figure (1) the pattern produced in WOOD SURFACE by irregular COLORATION and by annual growth rings, knots, and such deviations from regular GRAIN as interlocked and wavy grain;[153] (2) (numerical) an arithmetic value expressed by one or more digits.[36] (ASTM)

filament variety of FIBER characterized by extreme length which permits its use in yarn, having little or no twist and which usually does not require the spinning operation required for fibers.

filiform slender as a thread.

filiform corrosion a type of CORROSION that occurs under coatings on METAL SUBSTRATES characterized by a definite thread-like STRUCTURE and directional growth.[66]

filiform corrosion resistance the ability of a coating to RESIST that type of CORROSION of METAL SUBSTRATES characterized by a definite thread-like structure and directional growth that occurs under coatings.[72,66] See FILIFORM CORROSION.[66]

The FSCT Pictorial Standards of Coatings Defects contains photographic standards for the degree of filiform corrosion or organic coatings on metal from "Reference Standards for Reporting Filiform Corrosion," Detroit Society for Paint Technology (now "Coatings" Technology), Journal of Paint Technology, 44, No. 565 (1972).

ASTM Test Method D2803 covers two procedures for determining the susceptibility of organic-coated metal substrates to formation of filiform corrosion.

fill-type insulation loose INSULATING MATERIAL which is applied by hand or blown into wall spaces mechanically.[82]

filler (1) PIGMENTED composition used for filling fine CRACKS and indentations to obtain a smooth, even SURFACE PREPARATORY to painting;[147,158] cf. JOINT COMPOUND[147] (2) synonymous with EXTENDER;[58] (3) a thick coating applied to WOOD FURNITURE to fill large pores in the wood;[175] (4) a relatively nonadhesive substance added to an ADHESIVE to improve its working properties, permanence, strength, and other qualities;[79] (5) any COMPOUNDING ingredient, usually in dry, POWDER form, added to RUBBER in substantial amount to improve quality or lower cost. Fillers have various effects; some are relatively inert, like whiting or bartyes, and provide loading for cost reduction; others, like CARBON BLACKS, have a definite and desirable reinforcing effect with improvement in ABRASION RESISTANCE and other properties.[58,140] cf. JOINT COMPOUND[147]

filler coat a coat of paint, VARNISH, etc., used as a PRIMER.[71] (DAC)

filling of coated abrasives clogging of the ABRASIVE coat by SWARF.[59,158] Syn: loading

It can be reduced in many operations by using an open coat construction or a LUBRICANT.

filling up (1) covering the nonprinting areas (on a LITHOGRAPHIC plate) with a partially dried INK film. Effect caused by an excessive DRIER in the ink;[168] (2) a condition in the printing of HALFTONES where the ink fills areas between the dots, and produces a solid, rather sharp halftone print. This may also occur in the printing of type matter.[168]

film (1) a layer of one or more coats of paint or VARNISH covering an object or SURFACE;[55] (2) any supported or unsupported thin continuous covering or coating.[55,71] (DAC)

film applicator device to deposit a film of a specified THICKNESS, such as DOCTOR BLADE, WIRE WOUND ROD, DRAWDOWN BAR, etc.[77,59,157]

ASTM test methods for film applicators include: artists' paste paints, practice, D4941; blade applicators, D823; producing uniform films, D823; wire-wound (Meier) rods (bars), D4147.

film build the RHEOLOGICAL property which coatings possess of providing thickness in applied films.[166] See BUILD and HIGH BUILD.[166,133]

film former the part of a coating that remains on the SUBSTRATE after DRYING or CURING.[55]

Some film formers may be liquid but POLYMERIZE to form a solid when the coating is baked to the requisite curing temperature. (EPA)

film forming ability of a material to form a continuous dry film.[55]

film integrity continuity of a coating free of DEFECTS.[55]

film thickness THICKNESS of any applied coating, wet or dry.[128,157]

The thickness of the DRIED or CURED coating on the SUBSTRATE varies with the method of APPLICATION and the RHEOLOGY of the coating. Coating thickness on METAL generally ranges from 0.5 to 4 mils. Wet film thickness measurements aid in the prediction of dry film thickness. In instances where dry film thickness cannot be measured nondestructively, wet film thickness is frequently specified. Also, the ability to determine wet film thickness during application can provide the opportunity to correct the application procedures.

ASTM D4414 is a practice describing the use of thin rigid metal notched gages, also called step or comb gages, in the measurement of wet film thickness of organic coatings.

ASTM Test Method D1212 covers the determination of wet film thickness of organic coatings using the Interchemical wet thickness gage and the Pfund gage.

film thickness gage (gauge) device for measuring film THICKNESS; instruments for measuring either wet or dry films are available.[59,157]

ASTM test methods for film thickness gages include: dry films: by incision cut, Tooke gauge, D4138; on nonferrous metals (eddy), D1400; on steel, magnetic gauge, D1186; using micrometers, D1005; wet films: eccentric wheel, D1212; Interchemical and Pfund gages, D1212; notched gages, D4414.

filmogen FILM-FORMING materials.[55,71]

filter a porous material on which solid particles present in the air or in other fluid which flows through it are largely caught and retained.[59,131] (IUPAC)

Filters are made with a variety of materials: CELLULOSE and derivatives, GLASS FIBER, ceramic, SYNTHETIC PLASTICS, and fibers. Filters may be naturally porous or be made so by mechanical or other means. Membrane/ceramic filters are prepared with highly controlled pore size in a sheet of suitable material such as polyfluoroethylene, polycarbonate, or CELLULOSE ESTERS. Nylon mesh is sometimes used for reinforcement. The pores constitute 80 to 85% of the filter volume commonly, and several pore sizes are available for air sampling (0.45 to 0.8 μm are commonly employed). (IUPAC)

filter aid INERT, insoluble material, more or less finely divided, used as a filter medium or to assist in FILTRATION by maintaining adequate POROSITY of the FILTER CAKE.[131]

filter cake the solid MASS remaining on a FILTER after the liquid that contained it has passed through.[41,131]

filter, optical any uniform OPTICAL device which transmits RADIANT ENERGY of limited WAVELENGTHS and/or intensity.[43]

Filters may be made of GLASS, QUARTZ, PLASTIC, etc., containing absorbent material. They may consist of thin METALLIC coatings on glass or plastic; or they may consist of carefully made screens of uniform pore size. Optical filters may be used to modify the spectral distribution of radiant energy (selective filters) or the photometric intensity of radiant energy (neutral filters).

filter press apparatus for FILTERING, consisting of a number of flat chambers enclosed between sheets of filter cloth (PAPER or METAL) through which the liquid is forced by a pump, leaving the solid matter in the chambers.[59,131]

filter spectrophotometer ABRIDGED SPECTROPHOTOMETER, which uses filters of fixed, narrow band-pass transmissions of discrete WAVELENGTHS spaced across the spectrum, to measure the transmittance or reflectance of materials at these discrete wavelengths.[43,145]

The resulting spectral data arranged in order constitute an abridged spectrophotometric curve. Thus, the series of filters replaces the dispersion MONOCHROMATOR used in a continuous SPECTROPHOTOMETER.

filtration the PROCESS of SEGREGATION of phases; e.g., the SEPARATION of suspended solids from a liquid or gas, usually by forcing a carrier gas or liquid through a porous medium.[131] (IUPAC) See STRAINING.[131]

fin overflow material protruding from surface of CURED, molded articles, usually appearing at mold separation line or mold vent points.[140]

fine melt PROCESS of running COPALS at high temperatures as distinct from low temperature RUNNING, which is described as a slack melt.[131,167]

fine-etching a method of modifying a photomechanically prepared plate by controlled undercutting to change the size of the HALFTONE dots.[168]

fineness when applied to PIGMENTS, it describes the PARTICLE SIZE.[41,64]

When fineness is used in connection with the DISPERSION of pigments, it means the degree of dispersion of the pigment AGGLOMERATES.

fineness of dispersion a measure of the size and prevalence of oversize PARTICLES in the coating or INK.[64,128,157,168]

fineness of grind a numerical assessment of the degree of DISPERSION or of the presence of coarse PARTICLES in PIGMENTED coatings as determined by the protrusion of particles or AGGLOMERATES through the wet film at a given THICKNESS.[64,128,157]

ASTM test methods for dispersion (fineness of grind) include: coarse particles by sieve test, D185; dispersion stability, phthalo blue, D963; in pigment-vehicle systems, D1210; in printing inks, NPIRI grindometer, D1316; in titanium dioxide slurries, D3926.

fineness of grind gage a device to measure the FINENESS of DISPERSION of a PIGMENT, based on drawing a paint down with a steel blade over a channel machined in a stainless steel block of tapered depth and observing the minimum depth at which pigment particles are observed to interfere with the smooth wet SURFACE of the paint.[145,157]

In making pigmented products, the pigment is usually dispersed in a portion of the vehicle in some sort of mill. At this stage, it is necessary to be able to judge if the pigment agglomerates have been sufficiently broken up so not to interfere with the smoothness of the finished coating film.

Words presented in CAP/SMALL CAPS type indicate that the word is defined in another part of the *Coatings Encyclopedic Dictionary.*

Numerical superscripts classify terms in one or more of the categories listed in the second section of this volume.

ASTM Test Method D1210 covers the measurement of the degree of dispersion (commonly referred to as "fineness of grind") of the pigment in a pigment-vehicle system such as liquid coatings and their intermediates. It may also be used to assess the inclusion of particulates by a cleanliness rating.

fines portion of a POWDER composed of PARTICLES which are smaller than a specified size.[128]

These include finely crushed or powdered material or fibers, especially those smaller than the average in a mix of various sizes.

fingernail test gouging a dried film with fingernail to form a subjective, qualitative estimate of the relative HARDNESS and TOUGHNESS.[157]

fingers classification of fossil COPALS, especially of the CONGO type.[167]

The pieces of RESIN have long cylindrical shapes, similar to those of human fingers.

finish (1) final coat in a paint system; (2) sometimes refers to the entire coating system: the texture, color, and smoothness of a surface, and other properties affecting APPEARANCE.[54,71] cf. COATING and PAINT[71]

finish coat (1) the final layer of PLASTER applied over a BASECOAT or other SUBSTRATE. (ASTM) (2) See TOPCOAT.[54,71] See COATING.[71]

finn oil see TALL OIL.[44]

fire clay an earthy or stony mineral AGGREGATE which has the essential constituent hydrous silicates of ALUMINUM with or without free silica.[41,34]

It is PLASTIC when sufficiently pulverized and wetted, rigid when subsequently dried, and of suitable refractoriness for use in commercial refractory products.

Fire Diamond NFPA's hazard rating per NFPA 704 publication. Visual system that provides a general idea of the inherent hazards, and their severity, of materials relating to fire prevention, EXPOSURE, and control. Preferred reading order: health, flammability, reactivity, special.[172]

Position A—Health Hazard (Blue). Degree of hazard; level of short term protection.

0 = ordinary combustible hazards in a fire
1 = slightly hazardous
2 = hazardous
3 = extreme danger
4 = deadly

Position B—Flammability (Red). Susceptibility to burning.

0 = will not burn
1 = will ignite if preheated
2 = will ignite if moderately heated
3 = will ignite at most ambient conditions
4 = burns readily at ambient conditions

Position C—Reactivity, Instability (Yellow). Energy released if burned, decomposed, or mixed.

0 = stable and not reactive with water
1 = unstable if heated
2 = violent chemical change
3 = shock and heat may detonate
4 = may detonate

Position D—Special Hazard (White).
Position E—If two specific hazards exist.

OXY = oxidizer
ACID = acid
ALKALI = alkali
COR = corrosion
~~W~~ = use no water, reacts!
 *= radiation hazard.*

fire point the lowest temperature at which a liquid produces sufficient vapor to flash near its surface and continues to burn. Usually 10 to 30°C higher than the flash point.[172]

fire resistance the property of a material or assembly to withstand fire or give protection from it.[72,172]

As applied to elements of buildings, it is characterized by the ability to confine a fire or to continue to perform a given structural function or both.

fire resistive refers to properties of materials or designs to resist the effects of any fire to which the material or structure may be expected to be subjected.[172]

Fire resistive materials or structures are NONCOMBUSTIBLE, but noncombustible materials are not necessarily fire resistive. Fire resistive implies a higher degree of fire resistance than noncombustible.

fire retardant implies that the described product, under accepted methods of test, will significantly: (a) reduce the rate of FLAME SPREAD on the surface of a material to which it has been applied; (b) resist ignition when exposed to high temperatures; or (c) insulate a substrate to which it has been applied and prolong the time required to reach its ignition, melting, or structural-weakening temperature.[163,172] (ASTM)

fire-resisting finish the preferred term is "FIRE-RETARDANT COATING."[172,163]

fire-retardant chemical a chemical or chemical preparation used to reduce FLAMMABILITY or to retard the spread of flame.[83,172] (DAC)

fire-retardant coating a coating which will do one or more of the following: (1) reduce the FLAME SPREAD over which the coating is applied, sometimes at the sacrifice of the coating; See INTUMESCENT COATING;[71,163,172] (2) resist ignition when exposed to high temperature; or (3) INSULATE SUBSTRATE to which it is applied and thereby prolong time required to reach ignition, melting, or structural-weakening temperature.[71,163,172] (ASTM)

firewall any wall built for the purpose of restricting the spread of fire in a building.[82,172]

Such walls of solid MASONRY *or* CONCRETE *usually divide a building from the foundation to about a metre above the roof.*

first coat first coating applied in any painting schedule; in some cases, it would be the SEALING COAT; in others, the PRIMING COAT.[54,71]

fish eyes (1) paint DEFECT which manifests itself by the CRAWLING of wet paint into a recognized pattern resembling small "dimples" or "fish eyes";[56] (2) small globular MASS which has not blended completely into the surrounding material and is particularly evident in a TRANSPARENT or TRANSLUCENT MATERIAL.[56] (ASTM)

In RUBBER *technology, such globules are known as "cat eyes."*

fish oil a natural OIL extracted from fish and generally characterized by a rather large group of SATURATED FATTY ACIDS commonly associated with mixed TRIGLYCERIDES.[63]

The fatty acids derived from fish oils are three principal types: saturated, monounsaturated, and polyunsaturated with carbon chain lengths ranging from C_{12} *to* C_{24}*. Both the type and relative amounts of the fatty chains vary widely among different species of fish and different geographical areas. The fish oil commonly used in the coatings industry is* MENHADEN OIL, *produced from menhaden fish caught along the Atlantic Coast of the United States. This oil has an iodine value of about 175. Other common types of fish oils and their iodine values are: herring –134, anchovy –199. In addition to glycerides of stearic and the lesser unsaturated fatty acids, fish oils contain glycerides of clupanodonic acid, which appears to contain four* DOUBLE BONDS. *The iodine value varies over a wide range, approximately 130 to 190. The tendency of fish oil films to yellow considerably is due to the presence of highly* UNSATURATED *groups in the molecule.*

fitch long handled small BRUSH bound with METAL, with which nearly inaccessible areas are painted.[59]

fixative SOLUTION which can be SPRAYED onto drawings rendered in pencil, chalk, and other impermanent and easily removed materials so as to fix them and prevent SMUDGING.[54,71]

fixed oil see NONDRYING OIL.[63]

flag the end of a BRUSH BRISTLE which divides into two or more branches.[59]

Flagging provides the brush with the ability to hold a greater amount of liquid coatings.

flair a subjective term applied to the change in HUE of a COLORED material when the light source is changed. See METAMERISM.[43,69,35]

A color may appear blue in daylight but change to a purple-blue in incandescent light; ULTRAMARINE BLUE *is an example of such a blue, so it is said to "flair" red or to have a red "flair." Opposite of* COLOR CONSTANCY; *not to be confused with* METAMERISM, *which applies to a pair of colors.*

flake a flat, thin piece or layer; a chip.[133]

flake board same as PARTICLE BOARD.[75,153,159] (DAC)

flake metal see METALLIC PIGMENT.[41,45]

flake shellac ORANGE SHELLAC GUM in flake form.[146]

flake white (1) another name for WHITE LEAD of fine texture that was prepared electrolytically; (2) variety of white lead, that was used by artists.[41] (BSI)

flaking that phenomenon manifested in paint films by the actual detachment of pieces of the film itself either from its SUBSTRATE or from paint previously applied.[42,56] (ASTM) cf. PEELING[42] Syn: scaling

Flaking is generally preceded by CRACKING *or* CHECKING *or* BLISTERING *and is the result of loss of* ADHESION, *usually due to* STRESS-STRAIN *factors coming into play.*

flaking resistance the ability of a coating to resist the actual detachment of film fragments, either from its UNDERCOATING or SUBSTRATE.[56,72] Syn: scaling resistance

FLAKING *is generally preceded by* CRACKING, CHECKING, *or* BLISTERING *and is the result of loss of* ADHESION.

ASTM Test Method D772 covers the evaluation of the degree of flaking (scaling) of exterior paints by comparison with photographic standards. The FSCT Pictorial Standards of Coatings Defects *contains the same photographic standards.*

flamboyant finish a GLOSSY TRANSPARENT coating with or without COLORANT, over a bright UNDERCOAT, METALLIC SURFACE, or metallic FINISH.[54,163]

flame the visible heat rays which appear when the ignition of a material is reached.[172]

Hydrogen is one of the exceptions since the heat rays are not visible.

flame cleaning impingement of an intensely hot flame to the surface of structural steel resulting in the removal of mill scale and the dehydration of any remaining rust, leaving the surface in a

Words presented in CAP/SMALL CAPS type indicate that the word is defined in another part of the *Coatings Encyclopedic Dictionary.*

Numerical superscripts classify terms in one or more of the categories listed in the second section of this volume.

condition suitable for wire brushing followed by the immediate application of paint.[45,158] (BSI)

flame (flash) arrestor devices utilized on VENTS for FLAMMABLE LIQUID or gas tanks, storage containers, cans, gas lines, or flammable liquid pipelines to prevent flashback (movement of flame) through the line or into the container, when a flammable or explosive mixture is ignited.[172]

Wire screen of 40 mesh is utilized on smaller openings. On larger openings, parallel metal plates or tubes are more effective.

flame proofing see FIRE RETARDANT.[172]

flame spray any PROCESS whereby a material is brought to its MELTING POINT and SPRAYED onto a SURFACE to produce a coating. The process includes: (1) METALLIZING; (2) THERMOSPRAY; and (3) PLASMA FLAME.[77] cf. PLASMA DEPOSITION[77] and PLASMA COATING[77]

flame spread flaming combustion along a SURFACE; not to be confused with flame transfer by air currents.[135,172]

flammable any SUBSTANCE that is easily ignited, burns intensely, or has a rapid rate of FLAME SPREAD.[172]

Flammable and INFLAMMABLE are identical in meaning, however, the prefix "in" indicates negative in many words and can cause confusion. Flammable, therefore, is the preferred term.

flammable limits see EXPLOSIVE LIMITS.[172]

flammable liquid any liquid having an flp below 37.8°C (100°F), except a mixture having components with flp of 37.8°C (100°F) or higher, the volume of which make up 99% or more of the total volume of the mixture.[172] See COMBUSTIBLE LIQUID.[172]

Flammable liquids are known as Class I liquids, and are divided into three classes: Class IA—liquids having an flp below 22.8°C (73°F) and a bp below 37.8°C (100°F); Class IB—liquids having an flp below 22.8°C (73°F) and a bp at or above 37.8°C (100°F); and Class IC—liquids having an flp at or above 22.8°C (73°F) but below 37.8°C (100°F).

flash drying (1) rapid method of DRYING brought about by the EXPOSURE of a coating SURFACE to an elevated temperature or forced draft for a short period of time;[54,55] (BSI) (2) drying at ambient temperature, preparatory for a period of forced drying at elevated temperatures.[54,55] See FLASHOFF, FLASHOFF TIME, and FLASHOFF ZONE.[54]

flash point (flp) the lowest temperature of a liquid at which it gives off sufficient vapor to form an ignitible mixture with the air near the surface of the liquid or within the vessel used.

The flash point can be determined by the OPEN CUP or the CLOSED CUP method. The flash point determined by the open cup method is usually somewhat higher than the closed cup method. The latter is commonly used to determine the classification of liquids which flash in the ordinary temperature range. Open Cup Methods: Tagliabue, Cleveland Open Cup. Closed Cup Methods: Abel-Pensky Closed Cup, Tag Closed Cup, and Setaflash Tester are some of the methods used.

flash rusting is a form of CORROSION that occurs during the DRYING PROCESS of a WATER-BASED coating. Water-soluble corrosion products migrate to the surface of a coating and appear as a STAIN.[42,66]

An example of this is the appearance of a brown stain a few minutes after a plain steel nail is coated with a white LATEX paint. The currently accepted test method used to evaluate coatings as to their performance in flash rusting situations can be found in ASTM D2247-68. A more sophisticated method can be found in Journal of Coatings Technology, 65, No. 827, p. 45 (1983).

ADDITIVES used for the inhibition of flash rusting are basically of two types. The first is a liquid or soluble additive such as alkylamines, aminoalcohols, and others, which are used at 0.1 to 2% on the total weight of the formula. The second type is corrosion inhibiting pigments which must be ground into the paint and are only trace soluble in the water media. Both inhibitor types are also of direct benefit to the paint manufacturer in that they will inhibit the corrosion of the can.

flash-off time time allowed to elapse between the SPRAY APPLICATION of successive WET-ON-WET COATS or the time allowed for the EVAPORATION of the bulk of the SOLVENT before entering into a baking oven.[54] See FLASH DRYING,[54,55] FLASH OFF,[54] and FLASHOFF ZONE.[54]

flashing (1) a paint DEFECT in a paint film in which patches glossier than the general FINISH develop, especially at JOINTS or LAPS in the coating;[43,56] (2) the nonuniform APPEARANCE, including spotty differences in COLOR or gloss, usually due to improper or nonuniform sealing of a porous SUBSTRATE;[56] (3) NONCORROSIVE METAL used around angles or junctions in roofs and exterior walls to prevent leaks;[75] (4) the system used to seal MEMBRANE edges at walls, EXPANSION JOINTS, DRAINS, gravel stops, and other places where the membrane is interrupted or terminated.[82] (ASTM)

Base flashing covers the edges of the membrane. Cap or counterflashing shields the upper edges of the base flashing.

flashoff causing the greater part of the more VOLATILE SOLVENTS in a SPRAYED coat of LACQUER or ENAMEL to evaporate before proceeding with the APPLICATION of another coat.[54] See FLASH DRYING, FLASHOFF TIME and FLASHOFF ZONE.[54]

flashoff zone the area within the plant where SOLVENTS EVAPORATE from a coating during the interval between coats or before the painted object enters the bake oven.[164,172] (EPA) See FLASHOFF,[54] FLASHOFF TIME,[54,55] and FLASH DRYING.[54,55]

flat (verb) See FLATTING DOWN, and RUBBING.[54,158]

flat see GLOSS.[43] Syn: matte, velvet

flat coat coat of FILLER (British) also incorrectly used when meaning prime coat under ENAMEL.[71] (DAC)

The preferred term is "ENAMEL UNDERCOATER."[71] An INTERMEDIATE coat of paint used as a BASE for a TOPCOAT.

flat enamel PIGMENTED coating of low SPECULAR GLOSS, which has the leveling characteristics of a gloss ENAMEL.[43,71] See ENAMEL.[43,71]

Enamel has the connotation of both gloss and FLOW.

flat finish see FLAT PAINT (FINISH).[43,71]

flat grain WOOD or VENEER so sawed that the annual rings form an angle of less than 45° with the surface of the piece.[153] (ASTM)

flat lacquer a LACQUER with the appearance of having been rubbed after it has dried.[151]

flat paint brush see FLAT WALL BRUSH.[59]

flat paint (finish) paint which dries to a surface which scatters the light falling on it, so as to be substantially free from GLOSS or SHEEN.[43,71] See MATTE FINISH.[43,71,69] Syn: flat finish

flat screen printer a semicontinuous PROCESS for PRINTING on FABRIC.[131,148,77,81] (EPA) See SILK-SCREENING.[77,81,148]

The fabric is placed on a belt, the belt and fabric are moved into position under the flat screen, and the belt is stopped. The flat screen, on which print paste has been applied, is mechanically lowered onto the fabric. A squeegee moves across the screen forcing print paste through a pattern onto the fabric. The screen is then raised and the fabrics and belt are indexed to the next position to receive another screen and another color on the pattern.

flat spot an imperfection on a glossy painted surface; a spot lacking GLOSS, usually caused by a porous spot on the UNDERCOAT.[42,56] (DAC)

flat stone mill type of GRINDING MILL in which the material to be disintegrated is fed between the grinding surfaces of two flat stones, one of which is caused to rotate.[49,59,64]

The grinding surfaces are specially prepared with grooves or channels. These mills were used for either dry or wet grinding.

flat varnish VARNISH formulated with a FLATTING agent so as to dry with a dull FINISH. (Not to be confused with a flatting or rubbing varnish.)[58,71,152] Syn: matte varnish

flat wall brush a paint BRUSH, usually 4 to 6 in. (10 to 15 cm) in width, with long, stiff BRISTLES, usually made of synthetic fiber.[59] (DAC) Syn: flat paint brush

flattening agent see FLATTING AGENT.[58]

flatting agent any material which, when added to a coating or PLASTIC, lowers the GLOSS of the final surface.[58,83] See FLATTING PIGMENT.[58,41]

flatting down cutting or RUBBING down the SURFACE of a paint or VARNISH with fine ABRASIVES to produce a smooth, dull surface.[54,158] (BSI)

flatting oil a VARNISH-like composition made of heavy-bodied OIL dissolved in a THINNER, used to reduce paste paint to a FLAT paint.[63,152]

flatting or flattening (1) undesirable loss of GLOSS during DRYING; (2) addition of a FLATTING PIGMENT or agent to a paint or VARNISH; (3) addition of a FLATTING OIL to a paint or ENAMEL.[43,56]

flatting pigment any finely divided PARTICLE added to a paint formulation in order to decrease the GLOSS of the dried film. It may be an EXTENDER-type (low REFRACTIVE INDEX) or a HIDING pigment, and it is generally nonchromatic (neutral near-white), although not necessarily so.[41,58] See EXTENDER (PIGMENT), and DRY-HIDING.[4143,55]

The reduction in gloss comes about from surface-light scattering which occurs when light strikes the pigments protruding at the surface.

flatting (rubbing) varnish VARNISH containing a high proportion of hard RESIN which can be rubbed down after APPLICATION, to produce a smooth foundation for a finishing coat, or serve as the FINISH coat itself. (Not to be confused with a FLAT VARNISH.)[54,71,152,158] Syn: rubbing varnish, and polishing varnish

fleshing see FLUSHING.[41,131]

flex-cracking development of small CRACKS in FLEXIBLE articles or coatings when these articles are subjected to repeated flexing or bending.[42,47,56]

flexibility degree to which a coating after DRYING is able to conform to movement or deformation of its supporting surface, without CRACKING or FLAKING.[47,157] See IMPACT RESISTANCE.[47,157,72]

ASTM test methods for flexibility include: impact resistance, D2794; mandrel bend test, D522; on prepainted metal sheets, D4145. ASTM Test Method D2794 covers a procedure for rap-

Words presented in CAP/SMALL CAPS type indicate that the word is defined in another part of the *Coatings Encyclopedic Dictionary.*

Numerical superscripts classify terms in one or more of the categories listed in the second section of this volume.

idly deforming by impact of a coating film and its substrate and for evaluating the effect of such deformation.

flexibility test test applied to films to ascertain if they are able to accommodate ELONGATIONS without fracture.[47,157] Syn: bend test

FLEXIBILITY is usually determined by bending the film, applied to a suitable piece of thin METAL, around one or a series of rods or mandrels or a special conical mandrel in a specified time.

flexible wall coverings those which are pliable, such as PAPER, manmade vinyls, and FABRICS, as opposed to "liquid" wall coatings, such as paint, or "rigid" like WOOD paneling or other solid wall products, such as ceramic tile.[75,148]

flexographic ink quick DRYING, low VISCOSITY INK based on VOLATILE SOLVENTS that are used in the FLEXOGRAPHIC PRINTING PROCESS.[168]

flexography a typographic form of printing using RUBBER plates and relatively thin bodied, highly volatile RESIN-SOLVENT INKS.[168]

Formerly known as ANILINE PRINTING. In this method of printing, the image areas are raised above the non-image areas. The image carrier is made of rubber or other ELASTOMERIC material. (EPA)

flint abrasive not flint at all, but actually a natural quartz (silicon dioxide) which fractures into sharp-edged grains and is used on the common SANDPAPER for WOOD.[158]

flint paper an inexpensive PAPER using a natural quartz as the ABRASIVE, for small sanding jobs on WOOD.[59,158]

It tends to clog quickly and cannot be used for wet sanding.

flip-flop DECORATIVE effect with AUTOMOTIVE METALLIZED FINISHES in which there is a difference in color depth and flash when the FINISH is viewed at different angles.[43,69,35] See TRAVEL.[43,69]

Generally used in APPLICATION to automotive metallic finishes.

floating DEFECT which is sometimes apparent in COLORED paints containing one or more mixtures of different PIGMENTS. During DRYING or on STORAGE, one or more of the pigments separates or FLOATS apart from the others and concentrates in streaks or patches on the surface of the paint, producing a variegated effect.[42,56] cf. FLOODING[42,56]

This effect also occurs with single pigment type consisting of multiple particle sizes. This is believed by some to be the color consequence of BÉNARD (VORTEX) CELL circulation, observed as mottled, blotchy, or streaked APPEARANCE of a paint film.

floating coat see BROWN COAT.[73,75]

floc loose, open-structure mass formed in a SUSPENSION by the AGGREGATION of minute PARTICLES.[139]

flocculate cluster of PIGMENT PARTICLES formed in a paint after the pigment has been wetted or DISPERSED.[42,56,142] (ASTM)

The spaces between the particles are filled with VEHICLE, as distinguished from AGGLOMERATES, where the spaces between the particles are filled with air.[56] A group of two or more attached particles held together by physical forces such as surface tension, adsorption, or similar forces. Flocculates are much affected by the sample preparation and technique.

flocculation the PROCESS of contact and ADHESION whereby the PARTICLES of a DISPERSION form larger size clusters.[42,139,161] (IUPAC) cf. AGGLOMERATION and COAGULATION[42,139,161]

IUPAC defines it synonymously with agglomeration and coagulation; however, other references do not. In coatings it is generally the formation of clusters of pigment particles in a fluid medium which may occur after dispersion has been effected. The condition is usually reversible and particle clusters can be broken up by the application of relatively weak mechanical forces or by a change in the physical forces at the interface between the liquid and solid dispersed particles. Microscopically, flocculation appears as lacework or reticulum of loosely clustered particles. It often results in a more rapid settling and shows a loss in color strength. The use of an efficient dispersant at the optimized concentration can eliminate or minimize this problem.

flocculation of the vehicle in the PRINTING INK industry, sometimes used to mean LIVERING.[168]

floccule a small loosely AGGREGATED MASS of material suspended in or precipitated from a liquid; a cluster of PARTICLES.[64,139] (IUPAC)

flock (1) small disintegrated FIBERS or filaments; (2) WALLCOVERINGS imitating the surface of damask or cut velvet. Made by shaking finely chopped fibers over a pattern printed in VARNISH or some other sticky material.[148]

flock finish FINISH produced by the APPLICATION on a suitable ADHESIVE coat of short fiber cotton, WOOD, or other fiber, giving a soft suede-like feel and APPEARANCE.[54,77,163,35] (BSI)

flock gun special SPRAY GUN for applying FLOCK FINISHES.[59]

flocking a FINISHING technique in which small natural or synthetic fibers (flocks) are introduced as a SURFACE coating.[54,77,163] See FLOCK FINISH.[54,77,163]

flocking deposition an APPLICATION method of applying POWDER by SPRAY to a SUBSTRATE heated above the MELTING POINT of the powder.[77]

flooding (1) a concentration at the SURFACE of a paint film of one of the ingredients of the PIGMENT portion, giving rise to a change in color at the surface; (2) a differential separation of pigments in a dispersed pigment mixture. FLOATING is a differential separation of pigments (a) by gravity separation in the bulk dispersion; and (b) where the flooding results in a nonuniform or mottled surface COLORATION. While the terms "flooding" and "floating" are often used interchangeably, they also have been defined as separate, though related, phenomena. "Flooding" refers to uniform color changes and "floating" refers to local excess of one color;[166] (3) PROCESS of color change which a paint or ENAMEL undergoes homogeneously, from the freshly applied material to the finished dried film in which one or more pigments appear to be "flooding" the surface color;[42] (4) an excess of INK on the PRINTING plate caused by the ink fountain being open too much. In the case of LITHOGRAPHIC or offset work, the use of too little water or the absence of an etching material in the water fountain.[168] See FLOATING.[42,56]

floor and deck enamel an ENAMEL designed for ABRASION RESISTANCE.[71]

floor paint see DECK PAINT.[71]

floor sealer composition of RESINS, with or without OILS, in a SOLVENT; designed to penetrate the WOOD rather than produce a SURFACE FINISH.[71,147]

floor sweep a vapor collection duct designed to capture vapors which are heavier than air and which collect along the floor.[172] (EPA)

floor tile (1) a resilient material such as ASPHALT, RUBBER, VINYL, CORK, or LINOLEUM manufactured in modular units; laid on a floor as the FINISH flooring;[75,82] (2) structural tile units for floor and roof slab construction.[75,82]

floor varnishes VARNISHES which are formulated for the coating of flooring. Their main properties include rapid drying, tack-free surface, TOUGHNESS, and RESISTANCE TO ABRASION, washability, and receptiveness to WAX and other floor polishes.[152]

flop where two different painted panels appear to be a good match for COLOR when viewed at a given angle, but appear different at all other angles.[35] Syn: geometric metamerism[43,69]

flop color COLOR observed on paints (particularly METALLIC paints) when they are viewed near to the grazing angle, i.e., nearly parallel to the surface. The term is generally used in contrast to the terms "FACE COLOR" or "TOP COLOR," the color observed when viewing the surface at an angle near to the perpendicular.[43,35]

florals any WALLCOVERING design featuring recognizable flowers and foliage.[148]

florentine lake see CRIMSON LAKE.[41]

flow (1) RESISTANCE to movement by a liquid material and divided RHEOLOGICALLY into four categories: NEWTONIAN (simple) FLOW, PLASTIC FLOW, PSEUDOPLASTIC FLOW, dilatant flow;[166] (2) movement of a coating during and after APPLICATION and before the film is formed; (3) property which a coating or INK possesses of LEVELING after application. Coatings may possess excessive flowing properties and may SAG or RUN to an undesirable degree from vertical or inclined surfaces;[166] (4) INKS of poor flow are classed as short or buttery in body, while inks of good flow are said to be long. [168] (PTM) See LEVELING.[166]

flow coating PROCESS of applying paint in which the paint is poured or is allowed to flow over the object to be painted, the excess, if any, being allowed to drain off.[77,131] (BSI)

flow curve a plot of applied force against the resulting rate of FLOW.[166]

flower of antimony see ANTIMONY OXIDE.[41]

flowing varnish a VARNISH designed to produce a smooth lustrous surface without RUBBING or POLISHING.[152]

flowmeter an instrument for measuring the rate of FLOW of a fluid moving through a tube, pipe, or duct system. The instrument is calibrated to furnish volume or mass rate of flow.[145,166] (IUPAC)

Several types of flowmeters are in use: (a) variable area flowmeter: a flowmeter consisting of a tapered tube and float. The gas enters at the bottom of the tube. The inner diameter of the tube increases from the bottom to the top, and the float is supported in the tube at a height which is a function of the rate of flow and the nature of the gas in the tube; this is sometimes referred to as a "rotameter" which is derived from a trademark; (b) soap film flowmeter: a flowmeter in which a soap film is generated on a cross section of a graduated tube (sometimes a calibrated buret is employed) in which the gas to be measured passes. The film moves along the tube at the same rate as the gas flow. The measured time required for the film to sweep through a measured volume can be used to calculate the gas flow rate. This is often employed in gas chromatography work; (c) critical orifice flowmeter: a device consisting of a plate having an orifice of small diameter and placed in a gas stream such that the total flow passing through the orifice has a velocity at least equal to the sonic velocity. A pressure measuring device is attached upstream of the orifice. The flow is proportional to the upstream pressure and independent from the downstream pressure; (d) spirometer: an absolute flow measuring device consisting of two cylinders each sealed at one end

Words presented in CAP/SMALL CAPS type indicate that the word is defined in another part of the *Coatings Encyclopedic Dictionary.*
Numerical superscripts classify terms in one or more of the categories listed in the second section of this volume.

and having slightly different diameters. The smaller is inverted into the larger which is filled with a confining liquid. The inverted cylinder is suspended with counterweights so that it moves freely within the larger cylinder. Gas may either be displaced from the smaller cylinder or admitted to it. The dimensions of the smaller cylinder must be known. The rate of rise or fall of the inverted cylinder can be used to calculate the rate of flow of gas either into or out of the spirometer. The device is used principally for calibrating other devices. Other types of flowmeters also are in wide use today (e.g., Hastings-type mass flowmeter).(IUPAC)

flue a passage for the conduction of combustion gases in an incinerator; synonymous with chimney.[82,171] (IUPAC)

The term also applies to structures and openings of any kind used to transport gases in operations other than incineration.

flue gas waste gas from the combustion process.[171,172] (IUPAC)

fluid flow the movement of air or other fluid in the open or in a duct, pipe, or passage.[166] (IUPAC)

The FLOW can be of several types: (a) uniform flow is steady in time, or the same at all points in space; (b) steady flow is that for which the velocity at a point fixed with respect to a fixed system or coordinates is independent of time; (c) rotational flow has appreciable vorticity and cannot be described mathematically by a velocity potential function; (d) turbulent flow is that in which the fluid velocity at a fixed point fluctuates with time in a nearly random fashion; (e) laminar flow is that in which the mass of fluid may be considered separate laminae (sheets) with simple shear existing at the surface of contact of laminae if there is any difference in mean speed of the separate laminae; and (f) streamline flow is that in which fluid particles move along the streamlines; this motion is characteristic of viscous flow at low Reynolds numbers. (IUPAC)

fluid resin synonym for tall oil.[63]

fluidity the property of a material which allows it to FLOW when subjected to a SHEARING FORCE.[166]

The coefficient of fluidity can be expressed as the rate of shear induced by a shearing force of 1 dyne/cm^2. It is calculated from the slope of the FLOW CURVE. It applies only to NEWTONIAN materials. The coefficient of fluidity is the reciprocal of the coefficient of VISCOSITY.

fluidized bed coating method of applying a coating in which a heated or electrostatically charged article is immersed or passed over a fluidized bed of POWDERED coating (the coating material adhering to the hot metal), then heated in an oven to provide a smooth continuous film.[77]

The bed of powdered RESIN may be "fluidized" by vibration or by blowing compressed air through the powder from the bottom of the container.

fluorescein dye see ACID DYES.[41]

fluorescence optical phenomenon characterized by the reemission of absorbed RADIANT ENERGY by certain chemicals or materials.[43,69,35] See LUMINESCENCE.[43]

The WAVELENGTH of maximum emission is generally longer than the wavelength of maximum ABSORPTION, although not necessarily so.

fluorescent dyes DYESTUFFS which exhibit the phenomenon of FLUORESCENCE in the visible region of the SPECTRUM.[41,43]

Dyes are characterized as fluorescent if: they absorb invisible RADIATION, such as ULTRAVIOLET and emit visible light; they also absorb primarily visible light and emit visible light; or they absorb visible light and emit invisible radiation, generally infrared. Some dyes fluoresce more strongly than others, i.e., some are only slightly fluorescent and some are strongly fluorescent. The former may go undetected as fluorescent dyes. In order to measure the color of materials containing fluorescent dyes, the material must be illuminated with light of defined spectral quality, and the light reflected or transmitted by the materials must be passed through the monochromating device after reflectance or transmission, using so-called "reversed optics." Strongly fluorescent pigments are frequently referred to as high visibility pigments; their combined emission and reflectance is generally much higher at certain WAVELENGTHS than the REFLECTANCE of the perfect DIFFUSER.

fluorescent paint see LUMINOUS PAINT, and HIGH VISIBILITY PAINTS.[41,43,69,71,163]

fluorescent pigments (1) INORGANIC, usually coarse, crystalline, materials which emit light when activated by ULTRAVIOLET RADIATION, e.g., zinc sulfide, combined zinc and cadmium sulfides; (2) ORGANIC DYES or PIGMENTS used to manufacture LUMINESCENT materials. These include spirit-soluble dyes, oil-soluble dyes, and METALLIC ORGANIC COMPOUNDS, e.g., rhodamine, eosine, rhodamine tungstate, and zinc salt of 8-hydroxyquinoline. By absorbing unwanted WAVELENGTHS of light and converting them into light of desired wavelength, these pigments have the appearance of possessing an actual glow of their own; (3) pigments which exhibit the phenomenon of fluorescence in the visible region of the SPECTRUM. Few commercial mineral pigments are fluorescent; most of the fluorescent pigments are made from fluorescent dyes incorporated into an insoluble matrix and ground to a small particle size suitable for use as a pigment.[41,43] See FLUORESCENT DYES.[41,43]

fluorocarbons group of COMPOUNDS containing fluorine atoms, FLUOROPLASTICS, or SOLVENTS.[130,164]

fluoroplastics see FLUOROCARBONS.[38]

flushed (flushing) colors PIGMENTS which have been FLUSHED.[41] See FLUSHING.[41,131]

flushed pigment a PIGMENT obtained by the direct transfer of pigment PARTICLES that have been PRECIPITATED, from an AQUEOUS PHASE to a NONAQUEOUS STAGE (no INTERMEDIATE DRYING or PULVERIZING stage being involved).[41,131]

In the PROCESS, the PRESS CAKE (aqueous or water-wet pigment phase) is mixed and agitated with a nonaqueous vehicle (OIL, SOLVENT, and/or RESIN phase) in a heavy duty mixer. The pigment particles preferentially transfer (FLUSH) to the nonaqueous phase, and the bulk of the essentially clear water is poured off. Any residual water is then vacuumed off at a temperature of about 50°C (122°F). The flushed pigment production process, unlike the dry powder process, allows the pigment to preserve its optimum form. Pigment AGGLOMERATES remain soft and breakdown more easily in the flushing process because they have never been dried and case-hardened. They are also easier and less hazardous to handle because they do not possess the dusting characteristics of dry powder.

flushing PROCESS by which a PIGMENT DISPERSED in water is converted directly to a pigment dispersed in a different medium in PASTE form.[41,131] See FLUSHED PIGMENT.[41,131] Syn: fleshing

fluting parallel concave grooves used as FURNITURE DECORATION.[54,148,175]

Vertical and spiral flutings are common.

flux the number of particles flowing through a given area per unit time.[139] (ASTM)

flux, luminous RADIANT FLUX weighted for the relative LUMINOUS (visual) efficiency.[43,142]

flux, radiant see RADIANT FLUX.[43]

fly speck small spots of paint that are sprayed onto WOOD FURNITURE to create an antique effect.[175,148] (EPA)

flying a condition wherein a fine mist or SPRAY of INK is thrown off rapidly moving ink ROLLERS.[168] Syn: misting and spraying (2)

FMC Color Difference Equations letters "FMC" are the initials of Friele, MacAdam, and Chickering.[69]

There are two different equations, referred to as FMC-1 and FMC-2: FMC-1 is the same as FMC-2, except that the terms K_1 and K_2 are not included; FMC-2 is:

$$\Delta E = [(\Delta C)^2 + (\Delta L)^2]^{1/2}$$

where $\Delta C = K_1 \Delta C_1$

$$\Delta L = K_2 \Delta L_2$$
$$\Delta C_1 = [(\Delta C_{rg}/a)^2 + \Delta C_{yb}/b)^2]^{1/2}$$
$$\Delta C_{rg} = (Q\Delta P - P\Delta Q)/(P^2 + Q^2)^{1/2}$$
$$\Delta C_{yb} = S\Delta L_1/(P^2 + Q^2)^{1/2} - \Delta S$$
$$\Delta L_1 = (P\Delta P + Q\Delta Q)/(P^2 + Q^2)^{1/2}$$
$$\Delta L_2 = 0.279\Delta L_1/a$$
$$K_1 = 5.5669 \cdot 10^{-1} + 4.9434 \cdot 10^{-2}Y - 8.2575 \cdot 10^{-4}Y^2 + 7.912 \cdot 10^{-6}Y^3 - 3.0087 \cdot 10^{-8}Y^4$$
$$K_2 = 1.7548 \cdot 10^{-1} + 2.755 \cdot 10^{-2}Y - 5.7262 \cdot 10^{-4}Y^2 + 6.3893 \cdot 10^{-6}Y^3 - 2.6731 \cdot 10^{-8}Y^4$$
$$a^2 = 1.73 \cdot 10^{-5}(P^2 + Q^2)/[1 + 2.73P^2Q^2/(P^4 + Q^4)]$$
$$b^2 = 3.098 \cdot 10^{-4}(S^2 + 2.015 \cdot 10^{-1}Y^2)$$
$$P = 0.724X + 0.328Y - 0.098Z$$
$$Q = 0.48X + 1.37Y + 0.1276Z$$
$$S = 0.686Z$$

X, Y, Z are tristimulus values for the standard
X_1, Y_1, Z_1 are tristimulus values for the sample
ΔP, ΔQ, ΔS are above formulae for P, Q, S substituting $\Delta X = X_1 - X$ for X, etc.

foam DISPERSION of a gas in a liquid or a solid.[64,139]

foam paint roller cover one in which the paint applicating material is a man-made flexible cellular material.[59]

foamed plastic PLASTIC, the APPARENT DENSITY of which is decreased substantially by the presence of numerous air or gas cells distributed throughout its MASS.[38]

foaming agents (1) materials that increase the STABILITY of a SUSPENSION of gas bubbles in a liquid medium;[83,127,139] (2) BLOWING AGENTS in RUBBER or PLASTICS.

fog a visible SUSPENSION of fine droplets in gas; e.g., water in air.[64] See BLOOM and BLUSH.[42,56] cf. MIST[64]

fogged coat noncontinuous SPRAY COAT. See MIST COAT.[71]

fogged metal METAL, the luster of which has been sharply reduced by a film of CORROSION products.[43,66]

foil in WALLCOVERINGS, a very thin sheet of flexible METAL on a PAPER or FABRIC back.[142]

They can be printed with TRANSPARENT or OPAQUE COLOR, and MOTTLED to resemble marble, tortoise shell, etc.

foliage paint coatings designed to color grass lawns in the winter.[71]

Food and Drug Administration See FDA.[172,174]

foot-candle the quantity of LIGHT at a point on a plane SURFACE one foot from a standard candle and perpendicular thereto.[128,43] See CANDLE.[128,43]

footing CONCRETE BASE on which a foundation sits.[82]

Words presented in CAP/SMALL CAPS type indicate that the word is defined in another part of the *Coatings Encyclopedic Dictionary.*
Numerical superscripts classify terms in one or more of the categories listed in the second section of this volume.

foots SEDIMENT which settles from an OIL on standing; chiefly ALBUMINOUS matter.[63] See TAILINGS.[49]

Originally used to describe those solid impurities which PRECIPITATE from raw LINSEED OIL during storage and then settle to the bottom or "foot" of a storage tank. Later used to denote material insoluble in a mixture of equal parts of acetone and linseed or other oil under test, and insoluble in calcium chloride solution under the specific conditions of test method. Also, the term is used to denote that material which is precipitated from the oil by phosphoric acid and which is insoluble in acetone under the specific conditions of test method.[42,63](ASTM)

force the capacity to do work or cause physical change; energy, strength, or active power.[142]

forced drying DRYING coatings at a temperature between room temperature and 65.6°C (150°F) as opposed to AIR DRYING or BAKING. [67]

forced drying temperature temperature between room temperature and 65.6°C (150°F).[67,157] (ASTM) See BAKING.[67,131]

Ford cup efflux cup for measuring the VISCOSITY of paint or VARNISHES.[59,157,166] See EFFLUX VISCOMETER and VISCOMETER.[166]

foreign matter (or material) INSOLUBLE PARTICLES such as sand, lint, DUST, and dirt that adventitiously get mixed with the coating before, during, or after APPLICATION—causing the formation of raised specks in the dried film.[42,56] cf. INTERCOAT CONTAMINATION

form in art, the PRESENTATION of various shapes, either recognizable or abstract; the visible results of an artist's conception of the subject, but not the subject matter itself.[81] See ABSTRACT ART.[81]

form lacquer thin LACQUER or VARNISH used to coat CONCRETE forms to prevent concrete from ADHERING to the forms.[151,152]

form roller the form ROLLER is that roller in the INK DISTRIBUTION system of a PRINTING PRESS that is in direct contact with the printing plate and transfers the ink to it.[168]

formaldehyde HCHO. A colorless, pungent, volatile readily POLYMERIZABLE gas, readily soluble in water, widely used in the manufacture of PLASTICS and RESINS and as a DISINFECTANT.[144] (DAC) See FORMALIN.[144] Syn: methylene oxide, methanal

formalin HCHO. FORMALDEHYDE gas commercially available as a 37% wt SOLUTION in water with a small amount of methanol for inhibiting POLYMERIZATION.[144]

fossil gum resins fossil GUMS are the so-called hard gums or COPALS which were exuded from living plants in the form of liquids or semisolids, which have lain in the ground for centuries and thus hardened or fossilized with time.[167] See GUM, NATURAL, and RESIN, NATURAL.[155,130,167]

Some of the fossil gum resins are: Amber, Zanzibar, Kauri, Manila, Pontianak, and Congo.

foster charts graphs for determining color differences in modified MacAdam units from TRISTIMULUS and ratio tristimulus measurements made on a tristimulus COLORIMETER.[69]

They are based on the SIMON-GOODWIN CHARTS and give results which are a close approximation to those which would be obtained with the S-G charts.

fouling sessile marine organisms on the hulls of ships.[42,65,72]

foundation lower parts of walls on which the structure is built.[82]

Foundation walls of MASONRY or CONCRETE are mainly below ground level.

fountain part of a PRINTING PRESS which contains the INK to be fed to the distributing system.[168]

In LITHOGRAPHY, it is also the part which feeds the water or fountain solution to the dampening rollers.

fountain roller the ROLLER that revolves in the INK FOUNTAIN.[168]

In LITHOGRAPHY, it is also the roller that revolves in the dampening SOLUTION.

fountain solution in LITHOGRAPHY, generally a mixture of water, acid, buffer, and a GUM to prevent the nonprinting areas of the plate from receiving INK.[168] Syn: dampening solution

foxy condition where the PIGMENT crystals are sometimes arranged in different order by method of APPLICATION, causing different shades of COLOR when viewed from different angles.[43]

FR abbreviation for FEDERAL REGISTER, THE.[172]

fraction SOLVENT of definite boiling range obtained by FRACTIONAL DISTILLATION.[131]

fractional distillation separation of a liquid into its different components by collecting fractions or distillates of restricted boiling range.[131,164]

fractionation separation of a mixture into components or groups of components.[131]

This PROCESS may be carried out by DISTILLATION, partial freezing, SOLVENT segregation, etc.

frankincense see GUM THUS.[167]

free binder the volume of NONVOLATILE binder in a paint system that is in excess of its PIGMENTS' BINDER demand.[130]

free fatty acid unreacted FATTY ACID present in a coating VEHICLE.[144]

free radical see RADICAL.[169]

free silica SiO_2. SILICA generally present in small amounts in natural deposits of CLAY-like minerals and DIATOMACEOUS EARTH and usually considered to be a contaminant.[41]

freeze-thaw resistance extent to which WATER-BASED paints, utilizing SYNTHETIC LATEXES or synthetic RESIN EMULSIONS as VEHICLES: (1) retain their original properties, free from detrimental changes in CONSISTENCY, and (2) resist COAGULATION, or the formation of lumps and specks, when subjected to freezing and subsequent thawing.[72,157,161] (ASTM)

After leaving the plant, paints are sometimes shipped and/or stored at low temperatures where they can freeze. If the paint develops VISCOSITY instability or contains GRIT or gel after thawing, it may no longer be useable.

ASTM Test Method D2243 covers the determination of the extent to which waterborne coatings retain their original properties free of detrimental changes.

freezing point temperature at which a liquid material and solid are in equilibrium with one another, i.e., at a lower temperature the liquid will solidify.[142,157] cf. SOLIDIFICATION PAINT[142,157]

Amorphous materials, such as paints or paint vehicles, normally solidify over a temperature range referred to as freezing range. For test method see ASTM D1016.

French blue see ULTRAMARINE BLUE.[41]

French chalk $3MgO{\cdot}4SiO_2{\cdot}H_2O$. It is distinguished by its soapy or greasy feel.[41,58] Syn: talc, soapstone, or steatite

Chemically, it is substantially a hydrated magnesium silicate. It is supplied commercially as a white or gray powder with sp gr of 2.4 to 2.6.

French ochres pale yellow variety of OCHRE obtained from the Yonne and Vaucluse districts of France.[41]

French polish solution of SHELLAC in ETHYL ALCOHOL which is applied by a polishing cloth or pad instead of a BRUSH to produce an exceedingly high gloss to wood surfaces. The use of a little LINSEED OIL on the polishing cloth or pad is sometimes involved.[54,146,175] cf. FRENCH VARNISH[54,142]

This technique was developed early in the 18th century, but the tediousness and the skill it demanded hastened its near demise. It is also not as tough as the more conventional varnishes, urethanes, and lacquers. Also, unlike these conventional coatings, shellac solutions do not flow well because they evaporate quickly. Today, apart from restoration work, French polishing has largely been replaced in professional shops by padding LACQUER or by lacquer that has been sprayed on.

French process zinc oxide ZINC OXIDE PIGMENT made from zinc METAL.[41]

French ultramarine synthetic ULTRAMARINE BLUE.[41]

French varnish a hand-rubbed high-GLOSS FINISH, achieved by multiple APPLICATIONS of such varnish.[54,152] (DAC) cf. FRENCH POLISH[54,167]

French veronese green see HYDRATED CHROMIUM OXIDE.[41]

frequency number of vibrations per second, equal to the velocity of light (or any electromagnetic radiation) divided by the WAVELENGTH:

$f = \frac{v}{\lambda}$, where f is the frequency, v is the velocity, and x is the wavelength.

fresco method of painting in which PIGMENTS mixed only with water are applied to a freshly laid SURFACE of LIME PLASTER.[67,71,77,81] (BSI) cf. SECCO[71,81]

The painted surface is rendered durable by the pigment absorbing calcium hydroxide from the wet plaster, which is converted into insoluble CALCIUM CARBONATE as the surface dries. These, of course, require very ALKALI RESISTANT pigments. The term is sometimes used incorrectly to denote any form of mural painting.

fresco color water paints used for WALL DECORATIONS.[71,81]

Fresnel reflection phenomenon of REFLECTION at the SURFACE or interface where media or materials of different refractive indices join.[43]

Fresnel laws state that the angle of reflection is equal and opposite to the incident angle, and that the magnitude of the reflection depends on the angle of the incident light, on the relative refractive indices, and on the polarization of the incident light. When the surfaces are smooth and planar, SPECULAR (mirrorlike) reflection occurs, which is subjectively referred to as specular GLOSS. When the surfaces are rough, the reflectance at the air-material interface is diffuse, and the surface is observed to be matte-like.

fret a geometric band or border motif, consisting of interlacing or interlocking lines.[148] Syn: a key or meander pattern

friction top can can in which the lid is held in place by friction.[50]

Words presented in CAP/SMALL CAPS type indicate that the word is defined in another part of the *Coatings Encyclopedic Dictionary*.

Numerical superscripts classify terms in one or more of the categories listed in the second section of this volume.

Friele-MacAdam-Chickering Color Difference Equations see FMC COLOR DIFFERENCE EQUATIONS.[69]

frieze (1) any DECORATIVE horizontal band, as along the upper part of a wall in a room;[82] (2) a type of WALLPAPER popular in the early 1900s. Generally, a pictorial border which ran above door height, or, in dining rooms, above the plate rail.[148]

frisket device, usually of GUMMED PAPER placed on the working SURFACE, to block out subsequent painting.[54]

frosting (1) surface WHITENING of a painted surface due to leaching of SOLUBLE SALTS from the coating or SUBSTRATE. These salts are generally converted to an insoluble form due to the CO_2 and SO_2 in the air. Some of these mechanisms are similar to that of the formation of stalactites and stalagmites. Frosting is distinguished from EFFLORESCENCE and MOTTLING by knowledge of the paint film and underlying paints and substrates. The source of frosting is usually the TOPCOAT. If soluble salts are seen on a paint surface over WOOD, then it is called frosting; however, if the substrate were CEMENTITIOUS, then it could be efflorescence or frosting; (2) formation of a TRANSLUCENT finely wrinkled surface on a film of OIL, INK, or paint during DRYING, particularly when exposed to gas fumes, etc.; (3) the formation of an insoluble white crystalline haze on the surface of an exterior deeptone paint;[55] Frosting is not a problem in normal exterior weathering but is observed on protected areas of a home where CONDENSATION accumulates, i.e., horizontal face down areas such as overhangs and porch ceilings. Condensed moisture tends to leach out solubles, which recrystallize after drying into an insoluble form. The type of paint EXTENDER most responsible for this phenomenon is CALCIUM CARBONATE. (4) a CLOUDING of the surface of some rubber and synthetic RUBBER goods, appearing within a few hours or days after VULCANIZATION.[42,56,140] (DAC) See GAS CHECKING[56,63,157] and EFFLORESCENCE.[43,56]

The frosted appearance is different from BLOOM or BLUSH and cannot be readily removed by washing with a solvent. It may disappear if the article is heated moderately but will generally reappear on cooling. It is thought to be caused by ozone in the air which produces a maze of minute CRACKS. Some ANTIOXIDANTS have definite antifrosting effects. Frosting (3) is not a problem in normal exterior weathering, but is observed on protected areas of a house where CONDENSATION accumulates, i.e., the horizontal face down areas such as overhangs, soffits, and porch ceilings. It is believed that condensed moisture tends to leach out solubles, which recrystallize after drying into an insoluble form. The type of extender thought to be most responsible for this phenomenon is CALCIUM CARBONATE.

fruiting body the FUNGUS structure consisting of specialized hyphae where sexual reproduction occurs.[173]

FSCT abbreviation for the FEDERATION OF SOCIETIES FOR COATINGS TECHNOLOGY.[174]

fuchsin synthetic rosaniline DYESTUFF, a mixture of rosaniline and pararosaniline hydrochlorides.[41] Syn: basic fuchsin magenta

Dark green powder of greenish crystals with a bronze luster; faint odor. Soluble in water and ALCOHOL. Used in textile and leather industries, as a red dye, and as a pharmaceutical.

fugitive color (1) COLORING matter that exhibits color change or color transfer while wet or during subsequent DRYING;[41] (2) INKS made from PIGMENTS or DYES which are not permanent, and change or lose color rapidly when exposed to light, heat, moisture, or other conditions;[168,41] (3) colorant, PIGMENT, or DYESTUFF, which changes color (fades) rapidly following EXPOSURE to light.[69]

fugitive emission gas, liquid, solid, vapor, fume, mist, fog, or dust that escapes from process equipment or a product.[172,131]

fugitometer see FADEOMETER.[59,157]

full coat (1) maximum film THICKNESS of a particular coating which can be properly applied; (2) APPLICATION of a coating at a specified film thickness designed to achieve a desired effect.[54]

full gloss see GLOSS.[43]

full protective clothing fully protective gear that prevents skin contact with, INHALATION of, or ingestion of gases, vapor, liquids, and solids (dusts, etc.).[172]

This includes SCBA (self-contained breathing apparatus).

full-bodied see BODY.[166] (BSI)

Fuller's earth $3MgO \cdot 1.5Al_2O_3 \cdot 8SiO_2 \cdot 9H_2O$. Hydrated magnesium aluminum silicate (ATTAPULGITE) used for decolorizing solutions and OILS, and as a substitute for absorbent charcoal and as a dusting powder.[41,58] See HYDRATED MAGNESIUM ALUMINUM SILICATE.[41,58]

Properties: density, 2.36 g/cm³ (19.7 lb/gal). Refractive index, 1.50-1.55.

fumaric resin a hard, synthetic RESIN formed by the reaction of FUMARIC ACID and ROSIN.[130]

fumed silica see SILICA, SYNTHETIC (PYROGENIC).[41]

fumes gas-like emanation containing minute solid particles arising from the heating of a solid body such as lead, in distinction to a gas or vapor.[171,172] cf. SMOKE[171,172]

This physical change is often accompanied by a chemical reaction, such as OXIDATION. Fumes flocculate and sometimes coalesce. Odorous gases and vapors should not be called fumes. As distinguished from dusts, fumes are finely divided solids produced by other methods of subdividing, such as chemical processing, combustion, explosion, or DISTILLATION. Some solids, when heated to a liquid produce a vapor which, while arising from the molten mass, immediately condenses to a solid without returning to its liquid state. Fumes are much finer than dusts, containing particles from 0.1 to 1.0 μm in size.

functional group an atom or group of atoms, such as a CARBOXYL GROUP, that replaces hydrogen in an ORGANIC COMPOUND and that defines the STRUCTURE of a family of COMPOUNDS and determines the properties of the family.[144,133]

functionality effective number of bonds and points of intermolecular attachment in a chemical compound, the degree or order of reactivity.[134]

fungicidal paint paint which discourages the growth of fungi on its dry applied film.[71,163,173] (BSI)

The fungicidal properties are normally conferred by special ADDITIVES, although certain PIGMENTS such as ZINC OXIDE, commonly used in paints, may themselves contribute to the fungicidal properties of the paint.

fungicidal wash wash containing FUNGICIDES used before painting and designed to kill existing spores or germinations or to prevent their inception.[158] (BSI)

Many of these substances are toxic (to human beings) when used in concentrated form and therefore need careful handling. Use of the term "antiseptic wash" is discouraged.

fungicide chemical agent which destroys, retards, or prevents the growth of fungi and spores.[83,173] cf. FUNGISTAT[83,173]

fungistat COMPOUND which inhibits the growth of a fungus, or prevents the germination of its spores.[83,173] cf. FUNGICIDE[83,173]

fungistatic preventing the growth of a fungus by the presence of some chemical or physical agency but not FUNGICIDAL.[173]

fungus (fungi) multicellular plant or plants of the phylum, *thallophyta* that contain no chlorophyll and reproduce by sexual or asexual spore.[42,56,173] (ASTM)

Fungi cause discoloration and growth in or on a coating's surface, i.e., mildew, mold. Fungi are heterotrophs capable of reproducing both asexually and sexually. Asexual reproduction takes place either by the fragmentation of the MYCELIUM or by the production of SPORES. Sexual reproduction can occur by fusion of GAMETES, penetration of a gamete into a reproductive structure known as gametangium, or by fusion of gametangia.

fungus resistance see MILDEW (FUNGUS) RESISTANCE.[56,72,173]

furan resins RESINS in which the furan ring is an integral part of the POLYMER chain, made by the polymerization of POLYCONDENSATION of FURFURYL, furfuryl alcohol, or other compounds containing a furan ring, or by the reaction of these furan compounds with other compounds (not over 50%).[130]

furfural $C_4H_3O{\cdot}CHO$. Liquid ALDEHYDE with a bp of 161°C, mp of -36.5°C; sp gr of 1.159/22°C. Used in the preparation of aldehyde type RESINS, and as a SOLVENT for certain refining PROCESSES based on selective solvent action.[130] Syn: furfuraldehyde

furnace black CARBON BLACK obtained by burning natural gas or PETROLEUM OIL in a large furnace with a low percentage of the theoretical amount of air required for complete combustion.[41] See CARBON BLACK.[41]

furniture the movable articles in a room or an establishment that make it fit for living or working.[175]

fuse heating a coating component of DISPERSED RESIN(s) to a temperature at which it becomes homogenous.[167]

fused congo CONGO COPAL which has been subjected to a FUSING or RUNNING PROCESS, in order to induce OIL SOLUBILITY.[167]

fused driers DRIERS which are prepared by the direct interaction of the main constituents, stimulated by HEAT, in the absence of added water.[74,83]

The term is used to distinguish them from driers which are made by PRECIPITATION from aqueous solutions of METALLIC SALTS and ALKALI SOAPS.

fusel oil complex SOLVENT mixture obtained as the result of FERMENTATION.[164]

ALCOHOLS such as normal, iso, and tertiary butyl alcohols, propyl and isopropyl alcohols, and amyl and isoamyl alcohols may be present.

fusion point temperature of fusion or melting.[157]

fustics yellow coloring matters of natural origin.[41]

Old fustic is obtained from the species, Morus tinctoria, and consists of moric and morintannic acids. Young fustic is obtained from Rhus cotinus, and consists of fustin.

Words presented in CAP/SMALL CAPS type indicate that the word is defined in another part of the *Coatings Encyclopedic Dictionary.*

Numerical superscripts classify terms in one or more of the categories listed in the second section of this volume.

G

gable triangle formed by the edge of a ridged roof and the similarly shaped wall enclosed by the horizontal and raking cornices; called a pediment in classic architecture.[82]

gableboard see BARGEBOARD.[82]

gadoleic acid $CH_3(CH_2)_9CH = CH(CH_2)_7COOH$. Fatty acid from marine oils.[63]

gage length the initial length of the test specimen between the jaws of a tensile tester.[47,157] (ASTM)

Also spelled guage length.

Gahn's ultramarine see COBALT BLUE.[41]

galipot synonym for Bordeaux turpentine.[164]

galliolino old name for NAPLES YELLOW, which is substantially lead antimoniate.[41]

gallon, Imperial English gallon, equivalent to 277.77 cu in. or 4.54 litres. Equals 1.2 U.S. gallons. U.S. gallon = 0.833 Imperial gallons.[128]

gallon, U.S. (1) volume equal to 231 $in.^3$ For paint, VARNISH, LACQUER, and related products this is measured at 25°C (77°F); (2) volume of 8.33 lb of water; (3) equivalent to 3.785 litres.[128]

Gallows Birds, The Society of a FEDERATION OF SOCIETIES FOR COATINGS TECHNOLOGY's informal fellowship made up of selected members, who are usually well-known in the coatings industry.[174]

When initially formed, the candidates chosen were those who made an "error" when writing a Federal Specification. The choice of the name of the Society is attributed to Joseph Mattiello (of Mattiello Lecture fame). The gallows are for those who "stuck their necks out" writing vulnerable Federal Specifications, and who, in some cases, should have been "hung" for doing so.

The Society meets at the FSCT's Annual Meeting and Paint Industries' Show, where they have a social meeting and initiate new members. The chosen candidates are then blindfolded, their jackets are put on backwards, and they are then "tried" by a kangaroo court consisting of a "prosecutor," "defense attorney," and a "jury of their peers." They are always found guilty and are "hung" on a harmless gallows and noose with a trap door with less than a foot drop. This initiation rite is in the spirit of fun and good fellowship.

All new members are given a small gold key-like medallion inscribed with a gallows and noose on the front side; the obverse side is inscribed with the members' initials, the year they were initiated, and the Federal Specification number which contained the "error."

galvanic anode source of current for cathodic protection provided by a metal less noble than the one to be protected, e.g., magnesium, zinc, or aluminum as used for cathodic protection of steel.[61,66] See CATHODIC PROTECTION.[66]

galvanic cell (1) cell consisting of two or more dissimilar metals or alloys in contact with the same body of an electrolytic solution such as seawater. Upon electrically connecting the dissimilar metals, a current flows as the result of accelerated corrosion of the more active of the dissimilar metals or alloys; (2) electrolytic cell capable of producing electrical energy by electrochemical action.[61,66]

galvanic corrosion increased CORROSION above normal corrosion of a METAL that is associated with the FLOW of current to a less active metal in the same solution and in contact with the more active metal.[61,66]

galvanic protection see CATHODIC PROTECTION.[66]

galvanized steel zinc-coated steel.[45,158] See GALVANIZING.[158]

galvanizing APPLICATION of a coating of ZINC to STEEL by a variety of methods: barrel-galvanized dipped, electrogalvanized, electro-zinc plated, flake-galvanized, hot-dipped, hot-dip galvanized, mechanically plated, peer-coated galvanized, tumbler-galvanized, and wean-galvanized.[158]

gamboge $C_{30}H_{35}O_6$. Resinous exudation of intense yellow COLOR obtained from the species, *Guttifera*, which grows in India, Ceylon, Siam, and the East Indies.[167] Syn: cambogia

The product is a mixture consisting chiefly of a water insoluble resin (up to 80%), with a water soluble GUM *(up to 27%). The resin portion is chiefly gambogic acid.*

gamete a reproductive cell having the haploid number of chromosomes, especially a mature sperm or egg capable of fusing with a gamete of the opposite sex to produce the fertilized egg.[173] See FUNGUS (FUNGI).[173]

gamma protein protein derived from soya bean meal.[83]

The commercial product contains approximately 54% of actual protein. It disperses in water in the presence of a small amount of alkali and is used as an EXTENDER for CASEIN in water paints and distempers.

gamut, color total variety of COLORS that can be produced by any prescribed method, such as the mixtures of any particular sets of colorants.[69]

Gamuts are frequently described by plotting them on a chromaticity diagram.

garcia nutans oil VEGETABLE DRYING OIL obtained from the species, *Euphorbiaceae*, which grows in Mexico and Venezuela.[63]

It contains over 90% of eleostearic acid and gels in less than eight minutes at 282°C. It possesses the following constants: sp gr of 0.942/15°C, refractive index of 1.525, iodine value of 177.9, and saponification value of 189.

garden rocket oil oil with fair drying properties.[63]

It has an iodine value of 155, a sp gr of 0.928, and a saponification value of 192.

Gardner color standards liquid yellow to brown COLOR standards for describing the color of DRYING OILS, ASTM Test Method D1544.[69,157]

Gardner Mobilometer VISCOMETER used for determining the VISCOSITY of all types of VARNISHES, OILS, and pigmented compositions.[59,166,157]

The Mobilometer consists of a cylinder into which is poured the paint product of which the viscosity is required. Into this is fitted a closely fitting disc attached to the end of a guide rod. Provision is made at the other end of the rod for weighting it, if required, when dealing with viscous products, by including a shallow tray into which weights may be placed. The time is determined for the disc and rod to travel a definite distance into the paint product, and is a measure of its viscosity.

Gardner-Holdt viscosity tubes series of selected GLASS tubes of constant, standard diameter, which are filled with liquids of various viscosities, except for a small air space at the top.[59,157,166]

When these tubes are inverted, the air bubble travels through the liquid, and the rate of travel is a measure of the viscosity of the liquid. Empty tubes of similar standard dimensions are filled with liquids of which the viscosity is required, and the rates of travel of the bubbles compared with those of the tubes containing the liquids of known viscosities. Viscosities are compared under controlled temperature conditions.

garnet abrasive almandite, a type of garnet mineral occurring in New York State, is used in coated ABRASIVES.[158,34]

HARDNESS and toughness are increased by heat treating. It fractures along the cleavage planes of the crude crystals, and the resulting grains have very sharp edges.

garnet lac form of SHELLAC prepared from sticklac by a special refining PROCESS.[146]

The term "garnet" is merely an indication of the deep red COLOR of the material when dissolved in ALCOHOL.

garnet paper the paper generally favored for hand sanding on wood.[59,158]

The ABRASIVE is the same mineral as the semi-precious garnet gem, almandite.

gas a state of matter in which the material has a very low density and viscosity, can expand and contract greatly in response to changes in temperature and pressure, easily diffuses into other gases, and readily and uniformly distributes itself throughout any container.[171]

A gas can be changed to the liquid or solid state only by the combined effect of increased pressure and decreased temperature (below the critical temperature).

gas black synonym for CARBON BLACK, so called from its manufacture from PETROLEUM gas.[41]

gas blue variety of PRUSSIAN BLUE, so called from its origin in certain by-products of coal gas manufacture.[41]

gas checking film phenomenon associated with compositions containing VEGETABLE DRYING OILS, or derivatives therefrom, and demonstrated by the formation of crowsfoot-like wrinkles, or wrinkles which have the APPEARANCE of crystalline patterns.[56,63,157] Syn: crystallizing, crowsfooting, webbing, and frosting

The phenomenon occurs when susceptible films are allowed to dry in an atmosphere containing the products of combustion of coal gas or natural gas; hence the term, "gas checking."

gas crazing see GAS CHECKING.[56,63,157]

gasproof coating composition which does not gas check when exposed to an atmosphere containing the products of combustion of coal gas.[157,63] See GAS CHECKING.[157,63,56]

gauffrage PRINTING by pressure alone, without the use of a PIGMENT, producing an embossed effect on the paper.[81]

gavan Persian name for the plant which yields GUM tragacanth.[83,155]

gel (1) a semisolid material, somewhat elastic, composed of matter in a COLLOIDAL state that does not dissolve; remains suspended in a solvent;[139,166,161] (DAC) (2) semisolid system consisting of a net-

---

Words presented in CAP/SMALL CAPS type indicate that the word is defined in another part of the *Coatings Encyclopedic Dictionary.*

Numerical superscripts classify terms in one or more of the categories listed in the second section of this volume.

work of solid AGGREGATES in which liquid is held; (3) initial jellylike solid phase that develops during the formation of a RESIN from a liquid; (4) irreversible, massive increase in the resistance to FLOW of an oil, vehicle, or coating (clear or pigmented); characterized by a semi (jellylike) or fully rigid condition. In use, it may refer to a near-rigid state in which a very limited flow may be detected.[166,161,42]

gel coat (1) a thin, outer layer of RESIN, sometimes with PIGMENT, applied to a REINFORCED PLASTIC MOLDING as a cosmetic;[38,163] (2) high-build, chemical-resistant, thixotropic polyester coating.[163,71]

gel point the stage at which a liquid begins to exhibit ELASTIC properties and increased VISCOSITY.[142,166]

This stage may be observed from the inflection point on a viscosity-time plot.

gel strength very closely related to YIELD VALUE.[166]

It can be measured on an ICI Gel Strength Tester (also labeled "Rotothinner").

gel time (of a drying oil) time required for the OIL to form a solid GEL under specified conditions of temperature.[63,157] (ASTM)

gelatin protein type material extracted from animal skins, sinews, tendons, and from bones.[83,155]

Gelatin may be regarded as a pure glue. It dissolves in hot water to give a viscous solution with pronounced gelling characteristics. Inedible gelatin has useful applications in the production of sizes for various purposes, and in emulsions.

gelatinous having the consistency of a very soft ELASTIC solid, the nature and APPEARANCE of GELATIN.[142,35]

gelation conversion of a liquid to a GEL state.[166,139] See GEL.[42,161,166]

PROCESS by which a fluid system develops yield value. A sol is a colloidal suspension without yield value and a gel is a colloidal suspension with yield value. A gel (or jelly) is commonly considered a mass, retaining appreciable quantities of liquid which exhibits properties of a solid.

gelling apart from its general meaning (conversion of a liquid to a jelly-like consistency), this term is used specifically to denote the DETERIORATION of a coating owing to the partial or complete changing of the VEHICLE into an insoluble jelly-like condition that is unworkable even with the addition of SOLVENT.[42,161,166] cf. LIVERING[42,161,166,156]

general ventilation the removal of contaminated air and its replacement with clean air from the general workplace area as opposed to local ventilation, which is specific air changing in the immediate air of a contamination source.[172] Syn: dilution ventilation

An example of local ventilation is a laboratory fume hood.

generic name a designation or identification such as a code name, trade name, or brand name used to identify a chemical by other than its chemical name.[134,172]

genuine ultramarine natural ULTRAMARINE BLUE.[41]

geometric metamerism a phenomenon in which members of specimen pairs change relative APPEARANCE as the angle of illumination and viewing are changed, for example, because they incorporate flake metal or pearlescent colorants.[69,43,35] See METAMERISM.[69,43,35]

geranium lake type of colorant produced by PRECIPITATION of certain ACID DYESTUFFS with barium chloride followed by lead nitrate.[41]

gesso traditionally, a material made from chalk and gelatin or CASEIN GLUE, and painted on wooden panels as a surface for TEMPERA work.[71,81,158]

POLYMER gesso provides a flexible ground that can be applied to canvas and other materials.

ghatti gum GUM obtained from India and Sri Lanka, from *Anogeissus latifolia*, from which it exudes.[167] Syn: ghati gum, and Indian gum

ghosting (1) a visual defect in low SHEEN FINISHES where certain areas of the FINISH differ in APPEARANCE from the surrounding areas because they are slightly lighter in COLOR or because of slight variations in sheen but where there is no clear demarcation lines between the areas;[42,35] (2) with INKS, the presence of faint image of a printed design appearing in areas which are not intended to receive that portion of the image.[35,168]

gibbsite see ALUMINUM OXIDE, HYDRATED.[41]

gilder's liquor a mixture of ALCOHOL and water used in water gilding.[148,158]

It is mopped on an area to activate the glue before the gold leaf is tipped on with a BRUSH (a gilder's tip).

gilding art of covering substances, such as WOOD, with layers of gold leaf.[54,77,81]

gilsonite natural ASPHALTUM of American origin.[76] Syn: vintahite

It is hard, BITUMINOUS, and finds application in BLACK JAPANS, BRUNSWICK BLACKS, STAINS, BAKING ENAMELS, and electrical insulating VARNISHES of many types.

gingelli oil synonym for sesame oil.[63]

glance pitch synonym for Manjak.[76]

glass felt GLASS FIBERS bonded into a sheet with a

RESIN and suitable for impregnation in the manufacture of bituminous waterproofing, roofing membranes, and shingles.[76,75] (ASTM)

glass mat a thin mat of GLASS FIBERS with or without a BINDER.[75] (ASTM)

glass plate flow see PLATE FLOW.[166]

glass transition the transition temperature range (relatively small for most POLYMERS) within which an amorphous polymer changes from a RUBBERY or VISCOUS to a GLASS-like (brittle) state.[47] Syn: second order transition

This may be evidenced by a change in the slope of a stiffness—temperature curve, dielectric properties—temperature curve, etc.

glass transition temperature (T_g) temperature at which the noncrystalline portion of the POLYMER is transformed from a tough, RUBBERY material to a brittle, GLASS-like material.[47]

glaze (1) very thin coating of a paint product, usually a SEMITRANSPARENT tinted coating, applied on a previously painted surface to produce a decorative effect;[148,163,175] (2) small amount of COLOR coating added to WOOD furniture to highlight and give character to wood.[148,163,175] See STRIÉ.[148]

These are often manually wiped to remove most of the color. (EPA)

glaze coat (1) the top layer of ASPHALT in a smooth-surfaced BUILT-UP ROOF ASSEMBLY; (2) a thin protective coating of BITUMEN applied to the lower PLIES or top ply of a built-up membrane, when application of additional FELTS, or the flood coat and AGGREGATE surfacing are delayed.[76,71] (ASTM)

glazing (1) APPLICATION of TRANSPARENT or TRANSLUCENT PIGMENT on a painted surface to produce certain blended effects; (2) setting glass, as a window pane.[148,163,165]

glazing compound doughlike material consisting of PIGMENT and VEHICLE, used for sealing window GLASS in frames.[165] (ASTM) cf. PUTTY[165,158,147]

It differs from putty in that it retains its plasticity for an extended period.

GLC abbreviation of GAS-LIQUID CHROMATOGRAPHY; a technique by which the separation and identification of VOLATILE components of a mixture is possible.[157,37]

gloss subjective term used to describe the relative amount and nature of mirrorlike (SPECULAR) REFLECTION. Different types of gloss are arbitrarily differentiated, such as SHEEN, DISTINCTNESS-OF-IMAGE gloss, etc. Trade practice recognizes various stages of increasing gloss using the following terminology: FLAT (or MATTE)—practically free from sheen, even when viewed from oblique angles; EGGSHELL, semi-gloss, full gloss—smooth and mirrorlike surface when viewed from all angles.[43,35] See SILK, SATIN, MATTE, MAT, VELVET, PEARL, PLATINUM, and SUEDE.[43]

Gloss is measured quantitatively relative to the arbitrary value of 1000 for a perfect mirror. Polished plate GLASS, which on that basis has a gloss value of about 100, is used as the working standard. Gloss can vary greatly depending on the angle of observation. Meters for measuring the gloss of paints are generally set for angles of 85°, 60°, or 20° to normal, or combinations thereof. Measurements at 85° are significant primarily for paints of very low gloss ([sic], flat paints). Measurements at 60° are significant at all gloss levels. Measurements at 20° are useful for finer distinctions among very high gloss finishes, generally in the OEM category. No official agreement exists in the paint industry that relates instrumental gloss values to common gloss level terminology such as referred to in the preceding paragraph. Thus, in practice, paints marketed with those descriptive terms can vary widely in gloss. To promote understanding and standardization it has been proposed, in connection with architectural and industrial maintenance coatings, to establish eight numerical gloss categories, as shown in the following table:

Type[1]	*60° Gloss*	*85°Gloss*	*Description[2]*
I	*0-10*	*0-15*	*flat*
II	*0-10*	*15+*	*high sheen flat*
III	*10-25*	—	*eggshell*
IV	*25-40*	—	*low semigloss*
V	*40-55*	—	*(regular) semigloss*
VI	*55-70*	—	*high semigloss*
VII	*70-85*	—	*(regular) gloss*
VIII	*85+*	—	*high gloss*

(1) This numerical gloss category system was developed for consideration by ASTM Subcommittee D01.42 on Architectural Finishes.

(2) The descriptive terminology shown here is solely for guidance of the user of this volume, and does not claim to have any greater validity than the same terms used with commercial products that do not comply with the indicated gloss ranges.

It should be pointed out that the gloss of most air dried paint films decreases substantially on aging. Thus, strictly speaking, gloss level characterizations apply to gloss itself or to the gloss of the film at the particular time that it is being measured, rather than to the coating. In practice the gloss category of a coating is determined by standardizing the application and drying history of the film at the time it is measured; for example, 3.5 mils (90 µm) wet film thickness dried 7 days under normal room conditions of ventilation, temperature, and humidity.

ASTM test methods for gloss include: change, washability of coatings, D4828; effect of house-

Words presented in CAP/SMALL CAPS type indicate that the word is defined in another part of the *Coatings Encyclopedic Dictionary.*

Numerical superscripts classify terms in one or more of the categories listed in the second section of this volume.

hold chemicals, D1308; gloss differences, visual evaluation, D4449; haze of high gloss finishes; D4039; high gloss, goniophotometer test, E430; measurement of gloss and sheen, D523; tolerances, conformance evaluation, D3134; uniformity of brushouts, test, D3928.

gloss, distinctness-of-image the sharpness with which outlines are reflected by the surface of an object.[43]

gloss ink an INK that dries with a minimum of penetration into the stock and yields a high LUSTER.[168]

gloss oil solution of limed ROSIN or limed rosin acids in a volatile solvent, used chiefly in surface coatings. When made from tall oil, the source is usually indicated.[156] (ASTM)

gloss retention degree to which the original sheen of a coating is retained.[43,72]

gloss white a white mineral PIGMENT used as an INK EXTENDER, made by coprecipitation of alumina hydrate and blanc fixe.[41]

glossary a list of difficult or specialized words (jargon), with their DEFINITIONS, often placed at the back of book.[134]

glossing up undesirable BURNISHING.[42,56] See BURNISH, and BURNISH RESISTANCE.[56]

glossmeter an instrument for measuring the degree of GLOSS in relative terms.[145,43,157]

Such instruments measure the light reflected at a selected specular angle, such as 20° (from the perpendicular), 45°, 60°, 75°, or 85°. Results obtained are very dependent on the instrument design, calibration technique used, types of samples, etc. In essence, a glossmeter is an ABRIDGED GONIOPHOTOMETER.

glow discharge see CORONA DISCHARGE.[67]

glue (noun) originally, a hard GELATIN obtained from hides, tendons, cartilage, etc., of animals.[79] (ASTM) See ADHESIVE, BONE GLUE, MUCILAGE, and SIZE.[79]

Also an adhesive prepared from this substance by heating with water. Through general use the term is now synonymous with the term "adhesive."

glue (verb) see BOND.[79]

glycerides the ESTERS derived from GLYCEROL.[144]

Glycerol is a trihydroxy ALCOHOL and as such is capable of REACTING with three monobasic ACIDS. Such fully satisfied esters are described as the TRIGLYCERIDES, and the DRYING OILS are typical members of this class. If only one hydroxy group is reacted with an acid, the resultant product is a MONOGLYCERIDE, and if two are reacted a DIGLYCERIDE is obtained.

glycerin see GLYCEROL.[144,167]

glycerol $CH_2OH{\cdot}CHOH{\cdot}CH_2OH$. Trihydroxy ALCOHOL.[144] Syn: glycerin

Properties: a bp of 290°C, mp of 17°C, and sp gr of 1.260/15°C. It is of considerable value in the preparation of ALKYD RESINS, SYNTHETIC DRYING OILS, COPAL and similar ESTER type resins, esterified synthetic resins, and PLASTICIZERS.

glyceryl diacetate PLASTICIZER; SOLVENT for CELLULOSE derivatives, ALKYDS, and SHELLAC.[144] Syn: diacetin

Properties: a bp of 260°C, sp gr of 1.176/15°C, flp of 146°C (295°F), refractive index of 1.44.

glycol CH_2OHCH_2OH. Synonym for dihydric ALCOHOLS; ETHYLENE GLYCOL is the most simple of the glycols.[164]

ASTM test methods for glycols include: diethylene glycol, D2694; dipropylene glycol, D2696; ethylene glycol, ethylene glycol, D2693; hexylene glycol, D2636; propylene glycol, D2695.

glyptal glyptal was a proprietary name and is now applied to alkyd-type RESINS.[156]

It has been used as a European generic term for alkyd resins and as an alternative to the term "alkyd."

gold bronzes METALLIC POWDERS of gold COLOR produced from ALUMINUM or copper and its alloys.[41]

gold ink a PRINTING INK whose principal PIGMENT consists of BRONZE POWDER.[168]

gold lacquer VEHICLE of such COLOR that it will give the effect of gold when applied over ALUMINUM LEAF.[71,151,156]

gold leaf very thin leaves of metallic gold used for decorative purposes.[45,54,148]

Beaten gold (12-karat to 24-karat gold; those below 24-karat alloyed with silver, copper, or palladium). It is sold in books of leaves, each one 250 thousandth of an inch thick. Because the leaves are so thin, they are applied with BRUSH by oil gilding or water gilding. If the surface is properly prepared, gold leaf gives a satiny smooth FINISH that can last for centuries.

gold leafing in ANTIQUING, the art of applying a very thin sheet of gold over a "sized" area.[148,175]

It became fashionable in France in the 1700s.

gold number a measure of the protective ability of hydrophilic colloids.[139,157]

Specifically, the number of milligrams of colloids which, when added to a standard gold solution, is just insufficient to prevent a change in COLOR from red to blue.

gold size OLEORESINOUS VARNISH which dries rapidly to a tacky condition, but which hardens slowly.[79,148] (BSI)

It is used chiefly as an ADHESIVE for fixing GOLD LEAF to a SURFACE.

golden chrome see CHROME YELLOW, LIGHT and PRIMROSE.[41]

goniochromatic a colored material which exhibits GONIOCHROMATISM.[43,69]

goniochromatism the phenomenon where the COLOR of a material changes as the angle of ILLUMINATION and/or viewing is changed.[69,43]

gonioisochromatic a pair of colored materials exhibiting GONIOISOCHROMATISM.[43,69]

gonioisochromatism the phenomenon where the COLORS of two goniochromatic materials change color in exactly the same way when the ILLUMINATION and/or viewing angles are changed.[43,69]

goniometachromatic a pair of samples exhibiting GONIOMETACHROMATISM.[43,69]

goniometachromatism the phenomenon where the COLORS of two GONIOCHROMATIC materials match exactly at one set of ILLUMINATION-viewing angles but no longer appear to match if either the illumination or viewing angles, or both, are changed. Such pairs of samples are also called GEOMETRIC METAMERS.[43,69]

goniophotometer an instrument used to measure the amount of LIGHT REFLECTED as a function of the viewing angle when the incident angle is kept constant.[43,145]

Generally, the incident angle may be changed.

goniospectrophotometer an instrument used to measure a SPECTROPHOTOMETRIC curve at various angles of INCIDENCE and REFLECTANCE.[43,69]

The angles of incidence and reflectance may be changed. On GONIOCHROMATIC samples, the results obtained are dependent on the angles of incidence and reflectance, and on samples exhibiting specular reflectance (glossy materials), the results are also dependent on the total subtended angle of viewing relative to the specular angle.

gouache (1) method of painting with OPAQUE COLORS which have been GROUND in water and tempered with a preparation of GUM; (2) more generally, a term used to describe a method of water COLOR painting in which white is employed as a PIGMENT by contrast with the method of attaining whites by allowing the PAPER to show through.[77,81]

graffito last PLASTER coat of a FRESCO GROUND which is colored before APPLICATION.[71,73,81]

graft polymer or copolymer a high POLYMER, the molecules of which consist of two or more polymeric parts of different composition, chemically united.[130]

Such a graft copolymer may be produced, for example, by POLYMERIZATION of a given kind of MONOMER with subsequent polymerization of another kind of MONOMER onto the product of the first polymerization. The union of two different polymers by chemical reaction between their molecular end groups or by reaction producing CROSSLINKS between the different molecules would also produce a graft copolymer.

grahamite naturally occurring ASPHALTUM which varies considerably in composition.[76]

Some types are completely soluble in carbon disulfide, whereas others may contain as much as 50% of mineral matter.

grain the direction, size, arrangement, APPEARANCE, or quality of the fibers in WOOD or VENEER.[153,175] (ASTM)

To have a specific meaning the term must be qualified. See TYPES OF GRAIN, i.e., CROSS GRAIN, INTERLOCKED GRAIN, COARSE GRAIN, CLOSE GRAIN, DIAGONAL GRAIN, DIP GRAIN, etc.

grain a unit of mass in the U.S. Customary System, an avoirdupois unit equal to 0.002285 ounce (0.065 gram).[128]

grain alcohol see ETHYL ALCOHOL.[164]

grain crack resistance ability of a coating to resist breaks of the film that extend through to the SUBSTRATE when it is subjected to the effects of swelling and shrinking of the WOOD SUBSTRATE due to changes in temperature and/or humidity.[42,72,157]

The paint film is stressed most along the grain boundaries of the wood. Fluctuations of various substrates due to rapid temperature/humidity changes sometimes cause cracking of topcoats which, due to poor adhesion or inadequate elasticity, cannot adapt quickly enough.

grain raising swelling and standing up of short broken fibers of WOOD caused by absorbed liquids, especially water.[153]

Grain raising causes the surface of wood to look and feel rough.

graining (1) a PROCESS for simulating a grain such as that of WOOD or marble on a painted surface by applying a STAIN, either TRANSLUCENT or PIGMENTED, then working it into suitable patterns with tools such as graining combs, BRUSHES, and rags. Normally, it involves the APPLICATION of three coats: (a) ground COLOR—similar to a normal UNDERCOAT, drying to a flat FINISH; it is usually lighter in COLOR

Words presented in CAP/SMALL CAPS type indicate that the word is defined in another part of the *Coatings Encyclopedic Dictionary.*

Numerical superscripts classify terms in one or more of the categories listed in the second section of this volume.

than the graining color; (b) graining color—translucent or pigmented material, devoid of FLOW, by means of which the pattern or graining is achieved; (c) VARNISH COLOR—protective topcoat; (2) process of making the surface of a lithographic plate water-receptive by GRINDING with a muller and sand, or mechanically with pebbles and sand.[54,148,168,35]

gram subunit of mass in the metric system of weights and measures, equal to 1/1000 of a standard kilogram (453.6 grams = one pound).[128]

gram-molecule see MOLE OR MOL.[128]

grapeseed oil SEMIDRYING VEGETABLE OIL.[63]

Of the FATTY ACIDS present, up to 55% may be linoleic acid. Approximate constants for oil from various sources are: sp gr of 0.925/15°C, refractive index of 1.4899/15°C, saponification value of 190, iodine value of 125-157.

graphic arts coatings coatings which are marketed solely for APPLICATION to indoor or outdoor signs and include lettering enamels, POSTER COLORS, and bulletin colors.[71,81]

graphite Pigment Black 10. Soft, steel gray to black, hexagonally CRYSTALLIZED ALLOTROPE of CARBON.[41] Syn: black lead, and plumbago

When used as a PIGMENT, the main purpose is for its lubricating and conductive properties.

graphite paint a painting COMPOUND consisting of powdered graphite and oil, used to coat metallic structures to inhibit CORROSION.[71] (DAC)

graphitic mica see ALUMINUM POTASSIUM SILICATE.[41]

graphitization CORROSION of gray IRON in which metallic iron is converted into corrosion products, leaving a residue of intact graphite mixed with iron corrosion products and other insoluble constituents of cast iron and having approximately the original dimensions of the casting.[66] Syn: graphitic corrosion

grass cloth originally, a handmade product imported from Japan, made by gluing woven native grasses onto a PAPER backing.[148]

Printed wallpapers or dimensional wallcoverings, especially vinyl, that simulate the same.

grass sessile ALGAE on the side of a ship.[173]

grass-tree gum see ACCROIDES.[167]

gravel rounded or semirounded PARTICLES of rock that will pass a 3-in. (76.2 mm) sieve and be retained on a No. 4 (4.75 mm) U.S. standard sieve.[75] (ASTM)

gravel roofing see BUILT-UP ROOFING.[75,76,82]

gravure one of the methods of PRINTING using the INTAGLIO PROCESS, i.e., the INK is placed in cells below the plate surface.[168]

Gravure inks are quick drying, low VISCOSITY inks based on volatile solvents.

gravure tissue sheet of PAPER coated with dichromated GELATIN but containing no PIGMENT.[168]

gray blast (commercial blast) see NACE No. 3.[158]

gray scale an ACHROMATIC scale ranging from black through a series of successively lighter grays to white.[69,43]

Such a series may be made up of steps which appear to be equally distant from one another (such as the MUNSELL VALUE SCALE) or may be arranged according to some other criteria such as a geometric progression based on lightness. Such scales may be used to describe the relative amount of difference between two similar colors.

grazing angle angle of light very nearly parallel to the surface of a material; used to describe incident and/or reflected angles.[43]

grease a semisolid lubricant composed of emulsified PETROLEUM OILS and soluble HYDROCARBON SOAPS.

grease paint (1) nondrying and nonoxidizing coating for void spaces in ships which can be easily removed for control of the steel sheets;[163,71] (2) theatrical make-up.[71]

greaseproof inks and coatings INKS and coatings that are resistant to the action of fats, oils, and greases.[168]

greasiness see SPEWING.[42,56]

greasing sensitization of the nonprinting area of a LITHOGRAPHIC plate resulting in undesirable ADHESION of the INK to these areas.[168] Syn: scumming

green B, naphthol IRON LAKE of 1-nitroso-2-naphthol-3-carboxylic acid.[41]

green B, pigment PIGMENT produced by treating 1-nitroso-2-naphthol with a ferrous salt.[41]

green, brunswick see CHROME GREENS.[41]

green, chrome see CHROME GREENS.[41]

green, cinnabar see CHROMIUM OXIDE GREEN.[41]

green earth naturally occurring EARTH PIGMENT which is primarily an IRON SILICATE.[41] See GREEN, LIME.[41] Syn: terre verte

Used to make lime green.

green, emerald see EMERALD GREEN.[41]

green gold see NICKEL AZO YELLOW.[41]

green lake mixture of PRUSSIAN BLUE and YELLOW LAKE.[41]

green, lime green PIGMENTS which are employed for their fastness to ALKALI (LIME).[41]

For a considerable time it consisted of the green produced from GREEN EARTH but the term is now applied to other types, e.g., reduced Pigment Green B.

green lumber LUMBER which has not been dried or seasoned.[153] (DAC)

green, Paris PIGMENT consisting essentially of copper acetoarsenite.[41]

It is sometimes used in antifouling compositions, but its main use is in insecticides.

green rouge see CHROMIUM OXIDE GREEN.[41]

green stock raw or uncured PLASTIC or RUBBER stock, ready for VULCANIZATION.[140,38]

The term is not applied to crude rubber, synthetic rubber, or RESIN that has not been compounded.

green vitriol $FeSO_4 \cdot 7H_2O$. CRYSTALLINE FERROUS SULFATE used in the manufacture of IRON BLUE.[41]

green wood WOOD which has the original moisture present in the standing tree and has not been dried below the fiber saturation point.[153]

green, zinc range of greens made by mixing ZINC CHROMATE with IRON BLUE.[41]

They are less bright and weaker than the BRUNSWICK GREENS, but find application where lead-free greens are required.

greenockite naturally occurring CADMIUM SULFIDE MINERAL.[41,34]

greentone ink blue see IRON BLUE.[41]

grind (1) to reduce PIGMENT particle size mechanically; (2) misnomer used to describe the degree of DISPERSION.[64]

Its use to describe a MILL BASE is deprecated.

grinding (1) PROCESS by which pigment particles are reduced in size, mechanically;[41,131] (2) its use to describe pigment dispersion in MILL BASES is deprecated.

grinding Japan see JAPAN, GRINDING.[152]

grinding media charge see MILL CHARGE.[64]

grinding medium any medium which is used as a means of DISPERSING PIGMENTS or FILLERS by GRINDING or dispersing.[64,127]

grinding slab a flat piece, usually of GLASS or stone, on which COLOR is ground from a coarse to a finely divided state, frequently with the medium that is to blend it as a paint.[59,131]

grinning through showing through of the underlying SURFACE due to the inadequate OPACITY of a paint film which has been applied to it.[42,56] (BSI)

grisaille (Fr.) DECORATIVE PAINTING in which objects are rendered in tones of one COLOR, usually gray, often applied on GLASS, and intended to give the effect of sculptured relief.[81]

grit minute rough granules of foreign particles (sand, dust, etc.) causing lumps or texture in the coating.[42,56,161]

grit blasting ABRASIVE cleaning of a surface by blasting with angular chilled IRON GRIT, aluminum oxide, or any crushed or irregular abrasive.[45,158] See BLAST CLEANING.[158]

The grit is projected onto the surface either mechanically or by means of compressed air.

grit number see MESH NUMBER.[158]

gritty in a coating, roughness resembling grains of sand.[56,42]

ground (1) SURFACE applied to a canvas or other support, upon which a picture is painted; (2) base on which ORGANIC agents are precipitated to form LAKES; (3) any surface which is or will be painted.[81,158,159] See SUBSTRATE, and BASE (1).[168]

ground coat (1) PORCELAIN ENAMEL applied directly to the BASE METAL to function as an INTERMEDIATE layer between the metal and the cover coat; (2) coat of a paint with good OPACITY, applied before a GLAZE COAT. The final COLOR effect when glazed depends upon the mutual influence of the ground and glaze coats.[45,54,71,158] See GRAINING.[54,148]

ground oyster shells see CALCIUM CARBONATE, NATURAL.[41]

ground perlite see PERLITE, EXPANDED.[41]

ground silica see SILICA, CRYSTALLINE.[41,58]

ground wire a wire attached between apparatus, containers and instruments, and an electrical ground to dissipate an electric or electrostatic charge.[61,172]

groundnut oil another name for arachis or peanut oil.[63]

grout (1) a thin mortar used to fill cracks and crev-

Words presented in CAP/SMALL CAPS type indicate that the word is defined in another part of the *Coatings Encyclopedic Dictionary.*

Numerical superscripts classify terms in one or more of the categories listed in the second section of this volume.

ices in MASONRY.[73,165] *Mortar containing a considerable amount of water so that it has the consistency of a viscous liquid, permitting it to be poured or pumped into joints, spaces, and floor tile, and into the joints between pieces of ceramic, slate, and floor tile, and into the joints between preformed roof deck units. In foundation work, mixtures of CEMENT, cement-sand, CLAY, or chemicals, used to fill voids in granular soils, usually by a process of successive injection through drilled holes. Grout is sold either as a dry powder to be mixed with water or a ready mix;* (2) a thin plaster for finishing walls and ceilings;[73,165] (3) a mixture of BITUMINOUS material and fine sand that will FLOW into place without mechanical manipulation when heated.[76,165] (DAC)

guage length see GAGE LENGTH.[47,157]

guaiacol ortho methoxy PHENOL which occurs in WOOD TAR.[83]

Used for ANTISKINNING properties.

guanidine aldehyde resins reaction products of guanidine ALDEHYDES.[130]

guar gum the ground endosperms of *Cyanopsis tetra gonolobus*.[155]

The free-flowing POWDER is soluble in cold or hot water and can be gelled with BORAX. It is used as a THICKENING AGENT or PROTECTIVE COLLOID.

guide coat coat similar to the FINISH or COLOR coat but of a different color to assure good coverage.[71]

Guignet's green see HYDRATED CHROMIUM OXIDE.[41]

Guimet's blue synthetic ULTRAMARINE BLUE.[41]

gum any of a class of COLLOIDAL substances, exuded by or prepared from plants, sticky when moist, composed of complex carbohydrates and ORGANIC ACIDS, which are soluble or swell in water.[139,155] (ASTM) See GUM, NATURAL.[155,130,167]

gum Acacia see ACACIA GUM.[79,155]

gum accroides see ACCROIDES.[155]

gum animi see ANIMI.[167]

gum Arabic see ACACIA GUM.[79,155]

gum benzoin see BENZOIN.[167]

gum bloom a DEFECT in a painted SURFACE, appearing as a lack of GLOSS or a HAZE, resulting from the use of incorrect REDUCER.[56,35] (DAC)

gum karaya see KARAYA GUM.[155]

gum mastic a natural RESIN used for picture varnish to protect oil paintings, water colors, and the like.[81,167] (BSI)

gum, natural a CARBOHYDRATE HIGH POLYMER that is insoluble in ALCOHOL and other ORGANIC SOLVENTS, but generally soluble or dispersible in water.[155,130,167] See also RESIN, NATURAL; GUM.[155,130,167]

Natural gums are HYDROPHILIC polysaccharides composed of monosaccharide units joined by glycosidic bonds. They occur as EXUDATIONS from various trees and shrubs in tropical areas or as phycocolloids (ALGAE), and differ from natural resins in both chemical composition and solubility properties. Some contain acidic components and others are neutral. Their chief use is as protective colloids and emulsifying agents in food products and pharmaceuticals; as SIZING for textiles; and in electrolytic deposition of metals. Examples are: arabic, tragacanth, guar, karaya. The terminology of natural gums and resins is inconsistent and often confusing. The word "gum," often used as an adjective, seems to acquire a different meaning from the noun. For example, the resinous products obtained from PINE PITCH (produced by the parenchyma cells of softwoods) are conventionally called "gum turpentine" and "gum rosin." There are also such "gum resins" as gum benzoin, gum camphor, and others. The so-called ESTER GUM is a semisynthetic reaction product of rosin and a polyhydric alcohol. All these are actually resinous products having properties quite different from those of natural gums. Still further complicating the matter is the common application of the word "gum" to such plant latices as chicle and natural rubber, which are different from both carbohydrate gums and resins. It is probable that the confusion originated in the casual use of "gum" to refer to any soft, sticky product derived from trees. In view of this situation, any specific definition of the term is likely to be controversial.

gum rosin ROSIN which is produced from the GUM TURPENTINE obtained by tapping living pine trees, as distinguished from the WOOD rosin obtained from the stump and branches of dead trees.[44,167]

gum sandarac soft RESIN obtained from *Tetraclinis articulate*, used in the manufacture of paints, linoleum, and oil cloth.[167] Syn: juniper gum

gum spirits see TURPENTINE (1).[164]

gum thus botanically, the OLEORESIN from trees of *Boswellia* species native to Arabia and Somaliland.[44,167] (ASTM) Syn: olibanum or frankincense

As applied to the naval stores industry, the term refers to the crystallized pine oleoresin or "scrape" collected from scarified "faces" of trees being worked for turpentine.

gum turpentine see TURPENTINE.[44,164]

gumming (1) clogging of sandpaper by a coating; (2) addition of a GUM RESIN to a coating.

gummy British term for BRUSH DRAG.[42]

gun cotton highly nitrated form of CELLULOSE nitrate.[144,151]

gutta-percha product derived from the LATEX of Malayan and Brazilian trees, esp. *Palaquium gutta*.[140,167]

It resembles natural RUBBER in that it is a POLYMERIC COMPOUND obtained from ISOPRENE (C_5H_8), but contains other RESINOUS COMPOUNDS as well. Its SOLUBILITY characteristics resemble those of natural rubber.

gutter a channel at the EAVES for conveying away rain water.[82]

Gutzeit test method for testing and estimating small quantities of arsenic and antimony in other materials.[157]

Details are given in the British Pharmacopoeia.

gym finish composition of OILS and RESINS in a SOLVENT; designed to produce a high GLOSS, tough SURFACE FINISH on WOOD floors.[71,153]

gynocardia oil DRYING OIL, smelling like LINSEED OIL, and obtained from *Gynocardia odorata*, which grows in India and Pakistan.[63]

Its component acids include ISOLINOLENIC, LINOLENIC, LINOLEIC, OLEIC, and PALMITIC ACIDS. It has a sp gr of 0.925/25°C, iodine value of 153-160.

gypsum the mineral consisting primarily of fully hydrated calcium sulfate, $CaSO_4{\cdot}2H_2O$ or calcium sulfate dihydrate.[144,58,34] Syn: mineral white

gypsum, calcined $CaSO_4{\cdot}{}^1/_2\,H_2O$. GYPSUM partially DEHYDRATED by HEAT.[58] See PLASTER OF PARIS.[73,75]

gypsum cement see PLASTER OF PARIS.[73,75]

gypsum plaster see CEMENT PLASTER.[73,75]

gypsum wallboard a sheet or slab having an incombustible core, essentially GYPSUM, surfaced with paper suitable to receive DECORATION.[73,75] Syn: plasterboard

---

Words presented in CAP/SMALL CAPS type indicate that the word is defined in another part of the *Coatings Encyclopedic Dictionary.*

Numerical superscripts classify terms in one or more of the categories listed in the second section of this volume.

H

h abbreviation for hour or hours.[128]

hackberry tree seed oil obtained from the kernels of the tree, *Celtis occidentalis*. Its main constituent FATTY ACID is LINOLEIC ACID, present to the extent of over 70%.[63]

Properties: sp gr of 0.9204/25°C, refractive index of 1.4794/25°C, saponification value of 191, and an iodine value of 150.

hackles thin, needle-like, or sliver-like protrusions (ranging from 3 to 6 μm) found on steel plates which have been BLASTED with STEEL or GRIT.[158]

haco oil specially PROCESSED whale OIL, of German origin, with good drying properties.[63]

hair-cracking see CRACKING.[56]

hairline cracks very fine cracks which do not penetrate the TOPCOAT; they occur erratically and at random.[42,56,65] See CRACKING (1).[42,56,65]

halftone a PRINTED image composed of dots of varying frequency (number per sq. in.), size, or density, producing tonal gradations.[168]

The term is also applied to the process, plates, and INK used to produce this image.

halftone screens TRANSPARENT plates ruled diagonally with OPAQUE lines at right angles to one another, the distance between them and their thickness being approximately equal.[168]

halo effect piling up of INK at the edges of PRINTED letters and dots.[168]

Also, COLORED or sometimes uncolored areas adjacent to them, caused by the spread of colored or uncolored vehicles.

halogen member of the chemical group consisting of fluorine, chlorine, bromine, and iodine.[134,144]

halogenated solvents the SOLVENTS containing HALOGEN (usually chlorine) have improved SOLVENCY compared with the HYDROCARBONS from which they are derived and in addition flammability is reduced.[164]

Some of these are highly TOXIC, and precautions must be taken to avoid inhalation of their VAPORS.

halogenation PROCESS or reaction wherein any members of the halogen group are introduced into an ORGANIC COMPOUND, either by simple ADDITION or SUBSTITUTION.[131]

Halphen-Hicks test qualitative test for detecting ROSIN.[157]

The test involves dissolving the material under test in a solution of PHENOL in carbon tetrachloride. A drop of this mixture is then placed in close proximity to a drop of a solution of bromine in carbon tetrachloride. The bromine vapors diffuse into the droplet of the solution under test, and if rosin is present, a green COLOR, rapidly changing to blue or violet, is obtained.

ASTM Test Method D1542 covers procedures for the qualitative detection of rosin in varnishes by the Lieberman-Storch Test and the Halphen-Hicks Test. The rosin may be present as either free rosin (abietic acid), esterified rosin, or as metal salts.

hammer finish a paint FINISH which appears to have been applied over hammered METAL; produced by the use of NONLEAFING METALLIC PIGMENT, plus tinting pigments, which are mixed in a special BINDER.[71,163] (DAC)

hammer mill type of MILL used for PULVERIZING dry materials in which the disintegration is caused by the action of a number of small hammers.[64,131]

hand-blocking see BLOCK PRINTING.[145,168]

hand-held electrostatic spraying a method of SPRAY APPLICATION which uses a hand-held ELECTROSTATIC SPRAY GUN.[77] cf. ELECTROSTATIC SPRAYING[77]

This type of spraying results in a higher TRANSFER EFFICIENCY than conventional hand-held spraying.

hand-screening see SILK-SCREENING.[81,77]

hanger wire the wire employed to suspend the acoustical ceiling from the existing structure (WOOD JOISTS, steel bar joists, steel beams, CONCRETE slabs, etc.).[75] (ASTM)

hank a looped bundle, as of yarn.

hank shellac form of bleached SHELLAC which, instead of being GROUND and DRIED, is pulled into HANKS. Contains up to 25% MOISTURE.[146]

hansa yellow range of yellow PIGMENT DYESTUFFS which are characterized by clean tones and good fastness to light in MASSTONES but poor fastness to light in pale tints.[41]

HEAT RESISTANCE is also poor and therefore these pigments should not be used in baking finishes. They are produced by coupling diazotized bases with acetoacetanilide or its substitution products.

hanus iodine number method claimed to measure the complete UNSATURATION of both conjugated and nonconjugated substances.[157]

The method employs as a REAGENT a solution of iodine and bromine in glacial ACETIC ACID.

HAP abbreviation for HAZARDOUS AIR POLLUTANTS.[171,172]

hard dry see DRY-HARD.[55,67](BSI)

hard gloss paint a high GLOSS ENAMEL, formulated with a hard-drying RESIN VEHICLE.[71] (DAC)

hard gums RESINS exuded from living plants centuries ago.[167]

Some of the hard gums are Amber, Zanzibar, Kauri, Pontianak, Manila, and Congo.

hard lac resin RESIN resulting from SHELLAC when treated by fractional SOLVENT EXTRACTION to remove the soft resin component.[146,167]

Hard LAC resin is claimed to give films which are harder, and more durable, than those of ordinary shellac. In addition, hard lac resin POLYMERIZES more easily when HEATED and has better WATER RESISTANCE.

hard putty (1) LINSEED OIL putty which was hardened with litharge or basic lead carbonate; (2) a quick setting putty containing water and PLASTER OF PARIS.[165]

hard stopping see HARD PUTTY (2).[165]

hardboard a generic term for a smooth, grainless panel manufactured primarily from interfelted lignocellulosic fibers (usually WOOD), consolidated under HEAT and PRESSURE in a hot-press to a density of 31 lb per ft.^{3} (specific gravity 0.50) or greater, and to which other materials may have been added during manufacture to improve certain properties.[75,153,159]

In its most common form, it is smooth on one side and has a slight waffle-like texture on the reverse side. Used for interior panels or durable siding. It is also available with a smooth surface on both sides, and with a variety of embossed or textured finishes, as well as a perforated form that accepts hooks and fixtures, for converting any wall into a storage area.

hardener ADDITIVE (CROSSLINKING AGENT, RESIN, or other MODIFIER) used to promote or control the hardening or CURING reaction of a coating, PLASTIC, ADHESIVE, or resin system.[67,83] See CURING AGENT, CATALYST, and CROSSLINKING AGENT.[74,67,83]

hardinge mill conical type of BALL MILL used for GRINDING rock and similar materials.[49,59]

hardness ability of a coating FILM, as distinct from its SUBSTRATE, to resist cutting, INDENTATION, or PENETRATION by a hard object.[47] See KNOOP HARDNESS NUMBER, PENDULUM-ROCKER HARDNESS, MODULUS OF ELASTICITY.[47]

ASTM test methods for measuring hardness include: Knoop indentation tester, KHN, D1474; Koening pendulum test, D4336; pencil test, D3363; Persoz pendulum test, D4366; Pfund indentation tester, PHN, D1474; Sward rocker test, D2134.

hardness scale see MOHS SCALE, KNOOP HARDNESS NUMBER, PENCIL HARDNESS, ROCKER HARDNESS TESTER.[47,128]

hardwoods generally, the botanical group of trees that have broad leaves, in contrast to the conifers or SOFTWOODS.[153]

The term has no reference to the actual HARDNESS of the wood. Angiosperms is the botanical name for hardwoods.

HASE HYDROPHOBICALLY-MODIFIED ALKALI SOLUBLE EMULSION POLYMER.[83,166,130] See BRUSH DRAG, NATURAL SPREADING RATE, and ASSOCIATIVE THICKENER. cf. HEUR, HMHEC[83,166,130]

These anionic associative thickeners impart high SHEAR VISCOSITY, FLOW, FILM BUILD, GLOSS, and SPATTER RESISTANCE to LATEX paints. However, they can also reduce WATER RESISTANCE of EXTERIOR LATEX paints to the extent that early blister resistance over heavily CHALKED latex repaint surfaces is downgraded.

hashab a fine pale grade of GUM acacia.[167]

hatbanding synonym for PICTURE FRAMING (a).[42,56]

Hazard Communication Rule an OSHA law that requires chemical manufacturers and importers to assess the hazards associated with the materials in their workplace.[172]

MATERIAL SAFETY DATA SHEETS, labelling, and training are all a result of this law.

hazardous air pollutants (HAP) COMPOUNDS used in paints that have been identified as HAPs in the CLEAN AIR ACT of 1990.[171,172,144]

These include: acetamide, acrylamide, acrylic acid, acrylonitrile, allyl chloride, aniline, benzene, 1,3-butadiene, caprolactam, catechol, cumene, 1,2-dichloroethane, dichloroethyl ether, diethanolamine, dimethylamino-azobenzene, dimethylformamide, dimethyl phthalate, epichlorohydrin, ethyl acrylate, ethylbenzene, ethylene dibromide, ethylenimine, formaldehyde, hexachlorobenzene, n-hexane, hydroquinone, isophorone, maleic anhydride, methanol, methyl ethyl ketone, methyl isobutyl ketone, methylene chloride, naphthalene, nitrobenzene, 2-nitropropane, pentachlorophenol, PHENOL, *propylene oxide, styrene, 1,1,2,2-tetrachloroethane, toluene, 2,4-toluene diisocyanate, 1,1,1-trichloroethane, trichloroethylene, 2,4,6-trichlorophenol, vinyl* ACETATE, *vinyl chloride, xylenes, m-xylene, o-xylene, and p-xylene.*

The 1990 Clean Air Act (CAA), designed to eliminate VOLATILE ORGANIC COMPOUNDS *(VOCs),* OZONE *depleters, and other HAPs, has put limits on many organic compounds. It will require state and local governments to develop plans, if they have not already done so, to lower regional atmospheric pollution in compliance with federal law. An element in those plans will enforce limits on VOCs emitted from liquid coatings as they dry. CAA will restrict the use of solvents considered hazardous air pollutants.*

Hazardous Chemical, Material in a broad sense, any substance or mixture of substances having properties capable of producing adverse effects on the health or safety of a human.[172]

In 1971 OSHA adopted the following definition: "the term Hazardous Material means a material which has one or more of the following characteristics: (1) has a FLASH POINT *below 140°F, closed cup, or is subject to spontaneous heating; (2) has a* THRESHOLD LIMIT VALUE *below 500 ppm for gases and vapors, below 500 mg/m³ for fumes, and below 25 mppcf (million* PARTICLES *per cubic foot) for dusts; (3) has a single dose oral* LD_{50} *below 50 mg/kg; (4) is subject to* POLYMERIZATION *with the release of large amounts of energy; (5) is a strong* OXIDIZING *or* REDUCING AGENT; *(6) causes first degree burns to the skin from a short time* EXPOSURE, *or is systemically toxic by skin contact; or (7) in the course of normal operations, may produce dusts, gases, fumes, vapors, mists, or smokes which have one or more of the above characteristics." Also included are substances that are carcinogens, toxic, irritants, corrosives, sensitizers, and agents which damage the lungs, skin, eyes, or mucous membranes.*

hazardous decomposition a breaking down or separation of a substance into its constituent parts, elements, or into simpler COMPOUNDS accompanied by the release of HEAT, gas, or hazardous materials.[172]

hazardous decomposition products some materials give off hazardous materials when they decompose or burn.[172]

Hazardous Materials Identification System, The (HMIS) developed by the NPCA to provide information on acute health hazards, reactivity, and flammability encountered in the workplace at room temperatures.[172,128]

A number is assigned to a material indicating the degree of hazard, from 0 for the least up to 4 for the most severe. Letters designate personal protective equipment.

hazardous pollutant an AIR POLLUTANT to which no ambient air quality standard is applicable and which causes or contributes to air pollution which may reasonably be anticipated to result in an increase in serious irreversible or incapacitating reversible, illness.[172] (EPA)

This definition is given in Section 112 of the CLEAN AIR ACT. *Hazardous pollutants include* ASBESTOS, *mercury, beryllium,* BENZENE, *arsenic, vinyl chloride, and radionuclides. (EPA)*

hazardous substance a substance which, by reason of being explosive, flammable, poisonous, corrosive, oxidizing, or otherwise harmful, is likely to cause death or injury.[171,172,172] (DAC)

Element or COMPOUND *which, when discharged in any quantity, presents an imminent and substantial danger to the public health or welfare.*

Hazardous Waste Number an identification number assigned by the EPA, per the RCRA Law, to identify and track wastes.[172]

haze small amounts of LIGHT SCATTERING.[43,69,56]

When applied to TRANSPARENT *materials, it generally applies to the percentage of* TRANSMITTED LIGHT *which is scattered relative to that which is transmitted. When applied to glossy reflecting materials, it generally refers to the slight* SCATTERING *from the* SURFACE *which decreases the* DISTINCTNESS-OF-IMAGE GLOSS *(20° gloss).*

hazen color scale see APHA COLOR SCALE.[43,69,157,128]

hazy (1) a relatively small amount of nonsettling, finely DISPERSED matter which is not visibly homogeneous with the MASS of the liquid specified, even though the liquid is TRANSPARENT and transmits most of the light incident upon it;[43] (ASTM) (2) displaying haze.[43]

HDI abbreviation for HEXAMETHYLENE DIISOCYANATE.[141] cf. IPDI, MDI, TDI, and TMXDI[144]

headers double WOOD pieces supporting JOISTS in a floor, or double wood members placed on edge over windows and doors to transfer the roof and floor weight to the studs.[82]

heartwood the WOOD extending from the pith to the SAPWOOD, the cells of which no longer participate in the life processes of the tree.[153]

Heartwood may be infiltrated with GUMS, RESINS, *and other materials which usually make it darker and more decay resistant than sapwood.*

heat a form of energy associated with the motion of atoms or molecules and capable of being transmitted through solid and fluid media by conduction, through fluid media by convection, and through empty space by radiation.[176]

heat age stability, latex see LATEX HEAT AGE STABILITY.[161,157]

heat age stability test an accelerated test used to predict STABILITY of coatings with time (ARRHENIUS plot).[157,161,176]

These types of tests involve measuring paint properties (e.g., VISCOSITY, pH, GLOSS, BLOCK RESISTANCE, ADHESION) before and after the paint has been heat aged [typically a week to a month at 120°F (50°C) or 140°F (60°C)]. An excessive change in a property is unacceptable. These tests are meant to predict which paints may not be stable when stored by a retailer or consumer under elevated temperatures for a short period of time or at room temperatures for a longer period of time.

There is a general correlation between short term, high temperature oven stability and long term SHELF LIFE at AMBIENT TEMPERATURE. One exception is a paint containing a REACTIVE PIGMENT such as ZINC OXIDE. Another is the growth of some putrefying BACTERIA which is inhibited at these high temperatures.

ASTM Test Method D1849 covers the change in consistency and certain other properties that may take place when packaged paint of either solvent-reducible or water-reducible type is stored at a temperature above freezing.

heat capacity quantity of HEAT required to raise the temperature of a body 1°C.[128,176]

heat test test employed for assessing the purity of TUNG and similar rapidly POLYMERIZING OILS.[63,157] Syn: browne heat test. See GEL TIME.[63,157,176]

ASTM Test Method D1955 involves heating the oil at a specified temperature and noting the time required for GELATION to occur.

heat-bodied oil see OIL, HEAT-BODIED.[63,131,169]

heat-convertible resins THERMOSETTING RESINS which on controlled heating become infusible and insoluble.[67,130]

This phenomenon is associated with CROSSLINKING of the resin molecules.

heat-reactive phenolic resin PHENOL-FORMALDEHYDE RESIN which has been manufactured under alkaline conditions and which reacts with OILS, ROSIN, or other UNSATURATED COMPONENTS during VARNISH making, through the free methylol groups in its structure, along with further CONDENSATION of the methylol groups in the resin.[130]

heat-resistant finishes finishes designed to resist DETERIORATION on continuous or intermittent EXPOSURE to a predetermined elevated temperature.[71,176]

heat-sealing method of joining PLASTICS by simultaneous application of HEAT and PRESSURE to areas in contact.[38,176]

heat-sensitive paints coatings which change COLOR at a specific temperature.[71,163,176]

heat-set inks LETTERPRESS and LITHOGRAPHIC INKS which dry under the action of HEAT by EVAPORATION of their high boiling SOLVENT.[168,176]

heavy centered spray pattern an uneven SPRAY pattern having more coating in the center and less on the edges.[42,56] (ASTM)

heavy solvent naphtha high-boiling COAL-TAR NAPHTHA generally distilling below 190°C (90/190°C is a usual grade).[164]

heavy spar see BARIUM SULFATE, NATURAL.[58,41]

heavy-bodied having a thick consistency or high VISCOSITY.[166]

heavy-centered pattern spray pattern having most paint in center, less at edges.[56] (NACE)

hectorite a mineral $(Mg, Li)_3\ Si_4O_{10}\ (OH)_2$ consisting of a hydrous silicate of magnesium and lithium.[58] cf. KAOLINITE,[41,58] MONTMORILLONITE,[41,58,34] BENTONITE,[41,139] and ATTAPULGITE[41,58,34]

Hegman gauge see FINENESS OF GRIND GAGE.[145,157]

Helio Fast Rubine 4BL lake (58055). PIGMENT marketed as an ALUMINA HYDRATE LAKE; the parent toner is hard and horny when dried and, consequently, too difficult to disperse.[41]

It is black in MASSTONE but yields a purple tint on reduction with white. Its good LIGHTFASTNESS in dark shades, high COLOR intensity, and good chemical and BLEED RESISTANCE make Helio Fast Rubine 4BL lake particularly suitable for shading full shade durable maroon finishes. On the other hand, its poor lightfastness in very weak pastel shades plus its poor bake resistance at high temperatures make it unsuitable as a white toning agent for high bake exterior white finishes.

Hellige system or comparator standard instrument for COLOR determination of liquids such as VARNISHES, OILS, etc., based on comparisons which are made against a series of discs of colored GLASS, and which are carefully graded.[69,145,157]

hematite Fe_2O_3. Natural OXIDE of IRON.[41] See FERRIC OXIDE.[41]

hemiacetal groups FUNCTIONAL GROUPS derived from CARBONYL GROUPS by addition of one molecule of an ALCOHOL, of the general structure shown: –C–OH–OR–.[133,134] (ASTM)

hemicolloid COLLOIDAL PARTICLE of specified size, both as regards number of molecules in association, and length of chain.[139]

Words presented in CAP/SMALL CAPS type indicate that the word is defined in another part of the *Coatings Encyclopedic Dictionary*.

Numerical superscripts classify terms in one or more of the categories listed in the second section of this volume.

This includes substances involving between 20 and 100 molecules, and possessing an overall chain length not exceeding 25 nm.

hemihydrate a hydrate which contains one-half molecule of water to one molecule of the COMPOUND; the most common such material is partially dehydrated GYPSUM (PLASTER OF PARIS).[134,144] (DAC)

hemipolymer a readily soluble POLYMER with molecular weights between 1,000 and 10,000.[130]

hempseed oil a SEMIDRYING OIL obtained from the seeds of *Cannabis sativa*, which grows in India, Manchuria, China, Japan, and some parts of Europe.[63]

It is usually classed with soybean, poppyseed, sunflower, and walnut OILS, and has an iodine value of approx. 160, arising in the main from the presence of practically 50% of LINOLEIC ACID and 25% of linolenic acid. It has some application in manufacture of paints and VARNISHES.

heptanol-2 $CH_3(CH_2)_4CHOHCH_3$. Synonym for methyl amyl carbinol.[164]

Properties: bp of 160.4°C, vp of 0.9 mm Hg/ 20°C, and a sp gr of 0.8187/20°C.

heptyl acetate $CH_3COOC_7H_{15}$. Bp, 191°C; sp gr, 0.874.[164]

herbicide an agent used to destroy or inhibit plant growth. [173]

HEUR HYDROPHOBICALLY-MODIFIED ETHOXYLATED URETHANE POLYMER.[83,166,130] See BRUSH DRAG, NATURAL SPREADING RATE, and ASSOCIATIVE THICKENER. cf. HASE and HMHEC[83,166,130]

These are NONIONIC ASSOCIATIVE THICKENERS that impart HIGH SHEAR VISCOSITY, FLOW, FILM BUILD, GLOSS, and SPATTER RESISTANCE to LATEX paints. They do not appreciably downgrade water resistance as much as HASE thickeners.

hexabromide value HEXABROMIDE VALUE is a measure of the amount of acids of the LINOLENIC type which are present in a mixture of natural FATTY ACIDS.[63,157]

Bromine, under suitable conditions, forms addition compounds through addition at the unsaturated linkages. The hexabromides are separated from the dibromides and tetrabromides by reason of the insolubility of the former in cold ether. The hexabromide value is the percentage of hexabromide obtainable from a given material.

hexahydrophenol see CYCLOHEXANOL.[164,144]

hexalin see CYCLOHEXANOL.[164,144]

hexalin acetate see CYCLOHEXYL ACETATE.[164]

hexamethylene diisocyanate (HDI) an ALIPHATIC ISOCYANATE MONOMER used as an INTERMEDIATE in URETHANE coatings.[141] See ISOCYANATE RESINS, POLYURETHANES, POLYURETHANE FINISH, and URETHANE COATINGS.[156,71,130,163,67] cf. ISOPHORONE DIISOCYANATE (IPDI), DIPHENYLMETHANE DIISOCYANATE (MDI), TOLUENE DIISOCYANATE (TDI), and TETRAMETHYLXYLENE DIISOCYANATE (TMXDI)

hexamethalene tetramine $(CH_2)_6N_4$. Derived from FORMALDEHYDE and AMMONIA. It is used as a HARDENER for PHENOLIC RESINS.[67,83,144] Syn: HEXAMINE

hexamine see HEXAMETHYLENE TETRAMINE.[67,83,144]

hexane $CH_3(CH_2)_4CH_3$. Volatile, flammable, liquid PETROLEUM HYDROCARBON.[164]

Properties: sp gr, 0.660; bp, 69°C; solidification range—95° to -100°C; flp, –22°C (–70°F).

hexanol $CH_3(CH_2)_4CH_2OH$. High-boiling SOLVENT (144-156°C) with a vp of 4 mm Hg at 30°C; sp gr, 0.8186.[164]

hexone see METHYL ISOBUTYL KETONE.[164]

hexyl acetate $CH_3COO(CH_2)_5{\cdot}CH_3$. It has a bp of 169°C, sp gr of 0.890, and vp of 5 mm Hg/20°C.[164]

hexyl ether $C_6H_{13}OC_6H_{13}$. High boiling SOLVENT.[164]

Properties: bp of 226°C, sp gr of 0.794/20°C, vp of 0.07 mm Hg/20°C, and flp of 77°C (170°F).

hi-flash naphtha see NAPHTHAS.[164]

hickies DEFECTS in a PRINT appearing as specks surrounded by an unprinted "halo."[168]

hide finishes designed for use on leather.[71,151]

They are based on PIGMENTED NITROCELLULOSE, and completely cover the surface.

hiding, complete the state of uniform APPLICATION of paint at such a thickness that, when dry, the SUBSTRATE is no longer affecting the COLOR, i.e., a thickness that when increased even slightly does not change the perceived color.[43,69,157,35] See CONTRAST RATIO.[43,69,157]

COMPLETE HIDING is frequently judged by application of a uniform film over a white and black substrate such as a HIDING POWER chart. It is sometimes defined as the state when a contrast ratio of 0.98 is obtained (although this is seldom complete hiding as judged visually). Visual complete hiding is not necessarily SPECTROPHOTOMETRIC complete HIDING.

hiding, incomplete the state of APPLICATION of a coating at such a thickness that the COLOR of the substrate has an effect on the color which the coating would exhibit if it were applied at COMPLETE HIDING.[43]

Increasing the thickness of the coating thus changes the perceived color. The relative degree of incomplete hiding is frequently described by the contrast ratio. See CONTRAST RATIO, *and* HIDING, COMPLETE.[43]

hiding power (1) the ability of a paint, INK, etc. to hide or obscure a SURFACE over which it has been applied uniformly.[43,69,157] (ASTM) cf. COVERAGE[43,69,128] and SPREADING RATE[128]

When expressed numerically, it is generally in terms of the number of square feet over which a gallon of paint, or pound of pigment, as used, can be uniformly spread to produce a specified CONTRAST RATIO. HIDING POWER *is usually expressed in square feet per gallon or square metres per litre.*

ASTM test methods for hiding power include: brushouts, visual, relative, D344; drawdowns, reflectometry, D2805; and wet-to-dry change, visual, D5007.

hiding power chart (1) a PAPER chart, partially black and partially white. A paint film applied over such a chart gives a visual evaluation of the degree of HIDING. If a measurement of the REFLECTANCE of the film over the black portion is divided by the measurement of the reflectance over the white portion, the contrast ratio (opacity) is obtained; (2) a chart prepared by calculation of the KUBELKA-MUNK EQUATION for a white substrate of specified reflectance for various reflectances at complete hiding (R_∞) and for various reflectances over black, various contrast ratios, scattering-thickness values (SX). From such a chart, using measurements of reflectances over black and white, or measurements over black and of R_∞, changes in thickness, PIGMENT VOLUME CONCENTRATION, R_∞ for obtaining a desired contrast ratio (degree of hiding), or R_∞ can be determined. Such charts are frequently called JUDD CHARTS.[43,157,35] See TEST CHARTS.[157,59]

hiding, spectrophotometric, complete the state of uniform APPLICATION of a paint film which, when DRY, exhibits identical SPECTROPHOTOMETRIC CURVES over a perfect white of 100% REFLECTANCE and a perfect black of 0% reflectance.[69,43,157]

In the KUBELKA-MUNK EQUATIONS, *spectrophotometric* HIDING *is defined as R_∞ at all* WAVELENGTHS.

hiding, spectrophotometric, incomplete the state of a uniform paint film in which the SPECTROPHOTOMETRIC CURVES measured over white and black areas are not identical; addition of more paint is required to achieve complete spectrophotometric HIDING, or identity of REFLECTANCE curves of the film over white and black substrates.[43]

Films which appear visually to be at COMPLETE HIDING *are not necessarily at complete spectrophotometric hiding.*

high build producing thick, dry films per coat.[133,166] See BUILD.[133,166]

high gloss See GLOSS.[43] Syn: full gloss

high humidity gloss loss a loss in GLOSS of LATEX paints due to differences in RELATIVE HUMIDITY at the point of APPLICATION in the field (above the standard laboratory relative humidity for testing).[42,43]

This gloss loss may or may not be reversible when more favorable drying conditions become available. One problem encountered when coatings are painted outside of the laboratory is the considerable diversity in drying conditions.

high polymer a substance composed of POLYMER molecules of high molecular weight.[130] (IUPAC) See MACROMOLECULE.[130,139]

In general, a linear polymer of a given series is considered a high polymer if its physical properties (especially its viscoelastic properties) do not vary markedly with molecular weight. This term is customarily shortened to "polymer."

high sheen flat see GLOSS.[43]

high-boiling solvent a SOLVENT with an initial boiling point above 150°C (302°F).[164] (NACE)

high-build coating a coating composed of a series of uniform, tile-like films which are applied in THICKNESSES (minimum 5 mils) greater than those normally associated with paint films and thinner than those normally applied with a trowel.[71,166] (DAC) See TILE-LIKE COATING.[163]

high-density overlay an overlay consisting of PAPER that is impregnated with a THERMOSETTING RESIN and then applied to PLYWOOD; provides a smooth, hard, WEAR-RESISTANT SURFACE for high-quality concrete formwork and decking.[75,153] (DAC)

high-performance interior architectural wall coatings (HIPAC) differ from more conventional coatings in that they are tougher, more STAIN-RESISTANT, more ABRASION-RESISTANT, and, ordinarily, designed to be applied to wall surfaces of STEEL, MASONRY (poured CONCRETE, concrete block), and PLASTER or GYPSUM WALLBOARD.[71,82] (ASTM)

These coatings are tough, extra-durable ORGANIC *coating systems that are applied as a continuous (seamless) film and that* CURE *to a hard* FINISH. *The* FINISH *can be* HIGH GLOSS, SEMIGLOSS, *or* LOW GLOSS *as desired. These coatings are resistant to persistent* HEAT, HUMIDITY, ABRASION, STAINING, CHEMICALS, *and* FUNGUS *growth. They are used in areas where humidity, wear, or unusual chemical resistance requirements, particularly to soiling, are required and where strong detergents are used to maintain sanitary conditions.*

While they are excellent for walls, HIPAC are not usually intended for ceilings or floors. They would not ordinarily be used in homes, although parents with small children might want to use HIPAC on some walls.

Words presented in CAP/SMALL CAPS type indicate that the word is defined in another part of the *Coatings Encyclopedic Dictionary.*

Numerical superscripts classify terms in one or more of the categories listed in the second section of this volume.

ASTM D3730 is a guide that covers the selection and use of test methods for HIPAC.

high-solids coatings coatings containing about 70% or more solids by weight.[71]

high-visibility paints generally applied to paints containing FLUORESCENT PIGMENTS or to paints having high REFLECTIVITY because of the use of retroreflective pigments in them.[43,69,41]

high-visibility pigments see FLUORESCENT PIGMENTS.[41,43]

higher fatty acid See FATS.[63]

highlight the light or open areas of a HALFTONE PRINT.[168]

highlighting in ANTIQUING, accenting raised areas on FURNITURE with a lighter tone than the crevices.[54,148,175] (BSI)

Produced by rubbing harder. Emphasizing the impression of relief by making certain parts of a finished surface lighter than the general COLOR *of that surface.*

HIPAC abbreviation for HIGH PERFORMANCE INTERIOR ARCHITECTURAL COATINGS.[82,71]

hiroe grade of pale East India RESIN exported from Macassar.[167]

hitching post standards standards close to the COLOR of the specimens being measured.[69,35]

HLB value see HYDROPHILE-LIPOPHILE BALANCE VALUE.[128,127]

HMHEC abbreviation for HYDROPHOBICALLY MODIFIED HYDROXYETHYL CELLULOSE.[83,166] cf. HEUR and HASE[83,166,130]

HMIS abbreviation for the HAZARDOUS MATERIALS IDENTIFICATION SYSTEM.[172]

hold paint MARINE coating for holds of ships.[163,71]

holdout the ability of a PRIMER to seal porous SURFACES so that the FINISH COAT (TOPCOAT) is uniform; or the ability of the finish coat (topcoat) to dry to its normal uniform gloss when applied over a somewhat porous or nonuniform surface.[72,159] See ENAMEL HOLDOUT.[72,159] cf. ABSORPTION[142]

holiday detector device for detection of pinholes or voids.[145,157]

holidays APPLICATION DEFECT whereby small areas are left uncoated.[56] Syn: miss, skip, void, discontinuity, vacation

homogenizer MILL for making DISPERSIONS of EMULSION type.[59,60,64]

The term is not restricted to one type of machine, and is used to describe agitators and masticators as well as high-speed rotor-stator type precision machines with close clearances.

homopolymer HIGH POLYMERS consisting of molecules that contain (neglecting the ends, branch junctions, and other minor irregularities) either a single type of unit or two or more chemically different types in regular sequence.[130] (IUPAC) cf. COPOLYMER and POLYMER[130]

homopolymerization POLYMERIZATION in which a HOMOPOLYMER is formed.[169,130] (IUPAC)

honeycombing (1) checks often visible at the surface, that occur in the interior of a piece of WOOD, usually along the wood rays;[153] (2) lack of vertical FILM INTEGRITY; formation of cell structure; VOIDS.[56] (NACE)

hood a partial enclosure or canopy for capturing and exhausting, by means of a draft, the ORGANIC vapors or other fumes rising from a coating PROCESS or other source.[172] (EPA)

hot melt adhesive ADHESIVE which is applied in molten condition and cools rapidly to form a solid.[79,176] (EPA)

It contains little or no VOLATILE ORGANIC SOLVENTS.

hot oil expression method of obtaining VEGETABLE OILS from seeds by subjecting the oil-bearing seeds to pressure in a suitable press.[63,131,176]

To facilitate expression of the oil and/or increase the yield, the seeds are carefully and uniformly preheated before expression. Oils obtained from this process are known as hot-expressed oils.

hot spraying SPRAYING of hot LACQUERS or paints, the VISCOSITIES of which have been reduced to spraying CONSISTENCY by means of HEAT instead of by the addition of VOLATILE SOLVENTS.[77,176] (BSI)

By such a PROCESS *it is possible to apply materials with higher-solid contents and therefore better build.*

hot surface (1) an ALKALINE surface such as PLASTER, CONCRETE, etc; (2) an abnormally absorbent surface (British).[159]

hot-pressing the pressure-forming, between heated PLATENS, of PLYWOOD, LAMINATES, PARTICLEBOARD, FIBERBOARD, etc.; usually requires THERMOSETTING RESINS and heat for CURING.[75,67,153] (DAC)

hot-short inelastic, nonstretchable and easily broken by tension, when hot.[47,176]

hotmelt coatings compositions which liquify readily on heating and are applied to various surfaces in molten condition.[71,163,176]

house paint coating designed for use on large EXTERIOR surfaces of a building; generally of lower GLOSS than the coating used on TRIM areas.[71]

household chemicals, resistance to resistance of clear and PIGMENTED ORGANIC coatings to various liquids used in the home.[72,35]

ASTM Test Method D1308 covers the determination of the effect of household chemicals on these coatings, resulting in any objectionable alteration in the surface, such as discoloration, change in gloss, blistering, softening, swelling, loss in adhesion, or special phenomena.

housekeeping cleanliness, neatness, and orderliness of an area, with the designation of a proper place for everything and everything in its proper place.[172]

HPAM abbreviation for HYDROPHOBICALLY MODIFIED POLYACRYLAMIDE.[83,166]

Hubl solution reagent used for the determination of Hubl iodine values, involving iodine and mercuric chloride.[63,157]

hue the attribute of COLOR by means of which a color is perceived to be red, yellow, green, blue, purple, etc.[69]

White, black, and grays possess no hue.

hue, Munsell see MUNSELL HUE.[43,69]

Huemann's blue synthetic ULTRAMARINE BLUE.[41]

hull paint ships' hull paint, bottom paint.[163,71]

humidity the condition of the atmosphere in respect to water vapor.[128,154] (ASTM)

The amount of water vapor may be expressed on an absolute basis or on a relative basis.

humidity, absolute the mass of water vapor present in a volume of air. (ASTM) [128]

For example, grains per cubic foot, or grams per cubic metre.

humidity ratio in a mixture of water vapor and air, the weight of water vapor per unit weight of dry air. (ASTM) Syn: specific humidity

humidity, relative see RELATIVE HUMIDITY.[128]

humidity resistance the ability of a coating and the SUBSTRATE it is protecting to resist damage or DEGRADATION due to changes in humidity.[72,153,66]

As an example, WOOD SUBSTRATES and the coatings applied to them expand and contract to different degrees as the humidity changes, causing stresses that may produce CHECKS, CRACKS, SPLITS, BLISTERS, SWELLING, LOSS OF ADHESION, and various changes in surface APPEARANCE.

ASTM test methods for humidity resistance include: humid-dry cycling on wood, D3459; and humidity, on steel, D2247.

Some tests are run at a 100% RELATIVE HUMIDITY so that CONDENSATION forms on the test specimen and then it becomes a measure of WATER RESISTANCE.

humidity, specific see HUMIDITY RATIO.[128] (ASTM)

Hungary blue see COBALT BLUE.[41]

hungry surface surface, the absorptive powers of which have not been fully satisfied by the coats of paint applied to it, usually resulting in a patchy film.[42,56] (BSI) Syn: starved surface

Hunter Color Difference Equation generally refers to the COLOR DIFFERENCE EQUATION used for calculating color differences on certain color difference meters (especially those made by Hunter Labs and Gardner Instruments). It is a simplified equation to approximate the NBS EQUATION:

(1) $\Delta E = [(\Delta a_L)^2 + (\Delta b_L)^2 + (\Delta L)^2]^{1/2}$ where

$L = 10Y^{1/2}$

$a_L = 17.5\,(1.02X - Y)/Y^{1/2}$

$b_L = 7.0\,(Y - 0.847Z)/Y^{1/2}$ and the deltas are obtained as the difference of the sample minus the standard. Positive a or Δ a indicates redness, negative indicates greenness; plus b or Δ b indicates yellowness, minus indicates blueness. Other instruments may use the following version:

(2) $\Delta E = [(\Delta a_R)^2 + (\Delta b_R)^2 + (\Delta R)^2]^{1/2}$ where

$R = Y$

$a_R = 1.75f_y\,(1.02X - Y)$

$b_R = 0.70f_y\,(Y - 0.847Z)$

$f_y = 0.50\,(21 + 0.2Y)/(1 + 0.2Y)$. The particular equation used must be specified.[43,69]

hydrate COMPOUND in which molecules of water, as such, are present.[134]

Many CRYSTALLINE SALTS contain what is described as water of crystallization. Each molecule of salt is associated with one or more molecules of water. These salts are known as HYDRATED SALTS.

hydrated aluminum silicate see ALUMINUM SILICATE (CLAY).[41,58]

hydrated chromium oxide $Cr_2O_3{\cdot}2H_2O$. Pigment Green 18 (77289). Permanent green PIGMENT with excellent LIGHTFASTNESS but lower temperature resistance than chromium oxide and less TINTING STRENGTH.[41] Syn: Guignet's green, permanent green, viridian, vert emeraude (transparent), French veronese green, emerald oxide of chro-

Words presented in CAP/SMALL CAPS type indicate that the word is defined in another part of the *Coatings Encyclopedic Dictionary.*

Numerical superscripts classify terms in one or more of the categories listed in the second section of this volume.

mium, Mittler's green, native green, Pelletier's green, emerald green, smeraldino

Properties: density, 2.9-3.7 g/cm³ (24.2-30.8 lb/gal); O.A., 110.[41]

hydrated lime see CALCIUM HYDROXIDE.[73]

hydrated magnesium aluminum silicate $3MgO \cdot 1.5Al_2O_3 \cdot 8SiO_2 \cdot 9H_2O$. Natural ADSORPTIVE CLAY with a primary constituent of ATTAPULGITE CLAY which has rodlike particle shape.[41] Syn: Fuller's earth, attapulgite, palygorskite.

Properties: density, 2.36 g/cm³ (19.7 lb/gal); O.A., 80-190; particle size, 0.13 µm.

hydraulic cement a CEMENT that is capable of SETTING and HARDENING under water due to interaction of water and the constituents of the cement.[73,75] (ASTM)

hydraulic lime a LIME that contains more than 10% silicates; will HARDEN under water.[173] (DAC) Syn: quick-hardened lime

hydraulic spraying SPRAYING by hydraulic pressure.[77] See AIRLESS SPRAYING.[77]

hydric containing, or relating to, HYDROGEN in combination.[134]

hydroblasting cleaning with high-pressure water jet.[158]

hydrocarbon an ORGANIC COMPOUND consisting only of CARBON and HYDROGEN.[144] (EPA)

Sometimes the term is used loosely, but incorrectly, to refer to any organic compound.

hydrocarbon plastics PLASTICS based on RESINS composed of HYDROGEN and CARBON alone.[38,144]

Oxygen and other elements may be present as impurities in the resins, less than 1% by wt.

hydrocarbon resins POLYMERS consisting solely of HYDROGEN and CARBON, specifically denoting, within the trade, those RESINS which are hard and friable, which give good ADHESIVE strength, and lack the COHESIVE strength typical of ELASTOMERIC polymers.[130]

hydrogen ion concentration (pH) see PH.[128,157]

hydrogen value method of determining HYDROGEN ABSORPTION for UNSATURATED OILS or similar COMPOUNDS.[63,157]

Hydrogen is reacted with unsaturated OILS, and iodine values are calculated from the amount of hydrogen absorbed.

hydrogenated naphthalenes two commercial SOLVENTS are included in this description, namely tetrahydronaphthalene (TETRALIN) and decahydronaphthalene (DECALIN). The former is a partly HYDROGENATED NAPHTHALENE, whereas the latter is completely hydrogenated.[164]

hydrogenated naphthas series of SOLVENTS obtained by subjecting the sulfur dioxide extracts of PETROLEUM to HYDROGENATION.[164]

hydrogenated rosin modified ROSIN obtained by the HYDROGENATION of rosin at high pressures in the presence of a CATALYST.[44,167]

hydrogenated rubber RUBBER reacted in solution under PRESSURE, in the presence of a suitable CATALYST, to yield HYDROGENATED RUBBER, in which the unsaturation of natural rubber has been substantially satisfied.[140]

hydrogenation the chemical addition of HYDROGEN to a material.[131]

hydrogenation of oils method of treatment employed to convert UNSATURATED FATTY ACIDS or OILS into SATURATED types.[63,131]

hydrolysis disruptive reaction consisting of splitting a COMPOUND into two parts, one of which combines with the H ion of water and the other combines with the OH ion of water.[63]

For example, when VEGETABLE OILS or FATS are split, the ESTERS are broken down into an acid component and an ALCOHOL component, by taking the H and OH ions of water. When alkaline reagents are present, the acids do not appear as such at the end of the reaction, but as the corresponding ALKALI METAL SALTS.

hydrometer instrument for determining the SPECIFIC GRAVITY of a liquid.[145,157] See TWADDELL HYDROMETER.[59,157]

hydrophile see EMULSOID.[130,60]

hydrophile-lipophile balance (HLB) value an empirical value (arbitrary units of 0 to 20) which is based on the premise that all SURFACTANTS combine HYDROPHILIC and LIPOPHILIC groups in one molecule and that the proportion between the weight percentages of these two groups for NONIONIC SURFACTANTS is an indication of the behavior that may be expected from the product.[128,127]

hydrophilic substance which absorbs or exhibits affinity for water; wettable.[139,142,155,] See HYGROSCOPIC and DELIQUESCENCE.[142]

hydrophobic substance with does not absorb or exhibit affinity for water, nonwettable.[139,142,155]

hydrophobically modified hydroxy ethyl cellulose (HMHEC) hydrophobe-modified CELLULOSE ETHERS that provide some of the benefits of synthetic ASSOCIATIVE THICKENERS while retaining much of the predictability of the conventional cellulose ethers.[83,155] cf. HEUR and HASE

They have about half the molecular weight of conventional HEC with similar thickening efficiency.

hydrophobically modified polyacrylamide synthetic nonionic ASSOCIATIVE THICKENER.[83,166]

hydrosol COLLOIDAL DISPERSION with water as the dispersion medium.[64,139]

hydrostatic of or relating to fluids at rest and under pressure.[142]

hydroterpins HYDROGENATION products derived from TURPENTINE.[164]

They differ from the parent material in that they possess greater stability, i.e., less tendency to oxidation, higher boiling ranges, and slower EVAPORATION rates.

hydroxide HYDRATED METALLIC OXIDE; a BASE; a COMPOUND which will give hydroxyl (OH) ions in solution.[134]

hydroxyl group –OH. The monovalent group characteristic of HYDROXIDES and ALCOHOLS.[134] (ASTM)

(2-hydroxypropyl) methyl cellulose a mixed ether of CELLULOSE containing both hydroxypropyl and methyl groups.[83] (ASTM)

hygrometer an instrument that indicates directly or indirectly the RELATIVE HUMIDITY of the air.[145,157] (ASTM)

hygroscopic COMPOUND or substance which is capable of sorbing water. See HYDROPHILIC[139,142,159] and DELIQUESCENCE.

hypha any of the threadlike FILAMENTS forming the MYCELIUM of a FUNGUS.[173]

Hyphae (pl) may appear individually or have SPORES attached.

hypogoeic acid $C_{15}H_{29}COOH$. UNSATURATED FATTY ACID found in certain VEGETABLE OILS.

It has an mp of 33°C, and a bp of 236°C/15 mm Hg.[52]

hypohalous COMPOUND in which a hydroxyl group is combined with a HALOGEN atom.[134]

hysteresis the failure of a property that has been changed by an external agent to return to its original value when the cause of the change is removed.[166] See PERMANENT SET.[142]

hysteresis loop FLOW CURVE for a THIXOTROPIC material obtained by measurements on a rotational VISCOMETER showing for each value of RATE OF SHEAR, two values of SHEARING STRESS, one for an increasing rate of shear and the other for a decreasing rate of shear.[166]

A hysteresis loop characterizes a thixotropic material. If no hysteresis loop is obtained, the material is nonthixotropic. The area within the loop is a measure of the degree of THIXOTROPY. A hysteresis loop cannot be formed with an efflux viscometer because each point on the curve is measured on previously unagitated material.

Words presented in CAP/SMALL CAPS type indicate that the word is defined in another part of the *Coatings Encyclopedic Dictionary.*
Numerical superscripts classify terms in one or more of the categories listed in the second section of this volume.

I

IARC abbreviation for the International Agency for Research on Cancer.[172]

ICCATCI see COATINGS SOCIETIES INTERNATIONAL[174]

ice dam a MASS of ice formed at the transition from a warm to a cold roof surface.[76] (ASTM)

Frequently formed by the refreezing meltwater at the overhang of a steep roof, an ice dam may cause ice and water to back up under shingles or other roofing materials.

Iceland moss lichen grown in Iceland, Norway, and Sweden, which yields a MUCILAGINOUS extract, used in sizes and the like.[139,155] Syn: cetraria

Iceland spar natural, TRANSPARENT, double refracting, crystalline form of CALCIUM CARBONATE.[41,58]

ichnography (1) art of making DECORATIVE drawings by the use of compass and rule; in a sense, revived in "op" art techniques; (2) also refers to the art of tracing plans and illustrations.[81,82]

ICI (1) older abbreviation for International Commission on Illumination. (Use international abbreviation CIE);[174,43] (ASTM) (2) abbreviation for Imperial Chemical Industries.

ICI Cone/Plate viscometer a VISCOMETER with a range of 0 to 10 POISE and RATE OF SHEAR of 12,000 s^{-1} used to give information about the FLOW properties of coatings under APPLICATION conditions: BRUSHING, SPRAYING, ELECTROSTATIC DISK, or ROLL COATING.[145,166,157]

It provides a quick method of measuring small samples at high shear rates followed by a rapid clean up. It is suitable for use with materials having either NEWTONIAN or NON-NEWTONIAN flow properties, since the shear rate is fixed. It can be used with both WATER-BASED and SOLVENT-BASED systems.

BRUSH application of ARCHITECTURAL COATINGS imparts shear rates close to 10,000 s^{-1}. This viscosity measurement has good correlation to BRUSH DRAG and FILM BUILD.

ASTM Test Method D4958 on brush drag of LATEX paints reports that brush drag is directly related to paint high shear viscosity. There is generally good rank order agreement between this test method and ICI Cone/Plate Viscometer readings.

ASTM Test Method D4287 covers the viscosity of coatings at a shear rate of 12,000 s^{-1}. The viscosity value obtained gives information about the flow properties of the material under application conditions: brushing, spraying, electrostatic disk, or roll coating. Brush application of architectural coatings impart shear rates close to 10,000 s^{-1}.

IDLH abbreviation of immediately dangerous to life and health.[172]

Used to determine selection of a respirator. The maximum concentration from which one could escape within 30 minutes without any escape-impairing symptoms or irreversible health effects.

ignition temperature the lowest temperature at which a combustible material ignites in air and continues to burn independently of the source of heat.[172,176]

illuminance the LUMINOUS FLUX incident per unit area of a surface.[43]

illuminant mathematical description of the relative SPECTRAL power distribution of a real or imaginary light source, that is, the relative energy emitted by a source at each WAVELENGTH in its emission spectrum.[43]

Popularly used synonymously with the term "light source" or "lamp." Such usage is not recommended.

Illuminant A (CIE) incandescent illumination, yellow orange in COLOR, with a correlated color temperature of 2856K.[43,69] See CORRELATED COLOR TEMPERATURE.[43,69]

It is defined in the wavelength range of 380-770 nm.[43,69]

Illuminant B (CIE) an approximation of mean noon sunlight, approximately neutral in HUE, with a correlated COLOR temperature of 4874K.[69,43] See CORRELATED COLOR TEMPERATURE.[69,43]

It is defined in the WAVELENGTH range of 380-770 nm.[43,69]

Illuminant C (CIE) an approximation of average daylight, bluish, with a CORRELATED COLOR TEMPERATURE of 6774K.[43,69]

It is defined in the WAVELENGTH range of 380-770 nm.

Illuminants D (CIE) daylight ILLUMINANTS, defined from 300-830 nm, the UV portion 300-380 nm being necessary to describe correctly colors which contain FLUORESCENT DYES or PIGMENTS. They are designated as D with a subscript to describe the CORRELATED COLOR TEMPERATURE: D_{65} having a correlated color temperature of 6504K, close to that of ILLUMINANT C, is the most commonly used.[43,69]

They are based on actual measurements of the spectral distribution of daylight.

illuminating hand DECORATION of books, as done in medieval times, with drawings and miniature paintings.[81,148]

Also ornamentation and embellishments, usually in red, blue, and gold, that were sometimes added to initial letters and borders.

ilmenite $FeO \cdot TiO_2$. Natural IRON TITANATE.[41,34]

It is an iron-black mineral and is one of the main sources of TITANIUM DIOXIDE.

ilmenite black opaque black PIGMENT, prepared from the mineral of the same name.[41,34]

imbibition swelling of a solid, semisolid, or GEL by ABSORPTION of a liquid.[131]

IMCO abbreviation for International Maritime Consultative Organization, an intergovernmental organization in London charged with drawing up regulations for safety at sea, ship paints, and cargo.[174]

imitation gold ink a simulated GOLD INK that uses ALUMINUM POWDER to produce the metallic LUSTER.[168]

immiscible descriptive of the two or more fluids which are not mutually soluble.[164]

impact resistance ability of a coating to resist a sudden blow.[47,157,72] See FLEXIBILITY.[47,157]

Ability to resist DEFORMATION from impact. ASTM Test Method D2794 covers a procedure for rapidly deforming by impact of a coating film and its substrate and for evaluating the effect of such deformation.

impact test used to assess the ADHESION and FLEXIBILITY of applied coatings.[47,157]

A weighted plunger is allowed to fall from a specified height onto the front (direct) or back (reverse) of a coated panel. The extent of the damage caused by the impact is used as a basis for assessing the adhesion and flexibility.

impasto (Italian) literally, PASTE. Particularly thick or heavy APPLICATION of paint.[54,81]

It is the contrast of impasted areas with thinly painted passages which often impart character and robustness to an oil painting, e.g., the thickly painted highlights and thinly painted shadow areas in many of Rembrandt's paintings.

imperial gallon see GALLON, IMPERIAL.[128]

impregnated cloth a cloth impregnated with RESIN, VARNISH, SHELLAC, etc.[131] (DAC)

impregnating insulating varnish first coating INSULATING VARNISH applied to electrical equipment by means of vacuum-pressure impregnation, or by DIPPING.[61,152]

Its main purpose is to fill up the voids between the windings and elsewhere. The name distinguishes it from finishing insulating varnishes, which are used primarily to give an effective seal to the outside of windings, to prevent ingress of MOISTURE.

impression (1) the PRINTING pressure necessary for INK transfer; (2) also refers to a single print.[168]

impression cylinder the cylinder on a PRINTING PRESS that holds the material being printed against the printing plate, cylinder, or blanket.[168]

imprinting see PRINT RESISTANCE.[56,72,47,35]

impsonite naturally occurring ASPHALTUM, found in Oklahoma.[76]

in-can biodeterioration undesirable changes due to BACTERIAL growth in a can of paint (within the material mass), which can lead to gas production, foul odors, and marked changes in the properties of paints such as VISCOSITY loss.[42,173,83] See IN-CAN PRESERVATIVE.[173,83]

in-can preservative BIOCIDE used to prevent BACTERIAL growth in the paint that can cause odor or produce enzymes that can break down CELLULOSIC THICKENERS.[83,173,42] See IN-CAN BIODETERIORATION.[42] Syn: can preservative

inch-pound units units based upon the yard and the pound commonly used in the United States of America and defined by the National Institute of Standards and Technology.[128] (ASTM) cf. SI UNITS[128]

Note that units having the same names in other countries may differ in magnitude.

incident angle the angle between the INCIDENT RAY and a normal to the surface at the point of incidence.[43]

It is the same as "entrance angle" in SAE automotive nomenclature.

incident ray the center line of the light beam incident upon a surface.[43]

incinerator equipment in which solid, liquid, or gaseous combustible wastes are ignited and burned.[59,171] (IUPAC)

Types include flue-fed and multiple-chamber incinerators with several stages of combustion (IUPAC)

inclined plate flow see PLATE FLOW.[166]

inclusion presence of foreign material in the finished material.[42,56]

Words presented in CAP/SMALL CAPS type indicate that the word is defined in another part of the *Coatings Encyclopedic Dictionary.*

Numerical superscripts classify terms in one or more of the categories listed in the second section of this volume.

incompatibility effect when two or more components of a composition do not blend to produce a uniform or homogeneous mixture.[42,56,161]

It can be characterized by one or more of the following: SEPARATION, GELATION, CURDLING, PRECIPITATION, CLOUDINESS, SEEDINESS, LOSS IN GLOSS, or BLOOMING of RESULTING FILM.

incomplete hiding see HIDING, INCOMPLETE.[43]

indanthrone blue a member of a group of vat DYE PIGMENTS.[41]

Relative to PHTHALOCYANINE BLUE, it is considerably redder in HUE and superior in bronze resistance; LIGHTFASTNESS is excellent. It shows no tendency toward crystal growth in conventional paint thinners or solvents.

indelible inks compositions which are made water- and ALKALI-fast.[168]

Used on cloth to withstand laundering.

indene C_9H_8. Colorless liquid used for the preparation of SYNTHETIC RESINS.[144] See COUMARONE-INDENE RESINS.[167]

Insoluble in water, soluble in most ORGANIC SOLVENTS, rapidly absorbs oxygen from the air, forms POLYMERS by EXPOSURE to air and sunlight. Derivation: contained in the fraction of crude COAL TAR distillates which boils from 176 to 182°C. A sp gr of 1.006 (20/4°C), mp of 3.5°C, bp of 182°C, refractive index of 1.5726, and an flp of 78.3°C.

indene resins see COUMARONE-INDENE RESINS.[167]

indentation hardness resistance to PENETRATION by an indenter.[47] See KNOOP HARDNESS NUMBER, PFUND HARDNESS NUMBER.[47]

index of metamerism an index to describe the degree of METAMERISM existing between two samples.[43,69,128]

There is no standard accepted index. However, a frequently computed index is the COLOR DIFFERENCE, using a specified equation, which exists under a second ILLUMINANT when a pair of colors is an exact match under a primary or first illuminant.

index of refraction ratio of the sine of the ANGLE OF INCIDENCE of RADIATION traveling in a vacuum to the sine of the angle of REFRACTION in a medium, the perpendicular to the surface being 0°.[43,128]

The numerical ratio of the velocity of light in a vacuum to the velocity of light in a substance.

India ink a liquid INK made of LAMPBLACK mixed with a binding material and water.[168] Syn: Chinese ink

Indian ochre see IRON OXIDES, NATURAL.[41]

Indian red natural, Pigment Red 101 (77491). Synthetic, Pigment Red 101 (77491). Crystalline COPPERAS (ferrous sulfate in crystal form) is converted by CALCINATION to FERRIC OXIDE, producing a range of light to dark red oxide pigments. The dark shades are known as Indian red.[41]

Original Indian reds were a variety of natural red iron oxides obtained from India.[41]

Indian red ochre see IRON OXIDES, NATURAL.[41]

Indian yellow purree.[41]

A yellow PIGMENT containing the magnesium salt of euxanthic acid.

indigo $C_{16}H_{10}N_2O_2$ (molecular weight, 262.26) Dark blue, transparent DYE produced either by fermentation from the genus, *Indigofera*, indigenous to the Far East, or synthetically.[41]

induction period (of drying) that period prior to SET-TO-TOUCH in the DRYING process where there is but a small change in CONSISTENCY of the BINDER and the natural inhibitors of the OIL are removed by OXIDATION.[67]

induline range of blue or black DYESTUFFS belonging to the safranine group.[41]

They can be prepared in OIL, spirit, and water soluble forms. They are used as DYES for leather and in LAKE manufacture.

induline bases substances prepared from amidoazobenzene ANILINE and aniline hydrochloride.[41]

industrial finishes or coatings coatings which are applied to factory-made articles (before or after fabrication), usually with the help of special techniques for applying and drying—in contrast to ARCHITECTURAL PAINTS.[71]

ASTM D4712 is the Standard Guide for testing industrial water-reducible coatings. This guide covers the selection and use of procedures for testing water-reducible coatings, both pigmented and clear, utilizing synthetic latexes, synthetic resin emulsions, or water-reducible alkyds.

industrial maintenance paints high performance coatings which are formulated for the purpose of heavy ABRASION, water immersion, chemical, CORROSION, temperature, electrical, or SOLVENT RESISTANCE.[71] See MAINTENANCE PAINTS.[71]

industrial methylated spirits see METHYLATED SPIRIT.[64]

industrial talc $Mg_3Si_4O_{10}(OH)_2$. A mineral product varying in composition from that approaching the theoretical formula of TALC (which is nonfibrous), to mixtures of talc and other naturally associated minerals, some of which may be fibrous as defined in ASTM Test Method D2946, and/or ASBESTOS.[41,58,34] (ASTM) See MAGNESIUM SILICATE, FIBROUS.[41,83,34]

industrial talc, asbestos-free industrial TALC of which less than two PARTICLES per 100 particles are ASBESTOS, where "asbestos fiber" is defined as being both a fiber by ASTM Test Method D2946 and one of the asbestiform varieties of serpentine, riebeckite, cummingtonite (which are chrysotile, crocidolite, and amosite, respectively), anthophyllite, tremolite, or actinolite.[41,58] (ASTM)

The nonasbestiform varieties of these same materials are not asbestos.

industrial universal colorants the industrial coatings counterparts of TRADE SALES PAINT UNIVERSAL COLORANTS that can be used to shade ACRYLIC SOLVENT-BASED systems, both THERMOPLASTIC and THERMOSETTING, and epoxy systems, whether the simpler EPOXY ESTER type or the more complex two-component amine- and POLYAMIDE-CURED, including ORGANIC solvent-thinned and REACTIVE DILUENT HIGH-SOLIDS versions.[41,83] cf. UNIVERSAL COLORANTS[41,83]

These colorants can also be used with two-component and MOISTURE-CURED POLYURETHANES, POLYESTERS, and other industrial-type baking systems, as well as NITROCELLULOSE LACQUERS, VINYLS, and many others.

industrial waste water water discharged from an industrial PROCESS as a result of formation or utilization in that process.[131] (ASTM)

industrial water water (including its impurities) used directly or indirectly in industrial processes.[131] (ASTM)

inert not readily reactive with other elements; forming few or no COMPOUNDS.[144]

inert base a paint BASE which does not provide HIDING, COLOR, or DRYING properties.[71] (DAC)

Its main function is to provide solids, usually at low cost.

inert pigment (1) a PIGMENT which remains relatively inactive or chemically unchanged in paints under stated conditions. The term has little significance unless the conditions are stated; (2) frequently used for EXTENDER.[41,58] (BSI)

inflammable capable of being easily set on fire and continuing to burn, especially violently.[172] See FLAMMABLE.[172]

infrared (IR) electromagnetic RADIATION of WAVELENGTHS between 0.78 and 1.0 micrometre (780 to 1000 nanometres) sometimes used to CURE coatings.[43,67,176]

infrared drying method of HEATING for DRYING, CURING, and/or fusing a coating.[67,176]

The radiations involved in this form of heating have WAVELENGTHS greater than 780 nanometres, and they may be derived either from specially designed filament bulbs or special types of gas oven.

infusorial earth see DIATOMACEOUS SILICA.[41,58]

inhalator see RESPIRATOR.[172]

inhibitive pigment PIGMENT which assists in the prevention of CEDAR STAINING, or Redwood staining, MILDEW, CORROSION, YELLOWING, etc.[41,66] See CORROSION-INHIBITIVE PIGMENTS.[41,66]

inhibitor (1) COMPOUNDS or materials that have the effect of slowing down or stopping an undesired chemical change such as CORROSION, OXIDATION or POLYMERIZATION, DRYING, SKINNING, MILDEW GROWTH, etc.; (2) a negative CATALYST which prevents or retards VULCANIZATION or OXIDATION. A good inhibitor will retard vulcanization at room temperature to prevent precure but will not retard at normal CURING temperatures.[83,67,140] (DAC) See CORROSION INHIBITOR, INHIBITIVE PIGMENT.[83,41]

initial boiling point temperature indicated by the DISTILLATION thermometer at the instant the first drop of CONDENSATE leaves the condenser tube.[157,176] (ASTM Test Method D1078)

injection molding molding in which the RUBBER or PLASTIC stock is HEATED and, while in a flowable state, is forced or injected into the mold cavity.[38,49]

ink a fluid or VISCOUS substance used for writing or PRINTING.[168]

ink compounds ADDITIVES to remedy deficiencies in the INK vehicle or PIGMENT.[83,168]

ink fountain see FOUNTAIN.[168]

ink receptivity that property of a sheet of PAPER (or other material) which causes it to accept and/or absorb INK.[168]

ink-embossed wallpaper differs from regular embossed WALLPAPER in that the INK COLORS are applied at the time the PAPER is being embossed.[148]

inkometer an instrument indicating the TACK of an INK in terms of the torque developed by a system of rotating ink-wet rollers.[145,157,168]

Useful in determining tack differences in series of inks for multicolor wet printing and in the standardization of ink CONSISTENCY.

inks, cover generally OPAQUE, heavy bodied inks having a high PIGMENT concentration.[168]

inks, halftone INKS formulated for good reproduction of fine detail, such as HALFTONE dots on coated stock.[168]

Words presented in CAP/SMALL CAPS type indicate that the word is defined in another part of the *Coatings Encyclopedic Dictionary.*

Numerical superscripts classify terms in one or more of the categories listed in the second section of this volume.

They generally have high tinctorial strength and are finely dispersed.

inks, job heavy bodied INKS formulated to PRINT on uncoated stock using small sheet-fed presses.[168]

inks, quick-setting INKS for LETTERPRESS and OFFSET which dry by either FILTRATION, COAGULATION, SELECTIVE ABSORPTION or often a combination of these with some of the other drying methods.[168]

The vehicles are generally special resin-oil combinations which, after the ink has been PRINTED, separate into a solid material which remains on the surface as a dry film and an oily material which penetrates rapidly into the stock. This rapid separation gives the effect of very quick SETTING or DRYING.

inlay pattern or design formed by inserting cut shapes of contrasting materials into a surface, usually at the same level.[81,148]

inoculum that portion of the pathogen which is transferred to the plant or plant part for inoculation purposes.[173]

inorganic designation of COMPOUNDS that generally do not contain CARBON.[134]

Source: matter other than vegetable or animal. Examples: sulfuric acid and salt. Exceptions are carbon monoxide and carbon dioxide and their derivatives.

inorganic coatings coatings based on silicates or phosphates and usually used PIGMENTED with metallic zinc.[71]

inorganic polymers MACROMOLECULAR substances whose principal structural features are made up of homopolar interlinkages between multivalent elements other than CARBON.[58,130]

This does not preclude the presence of carbon-containing groups in side branches, or as interlinkages between principal structural members. Many examples of such POLYMERS are found in nature, e.g., MICA, CLAYS, TALC, etc.

inpaint to renew damaged areas on paintings or painted surfaces by repainting.[54,81] (DAC)

insect wax see CHINESE WAX.[137]

inside spray coating applied (SPRAYED) on the inside of cans to provide a protective film between the contents of the can and the can body.[163] (EPA)

inspection the PROCESS of measuring, examining, testing, or otherwise comparing the unit of product.[157] (ASTM)

ASTM D3276 is a Standard Guide for paint inspectors of coated METAL substrates. It includes the key elements of SURFACE PREPARATION, COATINGS APPLICATION, and final approval for both field and shop work.

insulate to prevent the passage of heat, electricity, or sound into or out of, especially by surrounding with a nonconducting material.[176]

insulating varnish VARNISH often used in combination with materials such as MICA, FABRIC, PAPER, etc., to provide electrical insulation for electrical equipment.[61,71,152] (BSI)

These are formulated to have a high resistance to electrical passage.

insulation resistance the electrical resistance of an insulating material to a direct voltage.[61]

It is determined by measuring the leakage of current which flows through the insulation.

intaglio form of PRINTING in which all the elements are below the surface, having been cut, scratched, engraved, or etched into the METAL to form INK-retaining grooves or cups.[81,168]

Surplus ink on the surface must be wiped off after each inking and before each printing impression.

integrating sphere a sphere coated inside with a highly REFLECTIVE, DIFFUSE material and used for the measurement of LUMINOUS FLUX.[43,145]

If the internal surface is perfectly diffuse, the intensity of any part of the sphere is the same. Many instruments used for reflectance measurements utilize such a device for measuring the diffuse or total reflectance from a sample material relative to a reference material.

integration (colorimetry) the PROCESS of summing the products of the three values: the ILLUMINANT, the observer response characteristics, and the REFLECTANCE or TRANSMITTANCE of an object at specified WAVELENGTHS in order to obtain the TRISTIMULUS VALUES.[43,69]

International Committee to Coordinate Activities of Technical Groups in the Coatings Industry (ICCATCI) see COATINGS SOCIETIES INTERNATIONAL.[174]

Inter-Society Color Council (ISCC) a U.S. council made up of delegates from about 30 "member-bodies," i.e., national ORGANIZATIONS and professional societies, and of individual members, all of whom have an interest in information about COLOR and in basic concepts of color. Member-body organizations and individual members represent: (1) creators of color effects in art and design; (2) manufacturers of colored materials; (3) producers of color reproduction processes; and (4) fields of science and technology.

The work of the Council is carried out in study groups, called "problem subcommittees," and is directed by specialists in the color problems under investigation. News of Council activities and newsworthy items in the world of color are published in the ISCC Newsletter, published bimonthly. Annual Council meetings provide a fo-

rum for discussion and resolution of color problems. Specialized interest meetings are held periodically. The ISCC is the official U.S. member of the International Color Association—Association Internationale de la Couleur, AIC.[174,69]

intercoat adhesion the ability of a coating to adhere to previously applied films.[79]

intercoat contamination the presence of foreign matter between successive coats.[42,56]

interface a SURFACE forming a common boundary between adjacent regions, bodies, substances, or phases.[133]

interfacial tension the molecular attractive FORCE between unlike molecules at an INTERFACE. It may be expressed in dynes per centimetre.[127,142] (ASTM)

interference color COLOR which appears in thin films ILLUMINATED with composite light as the result of the reinforcement of some colors (WAVELENGTHS) and weakening or elimination of others, depending on the phase differences and amplitudes of the light waves involved.[43,69]

The color produced depends on the ANGLE OF INCIDENCE, thickness of the film, and the REFRACTIVE INDEX. Interference color also appears when light strikes narrow slits, the principle of the diffraction grating.

interference filter OPTICAL filter made with extremely thin alternate layers of metals and dielectrics so that it passes only very narrow WAVELENGTH bands.[43,145]

Such filters are frequently used in ABRIDGED SPECTROPHOTOMETERS where they serve to isolate discrete wavelengths for measurement, thus effectively replacing a dispersion device in a spectrophotometer.

interference pigments materials used as PIGMENTS which depend on the interference of light for their COLOR.[41,43,69] See PEARLESCENT PIGMENTS.[41]

They are generally made from platelets coated with a thin film of higher refractive index. For example, thin films of TITANIUM DIOXIDE coated on MICA flakes are one type.

interlocked grain WOOD in which the FIBERS incline in one direction in a number of rings of annual growth, then gradually reverse and incline in an opposite direction in succeeding rings and then reverse again.[153] (ASTM)

intermediates (1) ORGANIC COMPOUNDS, either cyclic or acyclic. *These compounds may be considered as chemical stepping stones between the parent substance and the final product. The cyclic types still predominate as intermediates for synthetic DYES and have few other uses; the acyclic type in general have many independent uses. Exceptions are hexamethylenetetramine and butadiene. The distinction between an INTERMEDIATE and an end product is not always precise;* (2) materials which on reacting together yield dyes, either soluble or insoluble.[41,144]

intermolecular polymerization POLYMERIZATION which occurs as the result of association between molecules.[169]

internal mix spray gun SPRAY GUN in which the fluid and air are combined before they leave the gun.[77]

internal mixer enclosed machine for RUBBER or other suitable material, inside of which are two heavy mixing rotors which revolve in opposite directions with small clearance between themselves and the enclosing walls.[59,131]

The mixing chamber is jacketed or otherwise arranged for watercooling, and is provided with a feeding hopper which can be closed by means of a pneumatically operated vertical ram. Examples are the BANBURY MIXER and SIGMA BLADE MIXERS.

internal phase in an EMULSION, the DISCONTINUOUS PHASE. For example, in an OIL-IN-WATER EMULSION, the oil is the internal phase.[60]

internal plasticization PLASTICIZATION by means of internally combined groups, such as by COPOLYMERIZATION.[156,170]

internal plasticizer where the PLASTICIZING AGENT is chemically combined with the hard RESIN or POLYMER, the product is said to be internally plasticized.[170]

Common examples occur in OIL-MODIFIED ALKYDS, and such materials have considerable practical advantages over externally plasticized systems, where the plasticizer may be lost by EVAPORATION or leached out by water or other agents.

internally plasticized when a product is SYNTHESIZED from a reaction involving two or more raw materials, it may be said to be internally plasticized if one of the raw materials is able to confer PLASTICITY or FLEXIBILITY to it. In other words, the product is plasticized because it is built up from a component which is naturally PLASTIC.[130,170]

For example, an oil-modified alkyd, in which PHTHALIC ANHYDRIDE and GLYCEROL are combined with DRYING OIL FATTY ACIDS, is internally plasticized by reason of the presence of the fatty acid component. If a CONGO VARNISH were made by interaction of run congo and LINSEED OIL monoglyceride, it could be described as internally plasticized, in contrast to a varnish made by

---

Words presented in CAP/SMALL CAPS type indicate that the word is defined in another part of the *Coatings Encyclopedic Dictionary.*

Numerical superscripts classify terms in one or more of the categories listed in the second section of this volume.

simple dissolution of run congo in linseed oil, which would be externally plasticized. Also a POLYVINYL ACETATE-ACRYLIC COPOLYMER is internally plasticized while a polyvinyl acetate HOMOPOLYMER, into which a plasticizer has been stirred, is externally plasticized.

International Agency for Research on Cancer one of the three sources that OSHA refers to for data on a material's carcinogenicity.[172]

An agency of the World Health Organization, Geneva, Switzerland, distributed in the USA from 49 Sheridan Ave., Albany, NY 12210, (518) 436-9686.

intramolecular polymerization POLYMERIZATION which occurs within an individual molecule.[169]

For example, DRYING OILS consist of TRIGLYCERIDE ESTERS, involving up to three UNSATURATED FATTY ACIDS attached to a single GLYCEROL molecule. Under certain conditions, polymerization is believed to occur between neighboring fatty acid chains in a single molecule.

intrinsic viscosity the ratio of a solution's SPECIFIC VISCOSITY to the CONCENTRATION of the solute, extrapolated to zero concentration.[130,166] Syn: limiting viscosity number

intumesce to form a voluminous char on ignition. Foaming or swelling when exposed to heat.[172,176] See INTUMESCENCE,[72] INTUMESCENT COATING,[71,163,172] FIRE-RETARDANT COATING,[71,163,172] and CARBONIFIC.[172]

intumescence the puffing up of a coating on contact with high heat (fire) which results in a foam-like structure that insulates the SUBSTRATE.[72,176] See INTUMESCENT COATINGS,[71,163,172] FIRE-RETARDANT COATING,[71,163,172] CARBONIFIC,[172] and INTUMESCE.[172]

intumescent coating FIRE-RETARDANT coating which, when heated, becomes PLASTIC and produces nonflammable gases, such as CARBON DIOXIDE and AMMONIA.[71,163,172,176] See CARBONIFIC,[71,163,172,176] INTUMESCENCE,[72] and INTUMESCE.[172,176]

The gases are trapped by the film, converting it to a foam about fifty times as thick as the original paint film. At this stage, the film solidifies, resulting in a thick, highly insulating layer of carbon, which effectively protects the substrate from fire.

iodine number see IODINE VALUE.[157]

iodine test test for the detection of starches and dextrins.[157]

It involves the addition of a small amount of a solution of iodine in potassium iodide to an AQUEOUS SOLUTION of the suspected substance. Starches give purple or blue colors according to the degree of HYDROLYZATION, whereas erythrodextrin gives an orange or reddish-brown coloration.

iodine value measure of the UNSATURATION of OILS and FATTY ACIDS expressed in terms of the number of grams of iodine per 100 g of sample (percentage by wt of iodine absorbed).[157,128] (ASTM)

When iodine value is determined on oils or fatty acids having CONJUGATED SYSTEMS, the result is not a measure of total unsaturation, but rather is an empirical value that affords a comparison of total unsaturation. WIJS IODINE METHOD is not reliable for tall oil fatty acids containing appreciable quantities of ROSIN ACIDS.

ASTM test methods for iodine value include: drying oils, Rosenmund-Kuhnhenn, D1541; fatty amines, Wijs test, D2075; fatty quaternary ammonium chlorides, D2078; lac resins, test, D29; Wijs test, D1959.

ion an atom or a group of atoms that has acquired a net of electric charge by gaining or losing one or more electrons.[61,144]

ion exchange resins SYNTHETIC RESINS containing active groups (usually sulfonic, carboxylic, PHENOL, or substitute AMINO groups) that give the RESIN the property of combining with, or exchanging ions between, the resin and a solution.[130]

Used in water softeners for example.

ionization PROCESS by which neutral molecules or atoms dissociate to form positively and negatively charged PARTICLES.[61] (ASTM)

ionomer generic for that class of POLYMERS in which ionized carboxyl groups create ionic CROSSLINKS in the intermolecular STRUCTURE.[130]

The unique properties of ionomer resins are dependent on the presence of the ionic interchain forces clustered between the long chain molecules of the POLYMER. This interchain bonding develops properties normally associated with a crosslinked structure. However, these polymers are PROCESSED at conventional temperatures like other THERMOPLASTIC RESINS.

ionomer resins POLYMERS which have ETHYLENE as their major component, but containing both COVALENT and IONIC BONDS.[130]

The polymers exhibit very strong interchain ionic forces. The anions hang from the HYDROCARBON chain and the cations are metallic magnesium, zinc. These resins have many of the same features as POLYETHYLENE, plus high TRANSPARENCY, tenacity, resistance, and resilience, with increased resistance to OILS, greases, and solvents.

IPDI abbreviation for ISOPHORONE DIISOCYANATE.[141] cf. HDI, MDI, TDI, and TMXDI[141]

iridescence COLOR produced from light interference phenomenon.[43,69] See INTERFERENCE COLOR and OPALESCENCE.[43,69]

Irish gum see CARRAGEEN.[139,155]

Irish moss see CARRAGEEN.[139,155]

iron (Fe) a silvery-white, LUSTROUS, malleable, ductile, magnetic or magnetizable, metallic element occurring abundantly in combined forms, notably in HEMATITE, LIMONITE, MAGNETITE, and TACONITE, and used alloyed in a wide range of important structural materials.[45]

Atomic number, 26; atomic weight, 55.847; melting point, 1,535°C; boiling point, 2,750°C; specific gravity, 7.874 (at 20°C); valence, 2, 3, 4, 6.

iron blue $FeNH_4Fe(CN)_6$. Pigment Blue 27 (77510). Any of various PIGMENTS prepared by precipitating ferrous ferrocyanide from a soluble ferrocyanide and FERROUS SULFATE; subsequent oxidation produces a complex ferriferrocyanide, the shade and PIGMENT properties of which are dependent upon the oxidizing agent, reactant concentrations, pH, temperature, size of BATCH and other conditions of manufacture.[41] Syn: Brunswick blue, Chinese blue, greentone ink blue, milori blue, Prussian blue, Monthier's blue, paste blue, potash blue, redtone ink blue, steel blue, toning blue, Turnbull's blue, Wilkinson's blue

As a group, the pigments exhibit good SOLVENT RESISTANCE, and improved dispersibility can be obtained by resination. Density, 1.70-1.85g/cm^3 (14.2-15.4 lb/gal); O.A., 33-56; particle size, 0.05-0.2 µm.

ASTM test methods for iron blue include: analysis, D1135; specification, D261.

iron driers IRON SALTS of NAPHTHENIC ACID, 2-ethyl hexoic acids, or other acids which actively accelerate POLYMERIZATION at elevated temperatures, but are only feebly active at room temperature.[74,83] See DRIERS.[74,83,67]

The presence of the very dark FERRIC ION makes them applicable in dark or TINTED finishes. The British call them iron soaps.[67,74,83]

iron mill paint MILL consisting of a corrugated STEEL DISC revolving tightly against a stationary steel shell.[49,59]

As the PIGMENTED PARTICLES pass between these moving steel parts, the PIGMENT is DISPERSED into the VEHICLE.

iron oxides Fe_2O_3. PIGMENTS which are substantially OXIDES of IRON. (1) Natural: ochres, raw and burnt umbers; raw and burnt siennas; red oxides; metallic browns; maroon oxides; and black oxides. Range in Fe_2O_3 content: 20 to 99%; (2) synthetic: yellow, red, brown, black oxides. Red oxide—Fe_2O_3; yellow oxide—$Fe_2O_3 \cdot H_2O$; brown oxide—$Fe_2O_3 \cdot xFeO$; black oxide—Fe_3O_4.[41]

ASTM test methods for iron oxides include: black natural/synthetic analysis, D3872; synthetic, specification, D769; brown (natural) analysis, D50; specification, D3722; brown (synthetic) analysis, D3872; specification, D3724; red (natural) analysis, D50; specification, D3722; red (synthetic) analysis, D50; specification, D3721; yellow analysis, D50; specification, D768; iron oxides, ferrous iron content, D3872.

iron oxides, natural Pigment Yellow 42 (77492), Pigment Brown 6 (77499), Pigment Brown 7, Pigment Red 101 (77491). The COLORED NATURAL IRON OXIDES, HYDRATED iron oxides, and occasionally manganese oxide, in association and silicon oxide, aluminum oxide, and trace to minor amounts of calcium and magnesium oxides. In addition, CARBON and ORGANIC material may be present.[41] They include: limonite, natural yellow oxide, African ocher, Indian ocher, ocher, Spanish ocher, metallic brown, mineral brown, raw sienna, burnt sienna, terra de siena, raw umber, burnt umber, Turkish umber, hematite, natural red oxide, Persian gulf oxide, Persian oxide, Pompey red.[41]

iron oxides, synthetic Pigment Red 101 (77491). $Fe_2O_3 \cdot Fe_2O_3 \cdot H_2O$. IRON OXIDES manufactured by a SYNTHETIC PROCESS. There are four common types with varying compositions, physical properties, and particle shape: (a) copperas red, produced by two stage CALCINATION of FERROUS SULFATE to red iron oxide; (b) ferrite red, produced by the dehydration of yellow iron oxide; (c) iron oxide, slowly PRECIPITATED in solution in the presence of O and metallic iron; (d) synthetic red, produced by oxidation of synthetic black iron oxide.[41] They include: red iron oxide, Turkey red, rouge, Indian red, crocus, ferric oxide, Venetian red, ferrite red, ferrite yellow, brown iron oxide pigment, black iron oxide, jeweler's rouge.

iron phosphate coating see CHEMICAL CONVERSION COATING.[71,158]

iron soaps (British) See IRON DRIERS.[83,127]

iron titanate see ILMENITE.[41,34]

irradiance the quotient of the RADIANT FLUX INCIDENT on an infinitesimal surface element containing the point in question, by the area of that surface element.[43]

irradiation application of RADIATION to an object. (ASTM)

irritant a noncorrosive material that causes a reversible inflammatory effect on living tissue by chemical action at the site of contact as a function of concentration or duration of EXPOSURE.[172]

isano oil OIL that is obtained from the kernels of the tree, *Ongokea klaineana*, native to West Africa; the UNSATURATED OIL POLYMERIZES readily and is used in coatings.[63] See ISANOLIC ACID.[63]

isanolic acid eighteen-CARBON FATTY ACID with acetylene triple bonds at the 9 and 11 positions, and

---

Words presented in CAP/SMALL CAPS type indicate that the word is defined in another part of the *Coatings Encyclopedic Dictionary.*

Numerical superscripts classify terms in one or more of the categories listed in the second section of this volume.

having also an ETHYLENE BOND but no hydroxyl substituent.[63] Syn: erythrogenic acid

Occurs in ISANO OIL, along with closely related bolekic acid, which is a hydroxy acid.

ISCC abbreviation for INTER-SOCIETY COLOR COUNCIL.[174,69]

ISCC-NBS color designation a two-to-four-word descriptive phrase for a COLOR (such as "vivid orange" or "dark-grayish-reddish brown") defined according to sections of the MUNSELL color solid. (See NBS CIRCULAR 553.)[69] (ASTM) See ISCC-NBS COLOR NAMES DICTIONARY.[69]

ISCC-NBS Color Names Dictionary DICTIONARY of COLOR names and a color-naming system described specifically in terms of MUNSELL NOTATIONS.[69]

The system was devised by a committee of the INTER-SOCIETY COLOR COUNCIL in collaboration with members of the staff of the United States National Bureau of Standards. It is NBS Circular 553, issued in 1955. A supplement illustrating the Centroid Colors was issued as Standard Sample No. 2106 in 1965.

ISO abbreviation for International Organization for Standardization, Geneva.[174] (DAC)

ISO flow cups capillary FLOW cups used to measure the VISCOSITY of NEWTONIAN or near-newtonian coatings.[145,157,166]

The cup-orifice combination (ISO cup with orifice diameter of 3 mm, 4 mm, or 6 mm) is selected to provide an efflux time with the range of 20 to 200 s and viscosities up to 700 cSt (700 mm^2/s). The most commonly used cup is the one with the 4 mm orifice.

ASTM Test Method D5125 covers the determination of the flow time (viscosity) of coatings using ISO capillary flow cups.

isobutene $(CH_3)_2C{=}CH_2$. (2-methylpropene, isobutylene.) Colorless, highly volatile liquid or liquified gas.[141]

Properties: bp of –6.9°C, an fp of –139°C, an flp of –105°F, sp gr of 0.6 (20°C). POLYMERIZES easily. Uses: production of isooctane, butyl rubber, polyisobutene resins, tert-butanol, methacrylates, copolymer resins with butadiene, acrylonitrile.

isobutyl alcohol $(CH_3)_2CHCH_2OH$. Colorless flammable liquid used as a SOLVENT.[164] Syn: 2-methyl-1-propanol, and isopropyl carbinol

Properties: sp gr, 0.7982; bp, 108°C.

isocyanate a COMPOUND containing the functional group $-N{=}C{=}O$.[141] (EPA) See FUNCTIONALITY.[134]

These functional groups may be CROSSLINKED with a HYDROXYL containing material to form POLYURETHANES which are often used in coatings.

isocyanate resins RESINS synthesized from ISOCYANATES ($-N{=}C{=}O$) and ALCOHOLS ($-OH$).[130] See POLYURETHANES,[156,71,130,163,167] and URETHANE COATINGS.[156,71,163]

The REACTANTS are joined through the formation of the URETHANE linkage and hence this field of technology is generally known as urethane chemistry. Urethane coatings (SOLVENT or WATERBORNE) are noted for their overall balance of high performance properties. These include excellent toughness, FLEXIBILITY, HIGH IMPACT RESISTANCE, and TENSILE STRENGTH. They also offer excellent ABRASION RESISTANCE and resistance to a wide variety of chemicals, solvents, and stains.

Urethane coatings are generally divided into AROMATIC and ALIPHATIC types. Aromatic urethanes are generally based on TDI (toluene diisocyanate) or MDI (diphenylmethane diisocyanate). They are designed mainly for interior APPLICATIONS since they have a tendency to yellow when exposed to prolonged sunlight. Aliphatic urethanes offer excellent UV stability, but are more expensive than aromatic urethanes. Aliphatic urethanes are generally based on IPDI (ISOPHORONE DIISOCYANATE), HDI (HEXAMETHYLENE DIISOCYANATE), or TMXDI (TETRAMETHYLXYLENE DIISOCYANATE). In addition to excellent light stability, aliphatic urethanes have better flexibility than aromatic urethanes.

isoelectric heating form of HEATING by means of electrical energy, in which a special form of element is employed, and which is submerged in liquid raw materials.[59,176]

In practice a series of such elements is employed.

isoelectric point the PH value at which a substance or system (e.g., a PROTEIN SOLUTION) is electrically neutral; at this value, electrophoresis does not occur when a direct current is applied.[139]

isomeric polymers POLYMERS which have essentially the same percentage composition, but differ with regard to their molecular architecture.[130] (IUPAC)

isomerism the existence of two or more chemical COMPOUNDS with the same molecular formula but having different properties, owing to different arrangement of atoms within the molecule, e.g., ammonium cyanate NH_4CNO, and urea, $CO(NH_2)_2$, are ISOMERS.[133]

isomerized oils OILS which have been ISOMERIZED from a molecular rearrangement by several means, such as the CATALYTIC action of alkalis, sulfur dioxide, nickel, etc.[63]

Isomerized oils exhibit greater REACTIVITY, more rapid POLYMERIZATION, and greater DRYING RATE than their nonisomerized parent oils.

isomerized rubber RUBBER which has undergone cyclic rearrangement by heating in solution in the presence of a suitable CATALYST.[140] Syn: cyclized rubber

isomers chemical compounds with same molecular weight and atomic composition but differing molecular structure, e.g., n-pentane and 2-methylbutane.[133]

isophorone $COCH:C(CH_3)CH_2C(CH_3)_2CH_2$. A high boiling, cyclic ketonic solvent.[164] Syn: 3,5,5-trimethyl-2-cyclohexen-1-one; isoacetophenone

Poperties: mol wt, 138; bp, 215°C; flp, 84°C; vp, 0.1 mm Hg/20°C.

The ASTM Test Method for isophorone is D2619.

isophorone diisocyanate (IPDI) an ALIPHATIC ISOCYANATE MONOMER used as an INTERMEDIATE in urethane coatings.[141] See ISOCYANATE RESINS, POLYURETHANES, POLYURETHANE FINISH, AND URETHANE COATINGS. cf. HEXAMETHYLENE DIISOCYANATE (HDI), DIPHENYLMETHANE DIISOCYANATE (MDI), TOLUENE DIISOCYANATE (TDI), and TETRAMETHYLXYLENE DIISOCYANATE (TMXDI)

isophthalic acid $C_8H_6O_4$ m-phthalic acid. m-benzenedicarboxylic acid.[144]

Molecular weight, 166.13. CRYSTALLINE POWDER which melts at 345-348°C; sublimes without DECOMPOSITION.

isoprene $CH_2 = CHC(CH_3) = CH_2$. MONOMERIC HYDROCARBON constituent of natural RUBBER.[141] Syn: 2-methyl-1,3-butadiene, ß-methylbivinyl; hemiterpene

It has a molecular weight of 68.06, an mp of -120°C, and a bp of 34°C.

isopropyl acetate $(CH_3)_2CHCOOCCH_3$. It has a bp of 90°C, sp gr of 0.930, flp of 33°F, and a vp of 45 mm Hg/20°C.[144,164]

isopropyl alcohol $(CH_3)_2CHOH$. It has a bp of 82°C, sp gr of 0.791/15°C, an flp of 62°F, and a vp of 34 mm Hg/20°C.[164] Syn: propanol-2

isopropyl benzene see CUMENE.[164]

isopropyl butyrate $(CH_3)_2CHCOOC_3H_7$. It has a bp of 128°C and sp gr of 0.879/0°C.[144]

isopropyl cellosolve® $(CH_3)_2$ $CHOCH_2CH_2OH$. It has a bp of 141°C, sp gr of 0.906/20°C, and a vp of 8 mm Hg/30°C.[164]

isotactic pertaining to a type of POLYMERIC MOLECULAR STRUCTURE containing a sequence of regularly spaced asymmetric atoms arranged in like configuration in a POLYMER chain.[130,133]

isotactic polymer TACTIC POLYMER, the base unit of which possesses, as a component of the main chain, a CARBON (or similar) atom with two different lateral substituents, these atoms being so arranged that a hypothetical observer advancing along the bonds constituting the main chain, finds each of its substituents in the same steric order.[130,133] (IUPAC) cf. ATACTIC POLYMER, SYNDIOTACTIC POLYMER, and TACTIC POLYMER[130,133]

Here, hydrogen is counted a substituent.

isotropic applies to a material in which the velocity of propagation of RADIANT ENERGY (such as light) is identical for all directions.[43]

itaconic acid $CH_2C(COOH)CH_2COOH$. An mp of 167-168°C. Soluble in water, ALCOHOL, and acetone. It is used in COPOLYMERIZATIONS, RESINS, and PLASTICIZERS.[141] Syn: methylenebutanedioic acid; methylene-succinic acid

IUPAC abbreviation for INTERNATIONAL UNION OF PURE AND APPLIED CHEMISTRY.[174]

ivory black black PIGMENT prepared by CALCINATION of ivory chips and cuttings. In some respects it resembles BONE BLACK, to the better grades of which the name is also applied.[41]

It has a high calcium phosphate content.

J

j acid fusion product of beta-naphthylamine-3,6-bisulfonic acid with caustic.[41,144] Syn: 2-AMINO-5-naphthol-7-sulfonic acid; 6-amino-1-naphthol-3-sulfonic acid

Used in the manufacture of DYES.

jalousie window with movable, horizontal GLASS slats angled to admit ventilation and keep out rain.[82]

This term is also used for outside shutters of WOOD *constructed in this way.*

jamb vertical face inside an opening to the full thickness of a wall; vertical slide members of a window frame, door frame, or lining.[82]

jamba seed oil NONDRYING OIL, similar to rape oil, obtained from the seeds of *Eruca sativa*. Its main constituent acids are erucic, oleic, and linoleic.[63]

Approx. constants: sp gr, 0.916/15°C; iodine value, 100; saponification value, 173.

japan glossy black ENAMEL, either air drying or baking, based on ASPHALTUM and DRYING OIL.[71,76] Syn: black japan

Japan should not be confused with black enamels which are produced by PIGMENTATION *and do not have an asphaltum base.*

japan color PASTE containing PIGMENT and a GRINDING JAPAN VEHICLE, used for lettering and DECORATING.[81,148,163] (ASTM) See JAPAN, GRINDING.[146,152,156]

japan drier RESINATE-base liquid DRIER.[67,74,83] (ASTM) See RESINATES, METALLIC.[44,167]

japan, grinding rapid, hard-drying VARNISH suitable for use as a VEHICLE for JAPAN COLORS; frequently contains SHELLAC.[146,152,156] (ASTM)

Japan Society of Colour Material (JSCM) serves the Japanese coatings industry by promoting the technology of the paints, PIGMENTS, and PRINTING INKS industries through lectures, educational activities, and by practical and cooperative experimental work among its members.[174]

Founded in 1927 as the Association for Pigments, Paints and Printing Inks, the organization became JSCM in 1937. It is composed of three regional sections and publishes the monthly Journal of the Japan Society of Colour Material. JSCM is a member organization of the COATINGS SOCIETIES INTERNATIONAL.

japan wax product obtained from the berries of trees indigenous to Japan and China.[137]

Although described as a WAX, *it differs from waxes in composition, in that it is a mixture of* TRIGLYCERIDES. *Mp, 52°C; sp gr, 0.987/15°C; saponification value, 215; iodine value, 11.*

Japanese lacquer a glossy coating obtained by tapping the sap from the Japanese VARNISH tree (*Rhus vernicifera*) or sumac.[152,167]

Japanners' brown brown PIGMENT made by high temperature OXIDATION of FERROUS HYDROXIDE PRECIPITATED from a solution of an IRON SALT.[41]

japanning PROCESS of finishing with a baking BLACK JAPAN.[54] (BSI)

jar mill a small BALL MILL.[59,64]

jargon the specialized or technical language of a trade, profession, or similar group.[134]

jaspé paper WALLPAPER with a soft, irregular pattern giving the APPEARANCE of the SURFACE having been COMBED or BRUSHED.[148]

jaune brilliant artists' yellow COLORANT that was prepared from CADMIUM SULFIDE, yellow vermillion, and WHITE LEAD.[41]

The name is also applied to cadmium sulfide only.

JCT abbreviation for the JOURNAL OF COATINGS TECHNOLOGY.

jet the MASSTONE COLOR of black or near black SURFACES, and indicating blackness or intensity.[69]

It is frequently used to describe dark blues, blue-black, or black PIGMENTS.

jet abrader device used for measuring the ABRASION RESISTANCE of ORGANIC coating materials, in terms of the time required for a controlled jet of fine ABRASIVE PARTICLES to abrade through the coating to the SUBSTRATE.[47,157]

jet mill a doughnut-shaped tube GRINDING chamber, throughout which precrushed PIGMENT PARTICLES are carried by a stream of high-pressure steam or superheated compressed air.[59,64]

The steam or air is injected through jet nozzles located in the outer wall of the grinding chamber carrying the pigment particles in a circular path at high velocity. Grinding is done by attrition of the pigment particles against one another. Since high-pressure steam or superheated compressed air are the fluid media in jet milling, the pigment particles are subjected to considerable heat, which removes residual moisture from their surface. The fines are carried inwardly to a central outlet, and the large particles to the periphery for further attrition. The particle size is controlled by the rate of feed and fluid media pressure.

Jew's pitch ASPHALT.[76]

jewelers' rouge red iron oxide.[41]

JND abbreviation for JUST NOTICEABLE DIFFERENCE.[69]

joinery the art or craft of CABINETMAKING.[175]

joint the space between adjacent surfaces (as between MASONRY units), or the place where two members or components are held together by nails, fasteners, CEMENT, mortar, etc.[82]

joint compound a paste-like FILLER generally used for filling and finishing GYPSUM WALLBOARD.[147,158] cf. SPACKLE[147,158]

Primarily, it is used to cover nail heads and the tape used over seams between wallboard. It is also used as a texturing medium for walls.

joint filler SEALANT inserted between abutting ends of WALLBOARD.[165]

joint, lap (1) See LAP; (2) a joint in which one board, plank, METAL plate, etc., overlaps the edge of another piece; the overlapping part of each member is cut away to half thickness, resulting in flush surfaces;[82] (3) a joint formed by placing one piece partly over another and uniting the overlapped portions.[82]

joint roller ROLLER used to press down the edges of WALLPAPER and other WALLCOVERINGS in order to make the joints less conspicuous.[59]

joint, starved see STARVED JOINT.[79]

joint tape PAPER or paper-faced cotton tape, METAL, FABRIC, GLASS mesh, or other material, sometimes embossed or perforated, which is fixed over the joints between WALLBOARDS, to conceal the joints and provide a smooth surface for painting.[148,59,75]

jointless flooring see SEAMLESS FLOORING.[82]

joist a small rectangular sectional member arranged parallel from wall to wall in a building, or resting on beams or girders.[82]

They support a floor or the laths or furring strips of a ceiling.

jolly balance spring balance for determining the SPECIFIC GRAVITY of a solid by weighing it alternatively in air and water.[59,157]

joule, j and SI the unit of energy in the system of units.[128] (ASTM)

One joule is one watt-second.

Journal of Coatings Technology (JCT) the official journal of the FEDERATION OF SOCIETIES FOR COATINGS TECHNOLOGY.

journeyman painter one who has had at least three years' experience and schooling as an apprentice.[134,148]

JSCM see JAPAN SOCIETY OF COLOUR MATERIAL.[174]

Judd chart see JUDD HIDING POWER CHART.[43,157]

Judd color difference unit see NBS COLOR DIFFERENCE EQUATION and NBS UNIT.[69]

Judd graph convenient graph, devised by Judd at the National Bureau of Standards, showing simultaneously the variation in REFLECTANCE (LIGHTNESS) and OPACITY (CONTRAST RATIO) caused by variations in the THICKNESS of a film.[69] See HIDING POWER CHART (2).[43,157]

The curves are derived from the KUBELKA-MUNK EQUATION, as are also a second family of curves showing the variations from adding to a film of any given thickness varying amounts of black dye or CARBON BLACK.

Judd hiding power chart HIDING POWER chart containing lines obtained by computation from the KUBELKA-MUNK EQUATION.[43,157] See HIDING POWER CHART (2).[43,157]

juniper gum see GUM SANDARAC.[167]

just noticeable difference (JND) a psychological term used to describe the difference in perception which is noticeable statistically in 50% of the trials.[69,35] See MINIMUM PERCEPTIBLE DIFFERENCE.[43,69,35]

juta hycica resin Brazilian RESIN occasionally used in certain types of VARNISHES.[167]

It is a very hard and pale-colored material, insoluble in most of the common SOLVENTS. It is necessary to "run" this RESIN before it can be used in conjunction with a DRYING OIL.

jute (Burlap) coarse, heavy, plain-weave FABRIC of coarse single yarn.

jute seed oil obtained from the seed of *Corchorus capsularis*, grown chiefly in India, Egypt, and China.[63]

It contains LINOLEIC and OLEIC acids. Iodine value, 103; sp gr, 0.921/15°C; saponification value, 185; acetyl value, 27.[6]

K

K KUBELKA-MUNK ABSORPTIVITY COEFFICIENT.[43]

kakemono (Japanese) A painting mounted on a margin of brocade; hung by its top when in use, and rolled up when not in use.[81]

kalsomine see CALCIMINE.[71]

kaolin see CHINA CLAY.[41,58]

kaolinite $Al_2O_3{\cdot}SiO_2{\cdot}2H_2O$. A finely divided crystalline form of HYDRATED ALUMINUM SILICATE occurring as minute monoclinic crystals with a perfect basal cleavage, resulting chiefly from the alteration of FELDSPARS under conditions of hydrothermal or pneumatolytic metamorphism.[41,58,34] (ASTM)

It is an important CLAY mineral. Density, 2.58 g/cm³ (21.5 lb/gal); refractive index, 156; O.A., 32 to 55; Mohs hardness, 2.5.

karaya gum dry EXUDATION from *Sterculia urens*, which grows in India.[155] Syn: gum karaya

It swells in water, and has some resemblance to GUM TRAGACANTH.

karolith see PROTEIN ALDEHYDE RESINS.[38,130]

kauri fossil COPAL of coniferous origin, found in New Zealand.[167]

kauri-butanol value volume in ml at 25°C (77°F) of a SOLVENT, corrected to a defined standard, required to produce a defined degree of TURBIDITY when added to 20 g of a standard solution of KAURI RESIN in normal BUTYL ALCOHOL.[157,164,128] (ASTM) Abbreviation: KB value

For kauri-butanol values of 69 and over, the standard is toluene and has an assigned value of 105. For kauri-butanol values under 60, the standard is a blend of 75% n-heptane and 25% toluene and has an assigned value of 40.

kauri reduction test test for measuring the FLEXIBILITY of a VARNISH.[157]

kautschin see DIPENTINE.[44,164]

KB value see KAURI-BUTANOL VALUE.[157,164,128]

kelvin the SI UNIT of thermodynamic temperature defined as the fraction 1/273.16 of the thermodynamic temperature of the triple point of water, i.e., the triple point of water is exactly 273.16 kelvin.[128,176]

The units of kelvin and CELSIUS temperature interval are identical. A temperature expressed in degrees celsius is equal to the temperature in kelvins less 273.15°C. Symbol K; the name 'degree Kelvin' (Symbol °K) was discontinued by international agreement in 1967. Named after Lord Kelvin (1824-1907).

kelvin temperature scale temperature scale in which 0 is absolute zero, –273° Celsius.[128,] Symbol K. See KELVIN.[128]

This scale is used to describe the CORRELATED COLOR TEMPERATURE of light sources and ILLUMINANTS in color designations and color rendition.

kerosene a low VISCOSITY OIL distilled from PETROLEUM or shale oil, used as a fuel, paint thinner, and ALCOHOL denaturant.[164] Syn: coal oil and, in British usage, paraffin. Also "kerosine."

kerrolic acid $C_{15}H_{27}(OH)_4COOH$. Monobasic tetrahydroxy acid constituent of SHELLAC.[144,146]

ketone any of a class of ORGANIC compounds having a CARBONYL GROUP linked to an atom in each of two HYDROCARBON RADICALS and having a general formula R(CO)R′ where R may be the same as R′.[144,164]

ASTM test methods for ketone solvents include: acetone, D329; diacetone ALCOHOL, D2627; isophorone, D2619; methyl amyl ketone, D4360; methyl ethyl ketone, D740; methyl isoamyl ketone, D2917; methyl isobutyl ketone, D1153.

ketone condensation resins RESINS produced as the result of CONDENSATION of ketones and ALDEHYDES, for example, the RESIN obtained from METHYL ETHYL KETONE and FORMALDEHYDE.[130]

kettle reaction vessel for VARNISH or RESIN manufacture.[59,131]

kettle bodied oil see BODIED OIL.[156,63]

KHN see KNOOP HARDNESS NUMBER.[128,157]

kibbled broken into small lumps, about 1/4 in. diameter, e.g., kibbled glue.[79,131]

kick board see BASEBOARD.[82]

kidney oil fraction boiling between 250 and 270°C, obtained by the destructive DISTILLATION of ROSIN.[44,164]

Kienle's functionality theory fundamental postulate which covers, among other things, the likelihood of reactions between compounds to form products of high molecular weight; and the relation between size and shape of reacting molecules and physical properties of the reaction products derived therefrom.[130]

kieselguhr see DIATOMACEOUS SILICA.[58]

kiln dried LUMBER dried in chambers.[131,153] Syn: kiln seasoned

Heat is controlled to prevent cracking and warping.

kiln seasoned see KILN DRIED.[153,131]

kinematic viscosity of a NEWTONIAN liquid may be defined as the absolute or dynamic VISCOSITY divided by the density, both at the same temperature.[157,166] (ASTM)

The CGS UNIT of kinematic viscosity is the STOKES, which has the dimensions square centimetres per second. It is more common practice, for PETROLEUM products, to express kinematic viscosity in CENTISTOKES (cSt) or mm^2/s.

king's blue see COBALT BLUE.[41]

king's yellow ORPIMENT.[41]

kira-ye (Japanese) a PRINT with a MICA background.[81]

knife coater a coater used primarily for PAPER or FABRIC webs which employs an adjustable blade or "knife" to distribute a liquid coating evenly over a moving web surface.[77] (EPA)

knife mark see DOCTOR MARK.[66,77]

knife test for ADHESION BRITTLENESS, TOUGHNESS, and tendency to ribbon, by cutting a narrow strip of the coating from the test panel, with a knife.[47,157]

knifing filler FILLING composition suitable for application with a filling knife as distinct from one made for BRUSH APPLICATION.[147] (BSI)

Knoop hardness number, (KHN) the INDENTATION HARDNESS determined with a KNOOP INDENTER, and calculated as follows:

$$KHN = \frac{L}{Ap} = \frac{L}{l^2 Cp}$$

where:
L = load in kilograms applied to the indenter,
l = measured length of long diagonal of the indentation in millimetres,
Cp= indenter constant relating l^2 to Ap, and
Ap= projected area of indentation in square millimetres.[128,157] (ASTM)

Knoop hardness test an INDENTATION HARDNESS test using calibrated machines to FORCE a rhombic-based pyramidal diamond indenter having specified edge angles, under specified conditions, into the surface of the material under test and to measure the long diagonal after removal of the load.[47,157] (ASTM) cf. PENCIL HARDNESS[47,157]

ASTM Test Method D1474 covers the determination of indentation hardness of dried ORGANIC coatings when applied to acceptable plane rigid surfaces such as METAL or glass.

Knoop indenter pyramidal diamond of prescribed dimensions used for testing the indentation HARDNESS of ORGANIC coatings.[47,59,145,157] (ASTM)

In a more restricted sense, a type of diamond hardness indenter having edge angles of 172° 30' and 130°.

knot sealer solutions of various RESINS in ALCOHOL, used to seal knots in new WOOD.[153,71,42] See BLEEDING KNOTS.[42,153]

knotting fairly concentrated solution of SHELLAC in ALCOHOL, used for sealing knots in new WOOD.[146]

Koettstorfer number see SAPONIFICATION VALUE.[63,157,128]

König pendulum damping hardness test a pendulum resting on a dried coating surface is set into oscillation (rocking) and the time for the oscillation amplitude to decrease by a specified amount measured; the shorter the damping time, the lower the HARDNESS.[47,157,145] See PERSOZ PENDULUM HARDNESS TEST.[47,157,145]

Both pendulum hardness tests are covered by ASTM Test Method D4366.

Kordofan gum synonym for gum acacia, derived from the Sudanese province of Kordofan.[167]

Kraemer-Sarnow method (K & S) Softening point.[157]

Method of determining the softening points of resins and ALIPHATIC materials.

Krebs-Stormer Viscometer this is the most commonly used viscometer in the paint industry. Paint consistency is measured by its resistance to stirring by a two-vane paddle.[145,166] Syn: Stormer viscometer

The two vanes are offset to avoid channeling in viscous paints. Appropriate weights are added to a platform on a string attached to a pulley which is connected to the paddle until a speed of 200 rpm is reached, generally measured by a stroboscopic timer. The grams required to reach this speed are then converted, by using a conversion table, to Krebs units.

ASTM Test Method D562 covers the determination of the consistency of paints and related coatings using this viscometer, which is used as an indication of low-shear viscosity.

Krebs units (KU) values of a scale commonly used to express the CONSISTENCY of paints generally applied by BRUSH or roller.[128,166] (ASTM)

This scale is a log function of the "load to produce 200 r/min" scale. (ASTM)

Kremnitz white see CARBONATE WHITE LEAD.[41]

KU see KREBS UNITS.[166,157,128]

Kubelka-Munk Equation (Apply to a single wavelength of light)

$$(1) \quad R = \frac{1 - R_g(a - b\coth bSX)}{a - R_g + b\coth bSX}$$

coth = hyperbolic cotangent; abbreviation, $(e^u + 1/e^u)/e^u - 1/e^u)$

$$(2) \quad SX = \frac{1}{b}\left(\operatorname{arccoth}\frac{a-R}{b}\right) - \operatorname{arccoth}\frac{a-R_g}{b}$$

arc coth = inverse hyperbolic cotangent

where

R = reflectance over substrate of reflectance R_g

R_g = reflectance of substrate

$$a = \frac{S+K}{S} = \frac{1}{2}\left(\frac{1}{R_\infty} + R_\infty\right)$$

$$b = (a^2 - 1)^{1/2} = \frac{1}{2}\left(\frac{1}{R_\infty} - R_\infty\right)$$

S = unit scattering COEFFICIENT expressed in X^{-1}

X = thickness (or product of thickness times concentration)

R_∞ = reflectance at infinite thickness (complete hiding)

K = unit ABSORPTION coefficient expressed in X^{-1}

When hiding is complete (substrate is completely obscured), the first equation reduces to

$$\frac{(1-R_\infty)^2}{2R_\infty} = \frac{K}{S}$$

and for a mixture of pigments

$$(3) \quad \left(\frac{K}{S}\right)_{MIX} = \frac{C_1K_1 + C_2K_2 + C_3K_3 + \ldots}{C_1S_1 + C_2S_2 + C_3S_3 + \ldots}$$

where the C's refer to relative concentrations of pigments in the mixture (Ec's = 100%) and subscripts identify the pigments.

When all of the SCATTERING comes from a single component (such as white pigment in a pasted color or the FIBER in a textile), the denominator of the above equation is determined by the S of this component, and the equation becomes simply

$$(4) \quad \left(\frac{K}{S}\right)_{MIX} = C_1\left(\frac{K}{S}\right)_1 + C_2\left(\frac{K}{S}\right)_2 + C_3\left(\frac{K}{S}\right)_3 + \ldots$$

The latter is referred to as the single constant K-M equation compared to the third one using separate K's and S's, which is called the two-constant equation. The above equations are those most widely used for computer COLOR MATCHING.[43,69]

Kubelka-Munk Theory a theory describing the OPTICAL behavior of materials containing small PARTICLES which scatter and absorb radiant energy.[43,69]

It is widely used for COLOR matching calculations. The mathematical equation describes the reflectance or transmittance in terms of an ABSORPTION COEFFICIENT, K, and a scattering coefficient, S. The K-M theory is based on an assumption of multiple scattering, that is, reflectances from one particle on to other particles before the reflected radiant energy is observed. This assumed behavior is in contrast to the Mie Theory which is based on an assumption of single isolated scattering of individual particles.

kurenai-ye (Japanese) A hand-colored print in which beni is chiefly used.[81]

kusum one of the host trees to which the LAC insect affixes itself.[146]

L

L abbreviation for LIGHTNESS, generally used in COLOR ORDER SYSTEMS to describe the PERCEPTION of the amount of light REFLECTED or TRANSMITTED by materials.[43,69] See LIGHTNESS, LUMINANCE.[43,69]

L acid see CLEVE'S ACID.[144]

label any written, printed, or graphic sign or symbol displayed or affixed to containers of hazardous chemicals.[172]

A label should identify the hazardous material, appropriate hazard warnings, and name and address of the chemical manufacturer, importer, or other responsible party.

labradorite calcium sodium aluminum SILICATE, HARDNESS 6 Mohs scale, used as a wear resistant FILLER in DECK PAINTS.[41,58,34]

lac RESIN secreted by the female of the insect, *Laccifer iacca*, and deposited on the twigs of various species of trees in India and Indochina.[146,167]

After refining, the resin is known as SHELLAC.

lac dye red DYE obtained by the maceration of LAC.[41,146]

lac resins see SHELLAC.[146]

lac wax wax obtained from LAC consisting of myricyl and ceryl ALCOHOLS, free and combined with various FATTY ACIDS.[137,151]

lacquer (1) coating composition which is based on synthetic THERMOPLASTIC FILM-FORMING material dissolved in ORGANIC solvent and which dries primarily by solvent EVAPORATION. Typical lacquers include those based on NITROCELLULOSE, other CELLULOSE derivatives, VINYL RESINS, ACRYLIC RESINS, etc. They are resoluble in their original solvent; (2) FINISH on Chinese and Japanese lacquer ware.[71,151] (ASTM) See SPIRIT VARNISH[151] and URUSHIOL.[151,144]

ASTM test methods for lacquer include: cellulose nitrate content, D3133; ester value of lacquer solvents, D1617; freeze/thaw test, multicolored, D2337; guide for testing, D333; imprint resistance (of dried films), D2091; particle size analysis (multicolored), D2338; plasticizer migration, vinyl fabrics, D2199; selection of test methods, D333; stain removal (multicolored lacquer), D2198; temperature change resistance, test, D1211; testing wood FURNITURE lacquers, D2571; viscosity by dip type viscosity cups, D4212; viscosity by Ford viscosity cup, D1200.

lacrimator a material that produces tears.[172]

lactic acid $CH_3CHOHCOOH$. Colorless or yellowish, odorless, hygroscopic syrupy liquid.[144] Syn: alpha-hydroxypropionic acid and milk acid

It has a bp of 122°C/15 mm Hg, mp of 18°C, and sp gr of 1.2.

laitance milky white deposit on new CONCRETE; EFFLORESCENCE.[56,73]

lake special type of PIGMENT consisting essentially of an ORGANIC soluble coloring matter combined more or less definitely with an INORGANIC base or carrier.[41] (ASTM)

It is characterized generally by a bright COLOR and more or less pronounced TRANSLUCENCY when made into an OIL paint. Under this term are included two (and perhaps three) types of pigment: (a) the older original type composed of HYDRATE of ALUMINA DYED with a solution of the natural ORGANIC COLOR; (b) the more modern and far more extensive type made by PRECIPITATING from solution various COAL-TAR colors by means of a METALLIC SALT, TANNIN, or other suitable REAGENT, upon a base or carrier either previously prepared or coincidently formed; and (c) a number combining both types of varying degrees might be regarded as a third class.

lake Bordeaux B red DYESTUFF produced from 2, 3 hydroxy naphthoic acid and 2-naphthylamine-1-sulfonic acid.[41]

The calcium and manganese toners are known as Bordeaux R Toner, and Maroon Toner B.B. respectively. The latter has good LIGHTFASTNESS.

lake dyes DYES used for the making of LAKES by combination with, or adsorption on, salts of calcium, barium, chromium, aluminum, phosphotungstic acid, or phosphomolybdic acid.[41]

lake of acid yellow 1 acid yellow 1, LAKE (10316).[41]

A nitro DYE prepared by PRECIPITATION from an AQUEOUS SOLUTION by a METAL salt, i.e., aluminum or barium chloride.

lake red C Pigment Red 53 (15585). Red DYESTUFF derived from 6-chlor-3-toluidine-4-sulfonic acid, and ß naphthol. When laked with barium chloride, bronze scarlet is obtained.[41]

lake red P PIGMENT produced by coupling diazotized p-nitroaniline-o-sulfonic acid with betanaphthol.[41]

Words presented in CAP/SMALL CAPS type indicate that the word is defined in another part of the *Coatings Encyclopedic Dictionary.*

Numerical superscripts classify terms in one or more of the categories listed in the second section of this volume.

laketine a TRANSPARENT INK used for extending LETTERPRESS or LITHOGRAPHIC inks.[168]

lallemantia oil DRYING OIL obtained from the seeds of *Lallemantia iberica*, found in parts of Asia and Europe.[63]

It has a sp gr of 0.934/20°C, iodine value of 190, and saponification value of 190. Its main constituents are LINOLENIC and linolenic glycerides.

Lambert's law of absorption see BOUGUER'S LAW.[43]

Lambert's law of reflection the FLUX reflected per unit solid angle is proportional to the cosine of the angle measured from the normal (perpendicular) to the surface.[43]

If the reflected flux is isotropic, the surface is said to be a perfect Lambertian REFLECTOR or a perfect DIFFUSER.

laminar flow laminar FLOW is the requisite flow type for reliable measurements of VISCOSITY, and under the proper measuring conditions and limits is obtained in capillary-tube and rotational viscometers.[166,47] Syn: streamline flow or telescopic flow

If the rate of shear is too high, for both PLASTIC and NEWTONIAN systems, TURBULENT FLOW results; if the rate of flow for plastics is too low, PLUG FLOW results.

laminar scale RUST formation in heavy layers.[66,158]

laminate product made by bonding together two or more layers of material or materials.[79] (ASTM) See LAMINATION.[79,131]

laminate, high pressure LAMINATES MOLDED and CURED at pressures not lower than 1000 psi, and commonly in the range of 1200 to 2000 psi.[38,57]

laminated, cross LAMINATE in which some of the layers of material are oriented at right angles to the remaining layers with respect to the GRAIN or strongest direction in tension.[57,79] See LAMINATED, PARALLEL.[57,79]

Balanced construction of the laminations about the center line of the thickness of the laminate is normally assumed.

laminated, parallel LAMINATE in which all the layers of material are oriented approximately parallel with respect to the GRAIN or strongest direction in tension.[57,79] See LAMINATED, CROSS.[57,79]

lamination the PROCESS of adhering two WEB surfaces together to form a single composite.[79,131] (EPA)

PAPER, FABRIC, FOIL, and PLASTIC film may also be laminated to each other.

lampblack Pigment Black 6 (77266). Black PIGMENT made by burning low-grade heavy OILS or similar carbonaceous materials with insufficient air, in a closed system of a design that the SOOT can be collected in settling chambers.[41] Syn: Paris black

ASTM test methods for lampblack include: solvent extractable matter, test, D305; specification, D209; lampblack content, test, D305.

landfill (1) a method of solid waste disposal in which refuse is buried between layers of dirt so as to fill in or reclaim low lying ground; (2) a site used for or reclaimed by such disposal. See SANITARY LANDFILL.[171,172]

lanolin purified wool grease.[137] See ADEPS LANAE, ANHYDROUS.[137]

lap (1) noun. Region where a coat extends over an adjacent fresh coat. The object of the painter is usually to effect a joint between the two coats without showing the lap;[42,54,56] (BSI) (2) verb. To place one coat of FINISHING material alongside another, partly extending over it, causing increased thickness where the two coats are present;[42,54,56] (BSI) (3) to overlap or partly cover one SURFACE with another, as in shingling.[82] (BSI)

lap joint see JOINT, LAP.[54,82]

lap siding see CLAPBOARD.[75,82,153]

lapis lazuli a rich blue semiprecious stone; either used decoratively or ground and powdered for use as an ULTRAMARINE PIGMENT.[41,148] (DAC)

lapis lazuli blue natural ULTRAMARINE BLUE.[41]

lapping time see WET-EDGE TIME.[55]

larch turpentine see VENICE TURPENTINE.[167]

lasur (lasure) a SOLVENT or WATERBORNE interior or exterior CLEAR COATING for WOOD.[71]

This is a term used in parts of Europe. It generally refers to a clear coating with some film build as opposed to a penetrating stain type coating.

latent solvent see COSOLVENT.[164,83]

latex (1) stable DISPERSION of a POLYMERIC substance in an essentially aqueous medium; (2) fine dispersion of RUBBER or RESIN, natural or SYNTHETIC, in water; the synthetic is made by EMULSION POLYMERIZATION.[130,155] cf. EMULSION[60]

Strictly speaking, after polymerization, a LATEX is a solid dispersed in water and, therefore, is not an emulsion. Latex and emulsion are often used synonomously in the paint industry. Generally accepted plurals are latices or latexes.

ASTM test methods for latexes include: filter-retained solids content, D5097; guides to test procedures, D4143; nonvolatile content (solids), D4758; unreacted monomer content, D4827, D4747.

latex freeze-thaw resistance see FREEZE-THAW RESISTANCE.[157,161]

latex heat age stability the ability of LATEX to resist drastic change in PH and VISCOSITY, as well as SKIN formation, which could occur as the result of being subjected to a constant temperature of 140°F for 10 days or to occasional EXPOSURE to high temperatures over a longer period of time.[161,157] cf. LATEX ROOM TEMPERATURE STABILITY[161,157,176]

A generally accepted guideline is that 10 days at 140°F is roughly predictive of one year in a warehouse. Latex heat age stability is critical, because in shipment or in storage, latex can be subjected to high temperatures for extended time periods.

latex mechanical stability the ability of LATEX to resist COAGULATION under the influence of mechanical agitation.[161] (ASTM)

latex paint a paint containing a stable AQUEOUS DISPERSION of SYNTHETIC RESIN, produced by EMULSION POLYMERIZATION, as the principal constituent of the BINDER.[71] (ASTM)

Modifying resins may also be present.

latex room temperature stability the ability of LATEX to maintain its initial property/physical condition after allowing the sample to stand undisturbed for a long period of time (usually six months to a year).[161,157] cf. LATEX HEAT AGE STABILITY[161,157]

Property changes can be in PH, VISCOSITY, SEDIMENTATION, SKIN FORMATION, PHASE SEPARATION, SETTLING, etc. These are indicators of how sensitive a latex is to stresses encountered during shipping or extended STORAGE at ambient temperature.

lath one of a number of thin narrow strips of WOOD nailed to rafters, ceiling joists, wall studs, etc., to make a groundwork or key for slates, tiles, or plastering.[82]

latices see LATEX.[130,155]

laundry blue synthetic ULTRAMARINE BLUE.[41]

lauric acid $CH_3(CH_2)_{10}COOH$. FATTY ACID occurring in many vegetable fats as the glyceride, especially in coconut oil and laurel oil.[63] Syn: dodecanoic acid

Properties: a sp gr of 0.833, mp of 44°C, and refractive index of 1.4323.

laying off final light strokes of the BRUSH during a painting operation.[54] See FEATHERING.[54]

LC_{50} the concentration, in parts per million, of a substance in air that is lethal to 50% of the laboratory animals exposed to it.[172]

LD_{50} test (lethal dose 50%) a TOXICITY TEST based on the results of animal experiments, the results of which mean that half the animals die from a given dosage of the substance tested.[172] See ORAL DOSE (LD_{50}).[172]

leaching gradual release of toxins from ANTI-FOULING PAINTS.[172]

leaching rate milligrams of water-soluble material released per cm^2/day from an immersed ANTI-FOULING PAINT in seawater.[172]

lead acetate $Pb(C_2H_3O_2)_2 \cdot 3H_2O$. Crystalline salt, soluble in water, was used in the manufacture of lead pigments; also was used in production of VARNISHES.[144]

lead antimonate see ANTIMONY YELLOW.[41]

lead carbonate, basic see CARBONATE WHITE LEAD.[41]

lead chrome green see CHROME GREENS.[41]

lead drier one of many ORGANIC lead salts which are soluble in paints and VARNISHES; was used to speed the DRYING and HARDENING of the OIL VEHICLE.[67,74,83] (DAC)

lead in oil basic CARBONATE WHITE LEAD ground in LINSEED OIL; formerly in wide use, now replaced by TITANIUM DIOXIDE PIGMENTS.[64,71]

lead monoxide see LITHARGE.[41]

lead naphthenate lead soap of NAPHTHENIC ACIDS.[74,83,67]

Formerly used as a DRIER.

lead nitrate $Pb(NO_3)_2 \cdot 2H_2O$. Crystalline salt soluble in water and was used in the manufacture of lead chromes.[144]

lead octoate lead soap of 2-ethylhexanoic acid.[74,83]

Formerly used as a DRIER.

lead oxide see LITHARGE.[41]

lead phosphite, dibasic $2\ PbO \cdot PbHPO_3 \cdot {}^1/_2\ H_2O$. Fine white acicular crystals. PIGMENT grade is the dibasic lead salt of phosphorous acid.[41]

Formerly used as an ANTI-CORROSIVE white pigment, and for its ability to eliminate bleed-through of CEDAR and redwood STAINS in LATEX paints. Density of 55.5 lb/gal. refractive index of 2.25, O.A. of 14 lb/100 lb, and particle size of 2.5 µm.

lead primer see RED LEAD.[41]

lead sulfide PbS. Mineral which occurs as galena and is one of the chief sources of lead.[144,34]

Words presented in CAP/SMALL CAPS type indicate that the word is defined in another part of the *Coatings Encyclopedic Dictionary*.
Numerical superscripts classify terms in one or more of the categories listed in the second section of this volume.

It was prepared by passing hydrogen sulfide gas into an ACID solution of LEAD NITRATE. It is the COMPOUND responsible for darkening of paints when a lead-pigmented paint was exposed to sulfur-containing gases.

lead titanate $PbTiO_3$. A yellow buff-colored PIGMENT which possessed good RESISTANCE to weather and CHALKING.[41]

Pale yellow solid; insoluble in water. Its sp gr is 7.52. Derivation: Interaction of oxides of lead and titanium at a high temperature. It contained lead sulfate and lead oxide as impurities.

lead white see CARBONATE WHITE LEAD.[41]

leaded zinc oxide mixed PIGMENT comprising ZINC OXIDE and basic lead sulfate.[41]

Several grades are possible and vary in the proportions of the individual ingredients. Leaded zinc oxides are either mechanical mixtures of the two ingredients, or are made by heating together roasted zinc ore, lead sulfide ore, and coal. They are no longer used because of their toxicity.

leaf green see CHROMIUM OXIDE GREEN.[41]

leafing the action of PARTICLES of certain METALLIC and other PIGMENTS in the form of thin flat flakes in aligning themselves with or FLOATING on or near the surface of the vehicle.[55]

Leafing occurs when such pigments are mixed with a suitable vehicle and applied as a coating.

ASTM Test Method D480 covers procedures for sampling, qualitative analysis, and physical testing of flaked aluminum powders and pastes (leafing and nonleafing) for coatings.

lecithin a liquid, obtained in refinement of soya beans or cottonseed; used in paints to promote PIGMENT WETTING and to control pigment SETTLING and FLOW properties.[64,83,127] (DAC)

legs the stringy effect that is apparent when cemented surfaces are separated shortly after the bond is made.[79,140]

Long legs or strings are indicative often of a weak bond, whereas short legs indicate a strong bond.

leipsic yellow see CHROME YELLOW, LIGHT and PRIMROSE.[41]

Leithner's blue see COBALT BLUE.[41]

LEL see LOWER EXPLOSIVE LIMIT.[172]

LEL meter a device used to measure the concentration of combustible vapors in an air sample.[172,145] (EPA)

lemon chrome see CHROME YELLOW, LIGHT and PRIMROSE.[41]

lemon yellow see BARIUM YELLOW.[41]

lemongrass, oil of see CITRONELLA OIL.[83]

length property of an INK whereby it can be stretched out into a long thread without breaking.[168]

length out see PAINT BRUSH LENGTH CLEAR.[59](ASTM)

letdown (1) the PROCESS of paint manufacturing in which the PIGMENT PASTE (MILL BASE) is reduced (let down) by the addition of the remaining ingredients of the formula;[131,49] (2) reducing the intensity or depth of a colored pigment through the addition of a white, or sometimes colorless pigment.[41]

letterpress a PROCESS of TYPOGRAPHIC (raised type) PRINTING.[168]

In this method of printing, the image inked areas are raised above the background area. The material which transfers the image is made of METAL or other relatively hard material.

letterset see DRY OFFSET.[168]

leveler (1) group of steel ROLLERS that bends any edges that are not flat prior to coating;[59,70] (BSI) (2) a PROCESS in the ALUMINUM industry that flattens strip so that the THICKNESS is the same from one side to the other.[45,131]

leveling the measure of the ability of a wet coating to FLOW out to a smooth dry film after APPLICATION so as to obliterate any surface irregularities such as BRUSH marks, roller marks, ORANGE PEEL from spraying, peaks, or CRATERS which have been produced by the mechanical process of applying the film.[166,42] (ASTM) See FLOW.[166]

Leveling leads to uniformity of the surface of the coating. Uniformity favors HIDING, imaging, dirt release and DURABILITY with respect to CORROSION, CRACKING, and BLISTERING. BRUSH marks and other imperfections left in the film are normally undesirable particularly in SEMIGLOSS and GLOSS finishes where these irregularities are more readily apparent.

Leveling integrates the effects of gravity, film thickness, drying time, rheology, and surface tension. It also varies with humidity, air flow, temperature, porosity of the substrate, and on the orientation of the painted surface, viz., horizontal up, horizontal down, or vertical, as well as the direction of the brush strokes on a vertical surface.

ASTM Test Method D4062 covers the laboratory determination of the relative leveling or water and solvent-reducible architectural paints by comparing the ridges produced in a drawdown film to a series of plastic leveling standards.

levigation PROCESS of mixing a PIGMENT with water and washing.[41,131]

The percentage of SALTS and heavy PARTICLES is reduced or eliminated by controlled SEDIMENTATION.

leyden blue see COBALT BLUE.[41]

licanic acid $CH_3(CH_2)_3(CH{=}CH)_3(CH_2)_4COCH_2COOH$. Constitutes 75 to 80% of the FATTY ACIDS of OITICICA OIL; mp, 75°C.[63,144]

Lieberman-Storch Test procedure for the qualitative detection of ROSIN in VEHICLES. (ASTM)

The rosin may be present as either free rosin (ABIETIC ACID), esterified rosin, or METAL SALTS. In Europe, the Lieberman-Storch-Morawski test is more commonly used.

ASTM Test Method D1542 covers procedures for the qualitative detection of rosin in varnishes by the Lieberman-Storch Test and the Halphen-Hicks Test. The rosin may be present as either free rosin (abietic acid), esterified rosin, or as metal salts.

lifting a recoating and LAPPING APPLICATION problem sometimes encountered with ASSOCIATIVE THICKENERS, particularly the HEUR types, which result in tearing and pile-up at the lap.[42,56,77,] Syn: picking[42,56,77,]

Lifting occurs when the paint is applied over a previous coat of a different LATEX paint, one that contains a different THICKENER from that used in the TOPCOAT. The new coat applies evenly at first, but when an attempt is made to lap back into the freshly applied paint, it is wiped off, apparently having no adhesion to the SUBSTRATE. The effect is even more dramatic when the paint is applied by ROLLER; it goes on smoothly over the dry substrate, but when the roller is passed over the freshly painted area, as when FEATHERING in an adjacent section, the wet paint is picked up by the roller.

The lifting problem is due to an interaction between the thickener or other colloidal materials (such as latex stabilizers) in the dried film that is being recoated and the new topcoat that contains a different thickener. If the same thickener is used in both the old and new paints, no problem occurs. The dried paint that is being recoated must be a water-based paint; no problem has been observed in painting over solvent-based paints. The basecoat or surface beneath it must be impermeable to water; if the basecoat and surface are sufficiently porous, no appreciable lifting is seen.

The mechanism of lifting appears to be that phase separation takes place in the wet topcoat at the interface with the dry paint being recoated. The LATEX and pigment appear to flocculate and draw back from the interface, leaving a clear layer of fluid, which prevents adhesion to the topcoat. If the substrate is porous this clear liquid wicks into it, and adhesion is maintained.

light (1) ELECTROMAGNETIC RADIATION of which a human observer is aware through the visual sensations that arise from the stimulation of the retina of the eye. *This portion of the SPECTRUM includes WAVELENGTHS from about 380 nm to 770 nm. Thus, it is incorrect to speak of ultraviolet "light" because the human observer cannot see radiant energy in the ultraviolet region;*[43,69] (2) adjective meaning high reflectance, transmittance, or level of ILLUMINATION as contrasted to dark, or low level of intensity.[43,35]

light source an object which emits light or RADIANT ENERGY to which the human eye is sensitive.[43,69] See ILLUMINANT, and CORRELATED COLOR TEMPERATURE.[43,69]

The emission of a light source can be described by the relative amount of energy emitted at each WAVELENGTH in the visible SPECTRUM, thus defining the source as an ILLUMINANT, or the emission may be described in terms of its CORRELATED COLOR TEMPERATURE.

lightfastness (1) the relative degree of change or lack of change in COLOR of materials exposed to the same amount and character of light. Thus, LIGHTFASTNESS is a relative term dependent on conditions which must be defined;[43] (2) ability to withstand color change on EXPOSURE to light.[69,72] (ASTM) Syn: colorfastness, fastness, fast to light, and color retention

ASTM test methods for lightfastness (colorfastness, color retention) include: discoloration: microbiological discoloration, D3456; clear coatings in sunlight, D2620; white architectural enamels, D1543; fading: pigments in artists' paints, D4303; printed matter, D3424.

lightness (1) achromatic dimension necessary to describe the three-dimensionality of COLOR, the others being HUE and SATURATION. *Sometimes the lightness dimension is called "BRIGHTNESS." In the MUNSELL COLOR ORDER SYSTEM, the lightness dimension is called "VALUE."* See LUMINOSITY CURVE;[43,69] (2) perception by which white objectives are distinguished from gray, and light objects from dark ones.[43,69]

lignin chief noncarbohydrate constituent of WOOD, a POLYMER that functions as a natural BINDER and support for the CELLULOSE fibers of woody plants.[153,167]

It is one of the principal constituents of the woody cell walls, whose chemical composition is still unknown; the residue after treatment of solvent-extracted wood with strong mineral acids.

ligroine saturated volatile fraction of PETROLEUM boiling in the range 20-135°C (68-275°F).[164] (ASTM)

Words presented in CAP/SMALL CAPS type indicate that the word is defined in another part of the *Coatings Encyclopedic Dictionary.*

Numerical superscripts classify terms in one or more of the categories listed in the second section of this volume.

It is used as a SOLVENT, mostly in the laboratory. The term "ligroine" should be used in place of "benzine" or "petroleum ether."

lime CALCIUM OXIDE (CaO), or a mixture of calcium oxide and magnesium oxide (MgO); also, loosely, a general term for the various chemical and physical forms of QUICKLIME, HYDRATED LIME, and HYDRAULIC HYDRATED LIME.[144] (ASTM) cf. LIMESTONE[41,58] See CALCIUM OXIDE.[144]

lime blue mixture of ULTRAMARINE and TERRA ALBA.[41]

Another type of lime blue is made from METHYLENE BLUE by ADSORPTION on natural earth.

lime green see GREEN, LIME.[41,68]

lime putty see PUTTY.[158,165]

lime red LAKE produced by adsorbing magenta on a natural earth.[41]

lime yellow LAKE produced by adsorbing auramine or other yellow DYESTUFF on a natural earth.[41]

limed rosin commercial CALCIUM RESINATE made by the direct interaction of LIME and ROSIN.[156,167]

limekiln a furnace used to reduce naturally occurring forms of CALCIUM CARBONATE to LIME.[59]

limestone see CALCIUM CARBONATE, NATURAL.[41,58]

limewashing coating with limewash made from HYDRATED LIME or by slaking QUICKLIME, to which tallow is sometimes added.[71,163] (BSI) Syn: whitewash, whitening

limiting viscosity number see INTRINSIC VISCOSITY.[130,166]

limonite see IRON OXIDES, NATURAL.[41]

line etching a PRINT made up of lines or PIGMENTED areas and lighter spaces free from SHADING.[168]

liner see LINING TOOL.[59]

lining fabrics muslin or CANVAS used underneath fine WALLPAPERS to avoid small cracks possibly opening up in a PLASTER wall and showing through.[148]

lining paper plain PAPER applied before the WALLPAPER.[148]

Assures a smoother surface and better ADHESION.

lining tool (Brit.) A small flat FITCH with a slanting edge, used for painting lines with the help of a rule.[59] Syn: liner

linkrusta a permanent WALLCOVERING coated with a WOOD FLOUR and LINSEED OIL mixture on a PAPER backing.[148]

Linkrusta patterns are molded rather than printed.

linoleic acid $C_{18}H_{32}O_2$. cis-9, cis-12 octadecadienoic acid.[63]

A FATTY ACID constituent of LINSEED and other DRYING OILS, where it occurs as a glyceride; mol wt of 280.44, bp of 230°C, iodine value of 181.1.

linolein glyceride of LINOLEIC ACID. It is one of the constituents of LINSEED OIL which induces the DRYING properties.[63]

linolenate driers certain METAL SALTS of, and SOAPS of, LINSEED FATTY ACIDS.[67,83]

linolenic acid $CH_3CH_2CH{=}CHCH_2CH{=}CHCH_2CH{=}CH(CH_2)_7CO{-}OH$. Tripoly unsaturated fatty acid component of LINSEED and other DRYING OILS.[63]

Properties: bp of 230°C/17 mm Hg, an acid value of 201.6, and an iodine value of 273.7.

linoleum and oilcloth varnishes special highly flexible and elastic VARNISHES.[152,156]

linoleum, floor and wall covering made from oxidized LINSEED OIL or combinations of DRYING OILS, wood flour and/or ground cork, resins, and pigment, rolled out and compressed onto an ASPHALT saturated felt, burlap, or other backing.[75]

Heat, which fuses and sets the oils and resins to form strong binding agents, is applied to the mixture during compression.

linoxyn semisolid, highly oxidized LINSEED OIL; used in the manufacture of LINOLEUM.[63]

linseed oil DRYING OIL from seeds of the flax plant (*Linum usitatissimum*).[63] (PTM)

The oil is refined by treatments which remove water and MUCILAGINOUS material and is then described as refined oil, according to the method of treatment. Further processing produces BOILED OIL, BLOWN OIL, or BODIED OIL. This best known and most widely used oil in the paint industry is characterized by its relatively short DRYING TIME. Its high degree of UNSATURATION, to which its good drying characteristics can be partially ascribed, is due to the presence of large percentages of linolenic and linoleic triglycerides. Many years ago the oil was obtained from seed by mechanical pressure, including both hydraulic presses and later expellers. In recent years the more modern SOLVENT EXTRACTION is used. OILS thus obtained show lower percentages of impurities and better overall quality. Linseed oil responds very readily to a variety of refining techniques and is used in the paint industry both as a drying oil and as an ingredient in a wide array of modified RESINS of many varieties.

ASTM specification for boiled linseed oil is D260; for raw linseed oil, D234.

lintel horizontal beam supported at each end and spanning an opening.[82]

Usually supports the structure above it.

liquid state of matter in which a substance exhibits a characteristic readiness to flow, little or no tendency to disperse, and relatively high incompressibility.[142,166,164,133]

ASTM test methods for liquids include: clarity/cleanliness, visual, D2090; color, by Gardner scale, D1544; density and specific gravity, D3505; liquid/solid state, characterization, D4359.

liquid driers solutions of soluble DRIERS in VOLATILE ORGANIC SOLVENTS, usually HYDROCARBONS.[83,74] (BSI)

liquid emulsion polymer coating SYNTHETIC RESIN EMULSION (actually LATEX) which produces hard, tough coatings having a high LUSTER.[60,71,130]

Usually nonbuffable or semibuffable, and referred to and sold as self-polishing floor WAX.

liquid resin (deprecated) See TALL OIL.[63,44]

liquid rosin synonym for tall oil.[44]

liquid solvent wax mixture of WAXES and other ingredients in a SOLVENT BASE.[137] Sometimes also called SOLVENT WAX.

Must be polished for LUSTER. Occasionally colored.

liquid water emulsion wax DISPERSION of WAX and other modifying materials in water.[60,71] Sometimes called self-polishing wax or water wax.

liquifying stress the ease with which concentrated inks can be liquified by remilling after STORAGE.[168]

litharge PbO. Oxide of lead was made by the controlled heating of METALLIC LEAD.[41,74,83] Syn: lead oxide, lead monoxide, lead protoxide, massicot, plumbous oxide

Pure litharge has sp gr of 9.53, mp of 888°C, and mol wt of 223.21. It was used as a RAW MATERIAL in the manufacture of PIGMENTS and DRIERS and infrequently as a CATALYST in paints.

lithographed paper WALLPAPER made by the PRINTING PROCESS used for billboards and posters.[148]

Lithography exploits the affinity of COLOR for the greasy material in which the design is put on the ROLLER. A transfer process.

lithographic chalk a greasy CRAYON, composed of SOAP, WAX, OIL, and LAMPBLACK, for drawing designs on a lithographic plate prior to ETCHING.[168]

lithographic inks INKS used in the LITHOGRAPHIC PROCESS.[168]

The principal characteristic of a lithographic ink is its ability to resist excessive emulsification by the fountain solution.

lithographic stones slabs of LIMESTONE, three or four inches thick, the surfaces of which are smoothed or grained, for use in LITHOGRAPHIC printing.[168]

lithographic varnishes heat-bodied refined LINSEED OILS used as VEHICLES for PRINTING and LITHOGRAPHIC INKS.[168,152]

lithography reproducing graphic PROCESS in which the image configuration to be PRINTED is rendered on a flat surface, as on a stone or now chiefly on sheet zinc or aluminum, and treated so that it will retain ink while the nonimage areas are treated to repel ink and to accept FOUNTAIN SOLUTION.[54,77,81]

lithol red $C_{20}H_{13}N_2O_4SM^*$. (M^* = Na, Ba or Ca). Pigment Red 49 (15630). PIGMENT made by combining the INTERMEDIATES, TOBIAS ACID and BETA-NAPHTHOL. This type of red is available as sodium, BARIUM, and CALCIUM TONERS, also LAKES; the sodium is the lightest shade, the barium is what may be termed a medium shade, and the calcium lithols are deep reds and maroons.[41]

lithol rubine Pigment Red 57 (15850). AZO PIGMENT made by diazotizing paratoluidine-meta-sulfonic acid, and coupling with 3-hydroxy-2-naphthoic acid.[41]

Pigment is used as the calcium salt.[41] Most frequently encountered in the resinated form, it is characterized by its distinctly brilliant deep MASSTONE and its blue intense TINT. Because of its low HIDING POWER, it is generally used in combination with other pigments, such as molybdate orange, in bright, low-cost industrial red finishes which do not require good light and/or soap and ALKALI FASTNESS. Typical of the METAL precipitated azo pigments, it should not be used in paints which are either highly alkaline or highly acid; to do so invites trouble with BLEED, MIGRATION, and possibly, BLOOM.

lithopone $BaSO_4{\cdot}ZnS$. Pigment White 5 (77115). A mixed white PIGMENT with approximately equal molar parts of barium sulfate and zinc sulfide.[41]

Density, 4.3 g/cm³ (35.8 lb/gal); O.A., 11-17; particle size, 0.4-0.6 µm.

litre the volume of a kilogram of water at 4°C, equal to 1.06 quarts or 61.02 cu. in.[128]

live centers point at the very center at each end of the roll.[59,70]

This is usually a center point of the journal at which the roll turns. The roll is driven from this

Words presented in CAP/SMALL CAPS type indicate that the word is defined in another part of the *Coatings Encyclopedic Dictionary*.
Numerical superscripts classify terms in one or more of the categories listed in the second section of this volume.

point and is considered concentric from the same point.

live edge see WET EDGE.[54]

livering the progressive, irreversible increase in CONSISTENCY of a PIGMENT-VEHICLE combination.[166,156,161,42] British syn: feeding; cf. GELLING[166,161,42]

Livering in the majority of cases arises from a chemical reaction of the vehicle and the solid dispersed material, but it may also result from POLYMERIZATION of the vehicle. The irreversible character of the changes in the "livered" material distinguishes it from thixotropic "build-up," which is reversible.

lixivation to extract a soluble constituent from a solid mixture by washing or percolation.[131]

loading (of coated abrasives) see FILLING OF COATED ABRASIVES.[59,158]

logu sluggish, low snap or recovery.[67,140]

A condition formed in poorly CURED or overloaded VULCANIZERS.

long oil high ratio of OIL to RESIN in a medium; (1) long oil alkyd. An alkyd resin containing more than 60% of oil as a modifying agent; (2) long oil varnish. An OLEORESINOUS VARNISH, other than alkyd, containing more than 25 gal of oil per 100 lb of resin.[63,152,156] See SHORT OIL VARNISH, MEDIUM OIL VARNISH.

A long oil varnish is usually slower drying, tougher, and more elastic than a short oil varnish.

loss modulus "G"—the component of applied SHEAR STRESS which is 90° out of phase with the SHEAR STRAIN, divided by the strain.[46] (ASTM) Syn: viscous modulus, hysteretic modulus, imaginary modulus, out-of-phase modulus

loss of drier see DRIER DISSIPATION.[74,168,42,161,67]

loss of gloss a paint DEFECT in which a DRIED FILM of paint loses GLOSS, usually over a period of several weeks.[56,65] (DAC)

lot the quantity of liquid paint or coating produced in the final mixing operation after all the production processes are complete.[128,49] (ASTM) cf. BATCH[128,49]

lot number the number used by the manufacturer to identify an entity of production.[128,49] (ASTM) See BATCH.[128,49]

louvers slats placed at an angle, as in shutters.[82]

Lovibond color system a system of COLOR SPECIFICATION by means of numbers proportional to the OPTICAL DENSITIES of three GLASS filters (yellow, red, and blue) required to modify a standard light source (such as daylight or incandescent lamp light) to produce a COLOR MATCH.[43,69]

The carefully calibrated glasses should be used as described by the manufacturer, in order that results in different laboratories will agree. Lovibond specifications have long been used as the basis for describing the colors of edible OILS and vegetable oils used for paint vehicles.

Lovibond tintometer OPTICAL comparison instrument for the use of Lovibond glasses.[43,69,145] See LOVIBOND COLOR SYSTEM.[43,69]

low gloss see GLOSS.[43]

low lustre see GLOSS.[43]

low semigloss see GLOSS.[43]

low solvent coating a coating which contains a lower amount VOLATILE ORGANIC COMPOUND (VOC) than conventional organic SOLVENT-BORNE coatings.[172] (EPA)

Low solvent coatings usually fall into the three major groups of HIGH-SOLIDS, WATERBORNE, or POWDER coatings.

low temperature coalescence the ability of a coating to form a film at less than ideal temperature conditions.[55,157] See MINIMUM FILM FORMING TEMPERATURE (MFFT).[55,157,145]

lower explosive limit (LEL) the concentration of a COMPOUND in air below which a flame will not propagate if the mixture is ignited.[172] (EPA) See LEL METER.

lowest achievable emission rate (LAER) pursuant to section 171(1) of the CLEAN AIR ACT, LAER is that rate of emissions which reflects: (a) the most stringent emission limitation which is contained in the implementation plan of any state for such class or category of sources, unless the owner or operator of the proposed source demonstrates that such limitations are not achievable; or (b) the most stringent limitation which is achieved in practice by such class or category of source, whichever is more stringent.[172] (EPA)

In no event, however, shall the application of this term permit a proposed new or modified source to emit any pollutant in excess of the amount allowable under applicable new source standards of performance. Areas of the country that have not attained national ambient air quality standards may require LAER on new sources of pollution.

lubricant a substance, such as GREASE or OIL, that reduces FRICTION when applied to surface coating to moving parts.[83]

lumbang oil this OIL is obtained from the nuts of the tree, *Aleurites moluccana*.[63] (PTM) Syn: candlenut oil

Although a product of an Aleurites tree, it contains no elaeostearin. It dries somewhat better than SOYBEAN OIL.

lumber the product of the saw and planing mill not further manufactured than by sawing, resawing, passing lengthwise through a standard planing machine, crosscutting to length, and matching.[153,75] (ASTM)

lumber, edge-grained LUMBER so sawed that the annual rings form an angle of 45 to 90° with the surface of the piece.[153] (ASTM)

lumber, flat-grained LUMBER so sawed that the annual rings form an angle of less than 45° with the surface of the piece.[153] (ASTM)

lumber, matched LUMBER that is shaped to make a close tongued and grooved joint at its edges or ends.[153] (ASTM)

lumber, resawn rough or surfaced LUMBER that has been divided into two or more thicknesses by sawing.[153] (ASTM)

lumber, rough undressed LUMBER as it comes from the saw.[153] (ASTM)

luminance the LUMINOUS intensity (BRIGHTNESS) of light reflected or transmitted by a material in a given direction per unit of projected area of the material, as viewed from that direction.[43,69]

luminescence the emission of light by a molecule or ion which is a temperature below those required for incandescence. Luminescence can originate from electronically excited states of molecules or ions.[43] (IUPAC)

If the excited state is short lived, as is the case for excited singlet to ground state singlet electronic transitions, the emission follows excitation within a short time period (about 10^{-9} to 10^{-6} s commonly) and is called FLUORESCENCE. *The fluorescence from polycyclic aromatic compounds provides a useful means of identification and quantification of these species. If the emitting state undergoes a spectroscopically forbidden transition, for example, an excited triplet to ground state singlet electronic transition, then the emission may be delayed (greater than about 10^{-3} s) following the excitation; this emission is called* PHOSPHORESCENCE. *Some molecules (e.g., butanedione) give strong phosphorescence which is useful in identification.* CHEMILUMINESCENCE *is the emission which results from an excited state formed in a chemical reaction which does not involve light* ABSORPTION; *for example, luminol oxidation by H_2O_2 generates chemiluminescence which can be used for analytical purposes.*

luminescent emitting light not due to high temperature, usually caused by excitation by rays of a shorter WAVELENGTH.[43]

luminosity curve, CIE graph representing the LUMINOUS intensity of WAVELENGTHS in the visible SPECTRUM relative to the maximum intensity at the same wavelength.[43,69]

The relative luminosity curve is described by the Y color matching function in the CIE system. The shape of the curve depends on the area of the retina being stimulated and on the intensity of the incident light. Thus, the CIE LUMINOSITY CURVES *for the 2° and 10° observers are slightly different, and those for daylight vision, i.e., photopic vision, and for dark adapted vision, i.e., scotopic vision, are extremely different.*

luminous (1) adjective used to imply dependence on the spectral response characteristic of the STANDARD OBSERVER defined in the CIE SYSTEM. *Thus, the luminous reflectance or the luminous transmittance is described by the Y* TRISTIMULUS VALUE *in the CIE system. The adjective is applied to many measures of light, such as intensity, density, etc., and always indicates that the measures are weighted for the relative luminous sensitivity of the human observer*; (2) material emitting, or appearing to be emitting, visible radiant energy.[43,69]

luminous directional reflectance REFLECTANCE of a surface for specified directions of ILLUMINATION and view is the ratio of the BRIGHTNESS of the surface to the brightness that an ideally diffusing, perfectly white, surface would have if ILLUMINATED and viewed in the same manner.[43,69]

luminous flux see FLUX, LUMINOUS.[43]

luminous paint paint which exhibits FLUORESCENCE; (1) fluorescent paint: contains PIGMENTS which are capable of absorbing energy from the blue or ultraviolet end of the SPECTRUM and reemitting it in the form of light in the visible WAVELENGTHS. *A* FLUORESCENT PAINT *ceases to 'glow' if the activating source is removed*; (2) phosphorescent paint: contains pigments (phosphors) which absorb energy at one wavelength and emit it over a period of time, in the form of light at a longer wavelength in the visible spectrum. *It differs from a fluorescent paint in that it continues to glow after the stimulating source has been removed*; (3) radioactive or self-luminous paint: normally, this is a phosphorescent paint containing a portion of radioactive compounds, and in such a paint the phosphor is permanently activated by absorbing energy from the bombardment by the radioactive rays and emits light in the visible spectrum.[43,71,69] (BSI)

luminous pigments PIGMENTS used in paints to

make the surface visible in the dark and in coatings for electronic purposes.[41] See LUMINANCE[43,69] and LUMINOUS PAINT.[43,69,71]

luster (1) type of surface REFLECTANCE, or GLOSS, where the ratio of specular reflectance to diffuse reflectance is relatively high, but not so high as that from a perfect specular reflector (mirror);[43,69,35] (2) synonym for gloss.[43,35]

lusterless see DEAD FLAT.[43]

lyophilic characterizing a material which readily goes into COLLOIDAL SUSPENSION; if into water, it may be termed HYDROPHILIC.[139]

lyophobic characterizing a material which exists in the COLLOIDAL state without any significant affinity for the medium.[139]

M

m.k.s. system a system of units derived from the metre, kilogramme, and second.[128] cf. c.g.s. system[128]

Now superseded for scientific purposes by the SI units, which are based on the m.k.s. system.

MAC see maximum allowable concentration.[172]

MacAdam color difference equation a color difference equation developed by David MacAdam which is now used as modified by Hugh Davidson and Fred Simon to incorporate the effect of lightness on the chromaticity differences:

$$\Delta E = [1/K (g_{11} \overline{\Delta x}^2 + 2g_{12}\Delta x\Delta y + g_{22}\overline{\Delta y}^2 + G\overline{\Delta Y^2}]^{1/2}$$

where g_{11}, $2g_{12}$, and g_{22} are constants depending on the chromaticity coordinates, x and y, and K and G are constants depending on the luminous reflectance or transmittance, Y. This color difference is frequently calculated from charts prepared by Simon and Goodwin, which have the required constants built-in.[43,69]

MacAdam limits the theoretical limit or gamut of colors which can be obtained at various limits of luminance (Y).[43,69]

Thus, the gamut of colors which can be obtained theoretically decreases steadily as the luminance (Y) increases.

machine a device consisting of fixed and moving parts that modifies mechanical energy and transmits it to a more useful form.[59]

machine-printing the method by which the bulk of modern wallpapers are produced.[59,168]

Machine-printing employs a rotary press and a series of cylinders or rollers to turn out wallpaper at high speeds. Raw paper stock is first given a coating of the ground color by a special machine, after which the paper proceeds in a continuous web to the rotary press where the top colors are applied, and it is then festooned on specially heated drying racks. At the end of the line, the paper is rolled up and chopped into commercial rolls. Printing rollers are of three kinds: brass and felt, routed out, or engraved.

Mackey test test designed to measure the tendency of oxidizable oils to combust spontaneously.[157]

macromolecule high molecular weight molecule approaching the size of a colloidal particle.[130,139] See polymer and high polymer.[130]

madder lake lightfast, nonbleeding, red-colored pigment prepared from the coloring matter of madder root.[41]

Magdala red $C_{30}H_{21}N_4Cl$. Red dyestuff.[41] Syn: naphthalene red

magnesia MgO. (1) Magnesium oxide; (2) sometimes used incorrectly in the printing ink industry to mean magnesium carbonate.[41,168]

magnesite $MgCO_3$. Mineral, magnesium carbonate, principally used as a filler or extender.[41,58,34]

magnesite floors hard composition floors in which magnesium oxychloride is the binder.[75]

This binder is formed in laying the floor, when magnesium oxide is combined with a strong solution of magnesium chloride. Fillers which may be added to this binder are: asbestos, cork, sand, wood flour, marble dust, talc, leather, etc. This great variety of fillers produces magnesite floors having variable porosity, resiliency, appearance, and durability.

magnesium carbonate, precipitated chemically, this is the same as magnesite, but physically it has a much better color, in bulk being a very intense white.[41,58]

It is usually a very fine light powder of rather high oil absorption.

magnesium silicate, fibrous $3\,MgO{\cdot}2SiO_2{\cdot}2H_2O$. A fibrous chrysotile mineral white to gray powder chemically inert; high oil absorption; and has a high water demand.[41,34] Syn: asbestos, chrysotile, fibrous asbestos

Properties: density, 2.48-2.56 g/cm³ (20.7-21.3 lb/gal); O.A., 50-180.

magnesium silicate, nonfibrous $3MgO{\cdot}4SiO_2{\cdot}H_2O$. Pigment White 26 (77718). A hydrated magnesium silicate extender or filler of wide range of compositions. Soft white, gray, or yellow shade.[41,58] Syn: talc, asbestine

Natural product (talc) used in paint, rubber, ceramics, paper, and roofing compounds. Properties: density, 2.7-2.8 g/cm³ (22.5-23.3 lb/gal); O.A., 30-50; particle size, 0.5-2.5 µm.

ASTM test methods for magnesium silicate include: analysis, D717; specification, D605.

magnesium soaps saponification products of magnesium and various fatty acids.[83,127]

magnet wire wire used in equipment such as electric motors, generators, and transformers.[61,172] (EPA)

Words presented in cap/small caps type indicate that the word is defined in another part of the *Coatings Encyclopedic Dictionary.*

Numerical superscripts classify terms in one or more of the categories listed in the second section of this volume.

Magnet wire is coated electrically insulating VARNISH, ENAMEL, or very tough, SOLVENT RESISTANT, CROSSLINKED LATEXES.

magnet wire coatings those used to coat MAGNET WIRE.[71]

These coatings require an unusual combination of properties including high SOLVENT RESISTANCE, HEAT RESISTANCE, and ABRASION RESISTANCE, with TOUGHNESS and FLEXIBILITY. They are used as primary insulation on magnet wire to be used in the manufacture of core wound electric devices such as armatures, electromagnets, relays, coils, and the like.

magnetic inks INKS made with PIGMENTS which can be magnetized after printing and the printed characters later recognized by electronic reading equipment.[168]

magnetite black Fe_3O_4. Magnetic IRON OXIDE.[41] See BLACK IRON OXIDE.[41]

mahlstick long stick, padded at one end, on which a painter can rest his or her hand to steady it when working.[59,81]

mahogany sulfonates SOAPS, the sodium SALTS of SULFONIC ACIDS from PETROLEUM refining sludge; used in SYNTHETIC RESIN production, as are sorbitan oleates and laurates, polyoxyethylene esters.[127]

maintenance paints coatings used to maintain bridges, manufacturing plants, offices, stores, and other commercial structures, hospitals and nursing homes, schools and universities, government and public buildings, and both building and nonbuilding requirements in such areas as public utilities, railroads, roads, and highways, and including industrial paint, other than the original coating, the primary function of which is protection.[71] See INDUSTRIAL MAINTENANCE PAINTS.[71]

Residential maintenance is excluded.

makeup air fresh air drawn in from outside the plant to replace the dirty SOLVENT-laden air from a coating operation or drying oven that is exhausted from the building.[172] (EPA)

This makeup air must be heated to room temperature or even oven temperature, thus accounting for a large energy usage.

makeup solvent portion of SOLVENT required to compensate for the amount lost, normally via EVAPORATION, during the manufacturing process.[164,172] (EPA)

malachite $CuCO_3 \cdot Cu(OH)_2$. Basic carbonate of copper which occurs naturally.[41,34]

The COLOR varies from a bright emerald to a dark green.

malachite green bluish-green DYESTUFF made from dimethyl aniline and benzaldehyde.[41]

It is often sold in the form of its oxalate.

maleic acid $COOH(CH)_2COOH$. Dibasic acid used in the manufacture of SYNTHETIC RESINS.[130,144]

maleic anhydride unsaturated anhydride used in manufacturing SYNTHETIC RESINS and MALEINIZED OILS.[130,167]

Mp, 56°C; bp, 202°C; acid value, 1.143. It undergoes the Diels-Alder reaction.

ASTM test methods for maleic anhydride include: color, by platinum cobalt scale, D3366; maleic acid content, D2930; sampling and handling, practice, D3438; specification, D3504.

maleic anhydride value see DIENE VALUE OR NUMBER.[63,128]

maleic ester resin a SYNTHETIC RESIN made from MALEIC ACID or MALEIC ANHYDRIDE and a polyhydric ALCOHOL.[130]

maleic resin a RESIN made from a natural resin and MALEIC ANHYDRIDE or MALEIC ACID.[130,167] (ASTM)

maleic value synonym for diene value.[128]

maleinized oil OIL which has been reacted, through its double bonds, with MALEIC ANHYDRIDE.[63]

mandrel bend test test for determining the FLEXIBILITY and ADHESION of surface coatings, so named because it involves the bending of coated METAL panels around mandrels.[59,157,47]

Coatings attached to SUBSTRATES are elongated when substrates become elongated when they are bent during the manufacture of articles or when the articles are abused in service.

ASTM Test Method D522 covers the determination of the resistance to cracking (flexibility) of attached ORGANIC coatings on substrates of sheet METAL or RUBBER-type materials.

manganese black MnO_2. Manganese dioxide.[41,69]

A black PIGMENT. Principal uses are as a DRIER and as a COLORANT for ceramics.

manganese brown there are two types of manganese brown: (1) burnt turkey umber; (2) a brown oxide PIGMENT, made artificially as a by-product from chlorine manufacture.[41]

manganese driers these include manganese dioxide, the hydrated oxide, manganese acetate, sulfate, and borate. The ORGANIC DRIERS are SALTS of various organic acids such as naphthenic or 2-ethyl hexoic.[74,83,67]

Manganese driers are characterized by their reddish-brown colors and their surface drying activity.

ASTM Test Method D2375 covers the titrimetric determination of manganese in liquid paint driers that can be dissolved in a toluene-alcohol mixture and utilizes EDTA.

manganese green strong green PIGMENT, with good ALKALI RESISTANCE, prepared by roasting manganese dioxide and barium hydroxide together under oxidizing conditions.[41]

manganese violet see MINERAL VIOLET.[41]

Manila copal natural RESIN, two types of which are used in VARNISH manufacture, in which they are described as hard and soft manilas.[167]

The hard type requires running and is used to some extent in OIL VARNISHES. The soft type is readily soluble in industrial ALCOHOL and forms the basis of spirit PAPER and other air-drying varnishes; the native name for the soft type, obtained by tapping, is MELENGKET.

manjak intense black, naturally-occurring ASPHALTUM, obtained from Barbados. It differs from other asphaltums used in the trade by reason of its unusual STAINING POWER and difficult solubility.[76] Syn: glance pitch

Prolonged high temperature treatment is necessary in order to effect a reasonable solution in DRYING OILS, and a substantial amount of mineral matter always remains undissolved. It is used alone, or with GILSONITE, in black BITUMINOUS finishes of many types.

mannite see MANNITOL.[164,130]

mannitol $C_6H_8(OH)_6$. A hexahydric ALCOHOL which has been used in the production of SYNTHETIC OILS and ALKYD RESINS.[144,164,130] Syn: mannite

Properties: Bp, 278°C/1 mm Hg; mp, 166°C.

manufacture to make or process (a raw material) into a finished product, especially by means of a large-scale industrial operation.[49]

manufacturer's formulation a list of substances or component parts of coatings as described by the maker of the coatings.[172,49] (EPA)

This may be used, in many cases, to calculate the VOC of a coating.

mar resistance ability of a coating to resist damage caused by light abrasion, impact, or pressure.[47,142]

ASTM Test Method D5178 covers the determination of the mar resistance of coatings on smooth, flat surfaces. Results are expressed in terms of force-to-mar films of ORGANIC coatings.

marble flour see CALCIUM CARBONATE, NATURAL.[41]

marbling, marbleizing imitating with finishing materials, as in ANTIQUING, the figure and texture of polished marble or other DECORATIVE stones, usually by stippling or MOTTLING in conjunction with GRAINING, scratching, and SPATTERING.[54,148,175]

margaric acid see DATURIC ACID.[144]

marine borers mollusks and crustaceans that attack submerged WOOD in SALT and brackish water.[65,153]

marine coatings paints and VARNISHES used on ships and boats and specifically formulated to withstand water immersion and EXPOSURE to marine atmosphere.[71,152,163] See SPAR VARNISH.[152]

marine varnishes see MARINE COATINGS and SPAR VARNISH.[71,152,163]

marking nut oil OIL that resembles cashew nut shell liquid in that it is PHENOLIC and quite unlike the GLYCERIDE VEGETABLE OILS.[63,156] Syn: dhobi marking nut oil, or bhilawan oil

marouflage to glue a canvas to a wall which is to be covered by a MURAL PAINTING.[81,159]

marquetry inlaid work, as in FURNITURE; work inlaid with wood, shells, ivory, etc.[148,175]

martius yellow calcium derivative of NAPHTHALENE YELLOW.[41]

masking temporarily covering that part of a surface to which it is not desired to apply a coating.[54,158]

masking tape ADHESIVE backed PAPER tape used to mask or protect parts of a surface not to be finished.[59,158,54]

masonry the art of the mason in shaping, arranging, and uniting stone, brick, building blocks, tile, and similar materials, to form walls and other parts of a building.[73,75,82,159]

masonry conditioner a SOLVENT-BASED, PIGMENTED PRIMER coating formulated to have great PENETRATING power so as to prepare MASONRY (especially chalky STUCCO) to receive FINISH coats.[73,158]

Particularly important under LATEX paints.

masonry paint an ALKALI-RESISTANT coating, usually a LATEX paint, used for MASONRY SUBSTRATES.[71]

mass measure of the quantity of matter that a body or object contains.[128]

The mass of a body is not dependent on gravity and therefore is different but proportional to its weight.

mass spectrometry an instrument used in gas analysis which operates by ionizing gaseous atoms or molecules, generally with electrons of relatively low energy.[37,145] (IUPAC) See SPECTROMETRY, OPTICAL.[37,145]

Words presented in CAP/SMALL CAPS type indicate that the word is defined in another part of the *Coatings Encyclopedic Dictionary*.

Numerical superscripts classify terms in one or more of the categories listed in the second section of this volume.

The parent ions and ion fragments produced are accelerated electrically into a mass separator (involving electric and magnetic fields) which separates ions of specific mass to charge ratio.

masscolor the COLOR, when viewed by REFLECTED light, of a PIGMENT-vehicle mixture of such thickness as to obscure completely the background.[69,43] (ASTM) Syn: overtone or MASSTONE

massicot (massocot) synonym for lead monoxide.[41,144] See LITHARGE.[41]

masstone (1) a PIGMENT-VEHICLE mixture which contains a single pigment only; (2) occasionally, this term is used more loosely to describe a pigment-vehicle mixture which contains no white pigment.[69] See MASSCOLOR.[69]

masstone color the COLOR of a MASSTONE paint applied at complete HIDING.[69]

master batch a convenient means for handling small amounts of critical ingredients like COLOR PIGMENTS or CATALYSTS in higher CONCENTRATION (than that in which it occurs in normal mix) for subsequent LETDOWN (DILUTION) with the remainder of the coating ingredients.[49]

mastic (1) ADHESIVE composition. Loosely used to describe a PLASTIC FILLER, stopper, putty, or adhesive. Not to be confused with GUM MASTIC, which is a natural RESIN used for picture VARNISH;[79,165](BSI) (2) sometimes, high build coating.

masticate to work RUBBER on a mixing MILL or in an internal mixer until it becomes soft and PLASTIC.[140,131] Syn: breakdown

mastication of resins PROCESS of hot working of RESINS which is believed to reduce molecular complexity and to confer SOLUBILITY.[131,167,176]

It has been used on natural COPALS, and advantages of this treatment are that thermal cracking is avoided and pale colors maintained. Patented methods include treatment between rotating rollers, sometimes in the presence of solvents.

mat finish see GLOSS.[43]

match, conditional see CONDITIONAL MATCH.[69]

match, nonconditional see NONCONDITIONAL MATCH.[69]

matching act of hanging strips of WALLCOVERING so that the design will be in correct relation to the preceding strip.[148]

The types of match are as follows: (1) a random match—the simplest type of all. The pattern will look well no matter how one panel is placed in position in relation to the next one. Stripes, all over textures, and grasscloth patterns are usually random matches; (2) a straight match—one where the pattern on one strip must be joined with a portion of a pattern on another strip in order to complete the pattern across the wall. After being hung, the design will be seen repeated, from strip to strip, at the same height from the ceiling line; (3) a drop match—has a portion of the pattern on each strip like the straight match. However, the pattern will not repeat itself at the same distance from the ceiling line across the various strips on the wall. Instead, the matching point will apparently drop lower on each succeeding strip, but the pattern will be seen to match at the same horizontal points on alternate strips: the first and third, the second and fourth, and so on. Since there is a certain amount of waste with drop matching of large-scale patterns, professional paperhangers measure and cut adjacent strips from different rolls of the wallcovering, and plan to alternate them. Small, left-over pieces can often be matched to go above doorways or under windows. Elaborate patterns occasionally match at every third or fourth strip of paper but this, the multiple drop match, is quite rare.

matching, color act of making one material appear to match another COLOR.[69]

If the achieved match is dependent on the conditions of ILLUMINATION and viewing, the match is termed CONDITIONAL or METAMERIC. If the achieved match is independent of the quality of the illuminant viewer, or viewing conditions, the match is termed NONCONDITIONAL or NONMETAMERIC.

material balance a calculation based on conservation of mass, i.e., the amount of material going into a PROCESS is equal to the amount which leaves the process.[131,164] (EPA)

This relationship is often used to estimate SOLVENT losses from coating operations.

Material Safety Data Sheet (MSDS) OSHA's established guidelines for the descriptive data that should be concisely provided on a data sheet to serve as the basis for written hazard-communication programs. The thrust of the law is to have those who make, distribute, and use hazardous materials be responsible for effective communication.[172]

The purpose of an MSDS is to inform the user about: the material's physical properties or fast-acting health effects that make it dangerous to handle; the level of protective gear that you need; the first aid treatment to be provided when one is exposed to a hazard; the preplanning needed for safely handling spills, fires, and day-to-day operations; and how to respond to accidents.

A comprehensive MSDS access, storage, and retrieval system on CD-ROM called Panorama® is available from the FSCT.

matte See GLOSS. Also spelled mat.[43]

matte varnish see FLAT VARNISH.[152,163,58,71]

maturing PROCESS by which CLARITY, BRIGHTNESS, working properties, etc., are improved by storing VARNISH in tanks.[49,131] (BSI) See AGING.[157,161]

maximum allowable concentration (MAC) the maximum CONCENTRATION of a pollutant which is considered harmless to healthy adults during their working hours, assuming they breathe uncontaminated air at all other times.[172] (IUPAC)

MBK see METHYL BUTYL KETONE.[164]

MDI abbreviation for DIPHENYLMETHANE DIISOCYANATE.[141] cf. HDI, IPDI, TDI, and TMXDI

mechanical of or relating to machines or tools.[47,49]

mechanical adhesion see ADHESION, MECHANICAL, and ADHESION, SPECIFIC.[79]

mechanical properties those properties of a material that are associated with ELASTIC and inelastic reaction when FORCE is applied, or that involve the relationship between STRESS and STRAIN.[47] (ASTM)

It includes HARDNESS and TOUGHNESS.

mechanical stability (of latexes) the ability of a LATEX to withstand prolonged mechanical shear, without the formation of appreciable amounts of GEL or an unacceptable increase in VISCOSITY.[161]

Latexes may be sensitive to the shearing stress encountered during AGITATION, pumping, MILLING, SPRAYING, and shipping. Hence, they must have mechanical stability.

media AGGREGATE used to effect DISPERSION in certain types of production equipment, such as BALL, PEBBLE, and SAND MILLS.[49,59,131]

The media vary in size and composition. Some examples are: steel balls; natural stones or pebbles; synthetic ceramic balls; GLASS beads; and sand.

media mill any MILL using any one of the various types of GRINDING media, e.g., sand, steel ball, pebble, etc.[64,59]

medium in paints or ENAMELS, the CONTINUOUS PHASE in which the PIGMENT is DISPERSED; thus, in the liquid paint in the can, it is synonymous with VEHICLE, and in the dry film it is synonymous with BINDER.[156] (BSI)

medium (art) in a general sense, the particular material with which a work of art is executed: OILS, WATER COLOR, CHALKS, LITHOGRAPHIC STONE, pen and INK, etc.[156,81]

It may also refer to the liquid with which POWDERED PIGMENTS are ground to make ARTISTS' PAINT, and in a more restricted sense, to the liquid used to render such paint more fluid and workable.

medium oil varnish VARNISH of medium OIL content usually containing from 18 to 25 gal of oil per 100 lb of RESIN.[63,152,156] See LONG OIL and SHORT OIL.[63]

megapoise one million poises.[166,128] See VISCOSITY and POISE.[166,128]

This unit is used for materials of very high viscosity, e.g., ASPHALTS.

MEHQ abbreviation for monomethyl ether of hydroquinone.[83]

Meier bar see WIRE-WOUND ROD.[145,157,77]

MEK see METHYL ETHYL KETONE.[164]

MEK resistance see SOLVENT RUB[164,157] and DOUBLE RUB.[157]

melamine $C_3N_3(NH_2)_3$. Symmetrical triazine.[130] (ASTM)

It reacts with FORMALDEHYDE to give a series of heat reactive resins.

melamine resin a SYNTHETIC RESIN made from MELAMINE and an ALDEHYDE.[130] (ASTM)

melamine-formaldehyde resin (MF) an AMINO RESIN made by POLYCONDENSATION of MELAMINE with FORMALDEHYDE or a COMPOUND that is capable of providing methylene bridges.[130]

meldola blue METHYLENE BLUE type of DYE.[41] Syn: new blue or naphthol blue

melengket native name for soft MANILA COPAL obtained by tapping.[167]

melissic acid $CH_3(CH_2)_{28}COOH$. Monobasic FATTY ACID constituent of BEESWAX.[137,144]

Mp, 90°C.

melt viscosity measurement of the VISCOSITY of a molten POLYMER determined by any one of several methods.[130,142]

melting point temperature at which a solid changes to a liquid, and the liquid and solid phases are in equilibrium under a pressure of one atmosphere.[128,157,176]

membrane a flexible or semiflexible roof covering of WATERPROOFING, whose primary function is the exclusion of water.[76] (ASTM)

menaccanite old name for the TITANIUM MINERAL, ILMENITE.[41,34]

Words presented in CAP/SMALL CAPS type indicate that the word is defined in another part of the *Coatings Encyclopedic Dictionary.*

Numerical superscripts classify terms in one or more of the categories listed in the second section of this volume.

menhaden oil a DRYING OIL obtained by steaming or boiling the fish *Brevoortia tyrannus* caught along the Atlantic Coast of the United States.[63]

mer the smallest repeating structural unit of any high POLYMER.[130] See MONOMERIC UNIT.[130,141]

mercadium orange see CADMIUM-MERCURY SULFIDES.[41]

mercurials FUNGICIDES and BACTERICIDES containing mercury.[173]

No longer used in the U.S.

mercuric sulfide HgS. PIGMENT Red 106 (77766). (1) a naturally occurring mineral, CINNABAR; (2) synthetically produced by addition of Hg to alkali sulfides depending on temperature, the colors vary from red orange to bluish red.[41,34] Syn: vermillion, carmine vermillion, Chinese vermillion, English vermillion, patent vermillion, cinnabar, cenobrium, cinaper, cinoper, cynoper, vermiculus, zunsober, red cinnabar

Has good alkali resistance. Density, 8.0 g/cm^3 (66.6 lb/gal); O.A., 11-15.

mercury-cadmium lithopones see CADMIUM-MERCURY LITHOPONES.[41]

mesh (1) the square opening of a SIEVE;[128] (ASTM) (2) the number of apertures per linear inch in a woven or electroformed METAL screen or sieve, made especially for laboratory testing of high gravity DRY POWDERS or PIGMENTS for fineness and impurity content. Such screens are available to a mesh size of about 400.

mesh analysis see SIEVE ANALYSIS.[157]

mesh number the designation of size of an ABRASIVE grain.[59,158] Syn: grit number

Its name is derived from the openings per linear inch in the control sieving screen.

mesityl oxide $CH_3COCH=C(CH_3)_2$. Medium-boiling KETONIC SOLVENT.[164]

Properties: Bp, 130°C; flp, 25°C (78°F); vp, <10 mm Hg/30°C.

mesitylene $C_6H_3(CH_3)_3$. Powerful high-boiling HYDROCARBON SOLVENT.[164]

Bp, 165°C.

mesocolloid between hemicolloids and encolloids in size.[139]

COLLOID PARTICLES limited to aggregates of from 100 to 1,000 molecules and from 25 to 250 nm long.

mesopic vision vision at luminosities intermediate between luminosities required for completely photopic or completely scotopic vision; sometimes called twilight vision.[43]

metafiltration edge filtration through superimposed metallic strips with beveled edges, involving a change from coarse filtration (due to the strips) to fine filtration (due to the filter bed formed in their interstices).[49,131]

metal any of a category of electropositive elements that usually have a shiny surface, are generally good conductors of heat and electricity, and can be melted or fused, hammered into thin sheets, or drawn into wires.[45]

Typical metals form SALTS with nonmetals, basic OXIDES with oxygen, and ALLOYS with one another.

metal decorating LITHOGRAPHIC PRINTING on METAL.[45] See TEMPORARY COATINGS.[71]

It most often applies to the CAN COATING industries and includes the coating of can liners, outside BASECOATS, and OVERPRINT VARNISHES as well as LITHOGRAPHIC PRINTING. The coatings are generally applied by direct ROLLER COATING or by SPRAY onto TINPLATE or ALUMINUM, on individual metal sheets, or by continuous coil coating. This term is not to be confused with STRIPPABLE COATINGS, which are designed for temporary protection and subsequent easy removal.

metal leaf see DUTCH METAL.[148]

metal marking in the COIL COATING industry, the degree or amount of a DEFECT left on the SURFACE of a film when the edge of a piece of METAL is pulled across its surface.[70,56] Syn: coin marking or marring

Pencil line marks on a painted surface caused by scratching with metal.

metal powder pigment see METALLIC PIGMENT.[41,45]

metal primer the first coat of paint on METAL; a PRIMER, one coat.[71,158] (DAC)

metal spraying APPLICATION of a SPRAY coat of METAL (usually zinc or aluminum) onto a prepared surface (usually shotblasted mild steel).[45,71]

The metal to be sprayed is rendered molten by passing it, in wire or powder form, through a flame pistol which projects the semimolten metal onto the surface by means of a jet of compressed air.

metallic brown see BROWN IRON OXIDE PIGMENT.[41]

metallic inks INKS composed of ALUMINUM or BRONZE POWDERS in VARNISH to produce gold or silver COLOR effects.[168]

metallic paint paint which, on APPLICATION, gives a film with a METALLIC APPEARANCE. This effect is normally produced by the incorporation of fine flakes of such metals as COPPER, BRONZE, or ALUMINUM.[71,163] (BSI) See EFFECT FINISH.[71,163,43] Syn: metallic pigmented paint

The aluminum used may be LEAFING *or* NONLEAFING, *the former giving a far more brilliant metallic effect. These metals can be used in tinted or colored media to give polychromatic finishes, especially on automobiles because of their attractive appearance.*

metallic pigment PARTICLES or FLAKES of nonoxidized METALS or alloys used as PIGMENTS to modify the OPTICAL characteristics of a paint, to hide the SUBSTRATE, modify the COLOR, or adjust other properties.[41,69] See METALLIC PAINT.[41,45] Syn: flake metal and metal powder pigment

The metals most commonly used are ALUMINUM, COPPER, BRONZE, *and* ZINC.

metallic pigmented paint see METALLIC PAINT.[71,163]

metallic soaps a group of water-insoluble compounds containing alkaline earth or heavy metals combined with monobasic carboxylic acids of 7 to 22 CARBON atoms. They can be represented by the general formula $(RCOO)_x M$ (neutral SOAP), where R is an ALIPHATIC or ALICYCLIC RADICAL and M is a METAL with valence x.[83,67]

They differ from ordinary soap by composition and their insolubility in water. Their solubility or solvation in a variety of ORGANIC SOLVENTS *accounts for their many and varied uses.* DRIERS *are the most important group of metallic soaps promoting or accelerating the drying, curing, or hardening of oxidizable coating vehicles such as paints.*

metallizing (1) applying a thin coating of METAL to a nonmetallic surface. May be done by chemical DEPOSITION or by exposing the surface to vaporized metal in a vacuum chamber;[44,77,158] (2) see FLAME SPRAY;[77] (3) adding metallic pigments, such as aluminum, to a paint.

metamer one of a pair of COLORS which matches the other color when viewed in a described way but which does not match under all viewing conditions.[43,69] See METAMERIC PAIR.[43,69,35]

For example, if the light source, observer, or angle of viewing is changed, the color will no longer match the other color of the pair.

metameric match a conditional identity of COLOR exhibited by a pair of colors, each with different SPECTRAL DISTRIBUTION CURVES.[43,69] See METAMERISM.[43,69,35] Syn: conditional match

metameric pair a pair of COLORS which match when viewed in a described way but which do not match if the viewing conditions are changed.[43,69] See CONDITIONAL MATCH and METAMERISM.[43,69,35]

Thus, a metameric pair of samples exhibit the same TRISTIMULUS VALUES *for a described set of viewing conditions (observer, light source, geometry of the illumination, and viewing arrangement) but have different spectral distributions. Hence, they exhibit a match which is conditional only.*

metamerism property of two specimens that match under a specified illuminator and to a specific observer and whose SPECTRAL reflectances or transmittances differ in the visible WAVELENGTH. See METAMERIC PAIR[43,69,35] and CONDITIONAL MATCH.[43,69,35,157]

It is the phenomenon exhibited by a pair of colors which match under one or more set of conditions, be they real or calculated.

Metameric COLOR *matches result from the use of different colorants (*PIGMENTS, DYES, *and the like) in achieving the same color match. Usually, the purchaser requires that the color match be* NONMETAMERIC. *Where nonmetameric matches are not possible or practical, for cost or other reasons, some limitation of metamerism may be required. Metamerism should not be confused with* FLAIR *or* COLOR CONSTANCY, *which apply to the apparent color change exhibited by a single color when the spectral distribution of the light source is changed or when the angle of illumination or viewing is changed.*

ASTM Test Method D4086 describes visual methods for detecting metamerism and for estimating the magnitude of a metameric color difference. It is limited to the consideration of illuminant metamerism and observer metamerism. It is not designed to cover so-called geometric metamerism.

ASTM Test Method D1535 covers the visual evaluation of metamerism.

metamerism, degree of metamerism exists in varying amounts, depending on the magnitude of the differences in the SPECTRAL DISTRIBUTION CURVES of the two colors.[43,69,128]

Thus, the degree of metamerism may be slight to moderate to severe and is generally described by one of a number of types of metamerism (or metameric) indices.

metamerism index see INDEX OF METAMERISM.[43,69]

meter a general term for dial-type instruments used for measurement.[145] cf. METRE[128]

methanol CH_3OH. Colorless, flammable, poisonous liquid used as a SOLVENT, in the manufacture of FORMALDEHYDE, in ORGANIC SYNTHESIS, and for DENATURING (rendering impotable) ethanol for industrial use.[164] Syn: methyl alcohol, carbinol, wood alcohol and colonial spirits, MeOH

ASTM test methods for methanol include: acetone content of, D1612; content, in formaldehyde solutions, D2380; permanganate time, test, D1363; specification, D1152.

ASTM Test Method D1363 covers the detection in acetone and methanol of the presence of impurities that reduce potassium permanganate.

Words presented in CAP/SMALL CAPS type indicate that the word is defined in another part of the *Coatings Encyclopedic Dictionary.*

Numerical superscripts classify terms in one or more of the categories listed in the second section of this volume.

Method 18 an EPA test method which uses GAS CHROMATOGRAPHIC techniques to measure the concentration of individual VOLATILE ORGANIC COMPOUNDS in a gas stream.[157,172] (EPA)

Method 24 an EPA reference method to determine density, water content, and total volatile content (water and VOC) of coatings.[157,172] (EPA)

Method 25 an EPA reference method to determine the VOC concentration of a gas stream.[157,172] (EPA)

methoxybutyl acetate $CH_3OCH(CH_3)CH_2CH_2COOCH_3$. High-boiling SOLVENT.[164]

Properties: bp, 169°C; flp, 60°C (140°F).

methoxyl group the monovalent group, $-OCH_3$, characteristic of METHYL ALCOHOL and its ESTERS or ETHERS.[134] (ASTM)

methyl acetate CH_3COOCH_3. A low-boiling ESTER SOLVENT, exhibiting rapid EVAPORATION; bp, 58°C; vp, 173 mm Hg/20°C.[164]

methyl acrylate $CH_2{:}CHCOOCH_3$. Colorless liquid; MONOMER.[141]

Used in the manufacture of SYNTHETIC RESINS. Sp gr, 0.953; bp, 80.5°C.

ASTM D4709 is a specification for methyl acrylate.

methyl alcohol see METHANOL.[164]

methyl amyl carbinol see HEPTANOL-2.[164]

methyl amyl ketone a high boiling ketonic SOLVENT.[164] Syn: MAK, and 2-heptanone

Properties: boiling range, 147-153°C; flp, 106°F; vp, 4 mm Hg/20°C.

methyl butyl ketone $CH_3CO(CH_2)_3CH_3$. A high boiling KETONIC SOLVENT.[164] Syn: MBK and 2-hexanone

Properties: boiling range, 114-137°C; flp, 73°F; vp, 10 mm Hg/20°C.

methyl butyrate $C_5H_{10}O_2$. Medium-boiling SOLVENT. Bp, 102°C.[164]

methyl Cellosolve® acetate ETHER-ESTER SOLVENT.[164]

Properties: Bp, 143°C; vp, 6 mm Hg/30°C.

methyl Cellosolve®, methoxyethanol $CH_3OC_2H_4OH$. ETHER-ALCOHOL SOLVENT. Bp, 124°C.[164]

methyl chavicol ETHER constituent found in some TERPENE SOLVENTS. Boiling range, 215-216°C.[164]

methyl citrate (tri) PLASTICIZER with SOLVENT properties.[164,170]

Bp, 176°C/16 mm Hg.

methyl cyclohexane $CH_3C_6H_{11}$. HYDROCARBON SOLVENT. Bp, 101°C.[164]

methyl cyclohexanol $CH_3C_6H_{10}OH$. ALCOHOLIC SOLVENT.[164]

Because of its high boiling range, 160-180°C, it is used to improve flow properties. Small quantities added to weaker solvents often exert a very pronounced effect on the viscosity of a given solution. Flp, 60°C (140°F).

methyl cyclohexanone high-boiling SOLVENT. Boiling range, 160-170°C; flp, 49°C (120°F); vp, 4 mm Hg/30°C.[164]

It is often used in small amounts to improve FLOW or BRUSHING properties.

methyl cyclohexyl acetate boiling range, 175-190°C; flp, 68°C (155°F).[164] Syn: methyl hexalin acetate and sextate

methyl cyclohexyl stearate $CH_3(CH_2)_{16}COOC_6H_{10}CH_3$. Boiling range for the commercial product, 220-240°C/4 mm Hg; flp, 170°C (338°F).

methyl ethyl ketone $CH_3COCH_2CH_3$. A low-boiling SOLVENT, possessing all the properties of ACETONE without its extremely high VOLATILITY.[164] Syn: MEK and butanone

It is a solvent for some of the materials with lower solubility like VINYL RESINS and some CELLULOSE derivatives. Bp, 80°C; flp, 3°C (35°F); vp, 77 mm Hg/20°C.

methyl glycol phthalate (di) PLASTICIZER. Boiling range, 210-260°C/20 mm Hg; flp, 174°C (345°F).[170]

methyl hexyl ketone $CH_3COC_6H_{13}$. High-boiling KETONIC SOLVENT. [164]

Boiling range, 169-173°C; vp, 3 mm Hg/30°C.

methyl isoamyl ketone $CH_3COCH_2CH_2CH(CH_3)_2$ A medium boiling KETONIC SOLVENT.[164] Syn: 5-methyl-2-hexanone; MIAK

Properties: mol wt, 114; bp, 145°C; flp, 43°C; vp, 4 mm Hg/20°C.

The ASTM test method for methyl isoamyl ketone is D2917.

methyl isobutyl ketone $(CH_3)_2(CHCH_2)COCH_3$.[164] Syn: hexone, MIBK, isopropyl acetone, and 4-methyl-2-pentanone

Medium-boiling SOLVENT. Mol wt, 100; Bp, 118°C; flp, 74°F; vp, 13 mm Hg/20°C.

methyl lactate $CH_3CH(OH)COOCH_3$. Medium-boiling SOLVENT.[164]

Bp, 145°C; flp, 100°F.

methyl methacrylate $CH_2 = C(CH_3)COOCH_3$. Colorless liquid; MONOMER.[141]

ASTM Test Method D4747 covers the determination of free (unreacted) methyl methacrylate monomer content of acrylic latexes using gas-liq-

uid chromatography. ASTM Test Method D4827 does the same using capillary column gas chromatography. Sp gr, 0.936; bp, 100°C.

methyl methacrylate resins CH_2=$C(CH_3)$ $COOCH_3$. POLYMERIZATION products derived from METHYL METHACRYLATE.[130] Syn: acrylic resins

These RESIN SOLUTIONS of high VISCOSITY have restricted compatibility. Among their important properties are outstanding HEAT RESISTANCE and COLOR RETENTION.

methyl phosphate (tri) water-soluble, semipermanent PLASTICIZER. Bp, 196°C.[170]

methyl ricinoleate $CH_3(CH_2)_5CHOHCH_2CH$ $=CH(CH_2)_7COOCH_3$. PLASTICIZER and methyl ester of ricinoleic acid.[170]

Bp, 245°C/10 mm Hg. Sp gr, 0.9236.

methyl tartrate (di) a PLASTICIZER with a bp of 280°C.[170]

methylated spirit a mixture of ETHYL ALCOHOL and a small amount of methyl alcohol; used industrially as a SOLVENT for paints, LACQUERS, and VARNISHES.[164] (DAC)

methylcellulose any of several METHYL ETHERS of CELLULOSE.[83,155,166]

Commercially, a granular, white, flaky material which acts as a water-soluble THICKENER and STABILIZER; used in water-based paints.

methylene blue Blue 9 (52015). A thiazine dye used to make a type of LIME BLUE.[41]

methylene chloride CH_2Cl_2. Low-boiling CHLORINATED HYDROCARBON.[164] Syn: dichloromethane

It is a constituent of nonflammable paint removers. A SOLVENT, degreaser, and dewaxer. Properties: bp, 40°C; vp, 230 mm Hg/10°C. Sp gr, 1.34/4°C.

methylolureas $H_2NCONHCH_2OH$. First stage in the formation of UREA FORMALDEHYDE RESINS.[144]

Reaction products of urea and formaldehyde. Monomethylol urea is obtained by reacting one molecule of formaldehyde and one molecule of urea.

metre the basic unit of length in the INTERNATIONAL SYSTEM OF UNITS (SI), equal to 39.37 inches.[128]

metric perm see PERM.[128]

metric ton a unit of mass equivalent to 2,204.6 pounds or one megagram (10^6 grams).[128]

metrication any act tending to increase the use of the INTERNATIONAL SYSTEM OF UNITS (SI).[134]

mezzotints prints made from ground (e.g., crossed in several directions with knife-edge cuts) copper plates, the burred surface of which is scraped away to create the design.[81,168]

MFFT abbreviation of MINIMUM FILM FORMING TEMPERATURE.[55]

This is preferred over MFT.

MFT see MFFT.[55]

MIBK see METHYL ISOBUTYL KETONE.[164]

mica see ALUMINUM POTASSIUM SILICATE.[41,58]

ASTM Test Method D716 covers the analysis of mica. ASTM D607 is a specification for mica.

micaceous iron oxide naturally occurring IRON ORE which resembles MICA only in APPEARANCE and not chemically.[41]

It is converted to a grayish PIGMENT, which has value as a constituent of anti-corrosive paints and protective coatings. Density, 4.90 g/cm³ (40.8 lb/gal); O.A., 11 lb/100 lb. HARDNESS (mohs), 6.0; pH, 7.7.

micelle COLLOIDAL PARTICLE composed of many AGGREGATED small molecules having a layered structure.[139]

microbiological attack see BIODETERIORATION.[173]

microcrystalline silica see SILICA, MICROCRYSTALLINE.[41,58]

microcrystalline wax PLASTIC, high mp PETROLEUM WAX secured by removing most of the OIL from petrolatum by SOLVENTS or other means.[137]

microemulsion is a multicomponent (at least three) liquid dispersion system containing nanometer-size oil droplets in an aqueous continuous phase, or nanometer-size water droplets in an oil continuous phase. The first is called water-in-oil microemulsion and the latter, an oil-in-water microemulsion. They are thermodynamically stable, optically TRANSPARENT, and homogeneous.[60] cf. EMULSION[60]

microencapsulation PROCESS whereby a thin film of COLLOID, ELASTOMER, PLASTIC, or WAX is deposited on micron-sized solid or liquid bits of material and solidified in place to form microcontainers or capsules.[131]

The initial film may be applied in a gaseous, liquid, or solid state.

micrometer (1) instrument for measuring small lengths under the microscope; (2) micrometer caliper; instrument used in measuring the DRY FILM

Words presented in CAP/SMALL CAPS type indicate that the word is defined in another part of the *Coatings Encyclopedic Dictionary.*

Numerical superscripts classify terms in one or more of the categories listed in the second section of this volume.

THICKNESS of a coating or the thickness of METAL or other materials.[59,128]

micrometre (μm) a unit of length measure used to describe the WAVELENGTH of RADIANT ENERGY equal to one thousandth part of a millimetre or one millionth of a metre, 1000 nanometres. This term is replacing the term MICRON.

micron (μ) the use of this term is gradually being replaced by the term "MICROMETRE," designated by μm.[128]

micronized pigment PIGMENT with a narrow, PARTICLE-SIZE DISTRIBUTION.[41]

micronizer mill used for the dry GRINDING of PIGMENTS.[59,64]

Precrushed crudes or solids are fed into a relatively flat, cylindrical grinding chamber, the opening being in the bottom or top plate. The fluid medium, superheated compressed air or steam, enters the grinding chamber through nozzles located in the peripheral wall. The grinding action is done by the pigment PARTICLES hitting each other. The fines are carried out through an opening in the center of the MILL, the coarser, larger particles return for further grinding.

microorganisms in paint technology, BACTERIA and FUNGI which are harmful to liquid paint and dry paint films.[65,173] (DAC)

BACTERICIDES and FUNGICIDES are added to paints to inhibit the growth of these organisms.

microvoids small VOIDS or holes in a paint medium of such size that when filled with air (or some other material of different REFRACTIVE INDEX) they SCATTER light much as a PIGMENT does because of the difference in refractive index between the material in the void and the vehicle.[43] See BUBBLE COATING[71] and DRY HIDING.[43]

Such air-filled microvoids account for the white color of snow, for example.

middle chrome see CHROME YELLOW, LIGHT and PRIMROSE.[41]

Mie Theory relates the SCATTERING of a single spherical PARTICLE in a medium to the diameter of the particle, the difference in REFRACTIVE INDEX between the particle and the medium, and the WAVELENGTH of RADIANT ENERGY in the medium which is incident on the particle.[43]

This theory relates to the direct observation of the scattering of a single particle as compared to the KUBELKA-MUNK THEORY, which relates to multiple scattering between PARTICLES. MIE THEORY also takes into account the ABSORPTION which the particle may also exhibit.

migration movement of certain materials in a coating or PLASTIC (e.g., PLASTICIZERS in VINYL) to the surface or into an adjacent material.[56,170]

mil a unit of length equal to one thousandth of an inch (0.0254 mm).[128]

mildew (1) superficial growth produced on ORGANIC matter or living plants by fungi (as of the families *Erysiphaceae* and *Peronosporaceae*); (2) a fungi producing mildew; and (3) a discoloration caused by fungi.[42,56,65,173] cf. ALGAE[139]

Two types are common: (1) filamentous—thread-like structures, gray to black in COLOR, which make up the mycelium or vegetative form of fungi; (2) spores—spherical or nodular bodies, gray or black in color, which may or may not be associated with fungal mycelia. Spores are the reproductive form of fungi. (Mildew should not be confused with ALGAE which is usually green to brown in color and normally occurs in shaded damp areas.)

Mildew can be distinguished from dirt by using a bleach such as Clorox®. A drop of bleach is placed on the area in question and allowed to run down the area. If the discoloration turns lighter, then the discoloration is mildew. If it doesn't change, it is dirt. Mildew, unlike dirt, is very difficult to remove from the film. If the film is washed with water, dirt will be removed, but the mildew will remain imbedded in the film as a dark stain. The two can also be differentiated with a microscope.

mildew (fungus) resistance the ability of a coating to resist superficial FUNGUS growth that can cause discoloration and ultimate DECOMPOSITION of a coating's binding medium.[56,72,173] See MILDEW.[56,173,42,65]

Mildew collection is the phenomenon of a coating manifest by the presence of MOLD or fungus growth of a type familiarly known as mildew. Two types are common: (1) the SPORE type, which resembles caviar in appearance; *and (2) the MYCELIUM or filament type. A microscope is necessary to differentiate the spore type from dirt collection. The mycelium type is often recognized with the naked eye when it reaches an advanced stage, but a microscope is recommended for any mildew examination. The FSCT Pictorial Standards for Coatings Defects contains photographic standards for evaluating the degree of surface disfigurement of paint films by fungal growth or soil and dirt accumulation.*

ASTM Test Method D3274 also has these same photographic reference standards to provide a numerical basis for rating the degree of fungal and algal growth or soil and dirt accumulation on paint films.

mildewcide chemical agent which destroys, retards, or prevents the growth of MILDEW.[83,173] cf. MILDEWSTAT[83,173]

mildewstat a chemical agent that inhibits the growth of mildew.[83,173] (ASTM) cf. MILDEWCIDE[83,173]

milk acid see LACTIC ACID.[144]

milk paint is made from CASEIN, CLAY, earth PIGMENTS, and LIME.[71,175]

Almost a rarity, it is still used mainly by decorators, FURNITURE restorers, and history buffs. Its distinctive FINISH, which is flat and coarse, lends an authentic look to antique furniture, historic houses, and stencilled floors.

milkiness whitish or translucent APPEARANCE in an unpigmented liquid coating or film which should normally be TRANSPARENT.[56]

mill a machine or device that reduces a solid such as a PIGMENT or other coarse substance into minute grains by crushing, GRINDING, or pressing.[49,131,59]

mill base that portion of the coating formulation which is charged into the DISPERSION MILL.[49,131,64]

mill scale the black OXIDE of iron produced during the hot rolling, hot fabrication, or heat treatment of steel.[45,176]

The composition of the layer varies from Fe_3O_4 at the surface, to FeO at the METAL interface. Note: this term is sometimes used erroneously to include the concept of "CORROSION SCALE."

millilitre one thousandth of a LITRE.[128]

millimicron (deprecated) a unit of length used to describe the WAVELENGTH of electromagnetic RADIATION, particularly that in the visible region, equal to 10^{-9} metre. The use of this term is largely being replaced by the term NANOMETRE.[128]

millipoise one thousandth part of a POISE (measure of VISCOSITY) or 1/10 of a CENTIPOISE.[128,166]

millstone any stone employed for milling paint, CEMENT, grain, or minerals. They are made from sandstone, basalt, granite, or quartz conglomerate.[131,59]

milori blue a PIGMENT of the ferriferrocyanide family.[41] See IRON BLUE.[41]

milori green see CHROME GREEN.[41]

min (1) abbreviation for minimum (in figures and tables only); (2) abbreviation for minute (spell out when used with minimum).[128] (ASTM)

mineral any naturally occurring, homogeneous, INORGANIC substance having a definite chemical composition and characteristic CRYSTALLINE structure, COLOR, and HARDNESS.[34]

mineral acids strong INORGANIC acids, e.g., nitric, sulfuric, hydrochloric, etc.[144]

mineral black black PIGMENT made by GRINDING and/or heating black SLATE, SHALE, slaty coal, coke, or coal.[41,34] See BIDDIBLACK.[41]

It usually contains a high percentage of CARBON, mixed with substantial amounts of mineral matter, which may include aluminum silicate, and oxides of silicon, iron, calcium, and magnesium. By comparison with true carbon blacks, its staining strength is inferior.

mineral brown see BROWN IRON OXIDE PIGMENT.[41]

mineral oil any liquid product of PETROLEUM within the VISCOSITY range of products commonly called OILS and consisting of high molecular weight HYDROCARBONS.[63]

mineral orange see ORANGE MINERAL.[41]

mineral pigments see EARTH PIGMENT.[41,68]

mineral spirits a refined PETROLEUM DISTILLATE having a low AROMATIC HYDROCARBON content, with volatility, flash point, and other properties making it suitable as a THINNER and SOLVENT in paints, VARNISHES, and similar products. See ODORLESS SOLVENT.[164] Syn: petroleum spirits

mineral surfaced roofing FELT or FABRIC saturated with BITUMEN, coated on one or both sides with a bituminous coating and surfaced on its weather side with mineral granules.[75,76] (ASTM)

mineral violet (77742). A complex manganese ammonium phosphate PIGMENT.[41] Syn: manganese violet, permanent fast violet, Nurnberg violet

Its very low TINTING STRENGTH, low COLOR INTENSITY, and low HIDING POWER largely restrict its use to toning white finishes, in which service it offers low cost and excellent bake and BLEED RESISTANCE; however, tints fade severely in steam sterilizers. Because of its poor alkali resistance, it is not suitable for some household appliance finishes; in such cases, quinacridone violet can be used to advantage.

mineral white see GYPSUM.[41]

mini-bell an ELECTROSTATIC SPRAY APPLICATION device using a rapidly rotating bell to atomize paint.[77,172] (EPA) See BELL.

minimized spangle GALVANIZED sheet obtained by treating the regular galvanized sheet during the solidification of the zinc to restrict the normal spangle formation.[45,158] (ASTM)

This product usually has a dull APPEARANCE not characterized by a high degree of uniformity, and dissimilarity from COIL to coil is not unusual. This minimizes the CRYSTALLINE pattern from photographing through the applied coating, resulting in a smoother appearing FINISH.

Words presented in CAP/SMALL CAPS type indicate that the word is defined in another part of the *Coatings Encyclopedic Dictionary.*

Numerical superscripts classify terms in one or more of the categories listed in the second section of this volume.

minimum film forming temperature (MFFT) the minimum temperature at which a LATEX forms a continuous film, as evidenced by the visual lack of CRACKING and/or powdery APPEARANCE of film and the FILM INTEGRITY, by testing the film on a temperature gradient plate (MFFT apparatus) for film TOUGHNESS, by pushing against the film with a piece of WOOD or PLASTIC (so as not to scratch the METAL plate), until the film resists further cracking (remains continuous).[55,157,145]

MFFT measurement indicates POLYMER HARDNESS and is a useful measure of product suitability for a given end use and for coalescent requirements. Satisfactory film integrity of latex coatings requires that, as the water evaporates, the polymer portion of the vehicle coalesces into a continuous film. Low temperature reduces the ability of the film to adequately coalesce, hence, vehicles that can withstand low temperatures are required for latex paints. For a test method, see ASTM Test Method D2354.

minimum perceptible difference the minimum COLOR difference which can be observed between two colors.[43,69] See JUST NOTICEABLE DIFFERENCE.[43,69]

The magnitude is dependent on the conditions of viewing, i.e., size, illumination level and character, surrounding area, adaptation, etc.

minium naturally occurring and synthetically prepared RED LEAD OXIDE; was used as a PIGMENT.[41] (DAC)

minute value voltage which a unit thickness of INSULATOR (e.g., VARNISH) will withstand for one minute without breakdown.[61,152]

mirbane oil NITROBENZENE. Liquid with an almond-like smell, used as a deodorant, chiefly in polishes.[83,144]

mired a unit used to measure the reciprocal of COLOR temperature, equal to the reciprocal of a color temperature of 10^6 KELVINS.[128]

Derived from micro-reciprocal-degree.

mirror back coating used to prevent CORROSION and DISCOLORATION of the silver on the back of the mirror.[71,66]

An example of such a coating is an iron oxide pigmented SHELLAC.

miscellaneous metal parts as used in the *Environmental Protection Agency's Control Technique Guidelines*, any METAL parts excluding those produced in CAN, COIL COATING, magnet wire, METAL FURNITURE, and large appliance plants, and also excluding automotive, ship, and airplane bodies.[172,45] (EPA)

miscible liquids that can be mixed in all proportions (any ratio).[164] cf. COMPATIBILITY[161]

misses see HOLIDAYS.[56]

mist suspended liquid droplets in the air generated by CONDENSATION from the gaseous to the liquid state by mechanically breaking up a liquid by splashing or atomizing.[64]

mist coat (1) very thin sprayed coat; (2) a thin coat of VOLATILE THINNERS, with or without a small amount of LACQUER which is sprayed over a dry lacquer film to improve the smoothness and LUSTER.[54]

misting see FLYING.[168]

mitt see PAINTER'S MITT.[77]

Mittler's green see HYDRATED CHROMIUM OXIDE.[41]

mixed aniline point minimum equilibrium solution temperature of a mixture of two volumes of ANILINE, one volume of sample, and one volume of normal heptane of specified purity.[157] Refer to ASTM D1012.

mixing varnish a variety of VARNISHES, or paint VEHICLES, that are used by mixing with: (1) ALUMINUM PIGMENT (see ALUMINUM MIXING VARNISH);[152] (2) PIGMENT paste in OIL or COLORS in oil; (3) other vehicle types, such as gloss oil (if the mixing varnish is long in oil), or a spar varnish (if the mixing varnish is short in oil), to produce a desired oil length for the complete vehicle; (4) the MILL base discharged from pigment-dispersing equipment, as contrasted to the "GRINDING" varnish or vehicle in which the pigment has been dispersed; and (5) paint, to alter its properties.[152,156]

mixing white a white INK, either TRANSPARENT or OPAQUE, used in making tints.[168]

mixture a heterogeneous association of materials that cannot be represented by a chemical formula and that does not undergo chemical change as a result of interaction among the mixed materials.[172]

The constituent materials may or may not be uniformly dispersed and can usually be separated by mechanical means (as opposed to a chemical reaction). Uniform liquid mixtures are called SOLUTIONS. "If a hazardous chemical is present in a mixture in reportable quantities (i.e., 0.1% for carcinogens and 1.0% for other health hazards), it must be reported unless the mixture has been tested as a whole." (OSHA CPL 23-02, 38A)

mobility the property of a material which allows it to FLOW when a SHEARING FORCE larger than the yield value has been applied.[166]

The COEFFICIENT of mobility is the rate of shear induced by a shearing force per square cm of one dyne in excess of the YIELD VALUE. Mobility pertains to PLASTIC materials and is the analog of fluidity. It is calculated from the slope of the straight-line portion of the flow curve. The coefficient of mobility is the reciprocal of the coefficient of plastic viscosity.

Mobilometer RHEOLOGICAL instrument for measuring the CONSISTENCY of paints and similar products.[59,157] See GARDNER MOBILOMETER.[59,157]

The "mobility" is determined by the time in seconds required for a loaded perforated disk to pass through a specified depth of the sample contained in the cylinder.

model plant a description of a typical but theoretical plant used for developing economic, environmental impact, and energy impact analyses as support for regulations or regulatory guidelines.[49,172] (EPA)

It is an imaginary plant, the parameters of which are typical of existing or future plants. The parameters are used to estimate the cost of incorporating air pollution control technology as the first step in exploring the economic impact of a potential NSPS. Such analyses are a fundamental part of the development of a standard, as recorded in the Background Information Document.

modes of appearance various manners in which colors can be perceived, depending on spatial distributions and temporal variations of the light causing the sensation.[69,43,35]

Five modes are generally recognized: surface mode, volume mode, film (or aperture) mode, illumination mode, and illuminant mode. The first two of these, surface mode and volume mode, together make up the object mode. COLOR perceived as falling on an object, thus filling space around the object, describes the illumination mode. The illuminant mode is the APPEARANCE of the glow from a light source. The film or aperture mode describes color perceived in space, where no object is discernible.

modified alkyd see ALKYDS.[71,130]

modified phenolic resins RESINS in which the basic phenolaldehyde product has been modified by the introduction of ROSIN or other natural resin.[130,167]

The products are often ESTERIFIED.

modifier see ADDITIVE.[83]

modify to change in form or character; alter.[133]

modulus of elasticity the ratio of STRESS (nominal) to corresponding STRAIN below the proportional limit of a material.[47] (ASTM) See HARDNESS and STIFFNESS.[47] Syn: elastic modulus or Young's modulus

It is expressed in FORCE per area, usually kilograms-force per square centimetre (or pounds per square inch).

ASTM Test Method D2370 covers the determination of the modulus of elasticity (stiffness) as well as elongation, and tensile strength of ORGANIC coatings when tested as free films.

mohair paint roller cover a cover in which the paint applicating material is woven of short pile velour that contains wool or angora goathair.[77] (ASTM)

Mohr balance instrument used to measure the SPECIFIC GRAVITY of fluids.[59]

Mohs scale scale for determining the relative HARDNESS of a mineral, according to its RESISTANCE to SCRATCHING by one of the following minerals: (1) talc; (2) gypsum; (3) calcite; (4) fluorite; (5) apatite; (6) feldspar; (7) vitreous silica; (8) quartz; (9) topaz; (10) garnet; (11) fused zirconia; (12) fused alumina; (13) silicone carbide; (14) boron carbide; and (15) diamond. Other useful hardnesses are: fingernail, slightly more than (2); penny, about (3); pocket knife, slightly more than (5); window GLASS, (5.5); and a steel file, (6.5).[47,128,34] See HARDNESS.[47]

moiety a part, portion, or share.

moiré effect an OPTICAL effect which results from light interference, exhibiting a pattern of light and dark areas.[43,35]

The effect of superimposing a repetitive design, such as a grid, on the same or a different design to produce a pattern distinct from its components.

moiré papers WALLPAPERS having a watered silk sheen effect.[142] See MOIRÉ EFFECT.[43]

moisture diffuse wetness that can be felt as vapor in the atmosphere or condensed liquid on the surfaces of objects; dampness.[153]

moisture barrier treated PAPER or METAL that retards or bars water vapor, used to keep moisture from passing into walls or floors.[82]

moisture content the amount of moisture in a material determined under prescribed conditions and expressed as a percentage of the weight of the moist specimen, that is, the original weight comprising the dry substance plus any moisture present.[157] (ASTM)

ASTM test methods for moisture content include: blue pigment, D1135; capillary moisture in concrete, D4263; cellulose acetate, test, D871; cellulose, test, D1348; ethylcellulose, test, D914; hygroscopic moisture in pigments, D280; hydroxyethylcellulose, test, D2364; hydroxypropyl methylcellulose, D2363; Karl Fischer method, D4017; lac resins, test, D29; liquid naval stores, D890; methylcellulose, test, D1347; pine oil, D890; pine tars and pine tar oils, test, D856; sodium carboxymethylcellulose, D1439; tall oil, test, D803.

Words presented in CAP/SMALL CAPS type indicate that the word is defined in another part of the *Coatings Encyclopedic Dictionary.*

Numerical superscripts classify terms in one or more of the categories listed in the second section of this volume.

moisture equilibrium the condition reached by a sample when the net difference between the amount of moisture ABSORBED and the amount DESORBED, as shown by a change in weight, shows no trend and becomes insignificant.[157] (ASTM)

moisture vapor permeability (MVP) see SPECIFIC PERMEABILITY (of a film to moisture).[142,157]

ASTM Test Method D1653 covers the determination of the rate at which water vapor passes through films of paint, varnish, lacquer, and other organic coatings. The films may be free films or they may be applied to porous substrates.

moisture vapor transmission (MVT) rate of movement of moisture vapor through a membrane.[142,157]

moisture-set ink INKS that dry or set principally by PRECIPITATION.[168] See STEAM-SET INKS.[168]

The vehicle consists of a water-insoluble RESIN dissolved in HYGROSCOPIC SOLVENT. Drying occurs when the hygroscopic solvent has absorbed sufficient moisture either from the atmosphere, SUBSTRATE, or external APPLICATION to precipitate the BINDER. An important characteristic of these inks is their low odor.

mold lubricant see MOLD WASH.[83]

mold marks deformities caused during the molding operation.[38]

mold wash a material that will prevent ADHERENCE of CURED RUBBER to a mold and will facilitate removal of the cured product.[83,79,140] See ANTIBLOCKING AGENT.[83] Syn: mold lubricant, release agent, and stripping agent

The material is often a water solution or slurry of a suitable material sprayed or brushed on the hot mold prior to filling it with the rubber to be molded.

molding projecting or depressed surface, generally used at or around edges of walls, FURNITURE, etc., either plain or decorated, employed to ornament a wall surface, cornice, or capital.[82,148,175] Syn: mold or moulding

molds any of the small, multinucleate, possibly multicellular, filamentous fungi included in the *Eumycates*.[173]

mole or mol the amount of a substance that contains as many atoms, molecules, ions, or other elementary units as the number of atoms in 0.012 kilogram of CARBON 12. The number is 6.0225×10^{23}, or Avogadro's number.[128] Syn: gram-molecule

The mass in grams of this amount of a substance, numerically equal to the molecular weight of the substance. Syn: gram-molecular weight

molecular diffusion see DIFFUSION.[142]

molecular distillation of oils PROCESS of very high vacuum DISTILLATION involving vacua of the order 1/1000 mm of mercury.[131]

By means of this process, stand OILS can be separated into fractions of different molecular weight.

molecular sieve an ADSORBENT composed of porous aluminosilicates with pores of uniform molecular dimensions which will selectively adsorb molecules of the substance to be gathered.[58] (DAC)

molecular weight the mass in grams per MOLE of a substance.[128]

molecular weight(s) the sum of the ATOMIC WEIGHTS of all the atoms in a molecule.[144,128] (ASTM)

Atomic weights (and therefore molecular weights) are relative weights, arbitrarily referred to an assigned weight for oxygen. The molecular weight of a pure, unimolecular substance is as constant as the relative standard (oxygen). Theoretically, at least, pure dimers, tetramers, and other pure oligomers have constant weight.

molecule the smallest PARTICLE into which an element or compound can be divided without changing its chemical or physical properties; a group of like or different atoms held together by chemical forces.[144,133]

molybdate orange $25PbCrO_4 \cdot 4PbMoO_4 \cdot 1PbSO_4$. Pigment Red 104 (77605). Bright orange, INORGANIC PIGMENT which is a solid solution of lead chromate, lead molybdate, and lead sulfate. Syn: moly orange, molybdate red, chrome vermillion, scarlet chrome

Exhibits excellent HIDING and TINTING strength, brilliant HUE, good PERMANENCY, but only fair chemical resistance.[41] Properties: density, 5.41-6.34 g/cm^3 (4.5-5.8 lb/gal); O.A., 15.8-40 lb/100 lb; particle size, 0.1-1.0 μm.

mombassa gum see ANIMI.[167]

monk's cloth WALLCOVERINGS that simulate a basket-weave material of cotton, jute, or flax.[148]

monkey trade jargon for a BATCH of RESIN which has become THERMOSET and unusable during PROCESSING.[42,152] Syn: biff

monochromatic color scheme combining of COLORS in a room based on one COLOR used in various values and SATURATION or CHROMAS.[43,69,148]

monochromatic light light of a single WAVELENGTH.[43,69]

The degree of monochromaticness depends upon the DISPERSION device and accompanying OPTICAL scheme used to separate the different wavelengths from a continuous or "white" light source. EMISSION lines from elements which, when excited, emit light, such as mercury, sodium, etc., are most nearly true monochromatic light.

monochromator a device for DISPERSING white light into individual WAVELENGTHS, generally a prism or DIFFRACTION GRATING.[145,43,69]

monochrome painting in different VALUES and CHROMAS of one HUE.[71,81,69]

monocyclic terpenes heterogeneous mixture of monocyclic, bicyclic, and other related TERPENE HYDROCARBONS recovered or removed in the fractionation of certain terpenes or other essential OILS, or as a by-product in the chemical conversion of PINENES; generally sold under trade names.[44,164] (ASTM)

monodispersity a POLYMER system which is homogeneous in molecular weight, i.e., lacks MOLECULAR WEIGHT DISTRIBUTION.[169]

monoglyceride partial ester of GLYCEROL in which one of the three HYDROXYL groups is reacted with an ORGANIC acid.[130]

monolithic flooring see SEAMLESS FLOORING.[71,54,163]

monomer COMPOUNDS composed of low molecular weight molecules capable of reacting with like or unlike molecules to form a POLYMER.[130,141,169] See MONOMERIC UNIT (MER).[130,141]

monomeric unit (mer) unit of a POLYMER molecule containing the same kinds and numbers of atoms as the MONOMER (or as one of the monomers, in the case of a CONDENSATION polymer or a copolymer), allowing for atoms removed during the POLYMERIZATION PROCESS [e.g., POLYSTYRENE is composed of mers having the composition $-CH(C_6H_5)CH_2-$. Polyhexamethylene adipamide can be considered to be composed either of mers of the composition $-NH(CH_2)_6NHCO(CH_2)_4CO-$, or of equal numbers of these two types of mers: $-NH(CH_2)_6NH-$ and $-CO(CH_2)_4CO$].[130,141] See MER.[130]

monomethyl ether of hydroquinone (MEHQ) an INHIBITOR for ACRYLIC ACID and its ESTERS.[83] Syn: 4-methoxyphenol

ASTM Test Method D3125 covers the determination of MEHQ in colorless monomeric acrylate esters and acrylic acid.

monotype printing PROCESS of PRINTING from mechanically assembled pieces of type, each of which bears one character.[168]

montan wax hard WAX-like material obtained from lignite by SOLVENT EXTRACTION. It is sometimes described as a bitumen wax.[137]

The crude product is very dark in COLOR, almost black, but after refining it becomes pale yellow. Its mp varies from 72° to 82°C; its acid value, from 25 to 99; and its saponification value, from 58 to 104.

Monthier's blue see IRON BLUE.[41]

montmorillonite a group of CLAY minerals characterized by a weakly bonded sheet-like internal molecular structure; consisting of extremely finely divided hydrous aluminum or magnesium silicates that swell on WETTING, shrink on drying, and are subject to ion exchange.[41,58] (ASTM) See BENTONITE.[41,139] cf. KAOLINITE, ATTAPULGITE, and HECTORITE[41,58]

mop board see BASEBOARD.[82]

mop polishing (1) APPLICATION of a modified French polish with a camel hair "mop" to carved work; (2) POLISHING by friction with a rotary mop.[54] (BSI)

mopping the APPLICATION of hot BITUMEN with a mop or mechanical applicator to the SUBSTRATE or to the plies of a built-up roof.[77,76] (ASTM)

There are four types of mopping: (1) solid—a continuous coating; (2) spot—bitumen is applied in roughly circular areas, generally about 460 mm (18 in.) in diameter, leaving a grid of unmopped, perpendicular areas; (3) strip—bitumen is applied in parallel bands, generally 200 mm (8 in.) wide and 300 mm (12 in.) apart; and (4) sprinkle—bitumen is shaken onto the substrate from a broom or mop in random pattern.

mordant dyes DYES which develop their characteristic COLORS when PRECIPITATED on suitable BASES to form LAKES.[41]

It is often possible to obtain considerable variation in the COLOR of the final lake by altering the metallic RADICAL of the base. Mordant dyes in their original state are colorless.

mordants substances capable of uniting with both DYES and TEXTILE FIBERS so as to improve the bond between dye and textile and give improved textures and alter the colors.[41]

morphology the branch of biology that deals with the form and structure of organisms or one of its parts.[173,133]

mortar (1) material used in a PLASTIC state which can be troweled, and becomes hard in place, to bond units of masonry structures (the word "mortar" is used without regard to the composition of the material, and is defined only with reference to its use as a bonding material as contrasted with the words "STUCCO" and "PLASTER");[73] (2) a mixture of GYPSUM PLASTER with AGGREGATE or hydrated

---

Words presented in CAP/SMALL CAPS type indicate that the word is defined in another part of the *Coatings Encyclopedic Dictionary*.

Numerical superscripts classify terms in one or more of the categories listed in the second section of this volume.

LIME, or both, and water to produce a trowelable fluidity.[73] (ASTM)

mosaic (1) picture or surface DECORATION made by embedding small pieces of colored stone, marble, pottery, or GLASS in MORTAR. When made with stone and subsequently ground flat and smooth, it may be called TERRAZZO;[81] (2) APPEARANCE of a material which is nonhomogeneous in COLOR and approximates the appearance of a ceramic mosaic.[69]

mottle (1) irregular distribution or mixture of colorants or colored materials or GLAZES, giving a more or less distinct APPEARANCE of specks, spots, or streaks of COLOR. Mottling is often purposely achieved, although it may occur accidentally due to improper mixing;[148,69] (2) presence of irregularly shaped and randomly distributed areas of nonuniform appearance in color, GLOSS, or SHEEN. Syn: blotching; (3) figure which greatly increases the value of VENEER for cabinet work, for example, fiddleback.[153,81,35]

mottled spots of different tones and colors placed next to one another to obtain blotchy or patchy effects.[148,69]

mottler flat, thick BRUSH for GRAINING and MARBLING.[59]

mottling (1) the presence of irregularly shaped and randomly distributed areas of nonuniform APPEARANCE of COLOR, GLOSS, and SHEEN;[42,56](ASTM) (2) spotty round marks which appear as a DEFECT in a sprayed film of paint.[42,56] (DAC) Syn: blotching

ASTM Test Method D1848 covers descriptions of properties that may be observed in weathered latex paint films. Mottling/blotching is included.

moulding see MOLDING.[82]

mountain blue see AZURITE.[41]

MSDS abbreviation for MATERIAL SAFETY DATA SHEET.[172]

mu oil name sometimes given to TUNG OIL obtained from *Aleurites montana*.[63]

mucilage (1) FLOCCULANT or slimy deposit which separates from unrefined VEGETABLE OILS on heating;[63] (2) ADHESIVE prepared from a GUM and water. Also, in a more general sense, a liquid adhesive which has a low order of BONDING strength.[79]

mudcracking paint FILM DEFECT characterized by an irregular broken network of cracks in the film.[56,42]

muller an instrument usually of GLASS used for DISPERSING PIGMENTS in VARNISH or other VEHICLES for test purposes.[145,157]

The two glass GRINDING surfaces are generally ground glass. Pigment and vehicle are placed between them and the two rubbed together, either by hand or by machine power. Power driven automatic mullers have two circular glass grinding surfaces.

ASTM Test Method D387 is intended to be used to compare the color and strength of a pigment under test with a reference standard of the same type and grade. Pigments are dispersed in a suitable vehicle with a mechanical muller. Test and standard pigments are treated identically. Opaque drawdowns are made from the dispersions and compared, either visually or instrumentally, for color and strength differences.

mullion narrow dividers between windows or GLASS panels.[82]

multicolor finish a speckled coating containing flecks of small, individual colored PARTICLES different from the base COLOR.[54,71,163,69] Syn: speckled finish

multinotch applicator a DRAWDOWN BLADE with a series of notches of successively higher clearance used to measure sag resistance.[77,145] Refer to ASTM Test Method D4400.

multiple scattering see SCATTERING, MULTIPLE.[43]

Munsell book notation the MUNSELL COLOR NOTATION of a specimen obtained by visual or computational comparison with the MUNSELL HUE, VALUE, and CHROMA SCALES of the MUNSELL BOOK OF COLOR.[69] See MUNSELL RENOTATION.[69]

Munsell book of color a collection of COLOR chips which illustrate the MUNSELL COLOR SYSTEM, in both MATTE and GLOSSY finishes.[69,43]

It consists of pages of constant HUE at 2.5 hue step intervals, with darkest colors (of lowest value at the bottom) increasing in value upwards to the top, and colors of lowest CHROMA in the center fold increasing to highest chroma at the outside edge of the page.

Munsell chroma the COLOR term applied to the psychological sensation dimension of SATURATION.[69,43,35,128]

In MUNSELL COLOR NOTATION, neutrals, white, grays, etc., have a CHROMA of zero. Increasing deviation from gray to purest color in equally perceptible steps are given increasing numerical notations, with intermediate steps designated by decimals proportional to the perceptability differences.

Munsell color notation notation based on the concepts of the MUNSELL COLOR SYSTEM incorporating a number from 1 to 100 to describe the HUE, or a letter or letters with a number for hue description, a number from 0 to 10 to describe the VALUE (lightness scale) from perfect black to perfect white, and a number to describe the CHROMA (saturation) from zero for neutral to the highest possible chroma which varies in real colored materials up to about 15-16 for yellows and reds.[43,69,128] (ASTM)

A typical notation is hue value/chroma, such as 5.2R 4.2/10.8. The use of decimals allows for all possible colors to be defined. The letter hue circle is made up of 10 letters, R, YR, Y, GY, G, BG, B, PB, P, and RP, with each divided into 10 divisions plus any necessary decimals. Thus, 10R is the same as 0YR. This latter hue designation of a number combined with the appropriate letter(s) is most commonly used. A system of letters or numbers, or both, by which the color of any opaque object may be specified with respect to Munsell (1) hue, H, Munsell value, V, and Munsell chroma, C, written in the form, HV/C. The Munsell color notation of a specimen is obtained by visual comparison with the Munsell hue, value, and chroma scales of the Munsell Book of Color, or determined from its daylight REFLECTANCE, *Y, and* CHROMATICITY *coordinates, x, y, in the CIE system by reference to the scales of the ideal Munsell system.*

Munsell Color System a system of specifying colors in surfaces ILLUMINATED by daylight and viewed by an observer adapted to daylight, in terms of three attributes: HUE, VALUE, and CHROMA, using scales that are perceptually approximately uniform.[43,69,35] (ASTM)

ASTM Test Method D1535 provides a means of specifying the colors of objects of the Munsell color order system, a system based on the color-perception attributes. The test method is limited to opaque objects, such as painted surfaces viewed in daylight by an observer having normal color vision. This test method provides a simple visual method as an alternative to the more precise and more complex method based on spectrophotometry and the CIE system (ASTM Test Method E308 and Practice E1164). Provision is made for conversion of CIE data to Munsell notation.

Munsell hue that portion of Munsell notation that corresponds to the sensation attribute of HUE, the attribute which is described by the names of red, yellow, green, blue, and purple. See MUNSELL COLOR NOTATION.[69]

Munsell renotation revision of the Munsell Scales of attributes of surface colors based on studies by the Optical Society of America (OSA) Subcommittee on the Spacing of the Munsell Colors.[43,69]

Current samples and use of the Munsell System is based on this renotation in the CIE System as recommended by the OSA Subcommittee.

Munsell surface-color perception solid a spatial representation of COLORS in the form of a cylindrical coordinate system based on the three perceptual attributes: HUE, LIGHTNESS, and CHROMA.[43,69,35] (ASTM)

This solid forms the basis of the MUNSELL COLOR NOTATION *in which* MUNSELL HUE *corresponds to hue,* MUNSELL VALUE *corresponds to lightness, and* MUNSELL CHROMA *corresponds to chroma. The central vertical axis dimension represents neutral colors, ranging from black at the bottom, through a gradation of grays, to white on the top. The lightness of color perceived as* CHROMATIC *(not gray) is represented by the distance above the base plane. Hue is represented by the angular position about this axis. Chroma is represented by the perpendicular distance from the central axis. If the observer has normal color vision, is adapted to daylight, and views the specimen* ILLUMINATED *by CIE source c or D65, against a medium gray to white background, the Munsell value of the specimen correlates well with the observer's perception of the lightness of the color. Under the same conditions, the Munsell hue correlates well with the observer's perception of hue and the Munsell chroma with the perception of chroma.*

ASTM Test Method D1535 provides a means of specifying the colors of objects in terms of the Munsell color order system.

Munsell value (1) that portion of MUNSELL COLOR NOTATION that corresponds to the sensation of lightness. It is described by a number from zero for a perfect black to ten for a perfect white and is calculated from the equation:[43,69]

$$Y_0 = 1.2219V - 0.23111V^2 + 0.23951V^3 - 0.021009V^4 + 0.0008404V^5$$

(2) the daylight REFLECTANCE of a specimen expressed on a scale extending from 0 for ideal black to 10 for ideal white by steps of approximately equal visual importance. Value notations are within V/. Achromatic or neutral colors are designated N followed by the value notation, thus: N 5.0/.[43,69] (ASTM)

mural painting usually, large picture which is painted on a wall, or fastened to a wall surface.[81]

muscovite see ALUMINUM POTASSIUM SILICATE.[41]

muted colors COLORS whose CHROMA or SATURATION has been lessened or moderated by use of their complementary color or a neutral.[69] Syn: grayed colors or tones which, on losing most of their character as a color, approach "neutrals."

mutual solvent see COSOLVENT.[83,164]

MVP abbreviation for MOISTURE VAPOR PERMEABILITY.[142,157]

Words presented in CAP/SMALL CAPS type indicate that the word is defined in another part of the *Coatings Encyclopedic Dictionary.*

Numerical superscripts classify terms in one or more of the categories listed in the second section of this volume.

MVT abbreviation for MOISTURE VAPOR TRANSMISSION. See SPECIFIC PERMEABILITY (of a film to moisture).[142,157]

MW abbreviation for MOLECULAR WEIGHT.[128]

mycelium the mass of interwoven, threadlike FILAMENTS forming the vegetative part of a fungal plant.[173] See HYPHA and FUNGUS.[173]

mycology the science dealing with FUNGI.[173]

N

N abbreviation for Newton.[128]

N.S. abbreviation for North Standard, used as a dimension in the measurement of the fineness of grind.[64,157,128]

n'gart oil OIL obtained from *Plukenetia conophora*, which grows in the Cameroons.[63]

It polymerizes readily at normal stand-oil-making temperatures (300°C). Average reported constants for this oil are: iodine value, 200; sp gr, 0.937/15°C; and saponification value, 191.

n'gore oil see ISANO OIL.[63]

NAAQS abbreviation for NATIONAL AMBIENT AIR QUALITY STANDARDS.[172]

NACE abbreviation for National Association of Corrosion Engineers.[174]

NACE No. 1 white metal blast cleaned surface finish this FINISH is a surface with a gray-white, uniform METALLIC COLOR, slightly roughened to form a suitable anchor pattern for coatings; this surface shall be free of all OIL, GREASE, DIRT, visible MILL SCALE, RUST, CORROSION products, oxides, paint or any other foreign matter; the surface shall have color characteristic of the ABRASIVE media used; photographic or other visual standards of surface preparation may be used to further define the surface.[158] Syn: white blast

NACE No. 2 near-white blast cleaned surface finish this FINISH is one from which all OIL, GREASE, DIRT, MILL SCALE, RUST, CORROSION products, oxides, paint or other foreign matter have been removed from the surface except for very light shadows, very slight streaks or slight DISCOLORATIONS; at least 95% of the surface shall have the APPEARANCE of a surface blast cleaned to a white METAL surface finish and the remainder shall be limited to the light discoloration mentioned above; photographic or other visual standards of surface preparation may be used to modify or further define the surface.[158]

NACE No. 3 commercial blast cleaned surface finish this FINISH is defined as one from which all OIL, GREASE, DIRT, RUST SCALE, and foreign material have been completely removed from the surface and all rust, MILL SCALE, and old paint have been completely removed except for slight shadows, streaks, or DISCOLORATIONS; if the surface is pitted, slight residues of rust or paint may be found in the bottom of pits; at least two thirds of the surface area shall be free of all visible residues and the remainder shall be limited to light discoloration, slight staining, or light residues mentioned previously; photographic or other visual standards may be used to further define the surface.[158]

NACE No. 4 brush-off blast cleaned surface finish this FINISH is one from which OIL, GREASE, DIRT, RUST SCALE, loose MILL SCALE, loose rust, and loose paint or coatings are removed completely, but light mill scale and tightly adhered rust, paint, and coatings are permitted to remain provided they have been exposed to the ABRASIVE blast pattern sufficiently to expose numerous flecks of the underlying METAL fairly uniformly distributed over the entire surface; photographic or other visual standards of surface preparation may be used to further define the surface.[158] Syn: sweep blast

nacreous pearl-like. See PEARLESCENT.[43]

nacreous pigment TRANSLUCENT PIGMENT which, when added to TRANSPARENT coating material, creates a deep LUSTROUS and pearlescent FINISH.[41,69] See PEARLESCENT PIGMENT.[41,43]

NAD see NONAQUEOUS DISPERSION.[64,71]

nailhead rusting RUST from iron nails that penetrates or BLEEDS through the coating and stains the surrounding area.[42]

nanogram (ng) one billionth, 10^{-9}, of a gram.[128]

nanometre (nm) the unit of length used for describing the WAVELENGTH of light and equivalent to 10^{-9} metres or 10 ANGSTROMS; same as the older term millimicron.[128,43,69]

nap the dense, soft, or fuzzy FABRIC attached to a cylindrical cover used on paint ROLLERS.[59]

nap length see PAINT ROLLER COVER PILE HEIGHT.[59] See NAP.[59]

naphthalene red see MAGDALA RED.[41]

naphthas (1) AROMATIC HYDROCARBONS derived from COAL TAR, although the term "PETROLEUM NAPHTHA" is sometimes used for PETROLEUM SPIRITS of substantially ALIPHATIC type. The coal tar naphthas are the DISTILLATION fractions which contain a complex mixture of aromatic hydrocarbons, including XYLENES; (2) generally refers to hydrocarbon SOLVENTS, both AROMATIC and ALIPHATIC. Hi-flash naphtha is an example of the former and VM & P NAPHTHA an example of the latter.[164]

Words presented in CAP/SMALL CAPS type indicate that the word is defined in another part of the *Coatings Encyclopedic Dictionary*.

Numerical superscripts classify terms in one or more of the categories listed in the second section of this volume.

ASTM D3734 covers the specifications of two types of aromatic hydrocarbon solvents, normally petroleum distillates, having high flash points and moderately low volatility.

naphthenate (1) SALT of NAPHTHENIC ACID; (2) a DRIER used in paints; made with naphthenic acid and cobalt, calcium, iron, zinc, or manganese SALTS of naphthenic acid.[67,74,83] (DAC)

naphthenes C_nH_2n. Found in certain types of crude PETROLEUM.[164] Syn: cycloparaffins or hydrogenated benzenes

naphthenic acid ORGANIC ACID present as a minor constituent of certain grades of crude PETROLEUM.[144]

Commercial naphthenic acids are not pure COMPOUNDS, but consist of a mixture of acids based on cyclopentane rings.

naphthenic solvent HYDROCARBON SOLVENTS comprised wholly or partially of cycloparaffinic (naphthenic) hydrocarbon compounds.[164]

The only common commercial naphthenic solvent is CYCLOHEXANE.

naphthenic type of hydrocarbon CYCLOPARAFFINS.[164]

naphthol blue see MELDOLA BLUE.[41]

naphthol green B bluish-green PIGMENT usually prepared as a complex iron-barium LAKE on ALUMINUM HYDROXIDE.[41]

The starting material is 2-naphthol-6-sulfonic acid.

naphthol green Y bright green PIGMENT made by the interaction of a soluble iron SALT and nitroso-ß-naphthol.[41]

naphthol reds see ALKALI-RESISTANT RED.[41]

naphthol yellow S DYE made by the nitration of 4-nitroso-1-naphthol-2,7-disulfonic acid, or of 1-naphthol-2, 4, 7-trisulfonic acid. It is usually precipitated as the barium complex.[41]

NAPIM abbreviation for NATIONAL ASSOCIATION OF PRINTING INK MANUFACTURERS.[174]

Naples yellow a light yellow PIGMENT; the true pigment is a basic antimonate of lead, but is imitated by mixtures.[41] (DAC) Syn: galliolino, antimony yellow

NAPTAC abbreviation for the NATIONAL AIR POLLUTION CONTROL TECHNIQUES ADVISORY COMMITTEE.[172]

National Air Pollution Control Techniques Advisory Committee (NAPTAC) a group constituted pursuant to § 117 of the CAA consisting of representatives from industry, universities, local air pollution control agencies, and environmental groups.[172]

The committee meets at the request of the Agency to assist in the development of NSPS and NESHAPs. (EPA)

National Ambient Air Quality Standards (NAAQS) are developed by EPA pursuant to Section 109 of the CLEAN AIR ACT. They are standards which, in the judgement of the Administrator of EPA, must be attained and maintained in order to protect the public health, with allowance for an adequate margin of safety.[172] (EPA)

"Secondary" NAAQS specify a level necessary to protect the public welfare. NAAQS exist for nitrogen oxides, sulfur oxides, particulate matter, ozone, and carbon monoxide.

National Association of Corrosion Engineers (NACE) a professional engineering society dedicated to advancing the knowledge, understanding and practice of corrosion prevention and control. NACE International has over 15,000 members worldwide.[174]

Its address is: 1440 South Creek Dr., Houston TX 77084-4906; (713) 492-0535.

National Association of Printing Ink Manufacturers (NAPIM) a trade association whose purpose is to represent the printing ink industry in the USA and to provide direction to management in the areas of environmental issues, business management, government regulations, and regulatory compliance. All NAPIM affairs shall be conducted, without exception, in compliance with all governmental regulations and within well accepted business practices.[174]

Its address is Heights Plaza, 777 Terrace Ave., Hasbrouck Heights, NJ 07604; (201) 288-9454.

National Cancer Institute (NCI) part of the National Institutes of Health that studies cancer.[174,172]

National Decorating Products Association (NDPA) a not-for-profit retail trade association representing the independent retailers of paint, wallcovering, window coverings, floor covering, and related sundries and equipment. Its mission is to provide these retailers with sales training, product-related, and business operations information and programs to help them better operate their businesses. NDPA has a membership of approximately 3500 retailers who own approximately 6500 stores.[174]

Its address is 1050 N. Lindbergh, St. Louis, MO 63121-2994; (800) 737-0107.

National Fire Protection Association (NFPA) an international voluntary membership organization to promote/improve fire protection and prevention and establish safeguards against loss of life and property by fire.[174,172] See FIRE DIAMOND.[172]

It is best known for the National Fire Codes, sixteen volumes of standards, recommended prac-

tices, and manuals developed (and periodically updated) by NFPA committees. NFPA 704M publication is the code for showing hazards of materials using the familiar diamond-shaped label with appropriate numbers or symbols (NFPA hazard rating).

National Emission Standards for Hazardous Air Pollutants (NESHAP) regulations as defined in Section 112 of the CLEAN AIR ACT which regulate ASBESTOS, beryllium, BENZENE, arsenic, VINYL CHLORIDE, radionuclies, and any other substance for which no ambient air quality standards is applicable and which in the judgement of the EPA Administrator may cause an increase in mortality, or an increase in serious irreversible or incapacitating reversible illness.[172] (EPA)

National Institute of Occupational Safety and Health (NIOSH) the agency of the Public Health Service that tests and certifies respiratory and air-sampling devices. It recommends EXPOSURE LIMITS to OSHA for substances, investigates incidents, and researches occupational safety.[172,174]

National Paint and Coatings Association (NPCA) a nonprofit trade association representing some 500 paint and coatings manufacturers, raw materials suppliers, and distributors. NPCA's primary role is to serve as an advocate for the coatings industry on legislative, regulatory, and judicial issues at the federal, state, and local levels.[174]

Its address is: 1500 Rhode Island Ave., N.W., Washington, D.C. 20005-5597; (202) 462-6272.

National Toxicological Program (NTP) Federal activity overseen by the Department of Health and Human Services with resources from the National Institutes of Health, the Food and Drug Administration, and the Centers for Disease Control.[172,174]

Its address is: NTP Information Office, MD B2-04, Box 12233, Research Triangle Park, NC 27709.

native green see HYDRATED CHROMIUM OXIDE.[41]

natural finish any FINISH resulting from the APPLICATION of a TRANSPARENT substance (such as a VARNISH, WATER-REPELLENT PRESERVATIVE, SEALER, or OIL) which does not affect significantly the original COLOR or GRAIN.[71,163] (DAC)

natural gums fossilized or animal GUM RESINS.[167]

natural iron oxides natural mineral deposits consisting chiefly of ferric oxide.[41,34] See IRON OXIDES.[41,34]

COLOR range of yellow, red, brown, and black, consisting chiefly of ferric oxide.

natural pigments see EARTH PIGMENT.[41]

natural red 4 (75470) see CARMINE LAKE.[41]

natural resin see RESIN, NATURAL.[167]

natural spreading rate the SPREADING RATE that occurs when a coating is applied in a manner natural to the operator's technique, perceptions, and expectations, as they relate to coating tools, SUBSTRATE, and characteristics of the coating itself.[54,77]

Such a spreading rate can vary widely with the same paint applied under similar conditions by different operators, but a series of paints applied by different operators under the same conditions will tend to have approximately the same rank order. In BRUSH application the natural spreading rate is influenced strongly by viscous FLOW resistance referred to as brush drag, for which reason natural spreading rate generally ranks similarly to the high shear viscosity (about 12,000 s^{-1}) of the paints being applied. Paints applied by roller or other application procedures do not necessarily rank in the same order. ASSOCIATIVE THICKENERS are added to some paints to increase brush drag and thereby limit their spreading rate to obtain a thicker paint film.

natural yellow oxide see IRON OXIDES, NATURAL.[41]

naval stores chemically reactive OILS, RESINS, TARS, and PITCHES derived from the oleoresin contained in, EXUDED by, or EXTRACTED from trees chiefly of the pine species (*Genus pinus*), or from the WOOD of such trees.[44] (ASTM)

navy pitch see PITCH, NAVY.[44]

NBS Color Difference Equation COLOR DIFFERENCE EQUATION devised by Judd and modified by Hunter. The TOTAL COLOR DIFFERENCE, ΔE, is called an NBS unit.

$$\Delta E = f_g\left\{\left[221\,Y_m^{1/4}\left((\Delta\alpha)^2+(\Delta\beta)^2\right)^{1/2}\right]^2+\left[k(\Delta Y^{1/2})\right]^2\right\}^{1/2}$$

where

$$\alpha = \frac{2.4266x - 1.3631y - 0.3214}{1.0000x + 2.2633y + 1.1054}$$

$$\beta = \frac{0.5710x + 1.2447y - 0.5708}{1.0000x + 2.2633y + 1.1054}$$

$$Y_m = \frac{Y_1 + Y_2}{2} \qquad \Delta\alpha = \alpha_1 - \alpha_2,\ \Delta\beta = \beta_1 - \beta_2,\ \Delta Y = Y_1 - Y_2$$

(Y is expressed in percent)

k = constant expressing the relative importance of lightness and chromaticness in a particular viewing arrangement;

k = 12 for comparison made across a very narrow dividing line. With increasing separation of the two colors, the value of k decreases; k = 10 is frequently used.

f_g = factor which takes into account the mask-

Words presented in CAP/SMALL CAPS type indicate that the word is defined in another part of the *Coatings Encyclopedic Dictionary.*
Numerical superscripts classify terms in one or more of the categories listed in the second section of this volume.

ing effect of gloss on the detection of color difference; for normal observation

$$f_g = \frac{Y_m}{Y_m + 2.5}$$

(Y_m is measured with SPECULAR REFLECTANCE excluded.) If f_g is calculated in this way f_g decreases very rapidly when the Y_m becomes less than 10%, where f_g = 0.8. At 1% f_g = 0.287, for example. For ease in calculation k is frequently used as 10.000, especially where colors are not going to be used adjacently.[69]

NBS unit the TOTAL COLOR DIFFERENCE, ΔE, calculated by means of the NBS COLOR DIFFERENCE EQUATION, sometimes referred to as a Judd or a Judd Unit.[43,69,128] See NBS COLOR DIFFERENCE EQUATION, and HUNTER COLOR DIFFERENCE EQUATION.[43,69,134]

The term is frequently used erroneously to refer to the color difference calculated by other equations, normalized to agree in magnitude on the average to an NBS unit, or to the ΔE measured on tristimulus colorimeters with electrical circuits designed to give an approximation of the NBS Color Difference Equation, frequently one of the Hunter L, a, b equations. Such incorrect use of the term is to be discouraged. Description of the exact equation and method of measurement used will avoid confusion.

NCCA abbreviation for National Coil Coaters Association.[174]

NCI abbreviation for the NATIONAL CANCER INSTITUTE.[174,172]

NDPA abbreviation for NATIONAL DECORATING PRODUCTS ASSOCIATION.[174]

neatsfoot oil see ANIMAL OILS.[63]

negative catalyst see CATALYST.[74]

neoprene SYNTHETIC RUBBER POLYMER derived from 2-chloro-1, 3-butadiene.[140]

neoprene paint paint based upon PIGMENTED SOLUTIONS of the proprietary synthetic RUBBER neoprene (polychloroprene rubber), a VULCANIZING agent being added before use.[71] (BSI)

neou oil seed oil of the tree, *Parinarium macrophyllum*, which grows in several parts of West Africa.[63]

The main constituent acids of the oil are an ISOMER *of α-elaeostearic acid and* LINOLEIC ACID. *The oil has an iodine value of 135 and a saponification value of 190.*

nepheline a mineral of sodium or potassium-aluminum silicate, occurring worldwide in igneous rocks.[41,34] Syn: nephelite

nepheline syenite a mineral AGGREGATE consisting chiefly of ALBITE, microline, and NEPHELITE, each in significant amount.[41,58,34] (ASTM)

It is commonly used as an EXTENDER.

nephelite see NEPHELINE.[41]

nerve a condition difficult to define fully, but commonly used to denote the qualities of firmness, strength, and elasticity in crude RUBBER.[47,140]

In crude rubber, nerve is reduced or destroyed by milling or breakdown.

NESHAP abbreviation for the NATIONAL EMISSION STANDARDS FOR HAZARDOUS AIR POLLUTANTS.[172] (EPA)

neuberg chalk ALUMINUM SILICATE used to some extent as a FILLER or EXTENDER, but it has special properties as a constituent of METAL POLISHES.[58]

neutral (1) having no distinguishable HUE; ACHROMATIC; (2) not decided in COLOR; nearly achromatic; of low saturation.[69,43] See GRAY SCALE.[69,43]

neutral gray an ACHROMATIC gray representing a portion of the GRAY SCALE.[69,43]

neutral oil a light gravity mineral oil derived from PETROLEUM and used as a LUBRICANT in RUBBING finished work.[54,63]

neutral toner a STAIN of yellow-orange COLOR used to blend colored WOOD streaks in wood finishing.[54,71]

neutralization equivalent or number see ACID NUMBER.[157]

neutralizing (Brit.) SPRAY FINISH deliberately applied to produce an uneven or MOTTLED APPEARANCE.[54] (BSI)

new source any stationary source the construction or modification of which commences after publication of regulations or proposed regulations which prescribe a standard of performance which is intended to apply to that type of air pollution emission source.[172] (EPA)

new source performance standards (NSPS) standards for emission of air pollutants from new, modified, or reconstructed stationary emission sources which reflects the degree of emission limitation achievable through the application of the best system of emission reduction which (taking into account the cost of achieving such reduction) the administrator determines has been adequately demonstrated.[172] (EPA)

The CLEAN AIR ACT *usually refers to these as standards of performance for new stationary sources.*

newel large post at the termination of a stair rail.[82]

news inks PRINTING INKS designed to run on NEWSPRINT, consisting basically of CARBON BLACK or colored pigments dispersed in MINERAL OIL VEHICLES, which dry by ABSORPTION.[168]

Some news inks utilize EMULSION, OXIDATION, and heat-set systems.

newsprint the unsized, absorbent PAPER used in newspaper printing.[168]

newtonian flow type of FLOW characterized by a RATE OF SHEAR that is directly proportional to the SHEARING FORCE.[166] See NON-NEWTONIAN LIQUID.[166]

Newtonian flow can occur, at least ideally, under the influence of an infinitesimally small force. It is to be distinguished from PLASTIC flow which occurs only when a finite minimum force is exceeded. OILS, at sufficiently low rates of shear, exhibit newtonian flow. The curve obtained for a newtonian liquid on a rotational viscometer of the Couette type (RPM vs. deflection of torsion dial) passes through the origin.

newtonian liquid one in which the RATE OF SHEAR is proportional to the SHEARING STRESS.[166] (ASTM) Syn: simple liquid

The constant ratio of the shearing stress to the rate of shear is the VISCOSITY of the liquid. If the ratio is not constant, the liquid is NON-NEWTONIAN.

NFRA Hazard Rating see FIRE DIAMOND.[172]

NGR stain see NONGRAIN-RAISING STAIN.[71,54,163]

nibs small pieces of foreign material, pieces of SKIN, COAGULATED medium, etc., which project above the surface of an applied film, usually VARNISH or LACQUER.[56] See BITTY.[56,42]

nickel antimony titanate see NICKEL TITANATE.[41]

nickel azo yellow $C_{30}H_{18}O_4N_6Cl_2Ni$. Pigment Green 10 (12775). A nickel-chelated, AZO yellow PIGMENT derived from para chloraniline and 2,4-dihydroxyquinoline. It has excellent LIGHTFASTNESS in MASSTONE and TINT, FUME resistance, BAKE resistance, high CHROMA, and good TRANSPARENCY for metallized polychromatic finishes.[41] Syn: green gold

The major limitation of this pigment is its greenish HUE. When reduced with white, it produces an extremely lightfast but greenish yellow tone; in masstone, the pigment is green. The chemical resistance of finishes with this pigment is good; however, being a METAL chelate, its stability of pH extremes is somewhat lower than that of a true SALT.

nickel titanate $NiO{\cdot}Sb_2O_3$ 20 TiO_2. Pigment Yellow 53 (77788). An entire series of titanate PIGMENTS that range in HUE from yellow to brown and gray by substitution of nickel and antimony with other metals, e.g., chromium, cobalt, iron, zinc, and various combinations.[41] Syn: titanium yellow, nickel antimony titanate, nickel titanate yellow

Properties: density, 4.1-4.7 g/cm³ (34.2-39.2 lb/gal); O.A., 10-20; particle size, 0.5-1.2 μm.

nickel titanate yellow see NICKEL TITANATE.[41]

night vision see SCOTOPIC VISION.[69,43]

nigrosines deep blue or black ANILINE DYES.[41]

NIOSH abbreviation for NATIONAL INSTITUTE OF OCCUPATIONAL SAFETY AND HEALTH.[172,174]

nishiki-ye (Japanese) Brocade picture.[81]

The term was used at first to describe the brilliant COLOR invention of the Japanese artist, Harunobu, but is now loosely applied to all POLYCHROME prints.

nitrate green green PIGMENT made from nitrate lead chrome and ferriferro cyanide.[41]

nitration PROCESS of chemically adding nitrogen in combination with oxygen to any material.[131]

nitrobenzene $C_6H_5NO_2$. Mol wt, 123; bp, 210.85°C; mp, 5.7°C; sp gr, 1.1987; flp, 88°C (190°F).[144] Syn: mirbane oil

nitrocellulose another name for CELLULOSE NITRATE.[151,144] See DOPE COTTON, and PYROXYLIN.[130,151]

The product obtained by treating cellulose with a mixture of nitric and sulfuric acids. It is primarily used in the coatings industry as a base for lacquers and as a FILM-FORMING material widely used in flexographic and gravure inks.

ASTM test methods for nitrocellulose include: base SOLUTIONS, test methods, D365; content in alkyd lacquers, D3133; dilution ratio for solutions, D1720; testing, D301.

nitroethane $CH_3CH_2NO_2$. Used as SOLVENT in ORGANIC synthesis.[164]

Properties: bp, 114°C; flp, 41°C (106°F); vp, 16 mm Hg/20°C; sp gr, 1.052/20°C.

nitromethane CH_3NO_2. Used as SOLVENT in ORGANIC SYNTHESIS and for ZEIN.[164]

Properties: bp, 101°C; flp, 45°C (112°F); vp, 28 mm Hg/20°C.

nitroparaffins nitrated HYDROCARBONS, e.g., nitroethane, nitromethane, etc.[144]

1-nitropropane $CH_3CH_2CH_2NO_2$. Properties: bp, 132°C; flp, 50°C; vp, 7.5 mm Hg/20°C; sp gr, 1.003/20°C.

Words presented in CAP/SMALL CAPS type indicate that the word is defined in another part of the *Coatings Encyclopedic Dictionary.*
Numerical superscripts classify terms in one or more of the categories listed in the second section of this volume.

2-nitropropane $CH_3CHNO_2CH_3$. Properties: bp, 120°C; vp, 13 mm Hg/20°C; flp, 40°C; sp gr, 0.992/20°C.[164]

nm abbreviation for NANOMETRE.[128] cf. ANGSTROM UNIT[128,43]

NOC abbreviation for not otherwise classified.[172]

NOEL abbreviation for no effect level.[172]

nomenclature a systematically arranged list of terms and concepts in a particular field of knowledge, which may or may not include DEFINITIONS and other characteristics.[134] (ASTM)

nominal value a value assigned for the purpose of convenient designation; existing in name only.[128] (ASTM)

nonaqueous dispersion (NAD) the SOLVENT analog of a LATEX: the POLYMER is DISPERSED in a volatile ORGANIC liquid which is not a solvent for the polymer.[71]

NADs have a much higher solids content than conventional high molecular weight solvent coatings. Like latices, the VISCOSITY is independent of the molecular weight as opposed to solvent soluble resins.

nonattainment area an area, as defined under Section 107 of the CLEAN AIR ACT, which is in violation of one or more of the national ambient air quality standards.[172] (EPA)

nonbreak oil cf. BREAKFREE OIL[63,131]

noncombustible that which does not burn; preferred to "incombustible" by fire authorities.[172]

nonconditional match COLORS which match to all observers under all light sources.[69]

A NONMETAMERIC MATCH.

nonconvertible coating FILM-FORMER which, after being deposited from a solution, dries to give a film which can be redissolved in a SOLVENT from which it was originally deposited.[55,71] See CONVERTIBLE COATING.[55,71]

nondestructive test a test to determine the characteristics or properties of a material or substance that does not involve its destruction or DETERIORATION (e.g., X-ray examination, ultra high frequency sound, etc.).[157]

nondrying coating coating that does not dry in the course of a regular schedule.[42] See GREASE PAINT.[163]

nondrying oil OIL which does not of itself possess to a perceptible degree the power to take up oxygen from the air and lose its liquid characteristics.[63] (ASTM) Syn: fixed oil (DAC)

nonflammable incapable of being easily ignited or burning with extreme rapidity when lighted. Also, a DOT hazard class for any compressed gas other than a flammable one.[172] cf. INFLAMMABLE and FLAMMABLE[172]

nongrain-raising stain (NGR) one of many liquid wood stains, based on ALCOHOL or other SOLVENT; almost totally free of water.[71,163,54] (DAC)

These are usually ORGANIC solvent-borne because of the tendency of water to cause the grain of wood to swell.

nonionic surfactants those containing HYDROPHILIC groups which do not ionize appreciably in aqueous SOLUTIONS.[127]

nonmetameric match a pair of COLORS which appear to be identical to all observers under all conditions of ILLUMINATION and viewing; an unconditional match. cf. METAMERISM.[43,157,69] Syn: nonconditional match

Popularly, the term is frequently used to apply to a pair of colors which has identical SPECTROPHOTOMETRIC REFLECTANCE or TRANSMITTANCE CURVES, as measured in a prescribed geometric configuration, thus ignoring the effects of any differences in geometric characteristics between the samples in the pair. In practice, the term may be used to describe a pair of samples which does not appear to change in comparison to one another when viewed under several different light sources.

non-newtonian liquid any liquid that exhibits a VISCOSITY which varies with changing SHEAR STRESS or SHEAR RATE; thus, any liquid which does not satisfy the requirements for a newtonian liquid, i.e., that displays PLASTIC, PSEUDOPLASTIC, or DILATANT FLOW characteristics. Most paints are non-newtonian liquids.[166] (PTM)

non-newtonian oil or fluid a fluid that exhibits a VISCOSITY which varies with changing SHEAR STRESS or SHEAR RATE.[166] (ASTM)

nonpenetrating stain see OIL STAIN.[71]

nonpolar solvents AROMATIC and PETROLEUM HYDROCARBON groups characterized by low DIELECTRIC CONSTANTS.[164]

nonporous wood WOOD devoid of pores (vessels); same as SOFTWOOD or CONIFEROUS wood.[153]

nonscratch inks INKS which have high ABRASION and mar resistance when dry.[168]

nonskid paint see ANTISLIP PAINT.[71,163]

nonspecular reflectance REFLECTANCE other than the "mirror" reflectance that occurs at the angle equal and opposite to the incident angle; diffuse reflectance. See SPECULAR REFLECTANCE EXCLUDED, REFLECTANCE, DIFFUSE, and FRESNEL REFLECTION.[43,69]

nonvolatile matter (NVM) or content of a coating, the solid material remaining after the volatiles have been driven from the film under specified test conditions. The total percentage volatile present is obtained by subtracting the nonvolatile content from 100.[152] (ASTM) Syn: solids, total solids

ASTM test methods for nonvolatile/volatile content include: aluminum powders/pastes, D480; coatings, by volume, D2697; driers/naval stores, D4140; emulsions, resin solutions, D4209; guide to test methods, D2832; latices, test, D4758; pigment dispersions, D4139; printing inks/resin solutions, test, D4713; resin solutions, test, D1259; silane/siloxane masonry treatments, D5095; titanium dioxide slurries, D3926; varnishes, test, D1644; volatile solvents, test, D1353.

norelac resins particular type of NYLON RESIN in which the POLYMERIC ACIDS are those derived from DRYING OIL FATTY ACIDS.[130]

These products differ from the true nylon resins in that they are much more soluble in solvents, and can be used as constituents of VARNISHES which possess air-drying properties.

normal color vision vision of a normal observer who exhibits no symptoms of anomalous or defective COLOR response.[69,43]

normal conditions a qualitative term, dependent on the preference of the investigator; it often implies ambient pressure and "ROOM TEMPERATURE."[157] (IUPAC)

Preferably the variables of temperature and pressure should be quoted as values representative of the actual conditions (or range of conditions) employed in the study.

normal lead silico chromate normal lead chromate (medium shade) formed as a strongly adherent coating on a core of SILICA at a 50/50 weight ratio.[41] Syn: normal lead chromate coated/silica cored PIGMENT

Its major use was in traffic paints in which its lower density and lower cost were advantageous compared to pigmentations of pure lead chromate and extender. Density, 3.5 g/cm^3 (29.2 lb/gal); O.A., 14; particle size, 5 μm.

north light light from the north sky, generally through a window so located as to allow only light from the north to enter the room.[43,69] See DAYLIGHT.[43,69,157]

Because such light is the more constant throughout the day than light from other directions, it was and is preferred by artists, colorists, etc. for viewing COLOR. Artificial north light has been simulated with modern lamps and filters for the same purpose.

north standard see N.S.[64,157,128]

NOS abbreviation for not otherwise specified.[172]

novacite see SILICA, MICROCRYSTALLINE.[58,41]

novaculite see SILICA, MICROCRYSTALLINE.[58,41]

novelty siding see DROP SIDING.[82,153,75]

novolak PHENOLIC-ALDEHYDIC RESIN which, unless a source of methylene groups is added, remains permanently THERMOPLASTIC.[79,130] (ASTM) See RESINOID and THERMOPLASTIC.[130]

NO_x oxides of nitrogen (NO, NO_2, and N_2O_4).[144,172]

They react with the moisture in the respiratory tract to produce acids that corrode and irritate tissue, causing congestion and pulmonary edema.

NPCA abbreviation for NATIONAL PAINT AND COATINGS ASSOCIATION.[174]

NPIRI abbreviation for National Printing Ink Research Institute.[174]

nsa-sana oil see ESSANG OIL.[63]

NSPS abbreviation for NEW SOURCE PERFORMANCE STANDARDS.[172] (EPA)

NTP abbreviation for the NATIONAL TOXICOLOGICAL PROGRAM.[172,174]

nubs size grading of natural RESINS, about the size of the end of a finger.[167]

nuisance threshold that concentration of an air pollutant that is considered objectionable.[172] (IUPAC)

In the case of a substance with an objectionable odor, it is the smallest concentration of the substance which can be detected by a human being (nose).

Nurnberg violet see MINERAL VIOLET.[41]

NV abbreviation for NONVOLATILE.[142,156,157]

NVM abbreviation for NONVOLATILE MATTER.[142,156,157]

nylon resins POLYAMIDE RESINS made from the interaction of diamines and dicarboxylic acids.[130,156]

Hexamethylene diamine and adipic acid are typical reactants. These resins are composed principally of a long-chain synthetic polymeric amide which has recurring amide groups as an integral part of the main POLYMER chain.

Words presented in CAP/SMALL CAPS type indicate that the word is defined in another part of the *Coatings Encyclopedic Dictionary.*

Numerical superscripts classify terms in one or more of the categories listed in the second section of this volume.

O

O.A. abbreviation for OIL ABSORPTION.[63,64,157,128]

O.P. abbreviation for OVERPROOF.[164]

oak varnish short oil type of VARNISH for interior use, normally based on natural COPALS.[152]

OAQPS the Office of Air Quality Planning and Standards, a part of the U.S. Environmental Protection Agency.[172]

oatmeal paper rough-textured WALLPAPER in tan, EGGSHELL, or pale mixtures of COLOR which were very popular earlier in this century.[148]

object color COLOR of light reflected by an object as normally observed, as contrasted to other types of observed color.[43,69] See OBJECT MODE OF APPEARANCE.[43,69,35]

object mode of appearance mode of perceiving the APPEARANCE of tangible objects as located at a particular place; object mode includes two of the five recognized modes of appearance, that of surfaces and of bulk volume.[43,69,35] See MODES OF APPEARANCE.[43,69,35]

objet d'art (Fr.) any object of artistic merit; a curio.[81]

obliterating power see OPACITY.[43](BSI)

observation angle the angle between the INCIDENT RAY measured relative to the perpendicular or normal and the VIEWING ANGLE, also measured relative to the normal; therefore, the total subtended angle between incident and viewing. It is the same as "bend angle" in the literature relating to projection screens, and as "divergence angle" in some federal and military specifications.[43] (FTMS) Syn: viewing angle

observer the human viewer who receives a stimulus and experiences a sensation from it.[43,69]

In vision, the stimulus is a visual one and the sensation is an APPEARANCE.

observer, standard the SPECTRAL response characteristics of the average observer defined by the CIE.[43,69,134]

Two such sets of data are defined, the 1931 data for the 2° visual field (distance viewing) and the 1964 data for the annular 10° visual field (approximately arm's length viewing). By custom, the assumption is made that if the observer is not specified, the TRISTIMULUS data has been calculated for the 1931, or 2° field observer. The use of the 1964 data should be specified.

OCCA see OIL & COLOUR CHEMISTS' ASSOCIATION. (Brit.)[174]

occupational exposure limit (OEL) see EXPOSURE LIMITS.[172]

Occupational Safety and Health Administration (OSHA) part of the U.S. Department of Labor. The regulatory and enforcement agency for safety and health in most U.S. industrial sectors.[172,174]

When disagreements arise over the results of safety and health inspections performed by OSHA, employers have the right to dispute any alleged job safety or health violation found during the inspection, the penalties it proposed, and the time given by the agency to correct any hazardous situation. Employees and representatives of employees may initiate a case by challenging the propriety of the time OSHA has allowed for correction of any violative condition. Documents are available from OSHA Technical Data Center Docket Office, Rm N-3670, 200 Constitution Ave., N.W., Washington, D.C. 20210; (202) 523-7894.

ocher, ochre (yellow, golden, red) there are two types of ocher: (1) synthetic, and (2) natural (a naturally occurring yellow-brown hydrated iron oxide).[41]

Synthetic ocher possesses little advantage over the natural earth PIGMENTS, the difference being the method of manufacture. Natural ocher is found in France, Italy, Spain, Africa, Great Britain, and the United States.

ASTM Test Method D50 covers the procedure for the chemical analysis of these ochers as well as the analyses of yellow, orange, red, and brown pigments containing iron and manganese. ASTM D85 is the standard specification for ochre pigment.

octoates (1) SALTS of octoic, or 2-ethyl hexoic acid; (2) DRIERS used in paints; made with the cobalt, manganese, calcium, zinc, or iron salts of 2-ethyl hexoic acid.[67,74,83]

octyl acetate $CH_3COOCH_2CH(C_2H_5)C_4H_9$. High-boiling SOLVENT.[144,164] Syn: 2-ethyl hexyl ethanoate

Properties: bp, 199°C; sp gr, 0.873/20°C; flp, 83°C (180°F); vp, 0.3 mm Hg/20°C.

octyl alcohol high-boiling SOLVENT used for controlling FLOW and for coupling with HYDROCARBON solvents to reduce high VISCOSITIES.[164]

It is also used as an etherifying agent for UREA-FORMALDEHYDE resins to confer drying oil and ALIPHATIC hydrocarbon compatibility. Properties: bp, 194°C; sp gr, 0.834/20°C; flp, 81°C (178°F); vp, 0.3 mm Hg/20°C.

octyl phenol (tertiary) PHENOL used as an alternative to tertiary butyl phenol in the manufacture of pure oil-reactive PHENOLIC RESINS.[141]

odor that property of a substance which affects the sense of smell; any smell; scent; perfume.[42] (ASTM)

ASTM Test Method D1296 covers a comparative procedure for observing the characteristic and residual odors of volatile organic solvents and diluents to determine their odor acceptability in a SOLVENT system.

odor threshold the concentration of a COMPOUND which produces an odor which is detectable by a human being (nose).[172] (IUPAC)

For certain compounds this threshold is very low (e.g., 1 part in 10^9 for certain sulfides).

odorless mineral spirits see ODORLESS SOLVENT.[164]

odorless paint a paint such as a water-base LATEX paint or an OIL- or ALKYD-base paint which contains an odorless mineral spirit as a THINNER; produces a minimum amount of ODOR during application.[71] (DAC)

odorless solvent SOLVENTS generally of the mineral spirits type, that are SYNTHESIZED by the alkylation PROCESS and refined to remove odorous aromatics and sulfur compounds.[164]

OEL abbreviation for OCCUPATIONAL EXPOSURE LIMIT.[172]

OEM abbreviation for ORIGINAL EQUIPMENT MANUFACTURERS.[163]

OEM coatings original equipment manufacturers coatings, which include automotive, marine, furniture, and appliance, as well as many other miscellaneous consumer and industrial applications.[163]

oenanthal ALDEHYDIC DECOMPOSITION product derived during the DEHYDRATION of CASTOR OIL.[144] Syn: enanthaldehyde

off-color unacceptable COLOR difference.[69,157]

Nonmatching at different shade.

off-white COLOR which is obviously not white, but which is not sufficiently far removed from white to enable it to be called by a definite color name.[69,43] (BSI)

Official Digest one of the previous names of the present JOURNAL OF COATINGS TECHNOLOGY, or JCT, the official journal of the FEDERATION OF SOCIETIES FOR COATINGS TECHNOLOGY.

offset an indirect form of printing in which the INK is transferred from the printing plate to a RUBBER blanket and subsequently to the sheet.[168] See LITHOGRAPHY.[54,77,81]

offset lithography a method of printing in which the image area of the printing roll is essentially at the same level as the nonimage area.[168,81] See LITHOGRAPHY and OFFSET.[168,81]

The INK wets the image area only and the nonimage area is wetted by water. Ink is transferred from the printing roll to the RUBBER surface on a roll called the blanket cylinder which transfers the ink to the PAPER. When the WEB or continuous roll of paper is used, the process is called "web offset."

offset paper a variety of printing PAPER especially sized for offset lithography; it is noncurling and absorptive.[168]

offsetting of ink see SET OFF OF INK.[168]

oil any of numerous mineral, vegetable, and synthetic substances, and animal or vegetable fats that are generally unctuous, slippery, combustible, viscous, liquid or liquefiable at room temperatures, soluble in various ORGANIC solvents, but not in water.[63] See DRYING OIL, NONDRYING OIL, and SEMI-DRYING OIL.[63]

Oil & Colour Chemists' Association (OCCA) an organization composed of technical personnel of the surface coatings and allied industries of the United Kingdom.[174]

Founded in 1918, OCCA is composed of 11 sections within the U.K. and seven foreign sections under a General Secretariat located in London. OCCA promotes the technical advancement of paints, varnishes, printing inks, and allied industries through sponsorship of meetings, seminars, and exhibitions. It also publishes a monthly journal, Surface Coatings International. OCCA is a member organization of the COATINGS SOCIETIES INTERNATIONAL.

oil absorption (value) (O.A.) quantity of oil required to wet completely a definite weight of PIGMENT to form a stiff PASTE when mechanically mixed.[63,64,157,128] cf. BINDER DEMAND[157,156]

The oil ABSORPTION number or value of a pigment is the number of millilitres or grams of oil, usually acid refined LINSEED OIL, used to bind together 100 g of pigment under specified conditions of test. The unit used should be stated. The figure is not absolute, but depends on the operator as well as on the method of determination. The term "oil absorption" is frequently used as a measure of PLASTIC VISCOSITY. As such it can lead to unpredictable results, since oil absorption as measured bears little relation to a finished MILL-ground pigment dispersion. Oil absorption has meaning of a qualitative sort, however, in relation to a preliminary mixer operation.

Words presented in CAP/SMALL CAPS type indicate that the word is defined in another part of the *Coatings Encyclopedic Dictionary.*

Numerical superscripts classify terms in one or more of the categories listed in the second section of this volume.

ASTM test methods for oil absorption of pigments include: Gardner-Coleman method, D1483, and spatula rub-out test, D281.

oil, blown see BLOWN OIL.[63]

oil cake a cake or MASS of LINSEED, COTTONSEED, etc., from which the oil has been EXPRESSED.[63]

oil can in the COIL COATING industry, METAL irregularities of an otherwise flat looking metal sheet when viewed at near grazing angle.[43,70]

oil color OIL paint containing a high CONCENTRATION of colored PIGMENT, commonly used for tinting paint.[41,64] (ASTM) Syn: oil paste

oil content see OIL LENGTH.[156]

oil, debloomed mineral OILS which show a decided blue or purple cast when examined in a test tube or when spread in a thin film over a dark colored surface are said to possess a BLOOM.[63]

When an oil has been so treated as to remove this COLOR effect, it is said to be debloomed. Debloomed RUBBING oil gives a clearer, sharper, rubbed surface than that which shows a bloom.

oil, drying any UNSATURATED OIL which will react with oxygen from the air and dry to a relatively HARD, TOUGH, ELASTIC substance when spread out in a thin FILM.[63]

oil, fixed see NONDRYING OIL.[63]

oil gilding APPLICATION of METAL LEAF or GOLD LEAF onto a sealed surface, using OIL VARNISH as a SIZE.[148,158]

The FINISH is more MATTE than that produced by water gilding.

oil green see CHROMIUM OXIDE GREEN.[41]

oil, heat-bodied DRYING or SEMIDRYING OIL which has been heated to an elevated temperature and held at this temperature for the time required to produce a specific increase in VISCOSITY, as distinct from oil POLYMERIZED by a PROCESS involving OXIDATION.[63,131,169,176]

oil length ratio of OIL to RESIN in a medium.[156,176] See LONG OIL, MEDIUM OIL VARNISH, and SHORT OIL VARNISH.[156]

For an oleoresinous varnish, the oil length may be expressed in terms of parts by weight of oil to one part by weight of resin or, in American practice, in terms of U.S.A. gallons of oil per 100 lb of resin. Thus, a 25 gallon varnish would mean, in American usage, a varnish composed of 25 U.S.A. gallons of oil to 100 lb of resin. For an ALKYD RESIN, the oil length is expressed as the percentage of oil by weight in the resin.

ASTM test methods for oil content include: artists' paints, specification, D4302; and solvent-borne paints, D2245.

oil, long see LONG OIL.[63,152,156]

oil melt solution of RESIN in oil, obtained by heating.[63,131]

oil, nondrying see NONDRYING OIL.[63]

oil of lemon volatile OIL expressed from fresh peel of *Citrus limonum*.[63]

It contains limonene, terpinene, phellandrene, and pinene.

oil of mirbane see NITROBENZENE.[144]

oil of pine tar certain heavier fractions of the volatile OIL recovered by distilling pine-tar oil to convert it into pine TAR.[44,63] (ASTM)

oil of turpentine pharmaceutical name for spirits of TURPENTINE which conforms to the requirements of the National Formulary.[164] (ASTM)

oil paint (1) paint that contains DRYING OIL as the sole film forming ingredient; (2) a paint that contains drying oil, oil VARNISH, or oil-modified RESIN as the basic vehicle ingredient; (3) commonly (but technically incorrect), any paint soluble in ORGANIC solvents (deprecated).[71] (ASTM)

oil pastes very concentrated mixtures of PIGMENT and OIL, which are of paste-like CONSISTENCY.[41] Syn: colors in oil

They may be used for tinting purposes, or for preparation of oil paints by the simple addition of more vehicle and/or thinners, with the necessary DRIERS.

oil polish FINISH produced on WOOD by successive thin coats of LINSEED OIL, accompanied by RUBBING motion at the time of application.[54]

Also refers to any polishing material containing oil as one of the essential ingredients.

oil resistance ability of a FINISH to withstand softening when subjected to repeated or prolonged contact with oil.[72]

ASTM test methods for oil resistance include: electrical insulating varnish, D115; and wood furniture lacquers, D2571.

oil rubbing PROCESS of RUBBING the dried film of a finishing material with oil and pumice or some other suitable ABRASIVE.[54]

LINSEED OIL thinned with naphtha or turpentine is sometimes used, as are also light mineral OILS, such as neutral oil, straw oil, and paraffin oil.

oil, semidrying see SEMIDRYING OIL.[63]

oil softening the DETERIORATION of a paint film when the dried film becomes soft and excessively tacky due to EXPOSURE to various household or industrial OILS.[56,42]

Semigloss and gloss paints, when applied to kitchen cabinets or other high contact oil-prone areas, may eventually become soft and tacky. This may be the result of frequent contact and accumulation of body and/or household oils.

oil soluble materials capable of being dissolved in VEGETABLE or MINERAL OILS.[142,164]

oil stain solution of DYES and/or PIGMENT dispersions blended with OIL or VARNISH and reduced with HYDROCARBON SOLVENTS.[71] See STAIN.[41,43,64,69,163]

Those that are absorbed into the surface of the SUBSTRATE are called penetrating stains, while nonpenetrating stains contain higher levels of nonvolatile vehicle and/or pigments.

oil, stand see STAND OIL.[63]

oil varnish see OLEORESINOUS VEHICLE.[152,156,167]

oil, vegetable see VEGETABLE OIL.[63]

oil-bound description of a water paint, the medium of which contains a proportion of drying oil in the BINDER.[60,71] (BSI)

oil-bound distemper a DISTEMPER which contains a drying oil.[60,71,81] (DAC)

oil-modified alkyd see ALKYDS.[130]

oil-reactive phenolic resins PHENOL-FORMALDEHYDE resins which react with DRYING OILS on heating to yield products with special properties.[152,130]

oiticica oil drying oil obtained from the nuts of *Licania rigida*, a tree native to Brazil.[63]

It is unique by reason of the fact that its main constituent acid is LICANIC ACID. This acid is highly UNSATURATED because of its three conjugated double bonds and it also contains a KETONIC group. Oiticica oil is similar to TUNG OIL in that it has a high specific gravity, high refractive index, polymerizes rapidly on heating, dries quickly, and possesses good water resistance. Like tung oil, it webs, frosts and wrinkles unless adequately processed. It is used as a partial substitute for tung oil and special care is necessary when making VARNISHES from it. Sp gr, 0.982/20°C; iodine value, 145; saponification value, 190.

ASTM test methods for oiticica oil include: chloroform-insolubles, D1958; gel time, D1955; specification, D601.

olefins a class of unsaturated ALIPHATIC HYDROCARBONS having one or more double bonds.[133,134]

The double bond makes these compounds highly chemically reactive and useful as starting materials for the manufacture of other compounds. Examples are ethylene and propylene. (EPA)

oleic acid $CH_3(CH2)_7CH{=}CH(CH_2)_7COOH$. Fatty acid found in most vegetable OILS, whether drying or nondrying.[63]

Used to some extent in nondrying and nonyellowing ALKYD RESINS. Mp, 140°C; acid value, 199; iodine value, 90.1. A commercial grade is known as red oil.

oleoresin PINE GUM, the nonaqueous secretion of RESIN acids dissolved in a TERPENE HYDROCARBON OIL, which is: (1) produced or exuded from the intercellular resin ducts of a living tree; (2) accumulated, together with oxidation products, in the dead WOOD of weathered limbs and stumps.[44,167] See OLEORESINOUS VEHICLE.[156]

The term "oleoresinous," however, is not restricted to products of pine gum.

oleoresinous vehicle a vehicle prepared by the addition of a RESIN to a DRYING OIL.[156] See VARNISH.[156,63,152]

These two components may or may not be further processed to obtain specified properties. ALKYD RESINS are sometimes, but not generally, included in this category.

olibanum see GUM THUS.[167]

oligomer a POLYMER composed of molecules containing only two, three, or a few MERS.[130] (IUPAC)

one-compartment coating CROSSLINKING systems which can be stored in a single compartment, as opposed to a TWO-COMPARTMENT coating. Syn: one-pot coating

one-pot coating see ONE-COMPARTMENT COATING.

ongueka oil see ISANO OIL.[63]

opacity the degree to which a material obscures a SUBSTRATE, as opposed to transparency which is the degree to which a material does not obscure a substrate.[43,69] See CONTRAST RATIO and HIDING POWER.[43,69]

Opacity is sometimes described in terms of the contrast ratio.

opalescence exhibiting a milky IRIDESCENCE like that of an opal.[43,69]

opaque impervious to light rays.[43,69]

opaque ink one that does not allow the light to pass through it and has good HIDING POWER.[168]

It does not permit the PAPER or previous printing to show through.

opaque stain a pigmented composition intended for use on wooden surfaces to produce a uniform FINISH that obscures the grain and COLOR of the wood, but not its surface texture.[69,71] cf. STAIN and SEMITRANSPARENT STAIN [69,71] Syn: solid color or solid hide

opaque stains all stains that are not classified as semitransparent stains.[71]

opaqueness see OPAQUE and OPACITY.[43]

open coat in coated ABRASIVES, when the individual ABRASIVE grains are spaced at predetermined distance from one another the open coat covers about 50 to 70% of the coated surface with abrasive.[59,158]

Open coat has greater flexibility and resistance to filling or clogging than CLOSED COAT.

open grain common classification for woods with large pores such as oak, chestnut, and walnut.[153] (ASTM) Syn: coarse textured

open time length of time a coating remains wet enough to allow for BRUSHING-in at the LAPS.[42,54,55] See WET EDGE.[42,54,55] Syn: wet edge time

A major paint property. In LATEX paints, cosolvents such as propylene glycol as well as water soluble thickeners help increase open time. High temperatures and/or low humidity decrease open time.

optical brightener see BRIGHTENER, OPTICAL.[83]

optical filter see FILTER, OPTICAL.[43]

oral lethal dose (LD_{50}). The amount of a substance taken by mouth that would kill within 14 days half (50%) of those test animals exposed.[172] See LD_{50} TEST.[172]

The dose is measured in milligrams per kilogram of body weight of the test animal.

orange chrome see CHROME ORANGE, LIGHT and DEEP.[41]

orange lead LEAD OXIDE PIGMENT similar to RED LEAD but containing less litharge and more lead peroxide.[41] See RED LEAD, and RED LEAD (NONSETTING).[41]

It was used as a base for LAKES and to some extent as a pigment.

orange mineral LEAD OXIDE PIGMENT which is finer and of a brighter COLOR than RED LEAD. It was made by roasting WHITE LEAD CARBONATE or sublimed LITHARGE.[41] See RED LEAD.[41] Syn: mineral orange

orange peel (1) a paint surface APPEARANCE resembling an orange SKIN texture. Depending on the product, this may be desirable (some appliances) or highly undesirable (automobiles); (2) a type of mottle that occurs in FLEXOGRAPHIC, GRAVURE, and tin PRINTING.[56,42,168] See POCKMARKING and SPRAY MOTTLE.[56,42]

orange shellac a refined LAC which is soluble in ALCOHOL; contains some WAX and RESIN; used as a coating on floors and other wood surfaces.[146,71] (DAC) See SHELLAC.[146]

orbital pad sander a portable sanding machine consisting of a backup pad that moves in small circles, at high speed. See STRAIGHT-LINE PAD SANDER.[59,158]

Used for sanding METAL, the end grain of WOOD, PLASTICS, and the undercoats for paint and LACQUER finishes.

ore a mineral or an AGGREGATE of minerals from which a valuable constituent, especially a METAL, can be profitably mined or extracted.[34]

organic (1) designation of any chemical COMPOUND containing CARBON (some of the simple compounds of carbon, such as carbon dioxide, are frequently classified as INORGANIC compounds). To date, over one million ORGANIC compounds have been synthesized or isolated. Many occur in nature; others are produced by chemical synthesis; (2) in medicine, producing or involving alteration in the structure of an organ; opposed to functional.[134] cf. CARBONACEOUS[134,144]

organic aerosol fine ORGANIC PARTICLES or droplets which remain dispersed in the atmosphere.[172] (EPA)

organic yellow see HANSA YELLOW.[41]

organism an individual form of life, such as a plant, an animal, a bacterium, a protist, or a FUNGUS; a body made up of organs, organelles, or other parts that work together to carry on various processes of life.[173]

organization a group of persons organized for a particular purpose; an association.[174]

organometallic polymer a polymeric material, usually ACRYLIC, with BIOCIDAL ORGANIC MOIETIES chemically attached to their backbones, which when applied as a film and immersed, allow for controlled release of TOXICANT.[130,163]

organosol combination of DISPERSION RESINS and PLASTICIZER, dispersed in a mixture of VOLATILE ORGANIC SOLVENTS, which contain both polar and nonpolar types.[38,71,130] cf. PLASTISOL, and NONAQUEOUS DISPERSION[38,71,130]

It is similar to a plastisol except that an organosol contains more organic SOLVENT. (EPA)

Oriental blue synthetic ULTRAMARINE BLUE.[41]

Original Equipment Manufacturers coatings See OEM COATINGS.[163]

ormolu varnish a VARNISH having the APPEARANCE of gold or gilded bronze.[152,163] (DAC)

orpiment As_2S_3. Was used to a small extent as a PIGMENT, but was also used in SHELLAC to mask its natural COLOR.[41,146] cf. REALGAR.[41] Syn: king's yellow

ortho-chlor-para nitraniline Pigment Red 4 (12085). This PIGMENT has general properties similar to those of PARA RED, although it is slightly superior in tint LIGHTFASTNESS.[41] See CHLORINATED PARA REDS.[41] Syn: ortho chlor

OSHA (1) abbreviation for OCCUPATIONAL SAFETY AND HEALTH ADMINISTRATION, Department of Labor; (2) abbreviation for Occupational Safety and Health Act.[174,172] (DAC)

OSHRC abbreviation for Occupational Safety and Health Review Commission.[174,172]

osmosis DIFFUSION of molecules through a paint FILM or other semipermeable membrane caused by a difference in concentration.[139,142]

Ostwald color system system of classifying and designating COLORS in terms of their full COLOR content, their white content, and their black content.[69,43]

The system was devised by Wilhelm Ostwald, based on ADDITIVE color mixture, by means of spinning a Maxwell Disc and was used as the basic principle in making the COLOR HARMONY MANUAL.

Ostwald u-tube VISCOMETER in the form of a "U," in which the liquid flows from a bulb at a higher level on one side of the "U," through a CAPILLARY, to a receiving bulb on the other side.[59,166]

The time is measured for a given volume of liquid to pass from one side to the other.

otsu-ye (Japanese) A rough broadsheet painting, of small size on PAPER; the precursor of the print.[81] Also spelled "otsu-e."

ouricuri wax hard WAX obtained from the leaves of the Bahia palm, and used as a substitute for CARNAUBA WAX.[137]

Also spelled "ouricury."

ouropardo wax unusual hard WAX of exceptionally high sp gr (1.23), and possessing the property of resinifying on heating, without melting.[137]

outside finish the surface treatment or DECORATIVE trim on the exterior of a building.[71] (DAC)

oven a chamber which uses heat or irradiation to bake, CURE, POLYMERIZE, or dry a coating.[59,131,176]

overbaking an EXPOSURE of the coating to a temperature moderately higher or for a longer period of time, or both, than recommended by the manufacturer of the coating for normal curing.[42,67,176] (ASTM)

This condition is in contrast to "heat resistance" which is a parameter relating to the service life of a coating.

ASTM D2454 practice covers the determination of the time-temperature effect of overbaking on the physical and chemical properties of organic coatings.

overcure caused by an AFTERCURE or by being subjected to too high a temperature or too long a period at a proper temperature and resulting in a product less resistant to aging.[67,140,176]

overlap see LAP.[54]

overlapping seam method of hanging WALLPAPER in which only one SELVAGE is trimmed.[142]

overprint the printing of one impression over another.[168]

overprint varnish a clear VARNISH applied over a printed job to improve properties such as mar resistance.[168] cf. CLEARCOAT and BASECOAT/CLEARCOAT[71]

overproof (O.P.) containing a greater proportion of ALCOHOL than PROOF SPIRIT; especially, containing more than 50% alcohol by volume.[164]

overspray that solids portion of a coating sprayed from a SPRAY applicator which fails to ADHERE to the part being sprayed.[42,54] Syn: drift

Applied solids plus overspray solids equal total coating solids delivered by the spray application system. (EPA)

overtone see MASS COLOR.[69]

overvarnish coating applied over the INK on the outside of beverage cans to provide gloss and protect the can from corrosion and ABRASION.[168] (EPA) cf. OVERPRINT VARNISH, CLEARCOAT, and BASECOAT/CLEARCOAT[71]

oxalic acid $(COOH)_2 \cdot 2H_2O$. Dicarboxylic acid, used as a BLEACHING agent.[144]

Properties: mp, 99°C; sp gr, 1.65; mol wt, 126.07.

oxanilide $C_6H_5NHCOCONHC_6H_5$. A PLASTICIZER with bp, 320°C; mol wt, 240.11; mp, 245°C.[170] Syn: N. N-diphenyl-oxamide; oxalic acid dianilide, and camphol

oxidation (1) the chemical process by which the oxidation state of an atom in an element, ion, or compound is increased;[144] (IUPAC) (2) the formation of an oxide or, more generally, any increase in valence of an element;[144] and (3) in coatings, the introduction of oxygen into a molecule thereby producing a CURED (or degraded) film.[65,67]

Words presented in CAP/SMALL CAPS type indicate that the word is defined in another part of the *Coatings Encyclopedic Dictionary.*

Numerical superscripts classify terms in one or more of the categories listed in the second section of this volume.

oxidation inhibitor see ANTIOXIDANT.[83]

oxide a binary COMPOUND or element or RADICAL with oxygen.[144]

Oxides of nonmetals are covalent compounds having simple molecules.

oxide of chromium see CHROMIUM OXIDE GREEN.[41]

oxidized asphalt see BLOWN BITUMEN.[76]

oxidized bitumen see BLOWN BITUMEN.[76]

oxidized rosin ROSIN of much higher melting point than ordinary rosin and characterized by good solubility in industrial ALCOHOL.[167]

oxidized rubber RUBBER which has been subjected to OXIDATION by air, in the presence of CATALYSTS.[67,140]

Oxidized rubber differs considerably from the parent rubber in that it is a dark reddish-brown COLOR, and is much more compatible with solvents, DRYING OILS, etc. It has good heat resisting properties and yields useful films on baking.

oxidized turpentine turpentine which has been thickened by blowing or OXIDATION with air.[131,164] Syn: fat turpentine

oxidizer according to the DOT, it is a substance that yields oxygen readily to simulate the combustion (oxidation) of ORGANIC matter.[172] Syn: oxidizing material

Chlorate (CLO), permanganate (MnO), and nitrate (NO) compounds are examples of oxidizers.

oxidizing agent a chemical or substance that brings about an OXIDATION reaction.[144,172,83] cf. REDUCING AGENT[144,83]

The agent may (1) provide the oxygen to the substance being oxidized (in which case the agent has to be oxygen or contain oxygen), or (2) receive electrons being transferred from the substance undergoing oxidation. (Chlorine is a good oxidizing agent for electron-transfer purposes, even though it contains no oxygen.)

oxime COMPOUND of the general type characterized by the grouping >C = NOH.[83,161]

Butyraldoxime and methyl ethyl ketoxime are used as ANTISKINNING agents and as ANTIOXIDANTS for dip-tank stability, etc.

oxirane C_2H_4O. (1) oxygen atom of the epoxide ring; (2) ETHYLENE OXIDE.[67,133]

ASTM Test Method D1652 covers the quantitative determination of the epoxy content of epoxy resins. This is important in determining their reactivity and the properties of the coatings made from them.

oxygen, molecular (O_2) a reactive species which comprises about 21% by volume of the gases present in the earth's atmosphere.[144] (IUPAC) See OZONE.[144,171,172]

It is a necessary species for the existence of most forms of life on earth. It is a highly reactive gas which is capable of combining with most of the other elements.

oxygen (O) a nonmetallic element constituting 21 percent of the atmosphere by volume that occurs as a diatomic gas, O_2, and in many compounds such as water and iron ore.[144]

It combines with most elements, is essential for plant and animal respiration, and is required for nearly all combination. Atomic number 8; atomic weight 15.9994; melting point –218.4°C; boiling point –183.0°C; gas density at 0°C 1.429 grams per litre; valence 2. See Periodic Table of the Elements in this volume.

oxygenated solvent an ORGANIC SOLVENT containing oxygen as part of the molecular structure.[164,134] (EPA)

ALCOHOLS and KETONES are examples of oxygenated COMPOUNDS often used as paint solvents.

oxyns name given to the OXIDATION products of DRYING OILS, which are obtained by the addition of oxygen to the DOUBLE BONDS of the UNSATURATED FATTY ACIDS.[63,131]

ozokerite natural MINERAL WAX of variable properties, found in Germany, America, South Africa, and Austria. Its mp varies from 59° to 90°C and its sp gr from 0.900 to 0.960.[137]

ozone (O_3) a highly reactive COMPOUND which occurs in the atmosphere.[144,171,172] (IUPAC)

It is formed in the stratosphere through the photolysis of molecular oxygen at wavelengths less than 240 nm: (O_2 + $\underline{hv}$ → 2; O + O_2 + M(N_2/O_2) → O_3 + M). The stratospheric ozone acts as a protective shield which filters the wavelengths of sunlight which enter the lower atmosphere to those longer than about 290 nm. Ozone is a common secondary pollutant which is formed in a polluted air mass through the action of sunlight on NO_2 (NO_2 + hv → O + NO; O + O_2 + M(O_2/N_2) → O_3 + M). This is the most abundant of the oxidants formed in PHOTOCHEMICAL SMOG.

P

package stability the ability of a liquid, such as paint or VARNISH, to retain its original quality after prolonged STORAGE.[161] (DAC)

ASTM test methods for package stability include: coatings for ultraviolet curing, D4144; emulsion paints, biodegradation, D2574; freeze-thaw resistance, D2243; sampling liquid paints, D3925; settling of traffic paint, D869, and solvent and waterborne paints, D1849.

pad, padder see PAINT PAD.[77]

padding a method of applying a coating using a PAD or PADDER.[77]

paillon bright METAL FOIL, used to show through a thickness of ENAMEL or paint to alter its COLOR and give it brilliance.[148] (DAC)

paint (1) verb. To apply a thin layer of a coating to a SUBSTRATE by BRUSH, SPRAY, ROLLER, IMMERSION, or any other suitable means;[77](ASTM) (2) noun. Any pigmented liquid, liquefiable, or mastic composition designed for APPLICATION to a SUBSTRATE as a thin layer which is converted to an opaque solid film after application. Used for protection, decoration, or identification, or to serve some functional purpose such as the filling or concealing of surface irregularities, the modification of light and heat radiation characteristics, etc.[71] See COATING[71] and FINISH.[71]

paint and varnish remover liquid, principally SOLVENTS, sometimes with WAX or THICKENERS, which is applied to a coated surface in order to soften the old coating and bring it to such a condition that it can be easily removed.[158,164] Syn: stripper

Old clear or pigmented coatings may need to be removed from a surface before successful recoating can occur.

paint base the VEHICLE into which PIGMENT is mixed to form a paint; commonly ALKYD, LATEX, ACRYLIC.[64,156] (DAC)

paint brush a paint APPLICATION tool consisting of a flexible BRUSHING part composed of long filamentary material (brushing material) bound to a handle.[59,77] (ASTM) See BRUSH.[59,77]

Typical types include those designed for VARNISH, ENAMEL, sash, and wall painting. They are manufactured in a range of shapes and sizes. Until a paint brush is fully wetted with paint, only part of the paint loaded onto the brush can be transferred to the surface being painted. By properly preparing the brush before use, the amount of paint delivered to the surface can be made more uniform and reflect real use.

ASTM Standard Practice D5068 describes the preparation of paint brushes for evaluation. It is applicable to paint brushes 2 to 4 in. (50 to 100 mm) in width.

paint brush bristle hair of the swine (for example: pig, hog, boar), used in BRUSHING material.[59] (ASTM)

paint brush ferrule outer band that joins the BRUSHING material to the handle.[59] (ASTM)

paint brush filament a synthetic POLYMER extrusion used in BRUSHING material.[59] (ASTM)

paint brush head BRUSH without the handle.[59] (ASTM)

paint brush length clear also called "length out," the exposed length of the BRUSHING material from the ferrule to the tip end.[59] (ASTM)

paint brush thickness measurement of the BRUSHING material across the narrow opening of the ferrule.[59] (ASTM)

paint harling PROCESS of throwing paint-coated granite chips on to a sticky paint film previously applied to a surface, to give a roughcast effect.[54,77] (BSI)

This gives a thick durable FINISH which has been used, e.g., on steel-clad houses.

paint mitt see PAINTER'S MITT.[77]

paint pad a paint APPLICATION tool consisting of short filamentary material usually bonded to a flat, resilient backing connected to a handle designed to apply paint by a wiping action.[77] (ASTM)

The paint pad is dipped into a special tray containing the coating. The excess is wiped off the edge of the tray and the coating is wiped onto the SUBSTRATE. The pad comes in various shapes and sizes, generally with some type of handle. Some are small enough to coat window sashes.

paint remover see PAINT and VARNISH REMOVER.[158,164]

paint roller a complete paint APPLICATION tool consisting of a roller frame and a roller cover designed to apply paint by a rolling action.[77] (ASTM)

paint roller core a structural tube that forms the base of the roller cover to which applicating material is attached.[59] (ASTM)

Words presented in CAP/SMALL CAPS type indicate that the word is defined in another part of the *Coatings Encyclopedic Dictionary.*
Numerical superscripts classify terms in one or more of the categories listed in the second section of this volume.

paint roller cover a tubular sleeve consisting of a paint applicating material secured to a core.[59] (ASTM)

Until a paint roller cover is saturated with paint, only part of the paint loaded onto the paint roller cover can be transferred to the surface being painted. The remainder of the paint is absorbed by the paint roller cover. The amount of paint absorbed by the paint roller cover is inversely proportional to the amount already present within the paint roller cover. By saturation of the paint roller cover before testing, quantitative inaccuracies of the amount of paint delivered to a surface are eliminated.

ASTM Standard Practice D5069 describes the preparation or breaking in of paint roller covers for evaluation.

paint roller cover pile height also called "nap length"; the length of the roller cover paint applicating material from pile backing to pile face.[59] (ASTM)

paint roller frame a frame and handle assembly designed to hold a PAINT ROLLER COVER.[59] (ASTM)

paint system see COAT.[54] (BSI)

paint test charts see TEST CHARTS.[157,59]

paint thinner see THINNER.[164]

painter's mitt a glove-like device that slips over the hand.

Used for coating areas inaccessible with a BRUSH *or* ROLLER *such as fences and walls behind radiators.*[77]

painter's naphtha see VM & P NAPHTHA.[164]

painter's putty see PUTTY.[147,158,165]

Painting and Decorating Contractors of America (PDCA) a trade association of 3,000 painting and decorating contractors that promotes professional standards and practices. PDCA exists to advance the coating application industry by being a conduit for productive communication; by providing services and educational opportunities to members that cannot be obtained individually; by speaking as the voice of the coatings application industry; and by promulgating an attitude of ethical responsibility in business.[174]

Its address is: 3913 Old Lee Highway, Suite 33-B, Fairfax, VA 22030; 703-359-0826.

pale boiled oil lightly BLOWN LINSEED OIL containing a small amount of DRIERS.[63]

palette (1) surface on which an artist sets out, and mixes his pigments and range of colors which he uses; (2) the typical COLOR range of a school or group, e.g., the "Palette of the Impressionists (The Spectrum Palette)," "the Restricted Palette" of the painters of ancient Greece (generally speaking white, black, yellow ochre and red earth);[69,81] (3) the gamut of colors possible when mixing a prescribed group of COLORANTS.[69]

palette knife a spatula, usually smaller and slightly more flexible than the kind used for domestic and laboratory purposes, which serves to mix the ARTIST'S PAINT.[59,81]

palm oil pitch see FATTY ACID PITCHES.[144]

palmitic acid $CH_3(CH_2)_{14}COOH$. Properties: bp, 278°C; mp, 62°C; sp gr, 0.846/8°C; acid value, 218.8.[144]

pan in COIL COATING, the open container at the coater which holds the paint, and also the place where the pickup roller revolves in the paint.[59,70]

panel decoration a type of WALLPAPER which flourished in the second half of the 18th century, related to the wood-paneled walls then in use.[148]

Wallpaper panels of the present day are thought of more as "spot" decorations.

pantone printing planographic PRINTING PROCESS using plates having mercury- or amalgam-coated nonprinting areas.[168]

paper a material made of CELLULOSE pulp, derived from wood, rags, and certain grasses, PROCESSED into flexible sheets or rolls by deposit from AQUEOUS SUSPENSION, and used chiefly for writing, printing, drawing, wrapping, and WALLPAPER.[131,159,153]

paper, building see BUILDING PAPER.[76,82,75]

paper coating as used in the EPA's control technique guidelines, is the coating of PAPER, PLASTIC film, or metallic foil, usually with a roll, knife, or ROTOGRAVURE coater.[71] (EPA)

paperboard cardboard; pasteboard.[81] cf. CHIP BOARD[75]

papier-mâché (Fr.) PAPER which is wet, molded, and HARDENED into forms used for DECORATIVE purposes.[81,148]

para brown bright brown PIGMENT of good staining power, made by treating para red with copper SALTS.[41]

para nitraniline red see PARA REDS.[41]

para reds Pigment Red 1 (12070). Series of red PIGMENTS made by coupling REACTIONS involving diazotized p-nitroaniline and alkaline ß-naphthol.[41]

Adjustments in SHADE *are made by substituting part of the ß-naphthol.*

ASTM Test Method D970 covers para red and toluidine red pigments (toners) in the dry form commercially known as "pure." ASTM D475 is the standard specification for pure para red toner pigment. It is available in two types: Type I, light; and Type II, dark.

para toner see PARA REDS.[41]

para-chlor-ortho-nitraniline Pigment Red 6 (12090). The parachlor variety of chlorinated para red is redder (less yellow) in shade and more TRANSPARENT compared to the ortho-chlor but much superior in LIGHTFASTNESS. It is superior in tint lightfastness to toluidine red. Special precautions are required in drier addition to parachlor red paints since it discolors (dark and dull) with cobalt and/or iron DRIERS.[41] Syn: parachlor red

parachlor nitraniline red see PARACHLOR RED.[41]

parachlor red see PARA-CHLOR-ORTHO-NITRANILINE.[41]

paracoumarone-indene resins RESIN obtained from COUMARONE by POLYMERIZATION with sulfuric acid.[167] See COUMARONE-INDENE RESINS.[167]

paraffin oil an OIL either pressed or dry-distilled from paraffin distillate.[63,144]

paraffin wax INERT HYDROCARBON WAX derivative of crude PETROLEUM.[137]

Paraffin waxes are distinguished by their melting points. Their main uses include the conferring of WATER RESISTANCE, SLIP, or SOLVENT RETENTION in special types of compositions.

parallel laminated see LAMINATED, PARALLEL.[57,79]

paramerism phenomenon in which specimens having different SPECTROPHOTOMETRIC CURVES produce approximately the same COLOR sensation under the same illuminating and viewing conditions.[41,69,35] (ASTM) See METAMERISM.[69,43,35]

parameter a quantity or constant whose value varies with the circumstances of its application, as the radius line of a group of concentric circles, which varies with the circle under consideration.

An arbitrary constant or a variable in a mathematical expression which distinguishes various cases.

parameter of specular gloss a measurement of specular GLOSS dependent on GLOSSMETER geometry.[43] (ASTM)

parapet a low protective wall or railing along the edge of a raised structure such as a roof or balcony.[82]

parcel gilding selective GILDING to highlight details of WOOD or a painted surface.[148,158]

Paris black see LAMPBLACK.[41]

Paris blue, Parisian blue see PRUSSIAN BLUE and IRON BLUE.[41]

Paris green see EMERALD GREEN.[41]

Paris red see RED LEAD.[41]

Paris white see WHITING and CALCIUM CARBONATE, NATURAL.[41,58]

Paris yellow see CHROME YELLOW, LIGHT and PRIMROSE.[41]

Parkerized descriptive of IRON or STEEL which has received a rust-resisting treatment by being dipped in a boiling solution of manganese dihydrogen phosphate; this protective coating also improves the bonding (adhesion) of paints and lacquers.[45,158] (DAC)

parquet a wood floor INLAID with geometric patterns.[82,153,75,148]

A type of DECORATION dating from 17th century France.

parqueting method of reinforcing the backs of WOOD panels with battens to prevent them from warping.[75,82,148]

parquetry patterned WOOD inlay, especially for floors.[75,82,148]

partial aromatic solvents see AROMATIC SOLVENTS.[164]

particle a very small piece or part; a tiny portion or part; a tiny portion or SPECK.[43]

particulate small, separate pieces of an airborne material.[171,172]

DUST, fumes, smokes, mists, and fogs are examples. Generally, anything that is not a fiber and has an aspect ratio of 3 to 1.

particle board a panel material composed of small discrete pieces of WOOD or other lignocellulosic materials that are BONDED together in the presence of heat and pressure by a SYNTHETIC RESIN ADHESIVE.[75,153,159,175] (ASTM) cf. PLYWOOD[75,153,159,79] Syn: flake board

Particle boards are further defined by the method of pressing. When the pressure is applied in the direction perpendicular to the faces, as in a conventional multiplaten hot press, they are defined as flatplaten pressed, and when the applied pressure is parallel to the faces, they are defined as extruded.

Particle boards are manufactured from a range of materials, such as wood, flax, and hemp and are bonded with synthetic resins, usually UREA-FORMALDEHYDE. They are distinguished by their regularity of surface, thickness, size, and mois-

Words presented in CAP/SMALL CAPS type indicate that the word is defined in another part of the *Coatings Encyclopedic Dictionary.*

Numerical superscripts classify terms in one or more of the categories listed in the second section of this volume.

ture content, making them ideal for use in flooring, roof decking, partioning and wall linings, and FURNITURE. *Particle board can be supplied in several grades such as flooring grade, ready-to-paint, or ready faced with wood* VENEER, PLASTIC, *etc.*

particle size in paints, the diameter of a PIGMENT or LATEX particle; usually expressed in MILS or MICROMETRES.[41,60] (DAC)

ASTM test methods for particle size (analysis/distribution) include: coarse particle content, D185; pigments, reporting characteristics, D1366; polymeric powder properties, D3451; and white extender pigments, D3360.

particle size distribution the relative percentage by weight or number of each of the different size FRACTIONS of PARTICULATE MATTER.[128] (ASTM)

particulate small, separate pieces of an airborne material.[171,172]

Dusts, fumes, smokes, mists, and fogs are examples. Generally anything that is not a fiber and has an aspect ratio of 3 to 1.

particulate matter a general term used to describe airborne solids or liquid PARTICLES of all sizes. The term AEROSOL is recommended for use in describing airborne particulate matter.[172] (IUPAC)

parting agent a material applied to one or both surfaces of a sheet to prevent BLOCKING.[83,76] (ASTM) See SLIP AGENT (COMPOUND),[83] and SLIP SHEET.[168]

partition (1) any BUILDING component that divides space such as a wall, door, window, roof, or floor-ceiling assembly; or a combination of components such as a wall containing a door and a window; [82] (ASTM) (2) a SOLVENT system in which the solute is divided between two components of the liquid phase, the ratio of their CONCENTRATION (partition COEFFICIENT) being constant for a given substance.[164]

pass motion of SPRAY GUN in one direction only.[54]

passivation act of making inert or unreactive.[45]

paste ADHESIVE composition having a characteristic PLASTIC-type CONSISTENCY, that is, a high order of YIELD VALUE, such as that of a paste prepared by heating a mixture of starch and water and subsequently cooling the HYDROLYZED product.[79] (ASTM)

pasteboard synonym for paperboard and cardboard.[81]

paste blue see IRON BLUE.[41]

paste driers DRIERS made by GRINDING or mixing suitable INORGANIC DRYING COMPOUNDS of lead or manganese into drying oil, together with some cheap FILLER that were used to provide bulk.[83] Syn: patent driers

The resultant product is a stiff paste.

paste filler see FILLER.[147]

paste, pigment PIGMENT dispersion concentration to permit substantial reduction (let down) with SOLVENT, water, or VEHICLE.[41] cf. MILL BASE[49,131,64]

paste water emulsion wax similar to LIQUID WATER EMULSION WAX except furnished in paste form.[137] (CSMA)

Must be polished for LUSTER. *Occasionally pigmented. May contain some* SOLVENT.

pastel (1) light TINT; MASSTONE to which white has been added; (2) soft delicate HUE; (3) in art, a picture made with a crayon of PIGMENT with some BINDING MEDIUM such as GUM.[69,81]

pastel painting see PASTEL (3).[81]

pasteurizing varnishes VARNISHES for food containers which are capable of withstanding immersion in water at pasteurizing temperature for half an hour or more.[71]

Pasteurizing temperature is about 70°C, but food canners often require varnishes to withstand immersion in boiling water, or water boiling under pressure, without softening.

pastiche work of art composed in the style of another person.[81]

patent driers see PASTE DRIERS.[83]

patent vermillion see MERCURIC SULFIDE.[41]

pathogen, pathogene a parasite or virus capable of causing disease.[173]

pathogenicity the state or condition of producing disease or capable of doing so.[173]

patina (1) a thin layer of CORROSION, usually green or brown; (2) the GLOSS on wooden and other surfaces; (3) the softening of COLOR that develops with AGE.[81]

pat-out see TAP-OUT.[168]

pattern staining patterns on surfaces caused by deposits of DUST, the amount of which varies according to the relative THERMAL CONDUCTIVITY of different parts of the structural background.[56] (BSI)

The more highly conductive (colder) areas collect more dust than those of poor conductivity, thus the pattern of the underlying structure shows. The dust is often carried towards the surface by convection currents.

PCB see CHLORINATED DIPHENYLS.[170]

PDCA abbreviation for PAINTING AND DECORATING CONTRACTORS OF AMERICA.[174]

PE (1) abbreviation for POLYETHYLENE, polyethene;[144] (2) abbreviation for PENTAERYTHRITOL.[130]

peacock blue LAKE of acid glaucine blue DYE on ALUMINA HYDRATE. Structurally, the dye is alpha, alpha-bis [N–ethyl–N–(4–sulfobenzyl) AMINO phenyl]– alpha-hydroxy–ortho toluenesulfonic acid sodium SALT.[41]

peanut oil a fixed OIL.[63] Syn: arachis oil and groundnut oil

Yellow to greenish yellow. Typical NONDRYING oil of olive oil type. Soluble in ether, PETROLEUM ether, CARBON disulfide, and chloroform; insoluble in alkalies, but saponified by alkali hydroxides with formation of soaps; insoluble in water; slightly soluble in ALCOHOL. Principally glycerides of oleic and linoleic acids, with lesser amounts of the glycerides of palmitic, stearic, arachidic, behenic, and lignoceric acids. Made by pressing ground peanut meats or by extraction with hot or cold solvents. Latter method more common. Constants: Sp gr, 0.912-0.920 (25°C); solidifying point, –5 to + 3°C; saponification value, 186-194; iodine value, 88-98; refractive index, 1.4625-1.4645 (40°C).

pearl see GLOSS.[43]

pearlescent pigment used to obtain pearly LUSTER, iridescent effects, and METALLIC SHEEN. The PIGMENT PARTICLES are TRANSPARENT, thin PLATELETS of high REFRACTIVE INDEX which partially REFLECT and partially TRANSMIT INCIDENT LIGHT; simultaneous reflection from many layers of oriented platelets creates a sense of depth that is characteristic of nacreous luster.[41,43] See NACREOUS PIGMENT, INTERFERENCE PIGMENTS.[41,43]

pearl essence extract from fish scales used for obtaining the mother-of-pearl effect.[43,148]

pearlite same as perlite.[41] (DAC)

pearl lacquer see PEARLESCENT.[43,148]

pearl moss see CARRAGEEN.[139,155]

pearlescent an APPEARANCE resembling that of natural pearls or mother-of-pearl. See INTERFERENCE COLOR.[43,69,35] Syn: nacreous

It results from the specular REFLECTANCE of alternating thin layers of differing refractive indices.

pearlstone see PERLITE, EXPANDED.[41]

peat wax WAX, resembling montan wax, obtained by the SOLVENT EXTRACTION of peat. Mp, 70-100°C.[137]

pebble mill a rotating porcelain, buhrstone (or other nonmetallic lined) cylinder containing pebbles or porcelain balls or rods as the GRINDING media.[64,131,59]

In the manufacture of pigments or paints, it is used to grind and/or disperse.

peeling detachment of a coating film from the underlying surface or UNDERCOATING, in ribbons or sheets, generally accompanied by curling of the edges.[42,56] cf. FLAKING[42,56]

Peeling can be considered as an aggravated form of flaking (scaling). It is frequently due to collection of moisture beneath the film.

PEL abbreviation for PERMISSIBLE EXPOSURE LIMIT.[172]

pelletier's green see HYDRATED CHROMIUM OXIDE.[41]

pelletization processing of PIGMENTS or other chemical products into very small, free-flowing beads, which eliminate the DUST nuisance.[41,131]

pencil hardness a measure of coating HARDNESS based on the scratching of the film with pencil leads of known hardness.[47,157] cf. KNOOP HARDNESS TEST [47,157]

The result is reported as the hardest lead which will not scratch or cut through the film to the SUBSTRATE.

ASTM Test Method D3363 covers a procedure for a rapid, inexpensive determination of the film hardness of an organic coating on substrate in terms of drawing leads or pencil leads of known hardness.

pencilling painting MORTAR joints of brickwork with white paint to bring out the contrast between the joints and the brickwork.[148]

pendulum-rocker hardness HARDNESS measured by a pendulum or rocker, the mechanism being the damping of the oscillation due to the hardness of the coating.[47,157,145] See ROCKER HARDNESS TESTER.[47,157,145]

penetrating finish a low-viscosity OIL or VARNISH which penetrates WOOD, leaving a very thin film at the surface.[63,71,152] (DAC) See WOOD PRESERVATIVE COATING.

penetrating stain see OIL STAIN.[71]

penetration entering of an ADHESIVE into an ADHEREND.[79] (ASTM)

This property of a system is measured by the depth of PENETRATION of the adhesive into the adherend.

penetration number figure used, chiefly for BITUMINOUS type products, as a measure of their softness.[47,76,128]

It is obtained by allowing a weighted needle of specified dimensions to penetrate into the material under test, at a definite temperature. The PENETRATION figure is usually recorded as the number of units of depth which the needle penetrates

Words presented in CAP/SMALL CAPS type indicate that the word is defined in another part of the *Coatings Encyclopedic Dictionary.*
Numerical superscripts classify terms in one or more of the categories listed in the second section of this volume.

in a given time, e.g., 5 s. Thus, the softer a given material, the higher its penetration figure. Instruments used for determining this property are known as PENETROMETERS.

penetrometer apparatus for measuring PENETRATION number of a solid.[145,157] See PENETRATION NUMBER.[47,76]

Pensky-Martens closed-cup tester a device used in determining the FLASH POINT of liquids which have a viscosity of 45 SUS or more at 37.8°C (100°F), or contain suspended solids and require stirring to obtain uniform distribution of heat, or have a tendency to form a surface film under test conditions.[145,157]

Flash point measures the tendency of the sample to form a flammable mixture with air under controlled laboratory conditions. It is only one of a number of properties which must be considered in assessing the overall flammability hazard of a material.

ASTM Test Method D93 covers the determination of the flash point by Pensky-Martens closed-cup tester of fuel oils, lube oils, suspensions of solids, liquids that tend to form a surface film under test conditions, and other liquids of similar viscosities.

pentachloroethane $CHCl_2CCl$. Nonflammable SOLVENT. Bp, 161°C; vp, 5 mm Hg/20°C; sp gr, 1.685.[164]

pentachlorophenol C_6HCl_5O. A TOXIC, OIL-SOLUBLE chemical; widely used as a WOOD PRESERVATIVE for protection against decay and insects.[158,173,83,153] (DAC)

pentaerythritol (PE) $C_5H_{12}O_4$.2,2-bis(hydroxymethyl)-1,3-propanediol; tetramethylol methane; molecular weight, 136.15. A polyhydric ALCOHOL containing four primary alcohol groups. It is used in the manufacture of alkyds, copal ester, ROSIN ESTERS, SYNTHETIC RESINS, and LUBRICANTS. .[130,144,167]

Properties: bp, 276°C; mp, 262°C; sp gr, 1.399.

ASTM test methods for pentaerythritol/pentaerythritol content include: methods for testing, D2195; monopentaerythritol content, D2999; of alkyd resins, D1615.

penta resin ESTER GUM made from ROSIN and PENTAERYTHRITOL.[167,130] (ASTM)

peppery see BITTY.[42]

peptization PROCESS of bringing a solid into a COLLOIDAL SOLUTION.[64,139]

peptizing agents (1) materials which, when added in relatively small amounts, are capable of causing substantial reduction in VISCOSITY, as the result of DEPOLYMERIZATION, or DISPERSION. Peptization can occur in AQUEOUS MEDIA, such as when agents are added to prevent FLOCCULATION or AGGREGATION of PARTICLES;[64,83,139] (2) substances that act as a chemical PLASTICIZER for natural and synthetic rubbers. They act as catalysts for oxidation breakdown of RUBBER during the MILLING or MASTICATION period. Additional plasticization can be accomplished without further milling by heating the rubber containing the peptizing agent, thus reducing power consumption in breakdown.[83,74,140]

peracetylated rubber product prepared from raw RUBBER by treatment with ACETIC ACID and HYDROGEN PEROXIDE. Improved COMPATIBILITY is obtained.[140]

percent capture the ratio of SOLVENT vapors collected into the capture system to the amount of solvent vapors released by a process. It is expressed as a percentage.[172] (EPA)

percent volatile the percentage of a liquid or solid (by volume) that evaporates at an ambient temperature of 70°F (20°C) unless some other temperature is stated.[172]

perceptible capable of being perceived; discernible.[43]

perception the mode of integrated mental functioning that includes the combination of different sensations and the utilization of past experiences in recognizing the objects and "facts" from which the stimulation arises.[43]

perceptual adjective pertaining to or involving PERCEPTION.[43]

perchlorethylene used for VAPOR DEGREASING and dry cleaning.[164]

Properties: bp, 121°C; vp, 15 mm Hg/20°C.

perchloromethane see CARBON TETRACHLORIDE.[164]

perfect diffuser theoretical ideal white substance which reflects 100% of the INCIDENT LIGHT in a perfectly diffuse way; official CIE reference white to which colors being measured are compared; a Lambert surface, following the Lambert cosine law perfectly.[43] See LAMBERT'S LAW OF REFLECTION.[43]

perforated tape a type of tape used in finishing joints between GYPSUM BOARDS.[75] (DAC)

perilla oil a DRYING OIL obtained from the seed of the perilla plants, *Perilla ocymoides* and *Perilla nankinensis*, natives of the Orient.[63]

Its main constitutent acids are LINOLEIC *and* LINOLENIC, *and it has the highest* IODINE VALUE *of all known* VEGETABLE OILS *except* CHIA. *Like chia oil, raw perilla oil exhibits the property of* CRAWLING, *which is decreased by being heated at 260°C (500°F) for 15 minutes or more. It is superior to* LINSEED OIL, *both from the point of view of drying rate, especially in the form of stand oil, and also of* POLYMERIZATION *rate. Properties: sp gr, 0.933-*

0.937/15°C; iodine value, 194; saponification value, 192.

Periodic Table of the Elements a tabular arrangement of the ELEMENTS according to their ATOMIC numbers so that the elements with similar properties are in the same column.[144,142]

See the Periodic Table of the Elements contained in this volume.

perishing DEGRADATION of a FILM with age, resulting in loss of FILM STRENGTH, FLEXIBILITY, ADHESION, etc., and which may be shown by the development of CHALKING, CHECKING, CRACKING, and FLAKING, etc.[56,154] (BSI)

perlite, expanded 1 Na_2O·1 K_2O·2.5 Al_2O_3·19.5 SiO_2. A unique form of siliceous lava that is characterized by many spherical and convoluted cracks.[41] Syn: pearlstone, ground perlite, pearlite

The interior or internal structure causes it to break up into small spheres or "pebbles." Perlite expands when heated at a range of 720°C (1500°F) to 1090°C (2000°F) to 10 to 20 times its initial volume. Used as insulation and FILLER, also EXTENDER. Density, 2.2 g/cm³ (1.87 lb/gal); particle size, 44 μm.

perm a unit of measurement of WATER VAPOR PERMEANCE; a metric perm, or 1 g/24 h·m²·mm Hg. or U.S. unit, 1 grain/h·ft²·in. Hg.[128] (ASTM) See WATER VAPOR TRANSMISSION (WVT) and PERM CUP OR DISH.[145,157]

Perm is used to express the resistance of a material to the PENETRATION of moisture.

perm cup or dish a container made of noncorroding material, impermeable to water or water vapor. It is used to measure WATER VAPOR TRANSMISSION rate and/or WATER VAPOR PERMEANCE.[145,157]

permanence resistance of any object or material to change with age or EXPOSURE to deleterious conditions.[72]

permanent blue synthetic ULTRAMARINE BLUE.[41]

permanent blue toners these include VICTORIA BLUE and RHODULINE BLUE, both of which may be extended with a suitable base such as aluminum hydroxide.[41]

Victoria blue has good LIGHTFASTNESS. Rhoduline blue darkens before it bleaches.

permanent fast violet see MINERAL VIOLET.[41]

permanent green old name for HYDRATED CHROMIUM OXIDE reduced with BARYTES.[41]

permanent green toners these are based on MALACHITE GREEN, brilliant green, or mixtures of these with Thioflavin or Auramine.[41]

They are characterized by BRIGHTNESS of shade, good STAINING POWER, and fairly good LIGHTFASTNESS.

permanent inks INKS which do not readily fade or change COLOR when exposed to light and weather.[168]

permanent orange PIGMENT made from diazotized 2,4-dinitroaniline and ß-naphthol.[41]

permanent peacock blue see RHODULINE BLUE 66.[41]

permanent red 2B Pigment Red 48 (15865). These reds, also referred to as RUBINE REDS, are made from couplings of Red 2B acid with beta hydroxy naphthoic acid and PRECIPITATED with metals such as calcium, barium, and strontium. Those precipitated with manganese are referred to as BON reds and maroons. In view of their chemical similarity to LITHOL RUBINE, it is not surprising that the paint properties of the permanent red 2B reds are so similar to those of lithol rubine, but they are considerably lighter and yellower in MASSTONE and TINTS. Their chief application is in baked industrial enamels not requiring outdoor DURABILITY or FASTNESS TO SOAP and/or alkali. Their distinctive COLOR has led to their adoption for certain interior finishes.[41]

permanent red toners RHODAMINE and MAGENTA toners.[41]

permanent set increase in length, expressed as a percentage of the original length, by which an ELASTIC material fails to return to original length after being stretched for a standard period of time.[142] Syn: hysteresis

permanent violets a series of light resistant, tungstated, or molybdated methyl violet pigments.[41]

permanent white see BARIUM SULFATE, NATURAL.[41]

permanent yellow FGL Pigment Yellow 97 (U.S. Pat. 2,644,814). A monoazo PIGMENT that has gained significant use in TRADE SALES finishes. It is very similar in COLOR to DALAMAR YELLOW and somewhat more BLEED-RESISTANT.[41]

permanganate time involves a test that detects the presence of oxidizable materials, usually impurities, of certain solvents.[164,157]

Substances reacting with potassium permanganate in neutral SOLUTIONS reduce it to manganese dioxide which colors the solution yellow. In the permanganate test, the time required for the COLOR of the test solution to change to that of a standard solution is measured. The color of the test solution changes from pink-orange to yellow-orange.

Words presented in CAP/SMALL CAPS type indicate that the word is defined in another part of the *Coatings Encyclopedic Dictionary.*

Numerical superscripts classify terms in one or more of the categories listed in the second section of this volume.

ASTM Test Method D1363 covers the detection in acetone and methanol of the presence of impurities that reduce potassium permanganate.

ASTM Test Method D1721 covers the detection in tricresyl phosphate of the presence of impurities that reduce potassium permanganate.

permeability see WATER VAPOR PERMEANCE (WVP).[128,157]

permeance (water vapor) see WATER VAPOR PERMEANCE (WVP).[128,157]

permissible exposure limit (PEL) established by OSHA, this may be expressed as a time-weighted average (TWA) limit or as a ceiling EXPOSURE limit that legally must never be exceeded instantaneously even if the TWA exposure limit is not violated.[172]

OSHA PELs have the force of law. Note that the ACGIH TLVs and NIOSH RELs are recommended limits that may or may not be enacted into law by OSHA.

Persian Gulf red Fe_2O_3. Natural red IRON OXIDE (hematite) from the island of Ormus, in the Persian Gulf.[41] See IRON OXIDES, NATURAL, PIGMENT RED 101 (77491).[41]

Bright red COLOR. Range 70 to 80% Fe_2O_3.

Persian red this PIGMENT should not be confused with Persian Gulf red, from which it differs profoundly in being a BASIC LEAD CHROMATE.[41] See CHROME ORANGE, LIGHT and DEEP.[41] Syn: chrome red

personal protective equipment (PPE) devices or clothing worn to insulate a worker from direct EXPOSURE to hazardous materials.[172]

Examples include gloves and respirators.

Persoz pendulum hardness test similar to the KÖNIG PENDULUM DAMPING HARDNESS test, however, it has a greater degree of discrimination for measuring the hardness of soft coatings, but it may not be as suitable for testing hard, slippery coatings because of its tendency to skid on surfaces with a low COEFFICIENT of friction.[47,157,145]

Both pendulum hardness tests are covered by ASTM Test Method D4366.

perylene pigments a group of vat PIGMENTS, nearly all of which are N,N′-substituted peryzene-3,4,9,10-tetracarboxylic diimides.[41]

petrifying liquid usually a dilute emulsion of DRYING OIL and/or RESIN in water, used as a sealing coat before applying an oil-bound water paint to a POROUS SURFACE.[147,158] (BSI)

It may also be used in place of water to thin the first coat of water paint before application to a porous surface.

petrolatum purified mixture of semisolid HYDROCARBONS with unctuous nature derived from PETROLEUM.[137] Syn: petroleum jelly

petroleum a thick, flammable, yellow to black mixture of gaseous, liquid, and solid HYDROCARBONS that occurs naturally beneath the earth's surface, that can be separated into fractions including natural gas, gasoline, NAPHTHA, KEROSENE, fuel and lubricating oil, PARAFFIN WAX, and ASPHALT and is used as raw materials for a wide variety of derivative products.[164,137,131]

petroleum bitumens distillation residues derived from crude PETROLEUM.[76]

petroleum ethers low boiling ALIPHATIC fractions derived from crude PETROLEUM by FRACTIONAL DISTILLATION.[164]

The boiling ranges are fairly restricted, and products known as 40-60°C and 60-80°C petroleum ethers are common. The low boiling ranges and exceptionally high EVAPORATION rates restrict the use of the solvents in the paint trade considerably. The petroleum ethers bear no resemblance chemically to the true ethers. The only resemblance is the high evaporation rate.

petroleum hydrocarbons HYDROCARBONS of ALIPHATIC type obtained from crude PETROLEUM and including VM&P NAPHTHA, MINERAL SPIRITS, and PARAFFINS.[164]

petroleum jelly see PETROLATUM.[137]

petroleum naphtha a generic term applied to refined, partly refined, or unrefined PETROLEUM products and liquid products of natural gas, not less than 10% of which distills below 175°C (347°F) and not less than 95% of which distills below 240°C (464°F), when subjected to distillation in accordance with ASTM Method D86, Test for Distillation of Petroleum Products.[164] (ASTM)

petroleum spirits see MINERAL SPIRITS.[164]

PF resins see PHENOLIC RESINS.[130]

Pfund hardness number (PHN) the INDENTATION HARDNESS determined with a Pfund indentation tester, and calculated as follows:

$$\mathrm{PHN} = \frac{L}{A} = \frac{4L}{\pi d^2} = 1.27\frac{L}{d^2}$$

where L = load in kilograms applied to the indenter, A = area of projected indentation in square millimetres, and d = diameter of projected indentation in millimetres.[128,142,47] (ASTM)

Pfund indentation tester a hemispherical quartz or sapphire indenter of prescribed dimensions used for testing INDENTATION HARDNESS of ORGANIC coatings.[145,157]

pH the negative logarithm of the hydrogen ion CONCENTRATION, in mol/litre. Measure of the ACIDITY or ALKALINITY of an AQUEOUS SOLUTION. pH = 7 represents neutrality, i.e., the solution is neither acid nor alkaline. pH values from 0 to 7 are acidic, the lower the pH value, the higher the degree of acidity; pH values from 7 to 14 represent alkalinity. The higher the pH value above 7, the greater the degree of alkalinity.[128,157] cf. PK; PK_a

The stability and/or viscosity of many LATEX paint systems are dependent upon pH. In order to assess pH stability, the pH of the paint should be determined after heat aging, shelf aging, and freeze/thaw cycling.

phase a physically homogeneous, mechanically separable portion of a material system.[139] (ASTM) See CONTINUOUS PHASE and DISCONTINUOUS PHASE.[139]

phase, continuous see CONTINUOUS PHASE.[139]

phase, discontinuous see DISPERSED PHASE.[139]

phase separation the settling in a container of PIGMENTS, EXTENDERS, or other paint constituents in liquid paint exhibiting a clear or somewhat cloudy liquid on the surface.[42,161] cf. SYNERESIS Syn: settling

Phase separation or SYNERESIS sometimes occurs after LATEX paints stand for a long period of time undisturbed. All phase separations are not equivalent to with syneresis, which has a specific mechanism. Except in cases of very rapid (several hours) separation or severe pigment settling, phase separation does not affect paint performance and is merely unaesthetic.

phenol C_6H_5OH. The simplest PHENOL.[130,144] Syn: carbolic acid

Used for the preparation of PHENOL-FORMALDEHYDE RESINS. Bp, 182°C; mp, 42°C. Phenols are a class of AROMATIC COMPOUNDS containing OH groups attached directly to the BENZENE ring and which are also used in the manufacture of EPOXY RESINS, PLASTICIZERS, PLASTICS, and WOOD PRESERVATIVES.

phenol-aldehyde resins see PHENOLIC RESINS.[130]

phenol-formaldehyde resin a RESIN of the PHENOLIC type, made by the POLYCONDENSATION of PHENOL with FORMALDEHYDE.[130]

phenolic plastics PLASTICS based on RESINS made by the CONDENSATION of PHENOLS, such as PHENOL and CRESOL, with ALDEHYDES.[38] (ASTM)

phenolic resins RESINS made by the CONDENSATION of PHENOLS and ALDEHYDES.[130,152] Syn: phenol-formaldehyde or PF resins

ASTM test methods for phenolic resins include: apparent pH, D4613; free phenols content, D1312; methylol group determination, D4706; stroke cure time, D4640; volatile content, D4639.

phenoxy resin a high molecular weight, THERMOPLASTIC POLYETHER RESIN based on BISPHENOL A and EPICHLOROHYDRIN having bisphenol A terminal groups.[130]

phenyl acetamide see ACETANILIDE.[144]

phenyl benzoate $C_6H_5COOC_6H_5$. A PLASTICIZER. Bp, 314°C.[170]

phenyl ethyl alcohol $C_6H_5CH_2CH_2OH$. Semipermanent alcoholic PLASTICIZER. Bp, 220°C.[170]

phenyl methyl ketone see ACETOPHENONE.[164]

phenyl phosphate, di see DIPHENYL PHOSPHATE.[170]

phenyl phosphate, tri see TRIPHENYL PHOSPHATE.[170]

phloxine eosine lake.[41]

PHN abbreviation for PFUND HARDNESS NUMBER.[128,142,47]

phosphating PRETREATMENT of STEEL or certain other METAL surfaces by chemical SOLUTIONS containing metal phosphates and phosphoric acid as the main ingredients, to form a thin, inert, adherent, CORROSION-INHIBITING phosphate layer which serves as a good base for subsequent paint coats.[45,158]

phosphatize (phosphate coat) see PHOSPHATING.[45,158]

phosphomolybdic pigments a series of colored PIGMENTS derived from the interaction of BASIC DYES and complex ACIDS such as phosphomolybdic, phosphotungstic, and tungstomolybdic.[41]

They are characterized by good resistance to alkali, heat, light, and water.

phosphorescence the ability of certain substances to continue to emit LIGHT long after the source of excitation energy has been removed.[43] cf. LUMINESCENCE[43,69]

phosphorescent paint see LUMINOUS PAINT.[71]

phosphotungstic pigments see PHOSPHOMOLYBDIC PIGMENTS.[41]

photochemical catalysis CATALYSIS stimulated by action of light.[67,172]

photochemical oxidant ozone and smaller amounts of other irritating chemicals such as peroxyacetyl nitrate which are products of atmospheric reactions of VOLATILE ORGANIC COMPOUNDS, NO_x and sunlight.[172] (EPA)

Photochemical oxidants are a major portion of the air pollution commonly known as "smog."

photochemical reaction a reaction involving an electronically excited state of a molecule or atom

---

Words presented in CAP/SMALL CAPS type indicate that the word is defined in another part of the *Coatings Encyclopedic Dictionary.*

Numerical superscripts classify terms in one or more of the categories listed in the second section of this volume.

which can occur following light ABSORPTION of sufficient energy per quantum (usually ULTRAVIOLET or visible radiation).[67,43] (IUPAC)

In electronically excited molecules one of the several types of photochemical reactions may occur: photodecomposition into molecular fragments or RADICALS, *photodecomposition into stable smaller* PARTICLES, *photoisomerization, intersystem crossing (a singlet to triplet state, etc.), internal energy conversion (e.g., excited singlet to vibrationally excited singlet ground state),* FLUORESCENCE, PHOSPHORESCENCE, *electronic or vibrational energy quenching, etc.*

photochemical reactivity the measure of the rate at which an ORGANIC COMPOUND reacts in the presence of ULTRAVIOLET RADIATION to form PHOTOCHEMICAL OXIDANTS.[172] (EPA)

photochemical smog the resulting mix of ozone, peroxyacetyl nitrate, ALDEHYDES, and other oxidation products of HYDROCARBONS, oxides of nitrogen, AEROSOLS, etc., which form in highly polluted urban atmospheres under conditions of stagnation and high sunlight intensity in cities such as Los Angeles, California.[171,172] (IUPAC)

Most large cities in the world show some degree of photochemical smog generation today. Photochemical smog is a highly oxidized atmosphere in contrast with the SO_2-rich pollution and fogs which characterized many cities such as London, England, many years ago. The characteristics of photochemical smog also include eye-irritation, visibility reduction, damage to certain sensitive plants and trees, and possible pulmonary and other health-related problems in humans and animals.

photochemically-reactive organic material any ORGANIC material which will react with oxygen, excited oxygen, OZONE, and/or other FREE RADICALS generated by the action of sunlight on components in the atmosphere, giving rise to secondary contaminants and reaction INTERMEDIATES in the atmosphere which can have detrimental effects.[172]

photochemically-reactive organic substance see VOLATILE ORGANIC COMPOUND.[171]

photochemically-reactive solvent any SOLVENT with an AGGREGATE of more than 20% of its total volume composed of the chemical compounds classified below or which exceeds any of the following individual percentage composition limitations, referred to the total volume of solvent: (1) a combination of HYDROCARBONS, ALCOHOLS, ALDEHYDES, ESTERS, ETHERS, or KETONES having an olefinic or cyclo-olefinic type of unsaturation: 5%; (2) a combination of AROMATIC COMPOUNDS with eight or more CARBON atoms to the molecule except ethylbenzene: 8%; (3) a combination of ethylbenzene, ketones having branched hydrocarbon structures, trichloroethylene or toluene: 20%. (As defined by Rule 66.)[164,172]

photochromatic materials exhibiting the phenomenon of PHOTOCHROMATISM.[69,43]

photochromatism phenomenon of reversible change exhibited by some materials when exposed to light.[69,43]

photocurable coating a POLYMERIZABLE mixture that can be applied as a thin film to a SUBSTRATE and polymerized at a rapid rate by EXPOSURE to ACTINIC LIGHT.[67,55]

photoengraving PROCESS for producing a design on a sensitized METAL plate by placing a TRANSPARENT negative between the plate and a source of light. The areas not rendered water-insoluble by light are washed and etched with ACID SOLUTIONS.[168]

photogenic property the tendency of a PIGMENT to darken on EXPOSURE to sunlight and to be rebleached on being placed in the dark.[41]

photographing BRUSHMARKS or other irregularities in the previous coat or SUBSTRATE that show through the dried TOPCOAT.[42] Syn: telegraphing or showthrough

photogravure PROCESS for making prints from photochemically prepared INTAGLIO plates.[168]

photolithography PROCESS utilizing SURFACES which, under the selective action of LIGHT, are transformed from a water-receptive to a water-repellent and INK-receptive nature.[168]

photometer an instrument for the measurement of the INTENSITY of EMITTED, REFLECTED, or TRANSMITTED LIGHT.[43,145]

For the measurement of luminous intensity, a visual receptor element (the eye) may be used in the measuring device or a physical receptor element may be used which can be related to the calculated response of a standard observer.

photometric pertaining to the measurement of the INTENSITY of LIGHT.[43,69]

photopic vision mediated by the cone receptors in the retina of the eye, which give rise to the SENSATION of COLOR occurring at high and medium levels of LUMINANCE.[43,69]

phthalate esters a group of PLASTICIZERS, produced by the direct action of ALCOHOL on PHTHALIC ANHYDRIDE.[130]

They are the most widely used of all plasticizers and are generally characterized by moderate cost, good stability, and good all-round properties.

phthalic acid $C_8H_6O_4$; o-benzenedicarboxylic acid. mol wt, 166.13, crystals; mp, about 230°C when heated, forming phthalic anhydride and water.[144]

phthalic anhydride $C_6H_4(CO)_2O$. White, odorless, crystalline FLAKE ACID ANHYDRIDE used in the manufacture of ALKYDS, POLYESTER RESINS, PLASTICIZERS, SOLVENTS, DYES, and INTERMEDIATES.[156]

Properties: bp, 284°C; mp, 130°C; sp gr, 1.53; acid value, 758.0.

ASTM test methods for phthalic anhydride include: color in molten state, D3366; content in alkyd resins, D563, D1306; sampling and handling, D3438; specification, D2403.

phthalic anhydride test phthalic anhydride reacts with primary ALCOHOLS when the mixture is refluxed in BENZENE.[157]

Secondary alcohols react less readily, usually requiring a reaction temperature of 100-200°C, whereas the tertiary alcohols do not react.

phthalocyanine pigments series of ORGANIC PIGMENTS having as a structural unit four isoindole groups, $(C_6H_4)\ C_2N$, linked by four nitrogen atoms so as to form a conjugated chain.[41]

There are four commercially important modifications, including the basic compound: (1) phthalocyanine (METAL free), $(C_6H_4C_2N)_4\ N_4$, blue-green; (2) copper phthalocyanine, in which a copper atom is held by secondary valences of the isoindole nitrogen atoms; sp gr, 1.59; (3) chlorinated copper phthalocyanine, green, in which 15 to 16 hydrogen atoms are replaced by chlorine; (4) sulfonated copper phthalocyanine, water-soluble, green, in which two hydrogen atoms are replaced by sulfonic acid, HSO_4, groups.

ASTM Test Methods D1135 and D3256 cover the analysis of phthalocyanine (phthalo) blue and ASTM D963, its specification.

ASTM Test Method D3256 covers the chemical analysis of phthalocyanine (phthalo) green and ASTM D3021, its specification.

phycocolloid any of several polysaccharide hydrocolloids from brown to red seaweeds.[139,155] See GUM, NATURAL.[155,130,167]

physical of, or relating to matter and energy or the sciences dealing with them, especially physics.[142]

physical hazard a substance for which there is valid evidence that it is a combustible liquid, compressed gas, explosive, flammable, an ORGANIC peroxide, an OXIDIZER, PYROPHORIC, unstable (reactive), or water reactive.[172]

PIC test abbreviation for pseudoisochromatic test for defective COLOR vision.[43,157,69] See PSEUDOISOCHROMATIC PLATE TEST.[43,157,69]

pick see PICKING.[168]

pick-up roll (1) in the COIL COATING industry, the roll which revolves within the pan and is partially immersed in the paint. This roll picks up paint from the pan and applies it to the transfer or applicator roll;[59,70,79,77] (2) spreading device where the roll for picking up the ADHESIVE runs in a reservoir of adhesive. (ASTM)

picking (1) the adherence of a sheet of PAPER to the plate due to the tack of the INK; (2) the removal of the surface of the paper during printing. It occurs when the pulling FORCE (tack) of the ink is greater than the surface strength of the paper, whether coated or uncoated.[168] See LIFTING.[42,56,77]

picking up (1) the blending of a coat of freshly applied paint with another over which it is applied;[54] (2) the joining up of a WET EDGE.[54] cf. PULLING UP[42]

pickled pine a gray FINISH which duplicates the effect formerly produced by actually pickling the wood with nitric acid, but now obtained by using a gray STAIN.[148]

pickling (1) treatment for the removal of rust and MILL SCALE from STEEL by immersion in an acid solution containing an INHIBITOR. Pickling should be followed by thorough washing and drying before painting; (2) the process of removing paint and VARNISH with an alkaline preparation or strong solvents.[158] (BSI)

picture framing a perimeter thickness of COLOR difference (usually darker) relative to the rest of the painted surface.[42,56]

This problem can be due to various mechanisms: (a) architectural paints: the greater shear of a BRUSH used around the perimeter of a wall or ceiling compared to the lower shear of a roller used on the rest of the wall or ceiling; (b) industrial finishes: the FLOW of a FINISH during baking resulting in the build-up of the coating on the edge of the SUBSTRATE; (c) roofing: a rectangular pattern of ridges in a membrane over insulation or deck joints.(ASTM); CEMENT-ASBESTOS shingles: (on EXPOSURE fences) the darker color of the coating on the perimeter due to carbonation (from the CO_2 in the air) of the LIME in the shingle causing it to be more neutral compared to the rest of the shingle which is more alkaline and attacks some paints usually causing a lighter COLOR.

pig wrack see CARRAGEEN.[139,155]

pigging the cleaning out of pipelines, between processes, with a special contoured plug or pig.[49,131]

The pig is driven down the pipeline by compressed gas fluid or even product to produce a clean line ready for reuse.

pigment finely ground, natural or synthetic, INORGANIC or ORGANIC, insoluble DISPERSED PARTICLES (POWDER) which, when dispersed in a liquid VEHICLE to make paint, may provide, in addition to COLOR,

---

Words presented in CAP/SMALL CAPS type indicate that the word is defined in another part of the *Coatings Encyclopedic Dictionary*.

Numerical superscripts classify terms in one or more of the categories listed in the second section of this volume.

many of the essential properties of a paint—OPACITY, HARDNESS, DURABILITY, and CORROSION RESISTANCE.[41,69]

The term is used to include EXTENDERS, as well as white or color pigments. The distinction between powders which are pigments and those which are DYES is generally considered on the basis of SOLUBILITY—pigments being insoluble and dispersed in the material, dyes being soluble or in solution as used.

ASTM test methods for pigment content of paints and dispersions include: paint/traffic marking material, D4451; pigment pastes in oil, test, D1208; solvent paints, D2371; titanium dioxide slurries, D3926; water-based paints, D3723.

ASTM test methods for general properties and composition of pigments include: bleeding characteristics, D279; acidity/alkalinity, D1208; moisture content, D280 and D1208; volatile content, D4139; water soluble salts content, D2448; lightfastness in artist paints, D4303; oil absorption, Gardner-Coleman method, D281.

ASTM test methods covering pigment particle size include: reporting of characteristics, D1366; coarse particle content, D185; fineness of dispersion paint, D1210; fineness of grind, printing INK, D1316; particle size distribution, D3360; specific gravity (density), D153; tinting strength and color of colored pigments—with a mechanical muller, D387; with a miniature sand mill, D3022; tinting strength of white pigments—visual method, D332; instrumental method, D2745.

Pigment Black 6 (77266) see LAMPBLACK.[41]

Pigment Black 7 (77266) see CARBON BLACK.[41]

Pigment Black 10 see GRAPHITE.[41]

Pigment Black 11 (77499) see BLACK IRON OXIDE.[41]

pigment bleeding diffusing of coloring matter from a previously coated surface due to SOLVENT extraction of colorants.[41,42,56,69]

Pigment Blue 27 (77510) see IRON BLUE.[41]

Pigment Blue 28 (77346) see COBALT BLUE.[41]

Pigment Blue 29 (77007) see ULTRAMARINE BLUE.[41]

Pigment Brown 6 (77499) see BROWN IRON OXIDE PIGMENT.[41]

pigment dyes see DYE PIGMENTS.[41]

pigment extender see EXTENDER (PIGMENT).[41]

Pigment Green 8 (10006) see PIGMENT GREEN B.[41]

Pigment Green 10 (12775) see NICKEL AZO YELLOW.[41]

Pigment Green 15 (77510/77603) see CHROME GREENS.[41]

Pigment Green 17 (77288) see CHROMIUM OXIDE GREEN.[41]

Pigment Green 18 (77289) see HYDRATED CHROMIUM OXIDE.[41]

Pigment Green B $C_{30}H_{18}N_3O_6FeNa$. Pigment Green 8 (10006). A stable chelate PIGMENT based on nitrosation of 2-naphthol.[41]

Good ALKALI RESISTANCE and interior LIGHTFASTNESS; poor ACID RESISTANCE. Density, 1.4-1.58 g/cm³ (11.6-13.1 lb/gal); O.A., 75-100; particle size, 0.01-0.35 μm.

pigment grind (deprecated) Not a grind, but a DISPERSION of PIGMENT in VEHICLE.[64]

Preferred term is "MILL base."

pigment, metallic see METALLIC PIGMENT.[41]

pigment, nonhiding see EXTENDER (PIGMENT).[41,56]

Pigment Orange 21 (77601) see CHROME ORANGE, LIGHT and DEEP.[41]

Pigment Orange 23 (77201) see CADMIUM-MERCURY SULFIDES.[41]

pigment paste see PASTE, PIGMENT.[41]

Pigment Red 1 (12070) see PARA REDS.[41]

Pigment Red 3 (12120) see TOLUIDINE REDS.[41]

Pigment Red 4 (12085) see ORTHO-CHLOR-PARA-NITRANILINE.[41]

Pigment Red 6 (12090) see PARA-CHLOR-ORTHO-NITRANILINE.[41]

Pigment Red 38 (21120) see PYRAZOLONE RED.[41]

Pigment Red 48 (15865) see BON REDS AND MAROONS; PERMANENT RED 2B.[41]

Pigment Red 49 (15630) see LITHOL RED.[41]

Pigment Red 53 (15585) see LAKE RED C.[41]

Pigment Red 57 (15850) see LITHOL RUBINE.[41]

Pigment Red 81 (45160) see RHODAMINE 6G.[41]

Pigment Red 83 (58000) see ALIZARIN RED.[41]

Pigment Red 101 (77491) see INDIAN RED; IRON OXIDES, SYNTHETIC.[41]

Pigment Red 104 (77605) see MOLYBDATE ORANGE.[41]

Pigment Red 105 (77578) see RED LEAD.[41]

Pigment Red 106 (77766) see MERCURIC SULFIDE.[41]

Pigment Red 108 (77202) see CADMIUM-MERCURY SULFIDES.[41]

Pigment Violet 10 (45170) see RHODAMINE B.[41]

Pigment Violet 15 (77007) see ULTRAMARINE VIOLET.[41]

pigment volume see PIGMENT VOLUME CONCENTRATION.[41]

Pigment Volume Concentration (PVC) ratio of the volume of PIGMENT to the volume of total NONVOLATILE MATERIAL (i.e., pigment and BINDER) present in a coating.[41,128]

The figure is usually expressed as a percentage.

Pigment White 1 (77597) see CARBONATE WHITE LEAD.[41]

Pigment White 2 (77633) see BASIC LEAD SULFATE.[41]

Pigment White 4 (77947) see ZINC OXIDE.[41]

Pigment White 5 (77115) see LITHOPONE.[41]

Pigment White 6 (77891) see TITANIUM DIOXIDE, ANATASE (and RUTILE).[41]

Pigment White 7 (77975) see ZINC SULFIDE.[41]

Pigment White 11 (77052) see ANTIMONY OXIDE.[41]

Pigment White 12 (77990) see ZIRCONIUM OXIDE.[41]

Pigment White 16 (77625) see BASIC LEAD SILICATE.[41]

Pigment White 18 (77220) see CALCIUM CARBONATE, NATURAL AND CALCIUM CARBONATE, SYNTHETIC (WHITING).[41,58]

Pigment White 19 (77005) see ALUMINUM SILICATE (CLAY).[41,58]

Pigment White 20 (77019) see ALUMINUM POTASSIUM SILICATE (MICA).[41,58]

Pigment White 22 (77120) see BARIUM SULFATE, NATURAL (BARYTES) and SULFATE, SYNTHETIC (BLANC FIXE).[41,58]

Pigment White 26 (77718) see MAGNESIUM SILICATE, NONFIBROUS (TALC).[41,58]

Pigment White 27 (77811) see SILICA, MICROCRYSTALLINE and DIATOMACEOUS SILICA.[41,58]

Pigment Yellow 32 (77839) see STRONTIUM CHROMATE.[41]

Pigment Yellow 34 (77600) see CHROME YELLOW, MEDIUM.[41]

Pigment Yellow 37 (77199) see CADMIUM SULFIDE.[41]

Pigment Yellow 42 (77492) see IRON OXIDES, NATURAL.[41]

Pigment Yellow 53 (77788) see NICKEL TITANATE.[41]

Pigment Yellow 74 (11741) see DALAMAR YELLOW.[41]

pigment/binder ratio ratio (generally, by weight in the U.S.A.; by volume in Britain) of total PIGMENT to BINDER solids in paint.[41,128] Syn: binder ratio

pilaster flat vertical projection from a wall, with the proportions, capital shaft, and base of a column.[81,82]

pile burning see BIN CURE.[140]

piling (1) behavior of a very quick-drying paint which, during APPLICATION by BRUSH, becomes so sticky that the resulting film is thick and uneven;[42,56] (2) the buildup of INK on rollers, plate, or blanket.[168] See CAKING.[42]

pill flow see PLATE FLOW.[166]

pimple small, conical protrusion on the surface.[56]

May refer to small surface blister.

PIN abbreviation for PRODUCT IDENTIFICATION NUMBER.[172]

pinacolone $CH_3COC(CH_3)_3$. Methyl tert. butylketone. Medium-boiling ketonic SOLVENT. Bp, 106°C.[164] Syn: pinacolin

pinched coating in coated fabrics, a ridge or a wrinkle in the coating.[56]

pine needle oil ESSENTIAL OIL of typical fragrance obtained by steam distillation of the leaves (needles) of certain species of pine or other coniferous trees.[44] (ASTM)

pine oil colorless to amber colored VOLATILE OIL with characteristic pinaceous odor, consisting principally of isomeric tertiary and secondary cyclic TERPENE ALCOHOLS, with variable quantities of terpene HYDROCARBONS, ETHERS, KETONES, PHENOLS, and PHE-

Words presented in CAP/SMALL CAPS type indicate that the word is defined in another part of the *Coatings Encyclopedic Dictionary.*
Numerical superscripts classify terms in one or more of the categories listed in the second section of this volume.

NOLIC ETHERS, the amount and character of which depend on the source and method of manufacture. The four commercial kinds of pine oil are: (1) steam distilled pine oil: obtained from pine wood by steam DISTILLATION or by SOLVENT EXTRACTION followed by such distillation; (2) destructively distilled pine oil: obtained from the lighter distillate from the DESTRUCTIVE DISTILLATION (carbonization) of pine wood; (3) synthetic pine oil: obtained by chemical hydration of terpene hydrocarbons to form the terpene alcohols, or by dehydration of terpin hydrate; (4) sulfate pine oil: a high boiling fraction obtained in the refining and fractional distillation of the condensed vapors released during the digestion of wood by the sulfate process.[44,164]

ASTM Test Method D890 covers the measurement of moisture content of pine oil and ASTM D802, the sampling and testing of pine oil.

pine pitch dark-colored to black solidified material, somewhat pliant and tenacious, obtained by distilling off practically all the volatile oil from a RETORT PINE TAR; the genuine contains no added free ROSIN.[44] (ASTM)

pine tar, kiln burned heavy, oily liquid resulting from controlled carbonization (slow burning) of pine knots and stumpwood to charcoal in earth-covered piles or "kilns," with introduction of insufficient air to permit complete combustion; contains undecomposed RESIN ACIDS along with the DECOMPOSITION products.[44] (ASTM)

This product is sometimes called "country TAR."

ASTM Test Method D856 covers the sampling and testing of kiln burned pine tar, retort pine tar, and pine tar oil.

pine tar oil OIL obtained by condensing the vapors from the retorts in which resinous pine wood is destructively distilled (CARBONIZED).[44] (ASTM)

ASTM Test Method D856 covers the sampling and testing of kiln burned pine tar, retort pine tar, and pine tar oil.

pine tar, retort TAR produced by removal of volatile OILS from PINE TAR OIL by STEAM DISTILLATION.[44] (ASTM)

Several grades are marketed, namely: thin, medium, heavy, and extra heavy, so classified on the basis of viscosity, and depending upon the quantity of volatile oils removed.

ASTM Test Method D856 covers the sampling and testing of kiln burned pine tar, retort pine tar, and pine tar oil.

pine tar, Stockholm kiln-burned PINE TAR produced in Scandinavian countries from wood of the Northern European pine, *Pinus sylvestris*.[44] (ASTM)

Alpha-pinene: bp, 155°C; sp gr, 0.858/20°C; refractive index, 1.466.

Beta-pinene: bp, 163°C; sp gr, 0.869/20°C; refractive index, 1.476.

pinene $C_{10}H_{16}$. A bicyclic TERPENE HYDROCARBON, the principal constituent of all TURPENTINES, and existing therein in two ISOMERIC forms, alpha- and beta-pinene.[44,164]

The latter is found in appreciable quantity only in GUM SPIRITS and sulfate WOOD TURPENTINE. Pinene not otherwise described usually means alpha-pinene.(ASTM)

pinholes FILM DEFECT characterized by small pore-like flaws in a coating which extend entirely through the applied film and have the general APPEARANCE of pin pricks, fine holes, or voids when viewed by reflected light.[56] (ASTM) See PORES.[56]

The term is rather generally applied to holes caused by SOLVENT bubbling, moisture, other volatile products, or the presence of extraneous PARTICLES in the applied film.

ASTM Test Method D1848 recognizes pinholes as a paint film failure characteristic of exterior LATEX paints.

pinolin see ROSIN SPIRIT.[164,167]

pipe clay see ALUMINUM SILICATE (CLAY).[41,58]

pit circular depression, usually small.[56] See POCKMARKING.[56]

pitch the RESIN occurring in the WOOD of certain conifers.[44,76]

pitch, Archangel originally a genuine PINE PITCH made from pine TAR in the Archangel district of Russia; in this country a similar product is made from residues of pine origin, blended with various OILS to make a pitch for caulking boats.[44,76] (ASTM)

Its acidity is due mainly to ROSIN acids.

pitch, brewer's type of PITCH made by blending certain OILS, WAXES, or other ingredients with ROSIN for the coating of beer barrels.[44] (ASTM)

pitch, burgundy originally the solidified RESIN obtained by heating and straining the air-dried solid OLEORESIN exuded by the Norway spruce (*Picea excelsa*) and European silver fir (*Abies pectinata*); now denotes an artificial mixture made by heating ROSIN with certain FIXED OILS, the combination being used for ADHESIVE PLASTERS.[44] (ASTM)

pitch, navy pitch obtained by melting ROSIN with pine TAR, with or without rosin distillation residues.[44] (ASTM)

pitch oil see CREOSOTE.[76]

pitch pocket an opening between the growth rings containing RESIN in certain SOFTWOODS. Syn: resin pocket, pitch streak

pitch streak see PITCH POCKET.[153]

pitting formation of holes or pits in the surface of a METAL, by CORROSION, or in the surface of a coating while it CURES, often because of the escape of gas or SOLVENT.

pK the negative logarithm of the dissociation constant K or -log K that serves as a convenient measure of an acid [K for ACETIC ACID is 0.000018 or 1.8 $\times 10^{-5}$ from which pK is (5-0.25) or 4.75].[144,157] cf. PH, and PKA

pKa the PH at which the acid and its conjugate base exist at equal concentrations; where half the acid is neutralized.[144,157] cf. PK[144]

PKT see POTASSIUM TITANATE.[41]

Planckian radiator a body which ABSORBS all radiation falling on it and REFLECTS none. At each temperature to which it is heated, it radiates a precise magnitude per area of its surface at each WAVELENGTH throughout the spectrum. Thus, a family of SPECTRAL CURVES is formed for the family of temperatures. The CHROMATICITY COORDINATES of the integration of these individual curves trace a curve on a CIE CHROMATICITY DIAGRAM which is called the Planckian Locus and is the chromaticity reference for designating color temperature and CORRELATED COLOR TEMPERATURE. The temperatures are expressed in KELVIN (absolute) temperature scale.[43,69]

planography PRINTING PROCESSES which are dependent upon the fact that an oily INK will not ADHERE to water-moistened, nondesign areas of a level plate but will wet design areas treated with a greasy ink or VARNISH. [168]

plasma coating the formation of ORGANIC-INORGANIC coatings in plasma refers to the deposition of these coatings by subjecting vapors of the low molecular-weight starting materials to a FLOW discharge or plasma environment.[77] cf. PLASMA DEPOSITION and FLAME SPRAY[77]

A plasma state of a gas differs from other gas states in that it contains highly energetic ionized species. Plasmas are created by the collision of molecules and atoms with high-energy electrons, which are generated from electrical discharges associated with electric, radio, or microwave energy fields that create the glow discharge/plasma phenomena.

The nature of organic/inorganic molecules introduced into the plasma state strongly determines the nature of the final crosslinked film deposited on the surface of the SUBSTRATE.

plasma deposition technique a method of applying POWDER coatings using a compressed gas to melt the powder in a flame before the powder impinges on the SUBSTRATE to be coated.[77]

plasma flame see FLAME SPRAY.[77]

plaster a paste-like material, usually a mixture of PORTLAND CEMENT, LIME, or GYPSUM with water and sand; fiber or hair may be added as a BINDER; applied to surfaces such as walls or ceilings in the PLASTIC state; later it sets to form a hard surface.[75,73] (DAC)

plaster of Paris a group of CEMENTS which consist essentially of CALCIUM SULFATE and are produced by a partial dehydration of GYPSUM to the hemihydrate $CaSO_4 \cdot {}^1/{}_2H_2O$.[73,75] Syn: GYPSUM CEMENT

They usually contain ADDITIVES of various sorts.

plaster primer PRIMERS with a degree of resistance to ALKALI which are used for priming plasters and CEMENTS of varying degress of alkalinity.[71] (BSI)

The primers must not only resist SAPONIFICATION but must insulate succeeding coats of paint from attack.

plasterboard see GYPSUM WALLBOARD.[159,75]

plastic (1) noun. A HIGH POLYMER, usually synthetic, combined with other ingredients (e.g.) FILLERS, REINFORCING AGENTS, PLASTICIZERS, etc. The mixture is capable of being formed or molded under heat and pressure in its raw state. THERMOPLASTIC type can be resoftened by heat, while THERMOSETTING type cannot; (2) adjective. Capable of being shaped or molded with or without application of heat. WAXES, moist CLAY are examples.[47,166,38] See PLASTICITY.[47,166]

ELASTOMERIC materials, which are also shaped by FLOW, are not considered as plastics.

ASTM Test Method D3002 covers coatings for plastic substrates and ASTM D1652, the epoxy content.

plastic-bonded wallpapers PAPERS made with a PROTEIN size with a PLASTIC added.[148]

They are washable, but subject to staining by certain liquids.

plastic-coated wallpapers PAPERS that are more WASHABLE and STAIN-RESISTANT.[148] See WALLCOVERING.

A thick PLASTIC coating increases the resistance and permits vigorous washing.

plastic flow a liquid displays plastic flow when a YIELD STRESS must be overcome or exceeded before flow will take place. Plastic viscosity, U, is expressed as:

$$U = \frac{\text{shear stress} - \text{yield stress}}{\text{shear rate}}$$

Liquids which display plastic flow are called BINGHAM LIQUIDS.[166]

Words presented in CAP/SMALL CAPS type indicate that the word is defined in another part of the *Coatings Encyclopedic Dictionary.*

Numerical superscripts classify terms in one or more of the categories listed in the second section of this volume.

plastic paint a heavy-bodied, THIXOTROPIC paint which can be worked after APPLICATION, by STIPPLING or by PAINT ROLLERS having a TEXTURED pattern, to produce various textured or pattern surfaces.[71,163,54] (DAC) Syn: textured paint and texture-finished paint

plastic solid solid that DEFORMS continuously and permanently when subjected to a SHEARING STRESS in excess of its YIELD VALUE.[47,166] (ASTM)

plastic viscosity see VISCOSITY, PLASTIC.[166]

plastic wood a PASTE of WOOD FLOUR, PLASTICIZER, RESINS, and/or other materials dispersed in NITROCELLULOSE or other BINDERS and volatile SOLVENTS, used for repairing or filling holes in WOOD, etc.[147,153,158,165]

plasticity the complex property of a material involving a combination of the properties of MOBILITY and of YIELD VALUE, enabling it to be continuously deformed without rupture when acted on by a FORCE sufficient to cause FLOW and allowing it to retain its shape after the applied force has been removed.[166]

plasticization the PROCESS of increasing the FLEXIBILITY and ADHESION of an ORGANIC COATING film and decreasing its brittleness, cohesiveness, and GLASS TRANSITION TEMPERATURE. It can be achieved by chemical modification of the POLYMER (INTERNAL PLASTICIZER) or by adding COMPOUNDS (PLASTICIZERS) to the coating formulation (EXTERNAL PLASTICIZER).[170] See INTERNALLY PLASTICIZED.[170,130]

plasticizer (1) an ADDITIVE in a paint formulation to soften the film, thus giving it better FLEXIBILITY, CHIP RESISTANCE, and formability; (2) a chemical agent added to a PLASTIC composition to improve its FLOW and PROCESSABILITY and to reduce brittleness.[83,170] (DAC) See PLASTICIZATION, EXTERNAL PLASTICIZER and INTERNAL PLASTICIZER.[83,170]

plasticizer migration movement of PLASTICIZER in a coating or PLASTIC to the surface or into an adjacent material such as an undercoat or SUBSTRATE.[170,42,38]

plastigel a PLASTISOL to which a gelling agent has been added to increase its gel-like FLOW properties and YIELD VALUE.[71,83]

plastisol a SUSPENSION of a finely divided RESIN in a PLASTICIZER, that can be converted to a continuous film by the application of heat.[38,71,170] (ASTM) cf. ORGANOSOL[38,71,170] and PLASTIGEL[71,83]

Distinct from baking enamels, etc., in that substantially all the original mixture becomes a part of the film; there is no significant EVAPORATION of SOLVENT. The films are usually much thicker than obtainable from coatings which depend on the evaporation of a volatile solvent.

plastometer an instrument for determining the FLOW PROPERTIES of PLASTIC materials.[145,166]

The term "plastometer" is not actually necessary, for a plastometer is essentially any viscometer capable of measuring the flow properties of a material exhibiting YIELD VALUE.

plate cylinder that ROLLER of an offset PRINTING PRESS which bears the printing plate.[168]

plate flow the distance a POWDER COATING flows in the molten state.[166] Syn: inclined plate flow, glass plate flow, and pill flow

plateless engraving a TYPOGRAPHIC PRINTING PROCESS in which densely pigmented INK on printed sheets is dusted with a POWDERED RESIN of low melting point.[168]

After surplus powder has been blown from the PAPER, heat is applied to fuse the resin and yield characters which are raised above the paper.

platen (1) a flat plate in a PRINTING PRESS, which presses the PAPER against the inked type, thus securing an IMPRESSION;[168] (2) a plate of METAL, especially one that exerts or receives PRESSURE, as in a press used for GLUING PLYWOOD.[59] (ASTM) Syn: caul

platen press a PRESS which prints a single sheet by pressing the latter against an entire frame or chase of type at the same moment.[168]

platinum see GLOSS.[43]

pleochroism property of CRYSTALS or MINERALS of giving different ABSORPTION colors for different directions of VIBRATION of LIGHT within the crystal; in OPTICAL mineralogy this property is observed in crystals under a polarizing microscope with the lower nicol only and rotating the stage.[69,34,43] See DICHROISM.[69,43,34]

plessy's green $CrPO_4 \cdot nH_2O$. A bluish green PIGMENT consisting essentially of hydrated chromium phosphate.[41]

plochere color system a COLOR ORDER SYSTEM based on SUBTRACTIVE COLORANT MIXTURE developed by mixing a limited number of PIGMENTS in systematically varied proportions.[69]

plug flow movement of a material as a unit without SHEARING within the mass.[166]

plumbago synonym for GRAPHITE.[83]

ply (1) a layer of FELT in a BUILT-UP ROOFING membrane; four-ply membrane has at least four plies of felt at any vertical cross section cut through the membrane;[76] (ASTM) (2) one of the sheets of WOOD (VENEER) glued together to form PLYWOOD.[153]

plying cement any BITUMINOUS material used for adhering layers of FELTS, FABRICS, or MATS to structural surfaces and to each other.[76,79] (ASTM)

plywood cross-bonded assembly made of layers of VENEER or veneer in combination with a LUMBER core or plies joined with an ADHESIVE.[75,79,153,159] (ASTM) cf. PARTICLE BOARD[75,153,159]

Two types of plywood are recognized, namely; (1) veneer plywood; and (2) lumber core plywood. Generally, the grain of one or more plies is approximately at right angles to the other plies, and almost always an odd number of plies are used.

PMA (1) abbreviation for phosphomolybdic acid; (2) applied to PIGMENTS which have been PRECIPITATED with phosphomolybdic acid to give resistance to alkali, heat, light, and water.[41] See PHOSPHOMOLYBDIC PIGMENTS.[41]

PMA pigment see PRECIPITATED BASIC DYE BLUES.[41]

PMCC abbreviation for PENSKY-MARTENS CLOSED-CUP FLASH TESTER.[145,157]

pockmarking FILM DEFECT in the shape of irregular and unsightly depressions formed during the drying of a paint or VARNISH film.[56] See ORANGE PEEL.[56] Syn: pitting

pointing treatment of joints in masonry by filling with mortar to improve APPEARANCE or protect against weather.[82]

point source any discernible, confined, and discrete conveyance from which POLLUTANTS are or may be discharged.[171]

pointillism from the French, pointiller, meaning to dot, stipple. A system of late impressionist painting developed by Georges Seurat (1859-1891) and Paul Signac (1863-1935). [81]

Contrary to the practice of COLOR mixing on the PALETTE, pointillism consists of applying separated spots, or dots, of pure color, side by side on the canvas, e.g., red and yellow for orange, red and blue for mauve, etc. Theoretically, at the right distance, the spectator's eye automatically mixes the colors.

poise the C.G.S. UNIT OF VISCOSITY.[157] See VISCOSITY.[166,128]

The term "poise" is derived from the name of the man who discovered the laws of FLOW in CAPILLARIES —Poiseuille. Dimensions: dynes/cm^2. In SI units, P = 0.1 Pa.s.

Poison, Class A a DOT designation for an extremely dangerous poison such as a poisonous gas or liquid of such a nature that a very small amount of the gas or vapor of the liquid mixed with air is dangerous to life (e.g., phosgene, cyanogen, hydrocyanic acid, and nitrogen peroxide).

Poison, Class B a DOT designation for liquid, solid, paste, or semisolid substances other than class A poisons or irritating materials known or presumed, on the basis of animal tests, to be so toxic to man as to afford a hazard to health during transportation.

Poison Control Center provides medical information on a 24-hour basis for accidents involving ingestion of potentially poisonous materials.[172]

Call your area's largest hospital to find the one nearest you or look under the emergency page of your telephone directory (generally, the first few pages).

polar MOLECULE or RADICAL that has, or is capable of developing, ELECTRICAL charges.[133]

Thus, polar molecules ionize in solution and impart electrical conductivity. Water, ALCOHOL, and sulfuric acid are polar in nature; most HYDROCARBON liquids are not. Carboxyl and hydroxyl often exhibit an electrical charge. The formation of emulsions and the action of DETERGENTS are dependent on this behavior.

polimerisado oil a bodied liquid OITICICA OIL.[63]

polish (1) verb; see BURNISH.[158] (2) noun. A solid POWDER or a liquid or semi-liquid mixture that imparts smoothness, surface protection, or a DECORATIVE FINISH.[137,148]

polishing varnish very hard-drying, SHORT OIL VARNISH used for interior woodwork, FURNITURE, etc., and capable of being rubbed with ABRASIVE and MINERAL OIL LUBRICANTS to a very smooth surface for a desired degree of GLOSS.[71,152,54,163] See FLATTING VARNISH.[152] Syn: rubbing varnish

pollutant something that is unfit for or harmful to living things, especially a waste material that contaminates air, soil, or water.[171,172]

pollute to make unfit for or as harmful to living things, especially by the addition of waste.[171,172]

pollution the act or PROCESS of polluting or the state of being polluted, especially the contamination of soil, water, or the atmosphere by the discharge of harmful substances.[171,172]

pollution abatement ending POLLUTION.[171,172]

Distinguished from pollution control, which may only reduce pollution, and penalties, which principally punish violations.

pollution control reducing POLLUTION. Distinguished from POLLUTION ABATEMENT, which means ending pollution, and penalties, which principally punish violations.[171]

Words presented in CAP/SMALL CAPS type indicate that the word is defined in another part of the *Coatings Encyclopedic Dictionary.*

Numerical superscripts classify terms in one or more of the categories listed in the second section of this volume.

polyacetal POLYMER in which the repeated structural unit in the chain is of the ACETAL type.[130] See POLYOXYMETHYLENE.[130]

polyacrylate POLYMER of an ester of ACRYLIC ACID or of esters of acrylic acid homologues or substituted derivatives.[130] See ACRYLIC.[130]

polyacrylic plastics PLASTICS based on POLYMERS in which the repeated structural units in the chains are essentially all of the ACRYLIC type.[130] See ACRYLIC PLASTICS.[130]

polyaddition see ADDITION POLYMERIZATION.[130]

polyalcohol ALCOHOLIC or HYDROXY COMPOUND, containing more than one hydroxyl group.[144] Syn: polyol

Typical polyalcohols are ETHYLENE GLYCOL with two hydroxyl groups, GLYCEROL with three, and PENTAERYTHRITOL with four hydroxyl groups. The polyalcohols are sometimes known as polyhydric ALCOHOLS.

polyamide resins CONDENSATION RESINS of an AMINE and an ACID, the repeated structural unit in the chain being of the AMIDE type.[130]

polybasic acid ACID containing more than one REACTIVE HYDROGEN atom.[144]

With water-soluble ionizable acids which are polybasic, two or more hydrogen ions are made available, per molecule.

polycarbonate POLYMER in which the repeated structural unit in the chain is of the carbonate type.[130]

polychromatic finish (1) a multicolored paint FINISH;[69] (2) a paint containing reflecting METALLIC FLAKE and fine TRANSPARENT PIGMENTS which appear as a variety of colors when viewed from different angles.[71,163] (DAC)

polycondensate product of a CONDENSATION POLYMERIZATION.[169]

polycondensation see CONDENSATION.[169]

polycoumarone resins see COUMARONE-INDENE RESINS.[167]

polycyclization cyclization in which more than one internal ring formation occurs.[169]

polydicyclopentadiene resins friable, THERMOPLASTIC, unsaturated aromatic RESINS derived from PETROLEUM HYDROCARBONS whose principal component is dicyclopentadiene.[130]

Softening points from 10-140°C (50-285°F); sp gr, 0.99-1.11; iodine value, 170-200. Soluble in ALIPHATIC, AROMATIC, CHLORINATED SOLVENTS; insoluble in water, ALCOHOLS, and GLYCOLS.

polyelectrolyte high POLYMER substance, either natural or SYNTHETIC, containing ionic constituents; they may be either CATIONIC or ANIONIC.[130]

polyester a POLYMER in which the MONOMER units are linked by the group —COO—, used in the manufacture of RESINS, PLASTICS, and TEXTILE FIBERS.

polyester resins group of SYNTHETIC RESINS which are polycondensation products of dicarboxylic acids with dihydroxy ALCOHOLS.[130]

They are therefore a special type of ALKYD RESIN. Oil-free alkyds are a class by themselves. Often these resins are dispersed in a suitable MONOMER.

polyether glycol see POLYETHYLENE GLYCOL.[144]

polyether resins POLYMERS in which the repeating unit contains a CARBON-OXYGEN BOND derived from ALDEHYDES or EPOXIDES or similar materials.[130]

polyethylene glycol any of several CONDENSATION POLYMERS of ethylene glycol with the general formula: $HOCH_2(CH_2OCH_2)nCH_2OH$ or $H(OCH_2CH_2)\,nOH$.[144] Syn: polyether glycol and polyoxyethylene glycol

Average molecular weights range from about 200 to 6000.

polyethylene (PE) PLASTIC or RESIN prepared in the POLYMERIZATION of ETHYLENE as the sole MONOMER.[38,130]

polyformaldehyde see POLYOXYMETHYLENE.[130]

polyglycerols COMPOUNDS of ETHER-ALCOHOL type derived from the interaction of two or more molecules of glycerol.[156]

Thus, for example, two molecules of glycerol react to give diglycerol. The polyglycerols have some applications in ALKYD and ESTER RESIN manufacture.

polyglycol a polyhydric ALCOHOL of the monomeric glycol, of uncertain composition. Bp, 230-250°C.[144]

polyimide resins AROMATIC polyimides made by reacting pyromellitic dianhydride with aromatic diamines.[130,176]

Characterized by high resistance to thermal stress.

polyindene resins see COUMARONE-INDENE RESINS.[167]

polyisobutylene rubbery POLYMER of isobutylene (2-methyl-propene) which yields viscous SOLUTIONS in ALIPHATIC and AROMATIC HYDROCARBONS.[130,140] Syn: poly(2-methylpropene)

Films of great ELASTICITY can be obtained, and the product is characterized by excellent chemical and light RESISTANCE, and absorbs very little water.

polymer (1) substance, the molecules of which consist of one or more STRUCTURAL UNITS repeated any number of times; vinyl resins are examples of true POLYMERS. The name is also frequently applied to large molecules produced by any chemical process, e.g., CONDENSATION in which water or other products are produced; ALKYD RESINS are examples of these. Homopolymer—polymer of which the molecules consist of one kind of structural unit repeated any number of times; polyvinyl chloride and POLYVINYL ACETATE are examples. Copolymer—polymer of which the molecule consists of more than one kind of structural unit derived from more than one MONOMER; polyvinyl chloride-acetate, or polyvinyl acetate-acrylic copolymers are examples. See HIGH POLYMER;[130] (2) a substance consisting of molecules characterized by the repetition (neglecting ends, branch junctions, and other minor irregularities) of one or more types of MONOMERIC UNITS.[130] (IUPAC) Syn: macromolecule

polymer gels similar to liquid EMULSION POLYMER coatings, except furnished in paste form.[130]

Must be polished for LUSTER. May contain some SOLVENT. (CSMA)

polymerization chemical REACTION in which two or more small molecules (MONOMERS) combine to form large molecules (POLYMERS, MACROMOLECULES) that contain repeating STRUCTURAL UNITS of the original molecules and have the same percentage composition as the small molecules if the small ones were of the same kind. See MONOMER,[130,141,169] ADDITION POLYMERIZATION, and CONDENSATION POLYMERIZATION.[130,144]

This is the process by which PLASTICS, RESINS, and RUBBERS (ELASTOMERS) are synthesized. A hazardous polymerization is such a reaction that takes place at a rate that releases large amounts of energy that can cause fires or explosions or burst containers. Materials that can polymerize usually contain inhibitors that can delay the reactions.

ASTM test methods for polymerization include: cellulose nitrate, test, D1716 and unreacted monomer of latexes, D4747.

polymerized fatty acids polycarboxylic acids produced by POLYMERIZING ACIDS from animal or VEGETABLE FATS and OILS, in either an ester or free acid state, by means of heat alone or catalytically.[63,131,169] (ASTM) See DIMER ACIDS.[144]

poly(methyl methacrylate) plastics plastics based on POLYMERS made with methyl methacrylate as essentially the sole MONOMER.[130]

poly(methyl methacrylate) (PMMA) a POLYMER of METHYL METHACRYLATE.[130,38]

polyol see POLYALCOHOL.[144]

polyolefin a POLYMER derived by POLYMERIZATION from relatively simple olefins, including particularly POLYETHYLENE and POLYPROPYLENE from ethylene and propylene.[130]

polyolefin plastics PLASTICS based on POLYMERS made with an olefin (or olefins) as essentially the sole MONOMER (or MONOMERS).[130]

polyoxyethylene glycol see POLYETHYLENE GLYCOL.[144]

polyoxymethylene (POM) a POLYMER in which the repeated structural unit in the chain is oxymethylene.[130]

Polyoxymethylene is theoretically the simplest member of the generic class of polyacetals.

polypropylene (PP) a POLYMER of propylene (propene).[130] Syn: polypropene

polystyrene (PS) polymerized styrene. THERMOPLASTIC SYNTHETIC RESIN of variable molecular weight depending on degree of POLYMERIZATION.[38,130,169]

polyterpene resins friable, THERMOPLASTIC SATURATED cyclic resins produced by the CATALYTIC POLYMERIZATION of monomeric alpha- or beta- pinene which are principal constituents of TURPENTINE, with softening points, 10-135°C, Gardner COLOR, 3-6; sp gr, 0.93-0.97; iodine no., 48. Soluble in ALIPHATIC, AROMATIC, chlorinated solvents, insoluble in ALCOHOLS, glycols, or water.[167] See TERPENE RESINS.[167]

polyurethane finish an exceptionally hard and WEAR-RESISTANT PAINT or VARNISH made by the reaction of POLYOLS with a multifunctional ISOCYANATE.[71,152,163] (DAC) See POLYURETHANES,[156,71,130,161,67] URETHANE COATINGS and ISOCYANATE RESINS.[130]

URETHANE coatings (SOLVENT or WATERBORNE) are noted for their overall balance of high performance properties. These include excellent toughness, flexibility, high impact resistance, and tensile strength. They also offer excellent ABRASION RESISTANCE and resistance to a wide variety of chemicals, solvents, and stains.

Urethane coatings are generally divided into AROMATIC and ALIPHATIC types. Aromatic urethanes are generally based on TDI (TOLUENE DIISOCYANATE) or MDI (DIPHENYLMETHANE DIISOCYANATE). They are designed mainly for interior applications since they have a tendency to yellow when exposed to prolonged sunlight. Aliphatic urethanes offer excellent UV stability, but are more expensive than aromatic urethanes. Aliphatic urethanes are generally based on IPDI (isophorone diisocyanate), HDI (hexamethylene diisocyanate), or TMXDI (tetramethylxylene diisocyanate). In addition to excellent light stability, aliphatic urethanes have better flexibility than aromatic urethanes.

polyurethanes type I: one-package, prereacted-URETHANE COATINGS characterized by the absence of any significant quantity of free ISOCYANATE groups.

Words presented in CAP/SMALL CAPS type indicate that the word is defined in another part of the *Coatings Encyclopedic Dictionary.*

Numerical superscripts classify terms in one or more of the categories listed in the second section of this volume.

They are usually the REACTION product of a polyisocyanate and a polyhydric ALCOHOL ester of VEGETABLE OIL ACIDS and are hardened with the aid of METALLIC SOAP DRIERS. Type II: one-package, moisture-CURED urethane coatings characterized by the presence of free isocyanate groups and capable of conversion to useful films by the reaction of these isocyanate groups with ambient moisture. Type III: one-package, heat-cured urethane coatings that dry or CURE by thermal release of blocking agents and regeneration of active isocyanate groups that subsequently react with substances containing active hydrogen groups. Type IV: two-package catalyst-urethane coatings that comprise systems wherein one package contains a prepolymer or adduct having free isocyanate groups capable of forming useful films by combining with a relatively small quantity of CATALYST, ACCELERATOR, or CROSSLINKING AGENT such as a MONOMERIC POLYOL or polyamine contained in a second package. This type has limited pot life after the two components are mixed. Type V: two-package polyol-urethane coatings that comprise systems wherein one package contains a prepolymer or adduct or other polyisocyanate capable of forming useful films by combining with a substantial quantity of a second package containing a RESIN having active hydrogen groups, with or without the benefit of catalyst. This type has limited pot life after the two components are mixed. Type VI: one-package, nonreactive LACQUER-URETHANE SOLUTION coatings characterized by the absence of any significant quantity of free isocyanate or other FUNCTIONAL GROUPS. Such coatings convert to solid films primarily by SOLVENT EVAPORATION.[156,71,130,163,67,141] See URETHANE COATINGS and ISOCYANATE RESINS.[156,71,130,163,67]

Urethane coatings (solvent or waterborne) are noted for their overall balance of high performance properties. These include excellent toughness, flexibility, high impact resistance, and tensile strength. They also offer excellent abrasion resistance and resistance to a wide variety of chemicals, solvents, and stains.

Urethane coatings are generally divided into AROMATIC and ALIPHATIC types. Aromatic urethanes are generally based on TDI (TOLUENE DIISOCYANATE) or MDI (DIPHENYLMETHANE DIISOCYANATE). They are designed mainly for interior applications since they have a tendency to yellow when exposed to prolonged sunlight. Aliphatic urethanes offer excellent UV stability, but are more expensive than aromatic urethanes. Aliphatic urethanes are generally based on IPDI (isophorone diisocyanate), HDI (hexamethylene diisocyanate), or TMXDI (tetramethylxylene diisocyanate). In addition to excellent light stability, aliphatic urethanes have better flexibility than aromatic urethanes.

polyvinyl acetal (1) generically, a class of POLYMERS derived from polyvinyl ESTERS in which some or all of the acid groups have been replaced by hydroxyl groups and some or all of these hydroxyl groups have been reacted with ALDEHYDES to form ACETAL groups; (2) specifically, polyvinyl acetal made by the reaction of the hydroxyl group with acetaldehyde;[130] (3) a VINYL PLASTIC produced from the CONDENSATION of POLYVINYL ALCOHOL with an ALDEHYDE. There are three main groups: polyvinyl acetal, polyvinyl butyral, and polyvinyl formal; used in lacquers and adhesives. Polyvinyl acetal resins are THERMOPLASTICS which can be processed by casting, extruding, molding, and coating.[38,79,151] (DAC)

polyvinyl acetate (PVA and PVAc) a colorless, THERMOPLASTIC, water-insoluble, RESINOUS HIGH POLYMER derived from the POLYMERIZATION of VINYL ACETATE with a CATALYST; used as a LATEX BINDER in certain paints.[130]

polyvinyl alcohol (PVA) (1) water-soluble RESIN prepared by HYDROLYZING POLYVINYL ACETATE. Used in SYNTHETIC RESINS;[155] (2) oil resistant PLASTIC.[38]

polyvinyl butyral (PVB) RESIN formed by reacting polyvinyl ALCOHOLS with butyraldehyde.[38]

polyvinyl chloride co vinyl acetate a copolymer of VINYL CHLORIDE and VINYL ACETATE.[130] Abbreviation: PVC/VAc.

polyvinyl chloride (PVC) a hard and tough PLASTIC solid. STABILIZERS are necessary to prevent DISCOLORATION from EXPOSURE to light and heat. Used for plastics and coatings. Commonly known as VINYL.[38,130]

ASTM test methods for polyvinyl chloride include: chlorine content, test, D1156; residual vinyl chloride, D3680; test procedures, guide, D4368.

polyvinyl fluoride (PVF) $(-H_2CCHF-)_n$. POLYMER of VINYL FLUORIDE; in the film form it is characterized by superior WEATHER RESISTANCE, high DIELECTRIC CONSTANT, low permeability to air and water, as well as OIL, chemical, SOLVENT, and STAIN RESISTANCE.[130]

polyvinyl formal a POLYVINYL ACETAL made by the REACTION of the HYDROXYL GROUPS with FORMALDEHYDE.[130]

polyvinyl pyrollidone (PVP) $(C_6H_9NO)_n$. White, free-flowing AMORPHOUS POLYMER. It is soluble in water and ORGANIC solvents, and is COMPATIBLE with a wide range of HYDROPHILIC and HYDROPHOBIC resins.[130,155]

polyvinylidene chloride (PVDC) a POLYMER of vinylidene chloride.[38,130]

polyvinylidene fluoride (PVDF) $(-H_2CCF_2-)_n$. THERMOPLASTIC FLUOROCARBON POLYMER derived from vinylidene fluoride. In film form it is characterized by superior weather and UV RESISTANCE.[130]

pompey red synonym for FERRIC OXIDE.[41] See IRON OXIDES, NATURAL.[41]

pontianak manila type of semifossil copal obtained from Borneo.[167]

By reason of its ALCOHOL SOLUBILITY, it is used in SPIRIT VARNISHES, but after RUNNING it becomes oil-soluble and is sometimes used in OLEORESINOUS VARNISHES.

popcorn a name for nonuseful, hard, tough, insoluble POLYMER, resembling popcorn, formed by POLYMERIZATION in the manufacture of SYNTHETIC RUBBERS.[169,140]

popping (1) eruptions in a film of paint or VARNISH after it has become partially set so that craters remain in the film.[42,56] (BSI) Popping refers to the formation of BLISTERS in the surfaces of films that break but do not FLOW out before viscosity increase prevents LEVELING. Sometimes, the surface layer increases in viscosity to the extent that the blisters do not break; this condition is also called popping, but more usually blistering or bubbling;[42,56] (BSI) (2) of PLASTER. A mild form of blowing (BSI); (3) in the COIL COATING industry, a film DEFECT manifested as a PINHOLE completely through the film.[56]

Popping results from rapid loss of SOLVENT at the surface of a film during initial FLASH-OFF. The surface develops a high viscosity relative to solvent-rich lower layers of the film. When the coated object is put into an oven, solvent volatilizes in the lower layers of the film, creating bubbles that do not readily pass through the high-viscosity surface. As the temperature increases further, the bubbles expand, finally bursting through the top layer. The viscosity of the film meanwhile has increased so much that it is no longer possible for the coating to flow back together and heal the eruption.

poppyseed oil OIL with only fair drying properties, obtained from the seeds of *Papaver somniferum*, which grows in India, Russia, and France.[63]

Its main constituent ACID is LINOLEIC, which is present to approximately 62%. The oil is little used in the trade, its main applications being for artistic purposes. Properties: sp gr, 0.925/15°C, iodine value, 134; saponification value, 192.

porcelain enamel GLASS based materials that are melted onto a product.[163] See ENAMEL (2).

They have a broad range of uses from electronic circuits, chemical equipment, kitchen sink, bathtubs, etc.

porch paint see DECK PAINT.[71]

pore size the size of the openings of FILTERS or screens, usually expressed in micrometres.[128]

pores minute openings (holes) in surface of CURED goods.[56] See PINHOLES.[56]

May refer to minute bubbles within the article.

porosity presence of numerous minute voids in a CURED material. Also may refer to open cells in SKIN of cellular RUBBER, and in this case is not a DEFECT.[56]

ASTM Test Method D3258 covers the determination of the porosity of a white or near white, mineral spirits insoluble, paint film to indicate the degree to which a subsequent coat will penetrate.

Portland cement a hydraulic CEMENT produced by pulverizing clinker consisting essentially of HYDRAULIC CALCIUM SILICATES and usually containing one or more of the forms of calcium sulfate as an interground addition.[73] (ASTM)

Portland cement paint see CEMENT PAINT.[71,73]

post-cure heat or RADIATION treatment, or both, to which a CURED coating is subjected to enhance the level of one or more properties.[67,176]

poster color OPAQUE WATER COLOR (GOUACHE) obtainable in pots or tubes; often used by poster designers, these are water paints with a GUM BINDER.[71] Syn: tempera (a misnomer) and show card colors

postforming PROCESS in which CURED THERMOSETTING PLASTICS SHEETS, or other shapes, are heated to a high temperature and formed by stress into other configurations.[38,57]

pot life (1) the length of time a coating is useful after its original package is opened or other ingredients are added;[42,161] (2) the period of time after mixing during which a multiple-component coating remains usable with no decrease in desirable properties.[42,74,161] (DAC) Syn: usable life,[67,161,42,74] spreadable life.[67,161,42] See WORKING TIME.[74,161]

potash blue see IRON BLUE.[41]

potassium titanate (PKT) $(K_2O)_{1/x}(TiO_2)_4$. The value of x in the formula is greater than 1.0 because the commercial PIGMENT has the crystal structure of a leached tetratitanate.[41]

It is an ACICULAR WHITE HIDING pigment that has high SCATTERING POWER, and its ULTRAVIOLET REFLECTION is significantly higher than that of commercial TITANIUM DIOXIDE. It has been used in the PAPER industry and in vinyl plastics. Properties: density, 3.3 g/cm³ (27.5 lb/gal); O.A., approximately 80; particle size, diameter, 0.2 μm, length, 8-10 μm.

pounce (1) noun. Design on PAPER which has been formed by pricking it out with a sharp pointed instrument. The design is transferred to the sur-

---

Words presented in CAP/SMALL CAPS type indicate that the word is defined in another part of the *Coatings Encyclopedic Dictionary*.
Numerical superscripts classify terms in one or more of the categories listed in the second section of this volume.

face on which it is to be painted, by laying the paper on the latter and shaking on to it a dry powdered pigment through a linen bag called a pounce bag. Some of the PIGMENT passes through the pinholes in the paper and forms a replica of the design on the surface where it can then be rendered in paint;[54,81](BSI) (2) verb. To transfer the outlines of a drawing by applying a dry powder through small perforations made in the outline.[54,81]

pour point temperature at which materials possess a defined degree of fluidity.[47,166,176,128]

One particular method of determining pour point involves recording the times of efflux of a specified volume of the molten product through an orifice of standard dimensions. From the figures thus obtained, it is possible to arrive at a temperature which permits a standard volume of product to FLOW through the orifice in a specified time. This temperature is the pour point.

powder a substance consisting of ground, PULVERIZED, or otherwise finely DISPERSED solid particle.[64,41] See PIGMENT.[41,69]

powder coatings (1) a 100% solids coating applied as a dry powder which, when baked at a sufficiently high temperature, melts out to form a continuous film.[71] *For THERMOSETTING materials, a chemical reaction, either CONDENSATION or ADDITION, also takes place. This fused film has the uniformity, COLOR, toughness, and other properties generally associated with protective and DECORATIVE coatings;* (2) a coatings application method which utilizes a solid BINDER and PIGMENT. The solid binder melts upon heating and results in a pigmented coating upon cooling.[77,71] See FLUIDIZED BED COATING,[77] WRAP,[77] and COATING POWDER.[71]

ASTM Standard Practice D3451 covers the selection and use of procedures for testing polymeric powders and powder coatings.

powdered distemper see CALCIMINE.[71]

powdered quartz see SILICA, CRYSTALLINE.[41,158,58]

powdering of paints see CHALKING.[168]

powdering (of polishes) the partial or total disintegration of the polish film resulting in a fine, light-colored material.[137] (ASTM)

power cleaning see BLAST CLEANING.[158]

PO_x oxides of phosphorous.[144]

po-yok oil an oil derived from *Afrolicania elaeosperma* and *Parinarium sherbroense*, which occur in West Africa.[63] Also spelled "po-yoak" oil.

It is a DRYING OIL, since its main constituent acids are LICANIC and ELAEOSTEARIC, but it has never become available in quantity. It bodies rapidly on heating and behaves generally something like a mixture of OITICICA and TUNG OILS. Sp gr, 0.9612/15°C; iodine value, 150; saponification value, 188; refractive index, 1.5082.

ppb abbreviation of parts per billion.[128]

PRA abbreviation for PAINT RESEARCH ASSOCIATION. (British)[174]

precipitate substance separated from a solution in solid form by APPLICATION of cold or heat or by a chemical REACTION.[139]

precipitated barium sulfate see BARIUM SULFATE.[41,58]

precipitated basic dye blues the more commonly used pigments of this group are the so-called PTA, PMA, and PTMA pigments.[41,58]

They are so designated because they comprise the precipitated products of the reaction between basic DYES (e.g., Victoria Blue B) and complex INORGANIC acids such as phosphotungstic and phosphomolybdic acids or their mixture. These pigments as a group are nearly black in MASSTONE; accordingly, they are used chiefly for tinting purposes. In spite of certain deficiencies (e.g., poor BLEED RESISTANCE and poor LIGHTFASTNESS on outdoor EXPOSURE), the brilliance and high tinting strength of these pigments make them suitable for various interior paints of more-or-less special character (e.g., foil lacquers, poster paints, and some toy enamels). In general, PTA pigments are more lightfast than corresponding PMA pigments, although the latter are somewhat superior in tinting strength.

precipitated basic dye violets the general properties of the phosphotungstic or molybdic acid violet pigments in paint are essentially the same as those discussed under PRECIPITATED BASIC DYE BLUES in the section regarding deficiencies when subjected to EXTERIOR EXPOSURE.[41]

Its only application in paints might be lead-free finishes for interior use.

precipitated calcium carbonate see CALCIUM CARBONATE, SYNTHETIC.[41]

precipitated driers METALLIC SOAPS derived from the interaction of AQUEOUS SOLUTIONS of the alkali soaps of the drier acids, and of the METALLIC SALTS.[83]

The drying soaps are obtained as PRECIPITATES, and are usually characterized by much paler colors than those made by the fusion method.

precipitation the sedimentation of a solid material from a liquid solution in which the material is present in amounts greater than its solubility in the liquid.[164] (IUPAC)

precipitation number measure of the amount of solid matter PRECIPITATED from OIL in a test.[157,128]

The number of millilitres of solid matter formed in a certain amount of mixture of oil and SOLVENT.

precipitation scavenging see SCAVENGING.[171,172]

precision (reproducibility) the closeness of agreement between the results obtained by applying a given experimental procedure several times under prescribed conditions. The smaller the random part of the experimental errors which affects the results, the more precise is the method.[157,145,36] (IUPAC)

This term is not to be confused with accuracy which is a measure of the agreement between the true value and the measured value.

precure usually used for a definite controlled state of partial CURE to set up the article for intermediate handling before the final cure.[67]

Seldom used to refer to a defect.

precursor a chemical COMPOUND which is released into the atmosphere, undergoes chemical change, and leads to a new (secondary) pollutant, is called a precursor of that species.[171,172] (IUPAC)

prefabrication primer quick-drying material applied as a thin film to a METAL surface after cleaning, e.g., by a BLAST CLEANING PROCESS, to give protection during the period before and during FABRICATION.[71,158] (BSI) Syn: shop primer

Prefabrication primers should not interfere seriously with conventional welding operations or give off toxic fumes during such operations.

prehistoric art painting and sculpture produced by artists of the Old, Middle, and New Stone ages.[81] See CAVE PAINTING.[81]

The earliest known piece of prehistoric sculpture is the famous "Venus of Willendorf" (Natural History Museum, Vienna), a small fertility image of Paleolithic origin dating around 11,000 B.C.

premature vulcanization uncontrolled CURING or setting up of material before final CURE.[67,176] See BIN CURE.[67]

premix an admixture of several ingredients designed to be incorporated in a formulation or PROCESS as a group as opposed to individually.[49,131]

prepared linseed oil in the PRINTING INK industry, LINSEED OIL which has been treated with LITHARGE and other chemicals.[63,168]

prepasted ADHESIVE applied to the back of WALLCOVERING by the manufacturer.

DIPPING in water before hanging activates the paste.

prepolymer a POLYMER of degree of POLYMERIZATION intermediate between that of the MONOMER or MONOMERS and the final POLYMER.[130]

preservative a biocidal ADDITIVE used to prevent growth of MICROORGANISMS in LATEX and other waterborne paints that cause spoilage, foul-smelling metabolic products, gassing in the can, and paint viscosity reduction due to ENZYME digestion of cellulosic thickeners.[83,173] See WOOD PRESERVATIVE COATINGS[71] and IN-CAN PRESERVATIVES.[83]

press cake a PIGMENT dispersed in water (obtained directly from a FILTER PRESS) in which a water-insoluble RESIN is emulsified.[41,164]

A SOLVENT-based dispersion is made by breaking this emulsion and removing the water.

pressure FORCE applied over a surface, measured as force per unit area.[142]

pressure marking GLOSSY or DULL spots which become apparent as a STRIP is uncoiled.[56,70]

This is usually due to an undercured or soft film. Also can be caused by improper PLASTICIZER balance.

pressure mottling the film distortion or uneven pattern that causes a change of GLOSS and a nonuniform APPEARANCE in the coated surface, as opposed to BLOCKING.[56] See PRESSURE MARKING.[56]

pressure-sensitive adhesive see ADHESIVE, PRESSURE-SENSITIVE.[79]

When placed on a backing material, it adheres to another surface on contact without WETTING, heating, or adding a CURING AGENT.

pretreatment usually restricted to mean the chemical treatment of unpainted METAL surfaces before painting.[158] (BSI)

pretreatment primer see WASH PRIMER.[71,158,66]

pretrimmed papers rolls of WALLPAPER from which SELVAGE has been trimmed at factory.[148] Syn: trimmed papers

primary amine value the number of milligrams of potassium hydroxide equivalent to the primary AMINE basicity in 1 g of sample.[134,157,128] (ASTM)

primary colors three basic colors used to make most other colors by mixture, either ADDITIVE MIXTURE of lights or SUBTRACTIVE MIXTURE of colorants.[43,69] See PRIMARY COLORS, ADDITIVE; PRIMARY COLORS, CIE; and PRIMARY COLORS, SUBTRACTIVE.[69,43]

primary colors, additive three colored lights from which all other colors can be matched by ADDITIVE MIXTURE.[69,43]

The three must be selected so that no one of them can be MATCHED by mixture of the other two. Generally, a red, a green, and a blue are used.

Additive primaries are the complements of the subtractive primaries.

primary colors, CIE red, green, and blue (violet) primaries defined by the CIE in terms of SPECTRAL DISTRIBUTION CURVES.[43,69]

They are imaginary primary lights so selected that all possible real colors can be matched by computation with positive amounts, to avoid the use of negative amounts, which are required to match all colors when using red colored lights.

primary colors, subtractive COLORS of three COLORANTS or colored materials which, when mixed together subtractively, result in black or a very dark neutral color.[69,43]

Subtractive primaries are generally cyan, magenta, and yellow, the three basic colorants used in printing, for example.

primary high polymer one which is produced directly from small molecules, without chemical alteration subsequent to the POLYMERIZATION.[130] See DERIVED HIGH POLYMER.[130]

primary pollutant a pollutant emitted directly into the air from identifiable sources (e.g., SO_2, NO, HYDROCARBONS, etc.).[171,172] (IUPAC)

Secondary pollutants, such as ozone, are generated within the atmosphere through chemical changes which occur in primary pollutants.

prime coat see PRIMER.[71,158]

prime pigments PIGMENTS which possess colorant value and HIDING POWER.[41]

The refractive index of these pigments is 2.0 or higher in contrast to EXTENDER PIGMENTS.

primer (1) first complete coat of paint of a painting system applied to an uncoated surface prior to application of an INTERMEDIATE COAT or TOPCOAT.[71,158] *Such paints are designed to provide adequate ADHESION to new surfaces and are formulated to meet special requirements of the surfaces. The type of PRIMER varies with the surface, its condition, and the total painting system to be used. Thus, primers for new wood and certain other surfaces must provide for exceptional ABSORPTION of the medium. Primers for steelwork contain special anticorrosive pigments;*[71,158] (2) coating applied to a surface, prior to the application of an adhesive, to improve the performance of the bond;[79,158] and (3) a thin liquid BITUMEN applied to the surface to improve the adhesion of heavier applications of bitumen and to absorb DUST.[76,79] See METAL PRIMER[71,158] and PLASTER PRIMER.[71]

ASTM Standard Practice D3322 covers the selection and use of procedures for testing primers and primer surfaces.

ASTM Test Method D4752 describes a solvent rub technique for assessing the MEK resistance of ethyl silicate (inorganic) zinc-rich primers.

primer surfacer see SURFACER.[71,158]

priming the APPLICATION of a PRIMER.[158] (DAC)

priming paints see PRIMER.[71,158]

primrose chrome complex, primrose-colored chromates to which the formula $PbCrO_4 \cdot 4PbSO_4 \cdot Al_2(OH)_6$ has been given.[41]

print name for ETCHING, LITHOGRAPH, WOODCUT, etc.[81]

print paste a pigmented coating used to render designs on FABRIC, usually applied by the roller, rotary screen, or flat screen PRINTING PROCESS.[71] (EPA)

print resistance the ability of a coating to resist taking on the imprint of another surface placed on it under normal conditions of practical use or under defined conditions of experiment.[56,72,157] See PRINTING (2).[35,42,56,157]

The ability of a coating to resist printing is important because its APPEARANCE is adversely affected if the smoothness of the coating film is altered by contact with another surface, particularly one with texture. Interior paint systems, particularly GLOSS and SEMIGLOSS on window sills and other horizontal surfaces, often have objects such as flower pots placed on them that may tend to leave a permanent impression. This tendency for a paint film to "print" is a function of the HARDNESS of the coating, the pressure, temperature, humidity, and the duration of time that the object is in contact with the painted surface.

ASTM Test Method D2064 covers an accelerated procedure for evaluating the print resistance of architectural paints. It differs from print resistance Test Method D2091, which has reference pictorial standards, in that the latter is concerned with LACQUER finishes under packaging, shipping, and warehousing conditions, whereas Test Method D2064 is concerned with decorative coatings undergoing random on-site pressure contact.

Printing should not be confused with blocking, which is measured in ASTM Test Method D4946. The former relates to the indentation of a surface, and the latter, the sticking together of two surfaces.

print test see PRINT RESISTANCE.[47,72,157]

printability a collective term used to describe the properties required of all components in a PRINTING PROCESS.[168]

printing (1) the PROCESS, art, or business of producing PRINTED material by means of inked type and a PRINTING PRESS or similar means;[168] (2) forming a permanent impression in a semihardened paint film as a result of pressure from an object placed on it.[42,56,157] (DAC)

printing ink any FLUID or VISCOUS composition of materials, used in PRINTING, impressing, stamping, or transferring on PAPER or paper-like substances, WOOD, FABRICS, PLASTICS, FILMS, or METALS, by the recognized mechanical reproductive process employed in printing, publishing, and related services.[168] (ASTM)

ASTM test methods for printing ink include: apparent tack, inkometer test, D4361; fineness of grind, NIPIRI method, D1316; lightfastness, printed matter, D3424; nonvolatile content, D4731; viscosity, by falling-rod viscometer, D4040; water pick-up, D4942.

printing plate a surface carrying a design by which the INK is ultimately transferred to the material to be PRINTED.[59,168]

printing press a mechanical device to apply INK to a surface reproducing the pattern or design on the PRINTING PLATE.[59,168]

printing strength a relative value indicating how much INK is required to give an equal depth of tint to a definite amount of white ink as compared with the same amount of a standard ink of the same CONSISTENCY.[168]

process inks used in reproducing illustrations by the halftone COLOR SEPARATION PROCESS.[168]

The colors used are yellow, magenta (red), and cyan (blue); they are used with or without black.

processing aid a compounding material that improves the processability of a polymeric compound.[83,131](ASTM)

product coatings those used by ORIGINAL EQUIPMENT MANUFACTURERS. The products coated include: appliances, automotive, electrical insulation, factory-finished wood, film, PAPER and foil, machinery and equipment, marine, METAL containers, METAL FURNITURE and fixtures, nonautomotive transportation, pipe, sheet, strip and coil, toys and sporting goods, wood furniture and fixtures, miscellaneous consumer and industrial products.[71,159]

Product Identification Number (PIN) a four-digit number, prefaced by UN or NA, used in Canada under the Transportation of Dangerous Goods Regulation for use by emergency personnel to identify a material in the event of an accident.[172]

profile surface contour of a blast-cleaned or SUBSTRATE surface, viewed from the edge. (Cross section of the surface.)[45]

profile depth average distance between top of peaks and bottom of valleys on the surface of a coating.[45]

proof, apparent the proof of a liquid as calculated from its SPECIFIC GRAVITY at 60°F.[164,128]

It is equivalent to the proof of a solution of pure ALCOHOL and water having the same specific gravity at 60/60°F as the mixture in question. Since materials other than alcohol and water, such as DENATURANTS or other soluble ingredients, affect the specific gravity of the solution, the apparent proof is not necessarily the true "alcohol proof" of the solution.

proof gallon, U.S. the amount of ALCOHOL present in one wine gallon of 50% by volume of alcohol at 60°F.[164,128] See WINE GALLON.[128]

Proof gallons are calculated by multiplying the number of wine gallons at 60°F by the proof and dividing by 100. For example, one wine gallon of 190 proof alcohol contains 1.9 proof gallons.

proof spirit (Brit). This corresponds with a definite mixture of absolute ALCOHOL in water, and actually contains 49.24% by weight of alcohol. Sixty-four overproof (O.P.) industrial alcohol is commonly used for SPIRIT VARNISH manufacture, containing approximately 90% of alcohol by weight.[164,128]

proofing the PROCESS of RUBBERIZING FABRICS, to render them impervious to water.[131,140]

It is an operation most commonly done by spreading a RUBBER CEMENT of high viscosity or dough on the FABRIC, allowing the SOLVENT to EVAPORATE, and CURING in dry heat ovens or with sulfur chloride.

propanol see PROPYL ALCOHOL.[164]

2-propanone see ACETONE.[164]

propenoic acid see ACRYLIC ACID.[141]

proprietary solvents based on ETHYL ALCOHOL. Solvents containing more than 25% alcohol by volume which are manufactured from specially denatured alcohol, in accordance with authorized formulas.[164]

No permit is required to purchase proprietary solvents in the United States.

propyl acetate $C_3H_7COOCH_3$. Medium-boiling SOLVENT used for NITROCELLULOSE.[164]

Properties: bp, 102°C; sp gr, 0.897; flp, 12°C (53°F).

propyl alcohol $CH_3CH_2CH_2OH$. Used as a SOLVENT and as a DILUENT for NITROCELLULOSE LACQUERS.[164] Syn: n-propanol

Properties: bp, 97°C; sp gr, 0.804/20°C; flp, 25°C (67°F).

propyl benzoate $C_6H_5COOC_3H_7$. Semipermanent PLASTICIZER.[170]

Bp, 231°C.

Words presented in CAP/SMALL CAPS type indicate that the word is defined in another part of the *Coatings Encyclopedic Dictionary.*
Numerical superscripts classify terms in one or more of the categories listed in the second section of this volume.

propyl butyrate $CH_3CH_2CH_2COOC_3H_7$. Medium-boiling SOLVENT.[164]

Bp, 143°C.

propyl carbinol see N-BUTYL ALCOHOL.[164]

propyl propionate $CH_3CH_2COOC_3H_7$. Medium-boiling ESTER SOLVENT.[164]

Properties: bp, 122°C; sp gr, 0.883; mp, -76°C.

propyl ricinoleate $CH_3(CH_2)_5CHOHCH_2CH{=}CH(CH_2)_7COOC_3H_7$. Permanent PLASTICIZER.[170]

Properties: bp, 268°C/13 mm; sp gr, 9.908.

propylene $H_2C{=}CHCH_3$. A flammable gas obtained from PETROLEUM OILS during the refining of gasoline.[130,141] Syn: propene, methylethylene, methylethene

Used in the polymerized form as POLYPROPYLENE PLASTIC.

propylene dichloride $C_3H_6Cl_2$. Chlorinated hydrocarbon.[144]

Properties: bp, 96°C; flp, 21°C (70°F); vp, 38 mm Hg/20°C.

propylene glycol $CH_3CHOHCH_2OH$. Dihydric ALCOHOL used as an ESTERIFYING AGENT.[144] Syn: 1,2 propanediol

Also used as a WET-EDGE ADDITIVE. Properties: bp, 187°C; sp gr, 1.038/20°C; flp, 99°C (210°F); vp, 0.1 mm Hg/20°C.

protective coatings a thin layer of METAL or ORGANIC material, as paint APPLIED to a surface, primarily to protect it from OXIDATION, WEATHERING, and CORROSION.[71]

protective colloids materials such as gums, starches and proteins, POLYACRYLATES and CELLULOSE, and CELLULOSE DERIVATIVES, which are effective agents for protecting charged COLLOIDAL PARTICLES in AQUEOUS MEDIA against FLOCCULATION.[83,155]

protective factor (of an antioxidant) the proportion of millimoles of peroxide per kilogram in untreated oil to that in the oil containing an antioxidant.[168]

protein any of a group of complex nitrogenous ORGANIC COMPOUNDS of high molecular weight that contain AMINO ACIDS as their basic structural units and that occur in all living matter and are essential for the growth of animal tissue.[173] See CASEIN.[173]

protein-aldehyde resins PLASTIC derived from CASEIN and FORMALDEHYDE; was called karolith.[38,130]

Prussian blue brilliant deep blue PIGMENT of excellent staining power, good LIGHTFASTNESS, but unstable in the presence of alkalis.[41] See IRON BLUE.[41] Syn: erlangen blue, gas blue

It is usually obtained as a very fine POWDER. Its OIL ABSORPTION is about 90.

Prussian brown brown-colored, IRON OXIDE PIGMENT, obtained as a DECOMPOSITION product of Prussian blue subjected to heat.[41]

Prussiate of iron see IRON BLUE.[41]

PS abbreviation for POLYSTYRENE.[130]

pseudoisochromatic plate (PIC) test general term applied to a type of test plates used to determine defective or anomalous COLOR VISION.[69,157,43]

A PIC plate is a CHROMATIC figure formed by dots on a background of different chromatic dots, varying sometimes in lightness and size. The figure is either an Arabic numeral or some other identifiable pattern. The simplest version is a dichotomous test which involves the perception of chromatic dot patterns on backgrounds of different chromatic dots, used to test for red-green confusion. More complex tests are double number types (color defectives see one number, normals a different number), special camouflage type (Ishihara), those designed to detect qualitatively type and degree of defect, and quantitative diagnostic types (Hardy-Rand-Rittler PIC Plate Test).

pseudomonas a generic class of aerobic, mesophilic bacterium capable of releasing a variety of enzymes, including CELLULOSE-DECOMPOSING "CELLULASE" enzymes; these enzymes are a factor in VISCOSITY REDUCTION of LATEX paints modified with CELLULOSIC THICKENER, and may contribute to biodeterioration of paint films, enhancing their nutrient value for FUNGUS (MILDEW) growth.[173]

pseudoplastic flow type of FLOW characterized by a CONSISTENCY CURVE which shows no yield value (starts at the origin) and where the rate of flow increases faster than linearly with the shearing stress.[166] See VISCOSITY, STRUCTURAL.[166]

psychophysical adjective used to describe the sector of COLOR science which deals with the relationship between PHYSICAL description or specification of stimuli and the sensory perception arising from them.[43,69]

psychrometer see SLING PYSCHROMETER.[145,157]

psychrometry the use of a wet-and-dry-bulb thermometer for measurement of atmospheric humidity.[145] (IUPAC)

PTA (1) abbreviation for PHOSPHOTUNGSTIC ACID; (2) applied also to TONERS and PIGMENTS which have been precipitated with phosphotungstic acid to give permanence and insolubility.

PTA pigment see PRECIPITATED BASIC DYE BLUES.[41]

PTM abbreviation for Paint Testing Manual. No longer in print.[134] (ASTM)

It has been replaced by Paint and Coating Testing Manual.

PTMA pigment see PRECIPITATED BASIC DYE BLUES.[41]

PUD abbreviation for POLYURETHANE DISPERSION.[71,83] See POLYURETHANES.[156,71,130,163,67]

These are used mainly to enhance wear resistance of the FINISH. It also depresses the low shear viscosity and enhances mar resistance.

pug mill MILL used for the preliminary mixing of PIGMENTS into OILS or MEDIA to form stiff PASTES prior to GRINDING.[59]

This PROCESS of mixing is sometimes known as pugging.

pulldown a FILM of INK deposited on PAPER by a smooth-edged blade to evaluate the undertone and MASSTONE of the ink.[168] Syn: drawdown

pulling RESISTANCE of the movement of a BRUSH during the application of a material due to the viscous nature of the MEDIUM.[166] See DRAG.[166]

Such a material is sometimes referred to as being sticky under the brush.

pulling over PROCESS of LEVELING a CELLULOSE LACQUER film, usually on WOOD, by RUBBING it with a soft cloth pad soaked in a mixture of ORGANIC SOLVENTS which is only a partial solvent for the lacquer film.[54,151] (BSI) cf. LIFTING

pulling (under the brush) see DRAG.[166]

pulling up action of a coat of paint or VARNISH which softens a previous coat to such an extent as to make BRUSH APPLICATION difficult and, in extreme cases, causes an objectionable intermingling of the two coats.[42] (BSI)

pulp PRESS CAKE that has been further processed to yield a homogeneous paste of controlled solids content or controlled TINTING STRENGTH for commercial sale.[41,64]

pulverize to pound, crush, or grind, to a POWDER or DUST.[131]

pulverulent consisting of, or reducible to, a fine POWDER.[41,131]

Dusty, friable, crumbly.

pumacite see PUMICE.[41]

pumice a highly vesicular (frothy), glassy, volcanic lava, usually rhyolitic (granitic) in composition; composed of complex aluminum, calcium, magnesium, iron, sodium, and potassium silicates.[41] Syn: pumacite, pumice stone

Pumacite is the name for volcanic ash found in Kansas and Nebraska. Pumice is used as an ABRASIVE, polishing compound, and nonslip COMPOUND. Density, 2.2 g/cm³ (18.5 lb/gal).

pure black iron oxide see BLACK IRON OXIDE.[41]

purified stand oil see TEKAOL[63]

purity, colorimetric ratio of the LUMINANCE of the SPECTRUM LIGHT, in mixture with the specified ACHROMATIC LIGHT required to match the light being described, to the luminance of the COLOR itself.[69,43]

It is distinguished from excitation purity by the abbreviation P_c.

purity, excitation ratio of the straight line distance on a CIE CHROMATICITY DIAGRAM between the CHROMATICITY point of the sample and the ACHROMATIC or ILLUMINANT point on the diagram, to the linear distance between the point of intersection of this line with the SPECTRUM locus and the illuminant point.[43,69]

It is properly designated as P_e, *but is frequently abbreviated simply as P. The excitation purity, then, describes the relative distance from the neutral point and roughly corresponds in concept to the psychological description of saturation or chroma.*

purple brown artificial red oxide PIGMENT obtained by high-temperature treatment.[41,176]

purree see INDIAN YELLOW.[41]

putty (1) a heavy PASTE composed of PIGMENT, such as whiting, mixed with LINSEED OIL; used to fill holes and CRACKS in WOOD prior to painting, to secure and seal panes of GLASS in window frames; Syn: painter's putty, glazing compound; (2) in PLASTERING, a fine CEMENT consisting of lump LIME slaked with water; lime putty.[158,165] (DAC)

putty chaser see EDGE RUNNER MILL.[59,131]

PVA abbreviation for POLYVINYL ACETATE and POLYVINYL ALCOHOL.[130]

PVAc abbreviation for POLYVINYL ACETATE.[130]

PVC (1) abbreviation for PIGMENT VOLUME CONCENTRATION;[41] (2) POLYVINYL CHLORIDE.[38,71,130,128]

PVCL is also used as an abbreviation for polyvinyl chloride.

PVF abbreviation for POLYVINYL FLUORIDE.[130]

PVP abbreviation for POLYVINYL PYROLLIDONE.[130,155]

pycnometer a container whose volume is precisely known, used to determine the density of a liquid

Words presented in CAP/SMALL CAPS type indicate that the word is defined in another part of the *Coatings Encyclopedic Dictionary.*
Numerical superscripts classify terms in one or more of the categories listed in the second section of this volume.

by filling the container with liquid and then weighing it.[145] See WEIGHT-PER-GALLON CUP.[145]

pyrazolone red $C_{36}H_{28}N_8O_6Cl_2$. PIGMENT Red 38 (21120). A red POWDER used in the RUBBER, PLASTICS, and flour products industries.[41]

Properties: density, 1.35 to 1.58 g/cm³ (11.3 to 13.2 lb/gal); O.A., 41 to 70.

pyroabietic acids acidic products obtained from the ABIETIC ACID of ROSIN by heating.[144]

pyrolysis DECOMPOSITION or transformation of COMPOUND caused by heating.[131]

pyrometer instrument for measuring temperatures beyond the upper limit of the usual liquid thermometer.[145,176]

They may operate on the differential expansion of two metallic strips joined together, the measurement of changes of resistance, and the measurement of current flowing through two joined pieces of METAL. In addition, radiation pyrometers are based on the measurement of heat radiated from a hot body, and OPTICAL pyrometers on the measurement of the intensity of light emitted from a hot body.

pyrophoric describes materials that ignite spontaneously in air below 130°F (54°C).[131]

pyroxylin (1) soluble types of CELLULOSE NITRATE, and confined roughly to those containing less than 12.4% nitrogen;[151,130] and (2) synthetic medium more commonly known as LACQUER or by various trade names.[81,151] See NITROCELLULOSE.[144,151]

Q

QCT abbreviation for CLEVELAND CONDENSING HUMIDITY CABINENT.[59,157]

quality assurance the guarantee that the quality of a product (analytical data set, etc.) is actually what is claimed on the basis of the quality control applied in creating that product.[37,157] (IUPAC)

Quality assurance is not synonymous with QUALITY CONTROL. *Quality assurance is meant to protect against failures of quality control.*

quality control the maintenance and statement of the quality of a product (data set, etc.) specifically that it meets or exceeds some minimum standard based on known, testable criteria.[37,157]

quarry tile unglazed tile, usually six square inch or more in surface area and 0.5 to 0.75 inch (1.3 to 1.9 cm) in thickness made by the extrusion process from natural CLAY or shales.[75]

quartz CRYSTALLIZED SILICA which, when reduced to powder, is used as an EXTENDER.[41] See SILICA, CRYSTALLINE.[41,58]

quenching see CHILLING.[131]

quercitron $C_{15}H_{10}O_7C_{21}H_{20}O_{12}$. A natural yellow coloring matter obtained from the bark of Quercus nigra or Quercus tinctoria, and used for the production of yellow LAKES.[41] See BROWN PINK.[41]

The principal constituents are quercitrin and quercetin.

quick-hardening lime a hydraulic LIME.[73,75] (DAC)

quicklime see CALCIUM OXIDE.[73,144]

quinacridone golds these are very TRANSPARENT, red-shade yellows of particular interest in formulating the best quality metallic finishes; often in blends with transparent iron oxides and/or other, less transparent, ORGANIC pigments.[41] See QUINACRIDONE PIGMENTS.[41]

quinacridone pigments a family of PIGMENTS derived from quinacridone chemistry which span a HUE range including gold, scarlet, red, orange, magenta, violet, and maroon.[41]

These pigments most closely parallel the phthalocyanines in overall excellence of properties.

quinacridone violet a versatile member of the quinacridone PIGMENT family.[41]

It is used as a TONING *pigment; used to neutralize yellow tones in whites. It is also used in blends with* MOLYBDATE ORANGE *to obtain relatively low-cost, durable, nonbleeding, bright reds. Its high level of transparency is also very desirable for styling automotive metallized finishes. For maximum saturation when toning phthalocyanine blues to redder hues, the quinacridone reds or magentas may be preferred to the violet.*

QUV an accelerated WEATHERING machine that simulates the DETERIORATION, caused by water as rain and dew and the ULTRAVIOLET light energy in sunlight, of outdoor materials.[157,59,154] See ACCELERATED WEATHERING.[65,72,154,157] cf. WEATHER-OMETER[65,145,154,157]

The tester simulates the effects of sunlight by fluorescent UV lamps, and simulates rain and dew by direct CONDENSATION *of water on the test specimen. It is not intended to simulate the deterioration caused by localized weather phenomena, such as atmospheric pollution, biological attack, and* SALT *water* EXPOSURE. *For testing, see ASTM Test Method G53.*

Words presented in CAP/SMALL CAPS type indicate that the word is defined in another part of the *Coatings Encyclopedic Dictionary.*

Numerical superscripts classify terms in one or more of the categories listed in the second section of this volume.

R

r/min abbreviation for revolutions per minute.[128]

r/s abbreviation for revolutions per second.[128]

RACT abbreviation for REASONABLY AVAILABLE CONTROL TECHNOLOGY.[172] (EPA)

radiance quotient of the radiant intensity in a given direction of an infinitesimal surface element containing the point under consideration divided by the area of the orthogonal projection of this surface element on a plane perpendicular to a given direction.[43,69]

radiant energy that form of energy consisting of the electromagnetic SPECTRUM which travels at 115,890 kilometres/s (186,500 miles/s) through a vacuum, reducing this speed in denser media (air, water, GLASS, etc.).[43]

The nature of RADIANT ENERGY *is described by its* WAVELENGTH *or frequency, although it also behaves as distinct quanta ("corpuscular theory"). The various types of energy may be transformed into other forms of energy (electrical, chemical, mechanical, atomic, thermal, radiant) but the energy itself cannot be destroyed.*

radiant flux radiant power; RADIANT ENERGY emitted from, transferred to, or received through a surface per time interval.[43,69]

radiant heat baking CURING treatment in which heat is transferred to the paint surface mainly by RADIATION from a hot surface, e.g., electric lamps or gas heated panels.[67,176] Syn: infrared drying or BAKING

radiant heat drying see RADIANT HEAT BAKING.[67,176]

radiant heating form of heating applied to the BAKING of coatings; the temperature is attained by the ABSORPTION of radiant heat.[67,176] See INFRARED DRYING.[67,176]

radiation (1) EMISSION or transfer of energy in the form of electromagnetic waves or PARTICLES; (2) the electromagnetic waves or particles. (ASTM)

In general, nuclear radiations and radio waves are not considered in this vocabulary, only OPTICAL *radiations, that is, electromagnetic radiations (photons) of wavelengths between the region of transmission to X-ray (1 nm) and the region of transition to radio waves (1 mm).*

radiation cure method of CURING or DRYING coatings by EXPOSURE to ELECTROMAGNETIC WAVES or PARTICLES such as INFRARED, ULTRAVIOLET, or ELECTRON BEAM.[67] (EPA) See INFRARED DRYING, ULTRAVIOLET, and ELECTRON BEAM.[67]

Very ABRASION-*resistant coating systems can be formulated using radiation cure. They can be formulated using reactive* MONOMERS, OLIGOMERS, *and* POLYMERS, *which cure through the interaction of these* REACTIVE COMPOUNDS *with high-energy* ELECTRONS *or ultraviolet energies. Abrasion-resistant coatings produced through the use of high energy electrons are based on* ACRYLIC *monomers, multifunctional acrylic oligomers and polymers, which cure by a* FREE RADICAL *mechanism.*

radical (or free radical) a reactive atom or fragment of a stable molecule which contains an odd number of bonding electrons. These reactive species have very short lifetimes in the lower atmosphere and low concentrations, although they are responsible for much of the chemical transformations in the atmosphere.[171,172] (IUPAC)

Typical RADICAL *species encountered in atmospheric chemistry are: HO, HO_2, Cl, ClO, NO_3, CH_3O_2, $C_2H_5O_2$, etc. The concentrations show strong diurnal variations. NO and NO_2 are technically radicals as well (since they have an odd-number of bonding electrons), but they do not exhibit the high reactivity normally associated with typical radicals. Some chemists follow the practice of coupling with the formula of a radical, a dot placed near the atom in the radical which is judged to have the highest probability for the location of the odd electron.*

radioactive paint see LUMINOUS PAINT.[71]

rafaelite hard, natural ASPHALTUM mined in the Argentine.[76,167]

It is characterized by a high MELTING POINT, *and good* OPACITY, *but limited solubility in* DRYING OILS. *Special heating treatment is necessary to obtain stable* SOLUTIONS.

rag rolled finish see RAG ROLLING.[54,77,148] (DAC)

rag rolling (1) PROCESS of forming, in a SCUMBLE or GLAZE over a painted ground, a textured or variegated pattern by rolling a rag or washleather over the surface; (2) PRINTING a pattern on a dry painted surface by means of a rag or washleather or special paint roller charged with COLOR. The rag is crumpled or screwed up in the form of a rough roller to produce the pattern.[54,77,148] (BSI)

ragging is similar to SPONGING ON and SPONGING OFF. After applying a BASECOAT, GLAZES are either added (ragging on) or removed (ragging off) by patting or rolling with a lint-free cloth.[77,148] Syn: cloth distressing

To rag on, a rag is dipped in glaze, wrung nearly dry, then small areas of the surface are either patted or rolled over with the rag, bunched or rolled, over the surface to produce a smoother surface. To rag off, the glaze is applied as for sponging off, then lightly blotted or rolled from the surface with a rag dipped in the appropriate SOLVENT *and squeezed nearly dry. For a subtler effect, the surface is blotted again with cheesecloth before the glaze dries.*

railroad tracks in coated FABRIC, depressions in surface of a definite pattern as indicated by the name.[56]

railroading the horizontal, rather than vertical, APPLICATION of a WALLCOVERING or the use, horizontally, of an upholstery FABRIC.[148]

rain spotting particular case of WATER SPOTTING caused by rain.[56,42,65] (BSI)

raised grain (1) condition of WOOD surfaces where FIBERS from the wood structure have become prominent due to WETTING with water or materials containing water; (2) prominence of the harder portions of the GRAIN of wood when the softer portions have suffered shrinkage.[159] (BSI)

raising see LIFTING.[42,56]

RAL abbreviation for Reichsnormen Ausschuss fur Lieferbedingungen, issues of German Colors (Farbregisters RAL 840).[69]

rapeseed oil obtained from the seeds of the species, *Brassica*.[63] Syn: colza oil

It is a NONDRYING OIL, *the main constituent acid being* ERUCIC ACID, *with smaller amounts of* OLEIC *and* LINOLEIC ACIDS. *Properties: sp gr, 0.915/15°C; iodine value, 92; saponification value, 173.*

rate of shear in RHEOLOGY, rate of shear is often used interchangeably with VELOCITY GRADIENT.[166]

raw material an unprocessed natural product used in manufacture.[49]

raw sienna naturally occurring IRON OXIDE, limonite type ore.[41] See IRON OXIDES, NATURAL.[41]

COLOR *range, yellow to light brown, 40 to 70% Fe_2O_3. Somewhat translucent when dispersed in oil. General uses include tinting colors, artists' colors, and stains.*

raw umber see UMBER.[41]

raymond roller mill a type of MILL used for the dry-GRINDING of PIGMENTS or similar materials.[59,64]

The pre-crushed crudes or solids are fed into a circular grinding chamber in the base of the mill and by mechanical means are forced between rolls and a ring. Grinding of the pigment crude is obtained by the centrifugal FORCE *and pressure of the revolving rolls exerted against the ring. A stream of compressed air passes through the mill and carries out the pigment* PARTICLES *which are fine enough to "float" on it; the larger particles drop back into the mill grinding chamber for further grinding.*

RCRA abbreviation for RESOURCE CONSERVATION AND RECOVERY ACT.[172]

RCRA Hazardous Waste a material designated by RCRA as hazardous waste and assigned a number to be used in record keeping and reporting compliance (e.g., D003, F001, U169).[172]

reaction a change or transformation in which a substance DECOMPOSES, combines with other substances, or interchanges constituents with other substances.[144]

reactive compound a VOLATILE ORGANIC COMPOUND which is a PRECURSOR, i.e., reacts in the atmosphere to form PHOTOCHEMICAL OXIDANTS.[172,164] (EPA)

Almost all organic compounds are photoreactive, but a few have such low reactivity that they are usually considered nonreactive. Solvents that have been deemed of negligible photochemical reactivity by the EPA are: methane, ethane, 1,1,1-trichlorethane (methyl chloroform), methylene chloride, trichlorofluoromethane (CFC-11), dichlorodifluoromethane (CFC-12), chlorodifluoromethane (CFC-22), trifluoromethane (FC-23), trichlorotrifluoroethane (CFC-113), dichlorotetrafluoroethane (CFC-114), and chloropentafluoroethane (CFC-115).

reactive diluent a VISCOSITY REDUCER for coatings which has low volatility and will become a permanent part of the coating through chemical REACTION, usually under AMBIENT conditions.[164,171]

It is used in high-solids coatings to reduce the loss of ORGANIC SOLVENTS *into the atmosphere.*

reactive material a chemical substance or mixture that vigorously polymerizes, decomposes, condenses, or becomes self-reactive due to shock, pressure, or temperature.[172]

It includes materials or mixtures that fall within any of these categories: (1) <u>explosive material</u>—a substance or mixture that causes sudden, almost instantaneous release of pressure, gas, and heat when subjected to sudden adverse conditions; (2) <u>ORGANIC *peroxide*</u>*—an organic compound that contains the bivalent –O–O– structure, which can be considered a structural derivative of hydrogen peroxide, in which one or both of the hydrogen atoms has been replaced by an organic* RADICAL; *(3) <u>pressure-generated material</u>—a substance or mixture that may spontaneously*

Words presented in CAP/SMALL CAPS type indicate that the word is defined in another part of the *Coatings Encyclopedic Dictionary.*
Numerical superscripts classify terms in one or more of the categories listed in the second section of this volume.

polymerize with an increase in pressure unless protected by the addition of an inhibitor or by refrigeration or other thermal control; decompose to release gas in its container, or compromise the contents of a self-pressurized container; (4) water-reactive material—a substance or mixture that reacts with water to release heat or a flammable, toxic gas.

reactive pigments those PIGMENTS which REACT with the VEHICLE, as in the formation of zinc and lead soaps with DRYING OILS; also, pigments such as RED LEAD which react with acids formed at METAL surface to prevent rust.[41]

reactive resins (1) resins of PHENOL-FORMALDEHYDE type which are believed to REACT with DRYING OILS on heating, more particularly those of CONJUGATED types; (2) resins capable of CROSSLINKING with themselves or other resins; (3) resins with a high ACID NUMBER.[167]

ready-mixed aluminum paint see ALUMINUM MIXING VARNISH.[71]

ready-to-mix aluminum paint see ALUMINUM MIXING VARNISH.[71]

reagent a substance used in chemical reaction to detect, measure, examine, or produce other substances.[144]

realgar As_2S_2. Arsenic disulfide MINERAL, which was used as a PIGMENT under the name, "arsenic orange."[41,34]

Its chemical and physical properties are similar to those of ORPIMENT. Presently, it is not used as a pigment because of its toxicity.

reasonably available control technology (RACT) the lowest emission limit that a particular source is capable of meeting by the application of control technology that is reasonably available considering technological and economic feasibility.[172] (EPA)

RACT is usually applied to existing sources in nonattainment areas and in most cases is less stringent than new source performance standards.

recommended exposure limit (REL) the NIOSH REL is the highest allowable airborne concentration that is not expected to injure a worker.[172]

It may be expressed as a ceiling limit or as a time-weighted average (TWA), usually for 10-hour work shifts.

reconstituted oils DRYING OILS which are made by the reesterification of selected FATTY ACIDS, previously derived from a natural drying oil.[63]

recovered solvent SOLVENT which is extracted from a process or exhaust stream usually by ABSORPTION or CONDENSATION.[164] (EPA)

red cinnabar see MERCURIC SULFIDE.[41]

red iron oxide see IRON OXIDES, SYNTHETIC.[41]

red lake C pigments family of ORGANIC ACID AZO PIGMENTS prepared by coupling the diazonium SALT of ortho-chloro-meta-toluidine-para-sulfonic acid with ß-naphthol.[41]

red lake D strong red LAKE, sensitive to cobalt DRIERS, and made by coupling diazotized anthranilic acid with ß-naphthol.[41]

red lake P made by coupling diazotized p-nitro aniline-o-sulfonic acid with ß-naphthol.[41]

red lead Pb_3O_4. Pigment Red 105 (77578). Bright red to orange-red tetroxide; excellent opacity with good properties as a primary constituent of anticorrosive primers for iron and steel.[41,71] See ORANGE MINERAL.[41,66] Syn: minium and Paris red

Properties: density, 8.9 g/cm³ (74.2 lb/gal); O.A., 7-9.

red lead (nonsetting) special type of red lead which contained a minimum amount of free or reactive LITHARGE and a high content of LEAD PEROXIDE.[41]

The nonsetting types were used when it was desired to prevent undue THICKENING of oil mixtures or paints.

red ocher a mixture of HEMATITES; any of a number of NATURAL EARTHS used as red PIGMENTS.[41] (DAC) See IRON OXIDES, NATURAL.[41]

red oil commercial grade of OLEIC ACID.[144]

red oxide see IRON OXIDES, NATURAL.[41]

redtone ink blue see IRON BLUE.[41]

reduce to add a SOLVENT or THINNER to a coating, VARNISH, RESIN, LATEX, or EMULSION for the purpose of lowering its VISCOSITY and/or NONVOLATILE CONTENT.[164] See DILUTE.[164] cf. REDUCERS[168]

reduced viscosity the VISCOSITY of a coating or VEHICLE when THINNED (reduced) to a specified PERCENT SOLIDS with specified SOLVENTS.[166]

Often used as a portion of the specifications between buyer and seller.

reducers VARNISHES, SOLVENTS, OILS, or WAXY or GREASY COMPOUNDS that are employed to reduce TACK or CONSISTENCY of the INK for use on a press.[168,164] See THINNER.[164]

reducing agent a substance that chemically reduces other substances, especially by donating an ELECTRON or electrons.[144,83] cf. OXIDIZING AGENT[144,83]

reducing power the strength of a WHITE PIGMENT, i.e., the degree to which it is able to produce a very pale TINT when mixed with a defined proportion of colored pigment.[41,69]

The paler the tint produced, the greater the reducing power.

reducing solvent a SOLVENT added to dilute a coating usually for the purpose of lowering the coating's VISCOSITY.[164] (EPA)

reduction the chemical process by which the oxidation state of an atom in an element, ion, or compound is reduced.[144] (IUPAC)

reference standard the STANDARD to which a measurement is compared.[43,59,142,69]

refine, refining to reduce to a pure state; purify.[131]

refiner a machine similar to a two-roll mixing MILL, operated with rolls very close together to crush undispersed ingredients and hold them in the bite of the rolls for removal and discarding when the mass has passed through.[59,140]

Refiner rolls are shorter and have a much greater diameter than mixing rolls, and are operated at a higher surface speed ratio to provide more GRINDING effect and classification.

reflectance the ratio of the intensity of reflected RADIANT FLUX to that of the INCIDENT FLUX.[43,69]

In popular usage, it is considered as the ratio of the intensity of reflected RADIANT ENERGY to that reflected from a defined reference standard.

ASTM test methods for both reflectance and reflectivity include: gray scale, evaluating color change, D2616; hiding power by reflectometry, D2805; instrumental color difference, D2244; Munsell color system, D1535; of opaque specimens, E97; preparation of reflectance standards, E259; specific luminance, traffic coatings, D4061.

reflectance, absolute see ABSOLUTE REFLECTANCE.[43,69]

reflectance, diffuse REFLECTANCE over a wide range of angles.[43,69] See PERFECT DIFFUSER, and SPECULAR REFLECTANCE EXCLUDED.[43,69]

In popular usage, the diffuse reflectance is all of the reflected RADIANT ENERGY except that of the specular angle.

reflectance, directional measurement of RADIANT FLUX reflected for specified directions of ILLUMINATION and viewing, generally measured relative to that from a perfect diffuse reflector similarly illuminated and viewed. In practice, a reference white which approaches a PERFECT DIFFUSER, or a secondary calibrated reference standard, may be used.[43,69]

reflectance factor RADIANCE measured RELATIVE to the PERFECT DIFFUSER rather than to the IRRADIANCE.[43]

reflectance, Fresnel REFLECTANCE of RADIANT ENERGY at the surface separating media of different refractive indices.[43,69]

The magnitude of the light reflected may be calculated from FRESNEL'S LAW.

reflectance, hemispherical ratio of RADIANT FLUX reflected into a hemispherical collector to the incident flux.[43,69]

Normally, the portion reflected backward from a material is measured, although the portion reflected in the forward direction may be measured on translucent materials.

reflectance, luminous REFLECTANCE derived from RADIANT FLUX by evaluating the RADIANT ENERGY reflected according to its action upon a selective receptor (such as the eye), whose SPECTRAL sensitivity is defined by a standard relative luminous efficiency function.[43,69]

In popular usage, the term is used to describe the Y TRISTIMULUS VALUE in the CIE SYSTEM.

reflectance, nonspecular REFLECTANCE of RADIANT FLUX from a surface at angles other than that of the specular angle, i.e., DIFFUSE REFLECTANCE.[43,69]

reflectance, specular REFLECTANCE of a beam of RADIANT ENERGY at an angle equal but opposite to the INCIDENT ANGLE; the mirror-like reflectance.[43,69]

The magnitude of the SPECULAR REFLECTANCE on glossy materials depends on the angle and on the difference in refractive indices between two media at a surface and may be calculated from the Fresnel Law.

reflectance, total REFLECTANCE of radiant flux reflected at all angles from the surface, thus including both DIFFUSE and SPECULAR REFLECTANCES.[43,69]

reflection PROCESS or phenomenon of the return of RADIANT ENERGY from a surface.[43]

reflectivity ($R\infty$) (1) REFLECTANCE which would be attained if a material were completely OPAQUE; reflectance of a layer of material of such thickness that an increase in thickness will not change its reflectance; (2) the reflectance of a film so thick that a further increase in thickness does not change the reflectance.[43,69]

ASTM test methods for both reflectance and reflectivity include: gray scale, evaluating color change, D2616; hiding power by reflectometry, D2805; instrumental color difference, D2244; Munsell color system, D1535; of opaque specimens, E97; preparation of reflectance standards, E259; specific luminance, traffic coatings, D4061.

reflectometer instrument used for the measurement of REFLECTANCE.[43,145,69]

reflectorization incorporation of BALLOTINI in, for example, road signs or markings to give selective

REFLECTION in the general direction of the light source.[43,163] (BSI)

reflex reflector a reflecting material so designed that a beam of RADIANT ENERGY returns, or is REFLECTED back, along the line of INCIDENCE.[43,69] See RETROREFLECTIVE.[43,69]

Such reflectors are of importance for automobiles, traffic signals, and signs.

refraction deflection of RADIANT ENERGY from a straight path in one medium to a different path in another medium of different INDEX OF REFRACTION.[43]

refractive index see INDEX OF REFRACTION.[43]

refractometer instrument for measuring the REFRACTIVE INDEX of liquids or liquefiable products. Refractive indices are often taken as a guide to the purity of raw materials, such as DRYING OILS.[59] See ABBE REFRACTOMETER.[59]

Registry of Toxic Effects of Chemical Substances (RTECS) published by NIOSH, presents toxicity data on thousands of materials.[172]

Its objective is to identify "all known toxic substances" and to reference the original studies.

reglet a groove in a wall or other surface adjoining a roof surface for the attachment of COUNTERFLASHING.[76] (ASTM)

regulatory alternative any of the several air pollution control strategies or options that are considered during the development of a regulation.[172] (EPA)

Each such alternative is evaluated in terms of cost of control and economic impact on the industry.

Reichert-Meissl value the number of ml of N/10 potassium hydroxide solution required to neutralize the free volatile and soluble FATTY ACIDS in 5 g of the sample under test.[128]

reinforced membrane a roofing or waterproofing membrane reinforced with FELTS, mats, fabrics, or chopped fibers.[76] (ASTM)

REL abbreviation for RECOMMENDED EXPOSURE LIMIT.[172]

relative comparative; specifically applied to measurements made relative to a reference STANDARD, which values are dependent on the method of measurement used, as opposed to absolute measurements made in fundamental units and obtainable by different methods.[157]

relative density the ratio of the density of a given substance to the density of a reference material at specified conditions of temperature and pressure; in the older literature this is called SPECIFIC GRAVITY.[128] (IUPAC)

relative dry hiding power of a paint is the ability of that paint to reduce the contrast of a black and white surface to which it is applied and allowed to dry.[43] (ASTM)

It is quantitatively expressed in terms of the proportional SPREADING RATE of paint required to produce the same contrast reduction as obtained with the paint chosen as STANDARD.

relative humidity the ratio, often expressed as a percentage, of the partial pressure of water in the atmosphere at some observed temperature, to the saturation vapor pressure of pure water at this temperature.[128] (IUPAC)

relative leveling a measure of the ability of a coating to FLOW out after application so as to obliterate any surface irregularities, such as BRUSH marks or ORANGE PEEL or peaks and craters which have been produced by the mechanical PROCESS of applying the coating.[166,128]

The term is used with the "0" to "10" scale of leveling ratings of ASTM Test Method D2801, Leveling Characteristics of Paints by Drawdown Method.

release agent see MOLD WASH[83] and PARTING AGENT.[83,76]

release coating a coating applied to a SUBSTRATE (usually a flexible one) which limits the adherence of a pressure-sensitive coating.[71,79] (EPA)

This is used as a covering sheet for pressure-sensitive labels so that the labels can be pulled off and used. The back side of adhesive tapes are coated with release coating to allow the tape to be unrolled.

release paper a protective sheet having an ADHESIVE FILM on one side; it may be easily removed from the surface to which it is applied.[79] See TRANSFER COATING.[77]

releasing liquid a BASE for receiving PRINTED matter, which base later can be removed by treatment with a SOLVENT, e.g., water or ALCOHOL.[168]

relief DECORATION in which the design is given prominence by cutting away the background.[81,148]

relief offset PROCESS using a relief plate on an offset press.[168,131]

relief printing PROCESS utilizing those portions of the matrix which are in relief for taking INK from a supply roller and transferring it to the surface to be printed.[168,131]

remover see PAINT and VARNISH REMOVER.[164]

render to cover (brick, wood, or stone) with a first coat of PLASTER.[73]

rennet casein special type of CASEIN made from skimmed milk by PRECIPITATION with rennet.[155]

repainting cleaning and recoating (with similar materials) of extensive areas which are being redecorated or on which the existing coatings have deteriorated, or otherwise do not provide adequate protection.[71]

repeatability qualitatively, the closeness of agreement between successive results obtained with the same method of identical test material, under the same conditions (same operator, same apparatus, same laboratory, and short intervals of time). Quantitatively, the value below which the absolute difference between two single test results obtained in the above conditions may be expected to lie with a specified probability. In the absence of other indication, the probability is normally 95%.[37,157,36] (IUPAC)

reportable quantity (RQ) the amount of a material that when spilled must be reported to the DOT (Section 311 of the CLEAN WATER ACT).[172]

reproducibility qualitatively, the closeness of agreement between individual results obtained with the same method on identical test material but under different conditions (different operators, different apparatus, different laboratories, and/or different times). Quantitatively, the value below which the absolute difference between two single test results on identical material obtained by operators in different laboratories, using the standardized test method, may be expected to lie with specified probability. In the absence of other indication, the probability is 95%.[37,157,36] (IUPAC)

reproductive health hazard/toxin any agent that has a harmful effect on the adult male or female reproductive system or the developing fetus or child.[172]

Such hazards affect people in several ways, including loss of sexual drive, mental disorders, impotence, infertility, sterility, mutagenic effects on germ cells, teratogenic effects on the fetus, and transplacental carcinogenesis.

rescue orange special COLOR used for life belts and life rafts, etc., to improve visibility at sea.[69,35]

resene, ß DAMMAR is a mixture, and one of its constituents is ß resene, sometimes known as dammar wax.[137]

It is this material which is removed in the so-called dewaxing process, which is an essential operation to make the dammar suitable for use in NITROCELLULOSE LACQUERS.

resenes (1) ALKALI-RESISTANT constituents of certain NATURAL RESINS, e.g., DAMMAR, and characterized by the presence of oxygen. The resenes are not, however, acids, ALDEHYDES, esters, ketones, or lactones; (2) as applied to NAVAL STORES, those constituents of ROSIN which cannot be SAPONIFIED with ALCOHOLIC ALKALI, but which contain CARBON, HYDROGEN, and oxygen in the molecule.[44,167] (ASTM)

residual tack tackiness remaining in a film which, although SET, does not reach the really tack-free stage.[67] (BSI) See TACK.[67,79,166]

resin (1) general term applied to a wide variety of more or less TRANSPARENT and fusible products, which may be NATURAL or SYNTHETIC. They may vary widely in COLOR. Higher molecular weight synthetic resins are presently more generally referred to as POLYMERS;[130,167] (2) a solid, semisolid, or pseudosolid ORGANIC material that has an indefinite and often high molecular weight, exhibits a tendency to FLOW when subjected to STRESS, usually has a softening or melting range, and usually fractures conchoidally; (3) in a broad sense, the term is used to designate any POLYMER that is a basic material for coatings and plastics.[167] See RESIN, NATURAL; RESIN, SYNTHETIC; and POLYMER.[167,130]

ASTM test methods for the general properties of resins include: softening, ring and ball, E28; solubility/range test, D3132; volatile/nonvolatile, D4209; volatile resin acids, D3008; water content, Karl Fischer, test, D4017.

ASTM test methods for the general properties of resin solutions include: clarity/cleanliness, D2090; cloud point, D5062; dilutability (solvent tolerance), D5062; Gardner color scale, D1544; nonvolatile, ink vehicles, D4713; nonvolatile matter content, test, D1259; standard color solutions, D365 and D1209; unsaponifiable matter, D1397; viscosity, test, D1725.

resin emulsion paint a water paint consisting of a water EMULSION of an OIL-MODIFIED ALKYD or other RESIN; when dry, leaves a tough film of resin.[60,71,155] (DAC)

resin, natural a solid ORGANIC substance, originating in the secretion of certain plants or insects, which is THERMOPLASTIC, flammable, nonconductive of electricity; breaks with a conchoidal fracture (when hard); and dissolves in certain specific organic solvents but not water.[130,167] (ASTM) See RESIN; RESIN, SYNTHETIC.[167]

A resin is produced either in the metabolism of tree growth, (e.g., CONGO COPAL) or insects (e.g., SHELLAC). The former may be obtained as exudations from growing trees or dug up from the ground (FOSSIL GUM RESIN). With the exception of some fossil gum resins, natural resins are generally soluble in organic solvents but insoluble in water.

resin oil see TALL OIL.[44]

resin pocket see PITCH POCKET.[153]

resin, synthetic originally, a member of a group of SYNTHETIC substances which resemble and share

---

Words presented in CAP/SMALL CAPS type indicate that the word is defined in another part of the *Coatings Encyclopedic Dictionary.*

Numerical superscripts classify terms in one or more of the categories listed in the second section of this volume.

some of the properties of NATURAL RESINS, but now used for materials which bear little resemblance to natural resins.[130] (BSI) See POLYMER; RESIN, NATURAL.[167]

The term is generally understood to mean a member of the heterogeneous group of COMPOUNDS produced from simpler compounds by CONDENSATION and/or POLYMERIZATION. Chemically modified natural POLYMERS, such as CELLULOSE ACETATE and hardened CASEIN, are not considered to be synthetic resins.

resinamines RESINOUS basic products obtained from low temperature TAR.[167]

resinates, metallic ROSIN in which part or all of the rosin acids have been chemically REACTED with those METALS which give SOAPS or SALTS which are water insoluble.[44,167] (ASTM)

Limed rosin, zinc-treated rosin, and the resinates of lead, cobalt, copper, and manganese, are of the greatest importance.

resinoid any of the class of THERMOSETTING SYNTHETIC RESINS, either in their initial temporarily fusible state or in their final infusible state.[167,67] See NOVOLAK.[67,140]

resinous timber WOOD from certain trees which contain resinous material in the cells.[153,167]

The RESINS have high SOLVENT power for many paints and paint media. This resin frequently exudes through paint films applied to such wood, especially in sunny positions.

resist to strive to fend off or offset the actions, effects, or FORCE of.[72]

resist technique TAPING a fine line effect over a surface, and then blocking out the rest of the area and SPRAYING into the opening left by the tape.[148]

resistance the act or an instance of resisting or the capacity to RESIST.[72]

resistance to household chemicals see HOUSEHOLD CHEMICALS, RESISTANCE TO.[72,35]

resistivity see SPECIFIC RESISTANCE.[61]

resite a PHENOL-FORMALDEHYDE RESIN in the final stage of the CURING PROCESS. In this stage it is insoluble in ALCOHOL and acetone, and infusible.[67,130] See RESITOL and RESOL.[67,130,79] Syn: C-stage (ASTM)

resitol a PHENOL-FORMALDEHYDE RESIN in the transition stage of the CURING PROCESS.[67,130] See RESITE and RESOL.[67,130] Syn: B-stage

Under heating it softens to rubberlike consistency but without melting. It swells when it is immersed in ALCOHOL or acetone, but does not dissolve.

resol a fusible, soluble, PHENOLIC RESIN containing sufficient reactive methylol groups to enable the resin to become infusible on further reaction. See NOVOLAK.[67,130] Syn: A-stage

resorcinol test used to detect the presence of PHTHALIC ACID in phthalate PLASTICIZERS or ALKYD RESINS.[157]

It involves heating the product under test with concentrated sulfuric acid and resorcinol, and pouring the resulting mixture into excess of an alkaline solution. If phthalic acid is present, the fluorescein produced yields its characteristic yellow-green COLOR.

Resource Conservation and Recovery Act (RCRA) EPA has jurisdiction. Enacted November 21, 1976, and amended since.[172]

RCRA's major emphasis is the control of hazardous waste disposal. It controls all solid-waste disposal and encourages recycling and alternative energy sources.

respirator (1) a device that supplies oxygen or a mixture of oxygen and CARBON dioxide for breathing, used especially in artificial respiration;[172] (2) a screenlike device worn over the mouth or nose or both to protect the respiratory tract.[172] Syn: inhalator

There are quite a variety ranging from disposable DUST masks to self-contained breathing apparatus. All have specific uses and limitations.

retarder (1) generally, a component added to a composition to slow down a chemical or physical change. A slowly EVAPORATING SOLVENT may be added to a paint, VARNISH, or LACQUER to delay the SET of the film after application and so improve the application properties, or to give a better film, e.g., one with improved FLOW, or a retarder may be added to PLASTER to retard its setting;[67,83] (2) in flexography, gravure, and heatset PRINTING, high boiling solvent added to INK to slow the EVAPORATION rate;[168] (3) in letterpress and offset; (4) a material which in small amount added to a RUBBER COMPOUND retards VULCANIZATION or slows down the activity of the ACCELERATOR. Specifically, phthalic anhydride and salicylic acid are retarders. The most valuable retarder is one which slows the vulcanization at processing and early curing temperatures, but does not affect or may even activate the rate of CURE at full curing temperatures.[67,83,140] (BSI) See ANTIOXIDANT.[67,83,168]

reticulation the crepe-like APPEARANCE of the surface of the COLLOID on COLLOTYPE plates.[139,168]

retina the light-sensitive layer and highly organized neuron structure within the choroid, and lining the posterior two thirds of the eyeball.[43,69]

retinal of or pertaining to the RETINA.[43,69]

retort pine tar see PINE TAR, RETORT.[44]

retouching see SPOTTING IN.[54]

retraction see COLLING.[42,55,56]

retroreflection light that is reflected predominately in the direction of the source of that light.[43]

retroreflective a surface or device which reflects an INCIDENT RAY of light back along the direction of incidence with only limited spread, without requiring the incident ray to be normal to the surface.[43,69] See REFLEX REFLECTOR.[43,69]

Such a surface or device differs from both a perfect DIFFUSER and a perfect mirror. It may utilize small triple-REFLECTION cube corners, small TRANSPARENT spheroids, multiple hemispheric indentations, a multiple lens sheet backed by a reflective surface, or some similar device.

reverse impact test see IMPACT TEST.[157]

reverse roll coater a roll type coater for PAPER, film, foil, and METAL COIL which applies coating to the WEB by a roll which turns in a reverse direction of travel of the web.[77] (EPA) See ROLLER COATING.[71,77]

This procedure is said to reduce striations in the coating.

reversible colloid see COLLOID, REVERSIBLE.[139,64]

reversion the softening of VULCANIZED RUBBER (usually natural but sometimes synthetic rubber) when heated too long or at too high a temperature.[47,140]

Reversion is evidenced by increase in extensibility, decrease in TENSILE STRENGTH, and lowering of the STRESS to produce a given ELONGATION. Extreme reversion usually results in TACKINESS.

rewind in the COIL COATING industry, the apparatus used to recoil the STRIP after it is coated.[59,70]

rhe the C.G.S. UNIT of measurement of FLUIDITY.[166,128]

A material has a fluidity of one rhe when a SHEARING FORCE of 1 dyne/cm^2 induces a rate of shear of 1 cm/s.

rheological refers to RHEOLOGY.[166]

rheology science of the DEFORMATION and FLOW of matter.[166]

rheology modifier (1) an ADDITIVE used generally for changing the viscosity or FLOW of paints; (2) commonly used synonymously for ASSOCIATIVE THICKENERS.[83,166] cf. THICKENER[83,166]

rheopexy a special form of THIXOTROPIC gel which possesses the property of solidifying more rapidly when SHEARED (stirred) very slowly than when at complete rest.[166] cf. THIXOTROPY[47,166]

It should not be confused with DILATANCY. The equilibrium state of the former is a solid GEL, while the latter is a liquid.

rhodamine 6G Pigment Red 81 (45160). Xanthene type DYESTUFF made by CONDENSING monoethyl-*m*-AMINO-PHENOL with PHTHALIC ANHYDRIDE, and converting the product into the ethyl ester.[41]

rhodamine B Pigment Violet 10 (45170). Brilliant, strong DYESTUFF made by CONDENSING diethyl-*m*-AMINO-PHENOL and PHTHALIC ANHYDRIDE together, in the presence of zinc chloride.[41]

rhodamine reds a class of clean, blue shades of ORGANIC reds possessing good LIGHTFASTNESS; often called magenta in PROCESS PRINTING.[41]

rhoduline blue 66 a bright, greenish blue PIGMENT DYESTUFF, made by condensing o-chlorobenzaldehyde with dimethylaniline, which is then oxidized.[41] Syn: setoglaucine or permanent peacock blue

rhoduline heliotrope ANILINE DYE used in colored LACQUERS.[41]

Soluble in lacquer solvents, fast to light, and is not affected by EXPOSURE to air.

Rhus vernicifera $C_{22}H_{31}O_3$. Chinese tree which delivers a milky type of liquid.[151]

When the water is removed, a dark oily liquid is obtained, which yields a tough, flexible film on atmospheric OXIDATION. The chief constituent of the FILM-FORMING material is an UNSATURATED PHENOL-ACID-URUSHIOL. The liquid is used for the preparation of JAPANESE LACQUER.

ribbing in the coil coating industry, longitudinal streaks that do not FLOW out on painted STRIP.[70]

ribs raised ridge in the FINISH caused by heavy BRUSH marks which were not sanded or rubbed from underneath coats before finishing coats were applied.[54,55]

ricinoleic acid $CH_3(CH_2)_5CHOHCH_2CH = CH(CH_2)_7COOH$. Chief acid of CASTOR OIL, almost unique by reason of the presence of a HYDROXYL GROUP in the FATTY ACID chain.[144] See DCO.[63]

To exhibit drying properties, the ester must be dehydrated. Properties: bp, 250°C/15 mm Hg; mp, 17°C; sp gr, 0.945.

ring and ball method see BALL and RING METHOD.[157]

Rinmann's green see COBALT GREEN.[41]

ripple finish an intentional, uniformly WRINKLED FINISH which is obtained usually by BAKING.[71,163,176] Syn: crinkle finish, shrivel finish, wrinkle finish, suede finish

risk assessment a scientific process directed toward defining qualitatively and quantitatively, the

---

Words presented in CAP/SMALL CAPS type indicate that the word is defined in another part of the *Coatings Encyclopedic Dictionary.*

Numerical superscripts classify terms in one or more of the categories listed in the second section of this volume.

risks of adverse effects of chemicals on health and the environment under specified use conditions.[172]

The process is itself the product of two separate assessment processes: hazard assessment and EXPOSURE assessment. The former defines the intrinsic TOXICOLOGICAL and other properties of a material, while the latter is the process of defining the levels of exposure to the material under specified use conditions.

riveling, rivelling see WRINKLING.[54,56,42]

rock salt moss see CARRAGEEN.[139,155]

rocker hardness tester an instrument made up of two circular stainless runners connected through traverse bars spaced 25 mm apart.[47,157,145,128] Reference ASTM Test Method D2134.

Rocking frequency is calibrated by weights on a GLASS standard. It works on the principle that a "soft" surface dampens the oscillations of a rocking wheel more than a "hard" one.

Rockwell hardness a method of testing HARDNESS which is suitable for RESINS.[157,47]

It employs a loaded ball which is applied for a controlled period; the resulting indentation is measured.

rococo extremely ornate 18th Century style featuring scrolls, shells, and fanciful forms.[81]

roll coating see ROLLER COATING (2).[77]

roll of wallcovering a bolt, consisting of 36 square feet of WALLCOVERING, of which 30 square feet are estimated as usable.[148,128]

Bolts come in single, double, and triple rolls.

roll roofing coated FELTS; either smooth or mineral-surfaced.[76] (ASTM)

roll-out INK spread for TEST or SAMPLING purposes by use of a hand roller.[168]

roller see PAINT ROLLER.[77]

roller application see ROLLER COATING.[77]

roller cleaner see BRUSH or ROLLER CLEANER.[127]

roller coating (1) PROCESS by which a FILM is applied mechanically to sheet or strip material. The machine consists of a series of horizontal cylindrical rollers. The coating is picked up by one of the first rollers rotating in a trough containing the coating, is transferred to subsequent rollers until evenly distributed over the last roller, and thence to the flat surface of mild steel, tinned iron, or other suitable strips or sheets;[71,77] (2) APPLICATION of paint by means of a hand-operated ROLLER to wall surfaces, etc.[77]

roller coating enamel special type of enamel, designed for APPLICATION by a roller coating machine.[71] See ROLLER COATING.[71,77]

roller cover see PAINT ROLLER COVER.[59]

roller mill MILL involving one or more rolls; three and five rolls are commonly used.[59]

The rolls, which revolve at different speeds, are adjustable as to clearance between the milling surfaces, making it possible to obtain various FINENESS of PIGMENT DISPERSION.

roller spatter resistance the ability of a paint to resist the formation of various size droplets generated during paint roller application, which results in undesirable spots splashing on areas that are not being painted including the person applying the paint.[42,72] (ASTM)

Use of ASSOCIATIVE THICKENERS in LATEX paints has minimized roller spatter considerably.

rollformer in the COIL COATING industry, a machine that bends continuous STRIP into various shapes by a series of revolving METAL wheels or discs.[59,70]

rolling up the APPLICATION of INK to a plate by means of a ROLLER.[168]

roof the cover of a BUILDING, including the ROOFING and all other material and construction (such as supporting members) necessary to carry and maintain it on the walls or uprights.[82] (DAC)

roof cladding see ROOFING.[82]

roofing (1) any material used as a ROOF covering (such as shingles, slate, sheet METAL, or tile) to make it impervious to wind and water, and often to provide THERMAL INSULATION; (2) a roof.[75,82] (DAC) Syn: roof cladding

roofing system an assembly of interacting components designed to weatherproof, and normally to insulate, a building's top surface.[76,82] (ASTM)

room temperature temperature in the range of 20 to 30°C (68 to 85°F).[142,176] (ASTM)

room temperature stability, latex see LATEX ROOM TEMPERATURE STABILITY.[161,157]

R.O.P. colors abbreviation for run of press colors; a term used to describe the COLORED inks generally accepted as "STANDARD" colors for newspaper printing.[168]

ropiness see ROPY FINISH.[56,42]

ropy finish FINISH in which the BRUSH marks have not FLOWED out, this being the normal APPEARANCE of a paint or VARNISH having poor leveling properties.[56,35] (BSI)

A similar appearance may also be produced in a paint, which normally has good leveling prop-

erties, by continuing to brush the paint after the film has begun to SET.

rosin NATURAL RESIN obtained from pine OLEORESIN after removal of the VOLATILE FRACTIONS.[167] (BSI) Syn: colophonium and colophony

Two general kinds of ROSIN *are commercially available:* GUM ROSIN *obtained from living trees, and wood rosin obtained from dead wood, such as stumps and knots.* TALL OIL *rosin, a by-product of the* PAPER *industry, is a chemically similar material.*

rosin acids principally, monocarboxylic acids with the empirical formula $C_{19}H_{29}COOH$.[44] (ASTM)

They are classified into two groups: the ABIETIC *type and the pimaric type. Both types and their derivatives are found in* WOOD, GUM, *and* TALL OIL ROSINS.

rosin essence that portion of ROSIN distilled below 360°C.[167]

rosin esters a large family of RESINS produced from ROSIN ACIDS and POLYOLS.[167] See ESTER GUM.[44,130]

rosin, limed or zinc-treated ROSIN which has been chemically reacted with a CALCIUM or ZINC COMPOUND, or both, to form a product containing a METALLIC SALT of the rosin acids, along with unreacted rosin.[130,156] (ASTM) See RESINATES, METALLIC.[44,167]

The amount of calcium or zinc contained in the finished products of this category may vary over a wide range.

rosin, modified ROSIN that has been treated with heat or CATALYSTS, or both, with or without added chemical substances, so as to cause substantial change in the structure of the rosin acids, as ISOMERIZATION, HYDROGENATION, DEHYDROGENATION, or POLYMERIZATION, without substantial effect on the CARBOXYL GROUP.[44] (ASTM)

rosin oil relatively VISCOUS OILY portion of the CONDENSATE obtained when ROSIN is subjected to dry DESTRUCTIVE DISTILLATION; also used to describe specially compounded OILS having a rosin oil base.[44] (ASTM) See BLOOM OIL.[44]

rosin pitch dark, acidic residue obtained from the DESTRUCTIVE DISTILLATION of ROSIN.[44,167]

rosin, reclaimed ROSIN that has been recovered or reclaimed by any means from waste or deteriorated material, provided that the concentration of rosin acids is not below that normal for rosin, and any residual or contaminating component from the waste material itself or from any article used in the recovery process is not in sufficient quantity to cause the physical or chemical properties of the reclaimed product to differ materially from those of rosin.[44,167] (ASTM)

rosin size soda ROSIN SOAP containing varying amounts of free rosin; used for SIZING PAPER.[127,167]

rosin spirit relatively light, volatile portion of the CONDENSATE obtained in the first stages when ROSIN is subjected to dry DESTRUCTIVE DISTILLATION.[44] (ASTM) Syn: pinolin

rosin standards combinations of assembled colored glasses having the colors designated as representative of the established U.S. grades used in classifying ROSIN.[44,128] (ASTM)

The recognized official standards are those developed and issued by the U.S. Department of Agriculture, or similar standards made of Lovibond GLASS, *when certified by the same government agency. The official grades established by or under authority of the Federal Naval Stores Act, for which standards are provided, are as follows: X, WW, WG, N, M, K, I, H, G, F, E, D and FF (the latter grade is used only for wood rosin). Rosin darker in* COLOR *than the standard for grade D or FF is graded B. The designation* OPAQUE *with the grade letters OP is used to describe rosin which, because of a turbid, cloudy, or nontransparent condition due to occluded moisture, excessive* CRYSTALLIZATION, *or presence of foreign matter other than dirt, cannot be accurately graded by comparison with any of the above described rosin grade standards.*

rosin test (qualitative) see HALPHEN-HICKS TEST, and LIEBERMAN-STORCH METHOD.[157,167]

rosin type (sample) sample of ROSIN, or a mold of THERMOSETTING PLASTIC material, used as an unofficial STANDARD in grading rosin.[44] (ASTM)

Such sample shall be so selected, sized, and surface-finished that it will have the form of an approximate 7/8 in. cube with at least two opposite faces having smooth parallel surfaces and shall have a COLOR *when viewed through these faces which matches within rather narrow tolerances the color of the corresponding official government standard made of* GLASS.

rot DECOMPOSITION in WOOD by FUNGI and other microorganisms; reduces its STRENGTH, DENSITY, and HARDNESS.[153,173]

rotary press high-speed press which passes continuously running WEBS of PAPER between an impression cylinder and the curved form on the plate cylinder.[59,168]

rotary screen printer a widely used FABRIC printing technique in which a moving fabric web passes in contact with a series of rotating drums.[54,77] (EPA)

Words presented in CAP/SMALL CAPS type indicate that the word is defined in another part of the *Coatings Encyclopedic Dictionary*.

Numerical superscripts classify terms in one or more of the categories listed in the second section of this volume.

The cylindrical drums have holes in the drum surface in shape of a pattern. Liquid print paste is forced through the holes in the drum surface onto the fabric, leaving a pattern on the fabric surface.

rotational electrostatic spray a SPRAY paint device which uses a rotating BELL or DISK to shear the coating into a fine mist and which efficiently draws the coating onto the part being coated by electrostatic forces.[77] (EPA) See ELECTROSTATIC SPRAY.[77]

roto-dipping patented method of PRETREATING and PRIMING motor bodies in which the body is mounted on a longitudinal axle and is caused to pass through a series of SPRAY zones and DIP tanks.[77,158] (BSI)

At appropriate stages in the process the body is revolved on the axle.

rotogravure a method of PRINTING in which the image areas of the printing roll are recessed relative to the background. The image is in the form of cells or cups mechanically or chemically etched in the surface.[168] (EPA) See GRAVURE.[168]

Typically, a gravure cell is 35 microns deep by 125 microns square, with 22,500 cells in a square inch.

rotometer a device, based on STOKES LAW, for measuring rate of fluid FLOW.[145,166] (IUPAC)

It is a tapered vertical tube having a circular cross section in which a float moves in a vertical path to a height dependent upon the rate of fluid flow upward through the tube.

rottenstone brown, AMORPHOUS, siliceous LIMESTONE, similar in nature to pumice stone, but softer in TEXTURE.[158,58] Syn: tripoli and terra cariosa

Principal uses are as an ABRASIVE and FILTER medium.

rouge see IRON OXIDES, SYNTHETIC.[41]

round (round color) description of a FULL-BODIED paint, i.e., one which gives good BUILD.[166] (BSI)

round-robin the practice of planning, conducting, analyzing, and interpreting the results of interlaboratory tests.[157,36]

RQ abbreviation for REPORTABLE QUANTITY.[172]

RTECS abbreviation for REGISTRY OF TOXIC EFFECTS OF CHEMICAL SUBSTANCES.[172]

ru measure for surface roughness. 0.025 µm (10^{-6} in.) BSS 1134.[128]

rub up a procedure for determining how well a tintable paint has incorporated the COLORANT and how well white hiding pigments have been dispersed.[69,157,42]

A cast, wet paint is rubbed firmly with a finger to apply shear. Differences of COLOR depth in the dry film indicate that a color problem exists. Retail paint outlets stock tint-base paints which are custom colored for the customer by the addition of controlled volume of colorant. Good COLOR ACCEPTANCE is required in order to match color chips and give uniform color APPEARANCE. This rub up procedure has generally been replaced by ASTM Test Method D5326.

rub-off INK on printed sheets which, after sufficient DRYING TIME, smears or comes off on the fingers when handled.[168]

rubber (1) a highly resilient material, capable of recovering from large DEFORMATIONS quickly; manufactured from the juice of RUBBER trees as well as other trees and plants; (2) any of various synthetically produced materials having similar properties; an ELASTOMER.[140] (DAC)

rubber cement see CEMENT, RUBBER.[79]

rubber latex COLLOIDAL aqueous EMULSION of an ELASTOMER.[60,140] (ASTM)

rubber-emulsion paint see LATEX PAINT.[60,130]

rubber-seed oil OIL obtained from the seeds of *Hevea brasiliensis*.[63]

Properties: sp gr, 0.927/15 °C; saponification value, 190; iodine value, 138.

rubbing PROCESS of LEVELLING and lowering the GLOSS of a dried coating film by RUBBING it, either wet or dry, usually with a suitable ABRASIVE, to remove NIBS and other irregularities and to provide a suitable surface for subsequent coating.[54,158] See SANDING.[158] Syn: flatting down

rubbing down the fine tuning of a brushed-on VARNISH or POLYURETHANE FINISH to even out BRUSH marks or to eliminate DUST trapped in a hardened finish or to subdue a too slick look.[175,54]

It is generally the last stage of refinishing. Rubbing down merely covers the surface with a series of fine scratches that break up the reflection of light, but do not interfere with the clarity of the finish. Fine PUMICE powder in machine oil is used as the ABRASIVE.

rubbing flat PROCESS of FLATTING a coating means to subject it to ABRASIVE action to remove NIBS and other irregularities and to provide a surface with a suitable key for a subsequent coating.[158]

rubbing oil neutral, medium-heavy MINERAL OIL used as a lubricant for PUMICE stone in RUBBING VARNISH or LACQUER.[54,152,158]

rubbing test a procedure for testing the ease with which a LACQUER or ENAMEL can be rubbed to a smooth high-gloss FINISH.[157] cf. RUB UP[157]

rubbing varnish see POLISHING VARNISH, and FLATTING VARNISH.[54,152,163]

rubine pertaining to blue shade red, probably derived from the Latin word "rubia" for the madder plant from which MADDER LAKE (alizarin red), a blue shade red, was originally made.[41]

The name of the bluish red colored gem, the ruby, would appear to be similarly derived.

rubine reds see PERMANENT RED 2B.[41]

Rule 66 an air pollution regulation passed by the Los Angeles Air Pollution Control District in 1966.[172,168] See PHOTOCHEMICALLY-REACTIVE SOLVENT.[172,164]

This regulation was aimed at controlling HYDROCARBONS and defined certain types of compounds as unreacting and exempt from control. This was one of the first VOC regulations. It was very influential and was widely copied by other jurisdictions. (EPA)

rumbling see TUMBLING.[77]

runs narrow downward movement of a paint or VARNISH film; may be caused by the collection of excess quantities of paint at irregularities in the surface, e.g., CRACKS, HOLES, etc.; the excess material continuing to FLOW after the surrounding surface has SET.[56] (BSI) Sometimes called tear.[56]

Runs occur most frequently around detail areas such as nuts and bolts, where controlling film thickness while covering each surface is difficult, especially when SPRAY painting.

running PROCESS by which SOLUBILITY of a RESIN in an OIL is achieved. Fossil resins such as CONGO are "run" by heating the resin alone to elevated temperatures in incremental steps, thereby DISTILLING off unwanted volatile components and altering the resin so that it becomes compatible with additions of oil.[131,167,176]

Resins so treated are said to have been "run."

rust the reddish, brittle coating formed on IRON or its alloys resulting from EXPOSURE to humid atmosphere or chemical attack.[66]

This is not to be confused with WHITE RUST.

rust grade scale in evaluating the resistance to RUSTING, the linear, numerical rust grade scale is an exponential function of the area of rust so that slight amounts of first rusting have the greatest effect on lowering the rust grade. (ASTM D610).[66,157,128]

There is also a European rust scale and an ISO rust scale (ISO 4624).

rust resistance the ability of a coating to protect the SUBSTRATE of iron or its alloys from rusting.[72,66] See RUST.[66]

The amount of rusting beneath or through a paint film is a significant factor in determining whether a coating system should be repaired or replaced.

The FSCT Pictorial Standards of Coatings Defects contains photographic standards of the degree of rusting on painted steel surfaces.

ASTM Test Method D610 covers the evaluation of the degree of rusting on painted steel surfaces using the same visual standards as those in the FSCT Pictorial Standards of Coatings Defects but no photographic standards.

rust-inhibiting paint see ANTI-CORROSION PAINT.[71,66,72,163]

rust-inhibitive washes see CHEMICAL CONVERSION COATING.[158]

rustic siding see DROP SIDING.[82,153,75]

rusting CORROSION on the surface of IRON or ferrous metals resulting in the formation of products consisting largely of hydrous FERRIC OXIDE.[159,66]

rutile used to designate a crystal structure of TITANIUM DIOXIDE with a sp gr of 4.2 and a refractive index of 2.69.[41] cf. BROOKITE[41,34]

The rutile type is characterized by higher opacity, greater density, and greater inertness as compared to the ANATASE type.

Words presented in CAP/SMALL CAPS type indicate that the word is defined in another part of the *Coatings Encyclopedic Dictionary.*

Numerical superscripts classify terms in one or more of the categories listed in the second section of this volume.

S

S KUBELKA-MUNK SCATTERING COEFFICIENT.[43]

sa (0,1,2,2-1/2,3) Grades of cleanliness of BLASTED STEEL, Swedish Standard SIS 05 50 00, ASTM D-2200.[128,158]

sable pencil fine pointed BRUSHES of sable hair, set in quills, varying in size from lark's, through crow's, to duck's and goose's, or set in a METAL ferrule.[59,81]

Used for DECORATIVE detail. Because of the high cost of sable, squirrel hair may be used instead.

sable writer a BRUSH like a sable pencil, but longer, with a flat edge or a point, used for sign writing.[59,81]

sacrificial protection (1) the use of a METALLIC coating, such as a ZINC-RICH paint, to protect STEEL. In the presence of an electrolyte, such as salt water, a galvanic cell is set up and the metallic coating CORRODES instead of the steel;[66,163] (DAC) (2) zinc or aluminum anode.[66] (DAC)

saddle in roofing, a small structure that helps to channel surface water to drains.[76] (ASTM) See CRICKET.[76]

It is frequently located in a valley, a saddle is often constructed like a small hip roof, or like a pyramid with a diamond-shaped base.

safety inks PRINTING INKS which will change COLOR or BLEED when ink eradicator or water is applied to PRINTS.[168]

Also, inks which will erase easily. Usually, mixtures of a drying-oil ink with an AQUEOUS SOLUTION of a water-soluble color.

safflower oil SEMIDRYING to DRYING OIL, obtained from *Carthamus tinctorius*, a native of India.[63]

It is also available from seed grown in the United States. Its main constituent acid is LINOLEIC, with a small quantity of LINOLENIC acid. It DRIES and BODIES more slowly than LINSEED OIL. Properties: sp gr, 0.925/10°C; iodine value, 139; saponification value, 189. Its drying characteristics lie between those of linseed and soybean OILS. One of its main advantages for paint and VARNISHES is its extremely low after-yellowing. This is due to its very low linolenic acid content.

sag or sagging nonuniform downward FLOW of a wet paint film that occurs between the times of application and setting, resulting in an uneven coating having a thick lower edge.[56,42] (ASTM)

The sags usually occur at a thick local area of a vertical film and may have the characteristic APPEARANCE of a draped curtain, hence the synonym CURTAINING.

ASTM Test Method D4400 covers the laboratory determination of aqueous and nonaqueous liquid coatings at any level of sag resistance. It involves the use of a MULTINOTCH APPLICATOR.

salt a chemical COMPOUND formed by replacing all or part of the hydrogen ions of an acid with METAL ions or electropositive RADICALS.[144]

salt spray (fog) test a test to determine CORROSION RESISTANCE of coatings by exposing a coated standardized steel panel to a dilute SALT (sodium chloride) solution fog.[157,66] (ASTM, EPA)) ASTM Standard Test Method B117 can be used for testing.

This method is applicable to testing of ferrous and nonferrous metals, and is also used to test INORGANIC and ORGANIC coatings, etc., especially where such tests are the basis for material or product specifications.

Tests of this type may be used to evaluate the corrosion resistance of organic coatings on METAL products exposed to high salt environments such as coastal and road deicing. EXPOSURE conditions are complex and changeable. Important factors include climate, time of the year, presence of pollution, etc. Generally, it is difficult, if not impossible, to define or measure precisely all the factors which influence degradation.

sample (1) a portion, piece, or segment that is representative of the whole;[37] and (2) an entity that is representative of a class.

sampling the act, process, or technique of selecting an appropriate SAMPLE.[37,131]

sand small, loose grains of worn or disintegrated rock (chiefly of PARTICLES of weathered QUARTZ-bearing rock).[75]

sand finish rough FINISH PLASTER wall.[159]

sand grinder a type of PIGMENT DISPERSING equipment consisting of pumping pigment/vehicle slurry (the MILL BASE) through a cylindrical bank of sand which is being subjected to intense agitation.[59,64]

During passage upward through the agitated sand zone, the mill base is caught and dispersed between the sand PARTICLES; a strong SHEARING action which affects the dispersion of the pigment into the vehicle results. On emerging from the active sand zone, the dispersed mill base overflows through an exit screen sized to permit a free FLOW-THROUGH of the pigment dispersion while holding back the sand particles.

sand grinding PROCESS of DISPERSING PIGMENTS in the SAND GRINDER.[64,131]

sand-dry descriptive of a stage in DRYING of a paint FILM at which SAND will not ADHERE to the surface.[67,157] (DAC)

sandarac NATURAL RESIN which exudes from *Callitris quadrivalvis*, found in North Africa.[167]

sandblast to use SAND, flint, or similar ABRASIVE propelled by an air BLAST, on METAL, MASONRY, CONCRETE, etc., to remove DIRT, RUST, or paint, or to DECORATE the surface with a rough texture.[158,148] (DAC)

sanding ABRASIVE PROCESS used to prepare a surface prior to the APPLICATION of a protective or DECORATIVE coating.[158]

sanding sealer specially hard first coat that has the property of SEALING or FILLING, but not obscuring, the GRAIN of WOOD.[147,153,158]

The surface is then suitable for sanding before the application of subsequent coats.

sanding surfacer a heavily pigmented finishing material used for building the surface to a smooth condition. It is sanded after drying.[158,163]

sandpaper (1) a tough PAPER which is coated with an ABRASIVE material such as SILICA, garnet, silicon carbide, or aluminum oxide; used for smoothing and polishing; graded by a grit numbering system according to which the highest grit numbers (360 to 600) are used for fine polishing, and the lowest grit numbers (16 to 40) are used for coarse smoothing. Alternatively, sandpaper may be designated by the "0 grade" system, according to which "very fine" includes grades from 10/0 to 6/0, "fine," from 5/0 to 3/0, "medium," 2/0, 1/0, 1/2; "coarse," 1, 1 1/2, and 2; "very coarse," 2 1/2, 3, 3 1/2, and 4;[59,158] (DAC) (2) see COATED ABRASIVE. The general term of the Coated Abrasives Manufacturer's Institute for materials of this type.[59] (DAC)

sanitary landfill an engineered burial of refuse.[171,172] (IUPAC)

The refuse is dumped into trenches and compacted by bulldozer, where, it is hoped, aerobic metabolism by microorganisms decomposes the ORGANIC matter to stable compounds (H_2O, CO_2, etc.). Moisture is essential for the biological degradation and groundwater assists the process except when it fills voids and prevents the transport of oxygen to the refuse. Landfills of unsatisfactory design can be major sources of air, water, and soil POLLUTION.

saponification (1) alkaline HYDROLYSIS of FATS whereby a SOAP is formed; more generally, the hydrolysis of an ESTER by an alkali with the formation of an ALCOHOL and a SALT of the acid portion;[127,131] (2) the DECOMPOSITION of the medium of a paint or VARNISH film by alkali and moisture in a SUBSTRATE, e.g., new CONCRETE or fresh plaster.[56,65]

Saponified paint may become sticky and discolored. In severe cases, the film may be completely liquified by saponification. Loss of ADHESION may occur as a saponified layer develops next to the substrate.

saponification value a measure of the alkali reactive groups in OILS and FATTY ACIDS and is expressed as the number of milligrams of potassium hydroxide that react with one gram of sample.[63,157,128,127] (ASTM) Syn: Koettstorfer Number

The saponification value of oils and fatty acids is a measure of the content of ESTER linkages. For an oil, provided it is not significantly oxidized, the number of ester linkages per molecule (for example, three in triglyceride), can be used to calculate the molecular weight of the oil.

A saponification value higher than normal indicates that the oil has been oxidized (BLOWN) or chemically modified, with other acids such as MALEIC, FUMARIC, or PHTHALIC.

Saponification value alone is not a measure of the quality of the oil.

Chemically modified oils may require saponification times longer than one hour for complete reaction.

ASTM Test Method D1962 covers the determination of the saponification value of drying oils, bodied oils, fatty acids, and polymerized fatty acids.

sapwood the layers of WOOD next to the bark, usually lighter in COLOR than the HEARTWOOD and 1/2 to 3 inches or more wide.[153]

It is actively involved in the life processes of the tree. Under most conditions, sapwood is more susceptible to decay than heartwood.

SARA abbreviation for SUPERFUND AMENDMENTS AND REAUTHORIZATION ACT.[171,174]

sardine oils fish OILS, mainly of Japanese or American origin, which, when adequately refined, are useful DRYING OILS.[63]

Properties: sp gr, 0.930/15°C; iodine value, 187; saponification value, 190.

sash any framework of a window; may be movable or fixed; may slide in a vertical plane (as in a double-hung window) or may be pivoted as in a casement window; a pivoted sash is also called a ventilator. Syn: window sash (DAC)

sash tool a round BRUSH made in various sizes, bound with METAL or string and used for painting sashes, frames, and other small areas.[59]

satin See GLOSS.[43]

satin white FILLER made by PRECIPITATING together a mixture of CALCIUM SULFATE and ALUMINATE.[58]

Words presented in CAP/SMALL CAPS type indicate that the word is defined in another part of the *Coatings Encyclopedic Dictionary.*

Numerical superscripts classify terms in one or more of the categories listed in the second section of this volume.

The ingredients used for its preparation are calcium hydroxide and basic aluminum sulfate. It is used chiefly in aqueous pigmented coatings for PAPER, and also as a base for LAKE pigments.

saturated compound a COMPOUND in which all of the valences of the elements are satisified; in ORGANIC chemistry, having no double or triple bonds, i.e., incapable of absorbing substances by addition, such as hydrogen or iodine.[133,134]

saturated felt a FELT that has been immersed in hot BITUMEN; the felt adsorbs and absorbs as much bitumen as it can retain under processing conditions, but remains porous and contains voids.[76] (ASTM)

saturation the attribute of COLOR PERCEPTION that expresses the degree of departure from the GRAY of the same LIGHTNESS.[69,43,35] (ASTM)

All grays have zero saturation.

Saunderson correction equation for calculating a correction for the FRESNEL REFLECTANCES at the surface of a dielectric material for both the INCIDENT light (k_1) and exiting light (k_2). It is related to the INTERNAL REFLECTANCE, R_i, and to the TOTAL REFLECTANCE, R_T, by the following equation:

$$R_T = k_1 + \frac{(1-k_1)(1-k_2)R_i}{1-k_2 R_i}$$

(The equation was popularized by Saunderson in his work on computer COLOR MATCHING but actually dates back many years, at least to Walsh in the 1920s.)[43,69]

Saunderson-Milner Zeta Space Color Difference Equation a COLOR DIFFERENCE EQUATION designed to fit the spacing of the MUNSELL SYSTEM and based on Munsell value functions as described in the ADAMS CHROMATIC VALUE COLOR DIFFERENCE EQUATION.

$$\Delta E = [(\Delta\xi_1)^2 + (\Delta\xi_2)^2 + (\Delta\xi_3)^2]^{1/2}$$

where

$\xi_1 = (V_X - V_Y)(9.37 + 0.79 \cos\theta)$

$\xi_2 = kV_Y$

$\xi_3 = (V_Y - V_Y)(3.33 + 0.87 \sin\theta)$

$$\theta = \text{angle calculated from } \tan\theta = \frac{0.4(V_Z - V_Y)}{V_X - V_Y}$$

k = constant ranging from 1 to 5 depending on observation conditions such as fineness of line dividing the two colors being compared. Using a constant k of 2, the average magnitude of ΔE is one-sixth that of an NBS UNIT.[69]

Saybolt universal viscosity efflux time in seconds for 60 ml of fluid to FLOW through a calibrated orifice under specified conditions.[157,128]

s.b.p. spirits special boiling point spirits.[164]

A range of SOLVENTS obtained from PETROLEUM.

SBR abbreviation for STYRENE BUTADIENE RUBBER.[140]

SCAA see SURFACE COATINGS ASSOCIATION AUSTRALIA.[174]

scale (1) an adherent oxide coating that is thicker than the superficial film referred to as tarnish;[66] (2) a deposit formed from solution directly in place upon a confining surface.[66] *Scale is a deposit that usually will retain its physical shape when mechanical means are used to remove it from the surface on which it is deposited. Scale which may or may not adhere to the underlying surface is usually CRYSTALLINE and dense, frequently laminated, and occasionally columnar in structure;* (3) the oxide formed on the surface of the METAL during heating;[66] (4) RUST occurring in thin layers.[66,158]

scaler a hand-cleaning chisel.[59]

scaling (1) PROCESS of forming SCALE with or without acid fumes; sometimes refers to the spontaneous detachment of scale; [66,158] (2) see FLAKING.[42,56]

scaling resistance see FLAKING RESISTANCE.[56,72]

SCANZ see SURFACE COATINGS ASSOCIATION NEW ZEALAND.[174]

scarlet chrome mixed PIGMENT containing lead chromate, molybdate, and sulfate.[41] See MOLYBDATE ORANGE.[41]

scarlet lake $C_{18}H_{14}N_2O_7S_2Ba$. Lake, Acid Red 26 (16150). PIGMENT made by precipitating Scarlet 2R, which is the product of coupling diazotized m-xylidine with 2-naphthol-3, 6-disulfonic acid, on gloss white or hydrate of alumina.[41]

scatter, coefficient of the rate of increase of REFLECTANCE with thickness (weight per area) at infinitesimal thickness of material over an ideally black backing, $S = (dR_o/dX)x \rightarrow o$.[43] (ASTM)

scattering the process by which light or other ELECTROMAGNETIC radiant flux passing through matter is redirected over a range of angles.[43] (ASTM) See DRY HIDING.[43,55]

DIFFUSION of light passing through media with components of different REFRACTIVE INDICES (for example, PIGMENT and BINDER); scattering occurs at any such interface, at the surface, or inside a medium containing PARTICLES.

Kubelka-Munk parameters S and K are usually used to test scattering. S characterizes scattering and K characterizes the absorptivity.

scattering coefficient, Kubelka-Munk multiple (DIFFUSE) SCATTERING COEFFICIENT for a unit thickness and CONCENTRATION of scattering material in a medium of different refractive index as used in the KUBELKA-MUNK EQUATION.[43,69]

It is the rate of increase of REFLECTANCE of a layer over black as thickness is increased. Hence, the assumption is made that all of the scattering is in the backward or reverse direction from that of the incident diffuse light.

scattering coefficient, Mie single scattering COEFFICIENT of a PARTICLE in a medium of different REFRACTIVE INDEX, expressed as the ratio between scattering cross section and geometric cross section of the particle.[43,69] See MIE THEORY.[43,69]

It should properly be called scattering efficiency, but in popular usage, the term "scattering coefficient" is common.

scattering (light) an interaction of light with an object that causes the light to be redirected in its path.[43] (IUPAC)

scattering loss that part of transmitted energy lost due to roughness of REFLECTING SURFACE.[43] (ASTM)

scattering, multiple DIFFUSION or redirection of RADIANT ENERGY which results from the observation of the reflected energy after multiple reflections from one particle to others, as opposed to the observation of single scattering.[43] See KUBELKA-MUNK THEORY.[43,69]

scattering, single DIFFUSION or redirection of RADIANT ENERGY by a single particle spaced sufficiently distant from any other particle that the observed reflected energy from each particle has not impinged on any other particle.[43,69] See MIE THEORY.[43,69]

scavenging the removal of pollutants from the atmosphere by natural processes, including scavenging by cloud water, rainout, and washout.[171,172] (IUPAC)

This type of removal process is termed precipitation scavenging. Scavenging of airborne pollutants at surfaces of plant, soil, etc., is termed dry deposition.

SCE see SPECULAR REFLECTANCE EXCLUDED.[43,69] cf. SCI[43,69]

Schaffer's acid ACID used in the manufacture of DYESTUFFS.[41,144]

Chemically, it is 2-naphthol-6-sulfonic acid.

Scheiber oil one of the original DEHYDRATED CASTOR OILS.[63]

Schnitzer's green see CHROMIUM OXIDE GREEN.[41]

Schweinfurt green synonym for PARIS GREEN and emerald green.[41]

Chemically, cupric acetoarsenite.

SCI see SPECULAR REFLECTANCE INCLUDED.[43,69] cf. SCE[43,69]

science the observation, identification, description, experimental investigation, and theoretical explanation of a phenomena.[144,142,173]

scorching premature VULCANIZATION of a RUBBER COMPOUND, occurring during the mixing operation when the compound is calendered or EXTRUDED.[67,140] Syn: burning

Scorch is often controlled or prevented by selection of proper ACCELERATORS or by use of RETARDERS. Scorched or burnt stock is generally not processible.

scotopic vision vision mediated by rods alone at very low levels of LUMINANCE; night vision.[43,173,69]

scraper tool for SCRAPING off paint or other ADHERENT matter.[54,59]

scratch coat in three-coat PLASTERING, the first coat of plaster, which is then scratched to provide a bond for the second BROWN COAT.[73,75] (DAC)

screen in PRINTING, those areas of a plate or the print made from it in which tonal gradations are reproduced.[168]

screen analysis or test see SIEVE ANALYSIS.[131,157]

screen painting see SILK-SCREENING.[77,81]

screen print see SILK-SCREENING.[148,81]

screen process printing see SILK SCREENING.[71,81,148]

screening see SIEVING.[131]

scrub board see BASEBOARD.[82]

scrub resistance the ability of a coating to RESIST being worn away or to maintain its original APPEARANCE when RUBBED repeatedly with an ABRASIVE material.[72,157] Typically, the rubbing procedure can employ a BRUSH, SPONGE, or CLOTH, wetted with an abrasive SOAP solution, in which case it is more accurately referred to as wet scrub resistance. Syn: wet ABRASION resistance

Scrub (wet abrasion) resistance is a measure of FILM INTEGRITY. Interior wall paints often become soiled especially near doorways, windows, and in work and play areas.

ASTM Test Method D2486 covers the determination of the relative resistance of different wall paints to erosion when repeatedly scrubbed to remove stains during the life of the paint. Essentially it is a measure of film integrity. Results developed by use of this test method do not necessarily represent the scrub resistance of aged paint films. This test method measures scrub resistance by the traditional scrub-to-failure concept.

ASTM Test Method D4213 is a more precise, albeit more time consuming, method that measures essentially the same property by a weight-loss technique and reports it as a volumetric film erosion rate.

scrubber an apparatus used in sampling and in FLUE gas cleaning.[59,171,172] (IUPAC)

Words presented in CAP/SMALL CAPS type indicate that the word is defined in another part of the *Coatings Encyclopedic Dictionary.*

Numerical superscripts classify terms in one or more of the categories listed in the second section of this volume.

The gas is passed through a space containing wetted "packing" or SPRAY. In general, PARTICLES are collected in scrubbers by one or a combination of the following: impingement of particles on liquid medium; DIFFUSION of particles onto liquid medium; CONDENSATION of liquid medium; vapors on the particles; partitioning of the gas into extremely small elements to allow collection of the particles by BROWNIAN diffusion and gravitation settling on the gas-liquid interface. The devices include spray towers, jet scrubbers, Venturi scrubbers, cyclonic scrubbers, inertial scrubbers, mechanical scrubbers, and packed scrubbers. Normally the gas FLOW in the scrubber is counter to the liquid flow. Efficient scrubbers will collect particles as small as 1 to 2 μm in diameter.

scrubbing a PROCESS used in gas sampling or gas cleaning in which components in the gas stream are removed by contact with a liquid surface or a wetted packing, on SPRAY drops, droplets, or in a bubbler, etc.[131,171,172] (IUPAC)

sculptured wallcovering products molded of solid (usually SYNTHETIC) material in which the TEXTURE and shadows are real.[148]

scumble glaze TRANSPARENT preparation used in the SCUMBLING PROCESS.[54,71] (BSI)

scumble stain SEMITRANSPARENT STAIN for application over an OPAQUE groundwork of paint. (BSI)

BRUSH, STIPPLE, or sponge may be used for manipulating the scumble, or it may be COMBED, so that various effects, i.e., WOOD GRAINING and other more formal patterns, are possible. In this, the nonflowing property of the scumble greatly assists.

scumbling in painting, the operation of lightly RUBBING a BRUSH containing a small quantity of opaque or semiopaque COLOR over a surface to soften and blend TINTS that are too bright, or to produce a special effect; the coat may be so thin as to be SEMITRANSPARENT.[54,81,148] (DAC) cf. TIFFANY FINISH[54,148]

scumming see GREASING.[168]

scuttle a hatch that provides access to the roof from the interior of the building.[76,82] (ASTM)

SDC abbreviation of the Society of Dyers and Colourists (in the United Kingdom).[174]

seal coat, sealing coat see SEALER.[147]

sealant a mixture of POLYMERS, FILLERS, and PIGMENTS used to fill and seal joints where moderate movement is expected; unlike CAULKING, it cures to a resilient solid.[165] (ASTM)

These are soft enough to pour or EXTRUDE and are capable of subsequent hardening to form a permanent bond with the SUBSTRATE. Most sealants are synthetic polymers (POLYCHLOROPHENES, SILICONES, POLYURETHANES, ACRYLICS), which are semisolid before application and later become elastomeric. A few of the best known are natural products such as LINSEED OIL, ASPHALT, and various WAXES.

sealer (1) a liquid coat which SEALS WOOD, PLASTER, etc., and prevents the surface from absorbing paint or VARNISH; may be TRANSPARENT; may act as a PRIMER for a following coat or as a FINISH for the surface; (2) a coat, applied in liquid form, which is laid over a TAR-like substance to prevent its BLEEDING through an applied paint film.[147,158] (DAC) See SIZE, and FLOOR SEALER.[147,158] Syn: seal coat or sealing coat

sealing coat see SEALER.[147]

seamless flooring fluid or trowel-applied flooring without AGGREGATES.[71,54,163] Syn: monolithic flooring

seams in hanging WALLCOVERINGS, there are three methods of joining seams. All three are satisfactory, but the "BUTT" method produces the smoother finished job and is most often preferred; (1) in a "butt" the seam edges are fitted tightly together without any overlap. If the WALLCOVERING comes untrimmed, SELVAGES must be cut off at both sides of the roll. This leaves a flat, invisible seam, with no double thickness; (2) in the "LAP" method, one strip is lapped over the selvage of another. Wallcoverings are usually hung from left to right. In moving from left to right, the selvage is left intact on the right and trimmed on the left. The clean edge overlaps and covers the selvage edge. This procedure would be reversed in hanging from right to left; (3) the "wire-edge" method is a version of the "butt" technique, but will overlap about one sixteenth of an inch (1.5 mm) into the pattern, so that no wall shows through in case the butting is not perfect. Here, also, both selvages must be removed.[148]

sebacic acid $(CH_2)_8(COOH)_2$. Used for the preparation of sebacate PLASTICIZERS and as a STABILIZER for ALKYDS.[144]

Properties: mp, 133 °C; bp, 295 °C/100 mm Hg.

secco the art or an example of painting on dry PLASTER.[71,81]

Paints used for secco require a BINDER unlike fresco painting which is painted over moist setting LIME plaster, with water-based PIGMENTS.

secondary amine value the number of milligrams of potassium hydroxide (KOH) equivalent to the secondary AMINE basicity in 1 g of sample.[128,157] (ASTM)

secondary pollution (emission) the products of primary POLLUTANTS which form through photochemical and thermal reactions in the atmosphere (O_3, peroxyacetyl nitrate, etc.).[171,172] (IUPAC)

secondary reference standard a STANDARD which has been CALIBRATED relative to a primary standard and which, when used with the calibration values, essentially reproduces the primary standard.[43,69]

secondary treatment the second step in most waste-treatment systems, in which BACTERIA consume the ORGANIC parts of the wastes.[171]

sediment any solid which can settle or be centrifuged from the main portion of the liquid, for example, FOOTS, meal, grain, GUM.[161,64] (ASTM)

sedimentation (1) in the atmospheric sciences, the PROCESS of removal of an airborne particle from the atmosphere due to the effect of gravity;[171,172] (2) see SETTLING and PHASE SEPARATION.[42,61]

seed lac form of SHELLAC obtained from the original crude product, by removing twigs and other extraneous matter, and washing thoroughly with water.[146]

seediness (1) PARTICLES in a coating caused by REACTIVE PIGMENTS and acid components of the vehicle; (2) DEFECT in a CLEAR VARNISH or LACQUER caused by small PARTICLES sometimes VISIBLE when examined by TRANSMITTED LIGHT. On application, varnished or lacquered surfaces may present a BITTY, specky, or sandy APPEARANCE due to this defect.[56,42] (BSI)

seeding formation of small, undesirable PARTICLES or granules in a paint, VARNISH, or LACQUER.[42,56]

seeds see SEEDINESS.[56,42]

seedy descriptive of a paint FINISH that is not smooth owing to undispersed PIGMENT PARTICLES or insoluble gel particles in the paint.[42,56] (DAC) See BITTY.[42,56]

seeing into the film see DEPTH OF FINISH.[43]

segregated oils DRYING OILS which are obtained from the semidrying types by removal of nondrying constituents.[63]

This removal may be effected by selective CRYSTALLIZATION, SOLVENT EXTRACTION, or VACUUM DISTILLATION.

selected ordinate method method for determining the CIE TRISTIMULUS VALUES by adding the REFLECTANCES or TRANSMITTANCES at unequally spaced intervals of WAVELENGTH, so selected for each of the tristimulus values that the summation of the reflectances or transmittances give the same tristimulus values obtained by integration of weighted regularly spaced wavelength values.[43,157,69] See WEIGHTED ORDINATE METHOD.[43]

The accuracy of this technique is poor relative to that obtained with the weighted ordinate method unless an extremely large number of readings at different wavelengths is used (e.g., more than 100 for each tristimulus value).

selenium red see CADMIUM RED.[41]

self-adhesion see AUTOHESION.[79]

self-crosslinking see SELF-CURING, and AMBIENT CURE.[67,55]

self-curing undergoing CURE (crosslinking) without the application of heat.[67,55] See SELF-VULCANIZING, and AMBIENT CURE.[67]

self-etch primer see WASH PRIMER.[71]

self-lifting resistance the ability of dried films of ORGANIC coating materials to RESIST DEFORMATION upon application of an additional coat of the same material.[72,42]

self-luminous paint see LUMINOUS PAINT.[71]

self-polishing wax see LIQUID EMULSION POLYMER COATING, and LIQUID WATER EMULSION WAX.[137]

self-priming use of same coating for PRIMER and for subsequent coats.[54,158]

It may be thinned differently for the various coats.

self-sealing paint a paint which, when applied over a surface of varying POROSITY, SEALS the surface and yet dries with a uniform COLOR and SHEEN.[71] (DAC)

self-thermosetting see SELF-CURING.[67]

self-tone WALLCOVERING in which shades of one COLOR are featured.[148]

Habitually used for damasks and in FLOCKING.

self-vulcanizing pertaining to an ADHESIVE which undergoes VULCANIZATION without the application of heat.[67,69,79] (ASTM) See SELF-CURING.[67]

selvage (1) either edge of a roll of WALLPAPER, carrying no design, and intended to protect the latter, as well as give instructions for MATCHING or any other information; (2) an edge or edging which differs from the main part of: (a) a FABRIC; or (b) granule-surfaced ROLL ROOFING.[76] (ASTM) Also spelled "selvedge."

selvedge see SELVAGE.[148]

semicure a preliminary, incomplete CURE given to certain RUBBER articles to cause the rubber to set in some desired shape.[67,140]

Full VULCANIZATION is subsequently completed by a final cure.

semidrying oil oil which possesses the characteristics of a DRYING OIL but to a lesser degree; for example, upon being exposed to oxygen it only par-

Words presented in CAP/SMALL CAPS type indicate that the word is defined in another part of the *Coatings Encyclopedic Dictionary.*
Numerical superscripts classify terms in one or more of the categories listed in the second section of this volume.

tially hardens or is changed to a sticky mass.[63] (ASTM)

There is no definite line of demarcation between drying and semidrying OILS.

semigloss See GLOSS.[43,35]

semigloss lacquer LACQUER having the APPEARANCE of having been frictionally RUBBED.[43,151,35]

Usually arrived at through the introduction of a PIGMENT *(inert). A type of* SATIN FINISH.

semitransparent stain a pigmented composition intended for use on wooden surfaces to produce a uniform FINISH that does not fully obscure the GRAIN nor the texture of the WOOD.[69,71] cf. STAIN[69,71] and OPAQUE STAIN[69,71]

When applied to woods that differ in COLOR, *there is a color difference between the stained substrates.*

Senegal gum form of GUM ARABIC obtained from Senegal.[155] See ACACIA GUM.[155]

It is a natural EXUDATION *from Acacia senegal and is water-soluble.*

sensation mode of mental functioning that is directly associated with stimulation of the ORGANISM.[173,69]

sensitization a state of immune-response reaction in which further EXPOSURE elicits an immune or allergic response.[172]

A person previously exposed to a certain material is more sensitive when further exposed to it if that person had been sensitized to it originally.

sensitizer (1) an agent which cleanses the INK-repellant, nonprinting areas of LITHOGRAPHIC PRINTING plates;[168,83] (2) a material that on first EXPOSURE causes little or no reaction in man or test animals, but which on repeated exposure may cause a marked response not necessarily limited to contact site.[172]

SKIN SENSITIZATION *is the most common form. Respiratory sensitization to some chemicals is also known to occur.*

sepia dark brown artists' PIGMENT, which occurs as a natural secretion of a type of cuttle fish.[41]

sequestering agent see CHELATING AGENT.[83]

sequestration see CHELATION.[131]

sericite see ALUMINUM POTASSIUM SILICATE.[41,58]

serpentine see ASBESTOS.[41]

sesame oil pale yellow OIL, almost odorless, from the seeds of cultivated varieties of *Sesamum indium*.[63] Syn: benne, teel, or gingelli oil

Sp gr, 0.916. Saponification value, 188-193; iodine value, 103-122.

set (1) to convert an ADHESIVE into a fixed or hardened state by chemical or physical action, such as CONDENSATION, POLYMERIZATION, OXIDATION, VULCANIZATION, GELATION, hydration, or EVAPORATION of volatile constituents;[55] (2) condition of a paint or VARNISH film when it has dried to a point where, for all practical purposes, it ceases to FLOW.[67,79] See CURE and DRY.[67]

set off of ink the transferring or smearing of INK from freshly printed matter to another surface with which the undried print comes in contact, particularly from the face of one sheet onto the back of the sheet on top of it in the delivery pile or rewind roll.[168]

set-to-touch time the time required for the coating to reach a point where the adhesion to an external object is less than the internal cohesion of the film.[67,157,55] See DRYING TIME.[67,157,47]

Setaflash Tester a closed cup FLASH POINT tester in which a small sample of the material is heated in a crucible or cup, and at a definite temperature, the vapor-air mixture which forms is ignited by a DIPPING flame.[145,172,157]

ASTM Test Method D3278 covers procedures for determining whether a material does or does not flash at a specified temperature or for determining the lowest finite temperature at which a material does flash when using a Setaflash Tester. The test methods are applicable to paints, ENAMELS, LACQUERS, VARNISHES, *and related products having a flash point between 32 and 230°F (1 to 110°C) and viscosity lower than 150* STOKES *at 77°F (25°C).*

setoglaucine see RHODULINE BLUE 66.[41]

setting of ink the initial phase of the DRYING PROCESS, by slight ABSORPTION of the vehicle by the PAPER, wherein printed sheets, though not fully dry, can be handled without smudging.[168]

setting up (1) conversion of liquid paint during storage to a GEL-LIKE or pseudosolid condition. The process is usually reversible by agitation and thinning but may be permanent when chemically REACTIVE PIGMENTS or highly POLYMERIZED media are involved;[161,166] (BSI) (2) the THICKENING which occurs when paint stands in an open can;[161] (3) the increasing VISCOSITY of a paint film as it dries.[166] (DAC)

settling see PHASE SEPARATION.[41,42,161] cf. SYNERESIS

sgraffito painting form of MURAL DECORATION in which a design is produced by scratching through a WHITE or TINTED TOPCOAT of LIMEWASH to expose the darker colored MORTAR underneath.[54,71,81]

shade (1) a description of a particular HUE, or variation of a primary hue, such as a red shade blue;[69] (2) a description of the DEPTH of COLOR, such as a pale shade;[69] (3) a description of a mixture with black (or gray) as opposed to a TINT which is a mixture with WHITE;[69,43] (4) a color coat used on wood FURNITURE which gives character to the FINISH. It is often applied by hand wiping with a rag;[175,54] (EPA) (5) verb used to describe the PROCESS of making a small modification in a color by the addition of COLORANT(S);[69] (6) in INK manufacture, commonly used as a synonym for hue.[69,168] See SHADING AND BLENDING.[148]

shader see COLORIST.[69]

shading and blending altering the COLOR of a paint slightly by the addition of black tinting color to create a DECORATIVE effect of graduated colors when applied to adjacent areas.[148] (DAC) See SHADE.[69,168]

The LAP areas often are BRUSHED and ROLLED to achieve a subtle blending.

shale oil OIL obtained by DESTRUCTIVE DISTILLATION from SHALE TAR, which, in turn, is obtained by distilling shale.[76]

Shale is a BITUMINOUS type of MINERAL, mined similarly to bituminous coal, and is found in many places throughout the world. Shale oil varies in specific gravity from 0.750-0.850.

sharp paint rapid-drying paint which yields a flat film for use as a seal coat.[71]

It is usually highly pigmented and contains a minimum amount of BINDER. Sharp paints dry off to a TACK-FREE condition chiefly by EVAPORATION of SOLVENT. They are used as PRIMING or SEALING coats.

shear an action or STRESS, resulting from applied forces, which causes (or tends to cause) two contiguous parts of a body to slide, relative to each other, in a direction parallel to their plane of contact.[47,166] (ASTM)

shear modulus the ratio of the SHEAR STRESS to the corresponding shear strain for shear stresses below the proportional limit in shear of the material.[46] (ASTM)

shear rate, y the time rate of change of shear strain. For a one-dimensional shear FLOW, it is the velocity gradient.[47,166] (ASTM)

shear thickening (dilatancy) FLOW characterized by an increase in VISCOSITY as shear rate increases. True dilatancy involves an increase in the volume of the fluid.[166] Syn: dilatant flow

Dilatancy in a PIGMENT-vehicle system or pigment dispersions commonly results from high pigment loadings.

shear thinning FLOW characterized by a decrease in VISCOSITY as shear rate is increased. Usually due to PSEUDOPLASTIC FLOW and often considered as a synonym for that kind of behavior.[166] See THIXOTROPY.[166]

sheariness variation in GLOSS or SHEEN on a paint surface which should have been uniform in this respect.[43,56] (BSI)

This variation is due to differences in film thickness.

shearing strength the maximum SHEAR STRESS which a material is capable of sustaining.[47] (ASTM)

Shear strength is calculated from the maximum load during a shear or torsion test and is based on the original dimensions of the cross section of the specimen.

shearing stress the tangential SHEARING FORCE which is acting on a material to produce motion or FLOW in it.[47,166]

sheary see SHEARINESS.[43]

sheathing structural INSULATING board for use in housing and other building construction, which may be integrally treated, impregnated, or coated to give it additional water RESISTANCE.[75,159] (ASTM)

sheen an attribute of object mode of APPEARANCE which is similar to LUSTER; GLOSS with poor DISTINCTNESS-OF-IMAGE REFLECTANCE.[43,35] See GLOSS and LUSTER.[43]

In the paint industry the term sheen is generally used synonymously with gloss measured or observed at a grazing angle, such as 85° off the perpendicular. Sheen is therefore frequently evaluated in terms of gloss measurements made on an 85° glossmeter. Not infrequently there are coatings that are high in sheen, but have a low gloss when ILLUMINATED and viewed at or near the perpendicular.

sheepskin paint roller cover a cover in which the paint applicating material is wool fleece still attached to its tanned natural SKIN.[59] (ASTM)

shelf life the period of time in which a material may normally be stored and still be in a usable condition.[161,42] cf. STORAGE STABILITY.[161,42] Syn: staging life and pot life (1)

shellac ALCOHOL-soluble, orange-colored RESIN which is the secretion of the female of the insect (*Laccifer lacca*) found in great quantities in India and Indochina.[146]

Shellac appears as an INCRUSTATION on the twigs of certain trees. It forms the basis of FRENCH POLISH and several other types of SPIRIT VARNISH and is used for ADHESIVE purposes, because of its THERMOSETTING properties.

shellac, bleached see BLEACHED SHELLAC.[146]

shellolic acid very complex constituent of SHELLAC believed to consist of a tricyclic nucleus, with two carboxyl and two hydroxyl groups.[144,146]

Words presented in CAP/SMALL CAPS type indicate that the word is defined in another part of the *Coatings Encyclopedic Dictionary.*
Numerical superscripts classify terms in one or more of the categories listed in the second section of this volume.

sherardizing method of coating STEEL or cast IRON articles in intimate contact with ZINC.[58,71]

The articles are heated with zinc DUST for several hours. The zinc forms an alloy, at the interface of the ferrous surface, thus producing a thin, tightly adherent coating.

Sherardizing is used mainly for nuts, bolts, and other small parts. The coating has a gray matte APPEARANCE and consists mainly of zinc-iron alloy. It is made by treating the parts, after cleaning by PICKLING or GRIT BLASTING, in rotating steel drums or fixed containers containing ZINC DUST and sometimes sand, heated to about 400°C for three to five hours. The process is particularly suitable for threaded parts because the zinc penetrates deeply into recesses, and the parts mate accurately while retaining full coating thickness.

shingle (1) a small unit of prepared roofing designed for installation with similar units in overlapping rows on inclines normally exceeding 25%; (2) to cover with shingles; and (3) to apply any sheet material in overlapping rows like shingles.[75,76] (ASTM)

shingle stain a low-viscosity, pigmented, PENETRATING paint for use on WOOD SHINGLES to provide COLOR and protection against MOISTURE PENETRATION.[71,147] (DAC)

shingling (1) the procedure of laying parallel FELTS so that one longitudinal edge of each felt overlaps, and the other longitudinal edge underlaps, an adjacent felt. Normally, felts are shingled on SLOPE so that the water flows over rather than against each lap; (2) the application of SHINGLES to a sloped roof.[76] (ASTM) See PLY.[76]

shop coat one or more coats applied in a shop or plant prior to shipment to the site of erection or fabrication, where the field or finishing coat is applied.[54,158]

shop painting see SHOP COAT.[54,158]

shop primer see PREFABRICATION PRIMER.[71,158]

shop priming act of PRIMING new WOOD or METAL on the maker's premises prior to delivery to the purchaser.[71,158] (BSI)

short lacking TOUGHNESS or ELASTICITY.[166,47]

A dried film is referred to as being short when it is quite BRITTLE. The undried material is referred to as being short when it is crumbly and has no flowing properties; opposite of length.

short ink a highly pigmented INK which is stiff and cannot be drawn out to a thread.[168]

short oil alkyd an alkyd containing less than 40% oil in solids.[167,156] (DAC)

short oil varnish a VARNISH containing little OIL in comparison with the amount of RESIN present, less than 15 gal oil per 100 lb (1.25 litres oil per kg) resin.[167] (DAC)

shot blasting see SHOT PEENING.[158]

shot peening blasting with round iron shot, round steel shot, or any material which retains its spherical shape for peening purposes.[158] Syn: shot blasting

show card colors see POSTER COLOR.[71]

show-through (1) see PHOTOGRAPHING;[42] (2) the TRANSPARENCY of PRINTED sheets which permits printing to be seen from the other side of the sheet.[168]

shrinkage disruption of the level plane of the finished surface, resulting in a loss of GLOSS or WRINKLING.[56,142,159]

Contraction of the WOODEN SUBSTRATE, frequently resulting in CRACKING and/or CHECKING of the coated surface.

shrivel finish see RIPPLE FINISH.[163]

shutter a movable screen or cover used to cover an opening, especially a window.[82,75] (DAC)

SI units Systeme International d'Unités.[128]

An internationally agreed coherent system of UNITS, derived from the M.K.S. SYSTEM, now in use for all scientific purposes and thereby replacing the C.G.S. SYSTEM and the F.P.S. SYSTEM. The seven basic units are: the metre (symbol m), kilogram (kg), second (s), ampere (A), Kelvin (K), mole (mol), and candela (cd). The radian (rad) and steradian (sr) are supplementary units. Derived units include the hertz (Hz), newton (N), joule (J), watt (W), coulomb (C), volt (V), farad (F), ohm (Ω), weber (wb), tesla (T), henry (H), lumen (lm), and lux (lx).

siccative any REAGENT which CATALYZES or promotes OXIDATION of OILS; a DRIER.[74,83,67]

side seam sealer a COMPOUND for protecting the side seams of three piece cans.[147] (EPA)

siding the FINISH covering an exterior wall of a frame building; the siding may be a cladding material such as wood, aluminum, or vinyl (but not masonry); applied vertically or horizontally.[75,159] See CLAPBOARD, DROP SIDING.[75,159] (DAC) Syn: weatherboarding

siding shingle a shingle of any of a number of materials such as WOOD, aluminum vinyl, etc., used as a protective exterior wall covering over sheathing.[75] (DAC)

Sierra Leone copal fossil COPAL of African origin.[167]

It becomes soluble in DRYING OILS after RUNNING,

and because of its pale COLOR, *which is maintained after running, is used to some extent in* GRINDING *media for whites and tinted whites, silver baking* VARNISHES, *etc.*

sieve in laboratory work, an apparatus in which the apertures are square, for separating sizes of material.[145,157] (ASTM)

sieve analysis a determination of the proportions of PARTICLES lying within certain size ranges in a granular material by separation on SIEVES of different size openings.[157] (DAC) Syn: screen analysis

sieve number a number used to designate the size of a SIEVE, usually the approximate number of sieve cross wires per linear inch.[128] (DAC)

sieving separation of a mixture of various-sized PARTICLES, either dry or suspended in a liquid, into two or more portions, by passing through screens of specified MESH sizes.[131] See SIEVE ANALYSIS.[131] Syn: screening

sigma blade mixer see INTERNAL MIXER.[59,131]

significant digit any digit that is necessary to define a value or quantity.[36] (ASTM)

silanes group of compounds resembling the SATURATED HYDROCARBONS, in which the CARBON is replaced by silicon.[156]

For example, monosilane SiH_4, *cf. methane* CH_4; DISILANE Si_2H_6, *cf.* ETHANE C_2H_6.

silica, amorphous see AMORPHOUS SILICA.[41,58]

silica, crystalline SiO_2. Pigment White 27 (77811). PIGMENT grade CRYSTALLINE SILICA is quartz sand that has been crushed, ground, and/or air classified. It is a reinforcing FILLER for paints, elastomers, sealants.[41] cf. AMORPHOUS SILICA[41,58] Syn: powdered quartz, silica flour, ground silica

Chemically inert, heat resistant, nonconductive; imparts BURNISH RESISTANCE, SHEEN UNIFORMITY, *good* FLATTING *in* LATEX *paints. Properties: density, 2.65 g/cm³ (22.1 lb/gal); O.A., 24-36; particle size, 1.5-9.0 μm.*

silica, diatomaceous see DIATOMACEOUS SILICA.[58]

silica flour see SILICA, CRYSTALLINE.[41,58]

silica gel a regenerative ABSORBENT consisting of the AMORPHOUS SILICA manufactured by the action of HCl on sodium silicate.[74]

Hard, glossy, quartz-like in APPEARANCE. *Used in dehydrating and in drying and as a* CATALYST *carrier.*

silica, microcrystalline SiO_2. Pigment White 27 (77811). EXTENDER PIGMENT obtained from extensive geologic formations of hard, compact, homogeneous, microcrystalline SILICA located in the vicinity of Hot Springs, Arkansas.[41,58] Syn: novacite

Can be considered as a very finely divided or "micro" form of quartz in a bound state of subdivision. Excellent dielectric and low ABRASIVE *properties. Properties: density, 2.65 g/cm³ (22.1 lb/gal); O.A., 20.0; mean particle size, 4.0 μm. "Novaculite" is a generic name.*

silica, soft see AMORPHOUS SILICA.[41]

silica, synthetic (aerogel and hydrogel) SiO_2. Extremely POROUS and lightweight synthetic silicas which may contain up to 94% of "dead air" space that is enclosed by a tenuous webbing of microscopic SILICA FILAMENTS.[41,58] See SODIUM SILICOALUMINATE.[41,58]

silica, synthetic (pyrogenic) SiO_2. A COLLOIDAL SILICA used in the RUBBER, PLASTIC, and paint industry.[41] Syn: fumed silica

High purity, nonreactive PIGMENT *which imparts* THIXOTROPY, FLATTING, *and pigment* SUSPENSION. *Properties: density, 2.2 g/cm³; O.A., 280; particle size, 0.012 μm.*

silicate any member of the very widely occurring COMPOUNDS characterized by the presence of the elements, silicon, oxygen, and one or more metals with or without hydrogen (e.g., TALC).[41,58,34]

silicate paints water paints based on sodium, potassium, or lithium silicate. Used in ZINC-RICH paints.[71]

They are characterized by their nonflammability. Care must be exercised in the selection of pigments used with the silicate because of its alkalinity.

silicon carbide abrasive of the many CRYSTALLINE forms of silicon carbide produced in electric furnaces, the hexagonal, or alpha crystals made up of masses of interlocking crystals, is the major type used for coated ABRASIVES.[158]

It is greenish black iridescent, has high thermoconductivity, low thermoexpansion, and great chemical STABILITY, *being unaffected by nitric, sulfuric, hydrochloric acids.*

silicon carbide waterproof paper hard, sharp, coated ABRASIVE particularly suited for LACQUER, PLASTICS, and nonferrous metals, also for WET SANDING primers, undercoats, and between coats on wood and METAL using water, oil, or other lubricants.[59,158] See COATED ABRASIVE.[59,158]

silicone one of a class of COMPOUNDS comprising POLYMERIZABLE, high temperature-resistant RESINS, lubricant greases, and OILS, ORGANIC SOLVENT-soluble

---

Words presented in CAP/SMALL CAPS type indicate that the word is defined in another part of the *Coatings Encyclopedic Dictionary.*

Numerical superscripts classify terms in one or more of the categories listed in the second section of this volume.

water repellents, SURFACE TENSION MODIFIERS for ORGANIC SOLVENTS, etc.[83,130] (BSI)

Unlike organic compounds, which all contain CARBON, the basic structure of silicones consists of silicon-oxygen linkages, which are very stable to heat and other influences.

Silicone ADDITIVES offer MAR RESISTANCE, promote ADHESION and water RESISTANCE, help with LEVELING, and minimize FOAMING of paint.

silicone paint paint, based on SILICONE RESINS, that is RESISTANT to very high temperatures and therefore useful on smokestacks, heaters, stoves, and electrical insulation; requires heat to CURE or set; has a high resistance to chemical attack.[71,72,163] (DAC)

silicone plastics PLASTICS based on SILICONE RESINS.[38]

silicone resins group of RESINS containing a substantial amount of silicon, distinguished by their outstanding heat resistance, high water repellancy, and chemical resistance.[130,156]

They are made by preparing dialkyl dichlorsilanes from silicon tetrachloride and the corresponding alkyl magnesium bromide. The SILANES are converted to silanediols, which are POLYMERIZED into resinous products. Mixtures of silanediols and triols are COPOLYMERIZED to yield THERMOSETTING resins.

silicosis a chronic disease of the lungs caused by the continued inhalation of SILICA DUST.[171,172]

A disease due to breathing air containing silica characterized anatomically by generalized fibrous changes and the development of miliary modulation in both lungs, and clinically by shortness of breath, decreased chest expansion, lessened capacity for work, presence of fever, increased susceptibility to tuberculosis, and by characteristic X-ray findings.

silk See GLOSS.[43] British for SATIN.

silk-screen ink quick drying, full-bodied INK used in the silk-screen PRINTING PROCESS.[168]

silk-screen printing see SILK-SCREENING.[77,81,148]

silk-screening PRINTING PROCESS by which repeats of a motif or pattern can be made; the process involves the use of silk (or similar material of fine mesh) tightly stretched on frames; a separate frame is used for each COLOR.[77,81,148] See FLAT SCREEN PRINTER.[77,81,148]

Areas of the design to be printed in a given color are left open, while the rest of the silk is treated with an insoluble coating. The ground-coated PAPER, FABRIC, or other material on which the design is to be reproduced is laid out on long tables. The pattern is reproduced by drawing a suitable INK or paint across the screen with a RUBBER squeegee, which forces the color through the parts where the mesh is exposed, and the operator moves from one preset point to the next. By careful registering, a number of screens can be used in succession over the same design for multiple-color work.

silking surface DEFECT characterized by parallel irregularities left on or in the dried surface of a GLOSSY paint or VARNISH film, producing the APPEARANCE of silk.[56,35] (BSI)

In DIPPING or FLOW coating, the irregularities appear in the direction of the flow; in BRUSHING, in the direction in which the film was finally brushed. A stronger solvent or certain silicones can sometimes minimize or eliminate the problem.

sill the horizontal member that bears the upright portion of a frame; especially the base of a window.[82] See CILL.[82]

silver ink a PRINTING INK whose principal PIGMENT consists of aluminum powder.[168]

silver leaf beaten silver, used in silver leafing much in same manner as GOLD LEAFING.[148]

silver point method of drawing which consists of working with a silver tipped instrument on a specially prepared PAPER.[59,81]

The result is a drawing of great delicacy. Generally used by late medieval and renaissance artists, e.g., Leonardo da Vinci.

silver white (1) any of several white pigments used in paints; (2) a very pure variety of WHITE lead; FRENCH WHITE, CHINA WHITE.[41] (DAC)

Simon-Goodwin charts a series of charts for graphically calculating COLOR differences using the modified MACADAM COLOR DIFFERENCE EQUATION.[43,69] See FOSTER CHARTS.[43,69]

simple liquid see NEWTONIAN LIQUID.[166]

single roll see ROLL OF WALLCOVERING.[148]

single spread see SPREAD.[79]

sink (1) in thermodynamics, the part of a system from which heat, or more generally, energy is removed from the system; (2) in atmospheric chemistry, the sink is the receptor for material which is removed from the atmosphere.[171,172] (IUPAC)

Because of long range transport of many pollutants such as SO_2, sulfuric acid and its SALTS, the sink region can be many hundreds of kilometers from the source region of the pollutants.

sinkage the blotchy effect or loss of GLOSS due to ABSORPTION of the medium of a finishing coat by the UNDERCOAT.[43,56] Syn: sinking in

sinking in see SINKAGE.[43,56]

sintering (1) the COALESCENCE of solid PARTICLES;[131] (IUPAC) (2) the PROCESS by which fly ash produced in the combustion of fuels such as coal, is baked (sintered) at a very high temperature.[171,172,176]

The sintered material is used in the manufacture of CINDER BLOCKS and other CERAMIC products. This term is also relevant to the enrichment of low grade ores and preparation of the charge for, e.g., a blast furnace.

SIP abbreviation for STATE IMPLEMENTATION PLANS.[172] (EPA)

sissing see CISSING.[56,158]

size (1) usually a liquid composition to prevent excessive ABSORPTION of paints into PLASTER, old wall paint, and similar POROUS surfaces; also a liquid composition used as a first coat on METAL to improve ADHESION of succeeding coats (latter usage is limited to the metal decorating industry). The terms "SEALER" or "SIZE" are almost synonymous, but usage has established certain differences. A sealer is ordinarily a thin VARNISH or clear LACQUER and is usually applied on wood and metal surfaces. Ordinary painter's size is a thin solution of glue, starch, or other water-soluble substance and is usually applied on plaster surfaces, but size used in metal decorating is a thin varnish.[147,158] Syn: sizing; (2) an INK that dries with a sticky surface that will hold metallic or other powders;[168] (3) composition applied to PAPER to fill up the space between fibers and to produce a smooth surface;[168] (4) a sealer used to prepare the wall before WALLPAPER is applied;[148] (5) in ANTIQUING, a thin pasty varnish liquid used for adhering GOLD LEAF. It acts as a FILLER on porous surfaces and creates a more even effect with the gold leaf.[148] (ASTM)

sizing preparation of a surface for the application of paint.[147,158] Syn: SIZE

Skandinaviska Lackteknikers Förbund *[Federation of Scandinavian Paint and Varnish Technicians]* an organization composed of the technician members of the Scandinavian surface coatings associations.[174]

Founded in 1953 to promote cooperation and communication among the four Scandinavian surface coatings associations (Denmark, Finland, Norway, and Sweden), SLF publishes the monthly journal, Färg och lack Scandinavia. SLF is a member organization of the COATINGS SOCIETIES INTERNATIONAL.

skimmings (tall oil) CURD, not acidified or otherwise processed, skimmed from the black liquor of the alkaline PAPER pulp industry, from which TALL OIL is obtained.[44] (ASTM)

skin (1) film formed from a vehicle, liquid coating, or INK composition or formed during storage;[55,56,168] See SKINNING[42,67] and ANTISKINNING AGENTS;[83] (2) an ungrounded, nonwashable, low-priced grade of WALLPAPER.[148]

skinning the formation of an insoluble film (usually quite thick) on coating materials in the container.[42,67]

Skinning sometimes occurs after a paint is stored for a long period of time in a partially filled container.

skipping in COIL COATING, an irregular paint application usually occurring when improper contact is made between the applicator roll and the strip.[56,70] See STARVING OUT.[70]

skippy said of paint too heavy-bodied for uniform application, which causes the BRUSH to skip on the surface, leaving some spots insufficiently coated and others with too heavy a coating.[42,56]

skips see HOLIDAYS.[42]

skirting board see BASEBOARD.[82]

slack melt copal COPAL which has been run only to a limited extent, often at a comparatively low temperature.[167] cf. FINE MELT[131,167]

slake to combine (LIME) chemically with water or moist air.[131]

To undergo a slaking PROCESS; crumble or disintegrate, as lime.

slaked lime another name for CALCIUM HYDROXIDE, or QUICKLIME which has been slaked by addition of water.[144]

slate a hard, brittle metamorphic rock consisting mainly of CLAY minerals, characterized by good cleavage along parallel planes.[75]

Slate is used extensively as DIMENSIONAL STONE in thin sheets for flooring, roofing, panels (both DECORATIVE and electrical), chalkboard, and in granular form as surfacing on compositional roofing.

slate dust see SLATE POWDER.[58,158]

slate flour see SLATE POWDER.[158,58]

slate powder EXTENDER PIGMENT obtained from slate.[58,158] Syn: slate dust or slate flour

It is an aluminum silicate type of mineral, including a substantial amount of combined water. Used in wood and METAL FILLERS.

sleepy description of a recently applied glossy coating which has lost its initial GLOSS other than by BLOOM and has become dulled or lacking in LUSTER.[65] (BSI)

SLF see SKANDINAVISKA LACKTEKNIKERS FÖRBUND.[174]

Words presented in CAP/SMALL CAPS type indicate that the word is defined in another part of the *Coatings Encyclopedic Dictionary.*

Numerical superscripts classify terms in one or more of the categories listed in the second section of this volume.

slime a general name for any moist, sticky substance formed by FUNGI.[173]

sling psychrometer instrument containing independently matched dry- and wet-bulb thermometers suitably mounted for swinging through the atmosphere, for simultaneously indicating dry- and wet-bulb temperatures.[145,157] (ASTM)

slip (1) film of coating is said to have slip when it has negligible surface tackiness, and appears to be lubricated. *In actual fact, lubricants which are incompatible with the nonvolatile constituents of finishes are frequently added in order that they may be exuded to the surface and confer slip. Nongreasing gold and silver baking* VARNISHES *are typical products to which lubricants are added;*[47] (2) denoting the ease with which two surfaces slide in contact with each other. In a sense, it is the antithesis of friction, in that high COEFFICIENT of friction denotes poor slip and low coefficient of friction, good slip; (3) vaporized coating material emitted from SPRAY GUN.[54]

slip agent (compound) an ADDITIVE which exudes to the surface of a coating during and immediately after application and drying or baking.[83]

The additive coats the surface and provides the necessary lubricity to reduce the COEFFICIENT *of friction and improve* SLIP *characteristics.*

slip sheet a sheet of PAPER placed between two freshly printed sheets to prevent set off or BLOCKING.[168]

slitting in COIL COATING, a PROCESS by which a wide strip is slit into narrower size prior to recoiling.[70,131]

slope the tangent of the angle between the roof surface and the horizontal, expressed as a percentage, or in inches of rise per foot of horizontal distance.[76,82] (ASTM)

slow solvent SOLVENT with a slow EVAPORATION rate.[164]

slumping the PLUG FLOW of a surface layer of a coating on an inclined or vertical surface due to SAGGING of the under layer which arises from gravitational pull.[166]

slur a condition caused by slippage at the moment of impression between any two of the following: PAPER, printing plate, image carrier, or blanket.[168]

slurry a pourable mixture of liquid, generally water, and any finely divided insoluble solid such as PIGMENT, EXTENDER, or PORTLAND CEMENT.[64]

slushing PROCESS by which a coating is liberally applied to surfaces which require protection but which are more or less hidden from view and not readily accessible for painting by ordinary methods.[77] (BSI)

The paint or similar material is swilled on and the excess drained off.

slushing oil OIL used on metals to form a protective coating against RUST and tarnish.[145]

It should coat the surface completely, and yet be removable without undue labor.

SMA resins abbreviation for styrene-maleic anhydride RESINS.[130]

smalt a deep blue potassium-cobalt GLASS PIGMENT made by fusing pure sand and potassium carbonate with cobalt oxide, GRINDING and powdering.[41,148]

Smalt is sometimes applied to freshly coated surfaces to provide unusual DECORATING *effects.*

smearing the spreading of INK over areas of the plate and/or SUBSTRATE where it is not wanted.[168]

smectite dioctohedral (MONTMORILLONITE) and trioctahedral (SAPONITE) CLAY minerals, and their chemical varieties characterized by swelling properties and high cation-exchange properties.[41,58,83,34] cf. BENTONITE, ATTAPULGITE, KAOLINITE[41,58,83]

smeraldino see HYDRATED CHROMIUM OXIDE.[41]

smog a common name for the visible haze which results from air pollution containing PHOTOCHEMICAL OXIDANTS.[172] (EPA)

The word is a combination of SMOKE *and* FOG.

smoke a visible AEROSOL made up of small gasborne PARTICLES resulting from the incomplete combustion of ORGANIC matter.[171,172] (IUPAC)

It consists of CARBON, *carbon-rich products, and all other dispersible product particles from the incomplete combustion. This does not include steam (condensed water vapor).*

smoke abatement legal measures which a control agency of a community may take to enforce laws and regulations concerning SMOKE emission.[171,172] (IUPAC)

smoke alarm an instrument which can provide an objective method of continuous measurement of the smoke density.[145,171,172] (IUPAC)

Such devices are often dependent upon the attenuation of light, and they can actuate an alarm when the attenuation exceeds some predetermined amount.

smudge mixture of residues of paints to which thinners are sometimes added.[65] (BSI)

It is of unknown and variable quality and has no place in a normal paint system.

soaps METALLIC SALTS of FATTY ACIDS.[127] See SURFACTANTS[83,127] and DRIER.[83,74,67]

There is also a tendency to apply the term to metallic salts of all ORGANIC *acids.*

soapstone another name for the MINERAL from which TALC is obtained.[58,34]

sodium aluminosilicate see SODIUM SILICOALUMINATE.[41]

sodium borate $Na_2B_4O_7$. Powder or GLASS-like plates becoming OPAQUE when exposed to air. Slowly soluble in water.[144] See BORAX.[144]

sodium carboxymethyl cellulose see CARBOXYMETHYL CELLULOSE.[155]

sodium lithol red see LITHOL RED.[41]

sodium silicoaluminate $9Na_2O{\cdot}67SiO_2{\cdot}12Al_2O_3$. Ultrafine synthetic precipitate formed by reacting an aluminum SALT with an alkali silicate.[41,58] Syn: sodium aluminosilicate, synthetic silica pigment.

Properties: density, 2.1 g/cm³ (17.5 lb/gal); O.A., 125; particle size, 0.1-0.4 µm.

soffit (1) the exposed undersurface of any overhead component of a building, such as a roof overhang, an arch, balcony, beam, cornice, lintel, or vault;[75,82] (2) in WALLCOVERINGS, BORDER and SOFFIT PAPER are the same.[148] See EAVE.[75,82] See BORDER.[148]

soft distemper DISTEMPER, usually based on WHITING, lightly bound with glue, size, or other water-soluble BINDER so that the coating will withstand dry RUBBING without disturbance but can easily be removed by water washing.[71] (BSI)

soft ink PRINTING composition exhibiting an almost complete absence of tack.[168]

soft silica see AMORPHOUS SILICA.[41,58]

softener see BLENDER.[59]

softening point the temperature at which a disc of a sample held within a horizontal ring is forced downward a distance of 25.4 mm (1 inch) under the weight of a steel ball as the sample is heated at a prescribed rate in a water or GLYCEROL bath.[130,167,157,137] (ASTM) See BALL and RING METHOD.[157,137]

softwoods the botanical group of trees that have needles or scalelike leaves and are evergreen for the most part, cypress, larch, and tamarack being the exceptions.[153,173] cf. HARDWOODS[153,173]

The term has no reference to the HARDNESS of the wood. Softwoods are often referred to as conifers and, botanically, they are called gymnosperms.

soil see DIRT.[42]

soilant a discoloring substance that clings to the surface of a coating, whose dispersed COLOR component is not in solution and therefore does not penetrate the film.[42,157] (ASTM)

Soilant differs from STAIN in that the COLORANT of a stain is in solution, and therefore, can penetrate into the film.

ASTM Test Method D3450 covers the determination of the relative ease of removing soilant discolorations from the dried film of an interior coating by washing with either an ABRASIVE or nonabrasive cleaner.

soiling visible damage to materials by deposition of air pollutants.[171,172] (IUPAC)

sol a fluid colloidal system of two or more components, e.g., a PROTEIN sol, a gold sol, an EMULSION, a SURFACTANT solution above the CRITICAL MICELLE CONCENTRATION.[64,139] (IUPAC) See COLLOIDAL SOLUTION.[64,139]

solid (1) a substance having definite shape and volume; not liquid or gaseous.[133,142] (2) a printed area uniformly and completely covered with INK.[168]

solidification point an empirical constant defined as the temperature at which the liquid phase of a substance is in approximate equilibrium with a relatively small amount of the same substance in its solid phase.[157,142] (ASTM) For test method see ASTM D1493.

Solidification point is distinguished from FREEZING POINT, which is described in ASTM Test Method D1015. An interpretation of mole percent purity in terms of freezing point is given in ASTM Test Method D1016.

solids NONVOLATILE MATTER in a coating composition, i.e., the ingredients of a coating composition which, after drying, are left behind and constitute the dry film.[157] (BSI)

solids by volume the volume of the nonvolatile portion of a composition divided by the total volume, expressed as a percent. Syn: volume solids

solubility the maximum amount of material, pure solid, liquid, or gaseous COMPOUND (grams, mol, etc.) which dissolve at equilibrium in a given amount of SOLVENT (100 g, 1 kg, 1 litre of solution, etc.) at a given temperature.[164] (IUPAC)

The system is at equilibrium (at a fixed temperature) when the solution phase as well as the solid, liquid, or gas phase remain in contact indefinitely without further net change in amount of either phase. The solubility of gases in water is often considered in terms of Henry's law.

solubility parameter a measure of SOLVENCY whose major usefulness is in predicting whether a particular SOLVENT will dissolve a particular RESIN; an index of compatibility of coating components.[164]

Words presented in CAP/SMALL CAPS type indicate that the word is defined in another part of the *Coatings Encyclopedic Dictionary.*

Numerical superscripts classify terms in one or more of the categories listed in the second section of this volume.

soluble blue a PIGMENT dispersible in water that is made by treating an IRON BLUE with sodium ferrocyanide or OXALIC acid and used chiefly in permanent writing INKS and laundry blues.[41,168]

soluble drier see LIQUID DRIERS; DRIER.[83,74]

solute a substance dissolved in another substance, usually the component of a solution present in the lesser amount.[164]

solution a homogeneous mixture of two or more elements or COMPOUNDS.[164,142] (IUPAC) cf. COLLOIDAL SOLUTION

The term may be applied to mixtures of solids as well as liquids, but unless stated otherwise, solution normally refers to a liquid medium (e.g., H_2SO_4 in water). A homogeneous liquid or solid phase containing more than one substance, when for convenience one of the substances, which is called the SOLVENT and may itself be a mixture, is treated differently from the other substances, which are called solutes.

solution/solvent viscosity ratio see VISCOSITY RATIO.[166]

solvation ADSORPTION of a microlayer or film of water or other SOLVENT on individual dispersed PARTICLES of a SOLUTION or DISPERSION.[164]

It also applies to the action of PLASTICIZERS on RESIN dispersions in PLASTISOLS.

solvency the degree to which a SOLVENT holds a RESIN or other paint BINDER in SOLUTION.[164] (DAC) See KAURI-BUTANOL VALUE, ANILINE POINT, SOLUBILITY PARAMETER.[164] Syn: solvent power

solvent liquid, usually volatile, which is used in the manufacture of paint to dissolve or DISPERSE the FILM-FORMING constituents, and which EVAPORATES during DRYING and therefore does not become a part of the dried film.[164] (BSI)

Solvents are used to control the CONSISTENCY and character of the FINISH and to regulate application properties.

solvent balance the condition wherein a blend of SOLVENTS and/or DILUENTS produce the desired properties of SOLVENCY and SOLVENT EVAPORATION.[164] cf. SOLVENT IMBALANCE[164]

solvent density the weight per unit volume of a SOLVENT or solvent mixture.[168,172,128] (EPA)

This number is often used in calculating the VOC emissions from coatings. Densities of common ORGANIC solvents range from 6.6 lb/gal to 9.5 lb/gal. The EPA has chosen 7.36 lb/gal as an average density of a coating SOLVENT mixture to use in some calculations.

solvent imbalance the condition wherein the ratios of SOLVENTS and/or DILUENTS are such that inadequate solvency or improper EVAPORATION of volatiles results.[164] cf. SOLVENT BALANCE[164]

solvent paste wax similar to liquid SOLVENT WAX, except furnished in PASTE form. Occasionally colored.[137] (CSMA)

solvent pop BLISTERING caused by entrapped SOLVENT.[164,56]

solvent power see SOLVENCY.[164]

solvent release the ability of a RESIN to influence the rate at which SOLVENT EVAPORATES from a coating.[164]

solvent resistance the ability of a coating to resist SOLVENT attack, solution, or disfigurement.[72,164] See SOLVENT RUB TEST.[164,157]

solvent rub test a practice for assessing the SOLVENT RESISTANCE and/or CURE of an ORGANIC coating that chemically changes during the curing process.[164,157] (ASTM)

Coatings that chemically change during the curing process, such as EPOXIES, VINYL ESTERS, POLYESTERS, ALKYDS, and URETHANES, become more resistant to SOLVENTS as they cure. These coatings should reach a specific level of solvent resistance prior to being topcoated and prior to placing in service; the levels of solvent resistance necessary vary with the type of coating and the intended service.

RUBBING the coating with a cloth saturated with the appropriate solvent is one way to determine when a specific level of solvent resistance is reached. However, the level of solvent resistance by itself does not indicate full cure and some coatings become solvent resistant before they become sufficiently CURED for service. The time required to reach a specific level of solvent resistance can be influenced by temperature, film thickness, air movement and, for waterborne or water-reactive coatings, humidity. ASTM Test Method D4752 is the preferred solvent resistance test for ethyl silicate zinc-rich primer.

solvent shock the situation wherein some of the protective vehicle is washed off the fine PIGMENT PARTICLES, allowing them to pull together into clusters or FLOCS or when flocs of RESIN form due to DILUTION with a SOLVENT or diluent of insufficient strength.[139,42,164]

solvent wax see LIQUID SOLVENT WAX.[137]

solvent-borne coating coatings which contain only ORGANIC solvents.[71,168] (EPA)

If water is present, it is only in trace quantities.

solvolysis a chemical reaction in which the SOLUTE and SOLVENT react to form a new compound.[164,144]

soot the fine black PARTICLES, chiefly composed of CARBON, produced by incomplete combustion of coal, oil, wood, or other fuels.[171]

Soot gives smoke its COLOR.

sorbate see SORPTION.[139]

sorbent see SORPTION.[139]

sorbic acid $CH_3CH = CHCH = CHCOOH$. Naturally occurring UNSATURATED monobasic acid.[144]

Mp, 134°C. Used to improve characteristics of DRYING OILS and to improve ALKYDS.

sorbitol $C_6H_8(OH)_6$. Polyhydric ALCOHOL which has attracted some interest as a component of SYNTHETIC DRYING OILS and ALKYD-type RESINS and SURFACTANTS.[144]

Properties: sp gr, 1.47; mp, 97.5°C.

sorption process by which a substance (sorbate) is sorbed (adsorbed or absorbed) on or in another substance (sorbent).[139] (IUPAC) See ABSORPTION, ADSORPTION, and DESORPTION.[139]

It is sometimes difficult or impossible to discriminate experimentally between adsorption and absorption: in such cases it is convenient to use the noncommital term sorption (together with its derived terms, sorbent, sorbate, sorptive).

sorptive see SORPTION.[139]

SO_x oxides of sulfur where x equals the number of oxygen atoms.[172]

soya bean oil a pale yellow SEMIDRYING OIL EXTRACTED from the bean of the soybean plant, *Soja hispida*, a native of Asia, but also grown extensively throughout the world. Its DRYING properties are inferior to those of LINSEED but may be improved by suitable PROCESSING.[63,156] Syn: soybean oil

When refined, it finds wide use as a component in both exterior and interior paints, but its widest use is in the preparation of ALKYDS. Coatings made with soybean oil are less prone to YELLOWING than those based on LINSEED OIL.

spachtling see SPACKLE.[147,158]

Spackle® a PASTE, COMPOUND, or POWDER which can be mixed into a paste; used to fill holes, cracks, and defects in wood, plaster, wallboard, etc., to obtain a smooth surface prior to painting.[147,158] (DAC) cf. JOINT COMPOUND[147,158] Syn: spachtling, spackling, and sparkling

Spanish ocher see IRON OXIDES, NATURAL.[41]

Spanish red oxide FERRIC OXIDE MINERAL PIGMENT from the Malaga district of Spain.[41,34]

Of bright shade, with good HIDING POWER. Ferric oxide, 82-86%.

Spanish white synonym for PARIS WHITE.[41] See WHITING and CALCIUM CARBONATE, NATURAL.[41]

spanishing a method of depositing INK in the valleys of EMBOSSED PLASTIC FILM.[168]

spar varnish exterior VARNISH with good water RESISTANCE, capable of resisting WEATHERING.[152]

Obtained its name from use in coating ships' spars.

spark-testing method of detecting HOLIDAYS (flaws) on metallic substrates by means of spark-test tool.[157,56]

sparkling see SPACKLE.[147,158]

sparkproof tools bronze beryllium tools or other nonmetallic tools to prevent spark formation in potentially explosive atmospheres, as with concentrations of HYDROCARBON SOLVENT VAPORS within the EXPLOSIVE LIMITS.[172,59]

spatter small droplets of paint that are flung off a paint roller while rolling out a paint.[42,166]

These droplets are from the filaments of a paint (between roller and SUBSTRATE) which form while rolling. Spattering is an important issue in the Do-It-Yourself market, where lack of spatter is promoted as being beneficial. The use of ASSOCIATIVE THICKENERS in LATEX paints minimizes or eliminates this problem.

ASTM Test Method D4707 can be used to measure paint spatter resistance to roller application.

spatter finish a DECORATIVE effect wherein spots, globules, or spatters of contrasting COLOR appear on, or within the surface of, a white or differently colored background.[71,54]

The effect may be achieved through the use of a MULTICOLOR, multiphase single coat or by SPATTERING the contrasting colors onto a basecoat.

spatter resistance see ROLLER SPATTER RESISTANCE.[42,72]

spattering in ANTIQUING, applying small specks of COLOR to simulate textural qualities in WOOD.[148,175]

Usually done with a dry BRUSH and little paint, or by flicking a toothbrush.

SPCC abbreviation for Spill Prevention, Control, and Countermeasure plan.[172]

SPE abbreviation for Society of Plastics Engineers.[174,38]

special boiling point spirits see S.B.P. SPIRITS.[164]

special industrial solvents a special class of PROPRIETARY SOLVENTS manufactured in accordance with authorized formulas.[164]

They can be sold only for industrial or manufacturing use in quantities of 50 gallons or more, but otherwise no permit is required in the United States.

Words presented in CAP/SMALL CAPS type indicate that the word is defined in another part of the *Coatings Encyclopedic Dictionary.*

Numerical superscripts classify terms in one or more of the categories listed in the second section of this volume.

special-purpose coatings these generally include: AEROSOLS, arts and crafts, automotive and machinery refinishing, bridge maintenance, metallics, multicolored, roof, swimming pool, and traffic marking.[163,71,159] cf. SPECIALTY COATINGS[163] in the Categorical Listing section, ARCHITECTURAL COATINGS,[71] and PRODUCT COATINGS[71,159]

specialty printing PRINTING on such items as food packages and other than printing of publications.[168] (EPA)

species in chemistry, a given kind of atom, molecule, or RADICAL which has a characteristic chemical structure and composition.[144] (IUPAC)

A chemical species is a set of chemically identical atomic or molecular structural units in a solid array or of chemically identical molecular entities that can explore the same set of molecular energy levels on the time scale of the experiment.

specific conductance see CONDUCTIVITY.[61]

specific gravity weight of a given volume of any substance, compared with the weight of an equal volume of water.[128] See DENSITY and RELATIVE DENSITY.[128]

In the SI, the term "RELATIVE DENSITY" is preferred.

specific heat the number of JOULES required to raise the temperature of one gram of material one degree CELSIUS.[128,142]

Conversely, the amount of heat liberated in cooling one degree Celsius.

specific humidity see HUMIDITY RATIO.[128]

specific intensity specific intensity of a retroreflector is the ratio of the luminous intensity of the reflector at a given observation angle to the illumination of the projected area (perpendicular to the direction of illumination) of the reflector.[43,128]

The units may be expressed as candles/footcandle.

specific luminance specific luminance of a retroreflector is the ratio of the luminance of the reflector to the illumination of its actual area.[43]

specific permeability (of a film to moisture) the milligrams of water that permeate 1 sq. cm of film of 1 mm thickness each 24 hours after a constant rate has been attained under the preferred conditions of 25°C and using 100% relative humidity inside the cup and a phosphorous pentoxide desiccated atmosphere outside the cup.[142,157,128]

ASTM Test Method D1653 requires that a moisture differential be established across the test specimen, and an apparatus provided which will permit weighing the amount of moisture which has permeated through the film under the test conditions. It covers the determination of the rate at which water vapor passes through films of paint, VARNISH, LACQUER, and other ORGANIC coatings. The films may be free films or they may be applied to porous substrates.

specific resistance (resistivity) the ratio of the D.C. potential gradient in volts per centimeter paralleling the current FLOW within the specimen to the current density in amps per square centimeter at a given instant of time and under prescribed conditions.[61] (ASTM) cf. CONDUCTIVITY[61]

This is numerically equal to the resistance between opposite faces of a centimeter cube of the liquid. The units are ohm centimeters (or megohm centimeters). It is the reciprocal of conductivity.

specific volume reciprocal of specific gravity.[128,142] Syn: bulking value

specification a clear and accurate description of the technical requirement for materials, products, or services, which specifies the minimum requirement for quality and construction of materials and equipment necessary for an acceptable product.

In general, specifications are in the form of written descriptions, drawings, prints, commercial designations, industry standards, and other descriptive references.

speckled finish see MULTICOLOR FINISH.[54,71,163]

specks small PARTICLES of undispersed materials.[42]

spectra plural of SPECTRUM.[43,69]

spectral adjective referring to SPECTRUM.[43,69] See SPECTROPHOTOMETRIC.[43,69]

spectral match see COLOR MATCH.[69]

spectral power distribution curve intensity of RADIANT ENERGY as a function of WAVELENGTH, generally given in relative power terms.[43,69]

spectrogoniophotometer goniophotometer used to measure the geometric distribution of reflected or transmitted flux at individual wavelengths.[43,69,145]

spectrometry, optical (in trace component analysis of air): methods of identification of substances and determination of their concentration based upon the measurement of light transmittance at selected wavelengths or bands of wavelengths.[37] (IUPAC) See MASS SPECTROMETRY.[37]

spectrophotometer photometric device for the measurement of SPECTRAL TRANSMITTANCE, spectral REFLECTANCE, or relative spectral emittance.[43,69,145]

Spectrophotometers are normally equipped with dispersion optics (prism or grating) to give a continuous spectral curve.

spectrophotometer, abridged photometric device for measuring the TRANSMITTANCE, REFLECTANCE, or relative emittance at discrete wavelengths.[43,69,145]

The particular discrete wavelengths may be selected by the choice of a series of narrow bandpass filters (such as interference filters), or masks, etc. No continuous SPECTRAL curve can be obtained from such an instrument.

spectrophotometer, filter see FILTER SPECTROPHOTOMETER,[43,145] and SPECTROPHOTOMETER, ABRIDGED.[43,145,69]

spectrophotometric adjective referring to SPECTROPHOTOMETRY.[43,69]

The word "SPECTRAL" is frequently used in place of the longer "spectrophotometric," but they are not necessarily synonymous. Spectral is the adjective referring to spectrum and is the more general term, hence a SPECTROPHOTOMETRIC CURVE is also a spectral curve.

spectrophotometric curve a curve measured on a SPECTROPHOTOMETER; hence a graph of relative REFLECTANCE or TRANSMITTANCE (or ABSORPTION) as the ordinate, plotted versus WAVELENGTH or frequency as the abscissa.[43,69]

The most common curves in the visible region use wavelength units of nanometres, with the short WAVELENGTH units at the left of the scale.

spectrophotometric hiding see HIDING, SPECTROPHOTOMETRIC, COMPLETE and INCOMPLETE.[43]

spectrophotometry methods of using a photometer to measure light intensity as a function of WAVELENGTH.[145,43] (IUPAC)

spectrum spatial arrangement of components of RADIANT ENERGY in order of their WAVELENGTHS, wave number, or frequency.[43,69]

spectrum locus the curve, connecting points in a CHROMATICITY DIAGRAM, that represents various WAVELENGTHS of the visible SPECTRUM.[43,69]

specular refers to REFLECTION without distortions, for example, from a high quality first surface mirror.[43] (ASTM) See DISTINCTNESS-OF-IMAGE GLOSS.[43]

specular gloss relative luminous fractional REFLECTANCE from a surface in the mirror or SPECULAR DIRECTION.[43] See REFLECTANCE, SPECULAR and GLOSS.[43]

It is sometimes measured at 60° relative to a perfect mirror.

specular reflectance see REFLECTANCE, SPECULAR.[43]

specular reflectance excluded (SCE) measurement of REFLECTANCE made in such a way that the SPECULAR REFLECTANCE is excluded from the measurement; DIFFUSE REFLECTANCE.[43,69]

The exclusion may be accomplished by using 0° (perpendicular) incidence on the samples, thereby reflecting the specular component of the reflectance back into the instrument, by use of black absorbers or light traps at the specular angle when the incident angle is not perpendicular, or in directional measurements by measuring at an angle different from the specular angle.

specular reflectance included (SCI) measurement of the total REFLECTANCE from a surface, including the diffuse and specular reflectances.[43,69]

specular reflection light striking a surface, and being reflected or turned back at an angle equal to the ANGLE OF INCIDENCE.[43] See GLOSS.[43]

specular transmittance see TRANSMITTANCE, SPECULAR.[43]

sperm oil a yellow, WAXY OIL obtained from the head of the sperm whale and used as an industrial LUBRICANT.[63]

spermaceti wax wax obtained from certain fish OILS, e.g., SPERM OIL. It is characterized by a glistening white, CRYSTALLINE APPEARANCE. Properties: mp, 49°C; sp gr, 0.960; iodine value, 4; saponification value, 126.[137]

spewing the formation of a film, or the collection of PARTICLES, on a paint surface during the DRYING PROCESS; it results from the migration, to the surface, of the insoluble or incompatible portion(s) of the paint BINDER.[42,56,55] Syn: greasiness

spider legs film DEFECT where the coating material on an upright surface separates or breaks and the liquid runs down in long, crooked channels.[56]

spin coating SPRAY coating a panel while it is on a rotating table to obtain a uniform thick film.[77,157]

Generally the resulting coated panel is used for test purposes.

spinning method of APPLICATION which distributes the paint over a flat surface by CENTRIFUGAL ACTION.[77] (BSI)

spiral grain a form of CROSS GRAIN in which the FIBERS extend spirally about instead of vertically along the bole of a tree or axis of a TIMBER.[153,159] (ASTM)

spirit (1) in the paint industry, spirit is used somewhat loosely but generally refers to commercial ETHYL ALCOHOL normally sold as INDUSTRIAL METHYLATED SPIRIT; (2) the term "MINERAL SPIRITS" (in the U.K. as "WHITE SPIRIT") mainly defines a mixture of ALIPHATIC HYDROCARBONS with a proportion of AROMATIC HYDROCARBONS.[164]

spirit of wine see ETHYL ALCOHOL.[164]

Words presented in CAP/SMALL CAPS type indicate that the word is defined in another part of the *Coatings Encyclopedic Dictionary.*

Numerical superscripts classify terms in one or more of the categories listed in the second section of this volume.

spirit stains SOLUTIONS of suitable DYES in SOLVENT which may also contain a proportion of RESIN as BINDER.[162] See STAIN.[41,71,163]

spirit varnish (1) LACQUER based on a solution of RESIN or resins in INDUSTRIAL METHYLATED SPIRIT. The more correct term would be "spirit lacquer;"[151] (BSI) (2) a VARNISH which is converted to a solid film by SOLVENT EVAPORATION.[152] (ASTM) See LACQUER.[151] See SHELLAC.[146]

spirit-soluble dye dye soluble in ALCOHOL and insoluble in water; usually AZO COLORS lacking the group SO_2OH.[41]

spirit-soluble resin RESIN soluble in ALCOHOL and insoluble in water.[167]

spiriting off final operation in a FRENCH POLISHING PROCESS by which the last traces of oil are removed by drawing a rag, dampened with methylated SPIRIT, rapidly and repeatedly over the surface.[54] (BSI)

spirits of rosin see ROSIN SPIRIT.[164,167]

spirits of turpentine see TURPENTINE, SPIRITS OF.[164]

spitting see FLYING; MISTING.[168]

split spray asymmetrical SPRAY PATTERN resulting in the APPLICATION of bands of paint of uneven THICKNESS, caused either by a defective spray or nozzle or by partial blockage of the nozzle or air passages of a SPRAY GUN.[42] (BSI)

splitting a DEFECT in a painted surface; results from the PENETRATION of SOLVENTS, contained in a fresh coat of paint, into an older layer of paint over which it has been applied; likely to occur when the old layer has been SANDED too much.[56] (DAC)

sponging see SPONGING ON and SPONGING OFF.[77,148]

sponging off a DECORATIVE technique in which the BASECOAT is covered with GLAZE and patted with a sponge to remove it partly before it dries.[77,148] cf. SPONGING ON[77,148]

To sponge off, the glaze is applied with a ROLLER, BRUSH, or painting PAD. Before the glaze dries, it is patted with a sponge dipped in either water or paint thinner, depending on the type of paint, and squeezed nearly dry. The same hand motions are used as in sponging on.

sponging on a DECORATIVE technique in which the BASECOAT is dabbed with a sponge dipped in glaze.[77,148] cf. SPONGING OFF[77,148]

To sponge on, the flat sponge is dipped into the second COLOR, squeezed nearly dry and then dabbed onto the surface using a varying pattern of hand motion.

spore a simple reproductive body of lower plants. A spore is analogous to the seed of higher plants but lacks an embryo.[173]

It usually consists of one cell, but sometimes several spores are enclosed by a protective covering. It is an asexual reproductive cell capable of developing into a mature FUNGUS without fusion with another cell. It tends to be spherical in morphology and is observed singly or in clusters. Spores may or may not be associated with fungal MYCELIA and can range in a variety of colors. Although primarily gray or black, some spores are even green in COLOR and have been mistaken as terrestrial ALGAE.

spot finishing see SPOTTING IN.[54]

spotting development of small areas on a painted surface which differ in COLOR or GLOSS from the major portion of the work.[56] (BSI)

spotting in RUBBING down and refinishing small defective patches in a coating.[54] (BSI) Syn: retouching, spot finishing

spray booth an enclosed or semienclosed area used for the SPRAY painting of fabricated items; may be equipped with a source of filtered air to keep the atmosphere DUST-free, a waterfall backdrop to trap OVERSPRAY, and an exhaust system to VENT the FUMES of the EVAPORATING solvents.[59] (DAC)

spray coating see COATING, SPRAY.[77]

spray gun a tool, operated with compressed air or fluid pressure, which expels paint, mortar, etc., through a small orifice, onto the surface being coated.[59,77] (DAC) See CONCRETE GUN.[59,73] Syn: air BRUSH, SPRAYING pistol

spray mottle irregular surface of a sprayed film resembling the SKIN of an orange. See ORANGE PEEL.[56,42]

The DEFECT is due to the failure of the film to FLOW out to a level surface.

spray pattern configuration of SPRAY with gun held in a steady manner.[54]

spraying (1) method of APPLICATION in which the coating material is broken up into a fine mist that is directed onto the surface to be coated;[77] (2) see FLYING.[168] See AIRLESS SPRAYING and HOT SPRAYING.[77] cf. HAND-HELD ELECTROSTATIC SPRAY GUN[77]

Atomization is generally effected by compressed air, but other methods are direct pressure (airless SPRAY), centrifugal, and other mechanical or electrostatic forces.

spraying pistol see SPRAY GUN.[59]

spread quantity of ADHESIVE per joint area applied to an ADHEREND. It is preferably expressed in pounds of liquid or solid adhesive per thousand square feet of joint area; (1) single spread refers to application of adhesive to only one adherend of a joint; (2) double spread refers to application of adhesive to both adherends of a joint.[79] (ASTM)

spreadable life see POT LIFE.[161,42,67,74]

spreading a THICKENING or enlarging of PRINTED areas caused by BLEEDING of lateral PENETRATION of INK.[168]

spreading rate the average area covered by a specified volume of coating when applied over a normal surface in a manner appropriate for that coating.[77,128,54] See NATURAL SPREADING RATE.[77,54]

Spreading rate will vary with operator, the method of application, and the nature of the surface being coated.

Spreading rate is frequently expressed as ft^2/gal or m^2/litre.

springwood the portion of the annual growth ring that is formed during the early part of the season's growth.[153]

It is less DENSE and weaker mechanically than SUMMERWOOD.

square a roof area of 9.29 m^2 (100 ft^2), or enough material to cover 9.29 m^2 of DECK.[128,76] (ASTM)

squeegee coat SHELLAC, LACQUER, or similar materials applied in a heavy CONSISTENCY or body by means of a RUBBER gasket or squeegee.[77]

Broom handles and lead pencils are coated in this manner.

SSF abbreviation for seconds, Saybolt, furol.[166]

SSPC abbreviation for the STEEL STRUCTURES PAINTING COUNCIL. [174]

SSU abbreviation for seconds, Saybolt universal.[166]

stability resistance to change, DETERIORATION, or displacement.[161]

stabilizer (1) substance added, usually in small proportions, to retard undesirable chemical or physical changes, e.g., small quantities of stabilizers are added to retard the dechlorination of chlorinated RUBBER or the coagulation of an emulsion or LATEX;[83,161] (2) a substance used in the formulation of some plastics to assist in maintaining the properties of the material at or near their initial values during processing and service life.[38] (BSI)

stable tending toward DECOMPOSITION or other unwanted chemical change during normal handling or storage.[172] See STABILITY and REACTIVE MATERIAL.[172,161]

An unstable chemical in its pure state, or as commonly produced or transported, polymerizes vigorously, decomposes, condenses, or becomes self-reactive under conditions of shock, pressure, or temperature.

staging life see SHELF LIFE.[161,42]

stain (1) a TRANSPARENT or SEMITRANSPARENT SOLUTION or SUSPENSION of coloring matter (DYES or PIGMENTS or both) in a VEHICLE, designed to COLOR a surface (usually wood) by PENETRATION without hiding it or leaving a continuous film. True stains are classified water stains, nongrain-raising stains, oil stains, and spirit stains, according to the nature of the vehicle. The so-called "VARNISH stains" are varnishes colored with transparent material. These do not have the same power of penetration as the true stains and leave a colored coating on the surface; (2) an undesirable change of color resulting from FOREIGN MATTER;[56] and (3) to change the color by using a staining agent.[71,69] See OPAQUE STAIN and SEMITRANSPARENT STAIN.[71,69] cf. WOOD PRESERVATIVE COATINGS[71,153,65]

stain removal the ability of a paint to resist soiling by common household stains.[72,157]

Interior ARCHITECTURAL paints are subjected in use to soiling by dirt or other stains.

ASTM Test Method D4828 covers the determination of the relative ease of removal of common soil and stains from interior coatings by manual or mechanical washing with a sponge and a liquid or powder cleanser.

stainer see TINTER, and COLORIST.[41]

staining (1) the APPLICATION of a liquid DYE solution to a POROUS surface to impart COLOR;[148] (DAC) (2) DISCOLORATION produced by a RUBBER stock on ORGANIC FINISHES, LACQUERS, and FABRICS, owing to presence of discoloring type antioxidants in the compound. The condition is aggravated by EXPOSURE to heat, pressure, or sunlight. Also, discoloration of a surface due to migration of an ingredient from one rubber compound to an adjacent compound, even though they may not be in actual contact with each other.[65,140] cf. WATER STAINING[42,56]

staining power degree to which a colored PIGMENT imparts COLOR to a WHITE pigment under defined conditions of TEST.[43] (BSI) Syn: tinting strength[43,69]

The details of procedure for determining staining power, normally laid down in specifications for pigments, need to be carefully adhered to if consistent results are to be obtained. The corresponding property of a white pigment is "REDUCING POWER."

stand oil so-called because it was originally made by letting LINSEED OIL stand in the sun for a time, often as much as seven to ten years, during which time a natural POLYMERIZATION took place.[63,156] See OIL, HEAT-BODIED.[63,156]

The Dutch "Standolie" and German "Standöl" are counterparts of the English term which have been in use for hundreds of years, antedating the introduction and use of WOOD OIL.

Words presented in CAP/SMALL CAPS type indicate that the word is defined in another part of the *Coatings Encyclopedic Dictionary.*

Numerical superscripts classify terms in one or more of the categories listed in the second section of this volume.

standard (1) a reference point or practice established by general agreement;[134,157] (2) as used by AMERICAN SOCIETY FOR TESTING AND MATERIALS, a standard is a document that has been developed and established within the consensus principles of the Society and that meets the approval requirements of ASTM procedures and regulations. A concept that has been established by authority, custom, or agreement to serve as a model or rule in the measurement of quantity or the establishment of a practice or a procedure; (3) in air pollutant analysis, standard reference gases, liquids, or solids are used to calibrate equipment.[37,171,172] (IUPAC)

ASTM's six principal types of full consensus standards are: Standard Test Method—a definitive procedure for the identification, measurement, and evaluation of one or more qualities, characteristics, or properties of a material, product, system, or service that produces a test result; Standard Specification—a precise statement of a set of requirements to be satisifed by a material, product, system, or service that also indicates the procedures for determining whether each of the requirements is satisfied; Standard Practice—a definitive procedure for performing one or more specific operations or functions that does not produce a test result; Standard Terminology—a document comprising of terms, definitions, descriptions of terms, explanation of symbols, abbreviations, or acronyms; Standard Guide—a series of options or instructions that do not recommend a specific course of action; and Standard Classification—a systematic arrangement or division of materials, products, systems, or services into groups based on similar characteristics such as origin, composition, properties, or use.(ASTM)

The concentrations of standards specified by a government agency or private laboratory must be based upon accurate measurement techniques which are verifiable through crosschecks and the use of unambiguous methods.

standard conditions for testing temperature: 23 plus or minus 1.1°C (73.5 plus or minus 2°F); RELATIVE HUMIDITY: 50% plus or minus 4%.[157]

These conditions have been established for the use of all departments and establishments of the Federal Government, except where otherwise specified by an applicable specification.[157] (Fed. Stnd. No. 1) In ISO, standard conditions are 23±2°C and 50±5% relative humidity.

standard light source a reference light source whose SPECTRAL power distribution is known.[43,69] See LIGHT SOURCE, and ILLUMINANT.[43,69]

standard observer, 2° (1931) see OBSERVER, STANDARD.[43,69,134]

standard observer, 10° (1964) see OBSERVER, STANDARD.[43,69,134]

standard of performance the measure of POLLUTION control required by law, in the CLEAN AIR ACT of 1970.[171,172]

standard perfect white diffuser see PERFECT DIFFUSER.[43]

standard sand natural SILICA SAND from Ottawa, Illinois, graded to pass a No. 20 (850 μm) SIEVE and be retained on a No. 30 (600 μm) sieve.[131,157]

This sand shall be considered standard when not more than 15 g are retained on the No. 20 sieve and not more than 5 g pass the No. 30 sieve after five minutes of continuous sieving of a 100 g sample. Used in sand GRINDING and the FALLING SAND ABRASION TEST.

Refer to ASTM Test Method D968.

standard, white reflectance in general usage, may refer to a physical WHITE STANDARD which is nearly a perfect white DIFFUSER, such as pressed BARIUM SULFATE of high purity, pressed magnesium oxide, or freshly smoked magnesium oxide.[43,69]

Other less perfectly diffusing or reflective materials, such as white ceramics, may be CALIBRATED in reference to the perfect diffuser or to one of the near-perfect materials, and used in place of the material relative to which it is calibrated. Such a white reference standard should properly be called a secondary reference standard.

STAPPA abbreviation of STATE AND TERRITORIAL AIR POLLUTION PROGRAM ADMINISTRATORS.[172] (EPA)

starch coating surface coating for flat paints and/or WALLPAPER.[71]

Made with COLORS that do not SMEAR.

stardust imperfections in high gloss FINISHES caused by minute blemishes which give sparkling APPEARANCE.[55,56,35]

starter strip in the COIL COATING industry, a length of METAL threaded through a line after shutdown.[70]

This is used repeatedly to attach new strip to be coated.

starved joint a poorly BONDED GLUE joint resulting from an insufficient quantity of glue in the joint.[79] (DAC)

starved surface see HUNGRY SURFACE.[42,56,159]

starving out in the COIL COATING industry, similar to SKIPPING in that it does not apply an even coating to the strip.[70]

State Implementation Plans (SIP) are required by Section 110 of the CLEAN AIR ACT, as amended.[172] (EPA)

Each State is to submit a plan to the EPA Administrator which provides for implementation, maintenance, and enforcement of the national ambient air quality standards.

steam-distilled pine oil see PINE OIL.[44,164]

steam-distilled wood turpentine see TURPENTINE.[44,164]

steam-set inks INKS that dry by the APPLICATION of steam or moisture.[168] See MOISTURE-SET INK.[168]

stearic acid $CH_3(CH_2)_{16}COOH$. Hard, waxlike, SATURATED FATTY ACID.[144]

Properties: mp, 69°C; bp, 291°C/100 mm Hg.

stearin (stearine) $C_3H_5(C_{18}H_{35}O_2)_3$. Glycerol tristearate, tristearin.[144]

Properties: mol wt, 890.86; mp, 71.6°C; sp gr, 0.943.

stearin (stearine) pitch pitch obtained from stearin (stearine), from the residues of DISTILLATION of FATTY ACIDS, VEGETABLE OILS, etc.[76]

They differ from other pitches or BITUMENS by the possibility of possessing both acid and SAPONIFICATION values and having a tendency to OXIDIZE on EXPOSURE. Used in some inks to impart FLOW and good WETTING qualities.

steatite high purity TALC containing maximum allowable proportions of 1.5% CaO, 1.5% Fe_2O_3, and 4% Al_2O_3 as impurities. See MAGNESIUM SILICATE, NONFIBROUS.[41,58]

steel a generally hard, strong, durable, malleable alloy of iron and CARBON usually containing between 0.2 and 1.5% carbon, often with other constituents such as manganese, chrome, nickel, molybdenum, copper, tungsten, cobalt, or silicon depending on the desired alloy properties and widely used as a structural material.[75,159]

steel blue type of PRUSSIAN BLUE.[41] See IRON BLUE.[41]

steel plate printing an INTAGLIO type of PRINTING using METAL plates with the image etched or engraved below the surface.[168] Syn: steel die printing

Steel Structures Painting Council (SSPC) an association that publishes standards for cleaning and coating steel and concrete and for paints to protect steel and concrete in industrial environments. It holds educational meetings, exhibitions, and seminars related to the cleaning and painting of steel and concrete.[174]

Its address is: 40 24th St., Pittsburgh, PA 15222; (412) 281-2331.

steel wool a matted MASS of long, fine, STEEL FIBERS available in a variety of grades of coarseness.[59,158]

Used for cleaning and polishing surfaces, burnishing, removing film blemishes between coats, and the intentional dulling of coated surfaces.

stencilling method of applying a design by BRUSHING, dabbing, or spraying INK or paint through a cutout overlay placed on the SUBSTRATE and thus reproduce the pattern on the surface beneath.[77]

It is an expressive decorating technique especially suited to FURNITURE and surfaces like walls. The craft consists of applying COLOR (usually paint) onto a surface through a precisely cut pattern (stencil).

stepladder a ladder having flat steps, or treads, in place of rungs; usually provided with a supporting frame to steady it.[59] (DAC)

stereobase unit the BASE UNIT of a POLYMER, taking steric ISOMERISM into account.[130,133] (IUPAC)

stereochromy a comparatively new method of MURAL PAINTING in which water GLASS serves as the connecting medium between the COLOR and its substratum and also as a protective coating.[81]

stereospecific polymerization POLYMERIZATION in which a TACTIC POLYMER is formed.[169] (IUPAC)

stereotypes PRINTING plates cast in one piece from a heat-dried PAPIER-MACHÉ matrix containing the impression of the assembled type.[168]

stick shellac SHELLAC in a solid, sticklike form, which is manufactured in clear form and in numerous colors and shades to match all WOODS.[146] Syn: wood cement[146]

sticklac raw material obtained directly from the tree from which SHELLAC is derived.[146,167]

sticky see PULLING and DRAG.[166]

stiffness (modulus of elasticity) the load per unit area to elongate the film 1% from the first point in the STRESS-STRAIN CURVE where the slope becomes constant.[47,157] (ASTM)

Stiffness measurements are of value in studying the behavior of coatings subjected to environmental stresses, such as those produced by aging and WEATHERING. They may vary with specimen thickness, method of preparation, GAGE LENGTH, rate of load application, TENSILE TESTER response, and types of grips used.

ASTM Test Method D2370 covers the measurement of stiffness as well as tensile strength and elongation of ORGANIC coatings when tested as free films.

still life subject for painting composed of inanimate objects, e.g., fruit, flowers, etc.[81]

stillingia oil pale yellow, limpid, DRYING OIL; peculiar odor; mustard-like taste. Slightly soluble in

Words presented in CAP/SMALL CAPS type indicate that the word is defined in another part of the *Coatings Encyclopedic Dictionary.*

Numerical superscripts classify terms in one or more of the categories listed in the second section of this volume.

ALCOHOL. Sp gr, 0.943-0.945; iodine number, 160; saponification value, 210.[63] See CHINESE VEGETABLE TALLOW.[63]

stipple (1) to even out a coat of paint and remove BRUSH marks and other imperfections, immediately after application, by systematically dabbing the surface with a soft STIPPLING brush;[54,77] (2) to produce a BROKEN COLOR or TEXTURED effect, either by applying spots of a different color or by disturbing the surface of the paint coat, e.g., with a stippling brush or RUBBER STIPPLER.[54,77] (BSI)

stippler (1) a broad flat-based BRUSH having stiff BRISTLES for producing a TEXTURE on a surface such as soft PLASTER or paint; (2) any tool (as a RUBBER sponge or a textured or tufted ROLLER) used to create a stippled surface.[59] (DAC)

stippling (1) method of painting by means of small dots, touches of COLOR or pebbly-textured FINISH;[77] (2) ENGRAVING by means of dots; distinct from engraving in lines;[168] (3) in ANTIQUING, to obtain an irregular effect with a MOTTLED APPEARANCE. It usually entails the use of a sponge, newspaper, or points of dry BRUSH.[148] See POINTILLISM.[64,77,81]

stitcher stapling device used in COIL COATING to connect the end of a coil to the beginning of another coil.[59,70]

stochastic sampling random SAMPLING.[37]

Stockholm pine tar see PINE TAR, STOCKHOLM.[44]

stoddard solvent PETROLEUM SPIRITS with a minimum flash of 37.8°C (100°F) and low odor level.[164]

stoichiometric (1) involving chemical combination in simple integral ratios. Characterized by having no excess of reactants or products over that required to satisfy the balanced chemical equation representing the given chemical reaction.[144] (IUPAC); (2) characterized by, or being a proportion of, substances or energy in a specific chemical reaction in which there is no excess of any reactant or product. (ASTM)

Stokes Law the mathematical equation derived by G.G. Stokes which relates the velocity of fall v of a spherical body under gravity g through a fluid medium to the radius r of the body, to the viscosity η of the medium and to the difference between the density of the solid body (D_s) and that of the medium (D_l): [166]

$$v = \frac{2(D_s - D_l)\, r^2 g}{9\eta}$$

stokes (ST) the C.G.S. UNIT of kinematic viscosity.[166,128]

A liquid with a viscosity of one poise has a kinematic viscosity of one stokes if its density is one. A measure of kinematic viscosity equal to poises divided by density.

stopper pigmented composition used for filling fine CRACKS and indentations to obtain a smooth, even surface preparatory to painting.[54,58,158] Syn: FILLER

storage the act of storing goods or the state of being stored.[50]

storage life see SHELF LIFE.[161,42]

storage stability general composite property of RESISTANCE to any change, generally in a closed container, over a period of time; COLOR and liquid SEPARATION, formation of LUMPS, hard PIGMENT SETTLING, substantial changes in VISCOSITY, PH, development of ODOR, etc., are examples of undesirable changes.[161,42] See SHELF LIFE.[161,42]

Stormer viscometer see KREBS-STORMER VISCOMETER.[145,166,157]

stoving (Brit.) See BAKING.[67,131]

STP abbreviation for standard temperature (273.15 K or 0°C) and pressure (10^5 Pa); usually employed in reporting gas volumes.[128] (IUPAC)

Note that FLOWMETERS calibrated in standard gas volumes per unit time often refer to volumes at 24°C, not 0°C.

straight-line pad sander a portable SANDING machine consisting of a backup pad that moves to and fro in a straight line, with a stroke of about 5/16 inch on an average of about 3,200 times per minute. cf. ORBITAL PAD SANDER[59,158]

The resulting scratch pattern is ideal when sanding with the GRAIN in WOOD and far less noticeable than the swirls from an ORBITAL PAD SANDER when the FINISH coat is applied on either wood or METAL.

strain change per unit of length in a linear dimension of a body, which change accompanies a STRESS. Strain is a dimensionless quantity which may be measured conveniently in %, in in. per in., in mm per mm, etc.[47] (ASTM)

strain relaxation see CREEP.[140,166]

strainer an EXTRUSION machine in which unvulcanized RUBBER or reclaimed rubber is forced through a SCREEN backed by a perforated plate, to rid the stock of contaminating PARTICLES.[59,140]

straining the mechanical separation of relatively coarse PARTICLES from a liquid as distinguished from the PROCESS of FILTRATION.[131]

Strasbourg turpentine OLEORESINOUS EXUDATION obtained from the white fir.[44,167]

streaking long, narrow, irregular lines or bands or

layer-edges, especially ones distinguished by COLOR, visible on a surface.[35,42]

streamline flow a streamline is a line in a fluid such that the tangent to it at every point is in the direction of the velocity of the fluid particle at that point, at the instant under consideration.[166] cf. TURBULENCE[166]

When the motion of the fluid is such that, at any instant, continuous streamlines can be drawn through the whole length of its course, the fluid is said to be in streamline flow.

strength, dry strength of an ADHESIVE joint determined immediately after drying under specified conditions or after a period of conditioning in the standard laboratory atmosphere.[47,79] (ASTM) See STRENGTH, WET.[47,79]

strength, tensile see TENSILE STRENGTH.[47,157]

strength, wet strength of an ADHESIVE joint determined immediately after removal from a liquid in which it has been immersed under specified conditions of time, temperature, and pressure.[47,79] (ASTM) See STRENGTH, DRY.[47,79]

stress RESISTANCE to DEFORMATION developed within a specimen subjected to an external FORCE.[47] (ASTM)

Typical examples are TENSILE STRESS, SHEAR stress, and COMPRESSIVE stress. Stress usually reaches a maximum at the time of rupture. The load from the test output is divided by the sample cross section area. Units are in lb/in.^{2} or N/m^2.

stress-strain curve a graphical representation of the STRESS and STRAIN relationship of a material under conditions of compression, SHEAR, tension, or torsion.[47,157] (ASTM)

stress-strain diagram a diagram in which corresponding values of stress and strain are plotted against each other.[47,157] (ASTM)

Values of stress are usually plotted as ordinates (vertically) and values of strain are abscissas (horizontally).

strié a decorating process that gives a three-dimensional striped look using GLAZES.[148]

strike-through the PENETRATION of the VEHICLE of a PRINTING INK through the sheet so that it is apparent on the opposite side.[168]

string (varnish) VARNISH being POLYMERIZED will become VISCOUS and form strings which suspend from an object when DIPPED in it and then withdrawn.[152,161]

The length of the string is an indication of the degree of POLYMERIZATION.

stringiness property of an ADHESIVE that results in the formation of filaments or threads when adhesive transfer surfaces are separated.[79,166] (ASTM)

strip (1) a length of WALLPAPER cut to fit height of wall;[148] (2) a single section of the design, in a scenic wallpaper;[148] (3) a board of LUMBER less than 4 in. (10 cm) wide.[75,82,153]

strip coating strip coating or COIL COATING is the method of APPLICATION of a coating to raw METAL strip before its FABRICATION into finished products.[77]

strip flashing see STRIPPING (2).[76]

striping in ANTIQUING, painting a series of fine lines—in gold or a contrasting COLOR—to enhance some basic design characteristics on a piece of FURNITURE.[148,175] Syn: dutch gold

It may be accomplished by a resist technique (such as taping a fine line effect over a surface, and then blocking out the rest of the area and SPRAYING into the opening left by the tape). It may be painted free hand or with the use of a ruler or straight edge. It may be done with sizing and gold leaf.

strippable coating see TEMPORARY COATING.[71]

strippable paper a chemically treated stock TOUGH enough to resist tearing, with a special formulation which permits a release of the WALLCOVERING from the ADHESIVE.[148] Syn: dry strippable

This makes it possible to remove an entire strip from the wall without WETTING.

stripper see PAINT and VARNISH REMOVER.[158]

stripping (1) removing old paint, WALLPAPER, DISTEMPER, etc., by the use of a blowtorch, PAINT REMOVER, steam stripping appliance, stripping knife, or other SCRAPING tools; (2) SEALING the joint between a METAL sheet and a BUILT-UP ROOFING MEMBRANE;[76] (3) taping the joints between insulation boards;[158] (DAC) (4) a condition where the INK fails to adhere to and distribute uniformly on the metal rollers of the press.[168]

stripping agent see MOLD WASH.[83]

strontium chromate $SrCrO_4$. Pigment Yellow 32 (77839). A bright yellow PIGMENT of type similar to lead CHROMATE except that it is not blackened by hydrogen sulfide.[41] Syn: strontium yellow

It is used in CORROSION-RESISTANT PRIMERS. Properties: density, 3.67-3.82 g/cm^3 (30.6-31.9 lb/gal); O.A., 20-33; particle size, 10-30 µm.

strontium white synonym for the ground MINERAL, CELESTITE.[41,34]

strontium yellow see STRONTIUM CHROMATE.[41]

Words presented in CAP/SMALL CAPS type indicate that the word is defined in another part of the *Coatings Encyclopedic Dictionary.*

Numerical superscripts classify terms in one or more of the categories listed in the second section of this volume.

structure something made up of a number of parts that are held or put together in a particular way.[133]

structured paint paint having a GEL-like CONSISTENCY which breaks down under the kind of shear exerted during BRUSHING or roller application and re-forms when the shearing FORCE is removed.[166,163,71] See VISCOSITY, STRUCTURAL.[166]

When there is a time lag between the removal of the shearing force and the start of gel re-formation, the paint is said to be THIXOTROPIC. The thixotropic structure can vary from a very light gel to the fully gelled thixotropic paints of the nondrip type.

stucco (1) an exterior FINISH, usually TEXTURED; composed of PORTLAND CEMENT, LIME, and sand, which are mixed with water; (2) a fine PLASTER used for DECORATIVE work or MOLDINGS;[148] (3) simulated stucco containing other materials, such as EPOXY as a BINDER; (4) a partially or fully CALCINED GYPSUM that has not yet been PROCESSED into a finished product.[73,75,159] (DAC)

stud upright post in the framework of a wall for supporting LATH, sheets of wallboard, or the like.[82]

styrene $C_6H_5CH = CH_2$. Colorless to yellowish oily liquid. A MONOMER used to make SYNTHETIC RESINS and ELASTOMERS.[141] Syn: vinylbenzene, styrol, styrolene, cinnamene, phenylethylene

Properties: mol wt, 104.14; sp gr, 0.909; mp, -33°C; bp, 145-146°C.

styrene butadiene latex see BUTADIENE STYRENE LATEX.[130]

styrene resin SYNTHETIC RESIN made from VINYLBENZENE.[130] (ASTM)

sub spirits MINERAL SPIRITS. See TURPENTINE SUBSTITUTE.[164]

sub turps see TURPENTINE SUBSTITUTE.[164]

sublimed white lead synonym for basic lead sulfate.[41]

substrate (1) any surface to which a coating or printing INK is applied;[168] See PRODUCT COATING;[71,159] (2) the nonpainted material to which the first coat was applied; sometimes referred to as the original SUBSTRATE; (3) base on which ORGANIC coloring agents are PRECIPITATED to form LAKES.[41] See GROUND and BASE (3).[159,41]

ASTM test methods for coatings on metal substrates include: accelerated outdoor exposure, D4141; adhesion, by cut/tape test, D3359; blistering-evaluation of, D714; coil coating, D3794; flexibility/adhesion-deformed, D4145; mandrel bend test, D522; primers, testing practices, D3322.

subtractive colorant mixture see COLORANT MIXTURE.[43,69]

subtractive colorant mixture, complex COLORANT mixture which must take into account both the ABSORPTION and SCATTERING of two or more of the individual pigments used in the mixture.[43,69] See KUBELKA-MUNK THEORY.[43,69]

subtractive colorant mixture, simple COLORANT mixture which can be described by the single variable of ABSORPTION of RADIANT ENERGY.[43,69] See BEER'S LAW (LIQUIDS) and KUBELKA-MUNK THEORY.[43,69]

The term can be used to describe mixture of liquids which have no SCATTERING or of PIGMENT mixtures where almost all the scattering comes from one major component, frequently white pigment or a textile fiber, for example.

succinic acid $(CH_2)_2(COOH)_2$. A SATURATED dibasic acid.[144]

Used in the preparation of ALKYD type RESINS and PLASTICIZERS. Properties: mp, 184°C; bp, 235°C.

suction FORCE that causes a coating to be drawn into the pores of, or adhere to, a surface because of the difference between the external and internal pressures.[132]

suede see GLOSS.[43]

suede finish see RIPPLE FINISH.[71]

sulfate pine oil see PINE OIL.[44,164]

sulfate pitch see TALL OIL.[44,63,130]

sulfate resin see TALL OIL.[44,63,130]

sulfate rosin see TALL OIL.[44,63,130]

sulfate wood turpentine see TURPENTINE.[44,164]

sulfated oil same type of material as SULFONATED OIL.[127,144]

sulfation the introduction into an ORGANIC molecule of the sulfuric ester group (or its SALTS), $-O-SO_3H$, where the sulfur is linked through an oxygen atom to the parent molecule.[134] (ASTM)

sulfide staining the DISCOLORATION (generally gray or brownish yellow) of an exposed paint film resulting from sulfurous air pollutants reacting with METALLIC COMPOUNDS such as lead, mercury, or copper in the paint film.[42,56]

Sulfide staining of paint films generally occurs in areas where there are sulfurous gas air pollution from factories or marshes. This is an uncommon occurrence in most parts of the country.

sulfite turpentine is not in good usage, because the volatile oil recovered in the conversion of wood to pulp by the sulfite process consists chiefly of CYMENE, $C_{10}H_{14}$, rather than PINENE and other terpenes.[44,164] (ASTM)

sulfonated castor oil see TURKEY RED OIL.[83,127]

sulfonated oil a water-dispersible or soluble SURFACE ACTIVE material obtained by treating an unsaturated or hydroxylated fatty oil, acid, or ester with an agent capable of sulfating or sulfonating it at least partially.[83,127] (ASTM) Syn: sulfated oil

sulfonation the introduction into an ORGANIC molecule of the sulfonic acid group (or its SALTS), $-SO_3H$, where the sulfur atom is joined to a CARBON atom of the parent molecule.[134] (ASTM)

sulfurized oil OIL which has been reacted with sulfur or sulfur-containing compounds like sulfur chloride.[63,156]

The sulfur is believed to attach itself to the unsaturated bonds. Sulfur-treated OILS set or POLYMERIZE rapidly, and they are also rather unstable on storage.

sumi (Japanese). Black Chinese INK.[81]

sumi-ye (Japanese). A PRINT in black and white only.[81]

summerwood the portion of the annual growth ring that is formed during the latter part of the yearly growth period.[153,173]

It is usually more dense and stronger mechanically than SPRINGWOOD.

sunflower oil seed oil of *Helianthus annus*, which grows in abundance in many parts of the world.[63,156]

It can only be regarded as a SEMIDRYING OIL, but has the useful property of being nonyellowing. Properties: sp gr, 0.930/15°C; iodine value, 135.

Superfund Amendments and Reauthorization Act (SARA) Title III of SARA is known as the Emergency Planning and Community Right-to-Know Act of 1986. It is a revision and extension of CERCLA. SARA is intended to encourage and support local and state emergency planning efforts. It provides citizens and local governments with information about potential chemical hazards in their communities. SARA calls for facilities that store hazardous materials to provide officials and citizens with data on the types (flammable, corrosives, etc.); amounts on hand (daily, yearly); and their specific locations. Facilities are to prepare and submit inventory lists, MSDSs, and tier 1 and 2 inventory forms.[171]

The disaster in Bhopal, India, in 1987 added impetus to the passage of this law.

supersaturation an unstable system which has a greater concentration of material in solution than would exist at equilibrium is said to be supersaturated.[164] (IUPAC)

superstructure paint ship paint used on part of the ship's structure above the main deck.[71]

surface the exterior or upper boundary of an object or body.[159] (ASTM)

surface active agent see SURFACTANTS.[127,83]

surface area total area of the surface of all the PARTICLES in a mass of PIGMENT; this is usually expressed in terms of square metres per gram.[41,128]

surface area (B.E.T.) the total surface area of a solid calculated by the B.E.T. (Brunauer, Emmett, Teller) equation, from nitrogen adsorption or desorption data obtained under specified conditions.[128] (ASTM)

surface coating operation the system on an automotive, light truck, METAL FURNITURE, large appliance parts, or products surface coating line used to apply and dry or CURE an ORGANIC coating.[49]

The surface coating operation may be a prime coat or TOPCOAT operation and includes the coating application station(s), flash-off area, and curing oven.

Surface Coatings Association Australia (SCAA) an organization composed of technical personnel of the surface coatings and allied industries of Australia.[174]

Affiliated with the Oil & Colour Chemists' Association, SCAA was reformed under its present name in 1990. Its purpose is to further the development of the science, technology, and management of the coatings industries and to facilitate the exchange of information and ideas. Composed of five sections located in New South Wales, Victoria, Queensland, South Australia, and Western Australia, SCAA publishes a monthly journal, Surface Coatings Australia. SCAA is a member organization of the COATINGS SOCIETIES INTERNATIONAL.

Surface Coatings Association New Zealand (SCANZ) an organization composed of technical personnel of the surface coatings and allied industries of New Zealand.[174]

Affiliated with the Oil & Colour Chemists' Association. Its purpose is to further the development of the science, technology, and management of the New Zealand coatings industry and to facilitate the exchange of information and ideas. SCANZ is composed of two sections located in Auckland and Wellington. SCANZ is a member organization of the COATINGS SOCIETIES INTERNATIONAL.

surface conditioner preparatory coating applied to a chalked surface to bind chalk to the SUBSTRATE, prior to topcoating.[158]

---

surface drying the premature DRYING of the surface of a liquid coating FILM, so that the under portion is retarded in drying.[42,56,67] See DRYING (1) and DRYING TIME.[67,74,131]

surface preparation any mechanical or chemical method used to make a SUBSTRATE more receptive to good ADHESION and uniform APPEARANCE of the coating to be applied.[158,159] cf. SURFACE-TREATING[158]

ASTM test methods for surface preparation for painting include: abrading of concrete, D4259; acid etching of concrete, D4260; aluminum (hot-dip), D1731; aluminum/aluminum-alloy surfaces, D1730; architectural sandstone, D5107; concrete/masonry panels, D1734; galvanized steel, nonpassivated, D2201; glass panels, D3891; magnesium alloys, D1732; steel test panels, D609.

surface tensiometer an instrument used to measure surface and INTERFACIAL TENSIONS of liquids.[142,145]

surface tension property arising from molecular forces of the surface film of all liquids which tend to alter the contained volume of liquid into a form of minimum superficial area.[142] (ASTM)

Surface tension is numerically equal to the FORCE acting normal to an imaginary line of unit length in a surface. It is also numerically equal to the work required to enlarge the surface by unit area. The usual unit of measurement is dynes per cm (newtons per cm in SI UNITS).

surface-treating any method of treating a surface, in preparation of its reception to inks, paints, lacquers, adhesives and processes, such as chemical, flame, and electronic treating.[158]

surfacer pigmented composition for filling minor irregularities to obtain a smooth, uniform surface preparatory to applying FINISH coats; usually applied over a PRIMER and sandpapered for smoothness.[147,158] (ASTM)

Some types combine the properties and functions of both a primer and a surfacer and are called primer surfacers. It is intentionally thick enough to permit sanding without cutting through to the SUBSTRATE. A TOPCOAT (finish coat) is applied over a primer surfacer.

surfactants contracted from SURFACE ACTIVE AGENTS, these are ADDITIVES which reduce surface tension and may form MICELLES and thereby improve WETTING (wetting agents); help DISPERSE pigments (DISPERSANTS); inhibit foam (DEFOAMERS); or emulsify (EMULSIFIERS).[127,83]

Conventionally, they are classified as to their charge; ANIONIC (negative); CATIONIC (positive); NONIONIC (no charge); or AMPHOTERIC (both positive or negative).

surimono (Japanese). A PRINT, generally of small size and on thick soft PAPER, intended as a festival greeting or memento of some social occasion.[81]

SUS abbreviation for second, SAYBOLT UNIVERSAL.[157,166]

VISCOSITY is expressed in SUS, as determined by the standard method of test for Saybolt viscosity, ASTM D88, and may be determined by use of the SUS conversion tables specified in ASTM Method D2161, following determination of viscosity in accordance with the procedures specified in the standard method of test for viscosity of transparent and opaque liquids.

suspending agent see SUSPENSION AID.[64,83,161,127]

suspension relatively coarse, noncolloidal DISPERSION of solid PARTICLES in a liquid.[139,64] See COLLOID.[139,64]

suspension aid a substance incorporated in coatings to prevent hard SETTLING in the container during storage and to prevent PIGMENT separation during painting operations.[64,83,161,127] See ANTISETTLING AGENT.[83] Syn: suspending agent.

Sward rocker see ROCKER HARDNESS TESTER.[47,157,145]

swarf the workpiece cuttings, DUST, OIL, GRAIN PARTICLES, etc., created by abrading action of SANDPAPER, etc.[59,158]

sweating (1) EXUDATION of oily matter from a film of paint, VARNISH, or LACQUER after the film has apparently dried; (2) on a paint or varnish film, the development of GLOSS on a dull or MATTE FINISH; caused by RUBBING the film;[56,42] (3) development of gloss in a dry film of paint or varnish after it has been flatted down (sanding); (4) often incorrectly used to describe CONDENSATION of MOISTURE from humid atmospheres on relatively cold surfaces, e.g., walls.[56,42] (BSI)

Swedish oil see TALL OIL.[44,63]

Swedish olein see TALL OIL.[44,63]

Swedish pine oil see TALL OIL.[44,63]

Swedish resin see TALL OIL.[44,63]

Swedish rosin see TALL OIL.[44,63]

Swedish rosin oil see TALL OIL.[44,63]

sweep blast see NACE No. 4 BRUSH-OFF BLAST CLEANED SURFACE FINISH.[158]

swell to increase in size or volume as a result of internal pressure; expand; bulge; protrude; rise from within.[42,56]

syenite an igneous rock composed primarily of alkali FELDSPAR together with other MINERALS, such as hornblende.[41,34]

sylvic acid see ABIETIC ACID, COMMERCIAL GRADE.[144]

sylvic oil see TALL OIL.[44,63]

symbiont an organism which lives in a state of SYMBIOSIS.[173]

symbiosis a living together of dissimilar organisms.[173]

symbiotic living in that relation called SYMBIOSIS.[173]

sympathetic inks INKS that give markings which become invisible and can be made visible by the use of a developing solution.[168]

syndet a contraction for the term "SYNTHETIC DETERGENT" used loosely to signify synthetic detergents or compositions containing synthetic detergents.[127] (ASTM) See DETERGENT.[127]

syndiotactic polymer a TACTIC POLYMER in which the conventional base unit possesses, as a component of the main chain, a CARBON atom with two different lateral substituents, these substituents being so arranged that a hypothetical observer advancing along the bonds constituting the main chain finds opposite steric configurations around these chain atoms in successive conventional base units.[130,133] (IUPAC) See ATACTIC POLYMER and ISOTACTIC POLYMER.[130,133]

The true base unit is thus twice the size of the conventional base unit.

syneresis the separation of liquid from a GEL.[139,161] See CLEAR LIQUID SEPARATION (CLR) and PHASE SEPARATION.[161,42]

The spontaneous EXUDATION or squeezing out of SOLVENT or DILUENT as a separate PHASE which may occur when a gel stands undisturbed; the gel structure slowly densifies but there is no net volume change in the system.

synergism property by which two or more AGENTS act together more effectively than either alone. Syn: synergy

A phenomenon where the mixed effect of two influences is greater than the sum of the two influences acting separately, i.e., the unexpected interaction of parts in combination.

synergistic relating to the cooperative action of two or more discrete agencies such that their combined effect is different than the sum of the effects due to the individual agencies. (ASTM) See SYNERGISM.

synergy see SYNERGISM.

synthesis formation of a COMPOUND from simpler compounds or elements.[144,130,131,169]

synthetic produced from SYNTHESIS, especially not of natural origin.[144,130,131,169]

synthetic detergent a DETERGENT produced by chemical synthesis and comprising an ORGANIC composition other than SOAP.[127] (ASTM)

Often contracted to SYNDET.

synthetic magnetite see BLACK IRON OXIDE.[41]

synthetic paint a vague term which sometimes means paints containing SYNTHETIC RESINS rather than naturally occurring OILS or GUMS in the VEHICLE.[71]

The use of this term is deprecated.

synthetic paint roller cover a cover in which the paint applicating material is man-made.[59] (ASTM)

synthetic pine oil see PINE OIL.[44,164]

synthetic resin see RESIN, SYNTHETIC.[167,130]

synthetic rubber an ELASTOMER manufactured by a chemical PROCESS, as distinguished from natural RUBBER obtained from trees; rubberlike with respect to its degree of ELASTICITY.[140] (DAC)

synthetic yellow ocher SYNTHETIC YELLOW IRON OXIDE reduced with ALUMINUM SILICATE or other EXTENDER PIGMENTS.[41] See IRON OXIDES, SYNTHETIC.[41]

Syrian asphaltum natural ASPHALTUM mined near Damascus.[76] Syn: Egyptian asphaltum

It is not readily soluble in PETROLEUM HYDROCARBONS, and its value to the paint trade is limited.

system see ROOFING SYSTEM.[76]

Words presented in CAP/SMALL CAPS type indicate that the word is defined in another part of the *Coatings Encyclopedic Dictionary.*
Numerical superscripts classify terms in one or more of the categories listed in the second section of this volume.

T

Taber abraser an instrument used to measure ABRASION RESISTANCE.[157,145] See WEAR CYCLES.[157]

Specimen on a turntable rotates under a pair of weighted abrading wheels that produce abrasion through side slip.

tacciometer device for measuring the surface stickiness or tackiness of a dried paint or VARNISH film.[59,157]

It operates on the principle that a suitably weighted piece of PAPER will or will not adhere to the coated surface. The instrument consists of three steel balls which are placed on the paper, and these can be loaded further as required until on removing the load and the balls, the paper shows signs of adhering. The time during which the loaded paper is left in contact with the surface under test is usually specified, because the longer the time of contact, the greater the tendency for the paper to adhere.

tack (1) pull-resistance exerted by a material adhering completely to two separating surfaces. In liquids, tack is a function of VISCOSITY; in nonelastic PLASTIC materials, tack is a function of plastic viscosity and YIELD VALUE. COHESION becomes negligible since rupture occurs at very small areas;[166] (2) slight stickiness of the surface of a film of paint, VARNISH, or LACQUER, apparent when the film is pressed with the finger;[67,79,166] (BSI) (3) a relative measurement of the cohesion of an INK film which is responsible for its resistance to splitting between two rapidly separating surfaces;[168] (4) the natural adhesiveness of RUBBER in the raw state. Also, the property of raw or compounded rubber which causes layers of stock to cohere. It is a desirable property only when ADHESION or cohesion is desired.[79,140]

tack coat an application of BITUMINOUS material to an existing, relatively ABSORBENT surface, to prevent slippage planes and to provide bond between the existing surface and the new surfacing.[76]

tack, dry property of certain ADHESIVES, particularly NONVULCANIZING RUBBER adhesives, to adhere on contact to themselves at a stage in the EVAPORATION of volatile constituents, even though they seem DRY to the touch.[47,79] Sometimes called AGGRESSIVE TACK. (ASTM)

tack rag (cloth) FABRIC impregnated with a tacky substance such as a delayed drying VARNISH which is used to remove DUST from a surface after RUBBING down and prior to further painting.[59,158] (BSI)

Tack rag (cloth) should be stored in an air-tight container to conserve its tackiness.

tack range period of time in which an ADHESIVE will remain in the tacky-dry condition after application to an adherend, under specified conditions of temperature and humidity.[47,67,79] (ASTM)

tack-free freedom from TACK of a coating after suitable DRYING TIME.[67,55] See TACK and DRYING TIME.[67]

In some cases, coatings are tack-free after application; tack doesn't develop until a little later.

tackiness see TACKY.[67]

tacky (1) that stage in the drying of a paint at which the film appears sticky when lightly touched with the finger;[67] (2) having a TACK.[67] Syn: tacky dry

taconite a variety of chert containing MAGNETITE and HEMATITE, mined as an IRON ORE.[41]

Named after the Taconic Mountains.

tactic polymer a POLYMER in which there is an ordered structure with respect to the configurations around at least one main-chain site of steric ISOMERISM per conventional base unit.[130,133] (IUPAC) See SYNDIOTACTIC POLYMER, ISOTACTIC POLYMER, and ATACTIC POLYMER.[130,133]

The CARBON (or other) atom in the chain at the site of the steric isomerism need not, in a strict sense, be an asymmetric atom, since in a chain of infinite length the two chain portions may be considered as equivalent; however, for the purpose of this definition, such atoms are referred to as asymmetric atoms.

tacticity steric order in the main chain of a POLYMER.[130,133] (IUPAC)

tag closed tester an instrument for determining the FLASH POINT of a liquid which has a VISCOSITY of less than 45 SUS at 37.8°C (100°F), does not contain suspended solids, and does not have a tendency to form a surface film while under test.[157]

Reference ASTM Test Method D56 for flashpoint by tag closed tester.

tail (1) highest boiling SOLVENT fraction;[164] (2) elongated, somewhat pointed extension of the lower portion of the rising bubble in a BUBBLE TUBE VISCOMETER, characteristic of a VARNISH or RESIN solution that is near or approaching GELATION or which has a peculiar RHEOLOGICAL characteristic.[166]

tailings the remains, residues, or final products produced on refining any substance.[49] Syn: bottoms, foots

tails finger-like SPRAY pattern.[42,56]

talc see MAGNESIUM SILICATE, NONFIBROUS.[41,58]

tall oil generic name for a number of products obtained from the manufacture of WOOD pulp by the alkali (sulfate) PROCESS or more popularly known as the kraft process. To provide some distinction between the various products, designations are often applied in accordance with the process or composition, some of which are CRUDE TALL OIL, acid refined tall oil, distilled tall oil, tall oil fatty acids, and TALL OIL ROSIN.[63,130,44] (ASTM)

The following designations for tall oil shall be considered obsolete: crude resinous liquid, finn oil, liquid resin, liquid rosin, resin oil, sulfate pitch, sulfate resin, sulfate rosin, Swedish pine oil, Swedish resin, Swedish rosin, Swedish rosin oil, sylvic oil, talloel, tallol, Swedish oil, fluid resin, Swedish olein.

tall oil, acid refined product obtained by treating CRUDE TALL OIL in SOLVENT solution with sulfuric acid under controlled conditions to remove dark COLOR bodies and odoriferous materials.[44] (ASTM)

Removal of the solvent yields a product with lighter color and higher viscosity than crude tall oil with approximately the same fatty acids-to-rosin ratio.

tall oil, crude dark brown mixture of FATTY ACIDS, ROSIN, and neutral materials liberated by the acidification of SOAP skimmings.[44] (ASTM)

The fatty acids are a mixture of OLEIC and LINOLEIC acid with lesser amounts of SATURATED and other UNSATURATED FATTY ACIDS. The rosin is composed of RESIN acids similar to those found in GUM and WOOD rosin. The neutral materials are composed mostly of polycyclic HYDROCARBONS, sterols, and other high-molecular weight ALCOHOLS.

tall oil, distilled class of products obtained by distilling CRUDE TALL OIL in FRACTIONATING equipment under reduced pressure under such conditions that the ratio of ROSIN acids to FATTY ACIDS is varied over a wide range.[44] (ASTM)

The products which generally contain less than 90% of fatty acids are known as distilled tall oils. The fatty acids are a mixture of OLEIC and LINOLEIC acids with lesser amounts of SATURATED and other UNSATURATED fatty acids. The remainder consists of ROSIN and neutral materials.

tall oil fatty acids class of products generally containing 90% or more FATTY ACIDS obtained by FRACTIONATION of CRUDE TALL OIL.[44] (ASTM)

The fatty acids are a mixture of OLEIC and LINOLEIC acids with lesser amounts of SATURATED and other UNSATURATED fatty acids. The remainder consists of ROSIN and neutral materials.

tall oil heads (light ends) low-boiling fractions obtained by the FRACTIONAL DISTILLATION of CRUDE TALL OIL under reduced pressure.[44] (ASTM)

The composition of these products varies over a wide range but contains PALMITIC, OLEIC, LINOLEIC, and STEARIC acids together with lesser amounts of other SATURATED and UNSATURATED fatty acids. The neutral materials content is normally high.

tall oil pitch undistilled residue from the DISTILLATION of CRUDE TALL OIL.[44] (ASTM)

It is generally recognized that tall oil pitches contain some high-boiling ESTERS and neutral materials with lesser amounts of ROSIN and FATTY ACIDS.

tall oil rosin ROSIN remaining after the removal of substantially all of the FATTY ACIDS from tall oil by FRACTIONAL DISTILLATION or other suitable means.[44,167]

Such rosin shall have the characteristic form, APPEARANCE, and other physical and chemical properties normal for other kinds of rosin. The fatty acid content shall not exceed 5%.

tall oil soap product formed by the SAPONIFICATION or NEUTRALIZATION of tall oil with ORGANIC or INORGANIC bases.[127] (ASTM)

talloel see TALL OIL.[44,63]

tallol see TALL OIL.[44,63]

tan (1) a light or moderate yellowish brown to brownish orange;[69] (2) Japanese. A brick-red or orange COLOR, consisting of red oxide of lead.[69]

tan-ye (Japanese) A PRINT in which TAN is the only or chief COLOR used.[81]

Such prints, in which the tan was applied by hand, were among the earliest productions.

tandem line in the COIL COATING field, a rollercoat line with two coaters capable of applying and baking two coats (i.e., PRIMER and ENAMEL) prior to recoiling.[70]

tank coating paint used for the inside of tankers.[71]

tank white good HIDING, self-cleaning white paint for exterior METAL surfaces.[71]

tannin ORGANIC acid obtained in the form of brownish white scales from gall nuts, sumac, and other plants, used in dyeing, tanning, etc.[41]

tap-out spot of INK applied to PAPER by a finger, using tapping action to distribute the ink to approximately printing film thickness.[168] Syn: pat-out

With experience, one can determine whether ink has the proper TACK and working properties, and also can use the tap-out in COLOR MATCHING.

tape a continuous narrow, flexible, strip of cloth,

Words presented in CAP/SMALL CAPS type indicate that the word is defined in another part of the *Coatings Encyclopedic Dictionary.*

Numerical superscripts classify terms in one or more of the categories listed in the second section of this volume.

METAL, PAPER, or PLASTIC, such as ADHESIVE TAPE, etc.[79,75,45]

tape test a type of ADHESION TEST consisting of the APPLICATION of an adhesive tape to a dried coating and rapidly removing the tape with a swift, jerking motion.[157,79]

The coating can be either scribed or unscribed, depending on the specification. The "WET ADHESION" test for LATEX paints is performed by first WETTING the paint with a specified quantity of water for a specified time and blotting off the excess surface water. The tape test is made immediately after blotting.

tapered pattern elliptical-shaped SPRAY pattern; a spray pattern with converging lines.[54]

tar brown or black BITUMINOUS material, liquid or semisolid in consistency, in which the predominating constituents are bitumens obtained as condensates in the destructive distillation of coal, PETROLEUM, oilshale, WOOD, or other ORGANIC materials, and which yields substantial quantities of PITCH when distilled.[75,76] (ASTM)

tar acid PHENOL or its homologues either individually or blended together.[144] (ASTM)

tar and gravel roofing see BUILT-UP ROOFING.[75,76,82]

tar spirit powerful SOLVENT of varying composition and properties, obtained by distillation of wood TAR.[164]

Boiling range is extended and has been known to vary from about 70-80°C to practically 260°C. Tar spirits have sharp penetrating odors, dark colors, and a pronounced tendency to attack mild steel containers.

tarnish to dull the LUSTER of; DISCOLOR, especially by EXPOSURE to air or DIRT.[66]

tartaric acid COOHCH(OH)·CH(OH)COOH. Dihydroxydicarboxylic acid used in the preparation of PLASTICIZERS. Melting point, 169°C.[144]

T-bend a description of the severity of a bend in terms of thickness (T) of the sheet to which the coating has been applied; a measure of coating FLEXIBILITY.[47,157] (ASTM)

The T-bend rating is the minimum number of THICKNESSES of METAL around which the coated sheet is bent, or if bent around a die, the number of thicknesses of metal equivalent to the diameter of the die to achieve no fracture or removal of the coating.

ORGANIC coatings on precoated sheet are subjected to stresses when fabricated into products by roll forming, brake bending, or other deformation processes. These stresses can exceed the flexibility or adhesive strength of the coating, resulting in fracture of the coating, exposing the SUBSTRATE, or loss of adhesion to the substrate.

ASTM Test Method D4145 describes a procedure for determining the flexibility and adhesion of ORGANIC coatings (paints) on metallic substrates that are deformed by bending when the sheet is FABRICATED into building panels or other products. This test is a means of evaluating the ability of a coating system to withstand the stresses of fabrication.

T-bend flexibility test simple method for determining the FLEXIBILITY of coatings by bending a coated METAL test strip over itself.[47,157]

A panel is bent and pressed flat by means of a jig to achieve a 180° bend. Subsequent folds are equivalent to bending the panel around a rod of diameter equal to the thickness of the panel.

TBEP abbreviation for TRIBUTOXYETHYL PHOSPHATE.[144]

TBTF abbreviation for tributyltin fluoride.[144]

TBTO abbreviation for tributyltin oxide.[144]

TCP abbreviation for TRICRESYL PHOSPHATE.[170]

TDI abbreviation for TOLUENE DIISOCYANATE.[141] cf. HDI, IPDI, MDI, and TMXDI [141]

tea seed oil NONDRYING OIL. Iodine value, 90.[63]

tear see RUNS.[56]

teel oil see SESAME OIL.[63]

tekaol LINSEED STAND OIL from which the low-polymerized (SATURATED) highly dispersed phase has been removed.[63]

telechelic polymers those designed to contain END GROUP functional groups.[130]

telegraphing see PHOTOGRAPHING.[42]

telescopic flow see LAMINAR FLOW.[166]

telomer (1) low MOLECULAR WEIGHT POLYMER in which the terminal group on the end of the chain-like molecule is not the same as the side group;[130] (2) a polymer composed of molecules having terminal groups incapable of reacting with additional MONOMERS, under the conditions of the synthesis, to form larger polymer molecules of the same chemical type.[130] (IUPAC, ISO, ASTM)

tempera (1) a rapidly drying paint consisting of egg white (or egg yolk, or a mixture of egg white and yolk), GUM, PIGMENT, and water; especially used in painting MURALS;[71,81] (DAC) (2) a method of painting using tempera.[71,81]

temperature the THERMAL state of matter as measured on a definite scale.[176] (ASTM) See CELSIUS SCALE (°C), and FAHRENHEIT SCALE (°F).[176]

temperature checking tests of finishes as applied to FURNITURE.[56,157,175] Syn: cold checking

It is a film rupture at an angle to the run of the GRAIN.

temperature, color see COLOR TEMPERATURE.[43,69]

temperature, correlated color see CORRELATED COLOR TEMPERATURE.[43,69]

temperature, curing temperature to which an ADHESIVE or an assembly is subjected to CURE the adhesive.[79,67] (ASTM) See TEMPERATURE, DRYING and TEMPERATURE, SETTING.[176]

The temperature attained by the adhesive in the process of curing it (adhesive-curing temperature) may differ from the temperature of the atmosphere surrounding the assembly (assembly-curing temperature).

temperature, drying temperature to which an ADHESIVE on an ADHEREND or in an assembly or the assembly itself is subjected to dry the adhesive.[67,79] (ASTM) See TEMPERATURE, CURING and TEMPERATURE, SETTING.[176]

The temperature attained by the adhesive in the process of drying it (adhesive-drying temperature) may differ from the temperature of the atmosphere surrounding the assembly (assembly-drying temperature).

temperature, setting temperature to which an ADHESIVE or an assembly is subjected to SET the adhesive.[67,79,176] (ASTM) See TEMPERATURE, CURING and TEMPERATURE, DRYING.[176]

The temperature attained by the adhesive in the process of setting it (adhesive-setting temperature) may differ from the temperature of the atmosphere surrounding the assembly (assembly-setting temperature).

temporary coatings a coating designed to protect or decorate a SUBSTRATE for a limited time that can be readily removed either by mechanical or chemical means.[71] Syn: strippable coatings

tensile properties those mechanical properties measured in a test where the sample is elongated (pulled) at a constant rate, and the FORCE required is recorded.[47,157]

These include the ultimate TENSILE STRENGTH, ELONGATION, YOUNG'S MODULUS, YIELD STRENGTH, ELASTIC MODULUS, TOUGHNESS (also called ENERGY AT BREAK), and HARDNESS.

tensile strength (1) general: the strength shown by a specimen subjected to tension, as distinct from torsion, compression, or shear; (2) specific: the maximum tensile strength expressed in FORCE per unit cross-sectional area of the unstrained specimen; for example, kilogram-force per square millimetre (kgf/mm^2), pounds-force per square inch (psi) (ASTM); and (3) ability a material possesses of resisting deformation by the application of a force or load.[47,157] cf. ULTIMATE TENSILE STRENGTH[47,157]

Tensile properties are of value in studying the behavior of coatings subjected to environmental stresses, such as those produced by aging and WEATHERING. They may vary with specimen thickness, method of preparation, GAGE LENGTH, rate of load application, TENSILE TESTER response, and type of grips used.

ASTM Test Method D2370 covers the determination of tensile strength as well as ELONGATION and STIFFNESS (modulus of elasticity) of ORGANIC coatings when tested as free films.

tensile tester an apparatus designed to impart, or transmit, FORCE/extension, or stress/strain, to a material and to measure the effect of the action.[145,47,157] (ASTM) Syn: tensile testing machine

tensile testing machine see TENSILE TESTER.[145,47,157]

tensiometer see SURFACE TENSIOMETER.[145]

terebine form of liquid DRIERS, originally made by combining LINSEED OIL and natural RESIN with SALTS of lead and manganese and thinning with TURPENTINE.[67,83] (BSI)

terephthalic acid $C_8H_6O_4$. Forms POLYESTERS with GLYCOLS.[144] Syn: p-phthalic acid, p-BENZENE-dicarboxylic acid, tephthol

Mol wt, 166.13, sublimes above 300°C without melting.

term a word or expression that has a precise meaning, in some uses, or is peculiar to a science, art, profession, or subject.[134]

terpene resins neutral HYDROCARBON RESINS made by POLYMERIZATION of ß PINENE.[167]

They possess all the advantages of COUMARONE RESINS, e.g., neutrality, resistance to alkalis, etc., without the disadvantages of YELLOWING. They have excellent COMPATIBILITY.

terpenes a class of UNSATURATED ORGANIC COMPOUNDS having the empirical formula $C_{10}H_{16}$ occurring in most ESSENTIAL OILS and OLEORESINOUS plants.[44,164]

Structurally, the important terpenes and their derivatives are classified as monocyclic (DIPENTENE), bicyclic (PINENE), and acyclic (myrcene).

terpinenes $C_{10}H_{16}$. HYDROCARBONS of cyclic type, present in TURPENTINES.[44,164]

Properties: bp, about 179°C; sp gr, 0.865/15°C.

terpineol $C_{10}H_{17}OH$. Useful, powerful high-boiling SOLVENT PLASTICIZER.[164,170]

Words presented in CAP/SMALL CAPS type indicate that the word is defined in another part of the *Coatings Encyclopedic Dictionary.*

Numerical superscripts classify terms in one or more of the categories listed in the second section of this volume.

Bp, 210-220°C, depending upon the type, α, ß, or γ.

terpinolene $C_{10}H_{16}$. HYDROCARBON SOLVENT.[164]

Properties: bp, 183°C; mp, 15°C; flp, 44°C (111°F).

terpolymer a POLYMER composed of molecules containing three chemically different types of MERS.[130,141] (IUPAC)

terra alba synonym for GYPSUM or CALCIUM SULFATE.[58]

terra cariosa see ROTTENSTONE.[158]

terra cotta a fine reddish-brown CLAY mixed with sand and baked until hard, used for pottery, statues, and the like; also the reddish-brown COLOR of baked clay.[75,81]

terra di siena see RAW SIENNA and IRON OXIDES, NATURAL.[41]

terrazzo hard, marble-like flooring material consisting of 70% marble chips in CEMENT, that is cast in place or precast, and then ground down to a smooth surface; used as a DECORATIVE surfacing on walls as well as floors.[73,75,82] Syn: terrazzo concrete and Venetian mosaic

terrazzo seal a composition of ALKALI-RESISTANT RESINS or WAXES in SOLVENT or water.[147,137]

terre verte essentially an iron silicate. It is used as a base for LAKES.[41,58] Syn: green earth

tertiary amine value the number of milligrams of potassium hydroxide (KOH) equivalent to the tertiary amine basicity in 1 g of sample.[128,157] (ASTM)

test a procedure for critical evaluation; a means of determining the presence, quality, or truth of something; a trial.[157,37,145]

test charts smooth surface PAPER charts having adjacent black and white areas, and coated with a suitable VARNISH or LACQUER to render the surface impermeable to paint liquids.[157,59] (ASTM)

Charts of this type are frequently referred to as hiding power charts, and more generally as paint test charts. In addition to hiding power, they also serve as standard test surfaces for DRYING, GLOSS, LEVELING, SAGGING, HOLD OUT, and other paint characteristics. Their printed patterns are usually large juxtaposed areas of black and white, but traditional "checkerboard" and "zebra stripe" patterns are also used widely. Gray and white, black and gray, and multi-shade gray on white are commonly used variations. Unprinted plain white test charts are very widely used too, although of course not for hiding power purposes. Commercial sizes range from 3 × 5-1/2 in. (75 × 140 mm) up to 24 × 36 in. (6120 × 915 mm), with intermediate sizes of about 50 to 100 sq. in. (325 × 650 cm²) in area being the most common. The thickness of the paper stock is generally about 11 mils (280 mm).

test fence an apparatus consisting of a fence strategically located in a part of the country for specific weather conditions (temperature, humidity, sunlight duration, etc.) and facing a specific direction and angle.[154,157] See EXPOSURE RACK.[154,157]

It contains a series of exposure racks on which test panels are exposed. Coated test panels on various substrates (WOOD, METAL, PLASTIC, CEMENTITIOUS, etc.) are tested for exterior durability properties such as CHALK RESISTANCE, TINT RETENTION, ADHESION, CRACKING, PEELING, etc. The panels are exposed for various durations of time. Typical directions are north and south. Typical angles are vertical, horizontal, 5°, and 45°. Examples of U.S. test fence locations include: Arizona, Florida, and Pennsylvania.

test method a definitive, standardized set of instructions for the identification, measurement, or evaluation of one or more qualities, characteristics, or properties of a material.[157] (ASTM)

test pattern SPRAY pattern used in adjusting SPRAY GUN.[54]

tetrachloroethane $C_2H_2Cl_4$. SOLVENT, used at one time as the principal constituent of nonflammable PAINT REMOVER, but now largely discontinued because of its TOXICITY.[164]

Properties: bp, 147°C; vp, 11 mm Hg/20°C.

tetrachloromethane see CARBON TETRACHLORIDE.[164]

tetrahydrofurfuryl alcohol $C_4H_7OCH_2OH$. High-boiling SOLVENT. Used as a solvent for VINYL RESINS.[164] Syn: "THFA" and tetrahydrofuryl

Bp, 178°C; vp, < 1 mm Hg/30°C.

tetrahydronaphthalene $C_{10}H_{12}$. High-boiling SOLVENT.[164] Syn: tetralin

Properties: bp, 198-210°C; sp gr, 0.971/20°C; flp, 77°C (170°F); vp, < 1 mm Hg/30°C. It is used in ALKYD COMPOSITIONS, CHLORINATED RUBBER FINISHES, and in PAINT REMOVERS.

tetralin synonym for TETRAHYDRONAPHTHALENE.[164]

tetramethyxylene diisocyanate (TMXDI) an ALIPHATIC ISOCYANATE MONOMER used as an INTERMEDIATE in URETHANE COATINGS.[141] See ISOCYANATE RESINS,[141] POLYURETHANES,[141] POLYURETHANE FINISH,[141] and URETHANE COATINGS.[141] cf. HEXAMETHYLENE DIISOCYANATE (HDI),[141] ISOPHORONE DIISOCYANATE (IPDI),[141] DIPHENYLMETHANE DIISOCYANATE (MDI),[141] and TOLUENE DIISOCYANATE (TDI)[141]

Texanol® 2,2,4-trimethylpentanediol monoisobutyrate. A proprietary high boiling ESTER ALCOHOL water-insoluble COALESCENT.[164,83]

Properties: flash point, 118°C (Cleveland Open

Cup); density, 7.92 lb/gal; sp gr, 0.95.

textile printing the DECORATIVE enhancement of cloth by applying a pattern or colored design with INKS, DYES, or PRINT PASTES.[148,168] (EPA)

The design is usually applied by ROLLER, FLAT SCREEN, or ROTARY SCREEN techniques.

texture the structural quality of a surface.[158]

texture-finished paint see PLASTIC PAINT.[71,163,54]

textured paint see PLASTIC PAINT.[71,163,54]

T_g see GLASS TRANSITION TEMPERATURE.[47]

THEA see TETRAHYDROFURFURYL ALCOHOL.[164]

Thénard's blue synonym for COBALT BLUE.[41]

theoretical VOC content the quantity of VOC which will be released during APPLICATION, FLASHOFF, and CURE per unit volume of coating (or coating solids).[168,172] (EPA)

thermal of, relating to, using, producing, or caused by heat.[176]

thermal adhesive special type of ADHESIVE which only develops its adhesive properties when heat is applied to it, and partial liquefaction occurs.[79,176]

thermal discharge the introduction of water from a point source at a temperature different from the ambient temperature of the receiving waters.[171,176] See POINT SOURCE.[171]

thermal incinerator a device for oxidizing waste material via flame and heat.[172,176] (EPA)

This contrasts with a CATALYTIC INCINERATOR which incorporates a CATALYST to aid the combustion.

thermal paint coating containing small granules of METALLIC POWDER that create a resistance when current is applied and thereby heat the SUBSTRATE.[163,176]

thermography PRINTING PROCESS in which the INK, while still wet on the sheet, is dusted with a RESINOUS powder that adheres to the ink.[168,176]

The sheets are then put through a heating process which causes the PARTICLES of powder to fuse together with the ink, giving a raised effect to the letters which simulate steel-die ENGRAVING.

thermoplastic capable of being repeatedly softened by heat and hardened by cooling.[130,176] (ASTM)

thermoplastic resins RESINS which remain fluid on heating and do not become infusible.[130,176]

thermoset a material which will undergo or has undergone a chemical reaction by the action of heat, CATALYSTS, ULTRAVIOLET light, etc., leading to a relatively infusible state.[67,140,176] (ASTM)

thermosetting having the property of undergoing a chemical reaction by the action of heat, CATALYSTS, ULTRAVIOLET LIGHT, etc., leading to relatively infusible state.[67,140,176] (ASTM)

thermosetting ink INK which POLYMERIZES to a permanently solid and infusible state upon the application of heat.[67,168,176]

thermosetting resin a RESIN which POLYMERIZES to a permanently solid and infusible state upon the application of heat.[67,130,176]

thermospray see FLAME SPRAY.[77,176]

thesaurus usually a subject-based presentation of TERMS, where the content-based relationships (particularly hierarchical) between terms are delineated.[134] (ASTM)

thick having relatively great CONSISTENCY.[166]

thickener (1) any material used to THICKEN a liquid; (2) an ADDITIVE used to thicken (increase VISCOSITY) or modify the RHEOLOGY of a coating.[83,166] See THIXOTROPE.[83,166] Syn: bodying agent

thickening an increase in the CONSISTENCY of a coating.[166,161,42]

thinner (1) the portion of a paint, VARNISH, LACQUER, or PRINTING INK, or related product that VOLATILIZES during the DRYING process;[164] (ASTM) (2) any volatile liquid used for REDUCING the VISCOSITY of coating compositions or components; may consist of a simple SOLVENT or DILUENT or a mixture of solvents and diluents;[164] (3) solvents, diluents, low viscosity OILS, and vehicles added to inks to reduce their viscosity, CONSISTENCY, or TACK.[164,168]

thinning ratio the amount of THINNER that is recommended for a given quantity of paint.[164] (DAC)

thiocyanogen value measure of the number of single DOUBLE BONDS in a substance, for example, an OIL.[143,128]

Whereas thiocyanogen is selective in its action, adding on to isolated double bonds only, iodine is not selective, and thus a combination of both thiocyanogen and iodine values provides a means of assessing quantitatively the different types of UNSATURATED components in a material.

thioindigo pigments vat DYES covering the orange, red, violet, and brown shade ranges.[41]

Words presented in CAP/SMALL CAPS type indicate that the word is defined in another part of the *Coatings Encyclopedic Dictionary.*

Numerical superscripts classify terms in one or more of the categories listed in the second section of this volume.

They are noted for their BRIGHTNESS of shade and generally good FASTNESS properties.

thioindigoid maroons (73395). These are maroon in MASSTONE and violet in TINT.[41]

They are considerably bluer than most ORGANIC maroon pigments and find application in shading durable maroon finishes. Generally speaking, thioindigoid maroons are not bake resistant enough, nor sufficiently LIGHTFAST in weak tints to be used effectively as toning agents for whites. Some varieties BLEED slightly in AROMATIC and KETONE SOLVENTS. The nonbleeding types find extensive use in automotive finishes.

thiourea NH_2CSNH_2. Used in the preparation of THIOUREA-FORMALDEHYDE RESINS.[130] Syn: thiocarbamide

Properties: mp, 180°C; sp gr, 1.42.

thiourea resins resins made by the interaction of THIOUREA and ALDEHYDES.[130]

thiourea-formaldehyde resin an AMINO RESIN made by polycondensation of thiourea (thiocarbamide) with formaldehyde.[130]

thixotrope ADDITIVE used to impart THIXOTROPY to a coating.[83,166] See THICKENER.[83,166]

thixotropic adjective which describes full-bodied material which undergoes a reduction in VISCOSITY when shaken, stirred, or otherwise mechanically disturbed and which readily recovers the original full-bodied condition on standing.[166,133] See BODY; FALSE BODY.[166]

thixotropic paint paint which, while free-flowing and easy to manipulate under a BRUSH, sets to a GEL within a short time when it is allowed to remain at rest.[71,166] (BSI)

Because of these qualities a thixotropic paint is less likely to DRIP from a brush than other types and can be applied in rather thicker films without RUNNING or SAGGING.

thixotropy literally change by touch. FLOW behavior in which viscosity is reduced by touch (agitation, stirring). Thixotropy is time dependent. With a thixotropic material, viscosity decreases with time at a given shear rate, then rebuilds with time when the shearing stops. See FALSE BODY.[166] cf. RHEOPEXY[166]

Thixotropic does not necessarily mean shear thinning.

three-piece can a can made of three different pieces, a body, a top, and a bottom.[50] (EPA)

three-roll mill see TRIPLE ROLLER MILL.[59]

threshold limit value (TLV) refers to airborne CONCENTRATIONS of substances, and represents conditions under which it is believed that nearly all workers may be repeatedly exposed, day after day, without adverse effect.[172]

These values may be breathed continually for eight hours per day without harm. Because of wide variation in individual susceptibility, EXPOSURE of an occasional individual at or even below the threshold limit may not prevent discomfort, aggravation of a preexisting condition, or occupational illness. Threshold limits should be used as guides in the control of health hazards and should not be regarded as fine lines between safe and dangerous concentrations. The American Conference of Governmental Industrial Hygienists (ACGIH) adopts a list of threshold limit values (TLV) each year for more than 450 substances.

through-dry see DRY-THROUGH.[157]

through-drying uniform drying throughout the film as opposed to BOTTOM-DRYING or TOP-DRYING.[67]

throw out a PRECIPITATE or FLOC.

throwing DEFECT of impregnating insulating VARNISHES used on high-speed rotating equipment, and evidenced by the ejection of PARTICLES of VARNISH by the centrifugal FORCE generated.[56]

throwing out (1) see GAS CHECKING;[56,63,157] (2) formation of PRECIPITATE; (3) FLOCCULATION.[42,64,139,161]

throwing power the improvement of a coating (usually METAL) distribution over the primary current distribution on an electrode (usually cathode) in a given solution, under specified conditions.[61,77] (ASTM) See ELECTRODEPOSITION.[61,77]

The term may also be used for anionic processes for which the definition is analogous.

thus OLEORESINOUS EXUDATION of ROSIN type.[170]

Because of its relative softness it has been used to a small extent in SPIRIT VARNISHES as a PLASTICIZER.

tie coat intermediate coat used to BOND different types of paint coats.[71] Syn: block coat, transition PRIMER, barrier coat

Tiffany finish a wall decoration in which a number of TRANSLUCENT paints are applied in various HUES in an irregular manner and modified by wiping, blending, etc., a kind of SCUMBLING.[54,148]

tight cure sufficient VULCANIZATION to give a product good TENSILE STRENGTH and good snap or ELASTICITY.[67,140]

tile a CERAMIC surfacing unit, usually relatively thin in relation to facial area, made from CLAY or a mixture of clay and other ceramic materials, called the body of the tile, having either a "glazed" or "unglazed" face and fired above red heat in the course of manufacture to a temperature sufficiently high to produce specific physical properties and characteristics.[75,159] (ASTM)

tile-like coating (TLC) system applied by conventional means and intended to produce VITREOUS (tile-like) finishes on relatively rough MASONRY or other CEMENTITIOUS walls and ceilings.[163] (ASTM) See HIGH-BUILD COATING.[71,168]

They are generally thicker, harder, and more washable than conventional paints.

TLC see TILE-LIKE COATING.[163]

timber (1) trees on wooded land considered as a source of wood; (2) wood used as a building material; lumber.[153,75]

time, curing see CURING TIME.[67]

tinctorial strength see COLOR STRENGTH.[43,69,168]

tinge see CAST.[41,69]

tint (1) the COLOR produced by the mixture of white PIGMENT with absorbing (generally CHROMATIC) COLORANTS. The color of the resulting mixture is LIGHTER and less SATURATED than the color without the addition of the white;[69,168] (2) verb. See SHADE (4) and TINTING.[69]

tint base paint an architectural paint, generally containing less TITANIUM DIOXIDE than a full white paint, designed to be TINTED to a specific COLOR with one or more colorants.[71,69] See TINTABILITY,[69,42] COLOR DEVELOPMENT,[43,69] and COLOR UNIFORMITY.[69,42]

Retail paint outlets stock tint base paints which are custom colored for the customer by the addition of controlled volumes of colorant/s usually added by special equipment which controls the type and amount of colorants needed to match the specific color requested by the customer.

tint-tone COLOR obtained when a MASSTONE PIGMENT is mixed with a white pigment in a vehicle system.[43,69]

tintability the capability of a white or TINT BASE PAINT to accept various COLORANTS, as evidenced by COLOR DEVELOPMENT in the mixture.(ASTM)[69,42] Syn: color acceptance

Retail paint outlets stock tint base paints which are custom colored for the customer by the addition of controlled volume of colorant. Good color development is required in order to match color chips and give uniform color APPEARANCE.

ASTM Test Method D5326 covers a method for measuring color development in tinted latex paints for the purpose of determining the efficiency of colorants and the potential for poor color uniformity of applied paint films.

tinting (1) final adjustment of the COLOR of a paint to the exact color required;[41,43,69] (BSI) (2) in LITHOGRAPHY, a uniform DISCOLORATION of the background caused by the BLEEDING or washing of the PIGMENT in the FOUNTAIN SOLUTION.[168]

tinting color or colorant colored PIGMENT milled in media compatible with paint vehicles added in relatively small portions to already prepared paints to modify their color.[83,69,41] See TINT BASE PAINT.[71,69]

tinting (or tint) strength, scattering relative change in the scattering properties of a standard black material (with no white PIGMENT present) when a specified amount of a white or chromatic is added to it.[43,69] (ASTM) See TINTING STRENGTH.[43,69]

tinting strength measure of the effectiveness with which a unit quantity of COLORANT alters the COLOR of a material.[43,69] (ASTM)

For SCATTERING and ABSORBING colorants (pigments), both ABSORPTION and SCATTERING TINTING STRENGTH must be specified.

Tinting strength may be one factor in judging the relative economic value of paints, since PIGMENT concentration contributes to strength in a major way; other factors are formulation and COLOR DEVELOPMENT in GRINDING. The user may also select products for other properties, such as TRANSPARENCY, that are accompanied by different tinting strengths.

ASTM Test Method D4838 determines the relative tinting strength of chromatic paints.

tinting strength, absorption relative change in the ABSORPTION properties of a standard white material where a specified amount of an absorbing PIGMENT, BLACK or CHROMATIC, is added to it.[43,69] (ASTM)

This is the common definition of TINTING STRENGTH; however, this definition of the term can be misleading. For example, the tinting strength of a yellow colorant depends on its SCATTERING as well as its absorption. Its tinting strength as determined from a mixture with white provides no information about its behavior when mixed with low-scattering colorants, such as black.

tire pick-up resistance the ability of a LATEX paint to maintain ADHESION and resist imprint and black marks of tire treads when applied to a concrete surface.[79,72]

Water-based coatings do not penetrate into CEMENTITIOUS surfaces and therefore lack the adhesion of SOLVENT-based coatings. The alkalinity of the cementitious surface is a factor with certain systems.

titania see TITANIUM DIOXIDE.[41]

titanium brown brown PIGMENT made by precipitating a mixed solution of cobalt and ferrous SALTS,

Words presented in CAP/SMALL CAPS type indicate that the word is defined in another part of the *Coatings Encyclopedic Dictionary.*

Numerical superscripts classify terms in one or more of the categories listed in the second section of this volume.

containing a suspension of titanium hydroxide, with sodium carbonate.[41]

The mixed precipitate is washed and calcined.

titanium dioxide, anatase TiO_2. Pigment White 6 (77891). A high-opacity, bright PIGMENT of the chalking type, used as a prime pigment in paints, RUBBER, PLASTICS.[41] cf. TITANIUM DIOXIDE, RUTILE.[41] Syn: titania

Prepared from the mineral, ilmenite, or rutile ore. Density, 3.8-4.1 g/cm³ (32-34 lb/gal); O.A., 18-30; particle size, 0.3 µm; refractive index, 2.55.

titanium dioxide, rutile TiO_2. Pigment White 6 (77891). A high-opacity, bright white PIGMENT, nonchalking type, used as a prime pigment in paints, RUBBER, plastics.[41] cf. TITANIUM OXIDE, ANATASE[41] Syn: titania

Prepared from the mineral, ilmenite, or rutile ore. Properties: density, 3.9-4.2 g/cm³ (33-35 lb/gal); O.A., 16-48; particle size, 0.2-0.3 µm; refractive index, 2.76. Titanium dioxide in its rutile crystalline form has an exceptionally high REFRACTIVE INDEX and its OPACITY and TINTING STRENGTH, when finely dispersed, exceed all other white pigments. It is the most widely used white pigment in the paint and coatings industry.

In addition to giving opacity and whiteness to coatings, titanium dioxide absorbs or reflects harmful RADIATION, thereby protecting the SUBSTRATE. In its pure form it can also act as a photoactive CATALYST causing DEGRADATION of the BINDERS, such as ALKYDS, and eventual loss of pigment from the surface—one of the mechanisms of CHALKING. To prevent or reduce this, titanium oxide PARTICLES are coated with, for example, ALUMINA and SILICA. The coating of titanium dioxide is a highly developed process and has resulted in numerous grades intended for specific uses.

titanium greens complex PIGMENTS based on CALCINED mixtures of TITANIUM OXIDE or hydroxide with suitable other metallic oxides, carbonates, etc.[41]

The other METALLIC COMPOUNDS include those of zinc.

titanium lithopone this was made by mixing a minor proportion of TITANIUM DIOXIDE into LITHOPONE, or possibly by the COPRECIPITATION of the usual lithopone constituents in the presence of titanium hydroxide.[41] Syn: titanated lithopone

The resultant product in the latter case is subjected to controlled CALCINATION.

titanium yellow see NICKEL TITANATE.[41]

TLV see THRESHOLD LIMIT VALUE.[172]

TLV-TWA the allowable time-weighted average concentration for a normal eight-hour workday or 40-hour week.[172]

TMXDI abbreviation for TETRAMETHYXYLENE DIISOCYANATE.[141] cf. HDI, IPDI, MDI, and TDI[141]

TNO abbreviation for Dutch Organization for Applied Research[174] (Toegepast Natuurwetenschappelÿk Onderzock).

tobacco seed oil seed OIL obtained from *Nicotiana tabacum*.[63]

Considerable divergencies in composition of the oil have been reported. Some types contain as much as 70% LINOLEIC ACID, whereas others contain no linoleic acid and more than 54% of linolenic acid. In consequence, its constants as reported vary considerably. Certain types have excellent DRYING properties, and can replace LINSEED OIL without detriment.

tobias acid intermediate used in the manufacture of DYESTUFFS. 2-naphthylamine-1-sulfonic acid.[41]

TOC abbreviation for TOTAL ORGANIC COMPOUND.[172,152] (EPA)

tocopherols naturally occurring ANTIOXIDANTS in VEGETABLE OILS.[63,144]

tole (Fr.) painted tinware; today it is done on many surfaces.[81,148]

tolerance the total range of variation (usually bilateral) permitted for a size, position, or other required quantity; the upper and lower limits between which a DIMENSION must be held.[157] (ASTM)

toluene diisocyanate (TDI) an aromatic ISOCYANATE MONOMER used as an INTERMEDIATE in URETHANE COATINGS.[141] See ISOCYANATE RESINS,[167] POLYURETHANES, POLYURETHANE FINISH,[71] and URETHANE COATINGS.[71] cf. HEXAMETHYLENE DIISOCYANATE (HDI), ISOPHORONE DIISOCYANATE (IPDI), DIPHENYLMETHANE DIISOCYANATE (MDI), TETRAMETHYXYLENE DIISOCYANATE (TMXDI)[141]

toluene AROMATIC SOLVENT used in the manufacture of coatings.[141,164]

The commercial product has: boiling range, 105-112°C; flp, 50°F; vp, 26 mm Hg/30°C. The term "toluol" is still used commercially but is not preferred.

toluene-sulfonamide resins RESINS made by the interaction of toluene sulfonamide and FORMALDEHYDE.[130]

toluidine reds Pigment Red 3 (12120). Series of red DYESTUFFS made by diazotizing 2-nitro-p-toluidine and coupling this with ß naphthol under alkaline conditions.[41]

By altering the conditions of preparation, reds of different shade, brilliance, strength, etc., are obtained.

ASTM Test Method D970 covers para red and toluidine red pigments (toners) in the dry form commercially known as "pure."

ASTM D475 is the standard specification for

pure para red toner pigment. It is available in two types: Type I, light, and Type II, dark.

toluidine toner see TOLUIDINE REDS.[41]

toluidine yellow see HANSA YELLOW.[41]

tone use of this term is to be deprecated since it is variously employed in different senses.[43] (BSI)

According to the context, "HUE" or "UNDERTONE" are, respectively, the preferred terms.

toner (1) a single ORGANIC PIGMENT which does not contain INORGANIC pigment (EXTENDER), or a single inorganic pigment having maximum ABSORPTION strength for the given type of pigment; (2) also loosely applied to pure pigment DYESTUFFS, but this use of the term is deprecated.[41,43,69] See EXTENDED PIGMENT.[41,58]

toning blue see IRON BLUE.[41]

tooth in a dry paint film, a fine texture imparted either by a proportion of relatively COARSE or ABRASIVE PIGMENT or by the ABRASIVES used in SANDING; this texture improves the RUBBING properties and also provides a good base for the ADHESION of a subsequent coat of a paint or VARNISH.[47,158]

top color see FACE COLOR.[69]

top-drying drying of a film at the top only, e.g., COBALT NAPHTHENATE is used as a top DRIER.[67]

topcoat the coating intended to be the last coat applied in a coating system; usually applied over a PRIMER, UNDERCOATERS, or SURFACERS.[54,71] Syn: FINISH coat

topcoat holdout see ENAMEL HOLDOUT.[72]

torsion-braid analyzer an instrument which permits the measurement of thermomechanical properties of POLYMERS that are undergoing structural changes during CURE.[67,145]

total color difference the perceived difference between two colors, including the differences in HUE, SATURATION, and LIGHTNESS; generally used as a value calculated from a COLOR DIFFERENCE EQUATION and designated as ΔE.[69,35,43]

total enclosure an enclosure around the coating head of a web coating line or other application device so that all VOC from the coating application and FLASHOFF operations are collected and ducted through a stack or into the oven.[172,164]

With a total enclosure on a coating line there will be no fugitive emissions, only stack emissions. Some air pollution regulations may require a total enclosure so that all VOC emitted by the process can be measured.

total organic compound (TOC) those COMPOUNDS measured according to the procedures in an applicable test method.[172,157] (EPA)

total reflectance see REFLECTANCE, TOTAL.[43]

total solids see SOLIDS.[128]

total transmittance see TRANSMITTANCE, TOTAL.[43]

touch-dry see SET-TO-TOUCH TIME and DRYING TIME.[67]

touch-up coating often, after a METAL part such as an automobile body or a large appliance goes through a coating line, small imperfections will be present which are corrected later in a separate SPRAY BOOTH. This process is called "touch-up."[54,172] (EPA) cf. TOUCH-UP PAINTING[148,158,54]

VOLATILE ORGANIC COMPOUNDS from this repair process may be treated differently than emissions from the main coating line in regulations.

touch-up painting application of paint on small areas of painted surfaces to repair MARS, scratches, and small areas where the coating has DETERIORATED, in order to restore the coating to an unbroken condition.[148,158,54]

toughness that property of a material by virtue of which it can ABSORB work.[47,157] (ASTM) cf. BRITTLE[47,157]

This is a measure of the amount of energy that must be used to cause a sample in a TENSILE TESTER to rupture. It is measured from the area under the STRESS-STRAIN CURVE. Physically, RUBBER is "tougher" than GLASS, and will have a larger area under its stress-strain curve. Units are in lb-in/in² or N-m/m².

toxic poisonous.[172]

toxic pollutants those POLLUTANTS which, after discharge and upon contact with any organism, either directly from the environment or indirectly by ingestion through food chains, will cause death, disease, behavioral abnormalities, cancer, genetic mutations, physiological malfunctions, or physical deformities in such organisms or their offspring.[171,172]

toxic substance a substance that demonstrates the potential to induce cancer, to produce short- and long-term disease or bodily injury, to affect health adversely, to produce acute discomfort, or to endanger life of man or animal, resulting from EXPOSURE via the respiratory tract, SKIN, eye, mouth, or other routes in quantities which are reasonable for experimental animals or which have been reported to have produced toxic effects in man.[172] (OSHA)

Words presented in CAP/SMALL CAPS type indicate that the word is defined in another part of the *Coatings Encyclopedic Dictionary.*

Numerical superscripts classify terms in one or more of the categories listed in the second section of this volume.

Toxic Substances Control Act (TSCA) public law under EPA jurisdiction, it controls the EXPOSURE to and use of raw industrial chemicals not subject to other laws. Chemicals are to be evaluated prior to use and can be controlled based on risk. The act provides for a listing of all chemicals that are to be evaluated prior to manufacture or use in the U.S.[172]

toxicant a poison or poisonous agent.[83,172]

toxicity (1) the measure of the adverse effect exerted on the human body by a poisonous material;[172] (2) a relative property of a chemical agent with reference to a harmful effect on some biological mechanism and the condition under which this effect occurs. The quality of being poisonous.[172] (IUPAC)

It may be expressed as a fraction indicating the ratio of the smallest mass of the material which has some probability (often 50%) of causing death of some specific animal to the mass of the animal.

toxicology the study of the nature, effects, and detection of poisons and treatment of poisoning.[172]

Also, substances that are otherwise harmless but prove toxic under particular conditions. The basic assumption of toxicology is that there is a relationship among the dose (amount), the concentration at the affected site, and the resulting effects.

TPP abbreviation for TRIPHENYL PHOSPHATE.[170]

TPTF abbreviation for triphenyltinfluoride.[144,173]

Toxicant used in ANTIFOULING paints.

tracer a foreign substance mixed with or attached to a given substance to enable the distribution or location of the latter to be determined subsequently.[83,37] (IUPAC)

There are several types of tracers which are used: (a) a <u>physical tracer</u> is one that is attached by physical means to the object being traced; (b) a <u>chemical tracer</u> is a chemical with properties similar to those of the substance being traced with which it is mixed homogeneously; (c) an <u>isotropic tracer</u> is a unique isotope, either radioactive or an enriched, uncommon stable isotope, of the element to be traced; (d) a <u>radioactive tracer</u> is a physical or chemical tracer having radioactivity as its distinctive property which allows detection at small concentrations and hence after large transport distances. The composition of AEROSOLS in the troposphere has been used as a qualitative tracer of air masses. The elemental analyses (determined by neutron activation, X-ray fluorescence, etc.) of the aerosols transported from various source regions sometimes have characteristic patterns which are used to define qualitatively the origin of tropospheric aerosols collected in other geographic regions.

trade association an organization formed by members of an industry to promote the business of an industry as a whole or to encourage scientific and technical advances which will benefit the industry.[174] (EPA)

Government air pollution regulators often talk to trade associations who have an interest in volatile compound regulations. Prominent among these are: American Petroleum Institute, American Society for Testing and Materials, Can Manufacturing Institute, Chemical Manufacturers Association, Flexible Packaging Association, Graphic Arts Technical Foundation, International Gas Cleaning Institute, Motor Vehicle Manufacturers Association, National Coil Coaters Association, National Paint and Coatings Association, National Spray Equipment Manufacturers Association, Powder Coatings Institute, and Rubber Manufacturers Association.

trade journal a magazine devoted to the interests of a specific industry.[174] (EPA)

trade name a name used to identify a commercial product or service, which may or may not be registered as a trademark. Syn: brand name

It is not usually the chemical name, and the same or similar products can be marketed under different trade names by different companies.

trade sales paint sold as shelf goods through retail outlets to the general public and used mainly for coating of architectural structures.[71] (EPA) See ARCHITECTURAL COATINGS.[71]

trade sales coating see ARCHITECTURAL COATINGS.[71]

trade sales paints coatings applied on-site at ambient conditions by the consumer using application methods such as BRUSHING or ROLLER COATING.[71] See ARCHITECTURAL COATINGS.[71]

trade secret confidential information (formula, process, device, or compilation of data) that gives the owner an advantage over competitors.[172]

Manufacturers may choose to withhold data of a proprietary nature from an MSDS. Typically this would be the ingredients of a formulated product. OSHA permits this provided that (1) the trade secret claim can be substantiated; (2) the MSDS indicates that data is being withheld, and (3) the properties and health effects are included. State laws vary on this practice; some states require a trade secret registration number to be assigned to a material. There are procedures to obtain limited trade secret disclosure in emergency situations.

traffic marking MARRING or DISCOLORATION, or both, of a floor surface by traffic.[42] (ASTM)

traffic marking paint paint specially formulated to withstand wear of vehicular traffic and to be highly visible at night; used to mark center lines on roadways, highways, traffic lanes, parking lots, curb markings, crosswalks, etc.[71] (DAC) See BEADED PAINT and BALLOTINI.[71,163]

traffic paint abrasion resistance the ability of a coating to resist being worn away and to maintain its original APPEARANCE and structure when subjected to RUBBING, SCRAPING, or wear.[72] cf. TRAFFIC PAINT CHIPPING RESISTANCE[72]

The FSCT Pictorial Standards of Coatings Defects has a series of four photographs representing degrees of resistance to failure by ABRASION and/or EROSION to TRAFFIC PAINT providing a guide to evaluating failure.

traffic paint chipping resistance the ability of a coating or layers of coatings to resist removal, usually in small pieces, resulting from impact by hard objects or from wear during service.[72] cf. TRAFFIC PAINT ABRASION RESISTANCE[72]

The FSCT Pictorial Standards of Coatings Defects has a series of four photographs representing degrees of failure by chipping of traffic paint. ASTM Test Method D913 has these same photographs and covers the evaluation of degree of resistance to wear that may occur with traffic paints in road tests or in actual service.

tragacanth water-soluble GUM derived from *Astragalus* shrubs, and exported from Iran, Iraq, Turkey, India, Syria, and neighboring countries.[155]

It is obtained both as a natural and as an artificial EXUDATION. It resembles gum arabic in being composed of calcium, magnesium, and potassium SALTS. It is used as a STABILIZER for EMULSIONS.

transfer coating method of APPLICATION in which a coating is applied (cast or by other means) to a RELEASE PAPER, DRIED and/or CURED and then transferred onto the desired SUBSTRATE by pressure (nip roll or press) or ADHESIVE.[77]

transfer efficiency the ratio of the amount of coating solids deposited onto the surface of the coated part to the total amount of coating solids used.[49] (EPA)

transfer paper the temporary backing employed in DECALCOMANIA.[81,168,148]

transfer printing see DECALCOMANIA.[81,168,148]

transfer roll in the COIL COATING industry, the roll between the pick-up roll and applicator roll where three rolls are present in a coating operation.[70]

transition primer coating COMPATIBLE with PRIMER and also with a FINISH coat which is not compatible with the primer.[71,158] See BLOCK COAT, TIE COAT, and BARRIER COAT.[71]

translucency APPEARANCE state between complete OPACITY and complete TRANSPARENCY; partially opaque.[43,69,35]

translucent coating a liquid formulation (such as VARNISH, SHELLAC, or LACQUER) which when dry forms a TRANSLUCENT film.[71,146,151,152] (DAC)

transmission PROCESS by which RADIANT ENERGY is TRANSMITTED through a material or an object.[43]

transmittance the ratio of the transmitted SPECTRAL RADIANCE (intensity), I_t, of monochromatic light (single WAVELENGTH or small band of wavelengths) to that of the incident spectral radiance (intensity), I_o:I_t/I_o.[43] (IUPAC)

transmittance, diffuse ratio of RADIANT FLUX TRANSMITTED in all forward directions, except the undeviated direction (the SPECULAR TRANSMITTANCE), to incident flux.[43,69]

transmittance, internal ratio of RADIANT FLUX reaching the exit surface of material to flux which penetrated entry surface; sometimes referred to as transmittancy.[43,69]

transmittance, specular ratio of flux transmitted without change in image-forming state to incident flux; undeviated transmitted beam.[43,69]

transmittance, total ratio of total FLUX TRANSMITTED in all forward directions to INCIDENT flux; includes both DIFFUSE and SPECULAR transmittances.[43,69]

transparency state of being TRANSPARENT or completely nonscattering; attribute of located mode of APPEARANCE that permits PERCEPTION of object or space through or beyond a surface or within a volume.[43,69,35]

transparency, ink INK film of incomplete HIDING, exhibiting "SHOW-THROUGH."[43,168]

transparent adjective to describe a material which transmits light without DIFFUSION or SCATTERING.[43]

transparent coating a liquid formulation (such as VARNISH, SHELLAC, or LACQUER) which, when dry, forms a TRANSPARENT film.[71,146,151,152] (DAC)

transparent inks INKS which lack HIDING POWER and permit light to pass through.[168]

They permit previous PRINTING to show through, the two colors blending to produce a third, e.g., a TRANSPARENT yellow over a blue to produce a green, where the two colors are superimposed.

transparent painting PAINTING with TRANSPARENT PIGMENTS as opposed to OPAQUE ones, e.g., GLAZING, WATER COLOR PAINTING.[81]

Words presented in CAP/SMALL CAPS type indicate that the word is defined in another part of the *Coatings Encyclopedic Dictionary.*

Numerical superscripts classify terms in one or more of the categories listed in the second section of this volume.

Transparent PIGMENT is converted to opaque pigment by the addition of white.

trapping PRINTING of one INK film on another, in MULTICOLOR wet printing.[168] Syn: wet printing

Successful trapping depends upon the relative TACK and thickness of the ink film applied.

travel change in COLOR as the angle of viewing a GONIOCHROMATIC material, such as a metallic paint film, is changed from the perpendicular to near-grazing.[43,69] Sometimes called FLOP or FLIP-FLOP.

treated pigment PIGMENT that has been PROCESSED during manufacturing to impart specific properties.[41,131]

triad group of three colors harmoniously related to each other.[69]

tributoxyethyl phosphate (TBEP) ORGANIC phosphate with wide applications in FLOOR POLISHES, paints, SYNTHETIC RUBBERS, and as a DEFOAMER. It is probably best known in emulsion floor polishes as a LEVELING AGENT.[144]

tributyl citrate see BUTYL CITRATE.[144]

tricalcium silicate a COMPOUND which is a main constituent of PORTLAND CEMENT.[73] (DAC)

trichloroethyl phosphate $C_2H_4Cl_3PO_4$. PLASTICIZER of special value in the formulation of FIRE-RESISTING LACQUERS.[170]

Bp, 220°C/20 mm Hg.

trichromat person requiring mixtures of three component primaries to MATCH COLORS.[69,43]

Normal observers fall into this classification: the CIE STANDARD OBSERVER is defined in terms of the amounts of three PRIMARY COLORS required to match all SPECTRAL colors. Not all trichromats are normal, however. Some observers exhibit anomalous trichromatism, requiring abnormal proportions of three primary colors for COLOR matching.

trichromatic of or consisting of three COLORS.[43,69]

trichromatic coefficients synonym for CHROMATICITY COORDINATES or coefficients.[43,69] See CHROMATICITY COORDINATES, GENERAL and CHROMATICITY COORDINATES, CIE.[43,69]

tricresyl phosphate (TCP) $(CH_3C_6H_4)_3PO_4$. One of the most important of all PLASTICIZERS. Syn: tritolyl phosphate (TTP)[170]

Properties: bp, 430-440°C; flp, 216°C (420°F). ASTM Test Method D1721 covers the detection in tricresyl phosphate of the presence of impurities that reduce potassium permanganate.

trienol synthetic TUNG OIL of Swiss origin, practically indistinguishable from the natural oil.[63]

triethanolamine 2,2,2″-nitrilotriethanol. $N(CH_2CH_2OH)_3$. Very HYGROSCOPIC, VISCOUS liquid with a slight ammoniacal odor.[144]

Turns brown on EXPOSURE to air and light. Used in the manufacture of SURFACTANTS and EMULSIONS. Properties: mol wt, 149.19; sp gr, 1.124 (9.37 lb/gal); mp, 21.2°C.

triethylene glycol $(CH_2OCH_2CH_2OH)_2$. Polyhydric ALCOHOL.[144]

Properties: bp, 285°C; flp, 166°C (331°F); vp, <0.1 mm Hg/20°C.

triglyceride ESTER obtained by the interaction of the three reactive hydroxyl groups of GLYCEROL with FATTY or other acids.[144]

trim (1) the visible woodwork or MOLDINGS of a room, such as the BASEBOARDS, CORNICES, casings, etc.; (2) any visible element, usually of METAL or WOOD, which protects or covers joints, edges, or ends of another material; the finishings around fittings and openings, as a door trim, window trim, etc.[82] (DAC)

trimer a POLYMER composed of molecules containing three identical MERS.[130] (IUPAC)

trimmed papers see PRETRIMMED PAPERS.[148]

tri-n-propyl phosphate $(C_3H_7)_3PO_4$. A PLASTICIZER and SOLVENT.[170,144]

Properties: sp gr, 1.012; bp, 135-140°C.

triphenyl phosphate (TTP) $(C_6H_5)_3PO_4$. A PLASTICIZER.[170]

Properties: bp, 410°C; mp, 50°C; flp, 220°C (428°F).

triple-roller mill type of MILL used for the fine GRINDING of pigmented compositions.[59]

The PIGMENT AGGLOMERATES are DISPERSED by passage between accurately machined steel or granite cylinders, the clearance between the cylinders being capable of very delicate adjustment. The cylinders or rollers are made to operate at different speeds. The rough pigment-medium mix is fed between two rollers rotating towards each other at different speeds, and it is finally transferred to the third roller, which operates at the maximum speed. Pigment dispersion is effected by a reduction of agglomerates, controlled by the clearance between the rollers, and by a shearing action developed by the viscous medium on rollers which are rotating at different speeds. ROLLER MILLS are ideal for grinding stiff pigment PASTES.

tripoli see ROTTENSTONE.[158,58]

tripolite see DIATOMACEOUS SILICA.[41,58]

tristimulus of, or consisting of, three stimuli; generally used to describe components of ADDITIVE mixture required to evoke a particular COLOR sensation.[43,69]

tristimulus colorimeter instrument used to measure quantities which can be used to obtain an approximation of TRISTIMULUS VALUES.[145,43,69]

They are normally equipped with three (or four) special filters to obtain R, G, B values which must then be normalized to CIE magnitude. They were designed and are properly used only for measuring the COLOR DIFFERENCE between two similar, NONMETAMERIC samples.

tristimulus computation data products of relative SPECTRAL-energy distribution of an ILLUMINANT multiplied by each of three COLOR MATCHING functions; in the CIE SYSTEM designated as $E_c\bar{x}, E_c\bar{y}, E_c\bar{z}$, for example, for ILLUMINANT C and the color mixture data for the STANDARD OBSERVER at designated WAVELENGTHS.[43,69] See TRISTIMULUS VALUES.[43,69]

tristimulus integration use of TRISTIMULUS COMPUTATION DATA to determine TRISTIMULUS VALUES of a sample; in the CIE SYSTEM, the summation of the products of the REFLECTANCE or TRANSMITTANCE at regular WAVELENGTH intervals by the tristimulus computation data for the same wavelengths.[43,69] See TRISTIMULUS INTEGRATOR.[43,69]

tristimulus integrator device for performing the TRISTIMULUS INTEGRATION calculation, generally attached directly to a SPECTROPHOTOMETER.[43,69] See TRISTIMULUS VALUES.[43,69]

Several types have been used: mechanical analog type using specially cut cams giving a continuous integration, digital-mechanical type using selected ordinates (generally 100), and digital-computer type using weighted ordinates, most commonly at 10 nm intervals. The latter is most generally used today, because of the flexibility in obtaining CIE TRISTIMULUS VALUES for any OBSERVER whose COLOR MATCHING functions are known and any ILLUMINANT of defined SPECTRAL-power distribution.

tristimulus values, CIE amounts (in percent) of the three components necessary in a three-COLOR ADDITIVE mixture required for MATCHING a color; in the CIE System, they are designated as X, Y, and Z.[43,69]

The ILLUMINANT and STANDARD OBSERVER COLOR MATCHING functions used must be designated; if they are not, the assumption is made that the values are for the 1931 observer (2° field) and illuminant C. The values obtained depend on the method of integration used and on the relationship of the nature of the sample and on the instrument design used to measure the REFLECTANCE or TRANSMITTANCE. Tristimulus values are not, therefore, absolute values characteristic of a sample, but relative values dependent on the method used to obtain them. Approximations of CIE TRISTIMULUS VALUES may be obtained from measurements made on a TRISTIMULUS COLORIMETER, giving measurements generally normalized to 100, which must then be normalized to equivalent CIE values. The filter measurements should be properly designated as R, G, and B instead of X, Y, and Z. The calculation of tristimulus values in the CIE System is illustrated in the equation below for Y.

$Y = k \int E(\lambda)\, \bar{y}(\lambda) R$ or $T(\lambda) d(\lambda)$ where k is a normalizing factor to bring Y for a perfect white to 100.000%, and $d\lambda$ is the WAVELENGTH interval used. When a large number of wavelengths is not used, the integration should be designated as a summation.

$Y = k \sum E(\lambda)\bar{y}(\lambda)\ R$ or $T(\lambda)\Delta\lambda$.

ASTM Test Method D2244 covers the calculation, from instrumentally measured color coordinates based on daylight illumination, of small color differences between nonfluorescent, nonmetameric, opaque specimens such as painted panels.

tritolyl phosphate (TTP) see TRICRESYL PHOSPHATE.[170]

trivial name a name that is not produced by any systematic procedure of naming.[134] (ASTM)

Trivial names may be derived from geographic locations, names of people, or from descriptive words. Examples: "aragonite" after Aragon, Spain; "wollastonite" in honor of the English chemist, W.H. Wollaston; and "magnetite" alluding to its highly magnetic property.

trompe l'oeil (Fr.) Literally, "fool-the-eye."[81,148]

A design that creates a three-dimensional illusion by means of shadow and graphic textures.

trough in the COIL COATING field, synonymous with PAN.[70]

TRS abbreviation for total reduced sulfur.[172] (EPA) See DESIGNATED POLLUTANT.

truss framework of WOOD or METAL beams, stiffened by cross braces; used for roofs, bridges, etc.[82,153]

TSCA abbreviation for TOXIC SUBSTANCES CONTROL ACT.[172]

T-300 temperature at which the torsional MODULUS of an air-dried film is 300 kg/cm^2; a relative parameter of film STIFFNESS.[47]

Tukon see PENETROMETER.[145,157]

tumbling (1) PROCESS by which paint is applied to small articles, such as hairpins, children's building bricks, etc., which are unsuitable for coating by any of the normal methods.[77] *The articles are placed in a drum, together with a little more paint than will be sufficient to cover the total surface*

Words presented in CAP/SMALL CAPS type indicate that the word is defined in another part of the *Coatings Encyclopedic Dictionary.*

Numerical superscripts classify terms in one or more of the categories listed in the second section of this volume.

of all the articles, and the drum is rotated until the paint is evenly distributed. The articles are then emptied from the drum, generally onto wire trays, and the coating AIR-DRIED *or* BAKED. (BSI) Syn: rumbling, barrelling, barrel finishing, barrel polishing; (2) process used in some paint storehouses whereby containers are repeatedly upended to redisperse pigments which may have SETTLED or CAKED during STORAGE of the paint.[64]

tung oil DRYING OIL obtained from *Aleurites fordii* or *montana*, now grown in many different parts of the world.[63] Syn: China wood oil, wood oil, and mu oil

The Japanese tung oil is obtained from Aleurites cordata. Tung oil is distinguished by the presence of a very high proportion of elaeostearic acid, which contains three CONJUGATED DOUBLE BONDS. *It dries rapidly,* POLYMERIZES *readily on heating, "reacts" with certain types of pure* PHENOLIC RESINS, *possesses outstanding water resistance, and* WEBS *or* FROSTS *on drying unless properly heat-treated. Properties: sp gr, 0.939-0.943/15°C; refractive index, 1.486-1.522; iodine value, 145-175; saponification value, 189-195.*

tungates METALLIC soaps derived from TUNG OIL.[127]

tungstomolybdic pigments see PHOSPHOMOLYBDIC PIGMENTS.[41]

turbid characteristic of a liquid or sol having a relatively great amount of nonsettling FLOC, GELS, SUSPENDED MATTER, PARTICLES, droplets, or other insoluble or separated matter, even though the liquid is TRANSLUCENT and TRANSMITS at least a little light.[139]

turbidity term used to indicate the degree of cloudiness of water samples which is caused primarily by the presence of COLLOIDAL matter.[139] (IUPAC)

One unit of turbidity is defined as OPTICAL *density of 0.001 (yellow light, 580 nm) per cm of depth.*

turbulence (1) deviation from streamline or TELESCOPIC FLOW. TURBULENT flow occurs at relatively high RATES OF SHEAR and indicates the existence of EDDY currents in sheared material;[166] (2) atmospheric eddities, mechanical or thermal, which are produced primarily by shearing stress or by CONVECTION.[172]

Atmospheric turbulence is primarily responsible for the dilution and mixing of atmospheric pollutants. Turbulent fluctuations in stable air are mainly of high frequency with typical periods of seconds (mechanical turbulence). The fluctuations often constitute major deformations of flow and are capable of transporting momentum energy, and suspended matter, at rates far in excess of the rate of transport by molecular diffusion and conduction in a nonturbulent or LAMINAR FLOW. (IUPAC) cf. STREAMLINE FLOW[166]

turbulent violently agitated or disturbed.

turkey red artificial RED OXIDE of iron obtained by CALCINATION at relatively low temperatures.[41]

Hydroxide or carbonate of iron is PRECIPITATED *from a solution of an iron* SALT, *and the precipitate is washed before calcination. Soft, brilliant pale reddish brown pigments are obtained.* OPACITY *is high.*

turkey red oil SULFONATED CASTOR OIL.[83,127]

The term is sometimes applied to the sodium SALT *of this product. The oil is chiefly the reaction product of sulfuric acid with the hydroxyl groups of castor oil, but some secondary reactions also occur. Turkey red oil is used as a* WETTING AGENT.

turkey umber type of UMBER derived from Cyprus and originally exported via Turkey.[41] See UMBER.[41]

Turnbull's blue see IRON BLUE.[41]

turpentine a colorless, volatile OIL distilled from the products of certain pine trees and consisting primarily of a complex mixture of TERPENE HYDROCARBONS of the general formula $C_{10}H_{16}$.[44,164] (ASTM) Syn: turps

Turpentine was formerly extensively used in paints and VARNISHES *but has now been largely replaced by* MINERAL SPIRITS *or* WHITE SPIRIT. *Four kinds of turpentine are now recognized: (1)* <u>GUM turpentine or gum spirits</u>*: obtained by distilling the crude* EXUDED *gum or* OLEORESIN *collected from living trees; (2)* <u>steam-distilled wood turpentine</u>*: obtained from the oleoresin within the wood of pine stumps or cuttings, either by direct steaming of the mechanically disintegrated wood or after* SOLVENT EXTRACTION *of the oleoresin from the wood; (3)* <u>sulfate wood turpentine</u>*: recovered during the conversion of wood to paper pulp by the sulfate process; (4)* <u>destructively distilled wood turpentine</u>*: obtained by fractionation of certain* OILS *recovered by condensing the vapors formed during the* DESTRUCTIVE DISTILLATION *of pine wood.*

turpentine, oil of see OIL OF TURPENTINE.[44]

turpentine, spirits of see TURPENTINE.[44,164]

turpentine substitute this term is not recommended, because its use to describe a paint thinner of MINERAL SPIRITS origin is illegal under the Federal Naval Stores Act.[164] (ASTM) Syn: sub spirits and sub turps

turps see TURPENTINE.[164,44]

tuscan red red PIGMENT made from a mixture of IRON OXIDE and an alizarine dye.[41]

tusche a LITHOGRAPHIC drawing or painting material of the same nature as lithographic INK.[81,168]

Also used as a resistant in the ETCHING *process.*

TWA abbreviation for time-weighted average.

Twaddell hydrometer form of technical HYDROMETER used for measuring the SPECIFIC GRAVITY of liquids.[59,157]

It does not give a direct reading, but the specific gravity is calculated from the following simple equation: T° = 200(d-1), where T° is the reading in degrees Twaddell, and d is the required sp gr.

twilight vision rod vision by the dark-adapted eye in dim light.[173,43]

Twitchell reagent employed for the splitting or HYDROLYSIS of OILS and FATS, and made by the interaction of sulfuric acid, oleic acid, and naphthalene.[144]

two-component paint a coating which is manufactured in two components which must be maintained separately until shortly before use.[71,67,161] Syn: two-pot coating

When mixed, the two liquids rapidly CROSSLINK to form a solid composition.

two-compartment coating CROSSLINKING SYSTEMS which must be stored in separate containers before use.[67,161] Syn: two-pot coating. cf. ONE-COMPARTMENT COATING[67,161]

Otherwise, they would react and form a useless GEL.

two-component gun SPRAY GUN having two separate fluid sources leading to the spray head.[77,59]

two-piece can a can which consists of a body and bottom manufactured from a single piece of steel or aluminum and a separate top which is attached to the body.[50] (EPA)

two-pot coating see TWO-COMPONENT COATING.[67,161,71]

two-roller mills these are used for purposes similar to the TRIPLE-ROLLER MILLS, except that only two rollers are involved.[59]

tympan the roller opposite the PRINTING roller on a ROTARY PRESS over which the PAPER web passes.[168]

tympan sheet a sheet of PAPER or cloth placed between the IMPRESSION surface (PLATEN or cylinder press) and the paper to be printed.[168]

typography PROCESSES of making an impression using a raised, pigmented surface, from type, line etchings, halftones, or RUBBER stamps.[168] See LETTERPRESS.[168]

Words presented in CAP/SMALL CAPS type indicate that the word is defined in another part of the *Coatings Encyclopedic Dictionary.*
Numerical superscripts classify terms in one or more of the categories listed in the second section of this volume.

U

u, v CHROMATICITY COORDINATES in one of the CIE Uniform Color Spacings (1964).[43,69]

U*, V*, W* values used in a COLOR DIFFERENCE formula developed by Wyszecki in 1964, known as the CIE 1964 U*V*W* System.[43,69]

Ubbelohde viscometer suspended level apparatus for determining the VISCOSITY of a liquid.[157,166]

uchiwa-ye (Japanese). A PRINT in the shape of a fan.[81]

ultimate tensile strength the maximum stress necessary to break the sample in the grips of a TENSILE TESTER.[47,157] cf. TENSILE STRENGTH[47,157]

ultracentrifuge apparatus for applying centrifugal forces of many times gravity, up to 1,000,000 times, for causing very small PARTICLES to SETTLE out.[157]

The velocity of SEDIMENTATION can be determined according to STOKES' LAW, and the mean size and size distribution can also be determined. Molecular weights of particles can be calculated.

ultramarine blue $Na_6Al_6Si_6O_{24}S_2$ (Light); $Na_7Al_6Si_6O_{24}S_3$ (Medium); $Na_8Al_6Si_6O_{24}S_4$ (Dark). (1) A natural type of MINERAL origin.[41,34] Known as GENUINE ULTRAMARINE and LAPIS LAZULI; (2) the SYNTHETIC types are produced in several shades by heating sulfur, CLAY, soda ash, and a reducing agent. They have good LIGHTFASTNESS and good ALKALI RESISTANCE, but poor OPACITY and poor TINTING STRENGTH. Syn: brilliant ultramarine, factitious ultramarine, French blue, French ultramarine, Guimet's blue, Huemann's blue, laundry blue, Oriental blue, permanent blue

Properties: density, 2.2-2.7 g/cm^3 (18.6-20.2 lb/gal); O.A., 25-39; particle size, 0.8-2.0 μm.

ultramarine green ultramarines cover a COLOR range from pink to violet to blue and green.[41]

The pinks, violets, and greens are especially weak colors, and the green is no longer an item of commerce.

ultramarine violet $H_2Na_{4\text{-}6}Al_6Si_6O_{24}S_2$. Pigment Violet 15 (77007). Produced by mixing ULTRAMARINE BLUE with ammonium chloride and heating at about 150°C for several hours. Chemically and physically similar to ultramarine blue.[41]

ultraviolet absorber substance which absorbs ULTRAVIOLET RADIATION more readily than the coating material in which it is dissolved or dispersed, and which transforms the ultraviolet energy into longer WAVELENGTH energy which is relatively harmless to the coating.[83]

While some pigments are relatively nontransparent to ultraviolet radiation, the term is usually used to describe compounds which are physically dissolved in the coating and do not affect TRANSPARENCY to visible light. These UV absorbers are a class of stabilizers which have intense ABSORPTION up to 350 to 370 nm, but are TRANSPARENT in the visible. Examples of these are benzotriazoles and 2 hydroxy-benzophenones. They are widely used to stabilize exterior VARNISHES against UV degradation.

ultraviolet (UV) RADIANT ENERGY below about 380 nm; portion of the electromagnetic spectrum between about 10 and 400 nm.[43]

The term "ULTRAVIOLET LIGHT" is incorrect because light refers only to visible radiant energy.[43] (The term "light" is limited to the visible region of the SPECTRUM.) Sources of UV are used to CURE coatings which are specially designed for this purpose.

ultraviolet (UV) curing conversion of a coating from its application state to its final use state by means of a mechanism initiated by ULTRAVIOLET radiation generated by equipment designed for that purpose.[67,55] (ASTM) See CURE.[67]

ultraviolet light (deprecated) see ULTRAVIOLET.[43]

umber Pigment Brown 7. A naturally occurring brown earth containing FERRIC OXIDE, together with SILICA, ALUMINA, manganese oxides, and LIME.[41]

Raw umber PIGMENT is umber which has been ground; burnt umber pigment is umber calcined at a low temperature and ground.

undercoat (1) a coat of paint applied on new WOOD, or over a PRIMER, or over a previous coat of paint; improves the SEAL and serves as a BASE for the TOPCOAT, for which it provides better adhesion; (2) any paint which acts as a base for ENAMEL; (3) any primer which is colored.[71,147] (DAC)

An undercoat is applied in preparation for a FINISH coat to be applied over it. In the case of an unpainted surface, the undercoat is normally a middle coat, having been preceded by a priming coat; for previously painted surfaces, however, the undercoat is preceded by an older system.

undercoating a BITUMINOUS coating SPRAYED on the underside of automobiles to minimize RUSTING.[163,76,71]

undercure a condition or degree of CURE that is less than optimum, i.e., when insufficient time and/or temperature has been allowed for adequate cure;

may be evidenced by TACKINESS, longness, or inferior physical properties.[67,42]

undercutting the action of ETCHING solution on the sides of the resist of photomechanically prepared plates.[168]

underhue see UNDERTONE COLOR.[43,168,69]

undertone color COLOR of a thin layer of PIGMENT-vehicle mixture applied at incomplete HIDING on a white background; sometimes referred to as CAST or UNDERHUE.[43,69]

The term is sometimes used to refer to the color resulting when a pigment is mixed with white, but this should more properly be called TINT-TONE.

uniform chromaticity coordinates chromaticity coordinates yielding an approximately UNIFORM CHROMATICITY SCALE DIAGRAM for colors of equiluminosity.[43,69]

uniform chromaticity scale diagram any one of a variety of transformations of the CIE CHROMATICITY DIAGRAM to a diagram on which all pairs of just noticeably different COLORS of equal LUMINANCE are represented by pairs of points separated by nearly equal distances.[43,69]

uniform chromaticity spacing see UNIFORM CHROMATICITY SCALE DIAGRAM.[43,69]

uniform color scale a scale wherein the units of COLOR DIFFERENCE that are judged to be equal have equal scale differences.[43,69]

uniform color space three-dimensional space wherein all pairs of colors, judged to be equally different, are separated by nearly equal distances.[43,69]

uniform lightness scale a scale wherein the units of LIGHTNESS that are visually judged to be equal have equal scale values, and the lightness differences judged to be equal are represented by equal scale differences.[43,69,128]

unit, angstrom see ANGSTROM UNIT.[128](ASTM)

units in air pollution regulatory usage, units are the dimensional quantities such as length, weight, or volume in which an air pollution regulation or guideline is expressed.[128,172] (EPA)

In the series of reports known as CONTROL TECHNIQUE GUIDELINES, suggested RACT levels are given in units of "lb of VOC per litre of coating (less water)." The metric equivalent of this can also be used: "kg of VOC per litre of coating (less water)." For new source performance standard regulations, units of "kg of VOC per litre of coating solids" have been used. These latter units permit easier consideration of TRANSFER EFFICIENCY in any calculations that are required.

units, U.S. customary units based upon the yard and the pound commonly used in the United States of America and defined by the National Bureau of Standards.[128] (ASTM)

Some of these units have the same name as similar units in the United Kingdom (British, English, or U.K. units) but are not necessarily equal to them.

universal colorants colored PIGMENTS milled in media compatible with paint vehicles, added in small proportions to already prepared TRADE SALES PAINTS to modify their COLOR.[41,83] cf. INDUSTRIAL UNIVERSAL COLORANTS.[41,83] Syn: tinter, stainer

These colorants were designed for TINTING both LATEX (water-based) and ALKYD and LINSEED OIL/alkyd (SOLVENT-based) systems. Their composition was based primarily on pigments, ETHYLENE GLYCOL, water, and SURFACTANTS selected to provide COMPATIBILITY of colorants in both water- and solvent-based coating formulations, with minimal adverse effects.

It is these universal color formulations that are used for "in-store" tinting systems, allowing a paint company to make available to the homeowner a large selection of paint SHADES (many offer about 1500 different shades) by having in their inventory about 12 factory tinted most popular colors, and, in many cases, only two white tint bases for both water-based and solvent-based systems.

unsaponifiable matter that portion of FATS and RESINS which does not react with ALKALI to form a SOAP.[130,137,167]

unsaturated of or relating to a COMPOUND, especially of CARBON, containing atoms that share more than one VALENCE bond.[144]

unstable tending towards DECOMPOSITION or other unwanted chemical change during normal handling or storage.[172] See STABILITY and REACTIVE MATERIAL.[172,161]

An unstable chemical in its pure state, or as commonly produced or transported, POLYMERIZES vigorously, decomposes, condenses, or becomes self-reactive under conditions of shock, pressure, or temperature.

uralkyd see URETHANE ALKYD.[71,130]

urea NH_2CONH_2. Used in the production of UREA-FORMALDEHYDE RESINS.[130,144] Syn: carbamide

Properties: mp, 132°C; sp gr, 1.323.

urea resin a SYNTHETIC RESIN made from UREA and an ALDEHYDE.[130,156] (ASTM)

Words presented in CAP/SMALL CAPS type indicate that the word is defined in another part of the *Coatings Encyclopedic Dictionary.*

Numerical superscripts classify terms in one or more of the categories listed in the second section of this volume.

urea-formaldehyde resin (UF) an AMINO RESIN made by polycondensation of UREA (carbamide) with FORMALDEHYDE.[130,156]

urethane $NH_2COOC_2H_5$. Crystals. Used as an intermediate in ORGANIC synthesis, e.g., polyurethane resins.[141] Syn: ethyl carbamate, urethan, ethyl urethan

Properties: mp, 48-50°C; bp, 182-184°C.

urethane coatings coating vehicles containing a POLYISOCYANATE MONOMER reacted in such a manner as to yield POLYMERS containing any ratio, proportion, or combination of URETHANE linkages, active isocyanate groups, or polyisocyanate monomer.[71,156,163,130,141] See POLYURETHANES.[71,156,163,130]

The reaction products may contain excess isocyanate groups available for further reaction at time of application or may contain essentially no free isocyanate as supplied.

Urethane coatings (SOLVENT or WATERBORNE) are noted for their overall balance of high performance properties. These include excellent toughness, flexibility, high impact resistance, and tensile strength. They also offer excellent ABRASION RESISTANCE and resistance to a wide variety of chemicals, solvents, and stains.

Urethane coatings are generally divided into AROMATIC and ALIPHATIC types. Aromatic urethanes are generally based on TDI (TOLUENE DIISOCYANATE) or MDI (DIPHENYLMETHANE DIISOCYANATE). They are designed mainly for interior applications since they have a tendency to yellow when exposed to prolonged sunlight. Aliphatic urethanes offer excellent UV stability, but are more expensive than aromatic urethanes. Aliphatic urethanes are generally based on IPDI (isophorone diisocyanate), HDI (hexamethylene diisocyanate), or TMXDI (tetramethylxylene diisocyanate). In addition to excellent light stability, aliphatic urethanes have better flexibility than aromatic urethanes.

In a one- or two-part coating which contains polyisocyanate monomer and a hydroxyl-containing material, they react during CURE to form a polyurethane:

$$R_1NCO + R_2OH \longrightarrow R_1NHCOOR_2$$

urethane thickeners synonym for HEUR.[83]

urushi Japanese word for LACQUER.[81,151]

urushi-ye (Japanese) A print in which LACQUER is used to heighten the COLOR.[81]

The term is generally employed to describe only the early hand-colored prints in which lacquers, colors, and metallic DUST were applied to the printed black outline.

urushiol hydroxy acid of aromatic type present in *Rhus vernicifera*, the basis of Japanese LACQUER.[144,151]

usable life see POT LIFE.[67,161,42,74]

useful life the length of time the coating is expected to remain in service.[154,65]

U.S. gallon see GALLON, U.S.[128]

UV abbreviation for ULTRAVIOLET RADIATION.[43]

V

vacations see HOLIDAYS.[56]

vacuum filter filtering system in which filtrate is removed by application of a vacuum.[59]

vacuum forming PROCESS in which heated THERMOPLASTIC sheets are converted to other configurations by causing them to FLOW into molds with application of a vacuum.[49]

vacuum-pressure impregnation method for the commercial impregnation of electrical equipment with insulating VARNISH.[54]

It consists in first subjecting the equipment to a vacuum in order to remove as much occluded air as possible, permitting the varnish to enter also under vacuum to replace the air, and finally reducing the vacuum and applying pressure in order to assist the somewhat viscous varnish through the narrow interstices left between fine windings, etc.

valance (1) short drapery at top of a window;[148] (2) a DECORATIVE frame to conceal the top of curtains and fixtures.[148]

valence (valency) the combining capacity of an ATOM or RADICAL determined by the number of electrons that it will lose, add, or share when it reacts with other atoms.[144]

value, Munsell see MUNSELL VALUE.[43,69]

vanadium driers group of rather dark-colored DRIERS, based on soluble vanadium SALTS, which are fairly strong SURFACE-DRYING CATALYSTS.[74,83]

vandyke brown (1) a very dark, deep-brown NATURAL EARTH PIGMENT consisting essentially of ORGANIC matter obtained from peat or lignite found chiefly in parts of Germany; its principal uses are in ARTISTS' COLORS, STAINS, and GRAINING; (2) a synthetic pigment of similar COLOR.[41] Syn: cassel brown and cassel earth

vapor barrier a MOISTURE-impervious layer or coating (such as special paint, or a membrane on ROOFING FELT or on building PAPER) which prevents the passage of moisture or VAPOR into a material or structure.[75] (DAC)

vapor capture system any combination of HOODS and ventilation system that captures or contains ORGANIC vapors in order that they may be directed to an abatement or recovery device.[172] (EPA)

vapor degreaser an apparatus in which METAL surfaces are cleaned by SOLVENT VAPORS.[59,158]

vapor degreasing a cleaning PROCESS utilizing CONDENSING SOLVENT as the cleaning agent.[158,131]

vapor pressure pressure (usually expressed in millimetres of mercury) characteristic at any given temperature of a vapor in equilibrium with its liquid or solid form.[142]

vapor retarder see VAPOR BARRIER.[75]

variance a quantity, equal to the square of the STANDARD DEVIATION (σ^2), used to measure PRECISION of an AVERAGE of a number of measurements of a quantity.[36] (IUPAC) See STANDARD DEVIATION.[36]

varnish (1) a liquid composition which is converted to a TRANSPARENT solid film after application as a thin layer.[76] Some VARNISHES contain FLATTING AGENTS. BITUMINOUS—a dark colored VARNISH containing bituminous ingredients. The varnish may be either of the oil or spirit type.[76] OIL—a varnish which contains RESIN and drying oil as the basic FILM-FORMING ingredients and is converted to a solid film primarily by chemical reaction.[63] Spar—a varnish for exterior surfaces. The name originates from its use on spars of ships.[152] SPIRIT—a varnish which is converted to a solid film, primarily by SOLVENT EVAPORATION;[71,152] (ASTM) (2) in PRINTING INK technology, a broad term including fluid compositions comprising one or more of the following: OILS, RESINS, SOLVENTS, DRIERS, and WAXES; used either as a vehicle or to cover surfaces.[168] (ASTM) See OIL.[152,156,157]

varnish remover see PAINT AND VARNISH REMOVER.[164]

varnish stain a VARNISH which is colored with a TRANSPARENT material, leaving a colored coating on the surface; has less penetrating power than a true STAIN.[71,41] (DAC) See STAIN.[152]

varris Martin (Fr.) LACQUER work by a cabinet maker, named Martin, in 18th century France.[148,175,81]

It was an imitation of Oriental lacquer.

vat dyes those DYES which are insoluble in water, dilute acids, and alkalis but which, on being reduced, form a COLOR base soluble in ALKALINE SOLUTIONS.[41]

EXPOSURE of material dyed with such a color base produces the original colored COMPOUND by OXIDATION.

Words presented in CAP/SMALL CAPS type indicate that the word is defined in another part of the *Coatings Encyclopedic Dictionary.*
Numerical superscripts classify terms in one or more of the categories listed in the second section of this volume.

vegetable black originally, a sooty product, obtained by burning VEGETABLE OILS with restricted air.[41] See LAMPBLACK.[41]

Most vegetable black today is obtained by burning COAL-TAR OIL.

vegetable drying oil see DRYING OIL.[63]

vegetable oil an OIL extracted from vegetable matter, especially CASTOR, LINSEED, SAFFLOWER, SOYA, and TUNG oil; used in paints and PLASTICS.[63] (DAC)

vehicle the liquid portion of paint, in which the PIGMENT is dispersed; it is composed of BINDER and THINNER.[156]

veiling (1) formation of a cobweb pattern;[42,55] (2) CURTAINING or SAGGING.[42]

velocity gradient the change dv in relative velocity v between parallel planes with respect to the change dr in perpendicular distance r throughout the depth of material being sheared. Velocity gradient has the same dimensions as RATE OF SHEAR, which is reciprocal seconds.[166]

velvet see GLOSS.[43]

veneer (1) thin finishing or surface layer of fine WOOD, LAMINATED PLASTIC, formica, or the like, bonded to a SUBSTRATE;[82,148] (2) thin WOOD sheeting used to make PLYWOOD or to serve as the DECORATIVE surface layer on a laminate;[153] (3) an outside wall facing of brick, stone, etc.; provides a decorative, durable surface but is not load-bearing.[82,75,175]

Veneers are thin strips of wood cut in such a way as to reveal the wood's best natural features—GRAIN, COLOR, and pattern. They are produced either by rotary cutting, where the knife acts against the full length of the log turning on a lathe (veneers are so cut tangentially to the growth rings of the log) or, as in the case of most decorative veneers, sliced across the face of the log.

The advantages of veneers include a wide variety of choice from among the world's HARDWOODS and SOFTWOODS, and their economy when used as a decorative facing on other cheaper woods.

Veneered panels are supplied ready for use. The core may be plywood, particle board, or solid timber. The VENEER is laid at right angles to the grain on the face of the core. Balanced construction removes the possibility of shrinkage and warping and, when laid and polished, decorative veneers require little or no maintenance.

veneer plywood PLYWOOD which is faced with a DECORATIVE WOOD VENEER.[153,75] cf. PARTICLE BOARD[153,75,175]

Venetian mosaic see TERRAZZO.[73]

Venetian red chemically prepared oxide of iron, red PIGMENT.[41]

Made by CALCINING HYDRATED LIME and ferrous sulfate. Varies in TINTING STRENGTH according to the amount of FERRIC OXIDE.

Venetian window a window with one large fixed central pane and smaller panes at each side.[82]

Venice turpentine same as larch TURPENTINE, the OLEORESIN of the European larch tree (*Larix europea* or *Larix decidua*).[44,164,167] (ASTM)

The term is also now used widely to describe the clear yellow liquid portion of pine oleoresin, or a SYNTHETIC product of similar composition made by dissolving ROSIN in a TERPENE SOLVENT.

vent an opening designed to convey water vapor or other gas from inside a building or building component to the atmosphere.[82] (ASTM)

verdigris $Cu(C_2H_3O_2)_2 \cdot 2Cu(OH)_2$. (1) The dibasic ACETATE of copper which is a greenish blue, crystalline powder with an acetic odor.[41] *Used as a PIGMENT;* (2) the blue or green corrosion products which form the PATINA on copper, brass, or bronze upon WEATHERING.[66] Syn: aerugo

verditer basic copper carbonate.[41]

vergeboard see BARGEBOARD.[82]

vermiculite hydrated magnesium-aluminum-iron silicate capable of expanding six to 20 times the volume of the unexpanded MINERAL when heated to about 1100°C; the platelets exhibit an active curling movement when heated, hence the name.[58,34] See ALUMINUM POTASSIUM SILICATE.[58,41]

vermiculus see MERCURIC SULFIDE.[41]

vermillion, (vermilion) see MERCURIC SULFIDE.[41]

vernonia oil a low-viscosity epoxidized drying oil from the seed of an African plant, *Vernonia galamensis*, containing three reactive epoxy groups and three CARBON-carbon double bonds per triglyceride molecule and is characterized by its very low viscosity and melting point.[63] (ASTM)

It flows easily even at temperatures below 0°C and thus needs only a fraction of the volatile solvents usually used for other DRYING OILS. Thus it can be used as a reactive diluent for high solids alkyds and epoxy coating formulations.

vert emeraude (dull) see CHROMIUM OXIDE GREEN.[41]

vertical grain see EDGE GRAIN.[153]

vertical pattern a SPRAY pattern whose longest dimension is vertical.[54]

viable living; able to germinate or grow.[173]

Victoria Blue B strong, reddish blue DYESTUFF made from the interaction of Michler's KETONE and phenyl-α-naphthylamine.[41]

Vienna blue see COBALT BLUE.[41]

viewing angle the angle between the VIEWING RAY and a normal to the surface at the point of INCIDENCE.[43,69] Syn: OBSERVATION ANGLE

viewing booth an enclosure designed to exclude as much as possible all external light, and contain in it a light source or several light sources of standard SPECTRAL energy distribution, for the examination and comparison of the COLOR of materials.[69,145]

It is normally painted in a neutral color of average LUMINANCE.

viewing conditions the conditions under which a visual observation is made, including the angular subtense, directions of ILLUMINATION and viewing, character of the surroundings, etc.[43,69]

viewing geometry the geometry by which the INCIDENT ILLUMINATION is projected onto the sample and by which the observed light is collected and TRANSMITTED to the detector. See ANGLE OF INCIDENCE, ANGLE OF VIEWING, and ANGLE OF REFLECTION.[43,69]

viewing ray the line connecting the point of INCIDENCE on the surface and the center of the receptor entrance stop.[43]

vine black intense blue-black PIGMENT made by the partial burning of vine cuttings.[41]

vintahite see GILSONITE.[76]

vinyl (1) the UNSATURATED, univalent RADICAL CH_2:CH– derived from ETHYLENE; (2) any of the various compounds containing this group, typically highly reactive, easily POLYMERIZED, and used as a basic material for coatings and plastics; (3) any of the various plastics, typically TOUGH and FLEXIBLE.[130,134,38]

vinyl acetate plastics plastics based on resins made by the POLYMERIZATION of VINYL ACETATE or COPOLYMERIZATION of vinyl acetate with other UNSATURATED compounds, the vinyl acetate being in greatest amount by mass.[38] See POLYVINYL ACETATE.[130]

vinyl benzene see STYRENE.[141,164]

vinyl chloride CH_2 = CHCl. VINYL MONOMER. Slightly soluble in water; soluble in ALCOHOL and ether.[141]

Properties: sp gr, 0.9121; bp, –13.9°C.

vinyl chloride plastics PLASTICS based on RESINS made by the POLYMERIZATION of VINYL CHLORIDE or COPOLYMERIZATION of vinyl chloride with other UNSATURATED COMPOUNDS, the vinyl chloride being in greatest amount by mass.[38] (ASTM)

vinyl coating one in which the major portion of BINDER is of the vinyl RESIN family.[71]

vinyl cyanide see ACRYLONITRILE.[144]

vinyl plastics plastics based on RESINS made from MONOMERS containing the VINYL GROUP, CH_2=CH–.[38] (ASTM)

vinyl printing and topcoating the APPLICATION of a DECORATIVE pattern to the vinyl sheet or coated WEB with a ROTOGRAVURE type printing press. Frequently, a thin protective TOPCOAT is applied by the last print head of the press.[168,148] (EPA)

vinyl printing ink any mixture or solution containing polymeric RESIN and dye or PIGMENT which is applied in DECORATIVE patterns to vinyl sheets or coated WEBS using a ROTOGRAVURE PRINTING process.[168] (EPA)

vinyl resins these include products in which the UNSATURATED VINYL grouping is involved, i.e., CH_2=CH–.[130,156]

Vinyl resins include POLYVINYL ACETATE, POLYVINYL CHLORIDE, copolymers of these, the ACRYLIC and METHACRYLIC resins, the POLYSTYRENE resins, etc.

vinyl wallcoverings at present there are six types: (1) vinyl laminated to PAPER; (2) paper laminated to lightweight woven cloth and vinyl-coated; (3) vinyl laminated to lightweight woven cloth, natural or synthetic; (4) vinyl laminated to lightweight nonwoven cloth, natural or synthetic; (5) vinyl laminated to nonwoven paper/FABRIC web; (6) a man-made base impregnated with vinyl.[148] See VINYL.[130,134,38]

vinylformic acid see ACRYLIC ACID.[141]

vinylidene chloride CH_2:CCl_2. A MONOMER.[141] Bp, 37°C. Syn: 1,1-dichlorethylene

vinylidene chloride plastics PLASTICS based on POLYMER RESINS made by the POLYMERIZATION of VINYLIDENE CHLORIDE or COPOLYMERIZATION of vinylidene chloride with other UNSATURATED COMPOUNDS, the vinylidene chloride being in the greatest amount by mass.[38,130] (ASTM)

viridian green a complex derivative of hydrated chromium oxide.[41] See HYDRATED CHROMIUM OXIDE.[41] Syn: Guignet's green

viscoelasticity the property of a POLYMER that characterizes it as neither an ideal solid nor a viscous liquid, but seeming to have the character of both with the relative contribution of each dependent

Words presented in CAP/SMALL CAPS type indicate that the word is defined in another part of the *Coatings Encyclopedic Dictionary.*

Numerical superscripts classify terms in one or more of the categories listed in the second section of this volume.

upon time, temperature, stress, and strain rate.[47,166]

In addition to having some of the characteristics of elastic solids, they possess some of the characteristics of viscous liquids. Polymeric materials show creep under a certain load and stress relaxation if stretched.

viscometer (viscosimeter) an instrument for measuring FLOW properties.[166,145] See BROOKFIELD VISCOMETER,[166,157,59,145] BUBBLE TUBE VISCOMETER,[157,166,145] FORD CUP,[59,157,166] GARDNER-HOLDT VISCOSITY TUBES,[59,157,166] GARDNER MOBILOMETER,[59,157,166] ICI CONE and PLATE VISCOMETER,[145,157,166] ISO FLOW CUPS,[145,157,166] KREBS-STORMER VISCOMETER,[145,157,166] SAYBOLT,[145,157,166] and ZAHN CUP.[166,145,157]

The chief types of viscometers are as follows: capillary, rotational, outflow or efflux (Ford Cup, usually called orifice type), falling ball, tackmeter (radial flow type) bubble tube. The term "viscometer" is preferred to "viscosimeter."

viscose a thick, golden-brown viscous solution of CELLULOSE xanthate, used in the manufacture of rayon and CELLOPHANE.[39]

viscosimeter (deprecated) see VISCOMETER.[166,145]

viscosity (1) state or quality of being VISCOUS; (2) the property of a fluid whereby it tends to resist relative motion within itself. If different layers of fluid are moving with different velocities, viscous forces come into play, tending to slow down the faster-moving layers and to increase the velocity of the slower-moving layers.[166] cf. ABSOLUTE VISCOSITY, KINEMATIC VISCOSITY[166]

The constant of proportionality is called the COEFFICIENT of viscosity of the fluid. Measured in newton seconds per square metre (SI units) or poise (c.g.s. units). 1 centipoise = 10^{-3} Ns/m^2 or pascal-seconds (PAs) which are equivalent to 10 poise.

viscosity, absolute dynamic the FORCE per unit area that RESISTS the FLOW of two parallel fluid layers past one another when their differential velocity is 1 cm/(s)/cm separation. The viscous force is described by Newton's equation $f = \eta A(dv/dx)$ where A is the area (cm^2), (dv/dx) is the velocity gradient (s^{-1}), and η is the COEFFICIENT of absolute viscosity (poise).[166] (PTM)

viscosity, apparent the quantity obtained by dividing the SHEARING FORCE by the RATE OF SHEAR.[166]

This is a term applied only to NON-NEWTONIAN materials. It is not a constant for a given material, because its value depends on the rate of shear selected in making the measurement. Apparent viscosity is obtained by "one-point" methods. It has no general scientific value, and it is doubtful whether it has any real technical value.

viscosity coefficient the shearing stress necessary to induce a unit velocity FLOW gradient in a material.[166] (ASTM) See VISCOSITY.[166]

In actual measurement, the viscosity COEFFICIENT of a material is obtained from the ratio of SHEARING STRESS to SHEARING RATE. This assumes the ratio to be constant and independent of the shearing stress, a condition which is satisfied only by NEWTONIAN fluids. Consequently, in all other cases, values obtained are apparent and represent one point of the flow curve. In the C.G.S. SYSTEM, the viscosity coefficient is expressed in POISES, i.e., in dynes-seconds per square centimetre, and in the SI SYSTEM in NEWTON-SECONDS per square metre.

viscosity cup an efflux viscometer. See VISCOMETER.[145,166]

viscosity, intrinsic see INTRINSIC VISCOSITY.[130,166]

viscosity, kinematic VISCOSITY of a substance divided by the DENSITY of the substance at the temperature of measurement.[166]

Kinematic viscosity is commonly obtained from CAPILLARY and outflow VISCOMETER data. The unit is the stokes. A liquid having a relative density of one has a kinematic viscosity of one stokes if its viscosity is one POISE.

viscosity number $\frac{\eta - \eta_0}{\eta_0 c}$ where η is the VISCOSITY of the POLYMER solution; η_o is the viscosity of the pure SOLVENT; c is the concentration of the polymer solution in grams of solute per millilitre of solution.[166] (IUPAC)

viscosity, plastic resistance to FLOW in excess of the yield value in a PLASTIC material.[166] See PLASTIC FLOW, and PLASTICITY.[166]

Plastic viscosity (U) is proportional to (SHEARING STRESS-YIELD VALUE)/rate of shear. The COEFFICIENT of plastic viscosity is the FORCE in excess of the yield value, tangentially applied, that will induce a unit VELOCITY GRADIENT.

viscosity ratio the ratio of the VISCOSITIES of the POLYMER solution (of stated concentration) and of the pure SOLVENT at the same temperature.[166] (IUPAC) Syn: solution/solvent viscosity ratio

viscosity, relative (for liquids) see VISCOSITY, SPECIFIC (FOR LIQUIDS).[166]

viscosity, relative (for suspensions and solutions) the ratio between the VISCOSITY of a composition (η), divided by the viscosity of the liquid phase (η_o). It is designated η_{rel}.[166]

viscosity, specific (for liquids) the ratio between the VISCOSITY of a liquid and the viscosity of water at the same temperature. Specific viscosity is sometimes used interchangeably with RELATIVE VISCOSITY for liquids. Since water at 20°C has the viscosity of 1.005 centipoises (approximately 1 cP), the value obtained for the specific (or relative) viscosity of a liquid at 20°C is often given in CENTIPOISES.[166]

viscosity, specific (for suspensions and solutions) the VISCOSITY of a composition (η) minus the viscosity of the liquid (η_o), divided by the viscosity of the liquid phase. The symbol is η_{sp}.[166]

viscosity, structural materials which exhibit other than NEWTONIAN FLOW properties.[166] See PSEUDOPLASTIC FLOW, and STRUCTURED PAINTS.[166]

viscosity/density ratio $\frac{\eta}{\rho}$

where η is the VISCOSITY of the POLYMER solution; ρ is the DENSITY of the polymer solution.[166] (IUPAC)

viscous having relatively great VISCOSITY.[166]

visible region that portion of the electromagnetic spectrum between 380 and 770 nm.[43,69]

vision process of seeing; ocular perception.[43,173,35]

vision, defective color see COLOR VISION, DEFECTIVE.[43,69,173]

vision, normal color see COLOR VISION, NORMAL.[43,69,173,35]

visual of or pertaining to sight; ocular.[43,173,35]

vitreous enamel silicate GLASS fired on METAL. Syn: porcelain enamel

VM&P naphtha VARNISH maker's and PAINTER'S NAPHTHA.[164] Sometimes called simply painter's naphtha.

Any number of narrow-boiling-range fractions of PETROLEUM with boiling points of about 93 to 149°C according to specific use. The term "BENZINE" is still used for VM&P naphtha but is not preferred in modern nomenclature.

ASTM D3735 covers specifications for three types of moderately volatile hydrocarbon solvents, mainly ALIPHATIC in composition and normally petroleum distillates.

VOC abbreviation for VOLATILE ORGANIC COMPOUND.[172,164]

VOC-free paints see ZERO VOC PAINTS.[172,164]

vocabulary a list of selected words from the general language or a particular field of knowledge arranged in alphabetical or classified order.[134] (ASTM)

ISO uses TERM "terms and DEFINITIONS"; others use the term "GLOSSARY" for such a listing when definitions are included. A "controlled vocabulary" includes only those terms deemed standard terms as a result of a consensus process.

voids (1) See HOLIDAYS;[42] (2) See MICROVOIDS;[56] (3) interstitial space in media MILL. Knowledge of the void volume is necessary to calculate the optimum charge.[64]

volatile the easily evaporated components of any coating composition in contrast to the NONVOLATILE components.[164]

ASTM test methods for nonvolatile/volatile content include: aluminum powders/pastes, D480; coatings, by volume, D2697; driers/naval stores, D4140; emulsions, resin solutions, D4209; guide to test methods, D2832; latices, test, D4758; pigment dispersions, D4139; printing inks/resin solutions, test, D4713; resin solutions, test, D1259; silane/siloxane masonry treatments, D5095; titanium dioxide slurries, D3926; varnishes, test, D1644; volatile solvents, test, D1353.

volatile content the quantity, expressed as a weight percent of the powder coating, which is lost under specified conditions of temperature and time.[142]

volatile matter that portion of a coating that evaporates after application.[164]

volatile organic compound (VOC) any ORGANIC compound which participates in atmospheric photochemical reactions; that is, any organic compound other than those which the EPA designates as having negligible photochemical reactivity. (This is derived from EPA's official definition.)[171] Syn: volatile organic substances; photochemically reactive organic substance

VOC may be measured by a reference method, an equivalent method, an alternative method, or by procedures specified under any subpart. A reference method, an equivalent method, or an alternative method, however, may also measure nonreactive organic compounds. In such cases, an owner or operator may exclude the nonreactive organic compounds when determining compliance with a standard. The Administrator has designated the following organic compounds as negligibly reactive: methane, ethane, 1,1,1-trichloroethane (methyl chloroform), methylene chloride (dichloromethane), dichlorotrifluoroethane (HCF-123), tetrafluoroethane (HCFC-134a), dichlorofluoroethane (HCFC-141b) and chlorodifluoroethane (HCFC-142b), trichlorofluoromethane (CFC-11), dichlorodifluoromethane (CFC-12), chlorofluoromethane (CFC-22), trifluoromethane (CFC-23), trichlorotrifluoroethane (CFC-113), dichlorotetrafluoroethane (CFC-114), and chloropentafluoroethane (CFC-15).

volatile organic substance see VOLATILE ORGANIC COMPOUND.[171,164]

Words presented in CAP/SMALL CAPS type indicate that the word is defined in another part of the *Coatings Encyclopedic Dictionary.*

Numerical superscripts classify terms in one or more of the categories listed in the second section of this volume.

volatile thinner see THINNER.[164,168] (ASTM)

volume solids, percent see SOLIDS BY VOLUME.[128]

vortex cell see BÉNARD CELL.[164,42]

vulcanization physicochemical change resulting from CROSSLINKING of the unsaturated HYDROCARBON chain of polyisoprene (RUBBER) with sulfur, usually with application of heat.[67,140]

vulcanize to subject to VULCANIZATION.[67,79]

vulcanized fiber (1) RESIN-free CELLULOSIC PLASTIC.[38] *Made by immersing cotton waterleaf* PAPER *in a solution of zinc chloride, applying slight heat and pressure and subsequently leaching the zinc chloride from the newly created product. At the completion of this* PROCESS, VULCANIZED *fiber is no longer a paper, nor is it what is commonly known as chem-board; it is a completely new form of matter with unusual physical characteristics;*[38,75] (2) a nearly homogeneous material consisting of hydrated CELLULOSE, made by subjecting cellulose to a parchmentizing process.[38]

vulcanized oil VEGETABLE OIL which has been reacted with sulfur, or sulfur chloride.[63,156]

W

wainscot (1) wall paneling, usually of WOOD;[153,159] (2) a DECORATIVE or protective facing applied to the lower three or four feet of an interior partition or wall, making use of new material different from that of the upper wall, and often consisting of wood paneling or other facing material.[75,82,153] See DADO.[82,54,148]

wall an upright structure of MASONRY, WOOD, PLASTER, or other building material serving to enclose, divide, or protect an area, especially a vertical construction forming an inner PARTITION or exterior sides of a building.[82,75]

wallboard such BOARDS as pressed CELLULOSE fibers, plasterboard, CEMENT-ASBESTOS board, PLYWOOD, used in place of plaster interior surfaces.[75,159]

wallcovering any type of PAPER, vinyl, FABRIC, or specialty material that is pasted onto a wall or ceiling in a wide array of colors, patterns, textures, and performance characteristics, such as washability and ABRASION RESISTANCE.[148] (PDCA)

wallcovering adhesive (paste) adhesive material in POWDER or premixed form used to apply WALLCOVERING to a SUBSTRATE. Some wallcoverings are prepasted in the factory, and the adhesive is activated before application.[148, 470] (PDCA)

wallcovering, textile fabrics such as linen, cotton, velour, chintz, silk, and felt, which are used unbacked or PAPER-backed for WALLCOVERING.[148] (PDCA)

wallpaper PAPER, or paperlike material, usually decorated in colors, which is pasted or otherwise affixed to walls or ceilings of rooms.[148] (DAC) See WALLCOVERING.[148]

walnut oil VEGETABLE SEMIDRYING or DRYING OIL obtained from *Juglans regia*.[63]

It is used chiefly by artists, to whom its nonyellowing properties are attractive and in place of SOYA BEAN OIL, *in locations where walnut oil is economically available. It may contain as much as 63% of* LINOLEIC ACID *and 15% of* LINOLENIC *acid. Properties: sp gr, 0.926/15°C; iodine value, 140-150; saponification value, 195.*

warm-up a milling operation to PLASTICIZE uncured RUBBER COMPOUNDS before calendering, extruding, or molding.[131,140]

wash primer PRIMING paint usually supplied as one- or two-component systems. The paint contains carefully balanced proportions of an inhibiting chromate PIGMENT, phosphoric acid, and a synthetic RESIN BINDER mixed in an ALCOHOL. On clean, light alloy or ferrous surfaces, and on many nonferrous surfaces, such paints give excellent ADHESION, partly due to chemical reaction with the SUBSTRATE, and give a CORROSION-INHIBITING film which is a good basis for the application of subsequent coats of paint.[66,71] Syn: pretreatment primers, etch primers, and self-etch primers

Although these materials are referred to as primers, the films which they produce are so thin that it is more correct to consider them as etching SOLUTIONS *and to follow them with an ordinary* PRIMER *if maximum protection is required.*

washability ease with which the DIRT can be removed from a paint surface by washing; also refers to the ability of the coating to withstand washing without removal or substantial damage.[157]

washable distemper see WASHABLE WATER PAINT.[71]

washable water paint water paint which, in addition to a GLUE or CASEIN BINDER, contains an emulsified oil or similar fixing agent, thus rendering it washable.[71] (BSI)

Sometimes incorrectly described as washable distemper.

washboard see BASEBOARD.[82]

washcoat a WOOD FURNITURE coating which is applied after the body STAIN.[71,175,147] (EPA)

It seals the wood surface and stiffens the wood fibers for subsequent sanding.

washing see TINTING (2).[168]

washing out the removal of the original greasy INK on the design areas of etched LITHOGRAPHIC plates.[168]

washout inks INKS used on textiles which are easily removed by washing.[168]

water a clear, colorless, odorless, and tasteless liquid, H_2O, essential for most plant and animal life and the most widely used of all solvents.[144,164]

Properties: freezing point 0°C (32°F); boiling point 100°C (212°F); specific gravity (4°C) 1.0000; weight per gallon (15°C) 8.337 pounds (3.772 kilograms). It has many unique properties.

water beading the formation of beads of water (rain, etc.) on a film.[56] cf. WATER SPOTTING and RAIN SPOTTING[56]

Words presented in CAP/SMALL CAPS type indicate that the word is defined in another part of the *Coatings Encyclopedic Dictionary.*

Numerical superscripts classify terms in one or more of the categories listed in the second section of this volume.

This is of importance for *STAINS.*

water blasting BLAST CLEANING of METAL using high-velocity water.[158]

water break the APPEARANCE of a discontinuous film of water on a surface, signifying nonuniform WETTING and usually associated with a surface contamination.[158] (ASTM)

water gilding application of METAL LEAF or GOLD LEAF onto a SEALED surface, using an aqueous medium.[148,158]

It is a tedious process, but produces a brilliant FINISH. A wood surface is sanded and undercoated with layers of GESSO (in this case, a mixture of marble DUST and rabbit skin glue) and BOLE. GILDER'S LIQUOR is used to apply the leaf. Once dry, the gilding is burnished to the desired degree of brilliance with an agate stone.

water glass sodium or potassium silicate that is soluble in water.[144,155] (ASTM)

water paint (1) paint, the vehicle of which is a water EMULSION, water DISPERSION, or ingredients that react chemically with water[155] (ASTM); (2) a paint capable of being thinned or diluted with water, such as casein paint, LATEX paint, distemper, calcimine, and whitewash.[71,155] (DAC)

water repellent system an exterior coating system for above grade concrete or masonry which temporarily repels water but which is not intended to prevent the passage of moisture under HYDROSTATIC PRESSURE.[71,73,83,158] (ASTM)

There are two types: (1) film forming type—LATEX or oil based paints or similar materials that may require repair or renewal because of climatic action; (2) pore penetrating type—types that fill or coat the interior of the surface pores of concrete or masonry and are temporarily retained in these surface pores thus providing water repelling action; periodic repair or renewal may be required because of climatic action.

water resistance the ability of a coating and the SUBSTRATE it is protecting to resist damage or degradation due to water.[72] cf. HUMIDITY RESISTANCE[72]

It is also the measured ability to retard both PENETRATION and WETTING by water in liquid form. The method of test must be stated since the degree of resistance depends on the way it is determined or measured.

ASTM test methods for water resistance include: boiling water, of furniture lacquer, D2571; controlled condensation, D4585; of varnishes, D1647; water-fog test, D1735; water immersion test, D870.

water spotting see WATER STAINING.[42,56] cf. RAIN SPOTTING[42,56]

Typically, water spotting is caused by the redeposition of SURFACTANTS that have been dissolved in the water on a freshly dried film. These changes include COLOR or gloss resulting from the accumulation of water on or the contact of water with the painted surface apparent after drying.

water staining any nonuniform change in APPEARANCE, including COLOR, GLOSS, or SHEEN in the shape of spots and streaks resulting from the accumulation of water on or the contact of water with the painted surface. It is apparent after drying.[42,56,35] cf. STAINING[65,140] and RAIN SPOTTING.[42,56] Syn: water spotting

After a paint film has been exposed to rain, dew, sprinkling, etc. and the water dries, any water soluble materials (principally SURFACTANT) that have been extracted by the water will be concentrated on the surface and will leave a residue on the paint film. Water staining is more severe on fresher paint films.

water vapor permeance (WVP) the steady water vapor FLOW in unit time through unit area of body (WVT) induced by unit vapor pressure difference (Δ_p) between the two surfaces of a coating. Therefore, WVP = WVT/Δ_p.[128,157] See WATER VAPOR TRANSMISSION RATE (WVP),[128,157] PERM.[128] (ASTM)

Accepted inch-pound unit is grains per square foot per hour per inch of mercury (called a perm). Accepted SI UNIT is grams per square metre per 24 h per millimetre of mercury (called a METRIC PERM).

One of the factors affecting the performance provided by an ORGANIC coating is the degree of resistance to the passage of water vapor. Hence, the water vapor transmission characteristics of coatings are important in assessing their performance in practical use.

ASTM Test Method D1653 covers the determination of the rate at which water vapor passes through films of organic coatings. The films may be free or they may be applied to a porous surface.

water vapor transmission rate (WVT) the steady water vapor FLOW in unit time through unit area of a body, between two specific parallel surfaces, under specific conditions of temperature and humidity at each surface.[128,157] (ASTM) See WATER VAPOR PERMEANCE (WVP),[128,157] and PERM.[128]

Accepted inch-pound unit is grains per square foot per hour. Accepted SI unit is grams per square metre per 24 h.

One of the factors affecting the performance provided by an ORGANIC coating is the degree of resistance to the passage of water vapor. Hence, the water vapor transmission characteristics of coatings are important in assessing their performance in practical use. The water vapor transmission is not a linear function of film thickness, temperature, or relative humidity.

ASTM Test Method D1653 covers the determination of the rate at which water vapor passes through films of organic coatings. The films may be free or they may be applied to a porous surface.

water wax see LIQUID WATER EMULSION WAX.[137]

water white (WW) a material that approaches the colorless nature of water; generally applied to colorless TRANSPARENT liquids and solids.[69,157]

water-based coatings these include LATEX paints, EMULSION paints, and water paints. Coatings in which the volatile content is predominantly water.[71,155] Syn: WATERBORNE COATINGS and WATER-REDUCIBLE COATINGS

water-based inks inks containing a vehicle whose BINDER is water-soluble or water-dispersible.[168,155]

water-based paint see WATER PAINT and WATER-BASED COATINGS.[71,155]

water-break-free the ability of the rinse water to cover the entire surface in an unbroken film.[158,127] (ASTM)

water-reducible coatings see WATER-BASED COATINGS.[71]

water-reducible resins water-soluble types or latices or emulsions. Resins which can be diluted (reduced) with water, water-cosolvent mixtures, and sometimes with alkali (ALKALI-SOLUBLE RESINS).[155,156]

water-soluble resins in most cases, AMINES and/or cosolvents are required to solubilize these carboxyl-containing resins.[130,155]

The preferred term is "ALKALI-SOLUBLE RESIN." These systems are generally dispersions of micelles rather than true SOLUTIONS.

waterborne coatings see WATER-BASED COATINGS.[71]

watercolor (1) ARTISTS' COLORS in which the PIGMENT has been mixed with GUM as a BINDER;[41] (2) that form of TRANSPARENT water painting in which the white of the PAPER furnishes the lights and in which no white pigment is used during the painting of the picture.[71,155,81]

waterproofing treatment of a surface or structure to prevent the passage of water under HYDROSTATIC PRESSURE.[147] (ASTM)

wattle gum water-soluble or water-dispersible GUM obtained from the Australian acacia tree.[155]

wavelength distance between two successive points of a periodic wave in the direction of propagation in which the oscillation has the same phase; designated as λ in SPECTROPHOTOMETRY.[43,128,69] See FREQUENCY.[43,128]

wavy grain GRAIN in which the FIBERS and other longitudinal elements collectively take the form of waves or undulations.[153] (ASTM)

wax any of various unctuous, VISCOUS, or solid heat-sensitive substances, consisting essentially of high molecular weight HYDROCARBONS or ESTERS of FATTY ACIDS (C_{16} to C_{30}), characteristically insoluble in water but soluble in most ORGANIC SOLVENTS.[137]

Waxes can be classified as follows: (1) natural; (2) mineral; (3) synthetic; and (4) products of chemical modification of other waxes.

In coatings, waxes can lower the COEFFICIENT of friction, increase block resistance, improve mar resistance, and act as a water vapor barrier. However, they can downgrade intercoat adhesion.

wax-set ink a PRINTING INK designed to SET and DRY instantly upon immersion of the print in a bath of molten WAX.[168]

wear cycles in ABRASION RESISTANCE TESTS using the TABER ABRASER, the number of cycles of abrasion required to wear a film of specified thickness through to the test plate under a specified set of test conditions.[157]

wear index in ABRASION RESISTANCE TESTS using the TABER ABRASER, it is the loss in weight in milligrams per 1,000 cycles of abrasion under a specific set of test conditions.[157]

weather to age, DETERIORATE, DISCOLOR, etc., as a result of EXPOSURE to the weather.[65,154] (DAC)

weather resistance the ability of a material, paint film, or the like to withstand effects of wind, rain, sun, etc., and retain its APPEARANCE and integrity.[154,72] (DAC)

Most newly formulated coatings are tested by EXPOSURE to weather for several years before the paint is judged acceptable for specific use.

weather stripping METAL, WOOD, PLASTIC, or other material installed around door and window openings to prevent air infiltration.[82]

Weather-Ometer an apparatus in which specimen materials can be subjected to artificial and ACCELERATED WEATHERING tests which simulate natural weathering, by the use of controlled cycles of ULTRAVIOLET RADIATION, light, water, and heat.[65,154,157] See ACCELERATED WEATHERING[65,57,154,157] and ACCELERATED WEATHERING MACHINE.[154,65,157] cf. QUV[154,157,59]

Electric arcs, water spray, and heating elements are used to simulate the natural conditions of sun, rain, and temperature changes.

weatherboarding see SIDING.[75,179,159]

Words presented in CAP/SMALL CAPS type indicate that the word is defined in another part of the *Coatings Encyclopedic Dictionary.*

Numerical superscripts classify terms in one or more of the categories listed in the second section of this volume.

weathering behavior of paint films when exposed to natural weather or ACCELERATED WEATHERING equipment, characterized by changes in COLOR, texture, strength, chemical composition, or other properties.[42,56,65,154] See ACCELERATED WEATHERING.[65,154,57,145,157]

Natural outdoor weathering tests are normally carried out at selected EXPOSURE sites, on painted panels, generally exposed either vertically or at 45° facing south in the northern hemisphere.

web a roll of any SUBSTRATE that passes continuously through a PRINTING PRESS.[168]

webbing (Brit.) Development of WRINKLES, often in a well-defined pattern, in the surface of a paint or VARNISH during DRYING.[56] (BSI) See WRINKLING.[56,54,42]

This condition results from the irreversible swelling of a partially dried surface SKIN and may be aggravated by impure gas FUMES during stoving in a gas oven, in which case it is termed "GAS CHECKING." Webbing is generally regarded as a paint defect but is made use of in some paint finishes to give a TEXTURED coating which obscures minor faults and indentations in the surface to be coated.

weep hole a small hole in a WALL or storm windows which permits water to drain off.[82]

weight percent solids the portion of a coating that remains as part of the dry film expressed as weight.[128,157] (ASTM) See NONVOLATILE MATTER.[157] Syn: solids, total solids

Another convention of expressing solids content is by VOLUME PERCENT.

weight-per-gallon cup essentially, a METAL (less fragile than GLASS) PYCNOMETER to simplify computations; the popular size holds 83.2 g of water at 77°F (25°C) or 100 g for the IMPERIAL GALLON, so that multiplying the weight of the contents by 0.1 represents the DENSITY expressed in POUNDS PER GALLON.[145] (PTM)

weighted ordinate method method of calculating TRISTIMULUS VALUES from measurements made at regular WAVELENGTH intervals.[43,69] See TRISTIMULUS COMPUTATION DATA, TRISTIMULUS INTEGRATION, TRISTIMULUS VALUES, and SELECTED ORDINATE METHOD.[43,69]

weld a METALLIC BOND between like or unlike metals.[45]

weld slag AMORPHOUS deposits formed during welding.[45]

weld splatter beads of METAL left adjoining weld.[45]

Wenzel's blue see COBALT BLUE.[41]

Westphal balance special kind of balance used for the determination of the SPECIFIC GRAVITY of liquids or solids, by a direct weighing method.[59,157]

wet covered or soaked with a liquid, such as water.[142,164,35]

wet abrasion resistance see SCRUB RESISTANCE.[157,72]

wet adhesion the ability of a coating film to adhere tightly to a SUBSTRATE directly beneath it under wet conditions such as rain, dew, washing, etc.[72,79] (ASTM) See WET ADHESION TEST.[72,79,157]

wet adhesion test see TAPE TEST.[157,79]

wet and dry sandpaper sandpaper that can be used with water or other LUBRICANTS, making possible the sanding of some PLASTICS and metals that is not possible with dry sanding.[158]

The lubricant cools the workpiece, washes away swarf, reduces friction, produces finer finishes, resists rusting (with ADDITIVES), improves ABRASIVE life, and combats loading of the COATED ABRASIVE.

wet bulb temperature in PSYCHROMETRY, the temperature of the sensor or the bulb of the thermometer in which a constantly renewed film of water is evaporating.[145] (IUPAC)

The temperature of the water used to renew the film must be at the temperature of the gas.

wet edge the edge of a wet, painted area which remains workable.[54,42] (BSI) Syn: live edge

When painting large surfaces, it is generally necessary to join up to the edge of a paint film which has been left to dry for some period of time. When this can be done by blending this edge with free-working paint, without any lap mark between the drying and freshly painted area, the paint is said to present a wet edge.

wet edge extender high boiling liquids such as PROPYLENE GLYCOL added to LATEX or water-based paints to reduce the EVAPORATION RATE and thereby extend the (OPEN) TIME that they are wet enough to BRUSH into.[83,54] (ASTM) See WET EDGE.[42,54]

These wet edge extenders are used to minimize lapping problems.

wet edge time see OPEN TIME.[42,54,55]

wet film build see BUILD OF WET FILM.[128]

wet film gage device for measuring WET FILM THICKNESS of coatings.[145]

wet film thickness THICKNESS of the liquid coating film immediately after application.[128,157]

wet hiding the HIDING of freshly applied paint.[43] cf. DRY HIDING[43]

wet printing see TRAPPING.[168]

wet rot decay of TIMBER caused by FUNGI which flourish in alternate wet and dry conditions.[153,154]

wet sanding PROCESS of SANDING using water-resistant papers with liquids, such as water, soluble oil (emulsion), straight cutting OILS (mineral and fatty), or mineral lard oil (mixture of mineral and lard oils, sulfurized and chlorinated cutting oils, and wax).[59,158] See WET and DRY SANDPAPER.[59,158]

wet scrub resistance ability of a paint film to withstand SCRUBBING in contact with water.[72,47,157] (BSI) See SCRUB RESISTANCE.[72,157]

wet storage stain see WHITE RUST.[166,66]

wet-on-wet coating technique of painting whereby further coats are applied before the previous coats have dried, and the composite film then dries as a whole.[54,55] (BSI)

The PROCESS requires specially formulated paints.

wetting power a VEHICLE possesses of SPREADING UNIFORMLY and rapidly over the surface of PIGMENT PARTICLES.[127]

A vehicle with good WETTING properties assists in the GRINDING or DISPERSION of PIGMENTS and the ability to wet the surface to which the finished coating is applied.

wetting agent see SURFACTANTS.[127]

whipping arcing or waving a SPRAY GUN during SPRAYING.[54]

Considered poor technique.

whirling centrifugal removal of excess paint from articles which have been coated by DIPPING.[77] (BSI)

white most usually applied to neutral or near-neutral colors of high REFLECTANCE.[43,69]

white blast blast cleaning to WHITE METAL.[158] See NACE No. 1.[158]

white bole see ALUMINUM SILICATE.[41,58]

white earth nonspecific description applied to several different types of white FILLERS.[58]

white lac SHELLAC which has been chemically BLEACHED.[146]

white lead see CARBONATE WHITE LEAD.[41]

white metal blasting METAL to specified APPEARANCE such as SSPC-SP-10, NACE No. 1, or SA-3 (Swedish Standard SIS 05 59 00).[158]

white Portland cement a PORTLAND CEMENT, produced from raw materials low in iron, which hydrates to a white paste; used to yield a concrete of considerable WHITENESS.[73,75] (DAC)

white reference standard see STANDARD, WHITE REFLECTANCE.[43,69]

white reference standard, absolute see PERFECT DIFFUSER AND STANDARD, WHITE REFLECTANCE.[43,69]

white reference standard, primary see STANDARD, WHITE REFLECTANCE.[43,69]

white reference standard, secondary see SECONDARY REFERENCE STANDARD.[43,69]

white rust white CORROSION products (zinc hydroxide and zinc oxide) on zinc-coated articles.[66] Syn: wet storage stain

They form when the parts are stored so close together that CONDENSED MOISTURE is entrapped between them and the air circulation is inadequate to assist drying.

white spirit British name for MINERAL SPIRITS.[164]

white zinc synonym for zinc oxide.[41]

whiten, whitening to make or become white, especially by BLEACHING.[69]

whiteness perception of high LIGHTNESS, high DIFFUSION (scattering), and absence of HUE generally applied to OPAQUE or TRANSLUCENT solids or liquids.[43,69]

Whiteness in pigments results from the SCATTERING of light by multiple REFLECTION and REFRACTION; scattering is a function of the refractive index ratio between the PIGMENT and the surrounding BINDER and is strongly influenced by PARTICLE SIZE.

whiteness index any of several numerical indices used to indicate the degree of whiteness; examples are the following:

(1) $W = 4B-3G$ where B and G refer to tristimulus colorimeter readings.

(2) $W = 100 - [(100\text{-}L)^2 + (a^2 + b^2)]^{1/2}$ where L, a, and b refer to coordinates in HUNTER'S L, A, B COLOR DIFFERENCE EQUATION.

(3) Chemstrand Whiteness Scale $W = 10\ (Y - 2p^2)^{1/2}$ where Y is the CIE TRISTIMULUS VALUE and p is the CIE excitation purity.[43,69,128]

whitening (1) see LIMEWASHING; WHITEWASH;[71] (2) see WHITEN.[69]

whitening in the grain fault which sometimes develops in varnished or polished open-grained woods, filled or unfilled. It is manifested as a streaky white APPEARANCE.[56,159] (BSI) See BLEACHING (1).[43,65] Syn: grayness

---

Words presented in CAP/SMALL CAPS type indicate that the word is defined in another part of the *Coatings Encyclopedic Dictionary.*

Numerical superscripts classify terms in one or more of the categories listed in the second section of this volume.

whitewash cheap type of flat water paint based on LIME or WHITING loosely bound with GLUE, SIZE, CASEIN, or like water-dispersible BINDERS.[163] Syn: limewashing, whitening

whiting CALCIUM CARBONATE POWDER of high purity.[58,41] See CALCIUM CARBONATE, NATURAL.[58,41]

wicking ABSORPTION of liquid by CAPILLARY action.[142]

wide-belt sander a machine using the principle of a contact roll backup for a wide ABRASIVE belt.[59,158]

Wijs Test ANALYTICAL method for determining the degree of UNSATURATION of a material, which involves the addition of iodine to the existing DOUBLE BONDS.[157] See IODINE TEST, IODINE VALUE, and HANUS IODINE NUMBER.

The iodine is derived from a solution of iodine monochloride in glacial ACETIC ACID.

wilkinite see BENTONITE.[41,139,133]

Wilkinson's blue see IRON BLUE.[41]

winded the separation of PRINTED sheets so that they will be ventilated by air.[168]

window sash see SASH.[82]

wine gallon a STANDARD U.S. GALLON of 231 cubic inches (3.7854 litres).[128]

As commonly used, this term refers to ALCOHOL at 60°F, which is the standard temperature prescribed by the Federal Government for gauging alcohol.

winterized oils OILS (e.g., fish oils) which have been stored in a cold place for a long time to allow the solid (i.e., mostly SATURATED) components to SETTLE out.[63]

wire brush (1) a hand cleaning tool comprised of bundles of wires;[59] (2) the act of cleaning a surface with a wire BRUSH including power wire brushes.[158]

wire-bar applicator see WIRE-WOUND ROD.[145,157,77]

wire-wound rod METAL rod wound with a fine wire around its axis so that an INK or coating can be drawn down evenly and at a given thickness across a SUBSTRATE.[145,157,77] Syn: equalizer rod, Meier rod, wire-bar applicator

The thickness or gauge of the wire controls the depth of wet film applied by the rod. These are sets of calibrated stainless steel wire-wound drawdown rods (bars) at least 1/2 in. (13 mm) in diameter to prevent bowing during application. They are calibrated such that the diameter of the wire wound on the rod controls thickness of the wet film applied. The choice of specific drawdown rod is dependent on the dry film thickness required and the volume solids of the coating under test.

ASTM Test Method D4147 covers a procedure for applying a coating film of uniform thickness on a flat panel using the wire-wound drawdown rod (bar).

withering the loss of GLOSS caused by VARNISHING open-pore WOODS without filling pores, use of improper UNDERCOATING, or applying TOPCOAT before undercoat has dried.[42,56]

witherite see BARIUM CARBONATE.[41,58]

wollastonite see CALCIUM SILICATE, NATURAL.[41,58,34]

wood the hard, fibrous substance which composes the trunk and branches of a tree, lying between the pith and bark.[75,153] (DAC)

wood alcohol CH_3OH. METHANOL from the DESTRUCTIVE DISTILLATION of WOOD. When pure, its properties are identical with those of synthetic methanol.[164]

wood cement see STICK SHELLAC.[146]

wood chipboard see PARTICLE BOARD.[75,153]

wood filler heavily pigmented product used to fill the GRAIN of wood before the application of UNDERCOATS or FINISHES.[54,158]

wood finishing the planing, sanding, and subsequent staining, varnishing, waxing, or painting of a wood surface.[54,158,153] (DAC)

wood flour very fine wood PARTICLES generated from wood reduced by a ball or similar MILL until they resemble wheat flour in APPEARANCE, and of such size that the particles usually will pass through a 40-mesh screen.[153]

wood oil see TUNG OIL.[63,156]

wood preservative coatings coatings which are formulated to protect wood from decay and insect attack.[71,153,65] cf. PENETRATING FINISH[63,71,152] See STAIN and LASURE.[41,43,69,71,163]

Two basic types of finishes (or treatments) are used to protect wood surfaces during outdoor WEATHERING: (1) those that form a film, layer, or coating on the wood surface, and (2) those that penetrate the wood surface leaving no distinct layer or coating. FILM-FORMING materials include paints of all description, VARNISHES, LACQUERS, and also overlays (laminates) bonded to the wood surface. PENETRATING FINISHES include PRESERVATIVES, WATER REPELLENTS, pigmented semitransparent stains, and chemical treatments. Natural finishes are important in wood protection as are finishing practices.

wood rosin ROSIN obtained from pine stumps.[44,167]

wood tar a black, syruplike, viscous fluid that is a

by-product of the destructive distillation of wood and is used in pitch, wood preserving OILS, preservatives, and medicine.[153]

wood turpentine see TURPENTINE.[44,164]

wood veneer see VENEER.[82,148,159]

woodcut a wooden PRINTING plate, the image of which has been left in relief by cutting away the background.[81,168]

wool grease see LANOLIN.[137]

wool grease pitch see FATTY ACID PITCHES.

wool scale the comparative (textile) standard for assessing LIGHTFASTNESS.[157]

working time period of time during which an adhesive or coating, after mixing with CATALYST, SOLVENT, or other compounding ingredients, remains suitable for use.[161,74] See POT LIFE and USABLE LIFE.[161,74,42,67] Syn: working life

wrap a characteristic of POWDER COATINGS in electrostatic application to seek out and adhere to parts of the SUBSTRATE not in direct line of sight of the delivery system end point.[77]

wrinkle finish type of FINISH characterized by the presence of wrinkles of fairly uniform dimensions.[71,163]

The effect is obtained by inducing a pronounced but controlled tendency toward rapid surface DRY. This finish is generally BAKED rather than AIR DRIED. The size of the wrinkle can be adjusted by variation in the type and amount of drier, type and amount of PIGMENT, thickness of applied film, and baking temperature.

wrinkling the distortion in a paint film appearing as ripples; may be produced intentionally as a DECORATIVE effect or may be a DEFECT caused by drying conditions or an excessively thick film.[54,56,42] (DAC) See GAS CHECKING.[54,56,42] Syn: crinkling, riveling.

wurtzilite see ASPHALTIC PYROBITUMENS.[76]

WVP see WATER VAPOR PERMEANCE (WVP).[128,157]

WVT see WATER VAPOR TRANSMISSION RATE (WVT).[128,157]

WW initial letters meaning "WATER-WHITE" and applied to a pale grade in a COLOR scale of ROSIN.[142,43]

The term is also applied to other RESINS, or SOLUTIONS of the same, which are practically colorless.

Words presented in CAP/SMALL CAPS type indicate that the word is defined in another part of the *Coatings Encyclopedic Dictionary.*

Numerical superscripts classify terms in one or more of the categories listed in the second section of this volume.

X

X one of the three CIE TRISTIMULUS VALUES; the red primary.[43,69]

$\bar{x}$ SPECTRAL COLOR MATCHING functions of the CIE STANDARD OBSERVER, used for calculating the X TRISTIMULUS VALUE. See TRISTIMULUS COMPUTATION DATA, TRISTIMULUS VALUES, COLOR MATCHING FUNCTIONS.[43,69]

x one of the CIE CHROMATICITY COORDINATES calculated as the fraction of the sum of the three TRISTIMULUS VALUES attributable to the X value.[43,69]

$$x = \frac{X}{X + Y + Z}$$

xanthic containing yellow, or pertaining to yellow COLOR.[69]

xenon-arc light exposure apparatus an apparatus in which specimen materials can be subjected to artificial and accelerated WEATHERING tests which simulate natural weathering by using this light source EXPOSURE apparatus of water, or both.[65,145,154,157] See ACCELERATED WEATHERING,[65,57,154,157] ACCELERATED WEATHERING MACHINE,[154,65,157] and WEATHER-OMETER.[65,154,157]

ASTM Test Method G26 covers the basic principles and operating procedure for water- or light-exposure apparatus, or both, employing a xenon-arc light source.

xerography a dry method of photography or photocopying.[168]

xylene any of the three ISOMERS: ortho, meta, and para.[164] Syn: xylol

Commercially, a mixture of the three isomers in various proportions, with some ETHYLBENZENE.

m-xylene $C_6H_4(CH_3)_2$. Colorless liquid, insoluble in water, miscible with ALCOHOL, ETHER, and many other ORGANIC liquids.[164] See XYLENE.[164]

Properties: bp, 139.3°C; sp gr, 0.8684/15°C; mp, 47.8°C.

o-xylene $C_6H_4(CH_3)_2$. Colorless liquid, insoluble in water, MISCIBLE with ALCOHOL, ETHER, and many other ORGANIC liquids.[164] See XYLENE.[164]

Properties: mp, –25°C; sp gr, 0.897/20°C; bp, 144°C.

p-xylene $C_6H_4(CH_3)_2$. Colorless liquid, insoluble in water, MISCIBLE with ALCOHOL, ETHER, and many other ORGANIC liquids. CRYSTALLIZES at low temperature.[164] See XYLENE.[164]

Properties: mp, 13.2°C; bp, 137-138°C; sp gr, 0.8611/20°C.

xylenol resin a RESIN of the PHENOLIC type made by POLYCONDENSATION of a xylenol with an aldehyde or a ketone.[130]

xylography the art of PRINTING with wooden blocks.[81]

xylol the term "xylol" is still used commercially for XYLENE, but is not preferred in modern nomenclature.[164]

Y

Y one of the three CIE TRISTIMULUS VALUES, equal to the LUMINOUS REFLECTANCE or TRANSMITTANCE; the green primary.[43,69] See TRISTIMULUS VALUES, LUMINOUS DIRECTIONAL REFLECTANCE.[43,69]

$\bar{y}$ SPECTRAL COLOR MATCHING function of the CIE STANDARD OBSERVER used for calculating Y TRISTIMULUS VALUE.[43,69] See TRISTIMULUS COMPUTATION DATA, and COLOR MATCHING FUNCTIONS.[43,69]

y one of the CIE CHROMATICITY COORDINATES calculated as the fraction of the sum of the three TRISTIMULUS VALUES, attributable to the Y value:[43,69]

$$y = \frac{Y}{X + Y + Z}$$

yacca gum see ACCROIDES.[155,167]

Ye Olde Paint Stone see BOSTON STONE, THE.[59,64]

year ring an annual ring of growth in TIMBER.[153]

yellow accroides bright, yellow-colored RESIN obtained from Australia.[167] See ACCROIDES.[167]

It resembles red accroides, and is used to a limited extent in some types of SPIRIT VARNISHES.

yellow chrome see CHROME YELLOW.[41]

yellow iron oxide $Fe_2O_3 \cdot xH_2O$. Pigment Yellow 42 (77492). A natural or synthetic iron oxide.[41] Syn: ferrite yellow, ocher, monohydrate of ferric oxide, limonite

It has excellent BLEED RESISTANCE, chemical resistance (both acid and alkali), LIGHTFASTNESS, and DISPERSIBILITY, as well as excellent TOXICITY ratings. However, these yellow oxides have only fair heat resistance since they are subject to COLOR change above 177°C due to loss of water of hydration.

yellow limonite see IRON OXIDES, NATURAL.[41]

yellow ocher, synthetic synthetic yellow IRON OXIDE reduced with aluminum silicate or other EXTENDER PIGMENTS.[41] See OCHER.[41]

yellow pine a hard resinous wood of the longleaf pine tree, having dark bands of SUMMERWOOD alternating with lighter-colored SPRINGWOOD; used as flooring and in general construction.[75,153] (DAC)

yellowing development of a yellow COLOR on aging; most noticeable in the dried films of white paints or clear VARNISHES.[65,69,35] (BSI)

Primary cause of yellowing is the formation of COLOR bodies due to OXIDATION of the OIL or other unsaturated components of the BINDER.

yellowness the attribute by which an object COLOR is judged to depart from a preferred white toward yellow.[43,69,157] (ASTM)

yellowness index (YI) any of several numerical indices of the degree of departure from a preferred white REFLECTING material or colorless TRANSPARENT material towards yellow; an example is the following:

$$YI = 100\,(R - B)/G$$

where R, G, and B are TRISTIMULUS COLORIMETER readings, all normalized to 100. In CIE tristimulus terms, the equation becomes

$$YI = (128X - 106Z)/Y.$$ [43,69] (ASTM D1925)

YI abbreviation for YELLOWNESS INDEX.[43,69]

yield value minimum SHEARING STRESS which must be applied to a PLASTIC material to start LAMINAR FLOW.[166]

It may be measured in dynes/cm² or newtons/m² and is the shearing FORCE producing infinitely slow rate of flow between adjacent planes. The intercept on the force axis of the flow curve of a plastic material is a function of the yield value. It is the stress (either normal or shear) at which a marked increase in DEFORMATION occurs without an increase in load.

Young's Modulus the ratio of normal STRESS to corresponding STRAIN for TENSILE or compressive stresses less than the proportional limit of the material.[47] (ASTM) See COEFFICIENT OF ELASTICITY.[47]

Words presented in CAP/SMALL CAPS type indicate that the word is defined in another part of the *Coatings Encyclopedic Dictionary.*

Numerical superscripts classify terms in one or more of the categories listed in the second section of this volume.

Z

Z one of the three CIE TRISTIMULUS VALUES: the blue primary.[43,69]

$\bar{z}$ SPECTRAL COLOR matching function of the CIE STANDARD OBSERVER used for calculating the Z TRISTIMULUS VALUE.[43,69] See TRISTIMULUS COMPUTATION DATA, and COLOR MATCHING FUNCTIONS.[43,69]

z one of the CIE CHROMATICITY COORDINATES calculated as the fraction of the sum of the three TRISTIMULUS VALUES attributable to the Z primary:[43,69]

$$z = \frac{Z}{X + Y + Z}$$

zaffre see COBALT BLUE.[41]

Zahn cup efflux cup used for measuring the VISCOSITY of paints, VARNISHES, and PRINTING INKS.[145,157,166]

Zanzibar copal a very hard fossil RESIN obtained from the eastern side of Africa, and in the past from the island of Zanzibar.[167] Syn: ANIMI

zein ALCOHOL-soluble PROTEIN obtained from maize.

zeolite any one of a family of hydrous aluminum silicate minerals, whose molecules enclose cations of sodium, potassium, calcium, strontium, or barium, or a corresponding synthetic compound.[34,84]

They are used chiefly as molecular filters, ion exchange agents, and catalysts. The four oxygen anions surround silicon and aluminum cations to build a honeycomb of molecules which carries a negative charge. The excess charge is the key to the adsorbing and CATALYTIC powers of zeolites.

zero VOC paints those containing no PETROLEUM-BASED SOLVENTS.[71,163] See VOC.[171] Syn: VOC-free paints

These are environmentally friendly ORGANIC solvent-free coatings characterized by no lingering odor—generally LATEX paints.

zeta potential the potential across the INTERFACE of all solids and liquids.[61,139] Syn: electrokinetic potential

More specifically, it is the potential across the diffuse layer of ions surrounding a charged COLLOIDAL PARTICLE, and is largely responsible for colloidal STABILITY.

zeta space COLOR DIFFERENCE EQUATION devised by Saunderson and Milner.[43,69] See SAUNDERSON-MILNER ZETA SPACE COLOR DIFFERENCE EQUATION.[43,69]

zinc chromate $4ZnO{\cdot}K_2O{\cdot}4CrO_3{\cdot}3H_2O$. Pigment Yellow 36 (77955). Bright yellow PIGMENT which chemically is substantially zinc chromate, although its precise composition is rather complex.[41] Syn: zinc yellow

Its chief use is in ANTICORROSIVE paints and PRIMERS for STEEL. Properties: density, 3.36-3.49 g/cm³ (28.1-29.1 lb/gal); O.A., 28-31; particle size, 0.2-5.0 μm.

zinc drier zinc SALTS of acids generally used for DRIERS, such as naphthenic or 2-ethyl hexoic, are not driers. However, when used with driers such as cobalt they often prevent wrinkling and skinning. They act as auxiliary driers.[83,74,67]

zinc dust finely divided zinc METAL used as a PIGMENT in protective paints for iron and steel.[41,45] (BSI)

zinc ferrocyanide see ANTWERP BLUE.[41]

zinc green see COBALT GREEN.[41]

zinc oxide ZnO. Pigment White 4 (77947). A fine-particle, white PIGMENT used in RUBBER, paint, and PLASTIC industries for MILDEW resistance and film reinforcing properties.[41] Syn: zinc white, Chinese white, white zinc

Properties: density, 5.6 g/cm³ (46.7 lb/gal); O.A., 10-25; particle size, 0.2 μm; refractive index, 2.01.

zinc phosphate coating a thin, INORGANIC deposit formed on zinc treated with phosphoric acid.[158]

zinc resinate zinc SOAP of ROSIN with a melting point higher than rosin.[127,130]

zinc silicate primers INORGANIC ZINC-RICH PRIMERS that contain a silicate BINDER.[71]

zinc stearate insoluble zinc SOAP used to produce MATTENESS, FLATTING, or THICKENING in coatings.[83]

zinc stearate-coated finishing paper ZINC STEARATE-coated silicon carbide and garnet finishing papers used for sanding between SEALER coats.[59,158] See COATED ABRASIVE.[59,158]

The coating of zinc stearate reduces loading of the PAPER because of the lubricating properties of the stearate. Finer grades are also sometimes used in the final RUBBING operation of FURNITURE finishing.

zinc sulfide Pigment White 7 (77975). ZnS. Yellowish white PIGMENT, soluble in acids, insoluble in water.[41]

Certain specially treated grades of this pigment are used as FLUORESCENT PIGMENTS. Properties: density, 4.0 g/cm^3 (33.3 lb/gal); O.A., 13; particle size, 0.25 µm; refractive index, 2.37.

zinc tetroxy chromate see BASIC ZINC CHROMATE.[41,66]

zinc white synonym for ZINC OXIDE.[41]

zinc yellow see ZINC CHROMATE.[41]

zinc-cadmium sulfides combination of zinc sulfide and cadmium sulfide, along with special ADDITIVES, manufactured by highly specialized procedures and used as FLUORESCENT PIGMENTS.[41]

zinc-coated see GALVANIZING.[45,158]

zinc-rich primer ANTICORROSIVE PRIMER for iron and steel incorporating zinc DUST in a concentration sufficient to give electrical conductivity in the dried film, thus enabling the zinc METAL to corrode preferentially to the SUBSTRATE, i.e., to give CATHODIC PROTECTION.[66,71] (BSI)

zircon $ZrO_2 \cdot SiO_2$. Occurs as natural double oxide.[41,34] Syn: zirconium silicate

It is produced by physical separation of beach sand (Florida, Australia). Used in VITREOUS ENAMELS and CERAMICS. Properties: density, 4.56; particle size, 1.0-2.5 µm; refractive index, 2.0

zirconia see ZIRCONIUM OXIDE.[41,34]

zirconium driers zirconium SALTS of various ORGANIC ACIDS such as naphthenic or 2-ethyl hexoic.[74,83,67]

Used as a replacement for lead in FUME-proof paints or other lead-free paints where equivalent through drying can be obtained. It is also used as an auxiliary DRIER where COLOR and color retention are important.

zirconium oxide ZrO_2. Pigment White 12 (77990). A natural MINERAL, BADDELEYITE.[41,34] Syn: zirconium dioxide, zirconia, baddeleyite

Present commercial grades are unsuitable as PIGMENT grade in paint vehicles. Properties: density, 5.68 g/cm^3; particle size, 1.0-2.0 µm; refractive index, 2.1-2.2.

zunsober see MERCURIC SULFIDE.[41]

zwitterion a dipolar ion; a molecule containing both acidic and basic groups, which may be expected to neutralize itself with the production of a double SALT.[61,166] cf. ANION and CATION[61,166]

AMINO acids and proteins at their ISOELECTRIC POINTS exist in the form of zwitterions.

Words presented in CAP/SMALL CAPS type indicate that the word is defined in another part of the *Coatings Encyclopedic Dictionary.*

Numerical superscripts classify terms in one or more of the categories listed in the second section of this volume.

Categorical Listing of Terms

Throughout the *Coatings Encyclopedic Dictionary,* terms are classified into one or more of 80 categories. Superscripts appearing at the end of each definition identify the category listing for each term. These categories are designed to help the reader to gain a fuller understanding of the term and can serve as a check list for key words in the preparation of research papers and literature searches. The categories appear in this section in order by category number.

All of the categories and their page numbers are indexed in both numerical and alphabetical order.

Numerical List of Categories

Alphabetical List of Categories

Category 34
Minerals

albite
aluminum potassium silicate
amphibole
anhydrite
attapulgite
azurite
baddeleyite
barium sulfate
bauxite
brookite
celestite
crocidolite
crocoisite, crocolite, crocoite
cuprous (copper) oxide
dichroism
dichromism
dolomite
fire clay
garnet abrasive
greenockite
gypsum
ilmenite
ilmenite black
industrial talc
iron titanate
kaolinite
labradorite
lead sulfide
magnesite
magnesium silicate, fibrous
malachite
menaccanite
mercuric sulfide
mineral
mineral black
Mohs scale
natural iron oxides
nepheline
nepheline syenite
ore
pleochroism
realgar
rutile
silicate
soapstone
Spanish red oxide
strontium white
syenite
ultramarine blue
vermiculite
wollastonite
zeolite
zircon
zirconia
zirconium oxide

Category 35
Appearance

appearance
attribute
bronze
cast
chilling
chroma
chromatic
clarity
clean
color
color constancy
color uniformity
conditional match
crackle finish
dead flat
depth of finish
depth of shade
dullness
durability
eggshell
eggshell finish
end point
fading
finish
flair
flame spread
geometric metamerism
ghosting
graining
gum bloom
hiding power chart
hiding, complete
hitching post standards
household chemicals, resistance to
imprinting
just noticeable difference (JND)
light
luster
metamer
metameric match
metameric pair
metamerism
modes of appearance
moiré effect
mottle
Munsell chroma
Munsell color system
Munsell surface-color perception solid
object color
object mode of appearance
paramerism
print resistance
rescue orange
resistance to household chemicals
ropy finish
semigloss
sheen
silking
stardust
streaking
total color difference
vision
vision, normal color
visual
water staining
wet
yellowing

Category 36
Statistical

accuracy
approximate value
blistering
deviation
figure
precision
repeatability
reproducibility
round-robin
significant digit
variance

Category 37
Analytical

activated carbon
calibrate
chromatogram
chromatographic analysis (chromatography)
coulometry
diffraction analysis
dry weight
electrochemical methods of detection (analysis)
mass spectrometry
quality assurance
quality control
repeatability
reproducibility
sample
sampling
spectrometry, optical
standard
stochastic sampling
test
tracer

Category 38
Plastics

ablative coatings
ablative plastic
ABS
acetates
acrylate resins
acrylic plastics
acrylics
acrylonitrile-butadiene-styrene (ABS)
amino plastics
annealing
Bakelite®
base
binder
bloom
body putty (body solder)
camphor
cast coating
cast film
cellophane
celluloid
cellulose acetate butyrate (CAB)
cellulose acetate propionate (CAP)
cellulose ester
cellulose lacquer
cellulose nitrate
cellulose plastics
cellulose propionate
compacting
epoxy plastics
extrusion
extrusion mark
extrusion moldings
fluoroplastics
foamed plastic
gel coat
green stock
heat-sealing
hydrocarbon plastics
injection molding
karolith
laminate, high pressure
mold marks
organosol
phenolic plastics
plasticizer migration
plastisol
poly(methyl methacrylate) (PMMA)
polyethylene (PE)
polystyrene (PS)
polyvinyl acetal
polyvinyl alcohol (PVA)
polyvinyl butyral (PVB)
polyvinyl chloride (PVC)
polyvinylidene chloride (PVDC)
postforming
protein-aldehyde resins
silicone plastics
SPE
stabilizer
vinyl
vinyl acetate plastics
vinyl chloride plastics
vinyl plastics
vinyl wallcoverings
vinylidene chloride plastics
vulcanized fiber

Category 41
Pigments, Dyes

A-acid
abraum
absorption, oil
absorption tinting strength
acenaphthene
acetate chromes
acetate green
acetylene black
acicular
acid dyes
acrylide maroon
aerugo
African ochre
agglomerate
aggregate
albite
alizarin
alizarin madder lake
alizarin red
alkali blue
alkali-resistant red
alumina
alumina hydrate
alumina-blanc fixe
aluminum arsenite
aluminum hydrate
aluminum (in U.K., aluminium)
aluminum oxide
aluminum oxide cloth
aluminum oxide, hydrated
aluminum oxide paper
aluminum paste
aluminum phosphate
aluminum pigment
aluminum potassium silicate
aluminum powder
aluminum silicate (clay)
American process zinc oxide
American turpentine
American vermillion
amorphous silica
amphibole
anadonis green
anatase
aniline dyes
aniline pigments
animal blacks
anthracene
anthranilic acid
antimony orange
antimony oxide
antimony sesquioxide
antimony trioxide
antimony vermillion
antimony white
antimony yellow
antwerp blue
argillaceous
Arnaudon's green
arsenic orange
artificial ultramarine
asbestine
asbestos
attapulgite
aureolin
Austrian cinnabar
auxochrome
azo dyes
azo pigment
azure blue
azurite
baddeleyite
barium carbonate
barium chromate
barium chrome
barium lithol red
barium lithol toner
barium metaborate, modified
barium sulfate, natural
barium sulfate, synthetic
barium white
barium yellow
barium-extended titanium dioxide
barytes
base
basic carbonate of lead
basic dye
basic fuchsin magenta
basic lead acetate
basic lead carbonate
basic lead chromate
basic lead nitrate
basic lead silicate
basic lead silico chromate
basic lead sulfate
basic pigment
basic zinc chromate
bauxite
beni
bentonite
bentonite clay
benzidine yellow
Berlin blue
biddiblack
binder demand
binder ratio
binding power index
bismarck brown
black chalk
black iron oxide
black lead
black magnetic oxide
blanc fixe
bleed
bleed characteristics
bleeding pigment
blue basic lead sulfate (blue lead)
blue stone
blue toner
blue verditer
blue vitriol
blueing
BON
BON reds and maroons
bone ash
bone black
bone brown
bone white
BPI
Brazil-wood
brightener, optical
brightening agent
brilliant ultramarine
bronze blue
bronze gold powders
bronze paint
bronze paste
bronze powders
bronze scarlet
bronzing
brookite
brown iron oxide pigment
brown madder
brown pink
Brunswick blue
Brunswick green
burnt island red
burnt lime
burnt sienna
burnt umber
cadmium green
cadmium lithopone
cadmium/mercury lithopones
cadmium/mercury sulfides
cadmium orange
cadmium red
cadmium selenide
cadmium sulfide
cadmium yellow

cake, press
calcined clay
calcite
calcium carbonate, natural
calcium carbonate, synthetic
calcium lithol red
calcium metasilicate
calcium silicate, natural
calcium silicate, synthetic
calcium sulfate
calcium sulfate, anhydrous
carbazole dioxazine violet
carbon black
carbonate white lead
carmine
carmine lake
carmine vermillion
carminic acid
Carter lead
cassel brown
cassel earth
cassel yellow
celestial blue pigment
celestite
cement colorants
cenobrium
cerulean blue
cerussite
chalk
chalking
chalking resistance
chamber white lead
channel black
charcoal black
China clay
Chinese blue
Chinese red
Chinese vermillion
Chinese white
chips
chlorazol dyes
chlorinated para nitraniline red
chlorinated para reds
chromate of lead
chrome green
chrome ochre
chrome orange, light and deep
chrome oxide green
chrome red
chrome vermillion
chrome yellow, light and primrose
chrome yellow, medium
chromic hydrate
chromic oxide
chromium green
chromium hydrate
chromium hydroxide
chromium oxide green
chromium oxide, hydrated
chromium sesquioxide
chromophore
chrysotile
CI
cinaper
cinnabar
cinoper
claret red
clay
coal tar colors
cobalt aluminate
cobalt blue
cobalt green
cobalt ultramarine
cobalt yellow
cochineal
cologne yellow
color in oil
color in varnish
color wash
colorant
composite pigment
copper acetoarsenite
copper blue
copper bronze
copper powder
copper sub-oxide
copper sulfate
copperas
corrosion-inhibitive pigment
coupling
covellite
CPVC
cremnitz white
crimson antimony
crimson lake
crimson toner
critical pigment volume concentration (CPVC)
crocidolite
crocoisite. crocolite. crocoite
crystalline silica
cupric sulfide
cuprous (copper) oxide
custom color
cynoper
dalamar yellow
dead burned calcium sulfate
dehydrated gypsum
delaminated clay
deleafing
depth of shade
derby red
diarylide yellow
diatomaceous calcite
diatomaceous earth
diatomaceous silica
diazo
diazo solution
dihydroxy anthraquinone lake
dingler's green
dinitraniline
dinitraniline orange
diphenylamine
dispersion, degree of
dispersion, pigment
dolomite
drop black
dry color
dry hydrate
dumont's blue
Dutch pink
Dutch process
dye
dye, acid
dye, basic
dye, mordant
dye pigments
dye, spirit-soluble
dyes and dyestuffs
dyestuff
earth pigment
Egyptian blue
electrolytic white lead
emerald green
emerald oxide of chromium
enamel blue
English vermillion
eosin
erioglaucine dye
erlangen blue
extender
extender (pigment)
factitious ultramarine
fast to light
fastness, fast
feldspar
ferric oxide
ferriferro cyanide
ferrite yellow
ferrous sulfate
fibrous asbestos
fibrous magnesium silicate
filter cake
fineness
fire clay
flake metal
flake white
flatting agent
flatting pigment

fleshing
florentine lake
flower of antimony
fluorescein dye
fluorescent dyes
fluorescent paint
fluorescent pigments
flushed (flushing) colors
flushing
flushed pigment
free silica
French blue
French chalk
French ochres
French process zinc oxide
French ultramarine
French veronese green
fuchsin
fugitive color
Fuller's earth
fumed silica
furnace black
fustics
Gahn's ultramarine
galliolino
gas black
gas blue
genuine ultramarine
geranium lake
gibbsite
gloss white
gold bronzes
golden chrome
graphite
graphitic mica
green B, naphthol
green B, pigment
green, brunswick
green, chrome
green, cinnabar
green earth
green, emerald
green gold
green lake
green, lime
green, Paris
green rouge
green vitriol
green, zinc
greenockite
greentone ink blue
ground oyster shells
ground perlite
ground silica
Guignet's green
Guimet's blue
hansa yellow
heavy spar
hectorite
Helio fast rubine 4BL lake
hematite
high-visibility paints
high-visibility pigments
Huemann's blue
Hungary blue
hydrated aluminum silicate
hydrated chromium oxide
hydrated magnesium aluminum silicate
Iceland spar
ilmenite
ilmenite black
indanthrone blue
Indian ochre
Indian red
Indian red ochre
Indian yellow
indigo
induline
induline bases
industrial talc
industrial talc, asbestos-free
industrial universal colorants
inert pigment
infusorial earth
inhibitive pigment
inhibitor
interference pigments
intermediates
iron blue
iron oxides
iron oxides, natural
iron oxides, synthetic
iron titanate
ivory black
j acid
Japanners' brown
jaune brilliant
jewelers' rouge
kaolin
kaolinite
king's blue
king's yellow
Kremnitz white
labradorite
lac dye
lake
lake Bordeaux B
lake dyes
lake of acid yellow 1
lake red C
lake red P
lampblack
lapis lazuli
lapis lazuli blue
laundry blue
lead antimonate
lead carbonate, basic
lead chrome green
lead monoxide
lead oxide
lead phosphite, dibasic
lead primer
lead titanate
lead white
leaded zinc oxide
leaf green
leipsic yellow
Leithner's blue
lemon chrome
lemon yellow
letdown
levigation
leyden blue
lime blue
lime green
lime red
lime yellow
limestone
limonite
litharge
lithol red
lithol rubine
lithopone
luminous pigments
madder lake
Magdala red
magnesia
magnesite
magnesium carbonate, precipitated
magnesium silicate, fibrous
magnesium silicate, nonfibrous
magnetite black
malachite
malachite green
manganese black
manganese brown
manganese green
manganese violet
marble flour
martius yellow
massicot (massocot)
meldola blue
menaccanite
mercadium orange
mercuric sulfide
mercury-cadmium lithopones
metal powder pigment
metallic brown

metallic pigment
methylene blue
mica
micaceous iron oxide
microcrystalline silica
micronized pigment
middle chrome
milori blue
milori green
mineral black
mineral brown
mineral orange
mineral pigments
mineral violet
mineral white
minium
Mittler's green
molybdate orange
Monthier's blue
montmorillonite
mordant dyes
mordants
mountain blue
muscovite
nacreous pigment
naphthalene red
naphthol blue
naphthol green B
naphthol green Y
naphthol reds
naphthol yellow S
Naples yellow
native green
natural iron oxides
natural pigments
natural red 4 (75470)
natural yellow oxide
nepheline
nepheline syenite
nephelite
nickel antimony titanate
nickel titanate
nickel azo yellow
nickel titanate yellow
nigrosines
nitrate green
normal lead silico chromate
novacite
novaculite
Nurnberg violet
ocher, ochre (yellow, golden, red)
oil color
oil green
oil pastes
oil stain
orange chrome
orange lead
orange mineral
organic yellow
Oriental blue
orpiment
ortho-chlor-para nitraniline
oxide of chromium
para brown
para nitraniline red
para reds
para toner
para-chlor-ortho-nitraniline
parachlor nitraniline red
parachlor red
paramerism
Paris black
Paris blue, Parisian blue
Paris green
Paris red
Paris white
Paris yellow
particle size
paste blue
paste, pigment
patent vermillion
peacock blue
pearlescent pigment
pearlite
pearlstone
pelletier's green
pelletization
perlite, expanded
permanent blue
permanent blue toners
permanent fast violet
permanent green
permanent green toners
permanent orange
permanent peacock blue
permanent red 2B
permanent red toners
permanent violets
permanent white
permanent yellow FGL
Persian Gulf red
Persian red
perylene pigments
phloxine
phosphomolybdic pigments
photogenic property
phthalocyanine pigments
pigment
Pigment Black 10
Pigment Black 11 (77499)
Pigment Black 6 (77266)
Pigment Black 7 (77266)
pigment bleeding
Pigment Blue 27 (77510)
Pigment Blue 28 (77346)
Pigment Blue 29 (77007)
Pigment Brown 6 (77499)
pigment dyes
pigment extender
Pigment Green 10 (12775)
Pigment Green 15 (77510/77603)
Pigment Green 17 (77288)
Pigment Green 18 (77289)
Pigment Green 8 (10006)
Pigment Green B
pigment, metallic
pigment, nonhiding
Pigment Orange 21 (77601)
Pigment Orange 23 (77201)
pigment paste
Pigment Red 1 (12070)
pigment Red 101 (77491)
Pigment Red 104 (77605)
Pigment Red 105 (77578)
Pigment Red 106 (77766)
Pigment Red 108 (77202)
Pigment Red 3 (12120)
Pigment Red 38 (21120)
Pigment Red 4 (12085)
Pigment Red 48 (15865)
Pigment Red 49 (15630)
Pigment Red 53 (15585)
Pigment Red 57 (15850)
Pigment Red 6 (12090)
Pigment Red 81 (45160)
Pigment Red 83 (58000)
Pigment Violet 10 (45170)
Pigment Violet 15 (77007)
pigment volume
Pigment Volume Concentration (PVC)
Pigment White 1 (77597)
Pigment White 11 (77052)
Pigment White 12 (77990)
Pigment White 16 (77625)
Pigment White 18 (77220)
Pigment White 19 (77005)
Pigment White 2 (77633)
Pigment White 20 (77019)
Pigment White 22 (77120)
Pigment White 26 (77718)
Pigment White 27 (77811)
Pigment White 4 (77947)
Pigment White 5 (77115)
Pigment White 6 (77891)
Pigment White 7 (77975)
Pigment Yellow 32 (77839)
Pigment Yellow 34 (77600)
Pigment Yellow 37 (77199)

Pigment Yellow 42 (77492)
Pigment Yellow 53 (77788)
Pigment Yellow 74 (11741)
pigment/binder ratio
pipe clay
PKT
plessy's green
PMA
PMA pigment
pompey red
potash blue
potassium titanate (PKT)
powder
powdered quartz
precipitated barium sulfate
precipitated basic dye blues
precipitated basic dye violets
precipitated calcium carbonate
press cake
prime pigments
primrose chrome
Prussian blue
Prussian brown
Prussiate of iron
PTA pigment
PTMA pigment
pulp
pulverulent
pumacite
pumice
pure black iron oxide
purple brown
purree
PVC
pyrazolone red
quartz
quercitron
quinacridone golds
quinacridone pigments
quinacridone violet
raw sienna
raw umber
reactive pigments
realgar
red cinnabar
red iron oxide
red lake C pigments
red lake D
red lake P
red lead
red lead (nonsetting)
red ocher
red oxide
redtone ink blue
reducing power
rhodamine 6G
rhodamine B
rhodamine reds
rhoduline blue 66
rhoduline heliotrope
Rinmann's green
rouge
rubine
rubine reds
rutile
scarlet chrome
scarlet lake
Schaffer's acid
Schnitzer's green
Schweinfurt green
selenium red
sepia
sericite
serpentine
setoglaucine
settling
silica, amorphous
silica, crystalline
silica flour
silica, microcrystalline
silica, soft
silica, synthetic (aerogel and hydrogel)
silica, synthetic (pyrogenic)
silicate
silver white
smalt
smectite
smeraldino
sodium aluminosilicate
sodium lithol red
sodium silicoaluminate
soft silica
soluble blue
Spanish ocher
Spanish red oxide
Spanish white
spirit stains
spirit-soluble dye
stainer
steatite
steel blue
strontium chromate
strontium white
strontium yellow
sublimed white lead
substrate
surface area
syenite
synthetic magnetite
synthetic yellow ocher
taconite
talc
tannin
terra di siena
Thénard's blue
thioindigo pigments
thioindigoid maroons
tinge
tinting
tinting color or colorant
titania
titanium brown
titanium dioxide, anatase
titanium dioxide, rutile
titanium greens
titanium lithopone
titanium yellow
tobias acid
toluidine reds
toluidine toner
toluidine yellow
toner
toning blue
treated pigment
tripolite
tungstomolybdic pigments
turkey red
turkey umber
Turnbull's blue
tuscan red
ultramarine blue
ultramarine green
ultramarine violet
umber
universal colorants
vandyke brown
vat dyes
vegetable black
Venetian red
verdigris
verditer
vermiculite
vermiculus
vermillion, (vermilion)
vert emeraude (dull)
Victoria Blue B
Vienna blue
vine black
viridian green
watercolor
Wenzel's blue
white bole
white lead
white zinc
whiting
wilkinite
Wilkinson's blue
witherite
wollastonite
wood preservative coatings

yellow chrome
yellow iron oxide
yellow limonite
yellow ocher, synthetic
zaffre
zinc chromate
zinc dust
zinc ferrocyanide
zinc green
zinc oxide
zinc sulfide
zinc tetroxy chromate
zinc white
zinc yellow
zinc-cadmium sulfides
zircon
zirconia
zirconium oxide
zunsober

Category 42
Paint Problems

antilivering agent
barnacle
bead
Bénard (vortex) cell
bittiness
bitty
blanch
bleaching
bleed
bleed characteristics
bleeding
bleeding knots
bleeding pigment
blister
blister house
blister resistance
blistering
block resistance
blocking
bloom
blotching
blush, blushing
bottom
bounce back
broken color
bronzing
brush drag
brush mark
bubbling
burnish
caking
chalk masking
checking
checking resistance
cheesy
chilling
chipping
clear liquid separation (CLR)
cloudiness
CLR
coacervation
coagulation
coagulum
cobwebbing
coiling soup
cold-checking
color acceptance
color uniformity
copper staining
cracking
crawling
crinkling
crocking
crowsfooting
curing agent blush
curling
degree of efflorescence
dirt
dirt collection
dirt pick-up
dirt/soil resistance
discoloration
drier absorption
drier dissipation
drift
drips
dry spray
efflorescence
efflorescence/mottling
erosion
erosion resistance
exfoliate
exudation
fading
fatty paint
flaking
flash rusting
flat spot
flex-cracking
floating
flocculate
flocculation
flooding
fog
foots
foreign matter (or material)
fouling
frosting
fungus (fungi)
gel
gelation
gelling
ghosting
glossing up
grain crack resistance
greasiness
grinning through
grit
gritty
gummy
hairline cracks
hatbanding
heavy centered spray pattern
high humidity gloss loss
hungry surface
in-can biodeterioration
in-can preservative
inclusion
incompatibility
intercoat contamination
intercoat contamination
knot sealer
lap
leveling
lifting
livering
loss of drier
mildew
mildew (fungus) resistance
monkey
mottling
mudcracking
nailhead rusting
nibs
nondrying coating
odor
oil softening
open time
orange peel
overbaking

overspray
peeling
peppery
phase separation
photographing
picking
picking up
picture framing
pigment bleeding
piling
plasticizer migration
popping
pot life
print resistance
printing
pulling up
rain spotting
raising
retraction
riveling, rivelling
roller spatter resistance
rub up
sag or sagging
scaling
sedimentation
seeding
seedy
self-lifting resistance
settling
shelf life
show-through
skin
skinning
skippy
skips
soil
soilant
solvent shock
spatter
spatter resistance
specks
spewing
split spray
spray mottle
spreadable life
staging life
staining
starved surface
storage life
storage stability
streaking
sulfide staining
surface drying
sweating
swell
syneresis
tails
telegraphing
thickening
throwing out
tint base paint
tintability
traffic marking
usable life
veiling
voids
vortex cell
water spotting
water staining
weathering
webbing
wet edge
wet edge extender
wet edge time
withering
working time
wrinkling

Category 43
Optical Properties

"a," Kubelka-Munk equation
"b" or "ß"
"b," Kubelka-Munk
Å
Abbe Refractometer
abridged spectrophotometer
absolute reflectance
absorbance
absorption
absorption coefficient
absorption spectrophotometry
absorption tinting strength
absorptivity, a, n
achromatic
Adams Chromatic Value Color Difference Equation
additive color mixture
air entrapment
angle of incidence
angle of reflection
angle of viewing
Angstrom unit
anisotropy
appearance
artificial daylight
artificial illuminant
attribute
band spectrum
bead
beads, reflective
Beer's Law
Beer-Bouguer Law (Beer-Lambert Law)
bidirectional
bidirectional reflectance distribution function (BRDF)
birefringence
black
black light
blackbody
bleaching
blueing
blush, blushing
Bouguer's (Lambert's) Law
brightener, optical
brightening agent
brightness flop
brilliance
broken color
broken white
bronze
bronzing
bubble coating
burnish
burnish resistance
candle
cast
chemiluminescence
chroma
CIE 1976 (L*a*b*) Color Difference Equation
CIE 1976 (L*u*v*) Color Difference Equation
CIE Standard Observer
CIE tristimulus values
clarity
clean
clear
coefficient of scatter
cold
color

color constancy
color development
color floating
color of an object
color retention
color temperature
color uniformity
colorimeter, tristimulus
coloristic properties
complete hiding
conditional match
contrast ratio
correlated color temperature
coverage
coverage rate
covering power
cryptometer
daylight
dead flat
deep
deleafing
densitometer
density, optical
depth of finish
dichroism
dichromism
diffraction
diffraction grating
diffuse reflectance
diffuse transmittance
diffused light
direct lighting
directional reflectance
dispersion curve
dispersion, light
distinctness-of-image gloss
dominant wavelength, CIE
dry-hiding
dullness
effect finish
eggshell
eggshell finish
enamel
end point
excitation purity
extinction coefficient
fading
fast to light
filter spectrophotometer
filter, optical
flair
flashing
flat
flat enamel
flat finish
flat paint (finish)
flatting or flattening
flatting pigment
flip-flop
flop
flop color
fluorescence
fluorescent dyes
fluorescent paint
fluorescent pigments
flux, luminous
flux, radiant
fogged metal
foot-candle
foxy
Fresnel reflection
full gloss
geometric metamerism
gloss
gloss retention
gloss, distinctness-of-image
glossmeter
goniochromatic
goniochromatism
gonioisochromatic
gonioisochromatism
goniometachromatic
goniometachromatism
goniophotometer
goniospectrophotometer
gray scale
grazing angle
haze
hazen color scale
hazy
hiding power
hiding power chart
hiding, complete
hiding, incomplete
hiding, spectrophotometric, complete
hiding, spectrophotometric, incomplete
high gloss
high humidity gloss loss
high sheen flat
high-visibility paints
high-visibility pigments
hue, Munsell
Hunter Color Difference Equation
ICI
illuminance
illuminant
Illuminant A (CIE)
Illuminant B (CIE)
Illuminant C (CIE)
Illuminants D (CIE)
incident angle
incident ray
incomplete hiding
index of metamerism
index of refraction
infrared (IR)
integrating sphere
integration (colorimetry)
interference color
interference filter
interference pigments
iridescence
irradiance
isotropic
Judd chart
Judd graph
Judd hiding power chart
just noticeable difference (JND)
Kubelka-Munk Equation
Kubelka-Munk Theory
Lambert's law of absorption
Lambert's law of reflection
light
light source
lightfastness
lightness
Lovibond color system
Lovibond tintometer
low gloss
low lustre
low semigloss
luminance
luminescence
luminescent
luminosity curve, CIE
luminous
luminous directional reflectance
luminous flux
luminous paint
luminous pigments
luster
lusterless
MacAdam Color Difference Equation
MacAdam limits
mass color
matte finish
matte
mesopic vision
metallic paint
metamer
metameric match
metameric pair
metamerism
metamerism index
metamerism, degree of
microvoids

Mie theory
minimum perceptible difference
modes of appearance
moiré effect
moiré papers
monochromatic color scheme
monochromatic light
monochromator
multiple scattering
Munsell book of color
Munsell chroma
Munsell color notation
Munsell color system
Munsell renotation
Munsell surface-color perception solid
Munsell Value
nacreous
nacreous pigment
nanometre (nm)
NBS unit
neutral
neutral gray
night vision
nm
nonmetameric match
nonspecular reflectance
normal color vision
north light
object color
object mode of appearance
obliterating power
observation angle
observer
observer, standard
off-white
oil can
oil stain
opacity
opalescence
opaque
opaqueness
optical filter
Ostwald color system
paramerism
parameter of specular gloss
particle
pearl
pearl essence
pearl lacquer
pearlescent
pearlescent pigment
perceptible
perception
perceptual
perfect diffuser
phosphorescence
photochemical reaction
photochromatic
photochromatism
photometer
photometric
photopic
PIC test
Planckian radiator
platinum
pleochroism
primary colors
primary colors, additive
primary colors, CIE
primary colors, subtractive
pseudoisochromatic plate (PIC) test
psychophysical
purity, colorimetric
purity, excitation
radiance
radiant energy
radiant flux
reference standard
reflectance
reflectance factor
reflectance, absolute
reflectance, diffuse
reflectance, directional
reflectance, Fresnel
reflectance, hemispherical
reflectance, luminous
reflectance, nonspecular
reflectance, specular
reflectance, total
reflection
reflectivity ($R\infty$)
reflectometer
reflectorization
reflex reflector
refraction
refractive index
relative dry hiding power
retina
retinal
retroreflection
retroreflective
satin
saturation
Saunderson correction
scatter, coefficient of
scattering
scattering (light)
scattering coefficient, Kubelka-Munk
scattering coefficient, Mie
scattering loss
scattering, multiple
scattering, single
SCE
SCI
scotopic vision
secondary reference standard
seeing into the film
selected ordinate method
semigloss
semigloss lacquer
shade
sheariness
sheary
sheen
silk
Simon-Goodwin charts
sinkage
sinking in
specific intensity
specific luminance
spectra
spectral
spectral power distribution curve
spectrogoniophotometer
spectrophotometer
spectrophotometer, abridged
spectrophotometer, filter
spectrophotometric
spectrophotometric curve
spectrophotometric hiding
spectrophotometry
spectrum
spectrum locus
specular
specular gloss
specular reflectance
specular reflectance excluded (SCE)
specular reflectance included (SCI)
specular reflection
specular transmittance
staining power
standard light source
standard observer, 2° (1931)
standard observer, 10° (1964)
standard perfect white diffuser
standard, white reflectance
subtractive colorant mixture
subtractive colorant mixture, complex
subtractive colorant mixture, simple
suede
temperature, color
temperature, correlated color
thiocyanogen value

tinctorial strength
tint base paint
tint-tone
tinting
tinting (or tint) strength, absorption
tinting (or tint) strength, scattering
tinting strength
tinting strength, absorption
tinting strength, scattering
tone
toner
total reflectance
total transmittance
translucency
transmission
transmittance
transmittance, diffuse
transmittance, internal
transmittance, specular
transmittance, total
transparency
transparency, ink
transparent
travel
trichromat
trichromatic
trichromatic coefficients
tristimulus
tristimulus colorimeter
tristimulus computation data
tristimulus integration
tristimulus integrator
tristimulus values, CIE
twilight vision
U^*, V^*, W^*
u, v
ultraviolet (UV)
ultraviolet light
underhue
undertone color
uniform chromaticity coordinates
uniform chromaticity scale diagram
uniform chromaticity spacing
uniform color scale
uniform color space
uniform lightness scale
UV
value, Munsell
velvet
viewing angle
viewing conditions
viewing geometry
viewing ray
visible region
vision
vision, defective color
vision, normal color
visual
wavelength
weighted ordinate method
wet hiding
white
white reference standard
white reference standard, absolute
white reference standard, primary
white reference standard, secondary
whiteness
whiteness index
whitening in the grain
wood preservative coatings
$\bar{x}$
y
$\bar{y}$
yellowness
yellowness index (YI)
YI
z
$\bar{z}$
zeta space

Category 44
Naval Stores

abietates
abietic acid, commercial grade
alpha-pinene
American turpentine
archangel pitch
benzyl abietate
benzyl resinate
beta-pinene
bloom oil
bordeaux turpentine
borneol
brewer's pitch
burgundy pitch
cajeputene
cimene
country tar
destructively distilled pine oil
destructively distilled wood turpentine
dipentene
dipentene (pure compound)
dl-limonene
ester gum
fat turpentine
finn oil
gum rosin
gum thus
gum turpentine
hydrogenated rosin
Japan drier
kautschin
kidney oil
liquid resin
liquid rosin
metallizing
monocyclic terpenes
naval stores
navy pitch
oil of pine tar
oleoresin
pine needle oil
pine oil
pine pitch
pine tar, kiln burned
pine tar oil
pine tar, retort
pine tar, Stockholm
pinene
pitch, Archangel
pitch, brewer's
pitch, burgundy
pitch, navy
resenes
resin oil
resinates, metallic
retort pine tar
rosin acids
rosin esters
rosin, limed or zinc-treated
rosin, modified
rosin oil
rosin pitch
rosin, reclaimed
rosin spirit
rosin standards
rosin type (sample)

skimmings (tall oil)
steam-distilled pine oil
steam-distilled wood turpentine
Stockholm pine tar
Strasbourg turpentine
sulfate pine oil
sulfate pitch
sulfate resin
sulfate rosin
sulfate wood turpentine
sulfite turpentine
Swedish oil
Swedish olein
Swedish pine oil
Swedish resin
Swedish rosin
Swedish rosin oil
sylvic oil
synthetic pine oil
tall oil
tall oil, acid refined
tall oil, crude
tall oil, distilled
tall oil fatty acids
tall oil heads (light ends)
tall oil pitch
tall oil rosin
talloel
tallol
terpenes
terpinenes
turpentine
turpentine, oil of
turpentine, spirits of
turps
Venice turpentine
wood rosin
wood turpentine

Category 45
Metals, Metal Preparation

alloy
beam
bronze gold powders
bronze paint
bronze powders
buffing
cadmium plating
chemical conversion coating
chemical pretreatment
chemical vapor deposition
color constancy
conversion coating
CVD
degreaser
exfoliate
flake metal
flame cleaning
galvanized steel
gold leaf
grit blasting
ground coat
iron (Fe)
leveler
metal
metal decorating
metal powder pigment
metal spraying
mill scale
minimized spangle
miscellaneous metal parts
Parkerized
passivation
phosphating
phosphatize (phosphate coat)
profile
profile depth
tape
weld
weld slag
weld splatter
zinc dust
zinc-coated

Category 47
Mechanical Properties

abrasiometer
abrasion coefficient
abrasion cycle
abrasion resistance
after-tack
aggressive tack
alternating strain amplitude
alternating stress amplitude
anelasticity
bake hardness
ball punch impact test
Barcol hardness
bend test
Bingham liquid
block resistance
bond strength
bonding action
breaking extension
brittle
brittleness
bulk modulus
cheesy
coefficient of elasticity
cold stretch
complex Young's modulus
compliance
compounder's modulus
compression set
compressive strength
creep
distensibility
dry
dry-hard
drying time
ductility
dumbbell
durometer
dynamic fatigue test
elastic
elastic limit
elastic solid
elasticity
elastomer
elastomeric
elongation
energy at break
flex-cracking
flexibility
flexibility test
gage length
glass transition

glass transition temperature
guage length
hardness
hardness scale
hot-short
impact resistance
impact test
imprinting
indentation hardness
jet abrader
knife test
Knoop hardness test
Knoop indenter
König pendulum damping hardness test
laminar flow
mandrel bend test
mar resistance
mechanical
mechanical properties
modulus of elasticity
Mohs scale
nerve
pencil hardness
pendulum-rocker hardness
penetration number
penetrometer
Persoz pendulum hardness test
Pfund hardness number (PHN)
PHN
plastic
plastic solid
pour point
print resistance
print test
reversion
rheopexy
rocker hardness tester
Rockwell hardness
set-to-touch time
shear
shear rate, y
shearing strength
shearing stress
short
slip
stiffness (modulus of elasticity)
strain
strength, dry
strength, tensile
strength, wet
stress
stress-strain curve
stress-strain diagram
Sward rocker
T-300
T-bend
tack, dry
tack range
t-bend flexibility test
tensile properties
tensile strength
tensile tester
tensile testing machine
T_g
tooth
toughness
ultimate tensile strength
viscoelasticity
wet scrub resistance
Young's Modulus

Category 49
Manufacturing

aging (ageing)
agitation
air flotation
ball charge
ball mill
ball milling
banbury
batch
batch adjustment record
batch process
batch record
biff
blending
bottoms
Buhr mill
ceramic
clockspring
dilute
false body
flat stone mill
foots
hardinge mill
injection molding
iron mill
letdown
lot
lot number
manufacture
manufacturer's formulation
master batch
maturing
mechanical
media
metafiltration
mill
mill base
model plant
paste, pigment
pigging
premix
raw material
surface coating operation
tailings
transfer efficiency
vacuum forming

Category 50
Packaging

bail
bailing machine
bale
baling machine
barrel
bunghole
can coating
chimb
chime
chine
cleavage
container lining
cork
double seam can
friction top can
storage
three-piece can
two-piece can
zinc-rich primer

Category 54
Finishing Techniques

abrasive (noun)
abraum
airless spraying
antique finish
antiquing
arcing
back paint
back priming
ball
basecoat/clearcoat
bleaching
blender
bridging
bring forward
broadcast
broken color
broken-color work
brush graining
build
building coats
burned finish
burning-in
butt
chip resistance
chipping
cloth distressing
coat
coating
coating system
cobwebbing
comb
combing
cowtail
crackle finish
crackle varnish
crinkling
cross coating
crowsfooting
cutting in
dado
daub
decalcomania, decal
decorative painting
decoupage
distressing
drift
dry brush
dull rubbing
embossing
faux hois (Fr.)
feather edging
feathering
felting down
field coat
fiery finish
finish
finish coat
first coat
fixative
flamboyant finish
flash drying
flash-off time
flashoff
flashoff zone
flat
flatting down
flatting (rubbing) varnish
flock finish
flocking
fluting
French polish
French varnish
frisket
full coat
gilding
gold leaf
graining
ground coat
highlighting
impasto
inpaint
japanning
lap
lap joint
laying off
lithography
live edge
marbling, marbleizing
masking
masking tape
mist coat
monolithic flooring
mop polishing
multicolor finish
natural spreading rate
neutral oil
neutral toner
neutralizing
NGR stain
nongrain-raising stain (NGR)
offset
oil polish
oil rubbing
open time
overlap
overspray
paint harling
paint system
pass
picking up
plastic paint
polishing varnish
pounce
pulling over
rag rolled finish
rag rolling
retouching
ribs
riveling, rivelling
rotary screen printer
rubbing
rubbing down
rubbing oil
rubbing varnish
scraper
scumble glaze
scumbling
seamless flooring
self-priming
sgraffito painting
shade
shop coat
shop painting
slip
spatter finish
speckled finish
spiriting off
spot finishing
spotting in
spray pattern
spreading rate
stipple
stopper
tapered pattern
test pattern
texture-finished paint
textured paint
Tiffany finish
topcoat
touch-up coating
touch-up painting
vacuum-pressure impregnation
vertical pattern
wainscot
weather resistance
webbing
wet edge
wet edge extender
wet edge time
wet-on-wet coating
whipping
wood filler
wood finishing
wrinkling

Category 55
Film Formation

aluminum paste
ambient cure
bake system
baking schedule
baking temperature
baking time
Bénard (vortex) cell
blanch
blush, blushing
case hardening
close drying
coalescence
coalescent (coalescing agent)
coat
cocooning
convertible coating
crawling
curling
deformation
deleafing
dry-to-handle time
dry-to-recoat time
EB
electron beam (EB) curing
film
film former
film forming
film integrity
filmogen
flash drying
flash-off time
flashoff zone
flatting pigment
hard dry
lapping time
leafing
low temperature coalescence
MFFT
MFT
minimum film forming temperature (MFFT)
nonconvertible coating
retraction
ribs
scattering
skin
stardust
ultraviolet (UV) curing
veiling
wet-on-wet coating

Category 56
Film Defects

abrasion resistance
after-tack
agglomerate
aggregate
air bubble
air entrapment
air lock
alligatoring
antiwrinkling agent
bald spot
ball
bittiness
bitty
blanch
bleaching
bleed
bleeding
blemish
blister
blister resistance
blistering
block resistance
blocking
bloom
blotching
blush, blushing
brittle
brittleness
bronze
bronzing
brush mark
bubbling
carbonization
chalk masking
chalking resistance
checking
checking resistance
cheesy
chilling
chip resistance
chipping
cloudiness
coiling soup
coin marking
cold-checking
cold-cracking
color floating
cracking
crack resistance
cratering
craters
crawling
crazing
creeping
crinkling
crocking
crocodiling
crowsfooting
curing agent blush
curling
curtaining
decomposition
defect
degradation
degree of efflorescence
deleafing
deterioration
dirt
dirt/soil resistance
discoloration
discontinuities
dished
doctor mark or streak
dogskin
domed
drips
dry spray
efflorescence
efflorescence/mottling
erosion
erosion resistance
exudation
fading
fat edge
fish eyes
flaking
flaking resistance
flashing
flat spot
flatting or flattening
flex-cracking
floating
flocculate

flooding
fog
foreign matter (or material)
frosting
fungus (fungi)
fungus resistance
gas checking
gas crazing
gasproof
glossing up
greasiness
grinning through
grit
gritty
gum bloom
hair-cracking
hairline cracks
hatbanding
haze
heavy centered spray pattern
heavy-centered pattern
holidays
honeycombing
hungry surface
imprinting
inclusion
incompatibility
intercoat contamination
intercoat contamination
laitance
lap
lifting
loss of gloss
metal marking
migration
mildew
mildew (fungus) resistance
milkiness
misses
mottling
mudcracking
nibs
oil softening
orange peel
pattern staining
peeling
perishing
picking
picture framing
pigment bleeding
piling
pimple
pinched coating
pinholes
pit
pockmarking
popping
pores
porosity
pressure marking
pressure mottling
print resistance
printing
railroad tracks
rain spotting
raising
retraction
riveling, rivelling
ropiness
ropy finish
runs
sag or sagging
saponification
scaling
scaling resistance
seediness
seeding
seeds
seedy
shrinkage
silking
sinkage
sinking in
sissing
skin
skipping
skippy
solvent pop
spark-testing
spewing
spider legs
splitting
spotting
spray mottle
stain
staining
stardust
starved surface
sulfide staining
surface drying
sweating
swell
tails
tear
temperature checking
throwing
throwing out
vacations
voids
water beading
water spotting
water staining
weathering
webbing
whitening in the grain
withering
wrinkling

Category 57
Fabrication

accelerated weathering machine (or apparatus)
adhesive, assembly
adhesive, edge jointing
adhesive film
air assist thermoforming
air lock
bag molding
bank, rubber mill
blow molding
bond strength
bonding action
cast film extrusion
casting resin
chatter
chill roll
chill roll extrusion
clockspring
cold pressing
cold stretch
collet
compacting
compregnated wood
compression molding
edgewise
extrusion
extrusion moldings
fabricate
laminate, high pressure
laminated, cross
laminated, parallel
parallel laminated
postforming
weather-ometer
weathering
xenon-arc light exposure apparatus

Category 58
Extenders, Fillers, Flatting Agents

albite
alumina
alumina hydrate
alumina-blanc fixe
aluminum hydrate
aluminum oxide
aluminum oxide cloth
aluminum oxide paper
aluminum potassium silicate
aluminum silicate (clay)
aluminum stearate
amorphous silica
amphibole
asbestine
attapulgite
barium carbonate
barium sulfate, natural
barium sulfate, synthetic
barium white
barytes
base
bauxite
bentonite
blanc fixe
bleaching clay
burnt lime
calcined clay
calcite
calcium carbonate, natural
calcium carbonate, synthetic
calcium carbonate, syntheticcarbinol
calcium silicate, synthetic
calcium sulfate
calcium sulfate, anhydrous
calcium sulfate hemihydrate
casting plaster
celestite
chalk
chalk, precipitated
china clay
clay
constant white
dead burned calcium sulfate
dehydrated gypsum
delaminated clay
diatomaceous calcite
diatomaceous earth
diatomaceous silica
dolomite
dry hydrate
extender
extender (pigment)
feldspar
filler
flat varnish
flattening agent
flatting agent
flatting pigment
French chalk
Fuller's earth
ground silica
gypsum
gypsum, calcined
heavy spar
hectorite
hydrated aluminum silicate
Iceland spar
industrial talc
industrial talc, asbestos-free
inert pigment
infusorial earth
inorganic polymers
kaolin
kaolinite
kieselguhr
labradorite
limestone
magnesite
magnesium carbonate, precipitated
magnesium silicate, nonfibrous
matte varnish
mica
microcrystalline silica
molecular sieve
montmorillonite
nepheline syenite
neuberg chalk
novacite
novaculite
Paris white
Pigment White 18 (77220)
Pigment White 19 (77005)
Pigment White 20 (77019)
Pigment White 22 (77120)
Pigment White 26 (77718)
Pigment White 27 (77811)
pipe clay
precipitated barium sulfate
precipitated basic dye blues
quartz
satin white
sericite
sherardizing
silica, amorphous
silica, crystalline
silica, diatomaceous
silica flour
silica, microcrystalline
silica, synthetic (aerogel and hydrogel)
silicate
slate dust
slate flour
slate powder
smectite
soapstone
sodium silicoaluminate
soft silica
steatite
stopper
talc
terra alba
toner
tripolite
vermiculite
white bole
white earth
whiting
witherite
wollastonite

Category 59
Equipment, Apparatus, Machines

Abbe Refractometer
absorber
absorption hygrometer
accumulator
aerograph
agitator
air brush
air cap
air jet
air vent
alumina
aluminum oxide abrasive
aluminum oxide cloth
aluminum oxide paper
applicator
applicator roll
autoclave
automatic press
automatic turning sander
baffle
bailing machine
baling machine
ball charge
ball mill
ball milling
banbury
bird applicator
blender
booth
Boston Stone, The
brayer
breathing
bridle
bristle
Brookfield® Viscometer
brush
brush backed sander
Buhr mill
calender
camel hair
casing knife
cast film extrusion
caul
caulking cartridge
caulking gun
centrifugal clarifier
chaser
chatki
chill roll
chill roll extrusion
Cleveland condensing humidity cabinet (QCT)
coated abrasive
coater
collet
colloid mill
colorant dispenser
concrete gun
cone mill
crocus cloth
cup
curing oven
dabber
degreaser
deodorizer
desiccator
digester
disk
distemper brush
doctor blade
doctor roll
drop
drop cloth
drum (cylinder) sander
dust collector
edge runner mill
emery
emery abrasive
emery cloth
emission control equipment
external mix
fadeometer
ferrule
filling of coated abrasives
film applicator
film thickness gage (gauge)
filter
filter press
fitch
flag
flat paint brush
flat stone mill
flat wall brush
flint paper
flock gun
foam paint roller cover
Ford cup
fugitometer
Gardner Mobilometer
Gardner-Holdt viscosity tubes
garnet paper
grinding slab
hardinge mill
hiding power chart
hiding power chart
homogenizer
hydrometer
incinerator
internal mixer
iron mill
isoelectric heating
jar mill
jet mill
joint roller
joint tape
jolly balance
kettle
Knoop indenter
length out
leveler
limekiln
liner
lining tool
live centers
loading (of coated abrasives)
machine
machine-printing
mahlstick
mandrel bend test
masking tape
media
media mill
mesh number
micrometer
micronizer mill
mill
millstone
mobilometer
Mohr balance
mottler
nap
nap length
open coat
orbital pad sander
Ostwald u-tube
oven
paint brush
paint brush bristle
paint brush ferrule
paint brush filament
paint brush head
paint brush length clear
paint brush thickness
paint roller core
paint roller cover
paint roller cover pile height
paint roller frame
paint test charts
palette knife
pan
pebble mill
pick-up roll

platen
printing plate
printing press
pug mill
putty chaser
QCT
QUV
raymond roller mill
reference standard
refiner
refractometer
rewind
roller cover
roller mill
rollformer
rotary press
sable pencil
sable writer
sand grinder
sandpaper
sash tool
scaler
scraper
scrubber
sheepskin paint roller cover
sigma blade mixer
silicon carbide waterproof paper
silver point
softener
sparkproof tools
spray booth
spray gun
spraying pistol
steel wool
stepladder
stippler
stitcher
straight-line pad sander
strainer
swarf
synthetic paint roller cover
tacciometer
tack rag (cloth)
test charts
three-roll mill
triple-roller mill
Twaddell hydrometer
two-component gun
two-roller mills
vacuum filter
vapor degreaser
weather-ometer
Westphal balance
wet sanding
wide-belt sander
wire brush
Ye Olde Paint Stone
zinc stearate-coated finishing paper

Category 60 Emulsions

anionic emulsion
antifoaming agent
breaking
cationic emulsion
creaming of emulsion
discontinuous phase
dispersed phase
distemper, oil-bound
emulgator
emulsification
emulsifier
emulsifying agent
emulsion
emulsion polymerization
emulsion stabilizer
emulsoid
external phase of an emulsion
homogenizer
hydrophile
internal phase
latex
liquid emulsion polymer coating
liquid water emulsion wax
microemulsion
microemulsions
oil-bound
oil-bound distemper
particle size
resin emulsion paint
rubber latex
rubber-emulsion paint

Category 61 Electrical Properties

ampere
anion
anode
antistatic agents
band wire
boiled oil
breakdown voltage
cataphoresis
cathode
cathodic corrosion
cation
cell, electrolytic
complex dielectric constant
concentration cell
conductance
conductivity
corona discharge
corona, internal
corona resistance
coulomb
coulometry
couple
current density
dielectric constant
dielectric dissipation factor
dielectric strength
dipole
EDP
electric (electrical)
electric strength
electricity
electrode potential
electrodeposition
electrokinetic potential
electrolysis
electrolyte
electrolytic cell
electromagnetic waves
electron

electrophoresis
electrostatic
electrostatic detearing
electrostatic precipitation
Elpo
galvanic anode
galvanic cell
galvanic corrosion
ground wire
impregnating insulating varnish
insulating varnish
insulation resistance
ion
ionization
magnet wire
minute value
resistivity
specific conductance
specific resistance (resistivity)
throwing power
zeta potential
zwitterion

Category 63
Oils

absorption, oil
acid refined linseed oil
air blowing
aleurites
alkali-refined linseed oil
almond oil
amoora oil
animal oils
apricot kernel mixed fatty acids
apricot kernel oil
arachis oil
bagillumbang oil
basswood oil
baudouin test
Becchi-Millian test
beechnut oil
ben oil
benne oil
bleached oil
bloom
blown castor oil
blown oil
blown stand oil
blush, blushing
bodied oil
body gum (no. 8 varnish)
bone oil
break (of oils)
break test (oils)
breakfree oil
bunghole oil
cacahuananche oil
candlenut oil
cashew nut shell oil
castanha de cotia kernel oil
castor oil
castor oil, dehydrated
cedar nut oil
chia oil
China wood oil
clupanodonic acid
cochin oil
coconut oil
colza oil
congeal
corn oil
cottonseed oil
cottonseed pitch
crowsfooting
crude resinous liquid
crystallization
curcas oil
DCO
debloomed oil
dehydrated castor oil
dehydrogenated oils
diene value or number
dodecanoic acid
drying oil
elemi gum
essential oils
expression
fish oil
fixed oil
flatting oil
fluid resin
foots
frosting
gadoleic acid
garcia nutans oil
garden rocket oil
gas checking
gas crazing
gasproof
gel time (of a drying oil)
gingelli oil
grapeseed oil
groundnut oil
gynocardia oil
hackberry tree seed oil
haco oil
heat test
heat-bodied oil
hempseed oil
hexabromide value
higher fatty acid
hot oil expression
Hubl solution
hydrogen value
hydrogenation of oils
hydrolysis
isano oil
isanolic acid
isomerized oils
jamba seed oil
jute seed oil
kettle bodied oil
Koettstorfer number
lallemantia oil
lauric acid
licanic acid
linoleic acid
linolein
linolenic acid
linoxyn
linseed oil
liquid resin
long oil
lumbang oil
maleic anhydride value
maleinized oil
marking nut oil
medium oil varnish
menhaden oil
mineral oil
mu oil
mucilage
n'gart oil
n'gore oil
neatsfoot oil
neou oil
neutral oil
nonbreak oil
nondrying oil
nsa-sana oil

O.A.
oil
oil absorption (value) (O.A.)
oil, blown
oil cake
oil, debloomed
oil, drying
oil, fixed
oil, heat-bodied
oil, long
oil melt
oil, nondrying
oil of lemon
oil of pine tar
oil, semidrying
oil, stand
oil, vegetable
oiticica oil
oleic acid
oleoresinous vehicle
ongueka oil
oxyns
pale boiled oil
paraffin oil
peanut oil
penetrating finish
perilla oil
po-yok oil
polimerisado oil
polymerized fatty acids
poppyseed oil
prepared linseed oil
purified stand oil
rapeseed oil
reconstituted oils
ricinoleic acid
rubber-seed oil
safflower oil
saponification value
sardine oils
Scheiber oil
segregated oils
semidrying oil
sesame oil
soya bean oil
sperm oil
stand oil
stillingia oil
sulfate pitch
sulfate resin
sulfate rosin
sulfurized oil
sunflower oil
Swedish oil
Swedish olein
Swedish pine oil
Swedish resin
Swedish rosin
Swedish rosin oil
sylvic oil
tall oil
talloel
tallol
tea seed oil
teel oil
tekaol
throwing out
tobacco seed oil
tocopherols
trienol
tung oil
varnish
vegetable drying oil
vegetable oil
vernonia oil
vulcanized oil
walnut oil
winterized oils
wood oil

Category 64
Dispersion

absorption, oil
accretion
aerosol
agglomerate
agglomeration
aggregate
antisettling agent
ball charge
ball mill
ball milling
banbury
blender
Boston Stone, The
Buhrstone
calcium naphthenate
calcium octoate
calcium resinate
calcium stearate
calcium tallate
coagulation
cobalt naphthenate
cobalt octoate
colloid
colloid, linear
colloid, micellar
colloid, molecular
colloid, reversible
colloid, sphere
colloidal
colloidal particle
colloidal solution
colloidal state
cone mill
deflocculant
deflocculation
dispersant
disperse system
dispersed
dispersed phase
dispersion
dispersion, degree of
dispersion, pigment
dispersion, resin
dispersoid
DLVO theory
dust
fineness
fineness of dispersion
fineness of grind
flat stone mill
floccule
foam
fog
grind
grinding media charge
grinding medium
hammer mill
homogenizer
hydrosol
industrial methylated spirits
jar mill
jet mill
lead in oil
lecithin
media mill
micronizer mill
mill base
mist
NAD
north standard
O.A.
oil absorption (value) (O.A.)
oil color
oil stain
paint base

paste, pigment
pebble mill
peptization
peptizing agents
pigment grind
powder
pulp
raymond roller mill
reversible colloid
sand grinder
sand grinding
sediment
slurry
sol
stippling
suspending agent
suspension
suspension aid
throwing out
voids
Ye Olde Paint Stone

Category 65
Degradation of Coatings

ablation
abrasion
abrasion resistance
accelerated testing
accelerated weathering
accelerated weathering machine (or apparatus)
air loss
autooxidation
biodegradation
bleaching
brown rot
carbonization
chalk masking
chemical stability
chipping
corona
cracking
decomposition
degradation
depolymerization
deterioration
dirt collection
dirt pick-up
dry rot
erosion
erosion resistance
exfoliate
fouling
hairline cracks
loss of gloss
marine borers
microorganisms
mildew
mildew (fungus) resistance
oxidation
QUV
saponification
sleepy
smudge
stain
staining
useful life
water staining
weather
Weather-Ometer
weathering
whitening in the grain
wood preservative coatings
xenon-arc light exposure apparatus
yellowing

Category 66
Corrosion

active metal
aeration cell
anode
anticorrosive paint or composition
bacterial corrosion
basic lead silico chromate
basic zinc chromate
cathodic corrosion
cathodic polarization
cathodic protection
cation
cell, electrolytic
chemical conversion coating
chemical pretreatment
concentration cell
conversion coating
corrosion
corrosion fatigue
corrosion inhibitor
corrosion potential
corrosion rate
corrosion resistance
corrosion-inhibiting paint
corrosion-inhibitive pigment
crevice corrosion
deposit attack
descaling
electrolytic cell
erode
filiform corrosion
filiform corrosion resistance
flash rusting
fogged metal
galvanic anode
galvanic cell
galvanic corrosion
galvanic protection
graphitization
humidity resistance
inhibitive pigment
knife mark
laminar scale
mirror back coating
pretreatment primer
red lead
rust
rust grade scale
rust resistance
rust-inhibiting paint
rusting
sacrificial protection
salt spray (fog) test
scale
scaling
tarnish
verdigris
wash primer
white rust
zinc tetroxy chromate
zinc-rich primer

Category 67
Curing Chemistry, Cure

accelerator
acid catalysts
acid curing (hardening)
activator
adduct curing agent
after-bake
aftercure
air cure
air drying
ambient cure
ammonia cure
antioxidant
A-stage
bake hardness
bake system
baking
baking finish
baking schedule
baking temperature
baking time
bin cure
blowing
blown oil
bottom-drying
B-stage
calcium naphthenate
calcium octoate
calcium resinate
calcium tallate
catalyst
catalytic curing
cerium naphthenate
close drying
cobalt drier
cold-curing
cold-setting
cold-setting adhesive
corona discharge
cotton-free
cotton-free dry
crosslinking
crosslinking agent (or crosslinker)
C-stage
cure
cure cycle
cure stroke time
cure time
cure volatiles
curing
curing agent
curing oven
curing temperature
curing time
dielectric heating
diphenylmethane diisocyanate
drier
drier absorption
drier dissipation
dry-to-touch time
drying
drying time
dust-free
EB
electrocuring
electron beam
electron beam (EB) curing
epoxide
epoxy
forced drying
forced drying temperature
fresco
glow discharge
hard dry
hardener
heat-convertible resins
hexamethylene diisocyanate (HDI)
hexamine
hot-pressing
induction period (of drying)
infrared drying
infrared (IR)
inhibitor
iron driers
japan drier
lead drier
lead naphthenate
linolenate driers
logu
loss of drier
manganese driers
metallic soaps
naphthenate
octoates
one-compartment coating
one-pot coating
overbaking
overcure
oxidation
oxidized rubber
oxirane
photochemical catalysis
photochemical reaction
photocurable coating
polyurethane finish
polyurethanes
post-cure
pot life
precure
premature vulcanization
PUD
radiant heat baking
radiant heat drying
radiant heating
radiation cure
residual tack
resinoid
resite
resitol
resol
retarder
sand-dry
scorching
self-crosslinking
self-curing
self-thermosetting
self-vulcanizing
semicure
set
set-to-touch time
siccative
skin
skinning
soaps
spreadable life
stoving
surface drying
tack
tack range
tack-free
tackiness
tacky
temperature, curing
temperature, drying
temperature, setting
terebine
thermoset
thermosetting
thermosetting ink
thermosetting resin
through-drying
tight cure
time, curing
top-drying
torsion-braid analyzer

touch-dry
two component paint
two-compartment coating
two-pot coating
ultraviolet (UV) curing
undercure
usable life
vulcanization
vulcanize
working time
zinc drier
zirconium driers

Category 69
Color

"a," Kubelka-Munk equation
"a" or "α"
abbozzo
abridged spectrophotometer
absolute reflectance
absorbance
absorption coefficient
absorption tinting strength
achromatic
ACV
Adams Chromatic Value Color Difference Equation
Adams-Nickerson-Stultz Color Difference Equation
adaptation
additive color mixture
advancing colors
AIC
angle of incidence
angle of reflection
angle of viewing
ANLAB Color Difference Equation
ANS
APHA Color Scale
appearance
aptitude, color
artificial daylight
auxochrome
azo dyes
"b," Kubelka-Munk
"b" or "ß"
Beer's Law
Beer-Bouguer Law (Beer-Lambert Law)
bidirectional
bidirectional reflectance distribution function
birefringence
black
black light
blackbody
bleaching
bleed
bleeding
blueing
Bouguer's (Lambert's) Law
bozzo
brightener, optical
brightening agent
brightness
brilliance
broken white
bronze
bronzing
C or Δc
cast
CAT
chroma
chromatic
chromatic attributes
chromaticity
chromaticity coordinates, CIE
chromaticity coordinates, general
chromaticity diagram, CIE
chromaticity diagram, general
chromaticity difference diagrams
chromaticness
chrome orange, light and deep
chromophore
chrysotile
CIE 1976 (L*a*b*) Color Difference Equation
CIE 1976 (L*u*v*) Color Difference Equation
CIE chromaticity coordinates
CIE color difference equation
CIE Color Notation System
CIE Luminosity Curve
CIE Standard Observer
CIE tristimulus values
color
color acceptance
color aptitude
color aptitude test
color blindness
color constancy
color coordinates
color development
color difference
Color Difference Equations
color gamut
color harmony manual
Color Index Name
Color Index Number
color match
color matching
color matching functions
color measurement
color mixture, additive
color notation
color of an object
color order systems
color rendering index
color space
color specification
color strength
color temperature
color tolerance
color uniformity
color vision, anomalous
color vision, defective
color vision, normal
colorant
colorant match
colorant mixture
colorfastness
colorimeter
colorimeter, tristimulus
colorimetric
colorimetric purity
colorimetry
colorist
coloristic properties
coloristically
colour
Colour Index (C.I.)
Colour Index Name
Colour Index Number
complementary colors
complete hiding
complex subtractive colorant mixture
conditional match
contrast ratio
correlated color temperature
coverage
Cube Root Color Difference Equation
D6500 or D65
daylight
Daylight Illuminants, CIE

deadcoloring
deep
degree of metamerism
delta E, ΔE
density, optical
depth of shade
depth of shade, standard
dichroism
dichromism
diffraction
diffuse reflectance
DIN Color Difference Equation
DIN Color System
discolor
dispersion curve
dispersion, light
dominant wavelength, CIE
dullness
dye
excitation purity
extinction coefficient
exudation
face color
fading
Farnsworth-Munsell 100 hue test
flair
flat paint (finish)
flip-flop
flop
fluorescence
fluorescent paint
FMC Color Difference Equations
foster charts
Friele-MacAdam-Chickering Color Difference Equations
fugitive color
gamut, color
Gardner color standards
geometric metamerism
goniochromatic
goniochromatism
gonioisochromatic
gonioisochromatism
goniometachromatic
goniometachromatism
goniospectrophotometer
gray scale
haze
hazen color scale
hellige system or comparator
hiding, complete
hiding power
hiding, spectrophotometric, complete
high-visibility paints
hitching post standards
hue
hue, Munsell
Hunter Color Difference Equation
Illuminant A (CIE)
Illuminant B (CIE)
Illuminant C (CIE)
Illuminants D (CIE)
index of metamerism
integration (colorimetry)
Inter-Society Color Council (ISCC)
interference color
interference pigments
iridescence
ISCC
ISCC-NBS color designation
ISCC-NBS Color Names Dictionary
jet
JND
Judd color difference unit
Judd graph
just noticeable difference (JND)
Kubelka-Munk Equation
Kubelka-Munk Theory
L
light
light source
lightfastness
lightness
Lovibond color system
Lovibond tintometer
luminance
luminosity curve, CIE
luminous
luminous directional reflectance
luminous paint
luminous pigments
luster
MacAdam Color Difference Equation
MacAdam limits
manganese black
mass color
masstone
masstone color
matte finish
match, conditional
match, nonconditional
matching, color
metallic pigment
metamer
metameric match
metameric pair
metamerism
metamerism, degree of
metamerism index
minimum perceptible difference
modes of appearance
monochromatic color scheme
monochromatic light
monochromator
monochrome
mosaic
mottle
mottled
multicolor finish
Munsell book notation
Munsell book of color
Munsell chroma
Munsell color notation
Munsell color system
Munsell hue
Munsell renotation
Munsell surface-color perception solid
Munsell Value
muted colors
nacreous pigment
nanometre (nm)
NBS Color Difference Equation
NBS unit
neutral
neutral gray
night vision
nonconditional match
nonmetameric match
nonspecular reflectance
normal color vision
north light
object color
object mode of appearance
observer
observer, standard
off-color
off-white
oil stain
opacity
opalescenece
opaque
opaque stain
Ostwald color system
overtone
palette
paramerism
pastel
pearlescent
phosphorescence
photochromatic
photochromatism
photometric
photopic

PIC test
pigment
pigment bleeding
Planckian radiator
pleochroism
plochere color system
powder
primary colors
primary colors, additive
primary colors, CIE
primary colors, subtractive
pseudoisochromatic plate (PIC) test
psychophysical
purity, colorimetric
purity, excitation
radiant flux
RAL
reducing power
reference standard
reflectance
reflectance, absolute
reflectance, diffuse
reflectance, directional
reflectance, Fresnel
reflectance, hemispherical
reflectance, luminous
reflectance, nonspecular
reflectance, specular
reflectance, total
reflectivity ($R\infty$)
reflectometer
reflex reflector
rescue orange
retina
retinal
retroreflective
rub up
saturation
Saunderson correction
Saunderson-Milner Zeta Space Color Difference Equation
scattering coefficient, Kubelka-Munk
scattering coefficient, Mie
scattering, multiple
scattering, single
SCE
SCI
scotopic vision
secondary reference standard
selected ordinate method
self-vulcanizing
semitransparent stain
sensation
shade
shader
shading and blending
Simon-Goodwin charts
spectra
spectral
spectral match
spectral power distribution curve
spectrogoniophotometer
spectrophotometer
spectrophotometer, abridged
spectrophotometer, filter
spectrophotometric
spectrophotometric curve
spectrum
spectrum locus
specular reflectance excluded (SCE)
specular reflectance included (SCI)
stain
staining power
standard light source
standard observer, 2° (1931)
standard observer, 10° (1964)
standard, white reflectance
subtractive colorant mixture
subtractive colorant mixture, complex
subtractive colorant mixture, simple
tan
temperature, color
temperature, correlated color
tinctorial strength
tinge
tint
tint base paint
tint-tone
tintability
tinting
tinting color or colorant
tinting (or tint) strength, absorption
tinting (or tint) strength, scattering
tinting strength
tinting strength, absorption
tinting strength, scattering
toner
top color
total color difference
translucency
transmittance, diffuse
transmittance, internal
transmittance, specular
transmittance, total
transparency
travel
triad
trichromat
trichromatic
trichromatic coefficients
tristimulus
tristimulus colorimeter
tristimulus computation data
tristimulus integration
tristimulus integrator
tristimulus values, CIE
u, v
U^*, V^*, W^*
underhue
undertone color
uniform chromaticity coordinates
uniform chromaticity scale diagram
uniform chromaticity spacing
uniform color scale
uniform color space
uniform lightness scale
value, Munsell
viewing angle
viewing booth
viewing conditions
viewing geometry
visible region
vision, defective color
vision, normal color
water white (WW)
weighted ordinate method
white
white reference standard
white reference standard, absolute
white reference standard, primary
white reference standard, secondary
whiten, whitening
whiteness
whiteness index
whitening
wood preservative coatings
X
$\bar{x}$
xanthic
Y
$\bar{y}$
y
yellowing
yellowness
yellowness index (YI)
YI
Z
$\bar{z}$
z
zeta space

Category 70
Coil Coating

accumulator
applicator roll
backing coat
baffle
bridle
camber
clockspring
coater
coil coating
coin marking
deposition
leveler
live centers
metal marking
oil can
pan
pick-up roll
pressure marking
rewind
ribbing
rollformer
skipping
slitting
starter strip
starving out
stitcher
tandem line
transfer roll
trough

Category 71
Coating Types

ablative coatings
acoustic coating
acoustical paint
acrylate
acrylate resins
acrylics
aerosol coating
alkali-resistant paint
alkyds
aluminum paint
aluminum primer
anticondensation paint
anticorrosive paint or composition
antifouling paints
antinoise paint
antishatter composition
antiskid paint
antislip paint
antitracking varnish
appliance finish
architectural coatings
artists' colors (or paints)
asphalt paint
automotive finish
back paint
back priming
backing coat
bactericidal paint
baking finish
ballotini
barrier coat
base coat
basecoat/clearcoat
beaded paint
beads, reflective
bituminous paints
black japan
blackboard paint
blackout paint
bleeding knots
block coat
block filler
bond coat
bookbinders' varnish
breathing-type coating
bronze paint
bronzing fluid
bronzing lacquer
bronzing medium or liquid
bubble coating
buffer coat
cabinet finish
calcimine
calcicoater
can coating
cast coating
cellulosic coating
cement coating
cement paint
ceramic coating
chemical-resistant paint
Chinese white
clear coating
clear lacquer
clear water repellent coating
clearcoat
clearcole, clairecolle
coal tar-epoxy coating
coal tar-urethane coating
coating
coating powders
cobweb coating
cold water paint
color wash
compliance coating
container lining
convertible coating
corrosion-inhibiting paint
deck paint
diphenylmethane diisocyanate (MDI)
dispersion, resin
distemper
distemper, oil-bound
dope
effect finish
emulsion paint
enamel
encaustic
end sealing compound
epoxy paint
epoxy resins
etching primer
exterior basecoat
exterior finishes
exterior paints and varnishes
fabric coating
filler coat
film
filmogen
finish
finish coat
fire-retardant coating
first coat
fixative
flat coat
flat enamel
flat finish
flat paint (finish)
flat varnish

flatting (rubbing) varnish
floor and deck enamel
floor paint
floor sealer
fluorescent paint
fogged coat
foliage paint
fresco
fresco color
fungicidal paint
gel coat
gesso
glaze coat
gold lacquer
graffito
graphic arts coatings
graphite paint
grease paint
ground coat
guide coat
gym finish
hammer finish
hard gloss paint
heat-resistant finishes
heat-sensitive paints
hexamethylene diisocyanate (HDI)
hide finishes
high-build coating
high-performance interior architectural wall coatings (HIPAC)
high-solids coatings
HIPAC
hold paint
hotmelt coatings
house paint
hull paint
industrial finishes or coatings
industrial maintenance paints
inert base
inorganic coatings
insulating varnish
intumescence
intumescent coating
iron phosphate coating
isocyanate resins
japan
kalsomine
knot sealer
lacquer
lasur (lasure)
latex paint
lead in oil
limewashing
liquid emulsion polymer coating
liquid water emulsion wax
luminous paint
luminous pigments
magnet wire coatings
maintenance paints
marine coatings
marine varnishes
masonry paint
matte finish
matte varnish
metal decorating
metal primer
metal spraying
metallic paint
metallic pigmented paint
microvoids
milk paint
mirror back coating
modified alkyd
monochrome
monolithic flooring
multicolor finish
NAD
natural finish
neoprene paint
neutral toner
NGR stain
nonaqueous dispersion (NAD)
nonconvertible coating
nongrain-raising stain (NGR)
nonpenetrating stain
nonskid paint
odorless paint
oil paint
oil stain
oil-bound
oil-bound distemper
opaque stain
opaque stains
orange shellac
organosol
outside finish
overprint varnish
overvarnish
paint
paper coating
pasteurizing varnishes
penetrating finish
penetrating stain
phosphorescent paint
plaster primer
plastic paint
plastigel
plastisol
polishing varnish
polychromatic finish
polyurethane finish
polyurethanes
porch paint
Portland cement paint
poster color
powder coating
powdered distemper
prefabrication primer
pretreatment primer
prime coat
primer
primer surfacer
priming paints
print paste
product coatings
protective coatings
PUD
PVC
radioactive paint
ready-mixed aluminum paint
ready-to-mix aluminum paint
release coating
repainting
resin emulsion paint
reverse roll coater
ripple finish
roller coating
roller coating enamel
rust-inhibiting paint
screen process printing
scumble glaze
seamless flooring
secco
self-etch primer
self-luminous paint
self-sealing paint
semitransparent stain
sgraffito painting
sharp paint
sherardizing
shingle stain
shop primer
shop priming
show card colors
silicate paints
silicone paint
soft distemper
solvent-borne coating
spatter finish
special-purpose coatings
speckled finish
spirit stains
stain
starch coating
strippable coating
structured paint
suede finish
superstructure paint

synthetic paint
tank coating
tank white
tempera
temporary coatings
texture-finished paint
textured paint
thixotropic paint
tie coat
tile-like coating (TLC)
tint base paint
tinting color or colorant
topcoat
trade sales
trade sales coating
trade sales paints
traffic marking paint
transition primer
translucent coating
transparent coating
two component paint
two-pot coating
undercoat
undercoating
uralkyd
urethane coatings
varnish stain
vinyl coating
wash primer
washable distemper
washable water paint
washcoat
water paint
water repellent system
water-based coatings
water-based paint
water-reducible coatings
waterborne coatings
watercolor
whitening
wood preservative coatings
wrinkle finish
zero VOC paints
zinc silicate primers
zinc-rich primer

Category 72
Resistance Properties

ablation
abrasion resistance
accelerated weathering
acid resistance
age resistance
alcohol resistance
alkali resistance
alkali-resistant paint
anticorrosive paint or composition
barrier coat
blister resistance
blistering
block resistance
burnish resistance
chalking resistance
checking
checking resistance
chemical resistance
chemical stability
chip resistance
chipping
corrosion-inhibiting paint
crack resistance
dirt/soil resistance
enamel holdout
erosion resistance
filiform corrosion resistance
fire resistance
flaking resistance
flexibility
fouling
freeze-thaw resistance
fungus resistance
gloss retention
grain crack resistance
holdout
household chemicals, resistance to
humidity resistance
impact resistance
imprinting
intumescence
lightfastness
mildew (fungus) resistance
oil resistance
permanence
print resistance
print test
QUV
resist
resistance
resistance to household chemicals
roller spatter resistance
rust resistance
rust-inhibiting paint
scaling resistance
self-lifting resistance
silicone paint
solvent resistance
spatter resistance
stain removal
tire pick-up resistance
topcoat holdout
traffic paint abrasion resistance
traffic paint chipping resistance
water resistance
weather resistance
wet abrasion resistance
wet adhesion
wet scrub resistance

Category 73
Cementitious Products

ACB
ACI
acoustical plaster
adobe brick
agglomerate
aggregate
air entraining agents
anhydrite
artificial stone
asbestos board
asbestos-cement board (ACB), asbestos-cement wallboard, asbestos sheeting
beton
bond plaster, concrete bond plaster
brown coat
building block
building stone
cement
cement coating
cement colorants
cement paint
cement plaster
cement, Portland
cement-asbestos board
cinder block
colored cement
concrete
concrete block
concrete bond plaster
concrete gun
floating coat
graffito
grout
gypsum, calcined
gypsum cement
gypsum plaster
gypsum wallboard
hydrated lime
hydraulic cement
laitance
masonry
masonry conditioner
mortar
plaster
plaster of Paris
Portland cement
Portland cement paint
quick-hardening lime
quicklime
render
scratch coat
spray gun
stucco
terrazzo
tricalcium silicate
Venetian mosaic
water repellent system
white Portland cement

Category 74
Catalysts

accelerator
ACI
acid catalysts
activate
activated
activator
adduct curing agent
alkaline catalysts
aluminum naphthenate
benzoperoxide
benzoyl peroxide
butyl (tetra) titanate
calcium naphthenate
calcium octoate
calcium resinate
calcium tallate
catalysis
catalyst
catalytic curing
catalytic incinerator
cerium naphthenate
cobalt drier
cobalt naphthenate
cobalt octoate
crosslinking agent (or crosslinker)
curing agent
drier
drier absorption
drier dissipation
drying
enzyme
fused driers
hardener
iron driers
Japan drier
lead drier
lead naphthenate
lead octoate
liquid driers
litharge
loss of drier
manganese driers
naphthenate
negative catalyst
octoates
peptizing agents
pot life
siccative
silica gel
soaps
soluble drier
spreadable life
surface drying
usable life
vanadium driers
working time
zinc drier
zirconium driers

Category 75
Building Products; Materials

ACB
acoustical board
acoustical ceiling board
acoustical material
acoustical plaster
acoustical tile
adobe brick
agglomerate
aggregate
alabaster
artificial stone
asbestos board
asbestos plaster
asbestos-cement board (ACB), asbestos-cement wallboard, asbestos sheeting
asphalt paper
barium plaster
basecoat
bastard sawn
batt
bead and butt
bead, butt and square
beam
beam ceiling
bevel siding
board
board and batten
board and brace
board foot
bond plaster, concrete bond plaster
brick
brown coat
building block
building board
building paper
building stone
built-up roofing
caulk (calk)
caulking compound
cement
cement plaster
cement-asbestos board
ceramic
chip board
cinder block
clapboard
clinker block
cold-checking
cold-process roofing
composition roofing
composition siding
concrete
concrete block
concrete bond plaster
convertible coating
dado
dimension stone
doorframe
dressed lumber
drop siding
dry wall
eave
embossed hardboard
exterior type plywood
felt
felt and gravel roofing
fiberboard
flake board
flashing
flexible wall coverings
floating coat
floor tile
glass felt
glass mat
gravel
gravel roofing
gypsum, calcined
gypsum cement
gypsum plaster
gypsum wallboard
hanger wire
hardboard
high-density overlay
hot-pressing
hydraulic cement
joint tape
lap siding
linoleum, floor and wall covering
lumber
magnesite floors
masonry
mineral surfaced roofing
novelty siding
paper, building
paperboard
parquet
parqueting
parquetry
particle board
partition
perforated tape
plaster
plaster of Paris
plasterboard
plywood
quarry tile
quick-hardening lime
roofing
rustic siding
sand
scratch coat
sheathing
shingle
shutter
siding
siding shingle
slate
soffit
steel
strip
stucco
tape
tar
tar and gravel roofing
terra cotta
terrazzo
tile
timber
vapor barrier
vapor retarder
veneer
veneer plywood
vulcanized fiber
wainscot
wall
wallboard
weatherboarding
white Portland cement
wood
wood chipboard
yellow pine

Category 76
Bituminous Products; Roofing

air blowing
albino bitumen
alligatoring
asphalt
asphalt, air blown
asphalt cut back
asphalt emulsion
asphalt paint
asphalt paper
asphalt rock (rock asphalt)
asphalt seal coat
asphalt, steam blown
asphaltene
asphaltic bitumens
asphaltic pyrobitumens
asphaltite
asphaltum
Berlin blacks
bistre
bitumen
bituminous cement
bituminous coating
bituminous emulsion
bituminous grout
bituminous paints
black japan
black varnish
blown bitumen
Brunswick black
building paper
built-up roofing (BUR)
BUR
carbenes
coal tar
coal tar pitch
coking
cold-process roofing
composition roofing
counterflashing
creosote
cresol
cresylic acid
cutback
dead oil
dressing compound
Egyptian asphaltum
elaterite
emulsified asphalt
felt and gravel roofing
gilsonite
glance pitch
glass felt
glaze coat
grahamite
gravel roofing
grout
ice dam
impsonite
japan
jew's pitch
manjak
membrane
mineral surfaced roofing
mopping
oxidized asphalt
oxidized bitumen
paper, building
parting agent
penetration number
penetrometer
petroleum bitumens
pitch oil
ply
plying cement
primer
rafaelite
reglet
reinforced membrane
release agent
roll roofing
roofing system
saddle
saturated felt
scuttle
selvage
shale oil
shingle
shingling
slope
square
stearin (stearine) pitch
strip flashing
Syrian asphaltum
system
tack coat
tar
tar and gravel roofing
undercoating
varnish
vintahite
wurtzilite

Category 77
Application, Methods of

aerosol coating
air brush
air spray
airless spraying
anionic electrodeposition
application
application rate
applicator
autodeposition
automatic electrostatic spray application
barrel finishing
barrel polish
barrelling
bell
bird applicator
brayer
brush
brushing
casting
cathodic electrodeposition
cloth distressing
coating powders
coating, dip
coating, spray
cobwebbing
cocooning
coil coating
cold flow
crossing
curtain coating
dabber
daub
deposition
dip coating
dipcoat

dipping
disk
distressing
doctor blade
doctor roll
drawdown
drawdown bar (blade)
drawout
EDP
electrocoating
electrodeposition
electrostatic spraying
Elpo
equalizer rod
fadding
film applicator
flame spray
flat screen printer
flock finish
flocking
flocking deposition
flow coating
fluidized bed coating
fresco
gilding
gouache
hand-held electrostatic spraying
hand-screening
hot spraying
hydraulic spraying
internal mix spray gun
knife coater
knife mark
lifting
lithography
Meier bar
metallizing
mini-bell
mitt
mohair paint roller cover
mopping
multinotch applicator
natural spreading rate
offset
padding
paint
paint brush
paint harling
paint mitt
paint pad
paint roller
painter's mitt
picking
plasma coating
plasma deposition technique
plasma flame
powder coating
rag rolled finish
rag rolling
ragging
reverse roll coater
roll coating
roller
roller application
roller coating
roller coating enamel
rotary screen printer
rotational electrostatic spray
roto-dipping
rumbling
screen painting
silk-screen printing
silk-screening
slushing
spin coating
spinning
sponging
sponging off
sponging on
spray coating
spraying
spreading rate
squeegee coat
stencilling
stipple
stippling
strip coating
thermospray
throwing power
transfer coating
tumbling
two-component gun
whirling
wire-bar applicator
wire-wound rod
wrap

Category 79
Adhesion and Adhesives

A-stage
abhesive
acacia gum
adhere
adherend
adherometer
adhesion
adhesion promoters
adhesion, mechanical
adhesion, specific
adhesive
adhesive dispersion
adhesive film
adhesive, assembly
adhesive, cold-setting
adhesive, edge jointing
adhesive, hot-setting
adhesive, pressure sensitive
aggressive tack
amylaceous
anchorage
antiblocking agent
antishatter composition
arabic
Arco Microknife
assembly
assembly time
autohesion
B-stage
binder
blocking
bond
bond plaster, concrete bond plaster
bonding action
bone glue
brake lining medium
British gum
brush binder
C-stage
cat eyes
cement
cement, rubber
cohesion
cold-setting adhesive
concrete
concrete bond plaster
cross-cut (cross-hatch) tape test
delamination
dextrin
doctor roll
double spread
dry tack
filler

glue (noun)
glue (verb)
gold size
gum Acacia
gum Arabic
hot melt adhesive
intercoat adhesion
joint, starved
kibbled
laminate
laminated, cross
laminated, parallel
lamination
legs
mastic
mechanical adhesion
mold wash
mucilage
novolak
parallel laminated
particle board
paste
penetration
plying cement
plywood
polyvinyl acetal
pressure-sensitive adhesive
primer
release coating
release paper
residual tack
resite
resol
rubber cement
set
self-adhesion
single spread
spread
starved joint
strength, dry
strength, wet
stringiness
tack
tack range
tack, dry
tape
tape test
temperature, curing
temperature, drying
temperature, setting
thermal adhesive
tire pick-up resistance
vulcanize
weatherboarding
wet adhesion
wet adhesion test

Category 81
Art

abbozzo
abstract art
academy board
air brush
anamorphosis
applique
aquarelle
aquatint
artists' colors (or paints)
bas-relief (basso-rilieva)
beni
beni-ye
biocide
blender
block printing
bozzo
camel hair
canvas
canvas board
cardboard
cartoon
cassone painting
cave painting
chinoiserie
cloisonne
collage
craquelure
crayon
deadcoloring
decoupage
designers colors
distemper
distemper colors
encaustic
ferrule
flat screen printer
form
fresco
fresco color
gauffrage
gesso
gilding
gouache
graffito
graphic arts coatings
grisaille (Fr.)
ground
gum mastic
hand-screening
ichnography
illuminating
impasto
inlay
inpaint
intaglio
japan color
kakemono
kira-ye
kurenai-ye
lithography
mahlstick
marouflage
medium (art)
mezzotints
monochrome
mosaic
mottle
mural painting
nishiki-ye
objet d'art
offset
offset lithography
oil-bound distemper
otsu-ye
palette
palette knife
paperboard
papier-mâché
pastel
pasteboard
pastel painting
pastiche
patina
pilaster
pointillism
pounce
prehistoric art
print
pyroxylin
relief
rococo
sable pencil
sable writer
screen painting
screen print
screen process printing
scumbling
secco
sgraffito painting

silk-screen printing
silk-screening
silver point
stereochromy
still life
stippling
sumi
sumi-ye
surimono
tan-ye
tempera
terra cotta
Texanol®
tole
transfer paper
transfer printing
transparent painting
trompe l'oeil
tusche
uchiwa-ye
urushi
urushi-ye
varris Martin
watercolor
woodcut
xylography

Category 82
Architecture

abat-jour
abat-vent
abut
acoustical board
acoustical ceiling board
acoustical plaster
acoustical tile
AIA
AIMA
air brush
air ducts
alabaster
alcove
appurtenance
apron
architecture
bay
baluster
balustrade
banister
base
base plate
baseboard
batten
beam
bead and butt
bead, butt and square
beam ceiling
bevel siding
board
board and batten
board and brace
board foot
bond plaster, concrete bond plaster
bow window
brick
building
building block
building board
building code
building paper
building stone
built-up roofing
caisson
casement
casement window
ceiling
center of interest
chair rail
chimney
cill
clapboard
composition roofing
concrete bond plaster
coping
cornice
counterflashing
crawl space
cricket
dado
dead level
deck
detergent
detergent
doorframe
dormer
double-hung sash
drain
drop ceiling
drop siding
dry wall
eave
expansion joint
facade
felt and gravel roofing
fill-type insulation
firewall
flashing
floor tile
flue
footing
foundation
frieze
gable
gableboard
gravel roofing
gutter
headers
high-performance interior architectural wall coatings
HIPAC
jalousie
jamb
joint
joint, lap
jointless flooring
joist
kick board
lap
lap joint
lap siding
lath
lintel
louvers
masonry
moisture barrier
molding
mop board
moulding
mullion
newel
novelty siding
paper, building
parapet
parquet
parqueting
parquetry
pilaster
pointing
roof
roof cladding
roofing
roofing system
rustic siding
sash

scrub board
scuttle
sensitizer
shutter
sill
skirting board
slope
soffit
strip
stud
tar and gravel roofing
terrazzo
trim
truss
vent
veneer
Venetian window
vergeboard
wainscot
wall
washboard
weather stripping
weep hole
window sash
wood veneer

Category 83
Additives

acacia gum
accelerator
acid catalysts
activator
additive
adduct curing agent
adhesion promoters
air entraining agents
algicide
allotonic
aluminum naphthenate
aluminum resinate
aluminum stearate
2-amino-2-methyl-1-propanol
ammonium caseinate
ammonium soaps
AMP
antiblocking agent
antifloating agent
antiflooding agent
antifoaming agent
antilivering agent
antioxidant
antiozonant
antisag agent
antisettling agent
antisilking agent
antiskinning agent
antistatic agents
antiwrinkling agent
asbestos
associative thickener (AT)
AT
bactericide
bacteriostat
ballotini
barium naphthenate
barium oleate
barium stearate
bead
beads, reflective
blowing agents
bodying agent
brightener, optical
brightening agent
calcium chloride
calcium stearate
can preservative
carboxymethyl cellulose (CMC)
catalyst
cationic detergent
cellulose ethers
cellulose glycolate
cellulose gum
cellulosic thickeners
cerium naphthenate
chelating agent
chill-back
citronella oil
CMC
coalescent (coalescing agent)
cobalt drier
cobalt naphthenate
cobalt octoate
collector
copper carbonate
copper driers
copper resinate
copper soap
copper sub-oxide
cork
cork dust
corrosion inhibitor
cosolvent
coupling agent
crosslinking agent (or crosslinker)
cure
curing agent
deairentrainment
deflocculant
defoamer
denaturation
desiccant
dextrin
diphenyl
dispersant
dispersing agent
drier
drying agent
EDTA
emulgator
emulsifier
emulsifying agent
emulsion stabilizer
epichlorohydrin
ethylene glycol
ethylenediamine tetraacetic acid
fire-retardant chemical
flatting agent
foaming agents
fungicide
fungistat
fused driers
gamma protein
gavan
gelatin
guaiacol
hardener
HASE
HEUR
hexamine
(2-hydroxypropyl) methyl cellulose
HMHEC
HPAM
hydrophobically modified hydroxy ethyl cellulose (HMHEC)
hydrophobically modified polyacrylamide
in-can biodeterioration
in-can preservative
industrial talc
industrial universal colorants
inhibitor
ink compounds

iron driers
iron soaps
japan drier
latent solvent
lead drier
lead naphthenate
lead octoate
lecithin
lemongrass, oil of
linolenate driers
liquid driers
litharge
lubricant
magnesium soaps
manganese driers
MEHQ
metallic soaps
methylcellulose
mildewcide
mildewstat
mirbane oil
modifier
mold lubricant
mold wash
monomethyl ether of hydroquinone (MEHQ)
mutual solvent
naphthenate
octoates
optical brightener
oxidation inhibitor
oxidizing agent
oxime
parting agent
paste driers
patent driers
peptizing agents
plasticizer
plastigel
plumbago
precipitated driers
preservative
processing aid
protective colloids
PUD
release agent
retarder
rheology modifier
sensitizer
sequestering agent
siccative
silicone
skin
slip agent (compound)
soaps
soluble drier
stabilizer
stripping agent
sulfonated castor oil
sulfonated oil
surface active agent
surfactants
suspending agent
suspension aid
terebine
thickener
thixotrope
tinting color or colorant
toxicant
tracer
turkey red oil
ultraviolet absorber
universal colorants
urethane thickeners
vanadium driers
water repellent system
wet edge extender
zinc drier
zinc stearate
zirconium driers

Category 127
Surfactants

aluminum soaps
2-amino-2-methyl-1-propanol
ammonium caseinate
ammonium soaps
AMP
amphiphilic
angle of contact
anionic surfactants
barium naphthenate
barium oleate
biodegradable surfactant
brush (or roller) cleaner
calcium linoleate
calcium stearate
cationic detergent
cationic surfactant
cleaner
CMC
copper soap
critical micelle concentration (CMC)
detergent
detergent
DuNouy Ring
emulgator
emulsifier
emulsifying agent
foaming agents
grinding medium
HLB value
hydrophile-lipophile balance (HLB) value
interfacial tension
iron soaps
lecithin
magnesium soaps
mahogany sulfonates
nonionic surfactants
roller cleaner
rosin size
saponification
soaps
sulfated oil
sulfonated castor oil
sulfonated oil
surface active agent
surfactants
syndet
synthetic detergent
tall oil soap
tungates
turkey red oil
water-break-free
wetting
wetting agent
zinc resinate

Category 128
Dimensions, Scales, Values

Å
A.U.
absolute
absolute humidity
absolute temperature
absolute viscosity
absorption coefficient
acetyl number (value)
American gallon
amine equivalent weight
ampere
amu
Angstrom unit
aniline cloud point
API gravity
apparent density
approximate value
atmosphere
atomic mass unit (amu)
atomic weight
barrel
base box
batch
Beer-Bouguer Law (Beer-Lambert Law)
board foot
bold
British thermal unit (BTU)
BTU
build of wet film
bulk density
bulking
bulking value
C
°C
c.g.s.-emu system of units
c.g.s. system
calorie
candle
Celsius scale
centigrade
centimetre (cm)
centipoise (cP)
chips
coating solids
color rendering index
coulomb
coverage
cross-cut (cross-hatch) tape test
density
density, apparent
density, number
detection limit (threshold)
dew point
diene value or number
dimension
dry bulk
dust
dyne
English gallon
ester number
°F
F.P.S. system
Fahrenheit scale
film thickness
fineness of dispersion
fineness of grind
fines
foot-candle
gallon, imperial
gallon, U.S.
grain
gram
gram-molecule
h
Hazardous Materials Identification System, The (HMIS)
hazen color scale
heat capacity
hiding power
HLB value
humidity
humidity, absolute
humidity ratio
humidity, relative
humidity, specific
hydrogen ion concentration (pH)
hydrophile-lipophile balance (HLB) value
imperial gallon
inch-pound units
joule, j and SI
kauri-butanol value
KB value
kelvin
kelvin temperature scale
KHN
Knoop hardness number, (KHN)
Koettstorfer number
Krebs units (KU)
KU
litre
lot
lot number
m.k.s. system
maleic anhydride value
maleic value
mass
megapoise
melting point
mesh
metamerism, degree of
meter
metre
metric perm
metric ton
micrometer
micron (μ)
mil
millilitre
millimicron
millipoise
min
mired
Mohs scale
mole or mol
molecular weight
molecular weight(s)
Munsell chroma
Munsell color notation
MW
N
nanogram (ng.)
nanometre (nm)
NBS unit
nm
nominal value
north standard
O.A.
oil absorption (value)
particle size distribution
perm
permeability
permeance (water vapor)
Pfund hardness number (PHN)
pH
PHN
Pigment Volume Concentration (PVC)
pigment/binder ratio
poise
pore size
ppb
primary amine value
proof, apparent
proof gallon, U.S.
proof spirit
PVC
r/min

r/s
Reichert-Meissl value
relative density
relative humidity
relative leveling
rhe
roll of wallcovering
rosin standards
ru
sa
saponification value
Saybolt universal viscosity
secondary amine value
SI units
sieve number
specific gravity
specific heat
specific humidity
specific volume
spreading rate
square
stokes (ST)
STP
surface area
surface area (B.E.T.)
tertiary amine value
total solids
U.S. gallon
uniform lightness scale
unit, angstrom
units
units, U.S. customary
volume solids, percent
water vapor permeance (WVP)
water vapor transmission rate (WVT)
wavelength
weight percent solids
wet film build
wet film thickness
whiteness index
wine gallon
WVP
WVT

Category 130
Polymers (Synthetic Resins)

abietates
acetal resins
acetaldehyde resin
acetocopal
acetone resin
acid groups
acrolein polymers and resins
acrylate resins
acrylic latex
acrylic plastics
acrylics
acrylonitrile-butadiene-styrene (ABS)
addition polymer
addition polymerization
aldehyde resin
alkali-soluble resins (ASR)
alkyds
alkyl phenolic resin
allyl resin
alternating copolymer
amine resin
amine-furfural resin
amino plastics
amino resins
aniline-formaldehyde resin
aniline-furfural resin
anion exchange resins
aromatic petroleum residues
ASR
atactic polymer
azeotropic copolymer
Bakelite®
ball and ring method
base unit
benzoguanamine resins
bipolymer
bisphenol A and F
block copolymer
block polymer
branched
branched polymers
butadiene styrene latex
caprolactam
carbolic acid
cellulose acetate butyrate (CAB)
cellulose acetate (CA)
cellulose acetate propionate (CAP)
cellulose ester
chain length
chlorinated polyether
chlorinated rubber
condensation polymer
condensation polymerization
conventional base unit of a polymer
copolymer
copolymerization
cumar gum
cyclic diolefin resins
cyclic ketone resins
derived high polymer
diallyl esters
dimethyl urea
diolefin resins
diphenylmethane diisocyanate (MDI)
emulsion
emulsoid
end groups
epoxy adduct
epoxy ester
epoxy resins
ester gum
etherified urea resins
ethyl cellulose
eupolymer
fiber
fluorocarbons
fossil gum resins
free binder
fumaric resin
furan resins
furfural
graft polymer or copolymer
guanidine aldehyde resins
gum
gum, natural
HASE
heat-convertible resins
heat-reactive phenolic resin
hemipolymer
HEUR
hexamethylene diisocyanate (HDI)
high polymer
HMHEC
homopolymer
homopolymerization
hydrocarbon resins
hydrophile
inorganic polymers
internally plasticized
intrinsic viscosity
ion exchange resins
ionomer

ionomer resins
isocyanate resins
isomeric polymers
isotactic
isotactic polymer
karolith
ketone condensation resins
Kienle's functionality theory
latex
latices
limiting viscosity number
liquid emulsion polymer coating
macromolecule
maleic acid
maleic anhydride
maleic ester resin
maleic resin
mannite
mannitol
melamine
melamine resin
melamine-formaldehyde resin (MF)
melt viscosity
mer
methyl methacrylate resins
modified alkyd
modified phenolic resins
monoglyceride
monomer
monomeric unit (mer)
nitrocellulose
norelac resins
novolak
nylon resins
oil-modified alkyd
oil-reactive phenolic resins
oligomer
organometallic polymer
organosol
PE
penta resin
pentaerythritol (PE)
PF resins
phenol
phenol-aldehyde resins
phenol-formaldehyde resin
phenolic resins
phenoxy resin
phthalate esters
phycocolloid
plasticization
poly(methyl methacrylate) plastics
poly(methyl methacrylate) (PMMA)
polyacetal
polyacrylate
polyacrylic plastics
polyaddition
polyamide resins
polycarbonate
polydicyclopentadiene resins
polyelectrolyte
polyester resins
polyether resins
polyethylene (PE)
polyformaldehyde
polyimide resins
polyisobutylene
polymer
polymer gels
polymerization
polyolefin
polyolefin plastics
polyoxymethylene (POM)
polypropylene (PP)
polystyrene (PS)
polyurethane finish
polyurethanes
polyvinyl acetal
polyvinyl acetate (PVA and PVAc)
polyvinyl chloride co vinyl acetate
polyvinyl chloride (PVC)
polyvinyl fluoride (PVF)
polyvinyl formal
polyvinyl pyrollidone (PVP)
polyvinylidene chloride (PVDC)
polyvinylidene fluoride (PVDF)
prepolymer
primary high polymer
propylene
protein-aldehyde resins
PS
PUD
PVA
PVAc
PVC
PVF
PVP
pyroxylin
resin
resin, natural
resin, synthetic
resite
resitol
resol
rosin esters
rosin, limed or zinc-treated
rubber-emulsion paint
silicone
silicone resins
SMA resins
softening point
stereobase unit
styrene butadiene latex
styrene resin
sulfate pitch
sulfate resin
sulfate rosin
syndiotactic polymer
synthesis
synthetic
synthetic resin
tactic polymer
tacticity
tall oil
telechelic polymers
telomer
terpolymer
thermoplastic
thermoplastic resins
thermosetting resin
thiourea
thiourea resins
thiourea-formaldehyde resin
toluene-sulfonamide resins
trimer
unsaponifiable matter
uralkyd
urea
urea resin
urea-formaldehyde resin (UF)
urethane coatings
vinyl
vinyl acetate plastics
vinyl resins
vinyl wallcoverings
vinylidene chloride plastics
viscosity, intrinsic
water-soluble resins
xylenol resin
zinc resinate

Category 131
Processing, Process

abrasive (noun)
acetylation
acylation
aging (ageing)
agitation
agitator
air blowing
air flotation
air separation
alcoholysis
alkali-refined linseed oil
American process zinc oxide
application
aromatization
atomize
automation
baffle
bag molding
baking
balanced reaction
banbury
barrel finishing
barrel polish
batch process
biff
bleached oil
bleaching
bleaching clay
blow molding
blown bitumen
blown castor oil
blown stand oil
bodied oil
boiled oil
boxing
breakfree oil
buffing
cake, press
calcination
calcined clay
calender
calendered cloth
castor oil, dehydrated
ceramic
cermet
chaser
chatki
chelation
chilling
coacervation
coagulation
coil coating
coking
cold cut
cold pressing
collector
colloid mill
comminution
compacting
concentration
condensate
condensation
cone mill
copolymerization
counter-current
cracking
decantation
dehydration
dehydrogenation
deionization
delaminate
desorption
destructive distillation
dewaxing
digester
dilute
dispersion, pigment
distillation
dry blending
drying
Dutch process
edge runner mill
electrostatic detearing
elutriation
emulsification
ester interchange
esterification
expression
extraction
extrusion
festooning
filter
filter aid
filter cake
filter press
filtration
fine melt
flat screen printer
fleshing
flow coating
flushed (flushing) colors
flushing
flushed pigment
forced drying temperature
fraction
fractional distillation
fractionation
fugitive emission
grinding slab
halogenation
hammer mill
heat-bodied oil
hot oil expression
hydrogenation
hydrogenation of oils
imbibition
impregnated cloth
industrial waste water
industrial water
internal mixer
kettle
kibbled
kiln dried
kiln seasoned
laminate
lamination
letdown
leveler
levigation
lixivation
masticate
mastication of resins
material balance
maturing
media
metafiltration
microencapsulation
mill
mill base
millstone
molecular distillation of oils
nitration
nonbreak oil
oil, heat-bodied
oil melt
oven
oxidized turpentine
oxyns
paper
paste, pigment
pebble mill
pelletization
petroleum
pigging
polymerized fatty acids
premix
processing aid
proofing
pulverize
pulverulent
putty chaser
pyrolysis

pyrophoric
quenching
refine, refining
relief offset
relief printing
runs
running
sampling
sand grinding
saponification
screen analysis or test
screening
scrubbing
sequestration
sieving
sigma blade mixer
sintering
slack melt copal
slake
slitting
standard sand
stoving
straining
surface drying
synthesis
synthetic
treated pigment
vapor degreasing
warm-up

Category 133
Structure

acetyl
acicular
acyclic
acyl, acyl groups
additive reaction
allotropy
alternating copolymer
amorphous
amphoteric
anatase
anisotropic
anisotropy
aryl
atactic polymer
bentonite
bentonite clay
benzene ring
benzenoid
bond
branched
build
carbamide
composite
congeal
conjugated double bond
constitutional formula
di - compounds
dimorphous
diol
double bond
electron
enantiotropic
epoxide
epoxy
epoxy resins
film build
flake
functional group
hemiacetal groups
high build
interface
isomerism
isomers
isotactic
isotactic polymer
liquid
modify
molecule
morphology
olefins
oxirane
polar
saturated compound
solid
stereobase unit
structure
syndiotactic polymer
tactic polymer
tacticity
wilkinite

Category 134
Nomenclature

absolute
acetal
acid groups
acyclic
acyl, acyl groups
additive reaction
adduct
AIC
aldehyde
aliphatic amino group
aliphatic compounds
aliphatic solvents
alkyl groups
alternating copolymer
amides
amido
amine resin
amines
amino
anhydride
anhydrous
argillaceous
aromatic compounds
aromatic content
aryl
ASTM
basic
benzene ring
benzenoid
benzoyl
benzyl
bicyclic
calcareous
carbohydrates
carbonaceous
carbonium
carbonyl group
carbonyl value
carboxyl group
cementitious
Color Index Name
Color Index Number
color matching functions
Colour Index (C.I.)
Colour Index Name

Colour Index Number
combined water
combined-oxide formula
concordance
constitutional formula
cosanic acids
cyclic
decomposition
definition
degradation
di - compounds
dictionary
diol
DIY
double bond
eicosanic acid
empirical formula
encyclopedia
epoxy
ester
ether
ethyl
fatty acid
fatty nitrogen products
functionality
generic name
glossary
halogen
hemiacetal groups
hemihydrate
hydrate
hydric
hydroxide
hydroxyl group
hypohalous
inorganic
isocyanate
jargon
journeyman painter
methoxyl group
metrication
NBS unit
nomenclature
observer, standard
olefins
organic
oxygenated solvent
primary amine value
PTM
saturated compound
standard
standard observer, 2° (1931)
standard observer, 10° (1964)
sulfation
sulfonation
term
thesaurus
trivial name
vinyl
vinyl group
vinyl wallcoverings
vocabulary

Category 137
Waxes and Polishes

adeps lanae, anhydrous
animal waxes
ball and ring method
beeswax
blocking point
brazil wax
candelilla wax
carnauba wax
carnaubic acid
cera alba
cera flava
ceresine
cerotic acid
Chinese vegetable tallow
Chinese wax
chlorinated paraffin wax
dammar wax
degras
dewaxing
encaustic
esparto wax
insect wax
japan wax
lac wax
lanolin
liquid solvent wax
melissic acid
microcrystalline wax
montan wax
ouricuri wax
ouropardo wax
ozokerite
paraffin wax
paste water emulsion wax
peat wax
petrolatum
petroleum
petroleum jelly
polish
powdering (of polishes)
resene, ß
self-polishing wax
softening point
solvent paste wax
solvent wax
spermaceti wax
terrazzo seal
unsaponifiable matter
water wax
wax
wool grease

Category 139
Colloids

adsorbate
agar (or agar-agar)
algin
alginates
alginic acid
allyl starch
aluminum stearate
ammonium caseinate
attapulgite
bentonite
bentonite clay
breaking
Brownian motion
builder
carboxymethyl cellulose (CMC)
carrageen
cataphoresis
cetraria
chemisorption (or chemical adsorption)
CMC
coacervate
coacervation
coagulation
coagulum
colloid
colloid, linear
colloid, micellar
colloid, molecular
colloid, protective
colloid, reversible
colloid, sphere
colloidal
colloidal dispersion
colloidal particle
colloidal solution
continuous phase
desorption
discontinuous phase
dispersed
dust
electrokinetic potential
electrophoresis
emulsion stabilizer
floc
flocculation
floccule
flux
foam
foaming agents
gel
gelation
gold number
gum
hectorite
hemicolloid
high polymer
hydrophilic
hydrophobic
hydrosol
hygroscopic
Iceland moss
Irish gum
Irish moss
isoelectric point
lyophilic
lyophobic
macromolecule
mesocolloid
micelle
mildew
montmorillonite
osmosis
pearl moss
peptization
peptizing agents
phase
phase, continuous
phase, discontinuous
phycocolloid
pig wrack
precipitate
reticulation
reversible colloid
rock salt moss
sol
solvent shock
sorbate
sorbent
sorption
sorptive
suspension
syneresis
throwing out
turbid
turbidity
wilkinite
zeta potential

Category 140
Elastomers

accelerator
acid curing (hardening)
activator
aftercure
air cure
ammonia cure
anchorage
antiblocking agent
balata
bank, rubber mill
bin cure
bleeding
bloom
blowing
boot-legging
breakdown
butadiene styrene latex
butyl rubber
caoutchouc
cat eyes
chalking
checks
cracking
crack resistance
crawl
crazing
creep
crocking
curling
dumbbell
elastic
elastic limit
elastic solid
elasticity
elastomer
elastomeric
ethanite
factice
filler
fin
frosting
green stock

gutta-percha
hydrogenated rubber
inhibitor
isomerized rubber
legs
logu
masticate
mold wash
neoprene
nerve
overcure
oxidized rubber
peptizing agents
peracetylated rubber
pile burning
polyisobutylene
popcorn
proofing
refiner
resinoid
retarder
reversion
rubber
rubber latex
SBR
scorching
semicure
staining
strain relaxation
strainer
synthetic rubber
tack
thermoset
thermosetting
tight cure
vulcanization
warm-up
water staining

Category 141
Monomers

acrolein
acrylic acid
acrylonitrile
alpha-methylstyrene
anprolene
bengucopalic acid
benzofuran
butadiene
n-butyl acrylate
coumarone
cumarone
1,1-dichloroethylene
diphenylmethane diisocyanate (MDI)
diphenylmethane diisocyanate (MDI)
epichlorohydrin
erythrene
ethyl acrylate
ethyl carbamate
ethylene oxide
HDI
hexamethylene diisocyanate (HDI)
indene
IPDI
isobutene
isocyanate
isophorone diisocyanate (IPDI)
isoprene
itaconic acid
MDI
mer
methyl acrylate
methyl methacrylate
monomer
monomeric unit (mer)
octyl phenol (tertiary)
phosphotungstic pigments
polymerization
propenoic acid
propylene
styrene
TDI
terpolymer
TMXDI
toluene
toluene diisocyanate (TDI)
urethane
vinyl benzene
vinyl chloride
vinylformic acid
vinylidene chloride

Category 142
Physical Properties

absorption
activated
adsorbed water
adsorbent
adsorption
aeration
angle of contact
boiling point
boiling range
breathing
bulk density
burnish resistance
capillarity
CO_2
coefficient of cubical expansion
coefficient of expansion
critical point
deformation
deliquescence
diffusion
eddy diffusion
electric (electrical)
electricity
element
entrainment
evaporation
fastness, fast
feel
flocculate
flux, luminous
foil
force
freezing point
gel point
gelatinous

holdout
hydrophilic
hydrophobic
hydrostatic
hygroscopic
hysteresis
interfacial tension
liquid
mar resistance
melt viscosity
moiré papers
moisture vapor permeability (MVP)
moisture vapor transmission (MVT)
molecular diffusion
MVP
MVT
NV
NVM
oatmeal paper
oil soluble
osmosis
overlapping seam
Periodic Table of the Elements
permanent set
Pfund hardness number (PHN)
PHN
physical
prepasted
pressure
reference standard
room temperature
science
shrinkage
solid
solidification point
solution
specific heat
specific permeability (of a film to moisture)
specific volume
surface tensiometer
surface tension
vapor pressure
volatile content
wet
wicking
WW

Category 144
Chemicals, Chemistry

A-acid
abietic acid, commercial grade
absolute alcohol
acenaphthene
acetaldehyde
acetamide
acetanilide
acetates
acetic acid
acetic anhydride
acetic ester and acetic ether
acetonyl acetone
acetophenone
acetyl benzene
acid
acid sludge
acidity
acrolein
acrylic ester
acrylonitrile
adeps lanae, anhydrous
adipates
adipic acid
adronal
alcohol
aldehyde
alginates
aliphatic amino group
alkali
alpha-hydroxypropionic acid
aluminum arsenite
aluminum phosphate
American turpentine
amides
amido
amines
amu
amyl acetate
amyl alcohol
amyl citrate
anacardic acid
anacardol
angocopalolic acid
angocopaloresenes
anhydride
anilides
aniline, anilin
animal fats
anisole
anprolene
anthracene
anthracene oil
anthranilic acid
apricot kernel mixed fatty acids
apricot kernel oil
aspartic acid
atomic mass unit (amu)
atomic weight
azelaic acid
banana oil
base
basic lead acetate
bauxite
behenic acid
bengucopalresene
bengucopalolic acid
benzilic acid
benzoic acid
benzoperoxide
benzophenone
benzoyl peroxide
benzyl alcohol
bisphenol A and F
borax
boron resins
brine
buffer
butyl citrate
butyl lactate
butyraldehyde
butyrates
calcium chloride
calcium hydroxide
calcium oxide
carbamide
carbohydrates
carbolic acid
carbon dioxide (CO_2)
carbonaceous
carboxylic acid
cardol
carnaubyl alcohol
carotene (carotin)
carvone
caustic
caustic potash
caustic soda
cellulose nitrate

cerotic acid
ceryl alcohol
cetoleic acid
chemical
chemical family
chemical name
chemistry
chloroform
citrates
citric acid
Cleve's acid
compound
cyclohexadiene
cyclohexan-hexol
cyclohexane
cyclohexanol
cyclohexyl
daturic acid
diallyl esters
diazo solution
dibutyl phthalate
dimer acids
dimethyl phthalate
dimethylolurea
diphenyl
diphenylamine
dodecanoic acid
dope cotton
dry ice
effervescence
elaidic acid
electrolysis
element
eleomargaric acid
eleostearic acid
empirical formula
endothermic
epsom salts
equilibrium
erythritol
ester interchange
esterification
ethanolamine
ether
ethyl abietate
n-ethyl acetanilide
ethyl acetyl ricinoleate
ethyl acrylate
ethyl alcohol
ethyl lactate
ethyl succinate
ethylbenzene
ethylene (ethene)
ethylene oxide
exothermic
fatty acid pitches
formaldehyde
formalin
free fatty acid
functional group
glycerides
glycerin
glycerol
glyceryl diacetate
gun cotton
gypsum
halogen
hazardous air pollutants (HAP)
HDI
hemihydrate
hexahydrophenol
hexalin
hexamine
hydrocarbon
hydrocarbon plastics
inert
intermediates
ion
isophthalic acid
isopropyl acetate
isopropyl butyrate
j acid
kerrolic acid
ketone
L acid
lacquer
lactic acid
lead acetate
lead nitrate
lead sulfide
licanic acid
lime
maleic acid
mannitol
margaric acid
massicot (massocot)
melissic acid
methylolureas
milk acid
mineral acids
mirbane oil
molecular weight(s)
molecule
naphthenic acid
nitrobenzene
nitrocellulose
nitroparaffins
NO_x
octyl acetate
oenanthal
oil of mirbane
organic
oxalic acid
oxidation
oxide
oxidizing agent
oxygen, molecular (O_2)
oxygen (O)
ozone (O_3)
palm oil pitch
palmitic acid
paraffin oil
PE
pentaerythritol (PE)
Periodic Table of the Elements
phenol
phenyl acetamide
phthalic acid
pK
pKa
polyalcohol
polybasic acid
polyether glycol
polyethylene glycol
polyglycol
polymerization
polymerized fatty acids
polyol
polyoxyethylene glycol
POx
propylene dichloride
propylene glycol
pyroabietic acids
pyroxylin
quicklime
reaction
reagent
red oil
reduction
ricinoleic acid
salt
Schaffer's acid
science
sebacic acid
shellolic acid
slaked lime
sodium borate
solvolysis
sorbic acid
sorbitol
species
stearic acid
stearin (stearine)
stoichiometric
succinic acid
sulfated oil
sylvic acid
synthesis
synthetic
tar acid
tartaric acid

TBEP
TBTF
TBTO
terephthalic acid
tocopherols
TPTF
tributoxyethyl phosphate (TBEP)
tributyl citrate
triethanolamine
triethylene glycol
triglyceride
tri-n-propyl phosphate
Twitchell reagent
unsaturated
urea
urushiol
valence (valency)
vinyl cyanide
water
water glass
wool grease pitch

Category 145
Instruments

Abbe Refractometer
abrasiometer
abridged spectrophotometer
absorption hygrometer
accelerated weathering
accelerated weathering machine (or apparatus)
adherometer
air bubble viscometer
Arco Microknife
balance, Westphal type
Barcol hardness
baumé
beta gage (beta-ray gage)
bimodal distribution
bird applicator
blank reading or blank value
blocking test
boomerang
Brookfield® Viscometer
bubble tube viscometer
calibrate
capillary viscometer
cathode
centrifuge
Cleveland open cup
collector
colorimeter
colorimeter, tristimulus
colorimetry
comb
cryptometer
cup
cup, weight per gallon
densitometer
dip type viscosity cup
doctor blade
durometer
dust collector
equalizer rod
evaporometer
filter spectrophotometer
fineness of grind gage
flowmeter
glossmeter
goniophotometer
hand-blocking
Hegman gage
hellige system or comparator
holiday detector
hydrometer
hygrometer
ICI Cone/Plate viscometer
inkometer
integrating sphere
interference filter
ISO Flow Cups
knoop indenter
König pendulum damping hardness test
Krebs-Stormer Viscometer
LEL meter
low temperature coalescence
mass spectrometry
Meier bar
meter
minimum film forming temperature (MFFT)
monochromator
muller
multinotch applicator
pendulum-rocker hardness
penetrometer
Pensky-Martens closed-cup tester
perm
perm cup or dish
Persoz pendulum hardness test
Pfund indentation tester
photometer
plastometer
PMCC
precision
psychrometer
psychrometry
pycnometer
pyrometer
QUV
reflectometer
rocker hardness tester
rotometer
Setaflash Tester
sieve
sling psychrometer
slushing oil
smoke alarm
spectrogoniophotometer
spectrophotometer
spectrophotometer, abridged
spectrophotometer, filter
spectrophotometry
stormer viscometer
surface tensiometer
Sward rocker
Taber abraser
tensile tester
tensile testing machine
tensiometer
test
torsion-braid analyzer
tristimulus colorimeter
Tukon
viewing booth
viscometer (viscosimeter)
viscosimeter
viscosity cup
weathering
weight-per-gallon cup
wet bulb temperature
wet film gauge
wire-bar applicator
wire-wound rod
xenon-arc light exposure apparatus
Zahn cup

Category 146
Shellac

aleuritic acid
ber
bisacki
bleached shellac
bone dry bleached shellac
brown hard varnish
button lac
chatki
Chinese wax
cut
fadding
flake shellac
French polish
garnet lac
hank shellac
hard lac resin
japan color
japan, grinding
kerrolic acid
knotting
kusum
lac
lac dye
lac resins
orange shellac
orpiment
seed lac
shellac
shellac, bleached
shellolic acid
spirit varnish
stick shellac
sticklac
translucent coating
transparent coating
white lac
wood cement

Category 147
Sealers, Fillers

asphalt tile sealer
beaumontage
bituminous cement
clearcole, clairecolle
creep
filler
floor sealer
glazing compound
joint compound
knifing filler
painter's putty
paste filler
petrifying liquid
plastic wood
sanding sealer
seal coat, sealing coat
sealer
sealing coat
shingle stain
side seam sealer
size
sizing
spachtling
Spackle®
sparkling
surfacer
terrazzo seal
undercoat
waterproofing

Category 148
Decorating

all-over pattern
antique finish
antiquing
applique
block cutter
block printing
bole
border
broken-color work
bronzing
calendered papers
canopy ceiling
center of interest
chinoiserie
chintz
chip resistance
chipping
cloisonne
cloth distressing
collage
composition leaf
corduroy
cornice
correlated
cove ceiling
crystallization
dado
daub
decalcomania, decal
decapping
decor
decorated
decorative painting
decoupage
distressing
domino
drop ceiling
dry brush
dry strippable paper
duplex paper
Dutch gold
Dutch metal
ebonize
embossed paper
embossing
engraving
fabric printing
faux hois (Fr.)
flat screen printer
flexible wall coverings
flock

florals
fluting
fly speck
fret
frieze
gilder's liquor
glaze
glazing
gold leaf
gold leafing
gold size
graining
grass cloth
ground coat
highlighting
illuminating
ink-embossed wallpaper
inlay
japan color
jaspé paper
joint tape
journeyman painter
lapis lazuli
lining fabrics
lining paper
linkrusta
lithographed paper
marbling, marbleizing
marquetry
matching
metal leaf
molding
monk's cloth
monochromatic color scheme
mottle
mottled
oil gilding
paillon
panel decoration
papier-mâché
parcel gilding
parquet
parqueting
parquetry
pearl essence
pearl lacquer
pencilling
pickled pine
plastic-bonded wallpapers
plastic-coated wallpapers
pretrimmed papers
rag rolled finish
rag rolling
ragging
railroading
relief
resist technique
roll of wallcovering
sandblast
screen print
screen process printing
sculptured wallcovering
scumbling
seams
self-tone
selvedge
shade
shading and blending
silk-screen printing
silk-screening
silver leaf
single roll
size
skin
smalt
soffit
spattering
sponging
sponging off
sponging on
staining
stippling
strié
strip
striping
strippable paper
stucco
textile printing
Tiffany finish
tole
touch-up coating
touch-up painting
transfer paper
transfer printing
trimmed papers
valance
varris Martin
veneer
vinyl printing and topcoating
vinyl wallcoverings
wainscot
wallpaper
water gilding
wood veneer

Category 151
Lacquers

acetate dope
benzyl cellulose
bronzing lacquer
cable varnishes and lacquers
cellulose lacquer
cellulose nitrate
clear lacquer
compacting
crystallization
dilution ratio
dope
dope cotton
ethyl abietate
ethyl acetate
ethyl cellulose
ethyl propionate
extrusion
fadding
flat lacquer
form lacquer
gold lacquer
gun cotton
hide finishes
lac wax
lacquer
nitrocellulose
polyvinyl acetal
pulling over
pyroxylin
rhus vernicifera
semigloss lacquer
spirit varnish
translucent coating
transparent coating
urushi
urushiol

Category 152
Varnishes

antitracking varnish
baking vehicle (varnish)
biff
bituminous varnish
black japan
black varnish
bookbinders' varnish
bronzing fluid
bronzing medium or liquid
brown hard varnish
cable varnishes and lacquers
conversion varnish
copal varnish
depolymerized rubber
exterior finishes
exterior paints and varnishes
flat varnish
flatting oil
flatting (rubbing) varnish
floor varnishes
flowing varnish
form lacquer
French varnish
grinding japan
impregnating insulating varnish
insulating varnish
japan color
japan, grinding
japanese lacquer
linoleum and oilcloth varnishes
lithographic varnishes
long oil
marine coatings
marine varnishes
matte varnish
medium oil varnish
minute value
mixing varnish
monkey
nonvolatile matter (NVM) or content
oak varnish
oil, long
oil varnish
oil-reactive phenolic resins
oleoresinous vehicle
ormolu varnish
penetrating finish
phenolic resins
Poison, Class B
polishing varnish
polyurethane finish
rubbing oil
rubbing varnish
spar varnish
spirit varnish
string (varnish)
TOC
translucent coating
transparent coating
varnish
varnish stain

Category 153
Wood, Wood Products

alburnum
alpha cellulose
annual growth ring
bastard sawn
beam
beam ceiling
bevel siding
bird's-eye
bleeding
bleeding knots
block flooring
bloom
board
board and batten
board and brace
board foot
bow
boxed heart
brashness
burl
butt veneer
cedar
cellulose
charcoal
cherry
clapboard
close grain
coarse grain
coarse texture
composition siding
compregnated wood
coniferous wood
cross grain
dado
delamination
diagonal grain
diagonal-grained wood
dip-grained wood
dressed lumber
drop siding
edge grain
embossed hardboard
end grain
exterior type plywood
figure
flake board
flat grain
grain
grain raising
green lumber
green wood
gym finish
hardboard
hardwoods
heartwood
high-density overlay
honeycombing
hot-pressing
humidity resistance
interlocked grain
kiln dried
kiln seasoned
knot sealer
lap siding
lignin
lumber
lumber, edge-grained
lumber, flat-grained
lumber, matched
lumber, resawn
lumber, rough
marine borers
moisture
mottle
nonporous wood
novelty siding
open grain
paper

parquet
particle board
pitch streak
plastic wood
ply
plywood
resin pocket
resinous timber
rot
rustic siding
sapwood
softwoods
spiral grain
springwood
stain
strip
summerwood
timber
truss
veneer
veneer plywood
vertical grain
wainscot
wavy grain
wet rot
wood
wood chipboard
wood finishing
wood flour
wood preservative coatings
wood tar
year ring
yellow pine

Category 154
Weathering

accelerated testing
accelerated weathering
accelerated weathering machine (or apparatus)
alligatoring
artificial weathering
bleaching
brown rot
chalk
chalking
chalking resistance
checking
checking resistance
checks
climate cabinet
cold-checking
cracking
crack resistance
crazing
decomposition
degradation
deterioration
dry rot
durability
erosion
erosion resistance
exposure rack
exudation
fadeometer
fading
fast to light
fastness, fast
humidity
perishing
QUV
test fence
useful life
weather
weather resistance
Weather-Ometer
weathering
wet rot
xenon-arc light exposure apparatus

Category 155
Water-soluble Resins and Gums

acacia gum
agar (or agar-agar)
albumin
algin
alginates
alginic acid
alkali cellulose
alkali-soluble resins (ASR)
alpha-protein
ammonium caseinate
aqueous
arabic
ASR
British gum
carboxymethyl cellulose (CMC)
carrageen
cellulose ethers
cellulose glycolate
cellulose gum
cellulosic thickeners
cetraria
CMC
cold water paint
dextrin
distemper
emulsion
emulsion paint
fossil gum resins
gavan
gelatin
guar gum
gum
gum Acacia
gum accroides
gum Arabic
gum karaya
gum, natural
hydrophilic
hydrophobic
hydrophobically modified hydroxy ethyl cellulose (HMHEC)
Iceland moss
Irish gum
Irish moss
karaya gum
latex
latices
methylcellulose
pearl moss
phycocolloid
pig wrack
polyvinyl pyrollidone (PVP)
protective colloids
PVP
rennet casein

resin emulsion paint
rock salt moss
Senegal gum
sodium carboxymethyl cellulose
tragacanth
water glass
water paint
water-based inks
water-based paint
water-reducible coatings
water-reducible resins
water-soluble resins
watercolor
wattle gum
yacca gum

Category 156
Vehicles

acid refined linseed oil
acrylic latex
aluminum mixing varnish
amino plastics
amino resins
antilivering agent
baking vehicle (varnish)
Berlin blacks
binder
binder demand
binder ratio
binding power index
blown castor oil
blown oil
blown stand oil
bodied oil
bookbinders' varnish
BPI
butadiene styrene latex
calcicoater
casein
China wood oil
copal varnish
diphenylmethane diisocyanate (MDI)
epoxy adduct
epoxy ester
epoxy resins
ethyl cellulose
gelling
gloss oil
glyptal
gold lacquer
hexamethylene diisocyanate
internal plasticization
isocyanate resins
japan color
Japan, grinding
kettle bodied oil
limed rosin
linoleum and oilcloth varnishes
livering
long oil
marking nut oil
medium
medium (art)
medium oil varnish
mixing varnish
NV
NVM
nylon resins
oil absorption (value) (O.A.)
oil content
oil length
oil, long
oil varnish
oleoresin
oleoresinous vehicle
paint base
phthalic anhydride
polyglycerols
polyurethane finish
polyurethanes
PUD
rosin, limed or zinc-treated
silanes
soya bean oil
stand oil
sulfurized oil
sunflower oil
urea resin
urea-formaldehyde resin
urethane coatings
varnish
vehicle
vinyl resins
vulcanized oil
water-reducible resins
wood oil

Category 157
Test Methods

abrasiometer
abrasion coefficient
abrasion cycle
abrasion resistance
absorption hygrometer
absorption, oil
absorption spectrophotometry
accelerated aging
accelerated life
accelerated testing
accelerated weathering
accelerated weathering machine (or apparatus)
accuracy
acetyl number (value)
acid number (value)
acid value
acidity
adhesive tape test
aging (ageing)
air bubble viscometer
amine equivalent
amine value
aniline cloud point
aniline number
aniline point
applicator
Arco Microknife
Arrhenius equation
artificial weathering
ash
ball and ring method
ball punch impact test
Barcol hardness
baudouin test
baumé
bead
Becchi-Millian test
bend test
Benedict's test
BET
binder demand
binding power index
bird applicator
black light

blank
blank reading or blank value
blister house
blocking
blocking point
blocking test
boomerang
BPI
break test (oils)
breakdown voltage
breaking extension
brittle point, brittle temperature
Brookfield® Viscometer
browne heat test
brushability
brushout
bubble tube viscometer
calibrate
can stability
capillarity
capillary viscometer
chipping rating standards
clean
clear
Cleveland condensing humidity cabinet (QCT)
Cleveland open cup
climate cabinet
cloud point
cold cycle
cold flex temperature
cold test
cold-bend test
cold-cracking
color match
colorimetry
conditional match
consensus standard
contrast ratio
correlation
cotton-free
cotton-free dry
cottonseed oil
coverage
cross-cut (cross-hatch) tape test
crowsfooting
cryptometer
cumulative sample
cup
decantation
densitometer
deviation
dimension
DIN
dip type viscosity cup
dirt/soil resistance
discoloration
distillation range
doctor test
double rub
drawdown
drawout
dry
dry point
dry weight
dry-hard
dry-through
dry-to-handle time
dry-to-touch time
drying time
dumbbell
DuNouy Ring
dust-free
dust-free time
elongation
energy at break
equalizer rod
ester number
evaporation rate
evaporation rate, final
evaporation rate, initial
evaporometer
exposure rack
exposure tests
extraction
falling sand abrasion test
film applicator
film thickness
film thickness gage (gauge)
fineness of dispersion
fineness of grind
fineness of grind gage
fingernail test
flexibility
flexibility test
forced drying temperature
Ford cup
freeze-thaw resistance
freezing point
frosting
fugitometer
fusion point
gage length
Gardner color standards
Gardner mobilometer
Gardner-Holdt viscosity tubes
gas checking
gas crazing
gasproof
gel time (of a drying oil)
GLC
glossmeter
gold number
grain crack resistance
guage length
Gutzeit test
Halphen-Hicks test
hanus iodine number
hazen color scale
heat age stability, latex
heat age stability test
heat stability test
heat test
Hegman gage
hellige system or comparator
hexabromide value
hiding, complete
hiding power
hiding power chart
hiding, spectrophotometric, complete
holiday detector
Hubl solution
hydrogen ion concentration
hydrogen value
hydrometer
hygrometer
ICI Cone/Plate viscometer
impact resistance
impact test
initial boiling point
inkometer
inspection
iodine number
iodine test
iodine value
ISO Flow Cups
jet abrader
jolly balance
Judd chart
Judd graph
Judd hiding power chart
kauri reduction test
kauri-butanol value
KB value
KHN
kinematic viscosity
knife test
Knoop hardness number, (KHN)
Knoop hardness test
Knoop indenter
Koettstorfer number
König pendulum damping hardness test
Kraemer-Sarnow method
Krebs-Stormer Viscometer
KU
latex freeze-thaw resistance
latex heat age stability
latex room temperature stability

low temperature coalescence
Mackey test
mandrel bend test
maturing
Meier bar
MEK resistance
melting point
mesh analysis
metamerism
Method 18
Method 24
Method 25
minimum film forming temperature (MFFT)
mixed aniline point
mobilometer
moisture content
moisture equilibrium
moisture vapor permeability (MVP)
moisture vapor transmission (MVT)
muller
MVP
MVT
neutralization equivalent or number
nondestructive test
nonmetameric match
normal conditions
north light
north standard
NV
NVM
O.A.
off-color
oil absorption (value) (O.A.)
paint test charts
pencil hardness
pendulum-rocker hardness
penetrometer
Pensky-Martens closed-cup tester
perm
perm cup or dish
permanganate time
permeability
permeance (water vapor)
Persoz pendulum hardness test
Pfund indentation tester
pH
phthalic anhydride test
PIC test
PMCC
precipitation number
precision
primary amine value
print resistance
print test
printing
pseudoisochromatic plate (PIC) test
psychrometer
QCT
quality assurance
quality control
QUV
relative
repeatability
reproducibility
resorcinol test
reverse impact test
ring and ball method
rocker hardness tester
Rockwell hardness
room temperature stability, latex
rosin test (qualitative)
round-robin
rub up
rubbing test
rust grade scale
salt spray (fog) test
sand-dry
saponification value
Saybolt universal viscosity
screen analysis or test
secondary amine value
selected ordinate method
set-to-touch time
Setaflash Tester
sieve
sieve analysis
sling psychrometer
softening point
soilant
solidification point
solids
solvent resistance
solvent rub test
spark-testing
specific permeability (of a film to moisture)
spin coating
stain removal
standard
standard conditions for testing
standard sand
stiffness (modulus of elasticity)
stormer viscometer
strength, tensile
stress-strain curve
stress-strain diagram
SUS
Sward rocker
T-bend
Taber abraser
tacciometer
tag closed tester
tape test
t-bend flexibility test
temperature checking
tensile properties
tensile strength
tensile tester
tensile testing machine
tertiary amine value
test
test charts
test fence
test method
through-dry
throwing out
tolerance
total organic compound (TOC)
toughness
Tukon
Twaddell hydrometer
Ubbelohde viscometer
ultimate tensile strength
ultracentrifuge
varnish
viscometer (viscosimeter)
washability
water vapor permeance (WVP)
water vapor transmission rate (WVT)
water white (WW)
wear cycles
wear index
weather-ometer
weathering
weight percent solids
Westphal balance
wet abrasion resistance
wet adhesion
wet adhesion test
wet film thickness
wet scrub resistance
Wijs Test
wire-bar applicator
wire-wound rod
wool scale
WVP
WVT
xenon-arc light exposure apparatus
yellowness
Zahn cup

Category 158
Surface Preparation and Pretreatments

abhesive
abrasive (noun)
air jet
alumina
aluminum oxide abrasive
aluminum oxide cloth
aluminum oxide, hydrated
aluminum oxide paper
aluminum primer
anchorage
antiseptic wash
automatic turning sander
barrier coat
blast cleaning
bleaching
block coat
block filler
bole
bright blast
bring forward
brush backed sander
brush-off blast
buffer coat
buffing
burning off
cadmium plating
Carborundum
chemical conversion coating
cissing
CLA
closed coat
coatings remover
commercial blast
corrosion coating
corundum
crocus abrasive
crocus cloth
decapping
degreaser
degreasing
deposition
descaling
dirt
drum (cylinder) sander
dry-to-sand
dubbing out
emery
emery abrasive
emery cloth
etch
etching primer
field painting
filler
filling of coated abrasives
flame cleaning
flat
flatting down
flatting (rubbing) varnish
flint abrasive
flint paper
fungicidal wash
galvanized steel
galvanizing
garnet abrasive
garnet paper
gesso
gilder's liquor
glazing compound
gray blast (commercial blast)
grit blasting
grit number
ground
ground coat
hackles
hydroblasting
iron phosphate coating
joint compound
laminar scale
lime putty
loading (of coated abrasives)
masking
masking tape
masonry conditioner
mesh number
metal primer
metallizing
minimized spangle
NACE No. 1 white metal blast cleaned surface finish
NACE No. 2 near-white blast cleaned surface finish
NACE No. 3 commercial blast cleaned surface finish
NACE No. 4 brush-off blast cleaned surface finish
oil gilding
open coat
orbital pad sander
paint and varnish remover
paint remover
painter's putty
parcel gilding
Parkerized
pentachlorophenol
petrifying liquid
phosphating
phosphatize (phosphate coat)
pickling
plastic wood
polish
powdered quartz
power cleaning
prefabrication primer
pretreatment
pretreatment primer
prime coat
primer
primer surfacer
priming
priming paints
putty
roto-dipping
rottenstone
rubbing
rubbing flat
rubbing oil
rust-inhibitive washes
sa
sandblast
sanding
sanding surfacer
sandpaper
scale
scaling
sealer
self-priming
shop coat
shop painting
shop primer
shop priming
shot blasting
shot peening
silicon carbide abrasive
silicon carbide waterproof paper
size
sizing
slate dust
slate flour
slate powder
spachtling
Spackle®
sparkling
steel wool
stopper
straight-line pad sander
stripper
stripping
surface conditioner
surface preparation
surface-treating
surfacer

swarf
sweep blast
tack rag (cloth)
terra cariosa
texture
tooth
touch-up coating
touch-up painting
transition primer
tripoli
vapor degreaser
vapor degreasing
water blasting
water break
water gilding
water repellent system
water-break-free
wet sanding
white blast
white metal
wide-belt sander
wire brush
wood filler
wood finishing
zinc phosphate coating
zinc stearate-coated finishing paper
zinc-coated

Category 159
Substrates

academy board
ACB
acoustical board
acoustical material
architectural coatings
asbestos board
asbestos plaster
asbestos-cement board (ACB), asbestos-cement wallboard, asbestos sheeting
barium plaster
base
bottom
canvas
canvas board
cinder block
cold-rolled steel
concrete block
deck
dry wall
embossed hardboard
enamel holdout
end grain
exterior type plywood
fabric
flake board
ground
hardboard
holdout
hot surface
hygroscopic
marouflage
masonry
paper
particle board
plasterboard
plywood
product coatings
raised grain
rusting
sand finish
sheathing
shrinkage
siding
special-purpose coatings
spiral grain
steel
stucco
substrate
surface
surface preparation
tile
wallboard
whitening in the grain
wood veneer

Category 161
Stability

aging (ageing)
antilivering agent
antisettling agent
Arrhenius equation
bacteriostat
barium ion stability
bodying
calcium ion stability
can stability
clear liquid separation (CLR)
cloud point
clouding
CLR
coacervation
coagulation
coagulum
compatibility
curdle
dimensional stability
drier absorption
drier dissipation
emulsion stabilizer
fattening
feeding
flocculation
freeze-thaw resistance
gel
gelation
gelling
grit
heat age stability, latex
heat age stability test
heat stability test
incompatibility
latex freeze-thaw resistance
latex heat age stability
latex mechanical stability
latex room temperature stability
livering
loss of drier
maturing
mechanical stability (of latexes)
miscible
one-compartment coating
one-pot coating
oxime
package stability
phase separation
polyurethane finish

pot life
room temperature stability, latex
sediment
setting up
settling
shelf life
spreadable life
stability
stabilizer
stable
staging life
storage life
storage stability
string (varnish)
suspending agent
suspension aid
syneresis
thickening
throwing out
two component paint
two-compartment coating
two-pot coating
unstable
usable life
working time

Category 163
Specialty Coatings

ablative coatings
acoustic coating
acoustical paint
aerosol coating
antitracking varnish
anticondensation paint
anticorrosive paint or composition
antifouling paints
antinoise paint
antique finish
antiquing
antishatter composition
antislip paint
appliance finish
asphalt seal coat
asphalt tile sealer
automotive finish
bactericidal paint
ballotini
beaded paint
beads, reflective
Berlin blacks
bituminous paints
blackboard paint
blackout paint
broken-color work
bronze paint
bronzing lacquer
Brunswick black
brush graining
cable varnishes and lacquers
casein finish
casein paint
clearcoat
corrosion-inhibiting paint
corrosion-inhibiting paint
crinkle finish
designers colors
diphenylmethane diisocyanate (MDI)
dressing compound
effect finish
end sealing compound
exterior basecoat
fabric coating
facade paint
fire retardant
fire-resisting finish
fire-retardant coating
flamboyant finish
flock finish
flocking
fluorescent paint
fungicidal paint
gel coat
glaze
glazing
grease paint
hammer finish
heat-sensitive paints
hexamethylene diisocyanate (HDI)
high-build coating
hold paint
hotmelt coatings
hull paint
inside spray
intumesce
intumescence
intumescent coating
isocyanate resins
Japan color
limewashing
marine coatings
marine varnishes
matte varnish
metallic paint
metallic pigmented paint
monolithic flooring
multicolor finish
natural finish
NGR stain
nondrying coating
nongrain-raising stain (NGR)
nonskid paint
OEM
OEM coatings
oil stain
organometallic polymer
ormolu varnish
plastic paint
polishing varnish
polychromatic finish
polyurethane finish
polyurethanes
porcelain enamel
PUD
reflectorization
ripple finish
rubbing varnish
rust-inhibiting paint
sacrificial protection
sanding surfacer
seamless flooring
shrivel finish
silicone paint
special-purpose coatings
speckled finish
spirit stains
structured paint
texture-finished paint
textured paint
thermal paint
tile-like coating (TLC)
TLC
traffic marking paint
undercoating
urethane coatings
whitewash
wrinkle finish
zero VOC paints

Category 164
Solvents, Solvency, Thinners, Solubility

absolute alcohol
acetal
acetic ester and acetic ether
acetone
acetonyl acetone
acetophenone
acetyl benzene
alcohol
alcohol-ethers
aliphatic solvents
allyl alcohol
alpha-pinene
American turpentine
amyl acetate
amyl alcohol
amyl benzoate
amyl borate
amyl ether
amyl formate
amyl lactate
amyl propionate
anisole
aromatic solvents
azeotrope
balanced solvents
banana oil
Bénard (vortex) cell
benzene
benzine (deprecated)
benzol or benzole
benzoline
benzyl alcohol
benzyl benzoate
benzyl Cellosolve®
beta-pinene
bite
boilers
bordeaux turpentine
1-butanol
n-butyl acetate
n-butyl alcohol
Butyl Carbitol®
Butyl Carbitol® acetate
Butyl Cellosolve®
Butyl Cellosolve® acetate
butyl citrate
butyl lactate
butyric alcohol
cajeputene
carbinol
Carbitol®
carbon tetrachloride
carene
Cellosolve®
Cellosolve® acetate
chloroform
chlorinated hydrocarbons
cimene
cleaner
coal tar hydrocarbons
coalescent (coalescing agent)
coatings remover
cologne spirits
colonial spirit
commercial xylene
cosolvent
coupling agent
cracking
cumene
cut
cyclohexadiene
cyclohexane
cyclohexanol
cyclohexanone
cyclohexyl
cyclohexyl acetate
cycloparaffins
decahydronaphthalene
denaturation
denatured alcohol
destructively distilled pine oil
destructively distilled wood turpentine
diacetone alcohol
diethyl oxalate
diethylene glycol monobutyl ether
diethylene glycol monobutyl ether acetate
diluent
dilute
dilution ratio
dimethyl ketone
dioxan (dioxane)
dipentene
dipentene (pure compound)
diphenyl ketone
dipropyl ketone
dipropyl oxalate
dl-limonene
double rub
dry point
elute
ethanol
ethyl acetate
ethyl alcohol
ethyl hydroxide
ethyl lactate
ethyl oxalate
ethyl propionate
ethylbenzene
ethylene dichloride
ethylene glycol
EtOH
evaporation rate
evaporation rate, final
evaporation rate, initial
exempt solvent
fast solvent
fat turpentine
fenchol
fenchone
fenchyl alcohol
fermentation alcohol
flashoff zone
fluorocarbons
fractional distillation
fusel oil
galipot
glycol
grain alcohol
gum spirits
gum turpentine
halogenated solvents
heavy solvent naphtha
heptanol-2
heptyl acetate
hexahydrophenol
hexalin
hexalin acetate
hexane
hexanol
hexone
hexyl acetate
hexyl ether
hi-flash naphtha
high-boiling solvent
hydrogenated naphthalenes
hydrogenated naphthas
hydroterpins
immiscible
isobutyl alcohol
isophorone
isopropyl acetate
isopropyl alcohol
isopropyl benzene
isopropyl cellosolve®
kauri-butanol value
kautschin
KB value
kerosene
ketone

kidney oil
latent solvent
ligroine
liquid
makeup solvent
mannite
mannitol
material balance
MBK
MEK
MEK resistance
mesityl oxide
mesitylene
methanol
methoxybutyl acetate
methyl acetate
methyl alcohol
methyl amyl carbinol
methyl amyl ketone
methyl butyl ketone
methyl butyrate
methyl Cellosolve® acetate
methyl Cellosolve®, methoxyethanol
methyl chavicol
methyl citrate (tri)
methyl cyclohexane
methyl cyclohexanol
methyl cyclohexanone
methyl cyclohexyl acetate
methyl ethyl ketone
methyl hexyl ketone
methyl isoamyl ketone
methyl isobutyl ketone
methyl lactate
methylated spirit
methylene chloride
MIBK
mineral spirits
miscible
monocyclic terpenes
mutual solvent
naphthas
naphthenes
naphthenic solvent
naphthenic type of hydrocarbon
nitroethane
nitromethane
2-nitropropane
nonpolar solvents
O.P.
octyl acetate
octyl alcohol
odorless mineral spirits
odorless solvent
oil of turpentine
oil soluble
overproof (O.P.)
oxidized turpentine
oxygenated solvent
paint and varnish remover
paint remover
paint thinner
painter's naphtha
partial aromatic solvents
partition
pentachloroethane
perchlorethylene
perchloromethane
permanganate time
petroleum
petroleum ethers
petroleum hydrocarbons
petroleum naphtha
petroleum spirits
phenyl methyl ketone
photochemically-reactive solvent
pinacolone
pine oil
pinene
pinolin
precipitation
press cake
proof, apparent
proof gallon, U.S.
proof spirit
propanol
2-propanone
proprietary solvents
propyl acetate
propyl alcohol
propyl butyrate
propyl carbinol
propyl propionate
reactive compound
reactive diluent
recovered solvent
reduce
reducers
reducing solvent
remover
Rule 66
s.b.p. spirits
slow solvent
solubility
solubility parameter
solute
solution
solvation
solvency
solvent
solvent balance
solvent imbalance
solvent pop
solvent power
solvent release
solvent resistance
solvent rub test
solvent shock
solvolysis
special boiling point spirits
special industrial solvents
spirit
spirit of wine
spirits of rosin
spirits of turpentine
steam-distilled pine oil
steam-distilled wood turpentine
stoddard solvent
sub spirits
sub turps
sulfate pine oil
sulfate wood turpentine
sulfite turpentine
supersaturation
synthetic pine oil
tail
tar spirit
terpenes
terpinenes
terpineol
terpinolene
tetrachloroethane
tetrachloromethane
tetrahydrofurfuryl alcohol
tetrahydronaphthalene
tetralin
Texanol®
THEA
thinner
thinning ratio
total enclosure
turpentine
turpentine, spirits of
turpentine substitute
turps
varnish remover
Venice turpentine
vinyl benzene
VM&P naphtha
VOC
VOC-free paints
volatile
volatile matter
volatile organic substance
volatile thinner
vortex cell
water
wet
white spirit
wood alcohol
wood turpentine
xylene
m-xylene
o-xylene
p-xylene
xylol

Category 165
Sealants, Caulks

back putty
bed putty
bituminous grout
body putty (body solder)
calk
calking
caulk (calk)
caulking
caulking cartridge
caulking compound
caulking gun
clear water repellent coating
face putty
glazing
glazing compound
grout
hard putty
hard stopping
joint filler
lime putty
mastic
painter's putty
plastic wood
putty
sealant

Category 166
Rheology

absolute viscosity
antilivering agent
antisag agent
apparent viscosity
associative thickener (AT)
AT
average molecular weight (viscosity method)
Bingham body
Bingham liquid
body
bodying
bodying agent
Brookfield® Viscometer
brush drag
brushability
brushing consistency
bubble tube viscometer
build
capillary viscometer
centipoise (cP)
coefficient of plastic viscosity
coefficient of viscosity
cold flow
cold test
compression set
consistency
creep
cup
deflocculation
dilatancy
dilatant flow (shear thickening)
dip type viscosity cup
dispersion
drag (brush drag)
efflux viscometer
elastomer
eta (n)
false body
fattening
feeding
film build
flooding
flow
flow curve
flowmeter
fluid flow
fluidity
Ford cup
full-bodied
Gardner mobilometer
Gardner-Holdt viscosity tubes
gel
gel point
gel strength
gelation
gelling
glass plate flow
HASE
heavy-bodied
HEUR
high build
high-build coating
HMHEC
HPAM
hydrophobically modified polyacrylamide
hysteresis
hysteresis loop
ICI Cone/Plate viscometer
inclined plate flow
intrinsic viscosity
ISO Flow Cups
kinematic viscosity
Krebs units (KU)
Krebs-Stormer Viscometer
KU
laminar flow
leveling
limiting viscosity number
liquid
livering
megapoise
methylcellulose
millipoise
mobility
newtonian flow
newtonian liquid
non-newtonian liquid
non-newtonian oil or fluid
Ostwald u-tube
pill flow
plastic
plastic flow
plastic solid
plastic viscosity
plasticity
plastometer
plate flow
plug flow
poise
pour point
pseudoplastic flow
pulling
pulling (under the brush)
rate of shear
reduced viscosity
relative leveling
residual tack
rhe
rheological
rheology
rheology modifier
rheopexy
rotometer

round (round color)
setting up
shear
shear rate, y
shear thickening (dilatancy)
shear thinning
shearing stress
short
simple liquid
slumping
solution/solvent viscosity ratio
spatter
SSF
SSU
sticky
Stokes Law
stokes (ST)
Stormer viscometer
strain relaxation
streamline flow
stringiness
structured paint
SUS
tack
tail
telescopic flow
thick
thickener
thickening
thixotrope
thixotropic
thixotropic paint
thixotropy
turbulence
Ubbelohde viscometer
velocity gradient
viscoelasticity
viscometer (viscosimeter)
viscosimeter
viscosity
viscosity, absolute dynamic
viscosity, apparent
viscosity coefficient
viscosity cup
viscosity, intrinsic
viscosity, kinematic
viscosity number
viscosity, plastic
viscosity ratio
viscosity, relative (for liquids)
viscosity, specific (for liquids)
viscosity, relative (for suspensions and solutions)
viscosity, specific (for suspensions and solutions)
viscosity, structural
viscosity/density ratio
viscous
wet storage stain
yield value
Zahn cup
zwitterion

Category 167
Resins (Natural)

abietates
accra
accroides
acetylated congo
acetylated damar
almaciga
alpha cellulose
amber
amyrin
Angola copal
animi
ball and ring method
balsam
Batavia dammar or damar
batu
belly benzoin
bengucopalic acid
bengucopalresene
bengucopalolic acid
benguela copal
benzofuran
benzoin
black boy gum
black dammar
black East India
boea
bold
botany bay resin
bush kauri copal
cambogia
Canada balsam
Canada turpentine
cashew nut shell oil
chalk kauri
Chinese lac
coal tar resins
colophony, colophonium
congo copal
congo gum
congocopalic acid
Congolene
copal esters
copal oil
copals
coumarone
coumarone-indene resins
cyclized rubber
dagincoleic acid
dagingenoleic acid
dammar (damar)
dammarolic acid
dammarylic acid
dehydroabietic acid
dragon's blood
elaterite
elemi gum
ethyl abietate
fine melt
fingers
fossil gum resins
frankincense
French varnish
fuse
fused congo
gamboge
ghatti gum
glycerin
grass-tree gum
gum
gum animi
gum benzoin
gum mastic
gum, natural
gum rosin
gum sandarac
gum thus
gutta-percha
hard gums
hard lac resin
hashab
hiroe
hydrogenated rosin
indene
indene resins
isocyanate resins
japan drier
japanese lacquer

juniper gum
juta hycica resin
kauri
Kordofan gum
lac
larch turpentine
lignin
limed rosin
maleic anhydride
maleic resin
Manila copal
mastication of resins
melengket
modified phenolic resins
mombassa gum
natural gums
natural resin
nubs
oil varnish
oleoresin
olibanum
oxidized rosin
paracoumarone-indene resins
penta resin
pentaerythritol (PE)
phycocolloid
pinolin
polycoumarone resins
polyindene resins
polyterpene resins
pontianak
rafaelite
reactive resins
resenes
resin
resin, natural
resin, synthetic
resinamines
resinates, metallic
resinoid
resinous timber
rosin
rosin essence
rosin esters
rosin, limed or zinc-treated
rosin pitch
rosin, reclaimed
rosin size
rosin test (qualitative)
runs
running
sandarac
short oil alkyd
short oil varnish
Sierra Leone copal
slack melt copal
softening point
spirit-soluble resin
spirits of rosin
sticklac
Strasbourg turpentine
synthetic resin
tall oil rosin
terpene resins
unsaponifiable matter
wood rosin
yacca gum
yellow accroides
Zanzibar copal

Category 168
Printing Inks, Printing

after-tack
air brush
aluminum ink
aniline inks
aniline printing
backing away from fountain
base
binder
binding varnish
bituminous varnish
bleach
bleeding
blinding of lithographic plates
block printing
blocking test
blush, blushing
body gum (no. 8 varnish)
brayer
brilliance
bronze
bronze dusting
bronze powders
bronzing
build of wet film
caking
cast-coated paper
catch-up
caul
cell
chalking
Chinese ink
coated paper
cold-setting inks
collotype
color burn-out
color strength
comic inks
composition rollers
coverage
covering power
crawling
crocking
crystallization
cut
dampening solution
densitometer
desensitize
developing ink
die stamping
doctor blade
doubletone ink
drier absorption
drier dissipation
dry back
dry offset
drying of ink
drying time
ductor roller
dusting bronzes
end point
etching
extender
fastopake inks
feathering
filling up
fine-etching
fineness of dispersion
flexographic ink
flexography
flocculation of the vehicle
flooding
flow
flying
form roller
fountain
fountain roller
fountain solution
fugitive color
ghosting
gloss ink
gold ink
graining
gravure
gravure tissue

greaseproof inks and coatings
greasing
ground
halftone
halftone screens
halo effect
hand-blocking
heat-set inks
hickies
highlight
imitation gold ink
impression
impression cylinder
indelible inks
India ink
ink
ink compounds
ink fountain
ink receptivity
inkometer
inks, cover
inks, halftone
inks, job
inks, quick-setting
intaglio
laketine
length
letterpress
letterset
line etching
liquifying stress
lithographic chalk
lithographic inks
lithographic stones
lithographic varnishes
loss of drier
machine-printing
magnesia
magnetic inks
metallic inks
mezzotints
misting
mixing white
moisture-set ink
monotype printing
news inks
newsprint
nonscratch inks
offset
offset lithography
offset paper
offsetting of ink
opaque ink
orange peel
overprint
overprint varnish
overvarnish
pantone printing
parting agent
pat-out
permanent inks
photoengraving
photogravure
photolithography
pick
picking
piling
planography
plate cylinder
plateless engraving
platen
platen press
powdering of paints
prepared linseed oil
printability
printing
printing ink
printing plate
printing press
printing strength
process inks
protective factor (of an antioxidant)
pulldown
reduce
reducers
releasing liquid
relief offset
relief printing
retarder
reticulation
roll-out
R.O.P. colors
rolling up
rotary press
rotogravure
rub-off
Rule 66
safety inks
screen
scumming
sensitizer
set off of ink
setting of ink
shade
shading and blending
short ink
show-through
silk-screen ink
silver ink
size
skin
slip sheet
slur
smearing
soft ink
solid
soluble blue
solvent density
solvent-borne coating
spanishing
specialty printing
spitting
spraying
spreading
steam-set inks
steel plate printing
stereotypes
stippling
strike-through
stripping
substrate
sympathetic inks
tack
tap-out
textile printing
theoretical VOC content
thermography
thermosetting ink
thinner
tile-like coating (TLC)
tinctorial strength
tint
tinting
transfer paper
transfer printing
transparency, ink
transparent inks
trapping
tusche
tympan
tympan sheet
typography
undercutting
underhue
varnish
vinyl printing and topcoating
vinyl printing ink
volatile thinner
washing
washing out
washout inks
water-based inks
wax-set ink
web
wet printing
winded
woodcut
xerography

Category 169
Polymerization

acid curing (hardening)
addition polymer
addition polymerization
alpha-methylstyrene
alternating copolymer
atactic polymer
average molecular weight (viscosity method)
base unit
block copolymer
block polymer
blown oil
bodied oil
branched
branched polymers
browne heat test
casting resin
cationic polymerization
chain length
cold-setting
condensation
condensation polymer
condensation polymerization
congeal
conventional base unit of a polymer
copolymer
copolymerization
coumarone
crosslinking
degree of polymerization (DP)
depolymerization
depolymerized rubber
dimer
dimerization
emulsion polymerization
free radical
heat-bodied oil
homopolymerization
intermolecular polymerization
intramolecular polymerization
monodispersity
monomer
oil, heat-bodied
polycondensate
polycondensation
polycyclization
polymerization
polymerized fatty acids
polystyrene (PS)
popcorn
stereospecific polymerization
synthesis
synthetic

Category 170
Plasticizers

abietates
acetamide
acetyl ricinoleates
adipates
amyl benzoate
amyl borate
amyl stearate
amyl tartrate
benzyl abietate
benzyl benzoate
benzyl butyl phthalate
benzyl phthalate
benzyl resinate
bloom
butyl citrate
butyl phthalate
butyl ricinoleate
camphol
camphor
chlorinated diphenyls
chlorinated paraffin wax
citrates
cyclohexyl oxalate
cyclohexyl phthalate
cyclohexyl stearate
dibenzyl phthalate
dibutyl phthalate
dimethyl phthalate
diphenyl ether
diphenyl phosphate (DPP)
dipropyl succinate
DPP
ethyl abietate
n-ethyl acetanilide
ethyl acetyl ricinoleate
ethyl succinate
external plasticizer
internal plasticization
internal plasticizer
internally plasticized
methyl citrate (tri)
methyl glycol phthalate (di)
methyl phosphate (tri)
methyl ricinoleate
methyl tartrate (di)
migration
oxanilide
PCB
phenyl benzoate
phenyl ethyl alcohol
phenyl phosphate, di
phenyl phosphate, tri
plasticization
plasticizer
plasticizer migration
plastisol
propyl benzoate
propyl ricinoleate
TCP
terpineol
thus
TPP
trichloroethyl phosphate
tricresyl phosphate (TCP)
tri-n-propyl phosphate
triphenyl phosphate (TTP)
tritolyl phosphate (TTP)

Category 171
Ecology, Pollution

acid rain
air contamination
air pollutant
air pollutants, hazardous
air pollution
air pollution control
air quality control regions
air sampling
algae
ambient air
ambient air quality
building code
carbon dioxide (CO_2)
carbon monoxide (CO)
CERCLA
Comprehensive Environmental Response, Compensation, and Liability Act (CERCLA)
convection
diluent
dry deposition
dust
dust fall
ecology
eddy
effluent
effluent limitations
emission
emission control equipment
exempt solvent
fallout
flue
flue gas
fumes
gas
HAP
hazardous air pollutants (HAP)
hazardous substance
incinerator
landfill
oxygen, molecular (O_2)
ozone (O_3)
particulate
photochemical smog
photochemically reactive organic substance
point source
pollutant
pollute
pollution
pollution abatement
pollution control
precipitation scavenging
precursor
primary pollutant
radical (or free radical)
reactive diluent
sanitary landfill
SARA
scavenging
scrubber
scrubbing
secondary treatment
sedimentation
silicosis
sink
sintering
smoke
smoke abatement
smoke alarm
soiling
soot
standard
standard of performance
Superfund Amendments and Reauthorization Act (SARA)
thermal discharge
toxic pollutants
volatile organic compound

Category 172
Regulatory

abatement
acid
acid deposition
acid rain
action level
aerosol
air contamination
air pollutant
air pollution
air pollution control
air quality control regions
air sampling
ambient air
ambient air quality
asbestosis
autoignition temperature
band wire
bioavailability
biological monitoring
building code
CAA
carbon monoxide (CO)
carcinogen
CAS Registration Number (CAS; CAS RN.)
catalytic incinerator
CFR
chemical name
Chemical Transportation Emergency Center
CHEMTREC
class I, II, or III liquids
Clean Air Act (CAA)
Clean Water Act (CWA)
Code of Federal Regulations
collector
color code
combustible liquid
compliance coating
compliance schedule
concentration
control
control device
control device efficiency
control technique guidelines (CTG)
convection
corrosive
criteria pollutant
Department of Transportation, US (DOT)
dermatitis
designated pollutant
dilution ventilation
dioxin
dispersion
DOT

DOT Identification Numbers
dry deposition
dust collector
dust fall
eddy
effluent
emission
emission reduction
Environmental Protection Agency (EPA)
EPA
ergonomics
exempt solvent
explosive limits
exposure limits
f/cc
face velocity
fallout
FDA
Federal Register, The (FR)
Fire Diamond
fire point
fire resistance
fire resistive
fire retardant
fire-resisting finish
fire-retardant chemical
fire-retardant coating
firewall
flame
flame (flash) arrestor
flame proofing
flame spread
flammable
flammable limits
flammable liquid
flashoff zone
floor sweep
flue gas
Food and Drug Administration
FR
fugitive emission
fumes
general ventilation
generic name
ground wire
HAP
Hazard Communication Rule
hazardous air pollutants (HAP)
Hazardous Chemical, Material
hazardous decomposition
hazardous decomposition products
Hazardous Materials Identification System, The (HMIS)
hazardous pollutant
hazardous substance
Hazardous Waste Number
HMIS
hood
housekeeping
IARC
IDLH
ignition temperature
inflammable
inhalator
International Agency for Research on Cancer
intumesce
intumescence
intumescent coating
irritant
label
lacrimator
landfill
LC_{50}
LD_{50} test
leaching
LEL
LEL meter
low solvent coating
lower explosive limit
lower explosive limit (LEL)
lowest achievable emission rate (LAER)
MAC
magnet wire
makeup air
makeup solvent
manufacturer's formulation
Material Safety Data Sheet (MSDS)
maximum allowable concentration (MAC)
Method 18
Method 24
Method 25
mini-bell
miscellaneous metal parts
mixture
model plant
MSDS
NAAQS
NAPTAC
National Air Pollution Control Techniques Advisory Committee (NAPTAC)
National Ambient Air Quality Standards (NAAQS)
National Cancer Institute (NCI)
National Emission Standards for Hazardous Air Pollutants (NESHAP)
National Fire Protection Association (NFPA)
National Institute of Occupational Safety and Health (NIOSH)
National Toxilogical Program (NTP)
NCI
NESHAP
new source
new source performance standards (NSPS)
NFRA Hazard Rating
NIOSH
NOC
NOEL
nonattainment area
noncombustible
nonflammable
NOS
NOx
NSPS
NTP
nuisance threshold
OAQPS
occupational exposure limit (OEL)
Occupational Safety and Health Administration (OSHA)
odor threshold
OEL
oral lethal dose
organic aerosol
OSHA
OSHRC
oxidizer
oxidizing agent
oxygen, molecular (O_2)
ozone (O_3)
particulate
particulate matter
PEL
percent capture
percent volatile
permissible exposure limit (PEL)
personal protective equipment (PPE)
photochemical catalysis
photochemical oxidant
photochemical reactivity
photochemical smog
photochemically-reactive organic material
photochemically-reactive solvent
physical hazard
PIN

Poison Control Center
pollutant
pollute
pollution
pollution abatement
precipitation scavenging
precursor
primary pollutant
Product Identification Number (PIN)
RACT
radical (or free radical)
RCRA
RCRA Hazardous Waste
reactive compound
reactive material
reasonably available control technology (RACT)
recommended exposure limit (REL)
Registry of Toxic Effects of Chemical Substances (RTECS)
regulatory alternative
REL
reportable quantity (RQ)
reproductive health hazard/ toxin
Resource Conservation and Recovery Act (RCRA)
respirator
risk assessment
RQ
RTECS
Rule 66
sanitary landfill
scavenging
scrubber
scrubbing
sedimentation
sensitization
sensitizer
Setaflash Tester
silicosis
sink
sintering
SIP
smog
smoke
smoke abatement
smoke alarm
soiling
solvent density
SOx
sparkproof tools
SPCC
stable
standard
standard of performance
STAPPA
State Implementation Plans (SIP)
theoretical VOC content
thermal incinerator
threshold limit value (TLV)
TLV
TLV-TWA
TOC
total enclosure
total organic compound (TOC)
touch-up coating
toxic
toxic pollutants
toxic substance
Toxic Substances Control Act (TSCA)
toxicant
toxicity
toxicology
trade secret
TRS
TSCA
turbulence
units
unstable
vapor capture system
VOC
VOC-free paints

Category 173
Biological

adaptation
algae
algicide
algistatic
anaerobic
antibiotic
antifouling paints
antiseptic
A. pullulans
Aureobasidium pullulans
bacteria
bacterial corrosion
bacterial resistance
bactericidal
bactericidal paint
bactericide
bacteriostat
bacteriostatic
barnacle
bioavailability
biocide
biodegradable
biodegradable surfactant
biodegradation
biodeterioration
biological monitoring
biopolymer
brown rot
can preservative
casein
cellulase
color blindness
color vision, anomalous
color vision, defective
culture
dry rot
enzyme
ergonomics
essential fatty acids
fruiting body
fungicidal paint
fungicide
fungistat
fungistatic
fungus (fungi)
fungus resistance
gamete
grass
herbicide
hydraulic lime
hypha
in-can biodeterioration
in-can preservative
inoculum
mercurials
microbiological attack
microorganisms
mildew
mildew (fungus) resistance
mildewcide

mildewstat
molds
morphology
mycelium
mycology
organism
pathogen, pathogene
pathogenicity
pentachlorophenol
preservative
protein
pseudomonas
rot
science
scotopic vision
sensation
slime
softwoods
spore
summerwood
symbiont
symbiosis
symbiotic
TPTF
twilight vision
viable
vision
vision, defective color
vision, normal color
visual

Category 174
Organizations

AATCC
AIA
AICE
AIMA
American Chemical Society (ACS)
American Conference of Governmental Industrial Hygienists (ACGIH)
American National Standards Institute (ANSI)
American Petroleum Institute (API)
American Society for Testing and Materials (ASTM)
AOCS
BESA
BISRA
British Standards Institution (BSI)
BSCP
Coatings Societies International (CSI)
Commission Internationale d'Eclairage (CIE)
CSMA
Department of Transportation, U.S. (DOT)
Environmental Protection Agency (EPA)
Federation d'Association des Technicians des Peintures, Vernis, Emaux, et d'Imprimerie de l'Europe Continentale (FATIPEC)
Federation of Societies for Coatings Technology (FSCT)
Food and Drug Administration (FDA)
Gallows Birds, The Society of
IMCO
International Commission on Illumination (ICI)
International Committee to Coordinate Activities of Technical Groups in the Coatings Industry (ICCATCI)
International Organization for Standardization (ISO)
International Union of Pure and Applied Chemistry (IUPAC)
Inter-Society Colour Council (ISCC)
Japan Society of Colour Materials (JSCM)
National Association of Corrosion Engineers (NACE)
National Association of Printing Ink Manufacturers (NAPIM)
National Cancer Institute (NCI)
National Coil Coaters Association (NCCA)
National Decorating Products Association (NDPA)
National Fire Protection Association (NFPA)
National Institute of Occupational Safety and Health (NIOSH)
National Paint and Coatings Association (NPCA)
National Printing Ink Research Institute (NPIRI)
National Toxicological Program (NTP)
Occupational Safety and Health Administration (OSHA)
Occupational Safety and Health Review Commission (OSHRC)
Oil & Colour Chemists' Association (OCCA)
organization
Paint Research Association (PRA)
Painting and Decorating Contractors of America (PDCA)
Skandinaviska Lackteknikers Förbund (SLF)
Society of Dyers and Colourists (SDC)
Society of Plastics Engineers (SPE)
Steel Structures Painting Council (SSPC)
Superfund Amendments and Reauthorization Act (SARA)
Surface Coatings Association Australia (SCAA)
Surface Coatings Association New Zealand (SCANZ)
TNO
trade association
trade journal

Category 175
Furniture Finishing

adhesive, assembly
antique finish
antiquing
cabinet finish
cabinet making
cassone painting
cellulosic coating
cherry
cloth distressing
conversion varnish
cowtail
decapping
distressing
Dutch gold
filler
fluting
fly speck
French polish
furniture
glaze
gold leafing
grain
highlighting
joinery
marbling, marbleizing
marquetry
milk paint
molding
particle board
rubbing down
shade
spattering
striping
varris Martin
veneer
veneer plywood
washcoat

Category 176
Thermal Properties

baking schedule
baking temperature
baking time
baking vehicle (varnish)
heat
heat age stability, latex
heat age stability test
heat capacity
heat test
heat-resistant finishes
heat-sealing
heat-sensitive paints
heat-set inks
hot melt adhesive
hot oil expression
hot spraying
hot-short
hotmelt coatings
infrared drying
infrared (IR)
initial boiling point
insulate
intumesce
intumescence
mastication of resins
melting point
mill scale
oven
overbaking
overcure
polyimide resins
post-cure
pour point
premature vulcanization
pyrometer
radiant heat baking
radiant heat drying
radiant heating
ripple finish
room temperature
running
sintering
temperature
temperature, curing
temperature, drying
temperature, setting
thermal
thermal adhesive
thermal discharge
thermal incinerator
thermal paint
thermography
thermoplastic
thermoplastic resins
thermoset
thermosetting
thermosetting ink
thermosetting resin
thermospray

Appendix

Industry-Related Associations and Organizations

Air & Waste Management Association
1 Gateway Center, 3rd Floor
Pittsburgh, PA 15222
Phone: (412) 232-3444
FAX: (412) 232-3450

American Ceramic Society
735 Ceramic Place
Westerville, OH 43081
Phone: (614) 890-4700
FAX: (614) 899-6109

American Chemical Society
1155 Sixteenth St., NW
Washington, D.C. 20036
Phone: (202) 872-4600

American Conference of Governmental Industrial Hygienists
6500 Glenway Ave., Bldg. D-7
Cincinnati, OH 45211
Phone: (513) 661-7881

American Institute of Chemical Engineers
345 E. 47th St.
New York, NY 10017
Phone: (212) 705-8100
FAX: (212) 752-3294

American National Standards Institute
11 W. 42nd St., 13th Floor
New York, NY 10036
Phone: (212) 642-4900
FAX: (212) 302-1286

American Society for Testing and Materials
100 Barr Harbor Dr.
W. Conshohocken, PA 19428-2959
Phone: (610) 832-9500

American Society for Quality Control
611 E. Wisconsin Ave.
P.O. Box 3005
Milwaukee, WI 53201-3005
Phone: (414) 272-8575
FAX: (414) 272-1734

American Spray Coaters Association
230 W. Wells St., Ste. 311
Milwaukee, WI 53203
Phone: (414) 273-3430

ANAFAPYT, A.C.
Gabriel Mancera No. 309
Col. del Valle
03100 Mexico, D.F.
Mexico
Phone: 52-5-682-77-94
FAX: 52-5-543-64-88

Association for Finishing Processes/SME
One SME Dr.
P.O. Box 930
Dearborn, MI 48121
Phone: (313) 271-1500

Canadian Paint & Coatings Association
9900 Cavendish Blvd., Ste. 103
St. Laurent, Ont. H4M 2V2
Canada
Phone: (514) 745-2611

Chemical Manufacturers Association
2501 M St., NW
Washington, D.C. 20037
Phone: (202) 887-1100
FAX: (202) 887-1237

Dry Colors Manufacturers Association
206 N. Washington St., Ste. 202
Alexandria, VA 22314
Phone: (703) 684-4044

Federation of Societies for Coatings Technology
492 Norristown Rd.
Blue Bell, PA 19422
Phone: (610) 940-0777
FAX: (610) 940-0292

NACE International
National Association of Corrosion Engineers
1440 S. Creek Dr.
Houston, TX 77084-4906
Phone: (713) 492-0535
FAX: (713) 492-8254

National Association of Printing Ink Manufacturers
Heights Plaza
777 Terrace Ave.
Hasbrouck Heights, NJ 07604
Phone: (201) 288-9454
FAX: (201) 288-9453

National Coil Coaters Association
401 N. Michigan Ave.
Chicago, IL 60611-4267
Phone: (312) 644-6894

National Decorating Products Association
1050 N. Lindbergh St.
St. Louis, MO 63132-1994
Phone: (800) 737-0107
FAX: (314) 991-5039

National Paint and Coatings Association
1500 Rhode Island Ave., NW
Washington, D.C. 20005-5597
Phone: (202) 462-6272
FAX: (202) 462-8579

Occupational Safety and Health Administration
OSHA Technical Data Center
Docket Office
Rm. N-3670
200 Constitution Ave., NW
Washington, D.C. 20210
Phone: (202) 523-7894

Oil & Colour Chemists' Association
Priory House
967 Harrow Rd.
Wembley, Middlesex HA0 2SF
England
Phone: 44-181-908-1086
FAX: 44-181-908-1219

Painting & Decorating Contractors of America
3913 Old Lee Highway
Ste. 33-B
Fairfax, VA 22030
Phone: (800) 332-7322
FAX: (703) 359-2576

Paint Research Association
8 Waldegrave Rd.
Teddington, TW11 8LD
England
Phone: 44-181-943-4717
FAX: 44-181-943-4705

Powder Coatings Institute
2121 Eisenhower Ave., Ste. 401
Alexandria, VA 22314
Phone: (703) 684-1770
FAX: (703) 684-1771

RadTech International North America
60 Revere Dr., Ste. 500
Northbrook, IL 60062
Phone: (708) 480-9576

Roof Coatings Manufacturers Association
6000 Executive Blvd., Ste. 201
Rockville, MD 20852-3803
Phone: (301) 468-3233

Steel Structures Painting Council
40 24th St., 6th Fl.
Pittsburgh, PA 15222
Phone: (412) 281-2331
FAX: (412) 281-9992

Abbreviations

American Society for Testing and Materials ASTM
ampere ... A
angstrom ... Å
Association ... Assn.
atmosphere ... atm
average ... avg
barrel ... bbl
boiling point ... bp
British Standards Institution ... BSI
British thermal unit ... Btu
calorie ... cal
centimetre ... cm
centipoise ... cP
centistokes ... cSt
circa, about, approximate ... ca.
compare ... cf.
cubic centimetre ... cm^3
decigram ... d.g.
degree Celsius ... °C
degree Fahrenheit ... °F
Dictionary of Architecture and Construction ... DAC
edited, edition ... Ed.
editor ... Ed.
equivalent weight ... equiv wt
et alii (and others) ... et al.
et cetera ... etc.
exempli gratia (for example) ... e.g.
Federal ... Fed.
flash point ... flp
foot ... ft
freezing point ... fp
gallon ... gal
gram ... g
hour ... h
hydrogen ion concentration, negative logarithm of ... pH
ibidem (in the same place) ... ibid.
id est (that is) ... i.e.
inch ... in.
infrared ... IR
International Union of Pure and Applied Chemistry ... IUPAC
kelvin ... K
kilogram ... kg
kilometre ... km
Knoop hardness number ... KHN
logarithm (common) ... log
logarithm (natural) ... ln
maximum ... max
melting point ... mp
metre ... m
micrometre (formerly micron) ... μm
milliequivalent ... meq
milligram ... mg
millilitre ... ml
millimetre ... mm
millimetre of mercury (pressure) ... mm Hg
minimum ... min
minute ... min
molar ... M
mole ... mol
molecular weight ... mol wt
nanometre (formerly millimicron) ... nm
National Association of Corrosion Engineers ... NACE
newton ... N
normal ... N
Occupational Safety and Health Administration ... OSHA
oil absorption ... O.A.
ounce ... oz
Paint Testing Manual ... PTM
parts per billion ... ppb
parts per million ... ppm
pascal ... Pa
perdiagonal line in expressions with unit symbols
percent ... %
pint ... pt.
pound ... lb
pounds-force per square inch ... psi
quart ... qt
reference(s) ... ref
revolutions per minute ... rpm
revolutions per second ... rps
Saybolt Universal seconds ... SUS
second ... s
Society ... Soc.
specific gravity ... sp gr
specific heat ... sp ht
specific volume ... sp vol
square ... sq
Standard ... Stnd.
ultraviolet ... UV
United States ... U.S.
vapor pressure ... vp
versus ... vs.
volt ... V
volume ... vol.
volume (of a publication) ... Vol
watt ... W
weight ... wt
yard ... yd
Young's modulus ... E

ASTM system used for most abbreviations. All units in SI definitions.

Measures

Originally introduced in France after the Revolution of 1789, the metric system has been modified somewhat as it has come into nearly universal acceptance. Although use of the metric system has been legal in the United States since 1966, the United States is the only industrialized nation that still prefers the customary English units, except in scientific and technical usage. Scientists the world over now use the standards codified in 1981 by the International Standardization Organization (ISO). The International System of Units are known as SI units, after Système Internationale, the French spelling.

SI Base Units

Quantity	Unit	Symbol
length	metre	m
mass	kilogram	kg
time	second	s
electric current	ampere	A
thermodynamic temperature	kelvin	K
amount of substance	mole	mol
luminous intensity	candela	cd

SI Supplementary Units

plane angle	radian	rad
solid angle	steradian	sr

SI Decimal Prefixes

SI units are decimal: they are derived from multiples of the number ten. By scientific convention, the following set of prefixes is used when referring to multiples or fractions, thus a kilogram would be 1000 grams; a millimetre would be 1/1000 of a metre, and a picofarad would be one trillionth of a farad.

Value and Equivalent		Unit	Symbol
1 000 000 000 000 000 000	= 10^{18}	exa-	E
1 000 000 000 000 000	= 10^{15}	peta-	P
1 000 000 000 000	= 10^{12}	tera-	T
1 000 000 000	= 10^{9}	giga-	G
1 000 000	= 10^{6}	mega-	M
1 000	= 10^{3}	kilo-	K
100	= 10^{2}	*hecto-	h
10	= 10^{1}	*deka- or deca-	da
1	= 10^{0}	(unprefixed)	—
0.1	= 10^{-1}	*deci-	d
0.01	= 10^{-2}	*centi-	c
0.001	= 10^{-3}	milli-	m
0.000 001	= 10^{-6}	micro-	μ
0.000 000 001	= 10^{-9}	nano-	n
0.000 000 000 001	= 10^{-12}	pico-	p
0.000 000 000 000 001	= 10^{-15}	femto-	f
0.000 000 000 000 000 001	= 10^{-18}	atto-	a

*to be avoided in technical use.

Metric and U.S. Equivalents

Metric Conversion Tables

Length

Metric to U.S.

millimetres	x	0.04	=	inches
centimetres	x	0.39	=	inches
metres	x	3.28	=	feet
metres	x	1.09	=	yards
kilometres	x	0.6	=	miles

U.S. to Metric

inches	x	25.4	=	millimetres
inches	x	2.54	=	centimetres
feet	x	3.04	=	metres
yards	x	0.91	=	metres
miles	x	1.6	=	kilometres

Volume

millilitres	x	0.03	=	fluid ounces
millilitres	x	0.06	=	cubic inches
litres	x	2.1	=	pints
litres	x	1.06	=	quarts
litres	x	0.26	=	gallons
cubic metres	x	35.3	=	cubic feet
cubic metres	x	1.3	=	cubic yards

teaspoons	x	5	=	millilitres
tablespoons	x	15	=	millilitres
cubic inches	x	16	=	millilitres
fluid ounces	x	30	=	millilitres
cups	x	0.24	=	litres
pints	x	0.47	=	litres
quarts	x	0.95	=	litres
gallons	x	3.8	=	litres
cubic feet	x	0.03	=	cubic metres
cubic yards	x	0.76	=	cubic metres

Metric to U.S.					U.S. to Metric				
Mass									
grams	x	0.035	=	ounces	ounces	x	28	=	grams
kilograms	x	2.2	=	pounds	pounds	x	0.45	=	kilograms
short tons	x	0.9	=	metric tons	metric tons	x	1.1	=	short tons
Area									
square centimetres	x	0.16	=	square inches	square inches	x	6.5	=	square centimetres
square metres	x	1.2	=	square yards	square feet	x	0.09	=	square metres
square kilometres	x	0.4	=	square miles	square yards	x	0.8	=	square metres
hectares (ha)*	x	2.5	=	acres	square miles	x	2.6	=	square kilometres
					acres	x	0.4	=	hectares

*the hectare is not an official SI unit, but is permitted

Temperature

degrees Fahrenheit – 32 x 5/9 = degrees Celsius
degrees Celsius x 9/5 + 32 = degrees Fahrenheit

Linear Measure

U.S. Customary	Metric
1 inch	25.4 millimetres (mm)
	2.54 centimetres (cm)
1 foot (12 in.)	304.8 millimetres (mm)
	30.48 centimetres (cm)
	0.3048 metre (m)
1 yard (36 in.; 3 ft)	0.9144 metre (m)
1 rod (16.5 ft; 5.5 yds)	5.029 metre (m)
1 statute mile (5280 ft; 1760 yds)	1609.3 metres (m)
	1.6093 kilometres (km)

Metric	U.S. Customary
1 millimetre (mm)	0.03937 in.
1 centimetre (cm)	0.3937 in.
1 metre (m)	39.37 in.
	3.2808 ft
	1.0936 yds
1 kilometre (km)	3280.8 ft
	1093.6 yds
	0.62137 mi

Liquid Measure

U.S. Customary	Metric
1 fluid ounce (fl oz)	29.573 millilitres (ml)
1 pint (16 fl oz)	0.473 (litre) (l)
1 quart (2 pints; 32 fl oz)	9.4635 decilitres (di)
	0.94635 litre (l)
1 gallon (4 quarts; 128 fl oz)	3.7854 litres (l)

Metric	U.S. Customary
1 millilitre (ml)	0.033814 fl oz
1 decilitre (dl)	3.3814 fl oz
1 litre (l)	33.814 fl oz
	1.0567 qts
	0.26417 gal

Avoirdupois Weights

U.S. Customary	Metric
1 grain	0.064799 gram (g)
1 ounce (437.5 grains)	28.350 grams (g)
1 pound (16 oz)	0.45359 kilograms (kg)
1 short ton (2000 lb)	907.18 kilograms (kg)
	0.90718 metric ton
1 long ton (2240 lb)	1016 kilograms (kg)
	1.016 metric tons

Area Measure

U.S. Customary	Metric
square inch (0.007 sq ft)	6.452 square centimetres (cm^2) 645.16 square millimetres (mm^2)
square foot (144 sq in.)	929.03 square centimetres (cm^2) 0.092903 square metre (m^2)
square yard (9 sq yd)	0.83613 square metre (m^2)
square rod (30.25 sq yd)	
square mile (640 acres)	2.59 square kilometres (km^2)

Metric	U.S. Customary
1 square millimetre (mm^2)	0.00155 square inch (sq in.)
1 square centimetre (cm^2)	0.155 square inch (sq in.)
1 centiare	10.764 square feet (sq ft)
1 square kilometre (km^2)	0.38608 square mile (sq mi)

Capacity

U.S. Customary	Metric
cubic inch (0.00058 cu ft)	16.387 cubic centimetres (cc; cm^3) 0.016387 litre (l)
cubic foot (1728 cu in.)	0.028317 cubic metre (m^3)
cubic yard (27 cu ft)	0.76455 cubic metre (m^3)
cubic mile (cu mi)	4.16818 cubic kilometres (k^3)

Metric	U.S. Customary
1 cubic centimetre (cc; cm^3)	0.061023 cubic inch (cu in.)
1 cubic metre (m^3)	35.135 cubic feet (1.3079 cu yd)
1 cubic kilometre (km^3)	0.23990 cubic mile

Metric Units

Easy Estimation Guide (rounded off for rule-of-thumb estimations).

Prefix			Metric Unit		U.S. Equivalents				
milli-	=	1/1000	1 millimetre	=	0.039 inch				
centi-	=	1/100	1 centimetre	=	0.39 inch				
deci-	=	1/10	1 decimetre	=	3.937 inches	=	0.32 foot		
			1 metre	=	39.37 inches	=	3.2 feet	=	1.1 yard
deka-	=	10	1 dekametre	=	393.7 inches	=	32 feet	=	10 yards
hecto-	=	100	1 hectometre	=	3937 inches	=	328 feet	=	109 yards
kilo-	=	1000	1 kilometre	=	39300 inches	=	3280 feet	=	1090 yards

Temperature Conversions

°C ← °F			°C ← °F			°C ← °F			°C ← °F		
	°C → °F			°C → °F			°C → °F			°C → °F	
-17.8	0	32.0	10.6	51	123.8	38.9	102	215.6	73.8	165	329.0
-17.2	1	33.8	11.1	52	125.6	39.4	103	217.4	76.6	170	338.0
-16.7	2	35.6	11.7	53	127.4	40.0	104	219.2	79.3	175	347.0
-16.1	3	37.4	12.2	54	129.2	40.6	105	221.0	82.1	180	356.0
-15.6	4	39.2	12.8	55	131.0	41.1	106	222.8	85.0	185	365.0
-15.0	5	41.0	13.3	56	132.8	41.7	107	224.6	87.8	190	374.0
-14.4	6	42.8	13.9	57	134.6	42.2	108	226.4	90.5	195	383.0
-13.9	7	44.6	14.4	58	136.4	42.8	109	228.8	93.3	200	392.0
-13.3	8	46.4	15.0	59	138.2	43.3	110	230.6	96.1	205	401.0
-12.8	9	48.2	15.6	60	140.0	43.9	111	231.2	98.9	210	410.0
-12.2	10	50.0	16.1	61	141.8	44.4	112	233.0	101.6	215	419.0
-11.7	11	51.8	16.7	62	143.6	45.0	113	235.4	104.4	220	428.0
-11.1	12	53.6	17.2	63	145.4	45.6	114	237.2	107.2	225	437.0
-10.6	13	55.4	17.8	64	147.2	46.1	115	239.0	110.0	230	446.0
-10.0	14	57.2	18.3	65	149.0	46.7	116	240.8	112.7	235	455.0
-9.5	15	59.0	18.9	66	150.8	47.2	117	242.6	115.5	240	464.0
-8.9	16	60.8	19.4	67	152.6	47.8	118	244.4	118.2	245	473.0
-8.3	17	62.6	20.0	68	154.4	48.3	119	246.2	121.0	250	482.0
-7.8	18	64.4	20.6	69	156.2	48.9	120	248.0	123.8	255	491.0
-7.2	19	66.2	21.1	70	158.0	49.4	121	249.8	126.6	260	500.0
-6.7	20	68.0	21.6	71	159.8	50.0	122	251.6	129.4	265	509.0
-6.1	21	69.8	22.2	72	161.6	50.6	123	253.4	132.2	270	518.0
-5.6	22	71.6	22.7	73	163.4	51.1	124	255.2	135.0	275	527.0
-5.0	23	73.4	23.3	74	165.2	51.7	125	257.0	138.7	280	536.0
-4.5	24	75.2	23.9	75	167.0	52.2	126	258.8	140.5	285	545.0
-3.9	25	77.0	24.4	76	168.8	52.8	127	260.6	143.3	290	554.0
-3.4	26	78.8	25.0	77	170.6	53.3	128	262.4	146.1	295	563.0
-2.8	27	80.6	25.6	78	172.4	53.9	129	264.2	149.9	300	572.0
-2.3	28	82.4	26.1	79	174.2	54.4	130	266.0	151.7	305	581.0
-1.7	29	84.2	26.7	80	176.0	55.0	131	267.8	154.5	310	590.0
-1.1	30	86.0	27.2	81	177.8	55.5	132	269.6	157.2	315	599.0
-0.6	31	87.8	27.8	82	179.6	56.1	133	271.4	160.0	320	608.0
0.0	32	89.6	28.3	83	181.4	56.6	134	273.2	162.8	325	617.0
0.6	33	91.4	28.9	84	183.2	57.2	135	275.0	165.6	330	626.0
1.1	34	93.2	29.4	85	185.0	57.7	136	276.8	168.4	335	635.0
1.7	35	95.0	30.0	86	186.8	58.3	137	278.6	171.1	340	644.0
2.3	36	96.8	30.6	87	188.6	58.8	138	280.4	173.9	345	753.0
2.8	37	98.6	31.1	88	190.4	59.4	139	282.2	176.7	350	662.0
3.4	38	100.4	31.7	89	192.2	60.0	140	284.0	179.5	355	671.0
3.9	39	102.2	32.2	90	194.0	60.5	141	285.8	182.2	360	680.0
4.5	40	104.0	32.8	91	195.8	61.1	142	287.6	185.0	365	689.0
5.0	41	105.8	33.3	92	197.6	61.6	143	289.4	187.8	370	693.0
5.6	42	107.6	33.9	93	199.4	62.2	144	291.2	190.6	375	707.0
6.1	43	109.4	34.4	94	201.2	62.7	145	293.0	193.4	380	716.0
6.7	44	111.2	35.0	95	203.0	63.3	146	294.8	196.1	385	725.0
7.2	45	113.0	35.6	96	204.8	63.8	147	296.6	198.9	390	734.0
7.8	46	114.8	36.1	97	206.6	64.4	148	298.4	201.7	395	743.0
8.3	47	116.6	36.7	98	208.4	65.0	149	300.2	204.4	400	752.0
8.9	48	118.4	37.2	99	210.2	65.5	150	302.0	207.2	405	761.0
9.5	49	120.2	37.8	100	212.0	68.3	155	311.0	210.0	410	770.0
10.0	50	122.0	38.3	101	213.8	71.0	160	320.0	212.8	415	779.0

Periodic Table of the Elements

1 H Hydrogen 1.01								
3 Li Lithium 6.94	4 Be Beryllium 9.01							
11 Na Sodium 22.99	12 Mg Magnesium 24.31							
19 K Potassium 39.10	20 Ca Calcium 40.08	21 Sc Scandium 44.96	22 Ti Titanium 47.90	23 V Vanadium 50.94	24 Cr Chromium 52.00	25 Mn Manganese 54.94	26 Fe Iron 55.85	27 Co Cobalt 58.93
37 Rb Rubidium 85.47	38 Sr Strontium 87.62	39 Y Yttrium 88.91	40 Zr Zirconium 91.22	41 Nb Niobium 92.91	42 Mo Molybdenum 95.94	43 Tc Technetium (99)	44 Ru Ruthenium 101.07	45 Rh Rhodium 102.91
55 Cs Cesium 132.91	56 Ba Barium 137.34	57 La Lanthanum 138.91	72 Hf Hafnium 178.49	73 Ta Tantalum 180.95	74 W Tungsten 183.85	75 Re Rhenium 186.21	76 Os Osmium 190.2	77 Ir Iridium 192.22
87 Fr Francium (223)	88 Ra Radium (226)	89 Ac Actinium (227)	104 (261)	105 (262)	106 (263)	107 (262)	108 (265)	109 (266)
			Names not yet established for these elements					

Lanthanides	58 Ce Cerium 140.12	59 Pr Praseodymium 140.91	60 Nd Neodymium 144.24	61 Pm Promethium (147)	62 Sm Samarium 150.35	63 Eu Europium 151.96
Actinides	90 Th Thorium (232)	91 Pa Protactinium (231)	92 U Uranium (238)	93 Np Neptunium (237)	94 Pu Plutonium (242)	95 Am Americium (243)

								2 He Helium 4.00
			5 B Boron 10.81	6 C Carbon 12.01	7 N Nitrogen 14.01	8 O Oxygen 16.00	9 F Fluorine 19.00	10 Ne Neon 20.18
			13 Al Aluminum 26.98	14 Si Silicon 28.09	15 P Phosphorus 30.97	16 S Sulfur 32.06	17 Cl Chlorine 35.45	18 Ar Argon 39.95
28 Ni Nickel 58.71	29 Cu Copper 63.55	30 Zn Zinc 65.38	31 Ga Gallium 69.72	32 Ge Germanium 72.59	33 As Arsenic 74.92	34 Se Selenium 78.96	35 Br Bromine 79.90	36 Kr Krypton 83.80
46 Pd Palladium 106.42	47 Ag Silver 107.87	48 Cd Cadmium 112.41	49 In Indium 114.82	50 Sn Tin 118.69	51 Sb Antimony 121.75	52 Te Tellurium 127.60	53 I Iodine 126.90	54 Xe Xenon 131.30
78 Pt Platinum 195.09	79 Au Gold 196.97	80 Hg Mercury 200.59	81 Tl Thallium 204.37	82 Pb Lead 207.19	83 Bi Bismuth 208.98	84 Po Polonium (210)	85 At Astatine (210)	86 Rn Radon (222)

64 Gd Gadolinium 157.25	65 Tb Terbium 158.93	66 Dy Dysprosium 162.50	67 Ho Holmium 164.93	68 Er Erbium 167.26	69 Tm Thulium 168.93	70 Yb Ytterbium 173.04	71 Lu Lutetium 174.97
96 Cm Curium (247)	97 Bk Berkelium (247)	98 Cf Californium (251)	99 Es Einsteinium (254)	100 Fm Fermium (257)	101 Md Mendelevium (258)	102 No Nobelium (259)	103 Lr Lawrencium (260)

Order Extra Copies of the *Coatings Encyclopedic Dictionary* Today!

Number of Copies Hard Cover .... ____ Member ($105.00) ____ Non-Member ($135.00)
Soft Cover .... ____ Member ($80.00) ____ Non-Member ($105.00)

I am paying by: ❑ Check or Money Order ❑ Company Purchase Order (please attach)
❑ MasterCard ❑ Visa ❑ American Express ❑ Discover

Credit Card Number ______________________________ Exp. Date __________

Signature ____

Name ____

Title ____

Company____

Street Address ____

City/State/Zip ______________________ Country______________

❑ I would be interested in purchasing, if available, a CD-ROM version of the *Coatings Encyclopedic Dictionary*

❑ Please send me Membership Information for the FSCT

Mail to: Federation of Societies for Coatings Technology,
492 Norristown Rd., Blue Bell, PA 19422-2350.

Source Code —DC1095

Coatings Encyclopedic Dictionary

Please use this card to help us prepare for future updates. Send us your suggestions for terms to be added, definitions that should be expanded, clarified, or corrected, and any other technical contribution you care to make. Thank you for your assistance.

____

____

____

____

____

____

____

____

____

____

ATTACH
POSTAGE

Federation of Societies for Coatings Technology
492 Norristown Rd.
Blue Bell, PA 19422-2350

ATTACH
POSTAGE

Federation of Societies for Coatings Technology
492 Norristown Rd.
Blue Bell, PA 19422-2350